PRENTICE HALL
Algebra 1

Jan Fair

Sadie C. Bragg

Prentice Hall
Englewood Cliffs, New Jersey
Needham, Massachusetts

Prentice Hall Algebra 1

Student Text   Teacher's Edition   Teacher's Resource Book
Solution Manual • Computer Test Bank

AUTHORS

Jan Fair
Formerly Chairperson, Mathematics Department
Santa Maria High School
Santa Maria, California

Sadie C. Bragg
Associate Professor of Mathematics
Borough of Manhattan Community College
The City University of New York
New York, New York

REVIEWERS

William E. Cavanaugh
Mathematics Teacher
Elkhart Community Schools
Elkhart, Indiana

Nina L. Ronshausen
Graduate Faculty of the College of Education
Texas Tech University
Lubbock, Texas

Lee V. Stiff
Assistant Professor of Mathematics Education
North Carolina State University
Raleigh, North Carolina

Photo credits appear on page 718.

CONSULTANTS

Frederick Bell
Professor of Mathematics Education
University of Pittsburgh
Pittsburgh, Pennsylvania

John M. Erickson
District Mathematics and Science Coordinator
Hopkins Public Schools
Hopkins, Minnesota

Stephen Krulik
Professor of Mathematics Education
Temple University
Philadelphia, Pennsylvania

Mary Dell Morrison
Mathematics Instructor (Retired)
Columbia High School
Maplewood, New Jersey

Jesse A. Rudnick
Professor of Mathematics Education
Temple University
Philadelphia, Pennsylvania

Rex Schweers, Jr.
Professor of Mathematics
University of Colorado
Greeley, Colorado

Harris S. Schultz
Professor of Mathematics
California State University, Fullerton
Fullerton, California

© 1990 by Prentice-Hall, Inc., Englewood Cliffs, New Jersey 07632. All rights reserved. No part of this book may be reproduced in any form or by any means without permission in writing from the publisher.

Printed in the United States of America.

ISBN 0-13-021726-3

10 9 8 7 6 5 4 3

PRENTICE HALL
A Division of Simon & Schuster
Englewood Cliffs, New Jersey 07632

CONTENTS

1 REAL NUMBERS

1.1	Variables and Algebraic Expressions *Applications: Travel, Weather*	2
1.2	The Real Number Line *Applications: Computer*	7
1.3	Comparing and Ordering Numbers *Applications: Meteorology, Health, Navigation, Economics*	12
1.4	Addition on a Number Line *Applications: Meteorology, Number Problem, Sports, Economics*	17
1.5	Adding Real Numbers *Applications: Meteorology, Sports, Aviation, Finance, Calculator*	22
1.6	Subtracting Real Numbers *Applications: Computer*	27
1.7	Multiplying Real Numbers *Applications: Weather, Finance*	31
1.8	Dividing Real Numbers *Applications: Sports, Statistics, Meteorology, Finance, Calculator*	35
1.9	Problem Solving Strategy: Select Appropriate Notation	42

Featured Application: Meteorology 41
Chapter Summary and Review 46, Chapter Test 48
Preparing for Standardized Tests 49, Maintaining Skills 50

2 ALGEBRAIC EXPRESSIONS

2.1	Simplifying and Evaluating Expressions *Applications: Computer*	52
2.2	Exponents and Formulas *Applications: Calculator*	55
2.3	Properties of Real Numbers *Applications: Consumerism*	59
2.4	Combining Like Terms *Applications: Number Problems, Geometry*	63
2.5	Simplifying and Evaluating Expressions with Parentheses *Applications: Geometry, Calculator*	67
2.6	Translating Phrases to Algebraic Expressions *Applications: Sports*	74
2.7	Open Sentences and Solution Sets *Applications: Number Problems*	78

2.8	Problem Solving Strategy: Account for All Possibilities	82
2.9	Translating Word Statements to Equations	85
	Applications: Consumerism, Geometry	

Technology: Introduction to Spreadsheets 72
Chapter Summary and Review 90, Chapter Test 92
Preparing for Standardized Tests 93, Cumulative Review 94

3 EQUATIONS IN ONE VARIABLE

3.1	Solving Equations: Addition and Subtraction Properties	96
	Applications: Travel, Number Problems	
3.2	Solving Equations: Multiplication and Division Properties	101
	Applications: Calculator	
3.3	Solving Equations: More Than One Property	106
	Applications: Calculator	
3.4	Problem Solving Strategy: Make a Model	110
3.5	Algebraic Proof	113
	Applications: Language Arts	
3.6	Evaluating Formulas	117
	Applications: Travel, Geometry	
3.7	Problem Solving Strategy: Write an Equation	120
3.8	Problem Solving: Mixed Types	127

Technology: Using Spreadsheets 125
Chapter Summary and Review 130, Chapter Test 132
Preparing for Standardized Tests 133, Maintaining Skills 134

4 MORE EQUATIONS IN ONE VARIABLE

4.1	Solving Equations: Combining Like Terms	136
	Applications: Number Problem, Geometry, Hobbies	
4.2	Solving Equations: Variable on Both Sides	140
	Applications: Geometry, Finance	
4.3	Problem Solving: Consecutive Integers	144
4.4	Shortcuts in Solving Equations	148
	Applications: Number Problems, Geometry, Sales	
4.5	Solving Equations: Percents	152
	Applications: Demography, Finance	
4.6	Problem Solving: Percents	156

4.7	Problem Solving: Mixtures	161
4.8	Literal Equations	164
	Applications: Sports, Geometry	
4.9	Problem Solving Strategy: Make a Drawing or a Table	168
4.10	Problem Solving: Uniform Motion	172

Featured Application: Messenger Service 160
Chapter Summary and Review 178, Chapter Test 180
Preparing for Standardized Tests 181, Cumulative Review 182

5 INEQUALITIES IN ONE VARIABLE

5.1	Graphing Equations and Inequalities	186
	Applications: Computer	
5.2	Solving Inequalities: Addition and Subtraction Properties	190
	Applications: Number Problems	
5.3	Solving Inequalities: Multiplication and Division Properties	194
	Applications: Number Problems	
5.4	Solving Inequalities: More Than One Property	198
	Applications: Number Problems	
5.5	Combined Inequalities	202
	Applications: Sports	
5.6	Absolute Value Equations	206
	Applications: Number Problems	
5.7	Absolute Value Inequalities	210
	Applications: Meteorology, Sports	
5.8	Problem Solving Strategy: Write an Inequality	214

Featured Application: Marathon Running 218
Chapter Summary and Review 220, Chapter Test 222
Preparing for Standardized Tests 223, Maintaining Skills 224

6 POLYNOMIALS

6.1	Monomials and Multiplying Monomials	226
	Applications: Geometry	
6.2	Dividing Monomials	230
	Applications: Geometry	
6.3	Monomials and Exponents	234
	Applications: Computer, Calculator	

6.4	Scientific Notation		237
	Applications: Computer		
6.5	Polynomials		241
	Applications: Geometry		
6.6	Adding and Subtracting Polynomials		245
	Applications: Number Problem, Geometry		
6.7	Multiplying a Polynomial by a Monomial		248
	Applications: Number Problems, Geometry		
6.8	Multiplying Polynomials		251
	Applications: Number Problems, Geometry		
6.9	Multiplying Polynomials: Special Cases		256
	Applications: Geometry		
6.10	Problem Solving Strategy: Look for a Pattern		260

Featured Application: Fluid Motion 240
Chapter Summary and Review 264, Chapter Test 266
Preparing for Standardized Tests 267, Cumulative Review 268

7 FACTORING POLYNOMIALS

7.1	Factors and Exponents	270
	Applications: Computer	
7.2	Monomial Factors of Polynomials	274
	Applications: Geometry	
7.3	Factoring $x^2 + bx + c$, $c > 0$	278
	Applications: Geometry	
7.4	Factoring $x^2 + bx + c$, $c < 0$	282
	Applications: Gardening	
7.5	Factoring $ax^2 + bx + c$	285
	Applications: Geometry	
7.6	Factoring: Special Cases	289
	Applications: Geometry	
7.7	Factoring by Grouping	292
	Applications: Geometry	
7.8	Factoring Completely	296
	Applications: Geometry	
7.9	Solving Polynomial Equations by Factoring	300
	Applications: Number Problems	

7.10	Problem Solving Strategy: Use Polynomial Equations	304

Featured Application: Agriculture 308
Chapter Summary and Review 310, Chapter Test 312
Preparing for Standardized Tests 313, Maintaining Skills 314

8 RATIONAL EXPRESSIONS

8.1	Simplifying Rational Expressions	316
	Applications: Baking	
8.2	Multiplying Rational Expressions	320
	Applications: Calculator, Geometry	
8.3	Dividing Rational Expressions	324
	Applications: Physics, Manufacturing, Transportation	
8.4	Least Common Denominator (LCD)	328
	Applications: Photography, Construction	
8.5	Adding and Subtracting Rational Expressions	333
	Applications: Physics	
8.6	Mixed Expressions and Complex Rational Expressions	337
	Applications: Statistics, Electricity	
8.7	Dividing Polynomials	342
	Applications: Finance	
8.8	Ratios and Proportions	345
	Applications: Entertainment, Cartography, Travel	
8.9	Solving Rational Equations	350
	Applications: Geometry, Accounting	
8.10	Problem Solving: Work and Motion	354
8.11	Problem Solving Strategy: Solve a Simpler Problem	358

Featured Application: Interest 341
Chapter Summary and Review 362, Chapter Test 364
Preparing for Standardized Tests 365, Cumulative Review 366

9 LINEAR EQUATIONS

9.1	Graphing Ordered Pairs	370
	Applications: Chemistry, Transportation, Economics	
9.2	Graphs of Linear Equations	3⁻
	Applications: Finance	
9.3	Problem Solving Strategy: Estimate from Graphs	

9.4	Slope	384
	Applications: Geometry, Science	
9.5	Slope-Intercept Form of a Linear Equation	391
	Applications: Computer	
9.6	Equation of a Line	396
	Applications: Chemistry	
9.7	Linear Inequalities in Two Variables	401
	Applications: Computer, Nutrition	

Featured Application: Radio Waves 389
Chapter Summary and Review 406, Chapter Test 408
Preparing for Standardized Tests 409, Maintaining Skills 410

10 RELATIONS, FUNCTIONS, AND VARIATION

10.1	Relations	412
	Applications: Finance, Meteorology	
10.2	Functions and Function Notation	417
	Applications: Finance, Meteorology	
10.3	Linear, Constant, and Composite Functions	422
	Applications: Consumer	
10.4	Direct Variation	426
	Applications: Finance	
10.5	Inverse Variation	430
	Applications: Geometry	
10.6	Problem Solving Strategy: Use an Appropriate Formula	434

Featured Application: Cost Analysis 425
Chapter Summary and Review 438, Chapter Test 440
Preparing for Standardized Tests 441, Cumulative Review 442

11 SYSTEMS OF LINEAR EQUATIONS

11.1	Graphing Systems of Linear Equations	444
	Applications: Calculator, Business, Economics, Geometry	
11.2	The Substitution Method	451
	Applications: Business, Calculator	
11.3	Problem Solving Strategy: Write a Linear System to Solve a Problem	455
11.4	The Addition/Subtraction Method	459
	Applications: Agriculture, Science, Engineering	

11.5	The Multiplication-Addition/Subtraction Method	463
	Applications: Business, Food Preparation	
11.6	Problem Solving: Digit Problems	467
11.7	Problem Solving: Age Problems	470
11.8	Problem Solving: Money and Mixture Problems	474
	Applications: Computer	
11.9	Problem Solving: Wind and Water Current Problems	479
11.10	Graphing Systems of Linear Inequalities	484
	Applications: Office Administration, Agriculture	

Featured Application: Break-Even Point 450
Chapter Summary and Review 488, Chapter Test 490
Preparing for Standardized Tests 491, Maintaining Skills 492

12 RADICALS

12.1	Square Roots	494
	Applications: Physics	
12.2	Irrational Square Roots	497
	Applications: Geometry	
12.3	Decimal Forms of Rational Numbers	500
	Applications: Electricity	
12.4	Simplifying Square Roots	503
	Applications: Physics, Geometry	
12.5	Addition and Subtraction of Radicals	509
	Applications: Geometry	
12.6	Multiplication of Radicals	512
	Applications: Geometry, Number Problems	
12.7	Division of Radicals	516
	Applications: Chemistry, Physics, Number Problems	
12.8	Solving Radical Equations	520
	Applications: Number Problems, Physics	
12.9	The Pythagorean Theorem	524
	Applications: Construction, Physics, Geometry, Sports	
12.10	The Distance Formula	528
	Applications: Computer	
12.11	Problem Solving Strategy: Use Coordinate Geometry	531

Featured Application: Image Formation 507
Chapter Summary and Review 536, Chapter Test 538
Preparing for Standardized Tests 539, Cumulative Review 540

13 QUADRATIC EQUATIONS AND FUNCTIONS

13.1	Quadratic Equations with Perfect Squares	544
	Applications: Coordinate Geometry	
13.2	Completing the Square	547
	Applications: Number Problems	
13.3	The Quadratic Formula	551
	Applications: Geometry, Number Problem	
13.4	Mixed Practice: Solving Quadratic Equations by Any Method	554
	Applications: Physics, Construction, Number Problems, Landscaping	
13.5	Graphing Quadratic Functions	557
	Applications: Business, Physics	
13.6	The Discriminant	564
	Applications: Engineering, Finance	
13.7	Problem Solving Strategy: Use a Function	567
13.8	Sum and Product of the Solutions	572
	Applications: Engineering, Business	

Featured Application: Physics 562
Chapter Summary and Review 576, Chapter Test 578
Preparing for Standardized Tests 579, Maintaining Skills 580

14 STATISTICS AND PROBABILITY

14.1	Statistics: Measures of Central Tendency	582
	Applications: Geography, Business, Education	
14.2	Statistics: Graphing Data	586
	Applications: Geography, Demography	
14.3	Statistics: Measures of Variability	590
	Applications: Health, Business	
14.4	Simple Probability	596
	Applications: Testing, Finance	
14.5	Problem Solving Strategy: Draw a Diagram	600
14.6	Probability: Compound Events	603
	Applications: Marketing, Finance	

Featured Application: Scattergrams 594
Chapter Summary and Review 608, Chapter Test 610
Preparing for Standardized Tests 611, Cumulative Review 612

15 RIGHT TRIANGLE RELATIONSHIPS

15.1	Basic Geometric Figures	614
	Applications: Construction, Travel	
15.2	Triangles	618
	Applications: Hobbies, Forestry	
15.3	Congruence	623
	Applications: Construction	
15.4	Similar Figures	628
	Applications: Surveying, Gardening	
15.5	Trigonometric Ratios	633
	Applications: Navigation, Geometry, Architecture	
15.6	Using Trigonometric Ratios	639
	Applications: Surveying, Architecture, Engineering, Hobbies	
15.7	Problem Solving Strategy: Check for Hidden Assumptions	643

Featured Application: Networks 637
Chapter Summary and Review 648, Chapter Test 650
Preparing for Standardized Tests 651, Cumulative Review 652

Extra Practice	655
Table of Squares and Square Roots	670
Table of Trigonometric Ratios	671
Symbols	672
Answers to Selected Exercises	673
Glossary	698
Index	707

A Letter to the Student

Algebra 1 is going to be your first experience with secondary academic mathematics. This is just the beginning in a series of courses that will interest, excite, and challenge you.

Algebra is the study of numbers and their relationship to each other. The basic skills you learn this year will be your tools for success in other courses in mathematics, science, business, finance, engineering, and many more.

No matter what you do each day, you are involved in some sort of problem solving or decision making situation. Algebra will help you choose a successful approach to solving problems. To help you succeed in solving problems, this text provides completely worked-out examples to follow, strategies to guide you in choosing a successful approach, highlighted key concepts, and plenty of exercises, all written in a language you can read and understand. The applications at the end of every lesson underscore how useful this knowledge can be.

Algebra is different, not difficult. Read carefully, keep up with your assignments, enjoy the challenge, work hard, and above all take satisfaction in your accomplishments. Your hard work and accomplishments will certainly prepare you for your next course in mathematics.

THE AUTHORS

1 Real Numbers

Lightning is beautiful, but it can also be dangerous. The temperature inside the channel of a lightning flash is believed to reach 28,000°C. At this temperature, the air expands and produces a tremendous sound wave called thunder.

1.1 Variables and Algebraic Expressions

Objectives: To evaluate algebraic expressions for given value(s) of the variable(s)
To write algebraic expressions for word phrases

Suppose you work at a weather station where you spend $\frac{3}{4}$ of your time working in the office. The rest of the time you work in the field.

The hours worked per week may change, but the fraction of time charged to the office payroll is always $\frac{3}{4} \times n$, where n stands for the total number of hours worked. Out of 8 hours (h), your time in the office is:

$$\frac{3}{4} \times n = \frac{3}{4} \times 8$$

$$= \frac{3}{\cancel{4}_1} \times \frac{\cancel{8}^2}{1} = \frac{6}{1}, \text{ or } 6 \text{ h}$$

Capsule Review

If $6 is the hourly rate of pay and a is the number of hours worked, then $6 \times a$ is the total wages earned.

Use $6 \times a$ to find the pay for each number of hours (h) worked.

1. $\frac{3}{4}$ h
2. $3\frac{1}{2}$ h
3. 6.8 h
4. 10 h

5. How would $6 \times a$ change if the hourly rate became $8?

In $\frac{3}{4} \times n$, n is called a *variable*. $\frac{3}{4} \times n$ is a *variable expression*.

> A **variable** is a symbol used to represent one or more numbers. Any letter may be used as a variable.
>
> A **variable expression** is an expression that contains one or more variables.

You may not know the number(s) a variable stands for, but the mathematical meaning of a variable expression is clear. Here are different ways to read some expressions.

Addition	Subtraction
$n + 8$	$z - s$
Some number n plus 8	A number z minus another number s
n increased by 8	z decreased by s
The sum of n and 8	The difference between z and s
8 more than n	s less than z

Multiplication	Division
$\frac{2}{3} \times r$	$y \div 3$
$\frac{2}{3}$ times a number r	Some number y divided by 3
The product of $\frac{2}{3}$ and r	The quotient of y and 3
$\frac{2}{3}$ of r	The ratio of y and 3

To **evaluate** a variable expression, substitute a given number for each variable. The result is a **numerical expression** that may contain one or more of the operations of addition, subtraction, multiplication, or division. The number represented by the numerical expression is the value of the variable expression for the value(s) of the variable(s) used.

Both variable and numerical expressions are called **algebraic expressions.** Are 5 and $2xy$ algebraic expressions? Why or why not?

EXAMPLE 1 Evaluate each algebraic expression in the chart above if $n = 17$, $r = 9$, $y = 2.4$, $z = 7.89$, and $s = 0.92$.

Algebraic Expression	Numerical Expression (after substitution)	Value
$n + 8$	$17 + 8$	25
$z - s$	$7.89 - 0.92$	6.97
$\frac{2}{3} \times r$	$\frac{2}{3} \times 9$	$\frac{2}{3} \times 9 = \frac{2}{3} \times \frac{9}{1} = 6$
$y \div 3$	$2.4 \div 3$	$2.4 \div 3 = 0.8$

In algebra, a raised dot or parentheses are often used to indicate multiplication. In variable expressions, multiplication symbols may be omitted. The product of 6 and some number n is given in various ways below.

With a raised dot	With parentheses	With no symbol between
$6 \cdot n$	$(6)(n)$ or $6(n)$	$6n$

1.1 Variables and Algebraic Expressions

EXAMPLE 2 Evaluate the algebraic expression xyz if $x = \frac{3}{4}$, $y = 28$, and $z = \frac{3}{7}$.

$$xyz = x \cdot y \cdot z \qquad \textcolor{blue}{xyz \text{ means } x \text{ times } y \text{ times } z.}$$

$$= \frac{3}{4} \cdot \frac{28}{1} \cdot \frac{3}{7} \qquad \textcolor{blue}{\text{Substitute } \tfrac{3}{4} \text{ for } x, 28 \text{ for } y, \text{ and } \tfrac{3}{7} \text{ for } z.}$$

$$= \frac{3 \cdot 28 \cdot 3}{4 \cdot 1 \cdot 7} \qquad \textcolor{blue}{\text{Multiply.}}$$

$$= \frac{3 \cdot 3}{1} = 9$$

EXAMPLE 3 Evaluate $\frac{2}{3}(f + 11)$ if $f = 13$.

$$\frac{2}{3}(f + 11) = \frac{2}{3}(13 + 11) \qquad \textcolor{blue}{\text{Substitute 13 for } f.}$$

$$= \frac{2}{3}(24) \qquad \textcolor{blue}{\text{Add within the parentheses.}}$$

$$= 16 \qquad \textcolor{blue}{\text{Multiply.}}$$

Two operations—addition and multiplication—were needed to evaluate the algebraic expression in Example 3. In all such cases, *operations within the parentheses are done first*.

Division is most often shown in fraction form. $\frac{y}{5}$ means *y divided by 5*. In a fraction like $\frac{0.011s}{rt}$, the fraction bar creates invisible parentheses around the numerator and the denominator. This means that any operation(s) in the numerator and in the denominator are completed before the division.

EXAMPLE 4 Evaluate $\frac{k - m}{h}$ if $k = 7\frac{3}{4}$, $m = \frac{1}{4}$, and $h = \frac{1}{8}$.

$$\frac{k - m}{h} = \frac{7\frac{3}{4} - \frac{1}{4}}{\frac{1}{8}} \qquad \textcolor{blue}{\text{Substitute } 7\tfrac{3}{4} \text{ for } k, \tfrac{1}{4} \text{ for } m, \text{ and } \tfrac{1}{8} \text{ for } h.}$$
$$\textcolor{blue}{\text{Subtract.}}$$

$$= \frac{7\frac{1}{2}}{\frac{1}{8}} = \frac{15}{2} \cdot \frac{8}{1} \qquad \textcolor{blue}{\text{Recall the rule for dividing by a fraction.}}$$

$$= 15 \cdot 4 = 60$$

You will often translate word phrases into algebraic expressions.

EXAMPLE 5 Write an algebraic expression for each phrase.

    a. The cost of a meal c plus the tax t: $c + t$

    b. The cost of three movie tickets at d dollars per ticket: $3d$, or $3 \cdot d$, or $3(d)$

    c. The number of miles traveled m divided by the number of hours traveled h: $\frac{m}{h}$, or $m \div h$

CLASS EXERCISES

Identify the variable and describe the operation.

1. $4x$
2. $9 - a$
3. $m + 7$
4. $\frac{2}{g}$

Evaluate each expression if $a = 1$, $b = 5.4$, and $c = 0.5$.

5. $a + 89$
6. $b - c$
7. $\frac{b}{c}$
8. cab

PRACTICE EXERCISES

Evaluate each expression if $n = 12$.

1. $n + 15$
2. $30 - n$
3. $\frac{1}{2}n$
4. $\frac{5}{6}n$
5. $n \div 6$
6. $\frac{48}{n}$
7. $n + n$
8. $0 + n$

Evaluate each expression if $a = 15$, $b = 6$, and $c = 3$.

9. $\frac{ab}{2c}$
10. $5abc$
11. $\frac{c}{10ab}$
12. $\frac{abc}{25}$
13. $\frac{1}{3}(a + 3)$
14. $\frac{1}{5}(b + c)$
15. $\frac{2}{3}c + 5$
16. $\frac{5}{3}a + 1$

Write an algebraic expression for each phrase.

17. The sum of 3 and some number x
18. $\frac{3}{2}$ increased by some number y
19. A number t decreased by 1.07
20. The difference between a number m and 99

Evaluate each expression if $d = \frac{2}{3}$, $e = \frac{3}{4}$, and $f = \frac{1}{2}$.

21. $e \div d$
22. $e - d$
23. $e - f$
24. $\frac{de}{f}$
25. $4def$
26. $d - f$
27. $f - f$
28. $d + e + f$
29. $\frac{d + e}{f}$
30. $\frac{2e - d}{2f}$
31. $3d + 2e + 2f$
32. $\frac{3}{2}d + e + 4f$

Write an algebraic expression for each phrase.

33. The product of $\frac{3}{8}$ and some number j
34. 5 less than z
35. The quotient of 6.4 and some number b
36. 10 more than x.

Evaluate each expression if $p = 3\frac{5}{8}$, $q = \frac{3}{4}$, $r = 6$, $s = 4$, and $t = 1$.

37. $(p - q)8$
38. $(q + r)\frac{1}{3}$
39. $\frac{r + s}{r - s}$
40. $\frac{r - t + s}{r - q}$
41. $(8q) + \left(\frac{2}{3}r\right) - \left(\frac{1}{2}s\right)$
42. $\frac{7}{10}(r + s) - (4q)$

Applications

Write an algebraic expression for each phrase.

43. **Travel** The number of miles traveled m divided by the number of gallons used g
44. **Weather** The difference between the highest temperature h and the lowest temperature l

WRITING IN ALGEBRA

In Exercises 1–4, write one example of each expression.

1. An algebraic expression that involves multiplication and addition
2. An algebraic expression that involves division and subtraction
3. An algebraic expression that contains parentheses and the operations of addition and multiplication
4. In Exercise 3, what must be known before you can evaluate your algebraic expression? Give in order the arithmetic operations that must be performed. Is this still true if the operations are performed with a calculator?

1.2 The Real Number Line

Objectives: To graph real numbers on a number line
To classify numbers in subsets of the set of real numbers

The set of **real numbers** contains all positive numbers, all negative numbers, and zero. The real numbers can be represented as points on a number line.

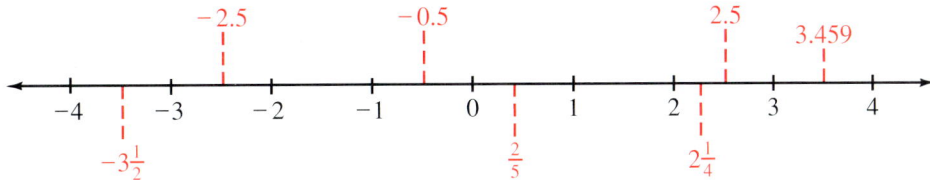

You have made and used number lines in earlier mathematics courses by marking off equal distances from a starting point, labeled 0. The zero point is also called the **origin.** Numbers to the right of 0 are called **positive numbers** and numbers to the left of 0 are called **negative numbers.** Zero (0) is neither positive nor negative.

Capsule Review

On a number line, numbers are associated with points of the line. The number associated with point J is $+4$, or 4.

Give the number associated with each point.

1. H 2. A 3. I 4. F

Name the point associated with each number.

5. -1 6. 5 7. -4 8. 0

Some sets of numbers, such as $\{-2, -1, 0, 1, 2, 3\}$, can be listed. The symbol $\{\ \}$ is used to enclose the members of a set. Three dots mean that the numbers of the set go on forever in the same pattern.

The set of all **natural** (counting) numbers: $\{1, 2, 3, 4, 5, \ldots\}$
The set of all **whole** numbers: $\{0, 1, 2, 3, 4, 5, \ldots\}$
The set of all **integers**: $\{\ldots, -3, -2, -1, 0, 1, 2, 3, \ldots\}$

EXAMPLE 1 Graph the integers −4, 2, and 5 on a number line. Label the points D, E, and F, respectively.

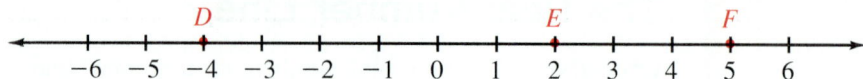

In Example 1, points D, E, and F are called the **graphs** of −4, 2, and 5, respectively. −4, 2, and 5 are called the **coordinates** of points D, E, and F, respectively. This relationship can be written as D(−4), E(2), and F(5).

There are points on the real number line that are *not* integers.

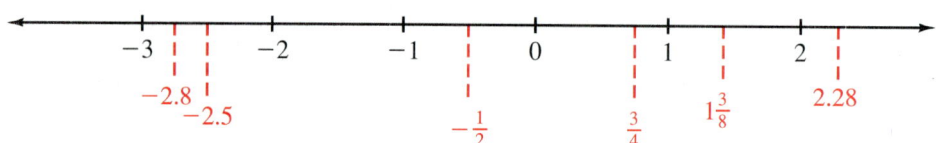

On the number line above, the fractions $-\frac{1}{2}$, $\frac{3}{4}$, and $1\frac{3}{8}$ and the decimals −2.5, 2.28, and −2.8 are graphed. These numbers are part of the set of *rational numbers*.

A **rational number** is a number that can be written in the form $\frac{m}{n}$, where m and n are integers and $n \neq 0$.

EXAMPLE 2 Show that the following numbers are rational numbers.

    a. $-3\frac{1}{2}$ **b.** 4   **c.** 5.02   **d.** 0

a. $-3\frac{1}{2} = \frac{-7}{2}$ $m = -7, n = 2$ **b.** $4 = \frac{4}{1}$ $m = 4, n = 1$

c. $5.02 = 5\frac{2}{100} = \frac{502}{100}$ $m = 502, n = 100$ **d.** $0 = \frac{0}{1}$ $m = 0, n = 1$

From Example 2, you can see that the set of rational numbers includes integers, decimals, and fractions.

Some numbers cannot be written in the form $\frac{m}{n}$, where m and n are integers and $n \neq 0$. These numbers are called **irrational numbers.** Examples of irrational numbers are:

    $\sqrt{2}$ The square root of 2, which is about 1.41
    $\sqrt{3}$ The square root of 3, which is about 1.73
    π Pi, the ratio of the circumference of a circle C and its diameter d, or $\frac{C}{d}$; π is approximately equal to 3.14.

Irrational numbers are real numbers and can be graphed on a number line, along with the rational numbers.

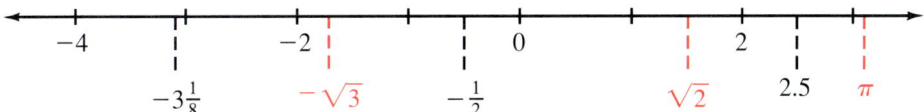

Until a later chapter of the book, most of the real numbers you will work with will be rational numbers.

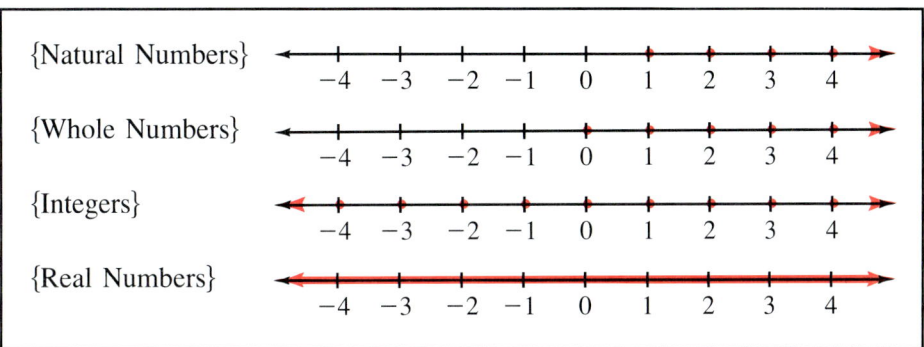

The entire number line is shaded to graph the set of real numbers. Each real number corresponds to exactly one point on the number line. Each point on the number line corresponds to a unique (one and only one) real number.

CLASS EXERCISES

For Exercises 1–8, refer to the number line below.

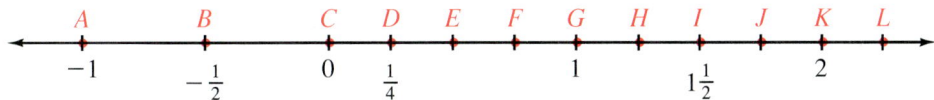

Give the coordinate of each point.

1. E 2. F 3. H 4. L

State the letter of each coordinate.

5. $\frac{1}{2}$ 6. $1\frac{3}{4}$ 7. $-\frac{1}{2}$ 8. $2\frac{1}{4}$

Graph these numbers on a number line. Label the points, as indicated.

9. $A(-4)$, $B(2)$, $C(-5)$

10. $D\left(-3\frac{1}{2}\right)$, $E(0)$, $F\left(3\frac{1}{2}\right)$

11. $G(-0.5)$, $H(-1.5)$, $J(0.5)$

12. $K\left(-2\frac{3}{4}\right)$, $L\left(1\frac{1}{2}\right)$, $M\left(-4\frac{1}{4}\right)$

1.2 The Real Number Line

PRACTICE EXERCISES

For Exercises 1–16, refer to the number line below.

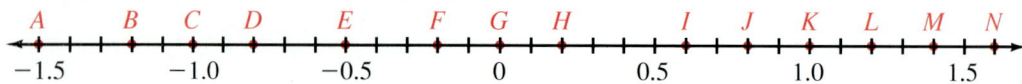

Give the coordinate of each point.

1. C
2. E
3. H
4. J
5. B
6. D
7. L
8. N

State the letter of each coordinate.

9. 0
10. 1.0
11. −1.5
12. −0.5
13. 0.6
14. 0.2
15. −0.8
16. −0.2

Show that each number is a rational number.

17. 7
18. −5
19. −3.012
20. 2.618
21. $11\frac{3}{4}$
22. $-6\frac{5}{6}$
23. −1
24. 0

For each set of numbers, draw a number line and graph the numbers. Use an appropriate scale.

25. $\{-1, 1, 2, 4\}$
26. $\{-3, 0, 2, 6\}$
27. $\{-4, -1.5, 2, 2.5\}$
28. $\{-6.5, -2, 3.5, 7\}$
29. $\{-1\frac{1}{4}, -\frac{1}{2}, \frac{3}{4}, 1\}$
30. $\{-2, -\frac{1}{4}, 0, 1\frac{3}{4}\}$

Draw this chart and indicate whether the given number belongs in the set.

	Natural Numbers	Whole Numbers	Integers	Rational Numbers	Irrational Numbers	Real Numbers
31. 8						
32. −11						
33. $\frac{0}{10}$						
34. 16.2						
35. $-3\frac{1}{6}$						
36. $-\sqrt{3}$						

37. On this number line, is there a point that corresponds to $-\frac{2}{3}$? That corresponds to 100? Explain.

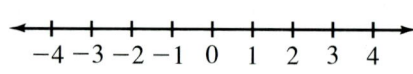

Chapter 1 Real Numbers

38. The graphs of the natural numbers, the whole numbers, and the integers are a series of points. The graph of the real numbers is a continuous line. Why is this so?

Applications

Computer The INT function on a computer, INT(X), always gives the greatest INTeger that is less than or equal to X. For example, INT(5.4) = 5, INT(7.9) = 7 and INT(−3.2) = −4. State the value of each expression.

39. INT(71) **40.** INT(−6.1) **41.** INT(4.87)

42. INT(1/4) **43.** INT(30/2) **44.** INT(17/2)

45. For what values of X does INT(X) = X?

46. For what values of X does INT(X/2) = X?

MATH CLUB ACTIVITY

Copy the number line on your own paper. Locate the points in Exercises 1–3.

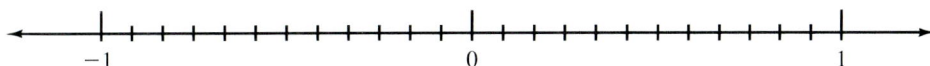

1. The temperature kept decreasing over a period of three hours. It decreased $\frac{1}{3}°$ the first hour and another $\frac{1}{2}°$ the second hour. The average decrease in temperature over a three-hour period was $\frac{1}{2}°$. Represent the temperature change in the third hour. Locate the corresponding point on the number line. Label it *W*.

2. Arrange the rational numbers $-\frac{5}{8}$, $-\frac{1}{4}$, $-\frac{5}{16}$ in order from least to greatest. Graph the greatest number on the number line. Label it *I*.

3. If water is poured down the mouth of a pipe system and divides evenly at each opening, what fraction of the water moves out of the opening labeled *N*? Graph the value on the number line. Label it *N*.

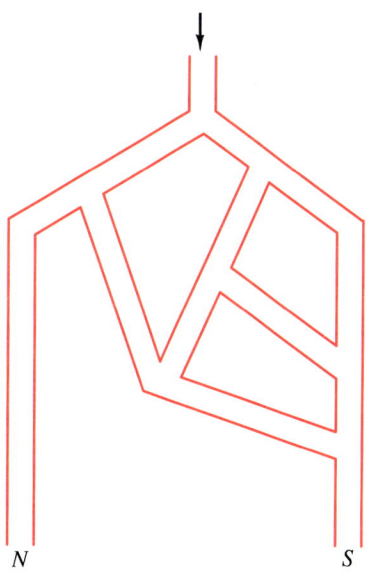

1.3 Comparing and Ordering Numbers

Objectives: To compare and order real numbers
To use the concepts of opposites and absolute value

Numbers are placed on a number line in increasing order from left to right. Given two numbers on a number line, the number to the left is less. The equal sign, =, and the inequality symbols, < and >, are used to compare numbers.

= means *is equal to*. $\frac{3}{2} = 1\frac{1}{2}$ is read "$\frac{3}{2}$ is equal to $1\frac{1}{2}$."
< means *is less than*. $0 < 2$ is read "0 is less than 2."
> means *is greater than*. $6 > 4$ is read "6 is greater than 4."

Capsule Review

On a horizontal number line:

$-1 < 0$ $\quad\quad$ -1 is to the left of 0.
$-8 > -9$ $\quad\quad$ -8 is to the right of -9.
$\frac{1}{2} = 0.5$ $\quad\quad$ $\frac{1}{2}$ and 0.5 correspond to the same point.

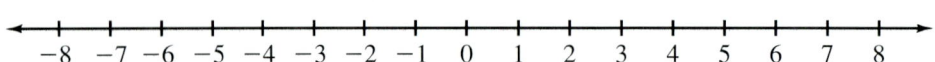

Replace each $\underline{?}$ with <, >, or = to make a true statement. Refer to the number line above.

1. $6 \underline{\ ?\ } -5$
2. $-4 \underline{\ ?\ } 3$
3. $-1.5 \underline{\ ?\ } 0$
4. $-2 \underline{\ ?\ } -2$
5. $-\frac{1}{2} \underline{\ ?\ } -3$
6. $-1\frac{1}{2} \underline{\ ?\ } -1$
7. $\frac{2}{3} \underline{\ ?\ } \frac{3}{4}$
8. $-1.2 \underline{\ ?\ } -1.25$

A number line can be used to compare or to show the order of two or more numbers.

EXAMPLE 1 Graph the numbers -1, $\frac{1}{2}$, -2, $-\frac{1}{3}$, and $\frac{1}{3}$. Then name them in order from least to greatest.

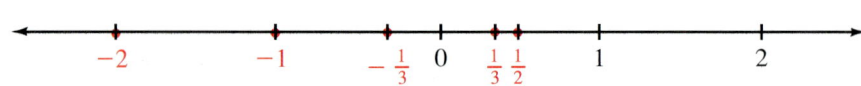

The numbers in order from least to greatest are -2, -1, $-\frac{1}{3}$, $\frac{1}{3}$, and $\frac{1}{2}$.

12 Chapter 1 Real Numbers

Every real number has an **opposite** whose graph is the same distance from the origin but in the opposite direction. To use the language of algebra, if n is a real number, then $-n$ is its opposite.

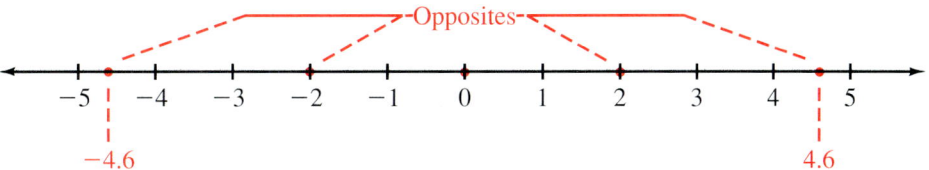

The opposite of 2 is -2; the opposite of -2 is 2.
-4.6 and 4.6 are opposites.
0 is its own opposite. That is, 0 is the opposite of 0.

EXAMPLE 2 Name the opposite of each number:
a. 13   b. $\frac{6}{7}$ c. -2.75

Number	13	$\frac{6}{7}$	-2.75
Opposite	-13	$-\frac{6}{7}$	2.75

To **simplify** a numerical expression such as $-(-33)$, you should replace the expression with its simplest name. $-(-33)$ means *the opposite of* -33. Therefore, $-(-33) = 33$.

EXAMPLE 3 Simplify: a. $-(-110)$ b. $-[-(-76)]$

a. $-(-110) = 110$ *The opposite of negative 110 is positive 110.*

b. $-[-(-76)] = -[76]$ *Remove the parentheses first.*
$ = -76$

Positive and negative numbers provide a useful way of representing opposite directions in real-life situations.

16 km above sea level $+16$, or 16   profit of $11,500 $+11,500$, or 11,500
3 km below sea level -3 loss of $160 -160

EXAMPLE 4 A temperature drop of 24.5° C is represented as -24.5. Write the opposite of this number and tell what it represents.

$$-(-24.5) = 24.5$$

This represents a temperature *rise* of 24.5° C.

By definition, the graphs of two numbers that are opposites are the same distance from the origin. When mathematicians are interested in distance only, and not direction, they use *absolute value*.

> The **absolute value** of a number is its distance from 0 on a number line. The absolute value of a number n is written $|n|$.

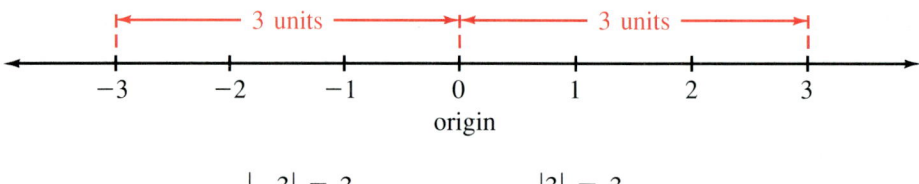

$|-3| = 3$ $|3| = 3$

EXAMPLE 5 Find each absolute value: **a.** $|15|$ **b.** $\left|-\frac{3}{4}\right|$ **c.** $|0|$

a. $|15| = 15$ **b.** $\left|-\frac{3}{4}\right| = \frac{3}{4}$ **c.** $|0| = 0$

Notice that the absolute value of a number is either positive or zero.

EXAMPLE 6 Simplify: $-|0.7|$

$-|0.7| = -(0.7)$ *Find the absolute value first.* $|0.7| = 0.7$
$ = -0.7$ *Then find the opposite. The opposite of 0.7 is -0.7.*

EXAMPLE 7 Simplify: $|6| + |-16|$

$|6| + |-16| = 6 + 16 = 22$

CLASS EXERCISES

1. How many rational numbers are there between 0 and 1?

Give the opposite of each number.

2. 19   **3.** -5 **4.** $-\frac{2}{3}$ **5.** $\frac{11}{8}$

Tell whether each statement is *true* or *false*.

6. $6 > 4$ **7.** $-3 < 5$ **8.** $-3 < -4$ **9.** $1 > 1$

10. $0 < -2$ **11.** $5.8 > 5.82$ **12.** $-\frac{7}{8} > -\frac{7}{9}$ **13.** $1.3 < 1.36$

14   Chapter 1 Real Numbers

For Discussion

14. Try the following definition of absolute value for some specific values of x. $|x| = x$ if $x > 0$, $|x| = -x$ if $x < 0$, or $|x| = 0$ if $x = 0$. Does this seem like a good definition of absolute value? Explain.

PRACTICE EXERCISES

Graph each group of numbers on a number line. Then list them in order from least to greatest.

1. $-4, 0, 3, 2, -1$
2. $5, 3, -3, -2, 1$
3. $2\frac{1}{2}, 2, -2\frac{1}{2}, -\frac{1}{2}, 0$
4. $-1.5, 1.5, -1, -0.5, 0$

Replace each $\underline{?}$ with $<$, $>$, or $=$ to make a true statement.

5. $-1 \underline{\ ?\ } 0$
6. $-3 \underline{\ ?\ } -4$
7. $\frac{1}{2} \underline{\ ?\ } \frac{1}{4}$
8. $\frac{2}{4} \underline{\ ?\ } \frac{3}{6}$
9. $\frac{3}{9} \underline{\ ?\ } \frac{2}{6}$
10. $-\frac{1}{4} \underline{\ ?\ } -\frac{1}{8}$
11. $-\frac{1}{3} \underline{\ ?\ } -\frac{1}{2}$
12. $-\frac{3}{5} \underline{\ ?\ } -\frac{3}{4}$

Simplify.

13. $-(-3.1)$
14. $-(-5.34)$
15. $-\left[-\left(-\frac{2}{5}\right)\right]$
16. $-\left[-\left(-\frac{1}{4}\right)\right]$
17. $|-6.2|$
18. $\left|\frac{1}{2}\right|$
19. $-\left|3\frac{1}{5}\right|$
20. $-\left|-\frac{5}{7}\right|$
21. $|-5| + |8|$
22. $|-15| + |29|$
23. $|-3| + |-2|$
24. $|9| + |-3|$

Write each set of numbers in order from least to greatest.

25. $\left\{\frac{1}{6}, \frac{1}{5}, \frac{1}{7}\right\}$
26. $\left\{-\frac{3}{5}, -\frac{3}{4}, -\frac{3}{8}\right\}$
27. $\left\{-1\frac{2}{3}, -\frac{5}{4}, -1\frac{3}{4}\right\}$
28. $\left\{1\frac{5}{6}, \frac{5}{3}, \frac{5}{2}\right\}$
29. $\{-0.104, -0.1, -0.1041\}$
30. $\{7.88, 7.885, 7.8809\}$

Simplify.

31. $-|-14|$
32. $\left|-4\frac{1}{5}\right|$
33. $-\left|-\left(-\frac{3}{8}\right)\right|$
34. $-[-(-5.23)]$
35. $-[21]$
36. $-[-50]$
37. $|-3| + |-9|$
38. $5 - |-4|$
39. $2 + |-(+5)|$

1.3 Comparing and Ordering Numbers **15**

Evaluate if $a = 7$, $b = 2\frac{3}{4}$, and $c = -6.1$.

40. $|c|$ **41.** $-|a|$ **42.** $|-c|$

43. $-|c|$ **44.** $|a + b|$ **45.** $|c| - |b|$

For all nonzero real numbers a, tell whether each statement is *true* or *false*.

46. $|a| > 0$ **47.** $a > -a$ **48.** $|a| = |-a|$ **49.** $-(-a) = a$

Applications

Use a positive or negative number to give the opposite of each situation. Then tell what the number represents.

50. Meteorology   16° C below freezing   **51. Meteorology**   20° C above freezing

52. Health   A gain of 6.5 lb   **53. Health**   A $3\frac{1}{4}$ lb weight loss

54. Navigation   120 ft above sea level   **55. Navigation**   55 ft below sea level

56. Economics   A $2560.38 debt   **57. Economics**   A profit of $785.95

PUZZLE

The following number pairs form a secret code. Read across the rows.

(19, 0)   (−10, 8)   (−6, −5)   (0, −16)
(−11, −9)   (−21, −19)   (−5, 0)   (−23, −21)
(0, 14)   (−4, 4)   (0, 5)   (−20, −18)
(−2, 0)   (20, −20)   (−8, −9)   (−16, 5)
(−17, 0)   (−8, 23)   (−7, −5)   (−20, 1)
(−21, 20)   (8, 0)   (−8, −5)   (−18, −19)

Why can't the meteorologist work? Here is the decoder. Use it to find the answer.

First Decoding
Use the greater of the two numbers in each pair. If the greater number is negative, then use its absolute value.

Second Decoding

0 = blank space   1 = A   2 = B   3 = C
4 = D   5 = E   6 = F   7 = G   8 = H
9 = I   10 = J   11 = K   12 = L   13 = M
14 = N   15 = O   16 = P   17 = Q   18 = R
19 = S   20 = T   21 = U   22 = V   23 = W
24 = X   25 = Y   26 = Z

1.4 Addition on a Number Line

Objective: To add numbers by "moves" on a number line

Positive and negative numbers can be added by using the concept of distance on a number line.

Capsule Review

On the number line below, segments are marked off in 1-unit lengths. The distance between points *A* and *B* is 1 unit, between *F* and *H* is 2 units, and between *J* and *C* is 7 units.

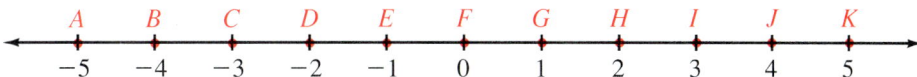

Find the distance between the given points.

1. *G* and *J*
2. *D* and *E*
3. *F* and *C*
4. *K* and *A*
5. *H* and *I*
6. *F* and *B*
7. *B* and *J*
8. *D* and *H*

Numbers can be represented by "moves" on a number line from one point to another. A move in the positive direction, from left to right, represents a **positive number.** A move in the negative direction, from right to left, represents a **negative number.**

A move of 4 units to the right from −5

A move of 2 units to the left from 4

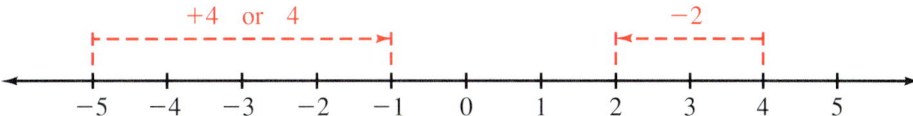

To add two numbers on a number line:

- Start at the origin, 0, and move the number of units and the direction indicated by the first number.
- From there, move the number of units and the direction indicated by the second number.
- The sum is the number that corresponds to the ending point.

EXAMPLE 1 Add: −3 + (−4)

Start at 0. Move left to −3. Then move four units to the left.

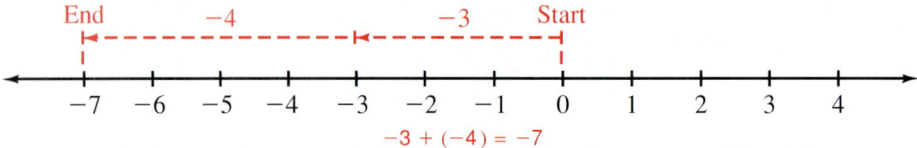

$-3 + (-4) = -7$

In Example 1, negative 4 is added to negative 3. Notice that parentheses are used so that the negative sign of the number to be added is not confused with the operational symbol (+ sign). When a positive number is added, parentheses are not necessary.

EXAMPLE 2 Add: −4 + 7

Start at 0. Move left to −4. Then move 7 units to the right.

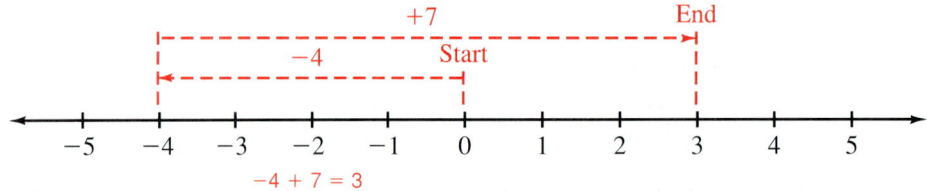

$-4 + 7 = 3$

EXAMPLE 3 Add: −2 + 0

Add 0 means *move no units*.

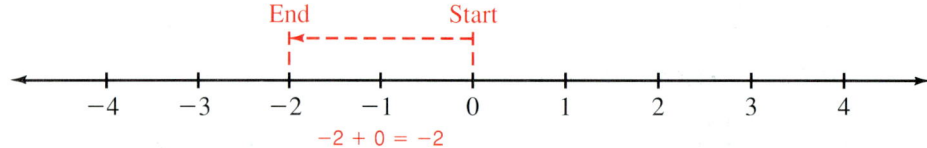

$-2 + 0 = -2$

Example 3 illustrates the special property of zero for the addition of real numbers.

Identity Property for Addition

For every real number n, there is exactly one real number 0 such that

$$n + 0 = n$$

and

$$0 + n = n$$

From this you can see that when zero is one of the addends, the sum is the other addend.

18   Chapter 1 Real Numbers

EXAMPLE 4 A weather balloon was reported to be 4.5 mi east of the weather station (+4.5). It then moved 7 mi west from this position. How far from the weather station was the balloon then?

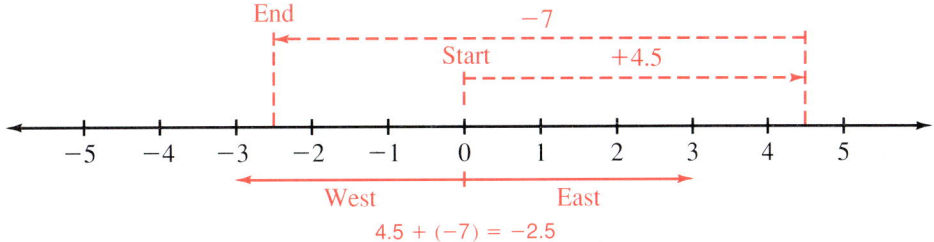

$4.5 + (-7) = -2.5$

The balloon was 2.5 mi west of the weather station.

It is possible to add more than two numbers on a number line.

EXAMPLE 5 Add: $5 + (-8) + (-2) + 7$

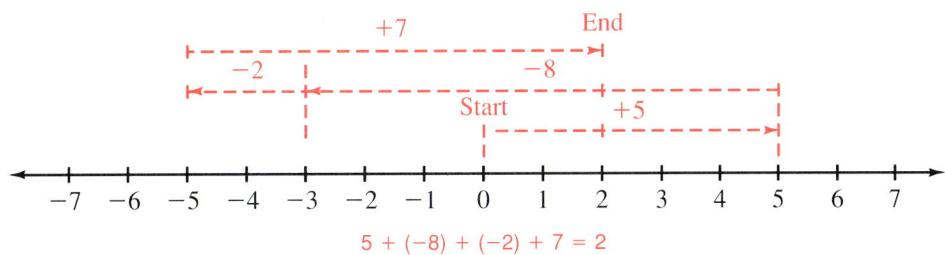

$5 + (-8) + (-2) + 7 = 2$

CLASS EXERCISES

Write an addition sentence for each diagram.

1.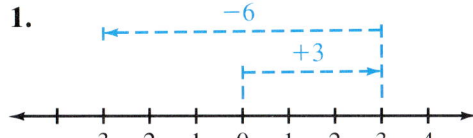

2.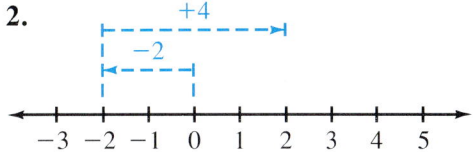

3.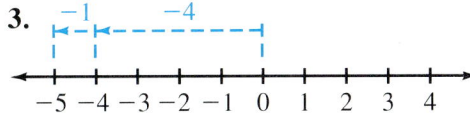

4.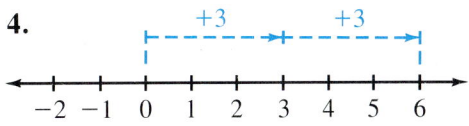

5.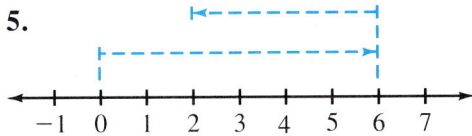

6.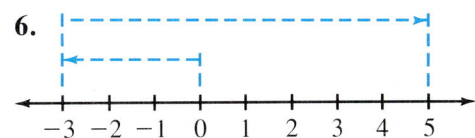

1.4 Addition on a Number Line

State whether the sum is positive, negative, or zero.

7. $-4 + (-2)$
8. $3 + (-9)$
9. $-1.5 + 2$
10. $-0.5 + 0.5$
11. $1\frac{1}{2} + \left(-\frac{1}{2}\right)$
12. $-\frac{2}{3} + \frac{1}{2}$

PRACTICE EXERCISES

Find each sum. Use the number line below to help you.

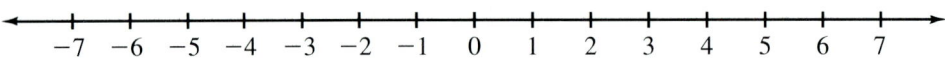

1. $-2 + (-3)$
2. $-4 + (-3)$
3. $-4 + (-2)$
4. $-7 + (-14)$
5. $-5 + 6$
6. $-7 + 4$
7. $-3 + (-4)$
8. $-9 + (-8)$
9. $-8 + 0$
10. $0 + (-6)$
11. $1.5 + (-5)$
12. $3.7 + (-8)$
13. $-4.5 + (-3.5)$
14. $5 + (-2) + (-4)$
15. $-3 + 2 + (-5)$
16. $-4 + 3 + (-8)$
17. $-3 + (-2) + 6$
18. $-8 + 2 + (-5)$
19. $8 + (-4) + 2$
20. $8 + (-9) + 3$
21. $-7 + 2 + (-1)$
22. $-5 + 7 + (-1)$
23. $-10 + (-2) + (-5)$
24. $-16 + (-3) + (-4)$
25. $-2.5 + (-5.5) + 1.5$
26. $1.5 + (-3.5) + 4.5$
27. $1\frac{1}{2} + \left(-2\frac{1}{2}\right) + \left(-1\frac{1}{2}\right)$
28. $3\frac{1}{4} + \left(-2\frac{1}{4}\right) + 1\frac{3}{4}$
29. $\frac{5}{3} + \left(-\frac{2}{3}\right) + \frac{4}{3}$
30. $\left(-\frac{9}{2}\right) + \left(-\frac{5}{2}\right) + \frac{3}{2}$

Replace $\underline{?}$ with $<$, $>$, or $=$ to make a true statement.

31. $-16.5 + (-4.3) + 4.3 \underline{} -16.3 + (-4.5) + 4.5$
32. $0.25 + (-0.9) + 16 \underline{} 0.9 + (-0.25) + (-16)$
33. $\left|-\frac{5}{8} + \frac{3}{7}\right| \underline{} \left|-\frac{5}{7} + \frac{3}{8}\right|$
34. $\left|\frac{2}{9} + (-1)\right| \underline{} \left|1 + \left(-\frac{2}{9}\right)\right|$

Applications

Write an addition expression for each situation. Then simplify.

35. **Meteorology** The temperature after a drop of 17° from 5° C

36. **Number Problem** The floor at which the elevator stopped if the detective followed the suspect into the elevator on the 24th floor and then the elevator went down 11 floors before the door opened

37. **Sports** The net loss if a football team lost 3 yd on each of 4 successive plays

20 Chapter 1 Real Numbers

38. Economics The net change if Maureen earned $25, spent $12 of it, and then earned $10 more

39. Economics The net change on the stock market for the SUP Corporation if its stock dropped $3\frac{1}{8}$ points on Monday and gained $4\frac{5}{8}$ points on Tuesday

BIOGRAPHY: Carl Friedrich Gauss

Carl Friedrich Gauss was born in Brunswick, Germany, on April 30, 1777. Before he was 3 years old, he was demonstrating his genius. At the age of 10, Gauss began to show an unusual ability in arithmetic. As a result of the encouragement and assistance of an able teacher, he was working with proofs in algebra before he was 11 years old.

There is a well-known story about Gauss at this age. His arithmetic teacher, hoping to keep the class busy for some time, told the students to find the sum of the first one hundred positive integers. To the teacher's astonishment, young Gauss found the correct answer within a few minutes. His method was something like this:

Write out the sum both forward and backward, lining up the terms.

$$1 + 2 + 3 + \cdots + 98 + 99 + 100$$
$$100 + 99 + 98 + \cdots + 3 + 2 + 1$$

Adding each vertical pair, you get a sum of 101. There are one hundred sums.

100(101) is two times the correct answer.

$$\frac{100(101)}{2} = 50(101) = 5050$$

The answer is 5050.

The system works just as well for adding a million numbers, provided the numbers are all equally spaced.

Use the Gauss method.

1. Find the sum of the first 50 positive integers.
2. Find the sum of the ten positive integers, 31 through 40.

1.5 Adding Real Numbers

Objectives: To add two or more real numbers
To evaluate algebraic expressions involving addition

Sums such as $-436 + 6347$ or $-2\frac{2}{7} + 4\frac{1}{6}$ are not easily found on a number line. If you know the specific rules for adding two or more numbers, a number line is not needed. These rules involve the concept of absolute value.

Capsule Review

The absolute value of a number is its distance from zero on a number line. So, $|15| = 15$, $|-15| = 15$, and $|0| = 0$.

Which absolute value is greater?

1. $|37|$ or $|-38|$ **2.** $|-6.2|$ or $|6.15|$ **3.** $|0.56|$ or $|0.07|$

A rule that can be used to add two positive numbers or two negative numbers is given below.

Addition Rule (Two numbers with the same sign)

To add two numbers that have the same sign, either both positive or both negative:
- Add their absolute values.
- Give the sum the same sign as the sign of each number.

EXAMPLE 1 Add: $-13 + (-26)$

Each number is negative. Add the absolute values.

$-13 + (-26)$ $|-13| + |-26| = 13 + 26$, or 39.
$= -39$ *The sum is negative.*

EXAMPLE 2 Add: $0.3 + 2.634$

Each number is positive. Add the absolute values.

$0.3 + 2.634$ $|0.3| + |2.634| = 0.3 + 2.634$, or 2.934.
$= 2.934$ *The sum is positive.*

Chapter 1 Real Numbers

Absolute value is also used to add two numbers that have different signs.

> **Addition Rule (Two numbers with different signs)**
>
> To add two numbers that have different signs:
> - Find their absolute values. Subtract the smaller absolute value from the larger.
> - Give the result the same sign as the sign of the number with the larger absolute value.

EXAMPLE 3 Add: $14 + (-72)$

Notice that one number is positive, the other is negative.

$14 + (-72)$ *Subtract the absolute values: $|-72| - |14| = 72 - 14$, or 58.*
$= -58$ *The sum is negative since $|-72| > |14|$.*

The procedure is the same for other rational numbers.

EXAMPLE 4 Add: $-\frac{3}{4} + \frac{7}{2}$

$-\frac{3}{4} + \frac{7}{2}$ *Subtract the absolute values: $\left|\frac{7}{2}\right| - \left|-\frac{3}{4}\right| = \frac{14}{4} - \frac{3}{4} = \frac{11}{4}$.*

$= \frac{11}{4}$, or $2\frac{3}{4}$ *The sum is positive since $\left|\frac{7}{2}\right| > \left|-\frac{3}{4}\right|$.*

To add several numbers that have different signs, you can add the positive and negative numbers separately, and then add the two sums.

EXAMPLE 5 Add: $-4 + 18 + (-40) + 2$

Add the positive numbers.	Add the negative numbers.
$\|18\| = 18 \qquad 18$	$\|-4\| = 4 \qquad 4$
$\|2\| = 2 \qquad +\ 2$	$\|-40\| = 40 \qquad +40$
$\qquad\qquad\quad 20$	$\qquad\qquad\quad 44$
Sum is positive: 20	Sum is negative: -44

Subtract the absolute values: $|-44| - |20| = 44 - 20 = 24$. The sum is negative since $|-44| > |20|$. So, $-4 + 18 + (-40) + 2 = -24$.

Another method to add more than two numbers when some of the numbers are positive and some are negative is to add the numbers in the order in which they occur.

1.5 Adding Real Numbers

EXAMPLE 6 Evaluate $-[5 + x] + y + z$ if $x = -\frac{1}{2}$, $y = 3\frac{1}{2}$, and $z = -4$.

$-[5 + x] + y + z$
$= -\left[5 + \left(-\frac{1}{2}\right)\right] + 3\frac{1}{2} + (-4)$ Substitute $-\frac{1}{2}$ for x, $3\frac{1}{2}$ for y, and -4 for z.
$= -\left[4\frac{1}{2}\right] + 3\frac{1}{2} + (-4)$ Add within the brackets.
$= -4\frac{1}{2} + 3\frac{1}{2} + (-4)$
$= -1 + (-4)$ $-4\frac{1}{2} + 3\frac{1}{2} = -1$
$= -5$

The sum of any two given real numbers is always another real number. Also, the sum is *unique*. That is, there is *one and only one* real number answer. These two ideas are summarized in a special property for real numbers.

Closure Property for Addition

For all real numbers m and n,

$m + n$ is a unique real number.

CLASS EXERCISES

Add, using the method that is easier for you.

1. $38 + (-28) + 15 + (-4)$
2. $-13 + 28 + (-14) + (-5)$
3. $-60 + 22 + (-40) + 72 + 6$
4. $-3\frac{1}{4} + 3\frac{1}{8} + \left(-2\frac{1}{2}\right) + (-6)$

For Discussion

5. Can any two real numbers be added by using a number line? Why are rules needed for addition?

PRACTICE EXERCISES

Add.

1. $-199 + (-301)$
2. $-86 + (-127)$
3. $649 + (-1102)$
4. $5851 + (-4998)$
5. $7.08 + (-2.47)$
6. $-25.31 + 75.07$

24 Chapter 1 Real Numbers

7. $-13.689 + 11.98$
8. $0.234 + (-1.07)$
9. $\frac{5}{7} + \left(-\frac{9}{7}\right)$
10. $-\frac{2}{9} + \frac{8}{9}$
11. $-\frac{3}{8} + \left(-\frac{1}{4}\right)$
12. $-\frac{5}{9} + \left(-\frac{1}{3}\right)$
13. $-\frac{3}{7} + \frac{1}{2}$
14. $\frac{5}{6} + \left(-\frac{2}{5}\right)$
15. $-17 + 27 + (-10)$
16. $55 + (-19) + 4$
17. $5.2 + (-6.7) + 10.1$
18. $-20.5 + 4.3 + (-7)$

Evaluate each expression if $a = -5$, $b = 2\frac{1}{2}$, $c = -\frac{1}{2}$, and $d = 15$.

19. $a + (-b) + c$
20. $b + (-d) + (-a)$
21. $-a + b + (-c)$
22. $-b + d + a$

Add.

23. $-\frac{3}{8} + \frac{7}{12} + \frac{1}{2}$
24. $\frac{5}{6} + \left(-\frac{2}{9}\right) + \left(-\frac{1}{3}\right)$
25. $2\frac{3}{25} + \left(-\frac{7}{15}\right)$
26. $\frac{4}{15} + \left(-3\frac{5}{12}\right)$
27. $7.3 + 12.12 + (-3.5) + 8.06$
28. $-28.5 + (-11.6) + 5.91 + 9.31$
29. $\frac{2}{3} + \left[3 + \left(-\frac{1}{3}\right)\right]$
30. $-\left[\frac{7}{8} + (-2)\right] + \frac{3}{8}$
31. $-(-3 + 11) + [9 + (-14)]$
32. $-[37 + (-15)] + (-8 + 14)$
33. $-|5 + (-7) + (-23)|$
34. $-|13 + (-3) + (-19)|$
35. $82 + [-|8 + (-31)|]$
36. $-|-17 + 28| + (-15)$
37. $|2 + (-6)| + |-30 + 19|$
38. $-|14 + (-30)| + |-17 + 3|$

Evaluate each expression if $r = -22$, $s = 7$, and $t = -8$.

39. $|-r| + [-|t + (-s)|]$
40. $-|s + (-t)| + |r|$
41. $-|t + (-r)| + |s + r|$
42. $|-s + (-r)| + [-|r + (-t)|]$

Applications

For each problem, state a sum. Then answer the question.

43. **Meteorology** The temperature at 8:00 AM was 19° C. By noon it had risen 6°, but by 4:00 PM it had fallen 8°. What was the temperature at 4:00 PM?

44. **Sports** On four successive plays, a football team gained 7 yd, lost 11 yd, lost 6 yd, and gained 15 yd. What was the total yardage?

45. Aviation A powered airplane towed a glider into the air to an altitude of 1000 m and then let it go. The glider dropped 60 m into a thermal (rising bubbles of warm air), which took it up 2200 m. Then it glided down 200 m into another thermal. Then the glider rose 1700 m. What was the altitude of the glider?

46. Finance Dee had $300.30 in her checking account. She wrote checks for $35 and $106.15 and deposited $20 and $43.75. What is the new balance in her account?

Calculator The sign change key, +/−, is used to enter negative numbers. To find the sum 25 + (−7) on a calculator, enter positive 7 first and then the sign change key to make it negative. Then add 25.

Find each sum using the sign change key on your calculator.

47. 23 + (−8)   **48.** 96 + (−33)   **49.** 5.83 + (−0.76)

50. 0.04 + (−98.9)   **51.** −29 + 101   **52.** −14.629 + (−9.36)

LOGICAL REASONING

Three key words in logical reasoning are *always* (every time), *sometimes* (at least once), and *never* (not even once).

Tell whether each of the following sentences is *always true*, *sometimes true*, or *never true*.

1. The sum of two positive numbers is positive.
2. The sum of two negative numbers is positive.
3. The sum of a positive number and a negative number is positive.
4. If two numbers are both positive or both negative, then the absolute value of their sum equals the sum of their absolute values.
5. If one number is positive and one number is negative, then the absolute value of their sum equals the sum of their absolute values.
6. If the absolute value of one number is equal to the absolute value of another, then the sum of the two numbers is 0.

1.6 Subtracting Real Numbers

Objectives: To subtract real numbers
To evaluate algebraic expressions involving subtraction of real numbers

You have learned to add real numbers and to find the opposite of a number. These two concepts together can be used to provide a rule for the subtraction of real numbers.

Capsule Review

Two numbers whose graphs are the same distance from the origin, one number positive and the other negative, are called *opposites*. The opposite of -4.5 is 4.5. The opposite of $2\frac{1}{2}$ is $-2\frac{1}{2}$.

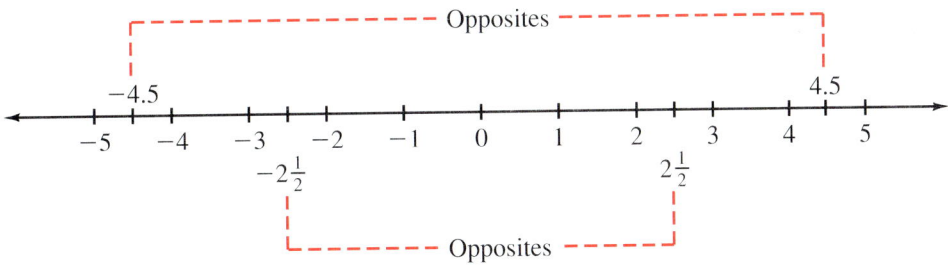

Name the opposite of each number.

1. -6
2. 15
3. $-\frac{1}{4}$
4. $-4\frac{1}{3}$
5. 13.09

Another name for opposite is **additive inverse**. Notice what happens when a number and its additive inverse are added.

Number	Additive Inverse	Sum
-5	5	$-5 + 5 = 0$
-0.6	0.6	$-0.6 + 0.6 = 0$

> ### Additive Inverse Property
> For every real number n, there is exactly one real number $-n$, such that
> $$n + (-n) = 0 \quad \text{and} \quad -n + n = 0$$

1.6 Subtracting Real Numbers

The additive inverse of a negative number is a positive number. For example, $-(-4)$ means "the additive inverse of negative 4," which is positive 4, or 4.

EXAMPLE 1 Simplify: **a.** $-(-2.1)$ **b.** $-\left(\frac{1}{3}\right)$

a. $-(-2.1) = 2.1$ **b.** $-\left(\frac{1}{3}\right) = -\frac{1}{3}$

The following definition may be used to subtract real numbers.

Definition of Subtraction

For all real numbers m and n,
$$m - n = m + (-n)$$

To subtract a number, add its opposite, or additive inverse. Any subtraction problem may be written as an addition problem. Then the rules for adding numbers are used.

EXAMPLE 2 Subtract: $3 - 9$

$3 - 9$ means "positive 3 *subtract* positive 9."
$3 - 9 = 3 + (-9)$ *Change to addition. The additive inverse of 9 is -9.*
$ = -6$

The difference is negative since $|-9| > |3|$.

EXAMPLE 3 Subtract: $5.6 - (-6.02)$

$5.6 - (-6.02)$ means "positive 5.6 *subtract* negative 6.02."
$5.6 - (-6.02) = 5.6 + 6.02$ *Change to addition. The additive inverse of -6.02 is 6.02.*
$ = 11.62$

The difference is positive since $|-6.02| > |5.6|$.

EXAMPLE 4 Subtract: $-(12 - 89) - (-9 - 23)$

Operations within parentheses are done first.
$-(12 - 89) - (-9 - 23) = -[12 + (-89)] - [-9 + (-23)]$
$ = -[-77] - [-32]$
$ = 77 + 32 = 109$

Explain why brackets are needed in Example 4.

Chapter 1 Real Numbers

EXAMPLE 5 Subtract: $-\frac{2}{3} - \left(-\frac{4}{5}\right)$

$-\frac{2}{3} - \left(-\frac{4}{5}\right) = -\frac{2}{3} + \frac{4}{5}$ *Change to addition.*

$\qquad = -\frac{10}{15} + \frac{12}{15}$ *Use a common denominator.*

$\qquad = \frac{2}{15}$ *The difference is positive since $\left|\frac{12}{15}\right| > \left|-\frac{10}{15}\right|$.*

EXAMPLE 6 Evaluate $m - (n - 7)$ if $m = 4.6$ and $n = -3.9$.

$m - (n - 7) = 4.6 - (-3.9 - 7)$ *Substitute 4.6 for m and -3.9 for n.*
$\qquad = 4.6 - [-3.9 + (-7)]$ *Within the brackets, change the*
$\qquad = 4.6 - [-10.9]$ *operation to addition: $-3.9 - 7$*
$\qquad = 4.6 + 10.9 = 15.5$ *means $-3.9 + (-7)$.*

CLASS EXERCISES

Write each subtraction expression as an addition expression. Then add.

1. $25 - 8$
2. $9 - 15$
3. $-6 - 19$
4. $17 - (-8)$

Subtract.

5. $-6 - 8 - 15$
6. $-14 - 7 - 5 - 6$
7. $-2 - (-8) - (+1)$

PRACTICE EXERCISES

Simplify.

1. $-(-4.1)$
2. $-\left(-\frac{3}{4}\right)$
3. $-(0.75)$
4. $-\left(1\frac{2}{3}\right)$

Subtract.

5. $10 - 18$
6. $110 - 125$
7. $6 - 19$
8. $23 - 30$
9. $1.4 - (-2.3)$
10. $2.5 - (-7.4)$
11. $1.5 - (-8)$
12. $4.2 - (-1.5)$
13. $-(10 - 15) - (-12)$
14. $-(7) - (-7 - 15)$
15. $-(2 - 7) - (-13)$
16. $-(13 - 90) - 10 - 24$
17. $-\frac{1}{9} - \left(-\frac{8}{9}\right)$
18. $\frac{2}{3} - \left(-\frac{1}{3}\right)$
19. $13 - (71 - 104)$
20. $-91 - (87 - 215)$
21. $-(25 - 88) - 36$

Evaluate each expression if $a = -6.7$, $b = 11.5$, $c = -4.9$, and $d = 15.2$.

22. $a - (b - 10)$
23. $b - (d - 4)$
24. $(c - a) - (d - b)$
25. $(a - c) - (b - d)$
26. $[-b - (-d)] - d$
27. $-c - [a - (-d)]$

Subtract. When several grouping symbols are used, simplify the expression in the innermost grouping first.

28. $-\frac{1}{2} - \frac{1}{3} - \left(-\frac{1}{3}\right)$
29. $-\left(\frac{3}{4}\right) - \frac{1}{4} - \frac{5}{8}$
30. $-0.26 - 5.3 - (-0.87)$
31. $-1.9 - 4 - (-0.25)$
32. $-5\frac{1}{2} - \left[6 - \left(-1\frac{1}{2}\right)\right]$
33. $\left(7\frac{1}{3} - 5\frac{2}{3}\right) - 16$
34. $-\left(9\frac{3}{8} - 1\right) - \left(-\frac{1}{4} - 2\frac{1}{8}\right)$
35. $\left[11\frac{1}{5} - (-14)\right] - \left(3\frac{1}{10} - 8\frac{4}{5}\right)$
36. $\{-7 - [3 - (-11 - 24)]\}$
37. $-41 - \{-(6 - [(-23) + 5])\}$
38. $|-8 - (-3 + 6)| - |17 + (-2)|$
39. $|28 - (-16)| + |-51 - (-31)|$

Evaluate each expression if $x = -3.2$, $y = 7.8$, and $z = -4.7$.

40. $3 - [(y - x) - (z + 1.4)]$
41. $-14.2 - [z - (-y) + (3.8 - x)]$
42. $z - |(1.2 - x) - [-(8.5 - y)]|$
43. $z - |[x - (-11)] - (13.1 - y)|$

Applications

Computer The ABS function on a computer, ABS(X), always gives the nonnegative value of X. For example, ABS(3.4) = 3.4, ABS(−7.05) = 7.05 and ABS(0) = 0. State the value of each expression.

44. ABS(−3)
45. ABS(6 − 9)
46. ABS(9 − 6)
47. −ABS(−5)
48. ABS(−6.1 − (−5.7))
49. ABS(−6.1) − ABS(−5.7)

TEST YOURSELF

Evaluate each expression if $a = 0.4$ and $b = 8$.

1. $21.9 + a$
2. $2ab$
3. $10(b - a)$ 1.1

Graph each set of numbers.

4. $\{-3, -2.5, 0, 2.5, 3.5\}$
5. {negative integers}   1.2

Replace each ? with <, >, or = to make a true statement.

6. $-27 \underline{\ ?\ } -28$
7. $\frac{5}{8} \underline{\ ?\ } \frac{5}{9}$
8. $-0.42 \underline{\ ?\ } -0.425$ 1.3

Add or subtract, as indicated.

9. $-5.5 + 4.5$
10. $\frac{1}{4} + (-2)$
11. $-8 + \left(-2\frac{1}{2}\right) + 15$ 1.4, 1.5
12. $-27 - (-30)$
13. $-0.82 - 3.6$
14. $[59 - (-23)] - (-24)$ 1.6

1.7 Multiplying Real Numbers

Objectives: To multiply real numbers
To evaluate algebraic expressions involving multiplication of real numbers

A stadium holds 80,000 people. If tickets to a concert in the stadium cost $12.50 each, what is the ticket sales maximum?

$$80,000 \cdot 12.50 = 1,000,000$$

The ticket sales maximum is $1,000,000.

Both 80,000 and 12.50 are positive numbers. The product of any two positive numbers is always a positive number. What happens when you multiply a positive and a negative number or two negative numbers?

Capsule Review

EXAMPLES   **a.** $12.50 \cdot 80,000 = 1,000,000$ **b.** $\frac{3}{5} \cdot \frac{6}{7} = \frac{18}{35}$

(positive · positive)   (positive · positive)

Find each product.

1. $56 \cdot 21$
2. $187 \cdot 100$
3. $(4.1)(1.5)$
4. $(0.49)2$
5. $1000(0.04)$
6. $\frac{1}{2} \cdot \frac{2}{3}$
7. $7 \cdot 1\frac{1}{7}$
8. $\frac{2}{3} \cdot 2\frac{2}{3}$
9. $\frac{5}{8} \cdot 5\frac{2}{6}$
10. $2\frac{1}{2} \cdot \frac{6}{9}$
11. $2\frac{4}{7} \cdot 4\frac{2}{3}$
12. $6\frac{3}{4} \cdot 2\frac{2}{3}$

Study the addition sentences below. What procedure do they suggest for multiplying a positive number and a negative number?

$$\underbrace{-3 + (-3) + (-3) = -9}_{3(-3)} \qquad \underbrace{-2 + (-2) + (-2) + (-2) = -8}_{4(-2)}$$

Multiplication Rule (Two numbers with different signs)

To multiply a positive and negative number, multiply their absolute values. The product is negative.

EXAMPLE 1 Multiply: $-\dfrac{1}{5} \cdot \dfrac{5}{9}$

$-\dfrac{1}{5} \cdot \dfrac{5}{9}$ Multiply the absolute values: $\left|-\dfrac{1}{5}\right| \cdot \left|\dfrac{5}{9}\right| = \dfrac{1}{\cancel{5}} \cdot \dfrac{\cancel{5}}{9} = \dfrac{1}{9}$

$= -\dfrac{1}{9}.$ The product is negative since the numbers have different signs.

A rule for multiplying two positive numbers or two negative numbers is stated below.

Multiplication Rule (Two numbers with the same sign)

To multiply two positive or two negative numbers, multiply their absolute values. The product is positive.

EXAMPLE 2 Multiply: $-13(-12)$

$-13(-12)$ The numbers have the same sign. Multiply the absolute values: $|-13| \cdot |-12| = 13 \cdot 12 = 156.$
$= 156$ The product is positive.

Multiplying a number by 1 does not change its value. The product of any number and 0 is 0.

Identity Property for Multiplication

For every real number n, $n \cdot 1 = n$ and $1 \cdot n = n$.

Property of Zero for Multiplication

For every real number n, $n \cdot 0 = 0$ and $0 \cdot n = 0$.

When multiplying more than two numbers, the multiplication can be done in any order.
- If there is an even number of negative numbers, the product is positive.
- If there is an odd number of negative numbers, the product is negative.

EXAMPLE 3 Multiply: $-7(-1)(4)(-3)(2)(-5)$

There are four negative numbers. The product is positive.
$-7(-1)(4)(-3)(2)(-5) = 840$

EXAMPLE 4 Evaluate $a \cdot b \cdot c$ if $a = -5$, $b = -2$, and $c = -3$.

$a \cdot b \cdot c = -5(-2)(-3)$ Substitute -5 for a, -2 for b, and -3 for c.
$= -30$ There are 3 negative numbers. The product is negative.

The product of any two given real numbers is always another real number. Also, the product is *unique*. There is *one and only one* real-number answer.

Closure Property for Multiplication

For all real numbers m and n, $m \cdot n$ is a unique real number.

CLASS EXERCISES

Multiply.

1. $-2 \cdot 1$
2. $4(-4)$
3. $0.5(-0.6)$
4. $-19{,}642 \cdot 0$
5. $-\frac{1}{2}\left(-\frac{5}{9}\right)$
6. $-\frac{4}{5} \cdot \frac{5}{8}$
7. $-\frac{2}{3} \cdot \frac{1}{9}$
8. $(-36)\left(-\frac{2}{9}\right)$

PRACTICE EXERCISES

Multiply.

1. $-15 \cdot 2$
2. $-12 \cdot 10$
3. $-2.34 \cdot 0.2$
4. $-0.005 \cdot 0.00$
5. $(-5)(-13)$
6. $-9(-11)$
7. $-0.6(-3.7)$
8. $-0.02(-0.08)$
9. $-6(10)\left(-\frac{1}{2}\right)$
10. $-\frac{3}{5}(-20)(-4)$
11. $(-50)\left(-\frac{3}{2}\right)\left(-\frac{3}{5}\right)$
12. $-24\left(\frac{0}{6}\right)\left(-\frac{1}{20}\right)$

Evaluate each expression if $a = -4$, $b = 3$, $c = -10$, $d = 5$, and $e = -6$.

13. $a \cdot b \cdot c$
14. $c \cdot d \cdot b$
15. $c \cdot (-a)(-d)$
16. $-b(-a)(-d)$
17. $-a(d)(-c)e$
18. $b(-a)(-c)(-d)$
19. $c(-b)ed$
20. $-e(-b)(-d)$

Multiply.

21. $-14\left[\frac{3}{7}\left(-\frac{1}{9}\right)(-27)\right]$
22. $\left[\frac{4}{5}(-15)\right]\left[\frac{3}{16}(-20)\right]$
23. $-8(-2)(0.5)(3)(-0.25)$
24. $32(-0.25)(-0.25)(20)(-10)$
25. $-37{,}037(-1.5)$
26. $-33.67(-0.0132)$
27. $-1.125(0.4)(24{,}691.358)$
28. $-5291(-2.1)(0.7)$
29. $-7(0.66)(-48.2)(-7)$
30. $0.0156(-3.4)(-5.5)(99)$

1.7 Multiplying Real Numbers

Evaluate each expression if $r = -6$, $s = -2$, $t = -3$, $v = 10$, and $w = -\frac{1}{2}$.

31. $(rs) - t$
32. $v + (sw)$
33. $(rs) - (vw)$
34. $(tr) - (vs)$
35. $-t(r - s)$
36. $(s - t)vw$
37. $-|rw| \cdot st$
38. $r \cdot |w - v|$
39. $|rw| - |st|$
40. $wv|-t - s|$
41. $|w(v + s)|$
42. $vs|(r - v)(-w)|$

Applications

43. **Weather** During a 1 hour and 30 minute period, a drop in temperature of 7° was recorded. Write a numerical expression to show a continuous drop in temperature at the same rate for 3 hours.

44. **Finance** A computer company lost $4.5 million each of the last three years. Use positive and negative numbers to express the total amount lost.

ALGEBRA IN METEOROLOGY

When you lose body heat, you feel cold. A low temperature can cause a loss of body heat. So can the wind. A combination of low temperature and wind makes it feel colder than the actual temperature. The table shows how cold you feel when a low temperature is combined with a wind.

Wind (mi/h)	Temperature in Degrees Fahrenheit										
	35	30	25	20	15	10	5	0	−5	−10	−15	−20
5	33	27	21	16	12	7	0	−5	−10	−15	−21	−26
10	22	16	10	3	−3	−9	−15	−22	−27	−34	−40	−46
15	16	9	2	−5	−11	−18	−25	−31	−38	−45	−51	−58
20	12	4	−3	−10	−17	−24	−31	−39	−46	−53	−60	−67
25	8	1	−7	−15	−22	−29	−36	−44	−51	−59	−66	−74

Example: The temperature is 25°F. The wind is blowing at 20 mi/h. How cold does it feel to your body?

Solution: Look across the top of the table to find the column for 25°F. Look down the left to find the row for 20 mi/h. At the intersection of the column and the row is the answer, −3°F.

Suppose the temperature drops from 20°F to −10°F while the wind speed remains at 25 mi/h. How many degrees does the actual temperature drop? How many degrees does your body feel that it has dropped?

1.8 Dividing Real Numbers

Objectives: To divide real numbers
To evaluate algebraic expressions involving multiplication and division of real numbers

Multiplication and division are opposite (or inverse) operations. You will see that the rules for determining the signs for the quotients of real numbers are the same as those for their products.

Capsule Review

The product of two positive numbers is positive. $(2.5) \cdot 3 = 7.5$
The product of two negative numbers is positive. $-3(-16) = 48$
The product of a positive number and a negative number is negative.

$$\left(-\frac{2}{3}\right)\left(\frac{1}{3}\right) = -\frac{2}{9}$$

The product of any number and 0 is 0. $-21 \cdot 0 = 0$

State whether the product is positive, negative, or zero.

1. $-5(-1)$
2. $40(0.256)$
3. $-9 \cdot \frac{1}{9}$
4. $0(3.6)$
5. $\left(\frac{2}{5}\right)\left(-\frac{1}{5}\right)$
6. $-25(-25)$
7. $\left(-\frac{5}{6}\right)\left(-\frac{6}{5}\right)$
8. $-1\frac{1}{3} \cdot 0$

The rules for dividing real numbers involve the mathematical concept of **reciprocals**. Two numbers whose product is 1 are called **reciprocals** or **multiplicative inverses** of each other.

Multiplicative Inverse Property (Reciprocal Property)

For every nonzero real number n, there is exactly one real number $\frac{1}{n}$, such that

$$n \cdot \frac{1}{n} = 1 \quad \text{and} \quad \frac{1}{n} \cdot n = 1$$

n and $\frac{1}{n}$ are reciprocals, or multiplicative inverses, of each other.

Zero does not have a reciprocal since the product of any number and 0 is 0, not 1.

EXAMPLE 1 Find the reciprocal of 9, $-\frac{1}{3}$, and -0.3.

Number	Reciprocal	
9, or $\frac{9}{1}$	$\frac{1}{9}$	$9 \cdot \frac{1}{9} = 1$
$-\frac{1}{3}$	$-\frac{3}{1}$, or -3	$-\frac{1}{3}(-3) = 1$
-0.3, or $-\frac{3}{10}$	$-\frac{10}{3}$	$-\frac{3}{10}\left(-\frac{10}{3}\right) = 1$

The following definition may be used to divide numbers.

Definition of Division

For all real numbers m and n, $n \neq 0$, $m \div n = m \cdot \frac{1}{n}$.

Since division by a number is defined as multiplying by the reciprocal or multiplicative inverse of the number, division can be rewritten as multiplication. Then the rules for multiplying real numbers can be used to decide if the quotient is positive or negative.

EXAMPLE 2 Divide: $\frac{-81}{3}$

$-81 \div 3 = -81 \cdot \frac{1}{3}$ *Use the reciprocal of 3 and rewrite as multiplication.*

$ = -\left(|81| \cdot \left|\frac{1}{3}\right|\right)$ *Multiply the absolute values. The quotient is negative.*

$ = -27$

EXAMPLE 3 Divide: $-1\frac{1}{2} \div \left(-\frac{9}{7}\right)$

$-1\frac{1}{2} \div \left(-\frac{9}{7}\right) = \left(-1\frac{1}{2}\right)\left(-\frac{7}{9}\right)$ *Rewrite as multiplication.*

$\phantom{-1\frac{1}{2} \div \left(-\frac{9}{7}\right)} = \left(\left|-1\frac{1}{2}\right| \cdot \left|-\frac{7}{9}\right|\right)$ *Multiply the absolute values. The quotient is positive.*

$\phantom{-1\frac{1}{2} \div \left(-\frac{9}{7}\right)} = \left(\frac{3}{2} \cdot \frac{7}{9}\right)$

$\phantom{-1\frac{1}{2} \div \left(-\frac{9}{7}\right)} = \frac{7}{6}$

The algebraic expression $r[2 \div (-s)]$ involves the operations of multiplication *and* division. After substituting given values for the variables, the operation within the brackets (division) is done first.

EXAMPLE 4 Evaluate $r[2 \div (-s)]$ if $r = 4$ and $s = -3$.

$$r[2 \div (-s)]$$
$$= 4[2 \div 3] \qquad \text{Substitute 4 for } r \text{ and } -3 \text{ for } s.\ -s = -(-3) = 3.$$
$$= 4\left[2 \cdot \frac{1}{3}\right]$$
$$= 4\left[\frac{2}{3}\right]$$
$$= \frac{8}{3}$$

Since 0 has no reciprocal, you cannot divide a number by 0. $\frac{1}{0}$ is undefined. However, it *is* possible to divide 0 by any real number other than 0. The quotient is always 0.

$$\frac{0}{5} = 0 \cdot \frac{1}{5} = 0 \qquad 0 \div \left(-\frac{1}{6}\right) = 0 \cdot (-6) = 0$$

EXAMPLE 5 Divide, if possible: a. $\frac{0}{3}$ b. $\frac{7}{0}$

a. $\frac{0}{3} = 0 \cdot \frac{1}{3}$ To divide by 3, multiply by the reciprocal of 3, or $\frac{1}{3}$.
     $= 0$

b. $\frac{7}{0} = 7 \cdot (?)$ Zero has no reciprocal.

It is *not possible* to divide by 0.

CLASS EXERCISES

State whether each quotient will be positive or negative.

1. $-7 \div 1$
2. $1.5 \div 0.5$
3. $\frac{119}{7}$
4. $-\frac{1}{2} \div \left(-\frac{1}{3}\right)$

Rewrite each division expression as multiplication.

5. $\frac{3}{4} \div \frac{7}{8}$
6. $-24 \div \frac{4}{5}$
7. $\frac{132}{-11}$
8. $-15 \div \left(-\frac{1}{4}\right)$
9. $\frac{0}{-10}$
10. $-\frac{2}{3} \div \frac{2}{3}$
11. $\frac{0.8}{5}$
12. $1\frac{1}{3} \div \left(-\frac{1}{2}\right)$

1.8 Dividing Real Numbers

PRACTICE EXERCISES

Find the reciprocal.

1. 5
2. 8
3. 75
4. 10
5. $-\frac{1}{5}$
6. $-\frac{1}{7}$
7. -0.9
8. -0.6

Divide, if possible.

9. $\frac{84}{-12}$
10. $\frac{76}{-19}$
11. $\frac{-36}{3}$
12. $\frac{-42}{14}$
13. $-1\frac{2}{3} \div \left(\frac{-2}{9}\right)$
14. $-\frac{1}{2} \div \left(\frac{-2}{3}\right)$
15. $-\frac{7}{12} \div \frac{7}{6}$
16. $-\frac{1}{3} \div \frac{5}{6}$
17. $\frac{0}{5}$
18. $\frac{17}{0}$
19. $9 \div 0$
20. $0 \div 15$

Evaluate each expression if $a = -8$, $b = -4$, $c = -3$, and $d = \frac{1}{2}$.

21. $\frac{a}{b}$
22. $\frac{b}{d}$
23. $d \div c$
24. $d \div a$
25. $\left(\frac{1}{2}d\right) \div b$
26. $\left(-\frac{3}{2}a\right) \div c$
27. $\frac{c - d}{d}$
28. $\frac{d}{b - a}$
29. $(-4c) \div (8d)$
30. $\left(-\frac{9}{2}b\right) \div (3c)$

Simplify. *Hint:* The numerator and the denominator of a fraction are each considered to be enclosed within parentheses.

31. $\frac{-3 + 5}{-2}$
32. $\frac{18}{-11 + 2}$
33. $(-13 - 3) \div (-2 - 6)$
34. $(100 - 16) \div (-14 + 2)$
35. $\frac{7(-3)}{-2 - 5}$
36. $\frac{-9(4)}{-1 - 2}$
37. $-36.3636 \div 12.5$
38. $789.62 \div 0.52$
39. $\frac{(-1.62)(97.87)}{0.27}$
40. $\frac{-90.81}{(6.25)(-0.025)}$
41. $\frac{718.3 - 84.7}{3.275 + (-2.923)}$
42. $\frac{7.061 - 33.925}{0.905 + (-0.321)}$
43. $\left(3 - 4\frac{1}{3}\right) \div \left(-\frac{2}{3} + \frac{5}{6}\right)$
44. $\left[3 - \left(-\frac{5}{4}\right)\right] \div \left(-5 + \frac{3}{4}\right)$
45. $\frac{|(-6)(5)|}{-4}$
46. $\frac{25 + (-4)}{|11 - 1 - 3|}$

If *a*, *b*, *c*, and *d* are any nonzero real numbers, with $c \neq d$, tell whether the statements are *true* or *false*.

47. $(a \div b) \div c = (a \div c) \div b$

48. $\dfrac{a}{bc} = (a \div b) \div c$

49. $\dfrac{|ab|}{|cd|} = \left|\dfrac{ab}{cd}\right|$

50. $\dfrac{a-b}{c-d} = \dfrac{b-a}{d-c}$

Applications

To find the **mean** or **average** of a group of measurements, divide their sum by the number of measurements.

Example $-5.5°$, $-8.2°$, $1.7°$, $3.9°$, $9.7°$, $10.1°$, and $3.0°$ are the 8:00 AM temperature readings in degrees Fahrenheit for a winter week in Minneapolis.
The average temperature at 8:00 AM is:

$$\frac{-5.5 + (-8.2) + 1.7 + 3.9 + 9.7 + 10.1 + 3.0}{7} = \frac{14.7}{7} = 2.1$$

2.1°F is the average 8:00 AM temperature for this week.

51. **Sports** A golfer shot 18-hole rounds with scores of 83, 79, 73, and 84. Find the average golf score for these rounds to the nearest whole number.

52. **Statistics** A student's algebra quiz scores are: 76, 58, 87, 80, and 82. Find the student's average quiz score to the nearest whole number.

53. **Sports** The heights of the members of the varsity basketball team are: 178 cm, 185 cm, 193 cm, 193 cm, and 201 cm. Find the average height of the players.

54. **Meteorology** One week the barometric readings were 28.90, 29.10, 30.20, 29.60, 30.10, 29.90, and 30.10. Find the average barometric reading for the week.

55. **Sports** Four friends ride to school on their bikes. The one-way distance for each is: 1.6 km, 0.8 km, 1.2 km, and 2.4 km. What is the average of these distances? Does any one of the riders travel the exact average number of kilometers?

56. **Finance** The stock market values of NJO stock over a particular week were $62\frac{1}{4}$, $61\frac{1}{2}$, $60\frac{3}{4}$, 61, and $62\frac{1}{2}$. Find NJO's average value for this week.

57. **Meteorology** One week the high temperature readings were 82.4°F, 94.1°F, 84.5°F, 89.5°F, 92.6°F, 90.8°F, and 87.9°F. Find the average high temperature reading for the week.

Calculator Most calculators have a reciprocal key, usually labeled 1/*x* or *x*$^{-1}$. This key is used to find the reciprocal of any number, expressed in decimal equivalent form. Find the reciprocal of each number using the reciprocal key on your calculator. Check your answer by division.

58. 17   **59.** 100   **60.** −19   **61.** −0.4

CRITICAL THINKING: ANALYSIS

You have seen expressions such as: $\frac{c}{b}$ ($b \neq 0$). This expression means that *b* can represent any number *except* 0. The following discussion shows logically why you cannot replace *b* with 0 in the expression $\frac{c}{b}$.

ASSUME   One way to define the division of *c* by *b* is to say that for any two numbers *c* and *b*, there is one number *a* such that

$$\frac{c}{b} = a \quad \text{provided that } a \times b = c.$$

For example, $\frac{6}{3} = 2$, since $2 \times 3 = 6$.

INFER   If division by 0 is possible, then you can divide 6 by 0 and get one number that is the quotient. This means that there is one number *a* that makes the following number sentence true.

$$\frac{6}{0} = a \quad \text{provided that } a \times 0 = 6.$$

ANALYZE   Try replacing *a* in $a \times 0 = 6$ with different numbers. Is there any number that makes the sentence true? Then, is there any number *a* that will make $\frac{6}{0} = a$ true?
Now try to divide 0 by 0.

Is there one number *a* such that $\frac{0}{0} = a$?

Try replacing *a* in $a \times 0 = 0$ with different numbers. Is there *only one* number that makes the sentence true?

CONCLUDE   Explain in your own words why 0 can never be a divisor (a denominator in a rational number). Explain also why $\frac{0}{0}$ is indeterminate (cannot be determined).

Chapter 1 Real Numbers

APPLICATION: Meteorology

Did you know that on April 12, 1934, the velocity of the wind reached 231 mi/h at Mount Washington, N.H.? This was the strongest wind ever recorded on the surface of the earth!

Meteorologists measure weather conditions and record and analyze the data. They are stationed around the world and use instruments as basic as the thermometer and as sophisticated as the computer. Meteorologists use the data they gather to prepare forecasts.

			Temperature Readings (°F)				
State	**Station**	**Date**	**6 AM**	**10 AM**	**2 PM**	**6 PM**	**10 PM**
N. Dak.	Bismarck	1/15	−1.8	0.5	1.5	−2	−2.7

You can use paper and pencil, a calculator, or a computer, to find the average temperature.

$$\frac{\text{Sum of the readings}}{\text{Number of readings}} = \text{Average}$$

$$\frac{(-1.8) + 0.5 + 1.5 + (-2) + (-2.7)}{5} = -0.9$$

```
10 INPUT "HOW MANY NUMBERS ARE YOU
   AVERAGING? ";N:PRINT
20 INPUT "ENTER THE FIRST NUMBER. ;S:
   PRINT
30 FOR I = 2 TO N
40 INPUT "NEXT NUMBER: ";X:PRINT
50 LET S = S + X
60 NEXT I
70 LET A = S / N
80 PRINT "THE AVERAGE IS";A
90 END
```

The average temperature for that day was −0.9°F. Is it possible to use that temperature to predict the temperature for the next day? Explain.

Solve.

1. The normal daily temperature for Bismarck in January is 8.2°F. How many degrees below normal is −0.9°F?

2. The lowest daily temperature for Bismarck in January was −2.8°F. How many degrees above the lowest temperature is −0.9°F?

3. What is the average of the following readings: 6.2°, −1.8°, 3.6°, 0°, −0.5°?

4. What is the average of the following readings: −2.3°, 0.6°, 5.2°, −4.7°, 3.2°?

1.9 Problem Solving Strategy: Select Appropriate Notation

A problem asks a question for which an answer is not known. The solution to a problem is the answer to the question. In arithmetic, problems are solved by using numbers. In algebra, the number or numbers may not be known. The unknown number (or numbers) is represented by a *variable* whose specific value is not known until the problem is solved.

The use of variables and other symbols (operation signs, grouping symbols, and equality and inequality symbols) is called **algebraic notation.**

Algebraic Notation

Letters of the alphabet from *a* to *z* are used as variables to represent one or more unknown numbers.	$a = -6 \quad m = 0.89 \quad x = 3\frac{1}{8}$
A raised dot or parentheses is used to represent the operation of multiplication.	$7 \cdot n \quad 4 \cdot x \quad 3(t) \quad -9(y)$
The product of a number and a variable or two or more variables can be expressed without a sign between the number and the variable.	$-3w \quad 16h \quad \frac{2}{3}b \quad 4yz$
The operations of addition, subtraction, and division are represented by the same symbols in algebra and arithmetic.	add: $x + y$ subtract: $5 - n$ divide: $18 \div b$
Division is also indicated by using the fraction form, as in arithmetic.	$\frac{3}{4} \quad \frac{n}{7} \quad \frac{a}{b} \quad \frac{25}{f} \quad \frac{t-6}{9}$
The symbol { } is used to enclose the members of a set. Three dots mean that the numbers in a set go on forever in the same pattern.	$\{1, 2, 3, 4, 5, 6, \ldots\}$
The equal sign and the inequality symbols are used to compare numbers.	$-5 = -5 \quad -3 < 0 \quad c > -2$
A negative sign before a variable means the opposite of whatever number the variable represents.	If $x = 7$, then $-x = -(7) = -7$. If $x = -4$, then $-x = -(-4) = 4$. If $x = 0$, then $-x = -(0) = 0$.
The absolute value of a number n is written as $\|n\|$.	$\|n\|$

Problem solving is a process in which several steps are applied sequentially. Each step consists of a set of skills.

Understand the Problem

Read the problem.
What facts are given?
What is the unknown, that is, what are you asked to find?
Review the definitions of all mathematical terms.

Plan Your Approach

Choose a strategy.
The strategy you choose is your plan of action for solving the problem. Many strategies will be developed in this book to help you plan your approach. In this lesson, the strategy is to use algebraic notation.

Complete the Work

Apply the strategy.
Use the algebra you know to apply the strategy to solving the problem. Very often, this means using the appropriate algebraic operations. Keep an open mind. Change your strategy if it does not work.

Interpret the Results

State your conclusion.

Check your conclusion.
What generalization(s) can you make?

EXAMPLE Maria asked Susan to list all the positive even numbers. Then Maria said she could name them by writing one algebraic expression. What is the algebraic expression Maria is thinking about?

Understand the Problem

What facts are given?
The problem is concerned with the set of positive even numbers.

What are you asked to find?
An algebraic expression that represents all positive even numbers.

Plan Your Approach

Choose a strategy.
Select a variable to use in writing the algebraic expression. Let n represent any counting number.

List the positive even numbers, 2, 4, 6, 8, 10, 12, ...

Observe that every even number is a multiple of 2.

$2 = 2 \cdot 1 \qquad 4 = 2 \cdot 2 \qquad 6 = 2 \cdot 3$
$8 = 2 \cdot 4 \qquad 10 = 2 \cdot 5 \qquad 12 = 2 \cdot 6$

Problem Solving Strategy: Select Appropriate Notation   **43**

- **Complete the Work** Study the relationships in the list above and observe that any positive even number is 2 times a counting number. Since *n* represents any counting number, $2n$ represents an even number.

- **Interpret the Results** **State your conclusion.**
  If *n* represents any counting number, the positive even numbers can be represented by the algebraic expression $2n$. You can check by substituting values of *n*.

CLASS EXERCISES

Use algebraic notation to represent each of the following.

1. An unknown number
2. The absolute value of an unknown number
3. An unknown number less than zero
4. The product of two unknown numbers
5. The sum of three unknown numbers
6. The absolute value of the difference of two unknown numbers
7. Two times the sum of two unknown numbers

PRACTICE EXERCISES

Use algebraic notation to represent each of the following.

1. Mary's age 2 years ago
2. One-half the current price
3. John's salary plus $500 commission
4. Martha's annual salary less $4500 in deductions
5. The price of an item plus 6% sales tax
6. Twenty-five percent of the boys
7. The average of salaries *a*, *b*, and *c*
8. List the positive multiples of 3. Then write an algebraic expression for any positive multiple of 3.
9. List the first 10 even natural numbers. Subtract 1 from each number. Are the resulting numbers odd or even? If *x* is any natural number, write an algebraic expression to represent the odd natural numbers.

10. If *p* represents the number of pears in a box, does 5*p* represent "5 pears" or "5 times the number of pears"?

11. If *x* and *y* are any two rational numbers, can *x* and *y* represent the same number? Explain.

12. Write an algebraic expression to represent the negative even integers.

13. Write an algebraic expression to represent the negative odd integers.

14. If *a*, *b*, *c*, and *d* represent any four integers, is the following equality true or false? Explain.
$$a + b + c = a + d + c$$

15. If *x* and *y* are any two rational numbers, is the following equality true or false? Explain.
$$|x - y| = |y - x|$$

16. If *a* and *b* are any two nonzero real numbers, and $a + b = c$, is it possible that $a = c$ or $b = c$? Explain.

PROJECT

This lesson shows that algebra is *more general* than arithmetic. Use a dictionary to look up the meaning of the word *general*. Write down the dictionary definition. Then explain in your own words what you think the dictionary definition of *general* means as it applies to algebra. Give two or three examples from this chapter to support your explanation.

TEST YOURSELF

Add or subtract, as indicated. 1.5–1.6

1. $-21 + (-13)$
2. $1.5 + 3.2$
3. $17 + (-9)$
4. $-\frac{2}{3} + \frac{1}{4}$
5. $-2 + 12 + -10$
6. $-\frac{1}{4} - \left(-\frac{3}{5}\right)$
7. $4 - 10$
8. $3.2 - (-4.1)$
9. $-(10 - 15) - (2 - 5)$

Multiply or divide, as indicated. 1.7–1.8

10. $(10)(-15)$
11. $-\frac{1}{2} \cdot \frac{2}{9}$
12. $(-2)(-3)(-7)(-2)$
13. $\frac{-36}{4}$
14. $-2\frac{1}{3} \div \left(-\frac{1}{2}\right)$
15. $\frac{3}{4} \div \frac{2}{4}$

Problem Solving Strategy: Select Appropriate Notation

CHAPTER 1 SUMMARY AND REVIEW

Vocabulary

absolute value (14)
additive inverse (27)
algebraic expression (3)
average (39)
coordinate of a point (8)
graph of a number (8)
integers (7)

irrational numbers (8)
multiplicative inverse (35)
natural numbers (7)
negative numbers (7)
numerical expression (3)
opposite (13)
origin (7)

positive numbers (7)
rational numbers (8)
real numbers (7)
reciprocal (35)
variable (2)
variable expression (2)
whole numbers (7)

Evaluating Algebraic Expressions To evaluate an algebraic expression, substitute the given numbers for the variables and then simplify. Operations within parentheses are done first.   **1.1**

Evaluate each expression if $w = 4$, $x = 0.5$, $y = 16$, and $z = \frac{2}{3}$.

1. $w - x$
2. $\frac{y}{w}$
3. $\frac{3}{4} y$
4. $w \div \frac{1}{2}$

5. $3yz$
6. $4(w - x)$
7. $\frac{8}{4y}$
8. $y - (wx)$

Translating Word Phrases To translate word phrases to algebraic expressions, look for key words that indicate operations:

addition (more than, sum)   multiplication (times, product)
subtraction (minus, difference)   division (quotient, ratio)

Write an algebraic expression for each phrase.

9. 18 more than some number x
10. The product of some number n and 22
11. The difference of a and b
12. The ratio of m and $\frac{2}{3}$

Graphing Numbers The graph of a number is a point on the number line.   **1.2, 1.3**

Graph each set of numbers.

13. $\left\{ -4\frac{1}{2}, -4, 0, 2.5, 3 \right\}$
14. {whole numbers}
15. {odd integers}

Replace each ? with <, >, or = to make a true statement.

16. $-11 \underline{\ ?\ } 10$
17. $0 \underline{\ ?\ } -3.5$
18. $-1\frac{1}{3} \underline{\ ?\ } -1.5$

46 Chapter 1 Real Numbers

Absolute Value and Opposites The absolute value of a number is the distance the number is from zero on a number line. -2 and 2, $4\frac{1}{3}$ and $-4\frac{1}{3}$ are called opposites. Their graphs are the same distance from the origin.

1.3

Simplify.

19. $-(4.31)$
20. $|-8|$
21. $-|6 + 3|$
22. $-\left|-\frac{8}{9}\right|$

Operations with Real Numbers A number line can be used to find the sum of real numbers. The rules for addition are given on pages 22–24 and the rules for multiplication are given on pages 31–33. Subtraction and division of real numbers are defined *in terms of* addition and multiplication, respectively.

1.4–1.8

$$m - n = m + (-n) \qquad m \div n = m \cdot \frac{1}{n} \; (n \neq 0)$$

Perform the indicated operations, if possible.

23. $-4.8 + 3.6$
24. $-\frac{2}{3} + \left(-\frac{2}{3}\right)$
25. $-11 + (-3.6) + 2.9$
26. $-18 - (-11)$
27. $-5.9 - 0.02$
28. $[-48 - (-48)] - (-9)$
29. $-6(-25)$
30. $-13 \cdot 3 \cdot 0$
31. $\frac{2}{3}(-20)(9)$
32. $\frac{-96}{16}$
33. $0 \div \left(-\frac{1}{3}\right)$
34. $\frac{-4.284}{-2.1}$

Problem Solving To solve problems of any type, the four-step problem-solving approach outlined below may be helpful.

1.9

Understand the Problem → Plan Your Approach → Complete the Work → Interpret the Results

35. The yards gained and lost on eight successive plays by the Lansing High School football team were as follows: $+6$, -10, $+4$, $+10.5$, -0.5, -9, -6, $+25$. What was the average yardage for the eight plays?

36. One carpenter charges $20 plus $25 per hour and another charges $30 per hour. Find how many hours a job must take for both to charge the same amount.

Summary and Review **47**

CHAPTER 1 TEST

Evaluate each expression if $a = 1.4$, $b = 25$, and $c = \frac{1}{5}$.   **1.1**

1. $0.9 + a$
2. abc
3. $(5c) - ab$

Write an algebraic expression for each phrase.

4. A number p decreased by 0.85
5. The quotient of a number b and 27
6. $\frac{2}{3}$ less than some number x
7. The product of w and z

Graph each set of numbers.   **1.2**

8. {positive integers}
9. {real numbers}

List the numbers in order from least to greatest.   **1.3**

10. $2.5, -1.5, 0, -2.25, -2, 2.25$
11. $0, -\frac{1}{3}, -1\frac{1}{3}, \frac{1}{2}, -1\frac{1}{2}, -3$

Simplify.

12. $-(-0.3)$
13. $-\left|\frac{2}{5}\right|$
14. $-[-|-9|]$

Perform the indicated operations, if possible.   **1.4–1.8**

15. $-45 + (-132)$
16. $(-8) + 2 + (-9) + (-11) + 6$
17. $3.2 - (-4.07)$
18. $(-6 - 5) + (-11)$
19. $0 \div \frac{1}{5}$
20. $-\frac{1}{4}(-5)(24)(-1)(-20)$
21. $-96 \div (-16)$
22. $-1.25 \div 0$

23. Find the value of each expression if $m = \frac{1}{2}$, $n = \frac{2}{3}$, and $p = \frac{3}{4}$.   **1.7, 1.8**
    a. $(np) + m$
    b. $-(np) - m$
    c. the average of m, n, and p

24. One house painter charges $15 plus $20 per hour. A second painter charges $25 per hour. Find out how many hours a job takes for the total charges of the second painter to be the same as the charges of the first painter.   **1.9**

Challenge

At 10:15 AM a train left Flatfoot, Washington, traveling east at 40 km/h (kilometers per hour). Two hours later, another train left the same station traveling west at 35 km/h. At what time will the trains be 230 km apart?

PREPARING FOR STANDARDIZED TESTS

Select the best answer for each and write the appropriate letter.

1. If $a = 3$, $b = 0$, and $c = 2$, then $2a + 3b - c$ has a value of
   A. 8   B. 6   C. 4   D. 2   E. 0

2. What is the result when the sum $[4 + (-5)]$ is subtracted from the sum $[9 + (-7)]$?
   A. 3   B. -3 C. 1   D. -1 E. 7

3. When simplified, $|7 + (3 - 4)|$ has a value of
   A. 8   B. 6   C. 0   D. -6 E. -8

4. If $\frac{4}{3} < k$ and $k < \frac{3}{2}$, then k could be:
   I. $\frac{11}{9}$
   II. $\frac{11}{8}$
   III. $\frac{11}{7}$
   A. I only   B. II only
   C. III only   D. I, II only
   E. II, III only

5. Which of the following is *not* a multiple of 9?
   A. 2637   B. 1008   C. 774
   D. 608   E. 540

6. $\left(\frac{2}{3}\right)\left(-\frac{3}{5}\right)\left(\frac{4}{5}\right)\left(-\frac{5}{3}\right)\left(-\frac{3}{2}\right)$ equals
   A. $\frac{4}{5}$ B. $-\frac{4}{5}$ C. $\frac{5}{4}$
   D. $-\frac{5}{4}$ E. 4

7. 18 is $\frac{2}{3}$ of what number?
   A. 6   B. 12   C. 24   D. 27   E. 36

8. Which of the following is less than $\frac{3}{4}$?
   A. $\frac{49}{64}$ B. $\frac{45}{60}$ C. $\frac{53}{70}$ D. $\frac{51}{67}$
   E. none of the above

9. When $a = 3$ and $b = \frac{2}{3}$, which of the following expressions represents the largest value?
   A. $(2a) + (3b)$ B. $2(a + b)$
   C. $a + (3b)$ D. $(6b) + a$
   E. $(3a) + (6b) - 7$

10. A group of boys pooled their money to buy a new baseball. They found they had 19 dimes, 18 pennies, 17 nickels, and 16 quarters. How much money did they have in all?
    A. $5.83   B. $5.93   C. $6.03
    D. $6.83   E. $6.93

11. A certain type of postage costs 56¢ for the first ounce and 17¢ for each additional $\frac{1}{2}$ oz or part. At this rate, what would be the amount of postage on a letter weighing 2.3 oz?
    A. 90¢   B. 97¢   C. $1.00
    D. $1.07   E. $1.24

Preparing for Standardized Tests

MAINTAINING SKILLS

Perform the indicated operation.

Example 1
```
    3.05   ← 2 decimal places
   ×4.8    ← 1 decimal place
   ─────
    2440
   1220
   ─────
   14.640  ← 3 decimal places
```

Example 2
```
         3.5
   2.4)8.4.0
       72
       ──
       120
       120
```

1. 3.81×0.87
2. 63.4×0.75
3. 7.51×5.7
4. 6.5×9.4
5. $5.76 \div 6.4$
6. $14.7 \div 0.98$
7. $28.12 \div 7.6$
8. $135 \div 0.25$
9. 249×0.06
10. 47.3×2.9
11. 82.5×3.4
12. 1.6×0.58
13. $14.56 \div 5.2$
14. $34.3 \div 9.8$
15. $62.32 \div 7.6$
16. $60.8 \div 6.4$
17. 0.0904×5
18. $0.0904 \div 5$

Example 3 $12 - (-17)$
$12 - (-17) = 12 + 17 = 29$

Example 4 $-36 \div (-9)$
$-36 \div (-9) = 4$

19. $8 - 15$
20. $-9 + 6$
21. -7×4
22. $56 \div (-8)$
23. $5 - (-9)$
24. $-6 \times (-8)$
25. $-7 + (-8)$
26. $-72 \div 9$
27. $-6 - (-2)$
28. $-42 \div (-7)$
29. $7 \times (-6)$
30. $12 + (-8)$

Solve.

Example 5

Gloria typed 24 pages in 2 hours (h). Jon typed 45 pages in 3 h. How many more pages per hour did Jon type than did Gloria?

Step 1 Gloria
$24 \div 2 = 12$

Step 2 Jon
$45 \div 3 = 15$

Step 3 $15 - 12 = 3$
Jon typed 3 more pages per hour.

31. The distance from Dominic's house to the office is 38 mi. How many miles does he travel going to and from the office in a 6-day workweek?

32. Sandi bought a stereo for $249 and 5 record albums for $7.98 each. How much did she spend in all?

33. Harold has 780 packages of pens to pack evenly into 12 boxes. The pens sell for $1.98 a package. How many packages are packed in a box?

2 Algebraic Expressions

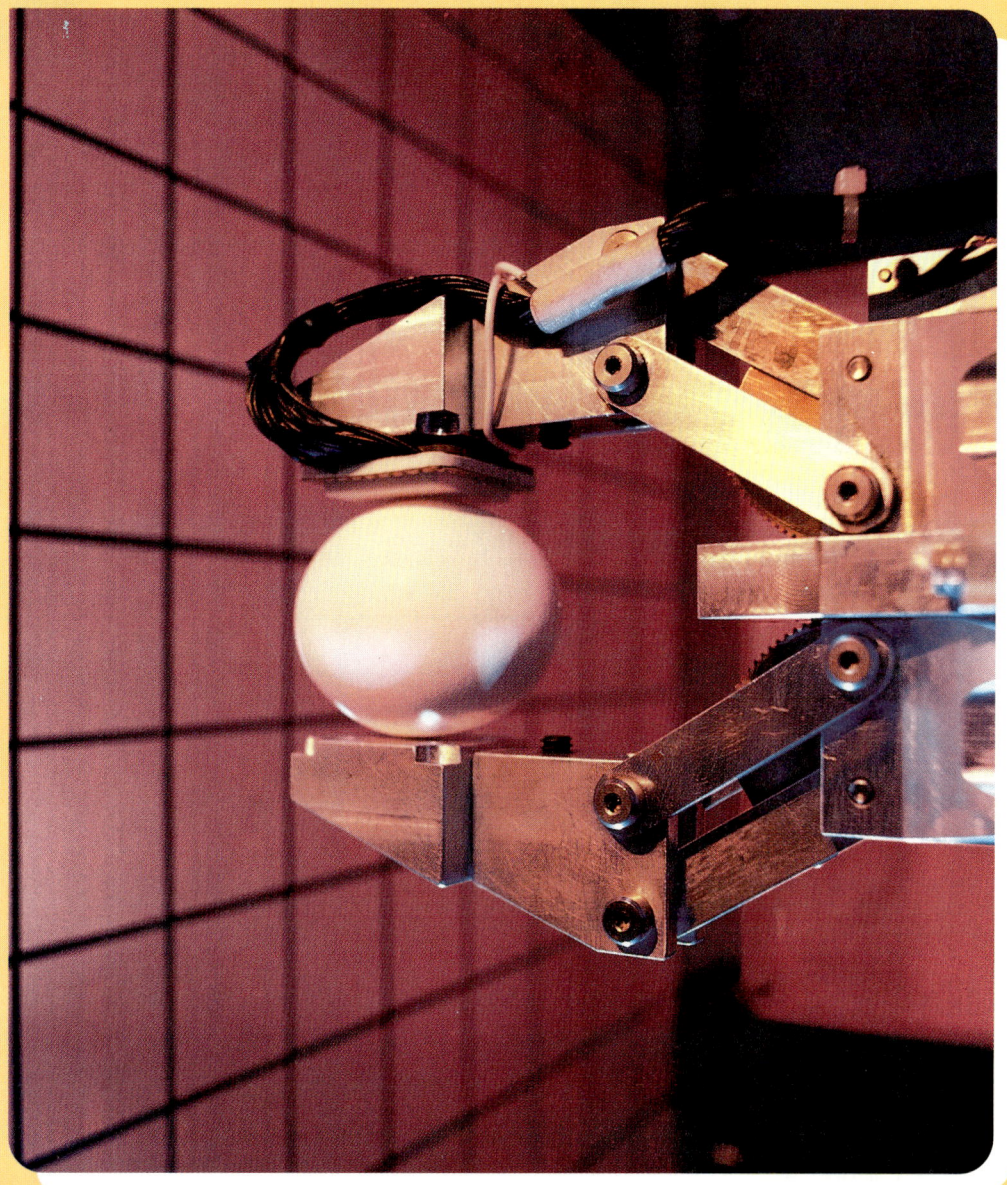

Although computers have been used in the business world for a long time, new advances in technology and research in such areas as artificial intelligence are contributing greatly to new knowledge and productivity.

2.1 Simplifying and Evaluating Expressions

Objectives: To simplify numerical expressions by using the rules for order of operations
To evaluate algebraic expressions

To *simplify* a numerical expression, replace the expression with its simplest name. To *evaluate* an algebraic expression, substitute given values for the variables and then simplify.

Capsule Review

EXAMPLE Evaluate: a. $-x + z$ if $x = -4$, $z = 2$
b. $3(y - 4)$ if $y = -5$

a. $-x + z = -(-4) + 2$ *Substitute -4 for x and 2 for z.*
            $= 4 + 2 = 6$ *Add.*

b. $3(y - 4) = 3(-5 - 4)$ *Substitute -5 for y.*
              $= 3(-9) = -27$ *Multiply.*

Evaluate each expression if $a = 2$, $b = -1$, $c = 13$, and $d = -4$.

1. $a + (-b) + c$ 2. cdb 3. $a \div b$ 4. $\dfrac{-b}{-d}$

Operations to be done first are shown within *grouping symbols*. Algebraic grouping symbols include:

    parentheses: $(4 + 3) \cdot (8 - 2) = 7 \cdot 6$, or 42
    brackets: $8 - [10 - 7] = 8 - 3$, or 5
    braces: $\{3 \cdot 2 \cdot 5\} + 3 = 30 + 3$, or 33
    fraction bar: $\dfrac{2 + 3}{13 - 3} = \dfrac{5}{10}$, or $\dfrac{1}{2}$

In the expression $14 - 12 \cdot 3 + 2 \div 2$, which operation should be done first? In order to obtain a unique (one and only one) value for this expression, mathematicians use the Rules for Order of Operations.

Rules for Order of Operations

- Perform any operation(s) within parentheses or other grouping symbols.
- Multiply and divide in order from left to right.
- Add and subtract in order from left to right.

Chapter 2 Algebraic Expressions

EXAMPLE 1 Simplify: $14 - 12 \cdot 3 + 2 \div 2$

$14 - \underline{12 \cdot 3} + \underline{2 \div 2}$
$= 14 - 36 + 1$ *Multiply and divide in order from left to right.*
$= -22 + 1$ *Add and subtract in order from left to right.*
$= -21$

When several grouping symbols are used in one expression, simplify the expression in the innermost grouping first. Then continue from the innermost to the outermost grouping symbols, using the Rules for Order of Operations.

EXAMPLE 2 Simplify: $\{40 - [3(2 + 4)] \cdot 2\}$

$\{40 - [3(2 + 4)] \cdot 2\}$
$= \{40 - [3(6)] \cdot 2\}$ *Simplify within parentheses.*
$= \{40 - [18] \cdot 2\}$ *Simplify within brackets.*
$= \{40 - 36\}$ *Simplify within braces. Multiply.*
$= 4$ *Subtract.*

EXAMPLE 3 Evaluate: $5a + 6b - \dfrac{2b}{c} + d$ if $a = 4$, $b = -6$, $c = -2$, and $d = 0$

$5a + 6b - \dfrac{2b}{c} + d$

$= 5 \cdot 4 + 6(-6) - \dfrac{2(-6)}{-2} + 0$ *Substitute 4 for a, -6 for b, -2 for c, and 0 for d.*

$= 5 \cdot 4 + 6(-6) - \dfrac{-12}{-2} + 0$ *Multiply in the numerator of the fraction.*

$= 20 + (-36) - 6$ *Multiply and divide.*
$= -16 - 6$ *Add.*
$= -22$ *Subtract.*

CLASS EXERCISES

Simplify each expression.

1. $2 + 6 \cdot 8 \div 4$ 2. $10 \div 5 \cdot 2 + 6$ 3. $6 + 8 \div 2 - 3$

Evaluate each expression if $a = 3$, $b = -4$, and $c = 1$.

4. $2a + 1$ 5. $3b + 1$ 6. $\dfrac{c}{6 - a}$ 7. $\dfrac{b + 1}{a}$

PRACTICE EXERCISES

Simplify each expression.

1. $16 - 11 \cdot 2 + 3 \div 3$ 2. $10 - 13 \cdot 3 + 4 \div 4$ 3. $21 + 16 \cdot 7 - 6 \div 2$

2.1 Simplifying and Evaluating Expressions

4. $42 + 21 \cdot 9 - 8 \div 4$ 5. $\{30 - [4(3 + 2)] \cdot 4\}$ 6. $\{40 - [6(4 + 3)] \cdot 7\}$

Evaluate each expression if $a = 5$, $b = -1$, $c = 0$, and $d = 2$.

7. $4a + 3$ 8. $2d - b$ 9. $a - \dfrac{d}{b}$ 10. $d - \dfrac{c}{b}$

11. $2a - 3b + 4c$ 12. $-4d + 2b - 3a$ 13. $\dfrac{7ab - 2d}{6a - b}$

14. $\dfrac{1 - 16bd}{15a + 1}$ 15. $8(a + b) - 11bd$ 16. $-5(2a + d) + 6b$

Simplify each expression.

17. $6 - [9 + 12(2 + 8) \div 4]$ 18. $5 - \{-4[8 - 11(-9 - 1)]\}$

Evaluate each expression if $m = 6$, $n = -2$, $p = 18$, $r = -3$, and $s = 9$.

19. $m + \dfrac{6n + p}{sr - 12}$ 20. $\dfrac{-r + 7n}{11 - rs}$

21. $m - (-n)[r - (-s)](m - n)$ 22. $(p + r)[n - (-m)] - mn$

Applications

Computer If you wanted to open a simple interest savings account, you could use the program below to help you determine which bank would give you the best return for your money (principal).

```
10 INPUT "ENTER THE PRINCIPAL:     ";P
20 INPUT "ENTER THE RATE AS A DECIMAL:    ";R
30 INPUT "ENTER THE TIME IN MONTHS:   ";T: PRINT : PRINT
40 LET I = P * R * (T / 12)
50 PRINT "YOUR RETURN ON $";P;" AT A RATE OF ";R * 100;"%
      FOR ";T;" MONTHS IS $"; INT ((I + 0.005) * 100) / 100
```

23. If bank ABC offers 5% annual interest for a deposit of $500 for 3 months, what interest would you earn?

LOGICAL REASONING

In each sentence a mathematical term was scrambled. Decode each term.

1. The CALROPICER of x is $\dfrac{1}{x}$, but x cannot be zero.
2. Addition and subtraction are mathematical ATOPERONIS.
3. Letters that represent numbers are called IBARLEVAS.
4. SESTERNAPEH are used as symbols of grouping.

2.2 Exponents and Formulas

Objectives: To simplify and evaluate expressions containing exponents
To evaluate formulas for given values of the variables

When an automobile makes a turn, it moves in a curved path. The acceleration (rate at which the car's velocity changes) can be found if the radius of the curve and the velocity or speed of the car are known.

Physicists have derived a formula for analyzing circular motion:

$$a = \frac{v^2}{r}$$

a = acceleration
v = velocity
r = radius of circle

The expression v^2 means $v \cdot v$, so $a = \frac{v^2}{r}$ is equivalent to $a = \frac{v \cdot v}{r}$.

This formula can be evaluated for a if the values of the variables v and r are known. (See Example 5.)

Capsule Review

EXAMPLE Evaluate: **a.** $m \cdot m \cdot m$ if $m = -3$
b. $2 \cdot 2 \cdot a \cdot a \cdot a \cdot b \cdot b$ if $a = 1$, $b = -4$

a. $m \cdot m \cdot m$ if $m = -3$
$= -3(-3)(-3) = -27$

b. $2 \cdot 2 \cdot a \cdot a \cdot a \cdot b \cdot b$ if $a = 1$, $b = -4$
$= 2 \cdot 2 \cdot 1 \cdot 1 \cdot 1 \cdot (-4)(-4) = 64$

Evaluate each expression if $x = -2$ and $y = 3$.

1. $y \cdot y \cdot y \cdot y$
2. $x \cdot x \cdot x \cdot x$
3. $-4 \cdot x \cdot x \cdot x \cdot y \cdot y$

When two or more numbers are multiplied, each number is called a **factor**. An **exponent** is used to show how many times the factor, or **base**, is multiplied.

If a is any real number and n is any positive integer, then:

$a^n = \underbrace{a \cdot a \cdot a \cdots \cdot a}_{n \text{ factors}}$
n is the exponent, a is the base.
a^n is read "a to the nth power."

EXAMPLE 1 Simplify: **a.** 5^2 **b.** $\left(\frac{1}{2}\right)^3$ **c.** 3^1 **d.** 1^5

	Read	Base	Exponent	Value
a. 5^2	5 squared	5	2	$5 \cdot 5 = 25$
b. $\left(\frac{1}{2}\right)^3$	$\frac{1}{2}$ cubed	$\frac{1}{2}$	3	$\frac{1}{2} \cdot \frac{1}{2} \cdot \frac{1}{2} = \frac{1}{8}$
c. 3^1	3 to the first	3	1	3
d. 1^5	1 to the fifth	1	5	$1 \cdot 1 \cdot 1 \cdot 1 \cdot 1 = 1$

When the exponent is 1, it is not necessary to write the exponent.

EXAMPLE 2 Simplify: **a.** -3^2 **b.** $(-3)^2$ **c.** $-(3)^2$ **d.** $-(-3)^2$

a. $-3^2 = -(3 \cdot 3) = -9$ **b.** $(-3)^2 = (-3)(-3) = 9$

c. $-(3)^2 = -(3 \cdot 3) = -9$ **d.** $-(-3)^2 = -[(-3)(-3)] = -9$

To simplify expressions with exponents, simplify the exponent first.

$$3 \cdot 2^4 = 3 \cdot (2 \cdot 2 \cdot 2 \cdot 2) = 48$$

The order of operations introduced on page 52 is thus extended to include exponents:

- Perform any operation(s) within grouping symbols.
- Simplify any exponents.
- Multiply and divide in order from left to right.
- Add and subtract in order from left to right.

EXAMPLE 3 Simplify: **a.** $2(3 + 4)^2$ **b.** $(4 - 2^3)^2$

a. $2(3 + 4)^2$
$= 2(7)^2$ *Simplify within parentheses.*
$= 2 \cdot 7 \cdot 7$ *Simplify the exponent.*
$= 98$ *Multiply.*

b. $(4 - 2^3)^2$
$= [4 - (2)(2)(2)]^2$ *Simplify $(4 - 2^3)$.*
$= [4 - 8]^2$
$= [-4]^2$ *Multiply.*
$= (-4)(-4) = 16$

EXAMPLE 4 Evaluate: **a.** $5x^3y^2$ if $x = 2$, $y = 3$
                   b. $2x^2 - y^3$ if $x = -1$, $y = -4$

Substitute values for x and y. Then simplify.

a. $5x^3y^2$ if $x = 2$, $y = 3$
$= 5 \cdot 2^3 \cdot 3^2$
$= 5 \cdot 2 \cdot 2 \cdot 2 \cdot 3 \cdot 3$
$= 360$

b. $2x^2 - y^3$ if $x = -1$, $y = -4$
$= 2(-1)^2 - (-4)^3$
$= 2(-1)(-1) - (-4)(-4)(-4)$
$= 2 - (-64)$
$= 2 + 64 = 66$

EXAMPLE 5 Evaluate $a = \dfrac{v^2}{r}$ for a if $v = 35$ and $r = 200$.

$$a = \dfrac{v^2}{r}$$
$$= \dfrac{35^2}{200} \quad \textit{Substitute 35 for v and 200 for r.}$$
$$= \dfrac{35 \cdot 35}{200} \quad \textit{Simplify the exponent.}$$
$$= \dfrac{1225}{200} = 6.125$$

CLASS EXERCISES

Simplify.

1. 10^2
2. $\left(\dfrac{2}{3}\right)^3$
3. -2^2
4. $2 \cdot 5^2$
5. $-(-2 - 3)^3$

Evaluate each expression if $x = 2$ and $y = 3$.

6. $x^2 y$
7. $-x^2 y$
8. $x^3 + y^2$
9. $(xy)^2$
10. $-(x - y)^2$

PRACTICE EXERCISES

Simplify each expression.

1. 5^2
2. 7^2
3. $\left(\dfrac{1}{2}\right)^2$
4. $\left(\dfrac{1}{5}\right)^3$
5. -4^2
6. -6^2
7. $(-4)^2$
8. $(-6)^2$
9. $-(4)^2$
10. $-(6)^2$
11. $-(-4)^2$
12. $-(-6)^2$
13. $2(4 + 3)^2$
14. $3(2 + 3)^2$
15. $(2 - 7^2)^2$
16. $(3 - 5^3)^2$
17. $3(7 + 4)^2$
18. $4(6 + 3)^2$
19. $(4 - 6^3)^2$
20. $(7 - 3^2)^2$

Evaluate each expression for the given value(s) of the variables.

21. $2m^2 n^3$ if $m = 2$, $n = 3$
22. $3r^3 s^2$ if $r = 2$, $s = 3$
23. $3b^3 - c^2$ if $b = -2$, $c = -3$
24. $6d^2 - e^3$ if $d = -2$, $e = -3$

Simplify.

25. $(0.3)^2$
26. $(0.5)^2$
27. $(0.001)^3$
28. $(0.001)^2$
29. $2^2(2 + 3)^2$
30. $3^2(4 + 2)^2$
31. $5^2(1 - 3)^2$
32. $4^2(1 - 5)^2$

Evaluate each expression if $c = 4$, $d = 3$, and $e = -2$.

33. $-c + d^2$
34. $-d + e^2$
35. $3c^2 - e^3$
36. $e^3 - 2c^2$
37. $c^3(d + e)^2$
38. $(e - d)^2 c^2$
39. $(2c + 4d)^2$
40. $(e - 2d)^3$

2.2 Exponents and Formulas

Evaluate each expression for the given value(s) of the variables.

41. $\dfrac{m^4 - n^4}{(m - n)^4}$ if $m = -1$, $n = -2$

42. $\dfrac{x^3 + y^3}{(x - y)^3}$ if $x = 1$, $y = -1$

43. $a^3 \div [b^2 - c^2 - (-d)^2]$ if $a = -3$, $b = 4$, $c = 3$, $d = 2$

44. $[r^3 - s^3 + (-t)^3]u^3$ if $r = 3$, $s = -2$, $t = 1$, $u = 4$

45. $-a^4 + b^2[c^2 - 2d^2] \div 2$ if $a = -1$, $b = 5$, $c = 0$, $d = 3$

Applications

Calculator The Power Key on your calculator, usually labeled y^x, can be used to simplify expressions with exponents.

Simplify the following expressions.

46. 9^4 47. 6^5 48. $(-7.1)^6$ 49. -3.9^4

BIOGRAPHY

Srinivasa Ramanujan was born in southern India in 1887. Although he had no formal schooling, he had an uncanny manipulative ability in arithmetic and algebra. When a mathematician said that he had just ridden in a taxi that was labeled with the "dull" number 1729, Ramanujan disagreed that the number was dull. He immediately pointed out that 1729 is the least number that can be represented in two different ways as the sum of two cubes:

$$1^3 + 12^3 = 1729 \text{ and } 9^3 + 10^3 = 1729$$

Ramanujan spent much of his life developing ways of calculating pi. The formulas found in his "Notebooks" never contained proofs. Also, computer algorithms were anticipated by Ramanujan, though he knew nothing about computer programming. Today, with our supercomputers, we are able to verify the accuracy of the work Ramanujan did by hand.

Use books about the history of mathematics or other reference books to answer the following questions.

1. Our system of numeration for integers is called the Hindu-Arabic system. What is the origin of the system?

2. Why is Srinivasa Ramanujan often called the "formula man"?

2.3 Properties of Real Numbers

Objectives: To identify the commutative, associative, distributive, and identity properties of real numbers
To simplify expressions using the properties of real numbers

Properties of real numbers are statements that are true for all real numbers. These properties help you perform arithmetic and algebraic calculations.

Capsule Review

Some of the statements below are examples of the properties of real numbers; others are just false statements.

In each case, tell whether the statement is *true* or *false*.

1. $-3 + 5 = 5 + (-3)$
2. $7 - 11 = 11 - 7$
3. $-4 + (12 + 9) = (-4 + 12) + 9$
4. $6 + (2 - 15) = (6 + 2) - 15$
5. $-\frac{3}{4}(16) = 16\left(-\frac{3}{4}\right)$
6. $18 \div \frac{3}{5} = \frac{3}{5} \div 18$

The sum of 0 and any real number a is equal to the number a. That is, $a + 0 = a$, so 0 is called the **additive identity**.

The product of 1 and any real number a is equal to the number a. That is, $a \cdot 1 = a$, so 1 is called the **multiplicative identity**.

The commutative properties state that the order in which you add or multiply two numbers does not change the sum or the product.

Commutative Property for Addition

For all real numbers a and b, $\quad a + b = b + a$.

Commutative Property for Multiplication

For all real numbers a and b, $\quad a \cdot b = b \cdot a$.

The associative properties state that when you add or multiply more than two numbers, the grouping of the numbers does not change the result.

Associative Property for Addition

For all real numbers a, b, and c, $(a + b) + c = a + (b + c)$.

Associative Property for Multiplication

For all real numbers a, b, and c, $(a \cdot b) \cdot c = a \cdot (b \cdot c)$.

EXAMPLE 1 Simplify: **a.** $(13 + 2) + 98$ **b.** $\dfrac{5}{7} \cdot \left(\dfrac{2}{3} \cdot \dfrac{7}{5}\right)$

a. $(13 + 2) + 98 = 13 + (2 + 98) = 13 + 100$, or 113

b. $\dfrac{5}{7} \cdot \left(\dfrac{2}{3} \cdot \dfrac{7}{5}\right) = \dfrac{5}{7} \cdot \left(\dfrac{7}{5} \cdot \dfrac{2}{3}\right) = \left(\dfrac{5}{7} \cdot \dfrac{7}{5}\right) \cdot \dfrac{2}{3} = 1 \cdot \dfrac{2}{3}$, or $\dfrac{2}{3}$

How did the *order* or the *grouping* of the numbers change? How did these changes make the computation easier?

Multiplication is said to be *distributive over addition* (or *subtraction*). The distributive property allows you to simplify an expression such as $5(11 + 7)$ in two different ways:

$$5(11 + 7) = 5(18) \qquad \text{or} \qquad 5(11 + 7) = 5(11) + 5(7)$$
$$= 90 \qquad\qquad\qquad\qquad\qquad = 55 + 35$$
$$\qquad\qquad\qquad\qquad\qquad\qquad\qquad = 90$$

Distributive Property

For all real numbers a, b, and c,

$a(b + c) = a \cdot b + a \cdot c$ and $(b + c)a = b \cdot a + c \cdot a$

$a(b - c) = a \cdot b - a \cdot c$ and $(b - c)a = b \cdot a - c \cdot a$

EXAMPLE 2 Simplify by using the distributive property:

a. $43(100 + 2)$ **b.** $\left(\dfrac{2}{5} - \dfrac{1}{9}\right)5$

a. $43(100 + 2) = 43 \cdot 100 + 43 \cdot 2 = 4300 + 86 = 4386$

b. $\left(\dfrac{2}{5} - \dfrac{1}{9}\right)5 = \dfrac{2}{5} \cdot 5 - \dfrac{1}{9} \cdot 5 = 2 - \dfrac{5}{9} = 1\dfrac{4}{9}$

Chapter 2 Algebraic Expressions

In your work in algebra, you will often use the following properties of equality.

> **Properties of Equality**
>
> For all real numbers a, b, and c:
> Reflexive Property $\quad\quad\quad a = a$
> Symmetric Property $\quad\quad$ If $a = b$, then $b = a$.
> Transitive Property $\quad\quad\;\;$ If $a = b$ and $b = c$, then $a = c$.

CLASS EXERCISES

Which property of real numbers justifies each statement?

1. $3 + [5 + (-2)] = (3 + 5) + (-2)$
2. $6 \cdot 13 = 13 \cdot 6$
3. $4(3 + 7) = 4 \cdot 3 + 4 \cdot 7$
4. $1.4 + 0 = 1.4$
5. $\frac{8}{7} \cdot \frac{4}{9} \cdot \frac{7}{8} = \frac{8}{7} \cdot \frac{7}{8} \cdot \frac{4}{9}$
6. $4 + (-1 + 9) = [4 + (-1)] + 9$

For Discussion

7. Is subtraction commutative? For all real numbers a and b, does $a - b = b - a$? Justify your answer by giving a numerical example.
8. Is division associative? For all real numbers a, b, and c, does $a \div (b \div c) = (a \div b) \div c$? Justify your answer by giving a numerical example.

PRACTICE EXERCISES

Simplify.

1. $(10 + 3) + 55$
2. $(16 + 7) + 32$
3. $98 + (27 + 9)$
4. $53 + (30 + 29)$
5. $\frac{2}{3} \cdot \left(\frac{1}{7} \cdot \frac{3}{2}\right)$
6. $\frac{7}{8} \cdot \left(\frac{1}{4} \cdot \frac{8}{7}\right)$

Simplify by using the distributive property.

7. $10(500 + 13)$
8. $12(200 + 10)$
9. $25(98 - 5)$
10. $36(53 - 12)$
11. $(1 - 100)0.25$
12. $(1000 + 30)1.5$
13. $\left(\frac{2}{3} - \frac{8}{21}\right)7$
14. $\left(\frac{2}{5} - \frac{7}{8}\right)5$

2.3 Properties of Real Numbers

Which property of real numbers justifies each statement?

15. $-3(a + 2) = -3(a) + (-3)(2)$
16. $(5 + y)(-2) = 5(-2) + y(-2)$
17. $8 - 7xy = 8 - 7yx$
18. $8(4b) = (8 \cdot 4)b$
19. If $12 = 3x$, then $3x = 12$.
20. If $y + 4 = 9$ and $9 = 5 + 4$, then $y + 4 = 5 + 4$.

Rewrite each expression, using the distributive property.

21. $2(3x - 11)$
22. $(6 + 2y)7$
23. $-\dfrac{3}{4}\left(\dfrac{2}{3}x - 8\right)$
24. $(-5j + 20)\left(-\dfrac{3}{5}\right)$
25. $12a - 3a$
26. $19b + 11b$

Is the property true with respect to the given operation(s) for all real numbers? If not, give a statement to show that it is false.

27. Associative; subtraction
28. Commutative; division
29. Distributive; division over multiplication
30. Distributive; multiplication over division
31. Distributive; subtraction over division
32. Distributive; multiplication over multiplication

Applications

Use the properties of real numbers to find the total cost.

33. **Consumerism** Flour—3 bags @ 99¢ per bag; Cereal—3 boxes @ $2.19 per box; Juice—3 bottles @ $1.31 per bottle
34. **Consumerism** Envelopes—10 boxes @ 89¢ per box; Ballpoint pens—5 boxes @ $1.59 per box; Staples—10 boxes @ $2.59 per box

WRITING IN ALGEBRA

On a separate piece of paper, write a brief description of each term. For example:

Variable: A letter or symbol used to represent unspecified number(s)

1. absolute value
2. additive inverse
3. grouping symbol
4. multiplicative inverse
5. rational number
6. real number
7. operation
8. zero
9. inequality symbol

2.4 Combining Like Terms

Objective: To simplify algebraic expressions by combining like terms

A *term* of an expression is a number, a variable, or a product or quotient of numbers and variables. Examples of terms are 13, x, $5y^2$, $6ab$, and $\frac{x}{7}$. The expression $6x^2 + x - 6$ consists of three terms: $6x^2$, x, and -6. The **numerical coefficient** or **coefficient** of the term $6x^2$ is 6. The coefficient of x is 1, since $x = 1x$.

Capsule Review

$5a + 11a = 11a + 5a$ by the commutative property for addition
$5a + 11a = (5 + 11)a$ by the distributive property

Use the property listed to write an expression equivalent to $4x^2 + 3x^2$.

1. commutative property for addition   **2.** distributive property

Use the property listed to write an expression equivalent to $9(3x - 7)$.

3. distributive property   **4.** commutative property for multiplication

Algebraic expressions such as $5a + 11a$ and $4x^2 + 3x^2$ contain *like terms*.

> **Like** (or **similar**) **terms** are terms that are exactly the same or differ only in their numerical coefficients. Terms that are not like terms are called **unlike terms**.

EXAMPLE 1 Name the terms and tell whether they are like or unlike:
a. $4x^2y + 3x^2y$ **b.** $-6a^2b + 3ab^2$ **c.** $m - 3m$ **d.** $3d + 6$

Expression	Terms	Like or Unlike
a. $4x^2y + 3x^2y$	$4x^2y$, $3x^2y$	Like terms
b. $-6a^2b + 3ab^2$	$-6a^2b$, $3ab^2$	Unlike terms
c. $m - 3m$	m, $-3m$	Like terms
d. $3d + 6$	$3d$, 6	Unlike terms

Explain why the terms in (a) are called like terms and the terms in (b) are called unlike terms.

2.4 Combining Like Terms

Some algebraic expressions can be simplified by using the distributive property to combine like terms, such as $4y + 7y$:

$$4 \cdot y + 7 \cdot y = (4 + 7)y, \text{ or } 11y$$

EXAMPLE 2 Simplify: **a.** $x + 6x$ **b.** $8z + 5z + 4z^2 - 3z^2$
c. $-4 + 3a - 2b + 5b - 6 - 7a$

a. $x + 6x$
$= 1x + 6x$
$= (1 + 6)x$
$= 7x$

b. $8z + 5z + 4z^2 - 3z^2$
$= (8z + 5z) + (4z^2 - 3z^2)$
$= (8 + 5)z + (4 - 3)z^2$
$= 13z + 1z^2$, or $13z + z^2$

c. $-4 + 3a - 2b + 5b - 6 - 7a$
$= (3a - 7a) + (-2b + 5b) + (-4 - 6)$
$= -4a + 3b + (-10)$
$= -4a + 3b - 10$

EXAMPLE 3 Simplify: **a.** $4 + 2(3x + 1) + 5x$ **b.** $4mn^2 + 3m^2n$

a. $4 + 2(3x + 1) + 5x$
$= 4 + 6x + 2 + 5x$
$= (6x + 5x) + (4 + 2)$
$= 11x + 6$

b. $4mn^2 + 3m^2n$
Unlike terms cannot be combined.

CLASS EXERCISES

For each expression; state the number of terms, name each term, and give its numerical coefficient.

1. $3x + y$
2. $-4xy^3$
3. $3a + a + 5a + 4$
4. $xy^2 + x^2y$

For each expression, tell whether the terms are like or unlike.

5. $6 + 5m$
6. $4xy - 5xy$
7. $3r^2s^3 - 6r^2s^3$
8. $u^3 + u^2$

Simplify.

9. $3x + 8x$
10. $5y + 4y$
11. $-4a^2 + 11a^2$
12. $16m^2 - 3m^2$

PRACTICE EXERCISES

For each expression, tell whether the terms are like or unlike.

1. $4xy^2z - xy^2z$
2. $r^2st - 12r^2st$
3. $7a^2b + 3ab^2$
4. $8m^2n^2 + 3mn^2$
5. $2x + 6x$
6. $8a + 3a$

7. $2 + 5m$
8. $7y - 6$
9. $t^2 - t$
10. $x - x^2$
11. $4a + a$
12. $b + 3b$

Simplify.

13. $3x + 5x$
14. $2y + 8y$
15. $-9t - 3t$
16. $-2q - 5q$
17. $9 + 2t + 6$
18. $8 + 3x + 7$
19. $15g^2 - 3g^2$
20. $12m^3 - 9m^3$
21. $5x + 2x + 6x^2 + x^2$
22. $3y^3 + 2y^3 + 8y - y$
23. $-2 + 5g - 4h + 3h + 4 + 7g$
24. $-1 - 4x + 6y + 3 - 4x - 6y$
25. $3ab + 6ab$
26. $10rs + 13rs$
27. $3ab^2 + 2ab^2$
28. $6m^2n + 3m^2n$
29. $1 + 6(m + 2) - 2m$
30. $-2 + 3(2b - 8) + 4b$
31. $2 + 3x - 4y + 11y - 6 - 9x$
32. $6m - 19 + 27 - 11n + 4m - 11n$
33. $3(x + 2) + 5$
34. $2(x + 1) + 3$
35. $3r + 2(r - 1)$
36. $6s + 3(s - 2)$
37. $4 + 3(2x + 1)$
38. $3 + 4(3x + 2)$
39. $3(5c) - 15c^2 + c$
40. $9b^2 + 4(5b) - 8b^2$
41. $4(2x^3 + x^2) - 5x^2$
42. $-8b^3 + 5(b^2 - 4b^3)$
43. $(10 + 5r)13 + (11 + 6r)15$
44. $(6p - 22)5 + (7p - 25)4$
45. $4(m + n) + 3(m + n)$
46. $6(a + b) + 7(a + b)$
47. $2(x^2 - y) + 4(x^2 - y)$
48. $7(g^3 - h) + 3(g^3 - h)$
49. $2(t - t^2) + 5(t + t^2)$
50. $5(b^2 + b) + 8(b^2 - b)$
51. $3(x^2y + xy^2) + 6(x^2y^2 - 2xy^2)$
52. $7(3mn - mn^2) + 5(mn^2 - m^2n)$
53. $\frac{1}{2}(4a + 2b) + \frac{1}{3}(6a - 3b)$
54. $\frac{1}{4}(8v + 4w) + \frac{1}{2}(10w - 12v)$
55. $\frac{1}{4}\left(y - \frac{1}{2}\right) + \left(y + \frac{1}{3}\right)\frac{3}{8}$
56. $\frac{3}{10}(a - 10) + \left(\frac{3}{4}a - 6\right)\frac{2}{5}$
57. $6[5c + 4(d - 2c)] - 5c$
58. $4[2a + 3(2b - a)] + 8b$
59. $0.69[b + 2(0.6b - 0.5)]$
60. $0.4[2.54(a + 0.2) - a]$

Applications

61. Number Problem Write an algebraic expression for the sum of a number n and the next two consecutive numbers, $n + 1$ and $n + 2$. Simplify the expression.

62. Number Problem Write an algebraic expression for the sum of an even number n and the next two consecutive even numbers, $n + 2$ and $n + 4$. Simplify the expression.

Geometry The area of a rectangle is found by multiplying length times width. Find the total area of each set of figures.

63.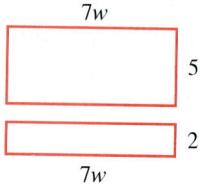

64.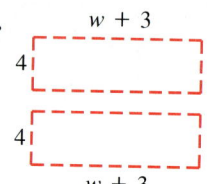

TEST YOURSELF

Simplify each expression. 2.1

1. $9 - 6 \cdot 3$
2. $-1 + 7(4 + 4)$
3. $8 \div 16 \cdot 2$

Evaluate each expression if $x = -2$, $y = 0$, and $z = 3$. 2.2

4. $5x + 3$
5. $2x - 3y + 4z$
6. $x + \dfrac{y + z}{y - z}$

7. Simplify $3 \cdot 2^4 \cdot 4$.
8. Evaluate $-m^2 + n$ if $m = -3$ and $n = -2$.
9. Evaluate $V = \pi r^2 h$ for V if $\pi \approx 3.14$, $r = 3$, and $h = 2$.

Which property of real numbers justifies each statement? 2.3

10. $m + (-n) = (-n) + m$
11. $[2(6.5)] \cdot 10 = 2[(6.5) \cdot 10]$
12. $8\left(\dfrac{3}{8}x - 1\right) = 3x - 8$
13. $\dfrac{2}{3} + \left(-\dfrac{4}{5}\right) + \dfrac{5}{6} = \dfrac{5}{6} + \dfrac{2}{3} + \left(-\dfrac{4}{5}\right)$

Simplify each expression. 2.4

14. $-4y^2 + y + 3y^2$
15. $5hk - 5hk^2 + 6hk - 5h^2k$
16. $3(a^2 - 1) + 4(a^2 - 2)$
17. $5(x^2 + 2) - 3(x^2 + 1)$

2.5 Simplifying and Evaluating Expressions with Parentheses

Objective: To simplify and evaluate algebraic expressions containing grouping symbols

In mathematics, science, and situations that arise daily, it is often necessary to find the value of an algebraic expression where the numerical values of the variables are given.

For example, when the local weather forecaster states, "The current temperature downtown is 5°C." You can estimate the Fahrenheit temperature by doubling the Celsius temperature and adding 32. The actual Fahrenheit temperature can be found by evaluating the algebraic expression $(1.8 \cdot C) + 32$.

$$(1.8 \cdot C) + 32$$
$$= (1.8 \cdot 5) + 32$$
$$= 9 + 32$$
$$= 41$$

The Fahrenheit temperature is 41°.

Capsule Review

To evaluate $x - 2y$ if $x = 2$, $y = -3$, substitute the given values for the variables, and then simplify the result.

$x - 2y$
$= 2 - 2(-3)$ *Substitute 2 for x and -3 for y.*
$= 2 + 6$ *Multiply.*
$= 8$ *Then add.*

Evaluate each expression if $x = 2$ and $y = -3$.

1. xy
2. $2xy$
3. $2x + y$
4. $x + 2y$
5. $x - y$
6. $y - x$
7. $3x - y$
8. $2y - 2x$

One way to simplify an expression such as $-(3x - 2)$ is to apply the property of -1 for multiplication.

Property of -1 for Multiplication

For every real number a,
$$-1 \cdot a = -a \text{ and } a \cdot -1 = -a$$

2.5 Simplifying and Evaluating Expressions with Parentheses

To apply this property to $-(3x - 2)$, think of $(3x - 2)$ as a real number a.

$$\overset{a}{\overbrace{-(3x - 2)}}$$
$$= -1(3x - 2) \qquad -a = -1 \cdot a$$
$$= -1(3x - 2) \qquad \text{Use the distributive property.}$$
$$= -1 \cdot 3x + (-1)(-2)$$
$$= -3x + 2$$

Because the terms in $-3x + 2$ are unlike, it is not possible to simplify the expression further.

EXAMPLE 1 Simplify: $2a - 3 - (a + 2)$

$$2a - 3 - (a + 2)$$
$$= 2a - 3 + (-1)(a + 2) \qquad -(a + 2) = -1(a + 2)$$
$$= 2a - 3 + (-1)(a) + (-1)(2)$$
$$= \underline{2a + (-1a)} + \underline{(-3) + (-2)} \qquad \text{Group like terms.}$$
$$= 1a + (-5) \qquad \text{Combine like terms.}$$
$$= a - 5$$

Here is another way to simplify $2a - 3 - (a + 2)$. Compare it with the steps shown in Example 1.

$$2a - 3 - (a + 2)$$
$$= 2a - 3 - 1(a + 2) \qquad \text{Omit the step using the + symbol.}$$
$$= 2a - 3 - a - 2 \qquad \text{Use the distributive property mentally.}$$
$$= a - 5 \qquad \text{Combine like terms without showing the grouping.}$$

EXAMPLE 2 Simplify: $(16x - 11) - 3(2x + 4)$

$$(16x - 11) - 3(2x + 4)$$
$$= 16x - 11 - 6x - 12 \qquad \text{Remove parentheses.}$$
$$= 16x - 6x - 11 - 12$$
$$= 10x - 23 \qquad \text{Combine like terms.}$$

In Example 2, $(16x - 11) = 1(16x - 11)$ by the identity property for multiplication. The parentheses are removed by distributing 1 over $(16x - 11)$.

EXAMPLE 3 Simplify: $-(2xy^2) - (3xy^2 + 4x^2y)$

$$-(2xy^2) - (3xy^2 + 4x^2y)$$
$$= -2xy^2 - 3xy^2 - 4x^2y$$
$$= -5xy^2 - 4x^2y$$

EXAMPLE 4 Simplify: $-(r + s^2)$. Then evaluate if $r = 3$ and $s = -2$.

$$\begin{aligned}&-(r + s^2)\\&= -r - s^2\\&= -(3) - (-2)^2 \quad \text{Substitute 3 for } r \text{ and } -2 \text{ for } s.\\&= -3 - (4)\\&= -3 - 4\\&= -7\end{aligned}$$

When an expression contains more than one set of grouping symbols, begin with the innermost grouping and work toward the outermost grouping.

EXAMPLE 5 Simplify: $3 + 2[-4(p - 4) + 8p]$. Then evaluate if $p = 0.5$.

$$\begin{aligned}&3 + 2[-4(p - 4) + 8p]\\&= 3 + 2[-4p + 16 + 8p] \quad &&\text{Use the distributive property.}\\&= 3 + 2[4p + 16] \quad &&\text{Combine like terms within brackets.}\\&= 3 + 8p + 32 \quad &&\text{Use the distributive property.}\\&= 8p + 35 \quad &&\text{Combine like terms.}\\&= 8 \cdot 0.5 + 35 \quad &&\text{To evaluate } 8p + 35, \text{ substitute 0.5 for } p.\\&= 4 + 35\\&= 39\end{aligned}$$

CLASS EXERCISES

Simplify.

1. $-(x + 3)$
2. $-(y - 7)$
3. $-(-a + 4)$
4. $-(-b - 3)$
5. $-(2m^2 + 3)$
6. $-(5n^2 - 6)$
7. $-(-f^2 + f)$
8. $-(-2g^2 - g)$
9. $-(ab^2) - (ab^2)$
10. $-(2rs) - (rs)$
11. $2x - 4 - (x - 5)$
12. $y - 7 - (4 + 2y)$

Simplify. Then evaluate if $a = 6$ and $b = -5$.

13. $-(b - a^2)$
14. $-4[2(b + 3) - a]$

PRACTICE EXERCISES

Simplify.

1. $2g - 4 - (g + 4)$
2. $4 + 5m - (3 + m)$

3. $(3a + 4) - (5a - 9)$
4. $(2h - 5) - (7h + 5)$
5. $(5t^2 - 2) - (2t^2 + 7)$
6. $(9p^2 - 1) - (3p^2 + 2)$
7. $(x + 5) - 3(x + 2)$
8. $(y + 3) - 2(y + 7)$
9. $-(ab^2c) - (3abc^2 - 2ab^2c)$
10. $-(a^2bc) - (-a^2bc^2 - a^2bc)$

Simplify each expression. Then evaluate for the given value(s) of the variables in Exercises 11–18 and 25–28.

11. $-(a + b^2)$ if $a = 2$ and $b = -5$
12. $-(m + n^2)$ if $m = 3$ and $n = -7$
13. $-(x - y^2)$ if $x = 4$ and $y = -1$
14. $-(r - s^2)$ if $r = 1$ and $s = -5$
15. $8 + 3[-2(p - 3) + 2p]$ if $p = 0.3$
16. $5 + 2[-1(z - 7) + 4z]$ if $z = 0.2$
17. $13 - 4[3(q + 1) - 2q]$ if $q = -1$
18. $10 - 3[4(a + 2) - 3a]$ if $a = -3$
19. $x^2yz - (-x^2yz^2 - x^2yz)$
20. $x^2yz^2 - (-xyz^2 - x^2yz^2)$
21. $2 + 2[-3(x - 2)]$
22. $7 + 3[-2(x - 1)]$
23. $(5r - 9) + (r - 9)16$
24. $(x - 11)12 + (x + 13)$
25. $-(m - n^2)$; $m = 4$, $n = -5$
26. $-(y - z^2)$; $y = -6$, $z = 7$
27. $5 - 6[10(x^2 + 2x) + 3x]$; $x = -2$
28. $11 - [4y + 3(y^2 - 5y)]8$; $y = 2$

Simplify.

29. $4 - \{m + [5(m + 1) - m]\}$
30. $2 - \{g + [3(g - 1) + g]\}$
31. $1 - \{a - [a + (a - 3)]\}$
32. $1 - \{q - [q - (q + 4)]\}$
33. $5 - 4\{13 + [3(2x - 7 + x)]\}$
34. $7 + 9\{11 - [4(3z + 6 - z)]\}$

Applications

The formula for finding the area A of a trapezoid is

$A = \frac{1}{2}h(b_1 + b_2)$

where h is the height and b_1 and b_2 are the lengths of the two parallel bases.

35. **Geometry** Find the area of a trapezoid if its height is 3 cm and the lengths of its bases are 5 cm and 7 cm.

36. **Geometry** Find the area of a trapezoid if its dimensions are: height, 14 ft, and bases, 10 ft and 12 ft.

Geometry Write an expression for the perimeter of each figure. Simplify the expression.

37.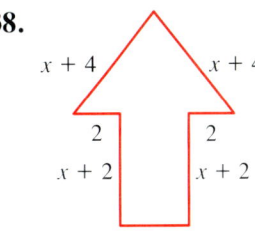

38.

Calculator One way to evaluate $4(p - 3) + p^2$ if $p = 0.3$ is to use the Memory Key on your calculator. Store the value 0.3 in Memory. Then evaluate as usual, except press Memory Recall Key when p is needed.

You may find that the following exercises can be worked more quickly by using a calculator.

Evaluate each expression for the given value of the variable.

39. $10 - (y - 2) + y^2$; $y = 1.7$
40. $9 + (2x - 3) + x^2$; $x = 3.1$
41. $(a - 4)12 + (a + 6)13$; $a = -1$
42. $10(b + 9) - 3(b - 8)$; $b = -2$
43. $(m - 6)^2 + 3(m - 6)$; $m = -2$
44. $\frac{1}{2}(x - x^2) + \frac{1}{3}(x^2 - 1)$; $x = 6$

ALGEBRA IN GEOMETRY

1. The perimeter of a figure is the sum of the lengths of its sides. If the sides measure n, $n + 1$, $n + 2$, and $n + 3$, find the perimeter of the figure in terms of n. What is the least integral sum that the perimeter can be?

2. The rectangle below measures 37 units by 13 units. Its area is 481 square units.

   13 units $A = l \cdot w$

   37 units

   The number $481 = (37)(13)$. Express the factor 37 as the sum of two squares. How many 6 by 6 squares can fit into the rectangle? How many 1-unit squares are left?

2.5 Simplifying and Evaluating Expressions with Parentheses

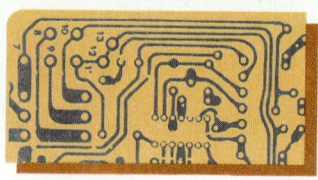

TECHNOLOGY:
Introduction to Spreadsheets

One of today's most widely used computer software applications programs is a spreadsheet. A **spreadsheet** is a table of information that is organized into rows and columns.

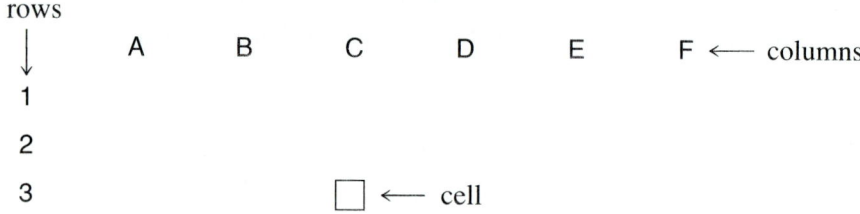

The intersection of a row and a column is a **cell**. A cell is named by the column-row (letter-number) that forms it. The cell indicated above is C3.

Each cell can contain two types of data:

- Labels: letters or words such as R or JOB
- Values: numbers such as 7 or $4.25 or expressions such as 28 + 72 or B3 + D3

The second expression above, B3 + D3, tells the computer to add the contents of cells B3 and D3. The sum will be displayed in the same cell where the expression is located.

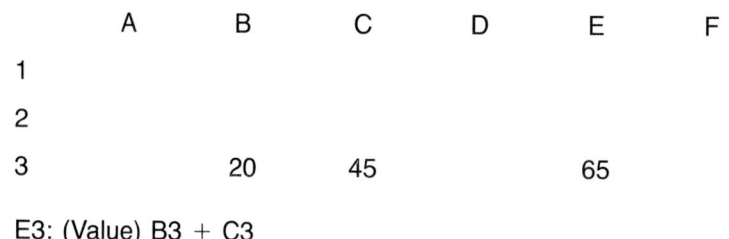

E3: (Value) B3 + C3

When you enter an expression or formula into a spreadsheet, the spreadsheet will calculate and display the results in various cells in certain ways. If you change the information in any cell, the program can recalculate every other cell that uses that information. A spreadsheet performs calculations using the same rules for order of operations that you use when making calculations in mathematics.

Chapter 2 Algebraic Expressions

A feature of spreadsheets is the "what if" speculating it allows you to do. Suppose you wanted to see how best to use your money. You might set up a spreadsheet like the one below.

```
File: Money                         REVIEW/ADD/CHANGE
========= A =========B=========C=========D=========E =========
 1:
 2: WEEKLY INCOME
 3:    BALANCE                 $15.45
 4:    ALLOWANCE
 5:    JOB
 6:    MISCELLANEOUS
 7:
 8:    TOTAL INCOME
 9:
10:
11: WEEKLY EXPENSES
12:    CLOTHES
13:    TAPES
14:    FOOD
15:    MOVIES:
16:    MISCELLANEOUS
17:
18:    TOTAL EXPENSES
19:
20:
21: MONEY FOR SAVINGS
C8: (Value)
```

If you enter values in the INCOME categories for a given week, you can then experiment by putting various values in the EXPENSES categories for the same time span.

Many functions available in a spreadsheet are shortcuts for certain kinds of arithmetic. The @SUM function, for example, totals a group of values and presents the result in a cell where the function appears. You will learn more about this function later.

Solve. Use cell names to state the formula that you would use for:

1. C8   **2.** C18   **3.** C21

4. Use a spreadsheet to project your monthly savings based on your weekly income and expenses.

5. If you want to save $100 a month toward the cost of a new stereo, how could you adjust your income and expenses to meet that goal?

6. If you planned on saving the same amount of money each week, what expression could you put in C23 if it represents how much money you save in a year?

7. If in a given week you received an additional $25 that you do not usually get, use the "what if," capability of the spreadsheet to see how you could spend the $25 and still save the amount you usually do.

2.6 Translating Phrases to Algebraic Expressions

Objectives: To translate a word phrase to an algebraic expression
To write word phrases for algebraic expressions

You can use algebra as a tool for applying mathematical logic to real-life situations and solving problems. In order to do so, you must be able to translate simple English word phrases into *algebraic expressions*.

Capsule Review

EXAMPLE Translate each word phrase to an algebraic expression.

x decreased by *y*	$x - y$	*a* less than *z*	$z - a$
p divided by *q*	$\dfrac{p}{q}$	*u* more than *y*	$y + u$

Translate each word phrase to an algebraic expression.

1. The sum of *a* and *b*
2. *x* decreased by *d*
3. The product of *e* and *f*
4. *g* divided by *h*
5. *i* less than *j*
6. *k* more than *l*
7. The quotient of *m* and *n*
8. The difference of *o* and *p*
9. Two minus *y*
10. *r* times 4

EXAMPLE 1 Translate each word phrase to an algebraic expression.

    a. Pedro's age in 4 years if *p* is his age now
    b. 3.14 times the square of the length of the radius *r*
    c. The sum of two consecutive integers if *n* is the first integer

    a. $p + 4$ **b.** $3.14 \cdot r^2$, or $3.14r^2$ **c.** $n + (n + 1)$, or $2n + 1$

When you need to translate a word phrase to an algebraic expression but the variable is not given, the following steps may apply:

1. Assign a variable to one of the unknown quantities.
2. Write an expression for any other unknown quantities in terms of the same variable.

EXAMPLE 2 Translate each word phrase to an algebraic expression.

    a. Twice the sum of a number and 3
    b. Woodrow's age 7 years ago
    c. A given length subtracted from 3 times that length

    a. Let x = the number. $2(x + 3)$

    b. Let w = Woodrow's age now. $w - 7$

    c. Let l = length.
        Then $3l$ = 3 times the length. $3l - l$

Explain why $2x + 3$ is not a correct translation of the word phrase in Example 2a.

EXAMPLE 3 Translate to an algebraic expression: The amount of money Della has (in cents) if she has 6 more dimes than Felix.

    Let d = number of dimes Felix has.
    Then $d + 6$ = number of dimes Della has.
    The value of $d + 6$ dimes is $10(d + 6)$ cents.

    So, Della has $10(d + 6)$ cents.

Working backwards can be a helpful strategy when translating word phrases to algebraic expressions.

EXAMPLE 4 Write two word phrases for each algebraic expression:
    a. $k - 5$ **b.** $5n + 25$

    a. A number k decreased by 5   **b.** 5 times a number n increased by 25
        5 less than some number k The sum of $5n$ and 25

CLASS EXERCISES

Translate each word phrase to an algebraic expression.

1. Six more than a number
2. A number decreased by 2
3. One-half of a number
4. The quotient of a number and 6
5. Two less than 5 times a number
6. Five more than 2 times a number
7. The product of 5 and a number squared
8. The quotient of a number squared and 3
9. Four times the sum of a number and 9
10. Twice the difference of a number and 7

2.6 Translating Phrases to Algebraic Expressions

Write two word phrases for each algebraic expression.

11. $2n + 3$ **12.** $9 - 3c$ **13.** $r^2 + 1$ **14.** $10d^2 + 60$

PRACTICE EXERCISES

Translate each word phrase to an algebraic expression.

1. Seven less than a number
2. A number decreased by 6
3. Five more than a number
4. A number increased by 6
5. One less than 8 times a number
6. Nine more than 1 times a number
7. The product of 5 and a number squared
8. The product of 6 and a number squared
9. The quotient of a number squared and 3
10. The quotient of a number squared and 2
11. Six times the sum of a number and 3
12. Twice the sum of 3 and a number

Write two word phrases for each algebraic expression.

13. $3d - 5$ **14.** $7d + 6$ **15.** $w(w + 3)$

Answer each question with an algebraic expression.

16. Let s be Sumi's age now.
    a. What was Sumi's age 4 years ago?
    b. Hiko is 2 years older than Sumi. Represent Hiko's age now.
    c. What was Hiko's age 4 years ago?

17. Let m be Mike's age now.
    a. What was Mike's age 4 years ago?
    b. Tina is 4 years younger than Mike. Represent Tina's age now.
    c. What was Tina's age 4 years ago?

Solve. Use $A = lw$ to find the area and $P = 2l + 2w$ to find the perimeter.

18. Let w be the width of a rectangle whose length is 4 units more than its width.
    a. What is the width?   b. What is the length?
    c. Find the area.   d. Find the perimeter.

19. Let l be the length of a rectangle whose width is 7 units less than its length.
    a. What is the length?   b. What is the width?
    c. Find the area.   d. Find the perimeter.

Chapter 2 Algebraic Expressions

20. Tony buys baseball pennants at 3 for a¢ and sells them all at 5 for b¢. Represent Tony's profit on each pennant.

21. Sadie bought several items at the record store. A tape costs $12 less than a disc, and the same amount as a record. If a tape costs t dollars, what is the cost of 4 tapes, 3 discs, and 2 records? How much change did Sadie receive if she paid for these items with $11t$ dollars?

Applications

Write an algebraic expression for one unknown in terms of the other.

22. **Sports** The Los Angeles Dodgers have won twice as many World Series as the Chicago Cubs.

23. **Sports** If the Pittsburgh Pirates win one more series, they will have won 3 times as many series as the Chicago Cubs.

24. **Sports** Baseball trivia.
    a. Joe DiMaggio hit 3 more home runs than Yogi Berra.
    b. If Babe Ruth had hit 2 more home runs, he would have hit twice as many as Yogi Berra.
    c. If Hank Aaron had hit 1 more home run, he would have hit 40 more than twice as many as Yogi Berra.

CAREER

For a computer to be useful, information must be input into the computer and processed information must be retrieved from it. Computer software is of two types: programs and data. Programs are sets of instructions; data are collections of organized information. Software is generally stored on disks.

If you are interested in computers, you may find a career as a **computer software developer** to be both rewarding and lucrative.

William Gates started a computer programming company in 1970 when he was in the 10th grade. One of his programs was written to schedule classes in schools.

In 1975 William Gates and a friend, Paul Allen, started a computer software company and they became multimillionaires from the company's profits.

Conduct research and list at least two other people who have been successful in the field of computers since 1975.

2.7 Open Sentences and Solution Sets

Objective: To find solution sets of open sentences from given replacement sets

Usually you see lightning flash and then listen for the thunder. The thunder often occurs within 10 seconds. Sometimes it is possible to tell the distance to the thunderstorm. A science reference book indicates that the formula $d = \frac{1}{5}s$ can be used to find the distance from the thunderstorm.

d = distance, in miles, of storm from observer
s = time in seconds

$$d = \frac{1}{5}s$$
$$= \frac{1}{5} \cdot 10 = 2 \text{ mi}$$

Capsule Review

Find the value of *A* for the given values of the variables.

EXAMPLE

$w = 3$, $l = 5$

$A = lw$
$= 5 \cdot 3$
$= 15$
$A = 15$ square units

1. $l = 8$ and $w = 6$
2. $l = 11.1$ and $w = 5$
3. $l = \frac{1}{2}$ and $w = 7$

Find the distance *d* for the given values of the rate *r* and the time *t*.

4. $r = 10$ and $t = 2$
5. $r = 50$ and $t = 3$
6. $r = 26$ and $t = \frac{1}{2}$

Some of the most powerful tools of mathematics are *equations* and *inequalities*. They are often used in solving problems in daily life.

> An **equation** is a mathematical sentence in which the symbol = (equals) connects two numerical or variable expressions.
>
> An **inequality** is a mathematical sentence in which the symbol < (less than), > (greater than), or ≠ (is not equal to) is used.

Chapter 2 Algebraic Expressions

An equation such as $x - 5 = 1$ or an inequality such as $x - 3 < 4$ that contains a variable is neither true nor false until you replace the variable with a number. Such equations and inequalities that contain variables are called **open sentences.** The symbols $\not<$ (not less than) and $\not>$ (not greater than) are also used in inequalities.

> A **solution** of an open sentence is any value of the variable that makes the open sentence a true statement.
>
> The **replacement set** of an open sentence is the set of numbers that may be substituted for the variable.
>
> The **solution set** of an open sentence is the set of all of the numbers from the replacement set that make the open sentence true.

EXAMPLE 1 Find the solution set of $1 - 6g = -1$. The replacement set is $\left\{-1, 0, \frac{1}{3}\right\}$.

Replace g with -1.

$$1 - 6g = -1$$
$$1 - 6(-1) \stackrel{?}{=} -1$$
$$1 + 6 \stackrel{?}{=} -1$$
$$7 \neq -1$$

Replace g with 0.

$$1 - 6g = -1$$
$$1 - 6(0) \stackrel{?}{=} -1$$
$$1 - 0 \stackrel{?}{=} -1$$
$$1 \neq -1$$

Replace g with $\frac{1}{3}$.

$$1 - 6g = -1$$
$$1 - 6\left(\frac{1}{3}\right) \stackrel{?}{=} -1$$
$$1 - 2 \stackrel{?}{=} -1$$
$$-1 = -1 \quad \checkmark \quad \text{True}$$

$\frac{1}{3}$ makes $1 - 6g = -1$ true. The solution set is $\left\{\frac{1}{3}\right\}$.

EXAMPLE 2 Find the solution set of $x - 5 < -4$. The replacement set is $\{-1, 0, 1\}$.

Replace x with -1.

$$x - 5 < -4$$
$$-1 - 5 \stackrel{?}{<} -4$$
$$-6 < -4 \quad \checkmark \quad \text{True}$$

Replace x with 0.

$$x - 5 < -4$$
$$0 - 5 \stackrel{?}{<} -4$$
$$-5 < -4 \quad \checkmark \quad \text{True}$$

Replace x with 1.

$$x - 5 < -4$$
$$1 - 5 \stackrel{?}{<} -4$$
$$-4 \not< -4$$

-1 and 0 make $x - 5 < -4$ true. The solution set is $\{-1, 0\}$.

When no number from the replacement set makes an open sentence true, the solution set is the **empty set,** { }. Another symbol for the empty set is $\emptyset$.

2.7 Open Sentences and Solution Sets

EXAMPLE 3 Find the solution set of $-2 > y - 3$. The replacement set is $\{1, 1\frac{1}{2}, 2\}$.

$$-2 > y - 3 \qquad -2 > y - 3 \qquad -2 > y - 3$$
$$-2 \stackrel{?}{>} 1 - 3 \qquad -2 \stackrel{?}{>} 1\frac{1}{2} - 3 \qquad -2 \stackrel{?}{>} 2 - 3$$
$$-2 \not> -2 \qquad -2 \not> -1\frac{1}{2} \qquad -2 \not> -1$$

The solution set is $\emptyset$.

EXAMPLE 4 Solve: $x \neq \frac{1}{x}$. The replacement set is $\{-1, 1, 2\}$.

$x \neq \frac{1}{x}$ is read as "x is not equal to $\frac{1}{x}$."

$$x \neq \frac{1}{x} \qquad x \neq \frac{1}{x} \qquad x \neq \frac{1}{x}$$
$$-1 \stackrel{?}{\neq} \frac{1}{-1} \qquad 1 \stackrel{?}{\neq} \frac{1}{1} \qquad 2 \stackrel{?}{\neq} \frac{1}{2}$$
$$-1 = -1 \qquad 1 = 1 \qquad 2 \neq \frac{1}{2} \quad \checkmark \quad \text{True}$$

The solution set is $\{2\}$.

CLASS EXERCISES

Classify each sentence as *true* or *false*.

1. $3 \cdot 4 - 8 = 10$
2. $-5 \cdot 6 < -7$
3. $-9 \cdot \frac{1}{9} \neq 9\left(-\frac{1}{9}\right)$

Find the solution set of each sentence. The replacement set is $\{-1, 0, 1\}$.

4. $a + 3 = 0$
5. $3m < 0$
6. $g - 2 \neq -1$
7. $q + q = 2$
8. $n + 1 > 0$
9. $x + 2 = 0$

PRACTICE EXERCISES

Find the solution set of each sentence. The replacement set is $\{-1, 0, 1\}$.

1. $5 - 4x = 1$
2. $2 - 3z = 5$
3. $n + 3 = 3$
4. $m + 2 = 1$
5. $y - 1 < 2$
6. $x + 1 > -2$
7. $a - 0 \neq 0$
8. $b - 1 \neq 0$
9. $3y + 5 = 5$
10. $2x + 1 = 3$
11. $2x + 3 > 0$
12. $1 - 6h < 7$
13. $3g = -3$
14. $-5d = -5$
15. $-4 > 2y - 5$

Find the solution set. The replacement set is $\{-2, -1, 0, \frac{1}{2}\}$.

16. $-6 < 3 - 3m$
17. $-x - 3 = 5x$
18. $-n - 4 = 6n$
19. $9c = 7c + 1$
20. $d^2 - 2 = 2$
21. $f^2 - 3 = -2$
22. $2d \cdot 2d = 0$
23. $1 = m \cdot m$
24. $-x = \frac{4}{x}$
25. $\frac{1}{a} = -1$
26. $5x - 2 \neq 2 - 5x$
27. $-7y - 6 \neq -6 - 7y$

Write two different open sentences for which each of the following is the solution set. The replacement set is $\{-2, -1, 0, 1, 2\}$.

28. $\{1\}$
29. $\{-2, 2\}$
30. $\{-2, -1, 0, 1, 2\}$
31. $\{0, 2\}$

If possible, find a number that can replace x in each open sentence to make it a true statement. If no such number can be found, explain why.

32. $\frac{7}{x} = 3$
33. $\frac{6}{x} < 0$
34. $x^2 + 2x \neq x + (x^2 + x)$

Applications

Number Problems Translate each open sentence into words.

35. $-5 > -7$
36. $4 > -11$
37. $a - 6 > 0$

DID YOU KNOW?

Did you know that the first successful electronic computer, Colossus 1, was built in England in 1943? It was designed for a single purpose—to crack secret codes during World War II. The main designer of Colossus 1 was an Englishman named Alan Turing (1912–1954). His theory of how a computer should work earned him the honor of being called "the father of computers."

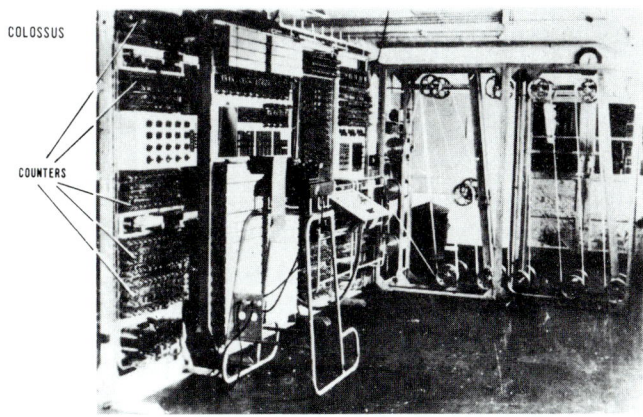

2.7 Open Sentences and Solution Sets

2.8 Problem Solving Strategy: Account for All Possibilities

The problem solving strategy of accounting for all possibilities means that every possible solution to a problem is identified and checked to see if it is actually a solution. This strategy can be used if the number of possible solutions is **finite,** meaning you can count the number of possible solutions.

The replacement set of the variable lists the possible solutions for a given open sentence. You can find the solution by replacing the variable with each number in the replacement set. When a number from the replacement set makes an open sentence a true statement, that number is a solution. A false statement indicates that the possible solution was not a solution.

If the replacement set of a variable is **infinite,** then the strategy of accounting for all possibilities cannot be used.

EXAMPLE Erik weighs w lb and his sister, Lauren, weighs 12 lb less. The sum of their weights is 158 lb. Find the weight of Erik and of Lauren.

Understand the Problem

Read the problem.

It is given that Erik weighs w lb and Lauren weighs 12 lb less. The sum of their weights is 158 lb.
You are asked to find the weights of Erik and of Lauren.

Plan Your Approach

Assign symbols.

Let w represent Erik's weight.
Let $w - 12$ represent Lauren's weight.

Write an equation.

$w + (w - 12) = 158$

The solution of the equation can be found by using the strategy of accounting for all possibilities. The number w must be less than 90, since $90 + (90 - 12) = 168$. Also, w is greater than 80, since $80 + (80 - 12) = 148$.
Thus, w is a number between 80 and 90.

Complete the Work Try the possible whole number solutions for *w*, starting with 81.

$81 + (81 - 12) = 150$ No
$82 + (82 - 12) = 152$ No
$83 + (83 - 12) = 154$ No
$84 + (84 - 12) = 156$ No
$85 + (85 - 12) = 158$ Yes

Interpret the Results **State your conclusion.**

Erik weighs 85 lb and Lauren weighs $85 - 12$, or 73 lb.

Check your conclusion.

$85 + 73 = 158$ ✔

CLASS EXERCISES

1. Explain the meaning of the word *finite*.

2. Is an open sentence true, false, or neither? Explain.

3. Does a solution to an equation or inequality make it true or false? Explain.

4. Explain the meaning of the word *infinite*.

PRACTICE EXERCISES

For each equation, find the solution set. In each case, the replacement set is the set of integers from −5 to +5, inclusive.

1. $x + 4 = 11$ **2.** $y - 2 = 3$ **3.** $n + 4 = 4$
4. $-t + 5 = 9$ **5.** $2k - 4 = -8$ **6.** $3(b - 2) = 0$
7. $m \div 3 = -1$ **8.** $z \div 5 = 0$ **9.** $2(a + 1) = 4$

Solve each equation by checking the possible solutions.

10. Lois has *d* dimes and $10(d + 6) = 140$.

11. Bob is *n* years old and $2n + 4 = 34$.

12. Ken worked *w* weeks and $\frac{w}{2} - 1 = 24$.

13. Suzanne edited *p* pages and $5 + \frac{1}{3}p = 16$.

Write an equation for each problem. Solve the equation by checking the possible solutions.

14. Helen weighs x pounds. Her sister weighs 9 lb more. The sum of their weights is 229 lb. How much do Helen and her sister weigh?

15. Carmen has 2 times as much money as Bill. She has $50 more than he has. How much money does each one have?

16. Harry and Beth each have some quarters. Harry has 5 more quarters than Beth. The combined value of their quarters is $4.75. How many quarters does each person have?

17. The sum of a number and six more than that number is -8. Find the number.

18. Is it possible to list the solutions of the inequality $x + 4 < -3$ if the replacement set for x is the set of all integers greater than -100? Explain.

19. The area of a rectangle is 252 ft^2. The perimeter of the rectangle is 64 ft. Find the length and width of the rectangle.

Mixed Problem Solving Review

1. A submarine made a dive of 419 ft and then came up 212 ft. What was the final depth of the submarine?

2. Write an equation for each word sentence. Solve the equation, using $\{-3, -2, -1, 0, 1, 2, 3\}$ as the replacement set.
   a. a number increased by 2 is -1.
   b. Twice a number less -3 is 7.
   c. The quotient of 9 and a number is -3.
   d. The difference between a number and 16 is -1.

PROJECT

Review the properties of real numbers discussed in Lesson 2.3. For each property, write its name and the replacement set for the variable or variables.

Are the replacement sets finite or infinite sets? Can you check if the properties are true by using the strategy of accounting for all possibilities? Explain your answer in writing.

Use a dictionary to look up the meaning of the words *assumption* and *postulate*. Do you think that every fact in arithmetic and algebra is a property? That is, do we just assume that all mathematical facts are true? If not, how can we tell true statements from false statements?

2.9 Translating Word Statements to Equations

Objectives: To translate a word statement to an equation
To write word statements for equations

The skills you acquire in your study of algebra have many useful applications. Various approaches can be used to solve application-type problems. One approach that is often useful is to translate a word statement into an equation and then solve the equation. The solution of the equation will lead to an answer to the question(s) asked in the problem.

To write an equation, you translate word phrases to algebraic expressions.

Capsule Review

Word Phrase	Algebraic Expression
Four times the length of a side s of a square	$4s$
The value in cents of a given number of dimes d	$10d$
The quotient of a number n and 2 less than the number	$\dfrac{n}{n-2}$

Translate each word phrase to an algebraic expression.

1. The sum of a number x and twice that number
2. The quotient of a number j and 5 times that number
3. The value in cents of a given number of half dollars
4. Zachary's age 9 years ago

Write a word phrase for each algebraic expression.

5. $r + 2$ 6. $30 - x$ 7. lw 8. $p + 8p$

EXAMPLE 1 Write an equation for each word sentence.
  a. The difference between x and y is 6.
  b. Nineteen is 7 more than twice z.
  c. The sum of 3 times a and 2 is 4.
  d. Three times the sum of a and 2 is 4.

The difference between x and y is 6.

a. $x - y = 6$ b. $19 = 2z + 7$ c. $3a + 2 = 4$ d. $3(a + 2) = 4$

Writing an equation becomes more challenging when the variable represents a specific number.

EXAMPLE 2 Use the first sentence to set up an algebraic expression. Write an equation for the second sentence.

John weighs p lb, and his brother weighs 17 lb more. The sum of their weights is 207 lb.

p = John's weight (lb)
$p + 17$ = brother's weight (lb)

Sum of their weights is 207 lb

$$p + (p + 17) = 207$$

By solving $p + (p + 17) = 207$, you would find John's weight, and then you would be able to find his brother's weight. In later chapters, you will solve equations of this type.

EXAMPLE 3 Write an equation using the first and second sentence.

Mineko has 4 times as much money as Tony. She has $15 more than he has.

m = amount of Tony's money (in dollars)
$4m$ = amount of Mineko's money

Mineko's money   15 more than Tony's

$$4m = 15 + m$$

If you solve the equation $4m = 15 + m$, you would know how much money Tony has, and you could find out how much money Mineko has (4 times as much).

EXAMPLE 4 Write an equation using the information given.

Loida and Rosita each have some quarters. Rosita has 6 more than Loida. The combined value of their quarters is $5.50.

q = number of quarters Loida has
$q + 6$ = number of quarters Rosita has

The value of Loida's quarters in cents is $25q$.
The value of Rosita's quarters in cents is $25(q + 6)$.

Combined value is $5.50, or 550¢.

$$25q + 25(q + 6) = 550$$

To improve your ability to translate word statements into equations, it is helpful to practice working backward. Note that there may be many ways to translate a given equation into a word statement.

EXAMPLE 5 Write a word statement for each equation.
  a. $w(w + 5) = 14$
  b. $x^2 = 144$

  a. The product of a number and five more than that number is 14.
  b. The square of a number is 144.

CLASS EXERCISES

Copy the sentence. Let $n =$ the unknown number and write an equation. Use arrows to show how the word statement and the equation are related.

1. Two times a number is 14.
2. A number decreased by 12 is 7.
3. One-half of a number is 34.
4. The quotient of a number and 6 is $\frac{2}{3}$.
5. Six less than twice a number is 5.
6. The product of 5 and a number squared is 45.

PRACTICE EXERCISES

Let $x =$ an unknown number. Write an equation for each sentence.

1. Three less than a number is 18.
2. A number decreased by 12 is 4.
3. Four more than a number is 27.
4. A number increased by -7 is 2.
5. Two less than 5 times a number is 18.
6. One more than twice a number is -13.
7. The product of a number squared and 8 is 56.
8. The product of 8 times a number and 7 is 79.
9. The quotient of 5 and 3 times a number is 10.
10. The quotient of 5 and a number squared is 100.
11. Six times the sum of a number and 9 is 132.
12. Three times the sum of a number and 3 is -18.

Write an algebraic expression. Then write an equation for the variable indicated for the last sentence.

13. Let b be Bonnie's age now.
    a. What was her age 6 years ago?
    b. If Clyde is 1 year younger than Bonnie, represent his age now.
    c. What was Clyde's age 6 years ago?
    d. If 6 years ago the sum of their ages was 35, write an equation to represent this.

14. Let *j* be Jesse's present age.
    a. What will Jesse's age be in 9 years?
    b. If James is 15 years younger than Jesse, represent his present age.
    c. What will James's age be in 9 years?
    d. If in 9 years Jesse will be 2.5 times as old as James, write an equation to represent this.

15. Let *n* be the number of $45 monthly payments made to purchase a VCR.
    a. What is the total value of these payments?
    b. If a $35 down payment was made, represent the total cost.
    c. If the VCR cost $485, write an equation to represent this.

16. Let *n* be the number of $55 monthly payments made to buy a stereo.
    a. What is the total value of the monthly payments?
    b. If there was a down payment of $65, represent the total cost.
    c. If the stereo cost $670, write an equation to represent this.

Write a word statement, or statements, for each equation.

17. $3n - 7 = 32$

18. $w - 3 = 4w$

19. $w(w + 6) = 48$

20. $2(l - 5) + 2l = 128$

Write an equation for each situation.

21. Joan's brother is now 3 times as old as she. In five years, his age will be twice hers. Let j = Joan's age now.

22. Paolo's mother is twice as old as he. Fourteen years ago, her age was 4 times his age. Let p = Paolo's age now.

23. Ernesto has a number of coins, all quarters. Laura has 6 more coins than Ernesto. Her coins are all dimes. The combined value of their coins is $2.35. Let e = number of coins Ernesto has.

24. Tony has a certain number of quarters and 11 more nickels than quarters. The total value of his coins is $2.65. Let q = number of quarters Tony has.

Applications

Write an equation for the underlined sentence. Tell what question(s) can be answered if the equation is solved.

25. **Consumerism** Tom's new car cost $150 more than twice as much as his old one. <u>The cost of the new car was $13,250.</u>

26. **Geometry** The length of a rectangle is 10 mm less than twice its width. <u>The area of the rectangle is 208 mm².</u>

Chapter 2 Algebraic Expressions

27. **Consumerism** Leila paid $35 down and made 5 equal monthly payments on her prom dress. The prom dress cost $110.

28. **Geometry** The length of a rectangle is 25 in. longer than three times its width. The perimeter of the rectangle is 130 in.

29. **Consumerism** Sheila paid $100 more than twice the amount Howard paid in tuition for college courses. Howard paid $1200 in tuition.

30. **Consumerism** Jim purchased a compact disc player. He paid $65 down and made 12 equal monthly payments. The total cost of his compact disc player was $245.

31. **Geometry** The height of a triangle is 2 cm more than 4 times the length of the base. The area of the triangle is 55 cm².

32. **Geometry** Two sides of a triangle have the same length. The third side is 12 cm less than 3 times the length of the equal sides. The perimeter of the triangle is 23 cm.

TEST YOURSELF

Simplify each expression.   2.5

1. $2r - 4(r + 3)$
2. $(y - 6) - 2(y + 2)$
3. Simplify $-(m^2 - n)$. Then evaluate if $m = -3$ and $n = -2$.

Translate each word phrase to an algebraic expression.   2.6

4. Seven times the sum of a number and 3
5. Eight less than the quotient of a number and 9

Write two word phrases for each algebraic expression.

6. $10d - 6$
7. $\frac{y}{4} + 6$

Find the solution set of each sentence. The replacement set is $\{-2, -1, 0, 1, 2\}$.   2.7

8. $m + 4 = 3$
9. $z - 2 > 1$
10. $2x + 1 \neq x$

Write an equation for each sentence.   2.8, 2.9

11. A number decreased by -2 is -11.
12. Two more than 5 times a number is 25.

CHAPTER 2 SUMMARY AND REVIEW

Vocabulary

additive identity (59)
base (55)
coefficient (63)
empty set (79)
equation (78)
exponent (55)
factor (55)
inequality (78)
like terms (63)
multiplicative identity (59)
open sentence (79)
properties of real numbers (59)
replacement set (79)
solution (79)
solution set (79)
unlike terms (63)

Rules for Order of Operations

- Perform any operation(s) within grouping symbols.
- Simplify exponents.
- Multiply and divide in order from left to right.
- Add and subtract in order from left to right.

Simplify each expression. 2.1, 2.2

1. $-2(5 - 8) + 18 \div 3$
2. $4\left(\dfrac{1}{2}\right)^4$
3. -0.01^2
4. $5(2^4 - 3^3)$

Evaluate each expression if $a = -1$, $b = 4$, and $c = 2$.

5. $c + \dfrac{b - a}{c + a}$
6. $(c - b)^3$

Properties of Real Numbers

Commutative: $-7 + 12 = 12 + (-7)$; $\quad r \cdot 5 = 5 \cdot r$
Associative: $(-3 + 8) + 14 = -3 + (8 + 14)$; $\quad (-9b)c = -9(bc)$
Distributive: $13(10 + 5) = 13 \cdot 10 + 13 \cdot 5$; $\quad a(21) - a(5) = a(21 - 5)$

Which property of real numbers justifies each statement? 2.3

7. $-5(29) + (-5)1 = -5(29 + 1)$
8. $4kl = 4lk$
9. $(2x + 3y) + 7y = 2x + (3y + 7y)$
10. $-2 + [5 + (-x)] = (-2 + 5) + (-x)$
11. $(13x)y = 13(xy)$
12. $(x + y)5 = 5x + 5y$

Combining Like Terms

Like terms may be combined by adding or subtracting their numerical coefficients: $-2t + 7t^2 - t^2 = -2t + 6t^2$.

Simplify. 2.4

13. $-3c + c + 5 + 4c^2$
14. $-8x + 3(x - 5)$
15. $mn^2 - mn + 3mn^2 - 2mn$
16. $4(5 - 3r) - 10r$

Chapter 2 Algebraic Expressions

Property of −1 $-(5a - 1) = -1(5a - 1) = -1(5a) - (-1)(1)$
$= -5a + 1$

Simplify. 2.5

17. $-(-2h^2 + h) + h$
18. $8t^3 - (4t^2 + t^3)$
19. Simplify $-3[2(4 - 5p) + p] - p$. Then evaluate if $p = 2$.

Translating Phrases to Algebraic Expressions

Word Phrase	Algebraic Expression
The product of −7 and a number	$-7 \cdot n$, or $-7n$
4 times the difference of a number and 6	$4(a - 6)$

Translate each word phrase to an algebraic expression. 2.6

20. The quotient of 11 and a number
21. Twice the sum of a number and −2
22. 8 less than the product of a number and 12
23. The number of cents in a given number of nickels

Finding Solution Sets The solution set of an open sentence is the set of all numbers from the replacement set that make the open sentence true.

Find the solution set. The replacement set is {−2, −1, 0, 1, 2}. 2.7

24. $5 + x = 5$
25. $2y - 1 = 3$
26. $v + 2 < -1$
27. $1 - 3h > 4$
28. $2w - 1 \neq -1$
29. $x^2 = 1$

Translating Statements to Equations

The product of some number and 5 reduced by 14 is 6.
    Let $n =$ some number.   Then: $n(5) - 14 = 6$, or $5n - 14 = 6$

The perimeter of an equilateral triangle is 96 cm.
    Let $s =$ length (cm) of a side.   Then: $3s = 96$

Let $x =$ an unknown number. Write an equation for each sentence. 2.8

30. Two times a number is −15.
31. Two less than twice some number is 1.
32. Carla has a number of $5 bills. Juan has 4 fewer $5 bills. The total value of their $5 bills is $160.

CHAPTER 2 TEST

Simplify each expression.

1. $-3 \cdot 2 + 27$
2. $24 \div (8 - 5) + (-2)$
3. 2^4
4. $9\left(\dfrac{1}{3}\right)^3$

Evaluate each expression if $a = -1$, $b = 4$, and $c = -2$.

5. $\dfrac{a + b}{c}$
6. $\dfrac{b - 2c}{a + b}$

Which property of real numbers justifies each statement?

7. $3(10t + u) = 3(10t) + 3u$
8. $4d + 11d = (4 + 11)d$
9. $(ha)t = h(at)$
10. $-6f + 5 = 5 + (-6f)$

Simplify.

11. $-7q + 5 - q + q^2$
12. $-9rt + r^2t - rt + 2rt^2$
13. $4(x - 2) + x$
14. $7 - (3 - 2j)$
15. $-5[2 - 3(2t + 7) - t]$
16. $y - [2y + 3(y - 1)]$

Translate each word phrase to an algebraic expression.

17. Five less than a number
18. The perimeter of a square if the length of one side is l

Find the solution set. The replacement set is $\{-2, -1, 0, 1\}$.

19. $x - 2 = -1$
20. $r^2 = 1$
21. $1(k) = k$
22. $2x + 1 = 0$
23. $z - 1 > -2$
24. $y + 0 \neq y$

Write an equation for each sentence.

25. Three more than half a number is 25.
26. Billy's father is twice Billy's age. Eleven years ago, Billy's father was 3 times Billy's age. (*Hint:* Translate the second sentence to an equation using algebraic expressions from the first sentence.)

Challenge

1. Simplify: $3 - \{x - [x - 3(x - 3)]\}$
2. Write an equation for this situation:

   Annie Oakley was 14 years younger than Buffalo Bill. When she became the star attraction in his Wild West Show, she was two-thirds his age.

PREPARING FOR STANDARDIZED TESTS

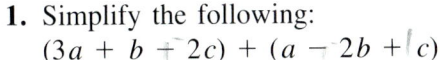

Select the best answer for each and write the appropriate letter.

1. Simplify the following:
   $(3a + b + 2c) + (a - 2b + c)$

   A. $2a - b - c$
   B. $2a - b - 2c$
   C. $3a - 2b - 2c$
   D. $4a - b - c$
   E. $4a - b - 2c$

2. Find the value of $a^2 + b^2$ when $a = 3$ and $b = 4$.

   A. 27 B. 25 C. 14
   D. 7   E. 5

3. An algebraic expression for "seven less than twice a certain number" is

   A. $2x - 7$ B. $2x + 7$ C. $7 - 2x$
   D. $x - 14$ E. $14 - x$

4. Which of the following would be the best *estimate* of 6832×19.81?

   A. 12,000 B. 14,000 C. 60,000
   D. 120,000 E. 140,000

5. Which of the following is a multiple of 4?

   A. 2874   B. 4315   C. 7982
   D. 13,410 E. 16,056

6. Find the value of $2[3(x^2 - 4) + 5y]$ when $x = 3$ and $y = 1$.

   A. 22 B. 40 C. 12
   D. 56 E. 55

7. Which of the following is less than 0.45 but more than 0.35?

   A. $\frac{1}{2}$ B. $\frac{1}{3}$ C. $\frac{2}{7}$
   D. $\frac{4}{9}$ E. $\frac{8}{25}$

8. If the replacement set is $\{-5, 5, 12, -12\}$, the solution set of the equation $x + 7 = 19 - 7$ is

   A. $\{-5\}$ B. $\{5\}$ C. $\{-12\}$
   D. $\{12\}$ E. $\{19\}$

9. If k is an integer, and $2k < 60$, then k cannot be

   A. 7   B. 12 C. 23
   D. 28 E. 32

10. Alex has one more math test to take before grades are averaged. So far, he has scores of 96, 85, 79, 91, and 89. What must his score be on this last test to maintain his present average?

    A. 86 B. 88 C. 91
    D. 93 E. 95

11. Alice and Betty spent their vacation doing yardwork. If their earnings were in the ratio 2:3, respectively, how much had Betty earned when Alice had earned $14.70?

    A. $29.40   B. $22.05
    C. $19.60   D. $17.15
    E. It cannot be determined from the information given.

Preparing for Standardized Tests **93**

CUMULATIVE REVIEW (CHAPTERS 1–2)

Write the numbers in order from least to greatest.

1. $2.5, -1, 0, -1.5, 2$
2. $-2, 1\frac{1}{2}, -\frac{1}{2}, 1, -1$
3. Graph $3, -2, -\frac{1}{2}, 0, 1$ on a number line.

Perform the indicated operation.

4. $-12 + 8$
5. $(-0.2)(-4)$
6. $9(-7)$
7. $\frac{3}{4} - \left(-\frac{3}{4}\right)$
8. $-36 \div (-4)$
9. $6 - (-8)$
10. $\frac{1}{2} \div \left(-\frac{3}{4}\right)$
11. $-5 - (-0.9)$
12. $-\frac{5}{6} \cdot \frac{3}{5}$
13. $-3.2 - 0.8$
14. $-5.4 \div 9$
15. $\frac{5}{8} + \left(-\frac{3}{8}\right)$

Simplify.

16. $-\left|-\left(-\frac{3}{4}\right)\right|$
17. $9 + 21 \div 3$
18. $6 + 24 \div (8 - 2)$
19. $-2[12 - 3(5 + 2)]$
20. $[6 - (-2)^2]^3$
21. $4x - 3y + 2x$
22. $m^2 - 2m^2 + m$
23. $3(a + 2) - 5a$
24. $p - (5 - 2p)$
25. $(5c + 2) - (3c - 6)$
26. $-2st^2 - (-3st^2 + 1)$

Evaluate each expression if $a = 3$, $b = \frac{2}{3}$, and $c = 4$.

27. $ab - 5$
28. $3b - ac$
29. abc
30. $|a - 2c|$
31. $(5a - 6b)c$
32. $\frac{2a - c}{b}$
33. $5[b \div (2a + c)]$
34. $\frac{3a + 1}{b + \frac{c}{a}}$
35. $2a^2 - c$

Translate each word phrase to an algebraic expression.

36. Karl's age in 5 years if x is his age now
37. Three less than twice a number
38. Five times the sum of a number and 3
39. The length of a rectangle is 1 cm more than twice the width
40. 14 lb more than 3 times Eva's weight

3 Equations in One Variable

Colleges use various criteria for selecting their students. For example, in addition to evaluating a student's high school academic record, many admissions officers look for involvement in school activities.

3.1 Solving Equations: Addition and Subtraction Properties

Objective: To solve equations using the Addition and Subtraction Properties for Equations

Suppose you are given directions for walking from the bus stop to the aquarium. To walk back from the aquarium to the bus stop, you would need to change a direction such as "Walk 10 blocks north" to its opposite, "Walk 10 blocks south."

Inverse (opposite) operations are used in mathematics as an efficient way to solve equations. Adding and subtracting the same number are inverse operations. One operation "undoes" the other.

Capsule Review

Recall what it means to add two numbers such as $5 + 4$.

$$\text{Start with } 5 \longrightarrow \text{add } 4 \longrightarrow 5 + 4.$$

To undo this, work backwards:

$$\text{Start with } 5 + 4 \longrightarrow \text{subtract } 4 \longrightarrow \text{to get back to } 5.$$

Simplify.

1. $4 - 4$
2. $-39 + 39$
3. $y - 1.5 + 1.5$
4. $2\frac{1}{8} + x - 2\frac{1}{8}$

Tell what must be done to get the variable x by itself.

5. $x + 6$
6. $x - 21$
7. $\frac{4}{5} + x$
8. $-2.1 + x$

In Chapter 2, you found the solutions of equations by substituting numbers from a given replacement set. From now on, you may assume that the replacement set is the set of real numbers unless otherwise stated.

Equivalent equations are equations that have the same solution(s) for the same replacement set. The equations $x + 2 = 10$ and $x = 8$ are equivalent; in each case, the solution is 8.

96 Chapter 3 Equations in One Variable

Adding or subtracting the same real number to each side of an equation produces an equivalent equation.

> **Addition Property for Equations**
>
> For all real numbers a, b, and c if $a = b$, then $a + c = b + c$.
>
> **Subtraction Property for Equations**
>
> For all real numbers a, b, and c if $a = b$, then $a - c = b - c$.

In the equation $x - 1.4 = 9.2$ below, 1.4 has been subtracted from some number x. The addition property for equations can be used to obtain an equivalent equation that has x alone on one side of the equation.

EXAMPLE 1 Solve and check: $x - 1.4 = 9.2$

$$x - 1.4 = 9.2$$
$$x - 1.4 + 1.4 = 9.2 + 1.4 \quad \text{Add 1.4 to each side of the equation.}$$
$$x + 0 = 10.6 \quad -1.4 + 1.4 = 0$$
$$x = 10.6 \quad \text{0 is the additive identity, so } x + 0 = x.$$

Check:
$$x - 1.4 = 9.2$$
$$10.6 - 1.4 \stackrel{?}{=} 9.2 \quad \text{Replace } x \text{ with 10.6.}$$
$$9.2 = 9.2 \checkmark \quad \text{True}$$

The solution is 10.6.

The solution of the equation is 10.6. Its *solution set* is {10.6}. When asked to solve an equation, you may give the solution without using set notation.

EXAMPLE 2 Solve: $8 = y + 4\frac{1}{5}$

$$8 = y + 4\frac{1}{5}$$
$$8 - 4\frac{1}{5} = y + 4\frac{1}{5} - 4\frac{1}{5} \quad \text{Subtract } 4\frac{1}{5} \text{ from each side of the equation.}$$
$$\frac{40}{5} - \frac{21}{5} = y + 0 \quad 8 = \frac{40}{5}; 4\frac{1}{5} = \frac{21}{5}$$
$$\frac{19}{5} = y$$
$$y = \frac{19}{5}, \text{ or } 3\frac{4}{5}$$

The solution is $\frac{19}{5}$, or $3\frac{4}{5}$. How do you *check* the solution?

3.1 Solving Equations: Addition and Subtraction Properties

The equation $a - (-52) = 25$ may be solved by removing parentheses and then using the subtraction property for equations.

EXAMPLE 3 Solve: $a - (-52) = 25$

$$a - (-52) = 25$$
$$a + 52 = 25 \quad \text{To subtract } -52, \text{ add its opposite, } 52.$$
$$a + 52 - 52 = 25 - 52 \quad \text{Subtract 52 from each side of the equation.}$$
$$a = -27$$

Check:
$$a - (-52) = 25$$
$$-27 - (-52) \stackrel{?}{=} 25 \quad \text{Replace } a \text{ with } -27.$$
$$-27 + 52 \stackrel{?}{=} 25$$
$$25 = 25 \checkmark \quad \text{True}$$

The solution is -27.

EXAMPLE 4 Solve: $-6.9 + z = -9.75$

$$-6.9 + z = -9.75$$
$$-6.9 + z + 6.9 = -9.75 + 6.9 \quad \text{Add 6.9 to each side of the equation.}$$
$$z = -2.85$$

Is -2.85 the solution? Check to be sure.

CLASS EXERCISES

Tell what operation you would perform on each side of the equation to get x alone on one side.

1. $x + 4 = -6$
2. $x - 2 = 10$
3. $-2 = -1 + x$
4. $x - 4 = -4$
5. $x - (-15) = 17$
6. $9 = x - 11$
7. $24 = x + 24$
8. $-35 + x = 50$

Solve and check.

9. $a - 7 = 24$
10. $x + 21 = -25$
11. $12 + d = -15$
12. $r - (-7) = 13$
13. $-2 = -15 + n$
14. $33 = -14 + c$

PRACTICE EXERCISES

Solve and check.

1. $x - 3.4 = 9.61$
2. $a - 12.5 = 13.9$
3. $y + 1.9 = 10.2$
4. $z + 2.4 = 5.3$
5. $m - 3.6 = 4.5$
6. $t - 5.3 = 2.3$
7. $9 = x + 2\frac{1}{3}$
8. $11 = z + 3\frac{2}{5}$
9. $4 = m + 1\frac{1}{3}$

Chapter 3 Equations in One Variable

10. $8 = s + 4\frac{1}{2}$
11. $3 = t - 1\frac{2}{3}$
12. $5 = y - 1\frac{1}{4}$
13. $a - (-60) = 30$
14. $b - (-25) = 24$
15. $x - (-5) = 10$
16. $m - (-2) = 12$
17. $z - (-6) = 50$
18. $t - (-9) = 41$
19. $-2.3 + x = -5.9$
20. $-5.3 + m = 10.2$
21. $-3.4 + s = -9.5$
22. $-8.2 + t = -12.4$
23. $-5.3 + r = -12.3$
24. $-2.5 + a = -5.5$
25. $y - 7.01 = 12.009$
26. $x - 3.12 = 5.23$
27. $z - 0.032 = 1.03$
28. $4 = n + 3\frac{1}{2}$
29. $5 = a + 2\frac{1}{3}$
30. $8 = x + 3\frac{1}{4}$
31. $45 - (-a) = 50$
32. $17 - (-x) = 22$
33. $52 = 25 - (-a)$
34. $-5.8 = a - 2.75$
35. $-8.3 = x - 2.5$
36. $-3.9 = b - 5.1$

Solve and check. The replacement set is the set of positive integers.

Hint: Solve the equation. If the result is not a positive integer, then write *no solution*.

37. $-15 + y = -15$
38. $x - 12 = -12$
39. $\frac{1}{2} + y = 5\frac{1}{2}$
40. $-\frac{4}{5} + x = -\frac{8}{5}$
41. $\frac{2}{3} + y = -\frac{1}{2}$
42. $-\frac{3}{4} + x = 1\frac{1}{4}$

Complete each sentence.

43. If $a - 12 = 15$, then $a - 1 = $ ___?___.
44. If $8 = t + 3$, then ___?___ $= t - 17$.
45. If $7 - 2x = 8$, then $11 - 2x = $ ___?___.
46. If $-3n - 5 = 17$, then $-3n + 5 = $ ___?___.

Find the solution set of each equation. Check your answer.

Example $y - 2 = |-5|$
$y - 2 = 5$
$y = 7$ The solution set is $\{7\}$.

Check: $y - 2 = |-5|$
$7 - 2 \stackrel{?}{=} |-5|$
$5 = 5$ ✓ True

47. $x = |3|$
48. $y = |-4|$
49. $x - 6 = |-3|$
50. $|z| = |-4| - |-6|$

3.1 Solving Equations: Addition and Subtraction Properties

Applications

The directions to go from school to the aquarium are given. State the directions from the aquarium to school.

51. Travel Go 7 blocks south, 3 blocks east, 2 blocks south, and 1 block west.

52. Travel Drive 6 mi west on Rt. 117 and 2 mi north on Rt. 9 to the third traffic light. Make a left at the third traffic light and go 2 blocks.

Write an equation for each statement. Then solve the equation.

53. Number Problem A number n increased by 25 equals 11. Find n.

54. Number Problem The sum of -43 and some number t is -18. Find t.

55. Number Problem Find a number x if x decreased by 29 equals -7.

56. Number Problem Find a number r if the difference between r and 11 is -58.

PUZZLE

Everett, Toni, Alice, and Enid are students at Lanier High School. Each takes part in one extracurricular activity. One plays the tuba in the school orchestra, one has the leading role in the school play, one is on the cycling team, and one is a cheerleader.

Everett and Alice were in the audience when the orchestra gave its concert. Enid and Toni tried out for the cycling team but did not make it. Everett is the cheerleader's best friend, but he does not know the cyclist. Enid and the cheerleader are cousins. Which student participates in which activity?

3.2 Solving Equations: Multiplication and Division Properties

Objective: To solve equations using the Multiplication and Division Properties for Equations

Laura gave Steve a copy of her favorite recipe. She told him that she had doubled the measure of each ingredient. To determine the original recipe, Steve had to "undo" Laura's doubling process. He had to divide each measure by 2.

Capsule Review

Multiplying and dividing by the same number are inverse operations. One operation "undoes" the other.

$$\text{Start with } x \longrightarrow \text{divide by 3} \longrightarrow \frac{x}{3}.$$

To undo this, work backwards:

$$\text{Start with } \frac{x}{3} \longrightarrow \text{multiply by 3} \longrightarrow \text{to get back to } x.$$

Simplify.

1. $\frac{1}{4} \cdot 4$
2. $\frac{-5}{-5}$
3. $\frac{r}{5} \cdot 5$
4. $-\frac{1}{8} a \cdot (-8)$

Tell what operation to perform to get the variable x by itself.

5. $\frac{x}{9}$
6. $-15x$
7. $\frac{x}{-3}$
8. $2.9x$

Multiplying or dividing each side of an equation by the same nonzero real number produces an equivalent equation.

Multiplication Property for Equations

For all real numbers a, b, and c, $c \neq 0$, if $a = b$, then $ac = bc$.

Division Property for Equations

For all real numbers a, b, and c, $c \neq 0$, if $a = b$, then $\frac{a}{c} = \frac{b}{c}$.

In Example 1, each side of the equation is divided by -16, since $\frac{-16}{-16} = 1$.

EXAMPLE 1 Solve and check: $-16x = 96$

$$-16x = 96$$
$$\frac{-16x}{-16} = \frac{96}{-16} \qquad \text{Divide each side by } -16.$$
$$x = -6$$

Check:
$$-16x = 96$$
$$-16(-6) \stackrel{?}{=} 96 \qquad \text{Replace } x \text{ with } -6.$$
$$96 = 96 \checkmark \qquad \text{True}$$

The solution of the equation is -6.

Example 2 illustrates a special case in which a property is used to solve an equation. Remember that you must obtain an equivalent equation in which the variable has a numerical coefficient of 1. The goal is an equation such as $x = 4$ or $-28 = x$, not $-x = 17$ or $-56 = -x$.

EXAMPLE 2 Solve: $5 = -x$

$$5 = -x$$
$$5 = -1x \qquad -x = -1x \text{ by the property of } -1 \text{ for multiplication.}$$
$$\frac{5}{-1} = \frac{-1x}{-1} \qquad \text{Divide each side of the equation by } -1.$$
$$-5 = x$$

The solution of the equation is -5. Check the solution.

To solve an equation, find an equivalent equation in which the variable (with a numerical coefficient of 1) is alone on one side of the equation. In Example 3, this is done by multiplying each side of the equation by 5.

EXAMPLE 3 Solve and check: $\frac{x}{5} = -7$

$$\frac{x}{5} = -7$$
$$5 \cdot \frac{x}{5} = 5 \cdot (-7) \qquad \text{Multiply each side of the equation by 5.}$$
$$x = -35 \qquad \qquad 5 \cdot \frac{x}{5} = \frac{\cancel{5}}{1} \cdot \frac{x}{\cancel{5}} = 1x, \text{ or } x$$

102 Chapter 3 Equations in One Variable

Check: $\frac{x}{5} = -7$

$\frac{-35}{5} \stackrel{?}{=} -7$ Replace x with −35.

$-7 = -7$ ✔   True

The solution of the equation is −35.

The equation $\frac{x}{5} = -7$ can also be solved by using reciprocals. Recall that the product of a number and its reciprocal is 1.

$\frac{x}{5} = -7$

$\frac{1}{5}x = -7$ $\frac{x}{5}$ means $\frac{1}{5} \cdot x$.

$5 \cdot \frac{1}{5}x = 5(-7)$ Multiply each side of the equation by 5, the reciprocal of $\frac{1}{5}$.

$x = -35$ $5 \cdot \frac{1}{5} = 1$

EXAMPLE 4   Solve: $\frac{3}{5}c = 6$

$\frac{3}{5}c = 6$

$\frac{5}{3} \cdot \frac{3}{5}c = \frac{5}{3} \cdot 6$ Multiply each side of the equation by $\frac{5}{3}$, the reciprocal of $\frac{3}{5}$.

$c = 10$ $\frac{5}{3} \cdot \frac{3}{5}c = 1 \cdot c$, or c

The solution of the equation is 10. How do you check the solution?

A calculator can be a very useful tool in solving equations. In order to use a calculator effectively, first write the equation in *calculation-ready form*. That is, write the equation with the variable alone on one side so that the computations are easily recognizable. Save the computations until the end.

EXAMPLE 5   Write the equation $\frac{3a}{17} = 11.3$ in calculation-ready form.

$\frac{3a}{17} = 11.3$

$\frac{3}{17}a = 11.3$ $\frac{3a}{17}$ means $\frac{3}{17}a$.

$\frac{17}{3} \cdot \frac{3}{17}a = 11.3 \cdot \frac{17}{3}$ Multiply each side of the equation by $\frac{17}{3}$, the reciprocal of $\frac{3}{17}$.

$a = 11.3 \cdot \frac{17}{3}$ Calculation-ready form

3.2 Solving Equations: Multiplication and Division Properties

CLASS EXERCISES

Tell what must be done to each side of the equation so the variable will be alone on one side. Then solve and check.

1. $5x = 30$
2. $-6y = 48$
3. $\frac{a}{5} = -4$
4. $6 = -\frac{1}{3}t$
5. $-24 = -y$
6. $\frac{2}{3}x = 1$
7. $-1.01b = 6.06$
8. $\frac{x}{-6} = 5$

PRACTICE EXERCISES

Solve and check.

1. $-9t = 72$
2. $-17t = 51$
3. $-8a = 56$
4. $-2x = 12$
5. $-24r = 120$
6. $-5p = 75$
7. $3 = -x$
8. $15 = -z$
9. $7 = -x$
10. $\frac{c}{25} = -1$
11. $\frac{x}{7} = -3$
12. $\frac{t}{3} = -12$
13. $\frac{a}{4} = -3$
14. $\frac{m}{2} = 10$
15. $\frac{s}{5} = 11$
16. $\frac{1}{2}e = 2$
17. $\frac{3}{5}a = 12$
18. $\frac{1}{4}x = 9$
19. $\frac{1}{8}n = 3$
20. $\frac{2}{3}m = 4$
21. $\frac{3}{7}t = 15$

Write the following equations in calculation-ready form.

22. $\frac{a}{12} = 11.5$
23. $\frac{m}{15} = 10.5$
24. $\frac{2x}{7} = 21.2$
25. $\frac{5t}{11} = 55.5$
26. $\frac{-2s}{3} = 33.5$
27. $\frac{-5r}{6} = 24.6$

Solve and check.

28. $\frac{x}{5} = -1\frac{1}{10}$
29. $\frac{m}{8} = -2\frac{1}{4}$
30. $\frac{y}{9} = -\frac{5}{3}$
31. $\frac{z}{3} = -\frac{9}{3}$
32. $\frac{7}{9}d = \frac{14}{3}$
33. $\frac{2}{3}a = \frac{4}{9}$
34. $-4.027m = 50.1$
35. $-7.850n = 0.929$
36. $-\frac{x}{5} = -2.5$
37. $-\frac{z}{2} = -4.6$

Chapter 3 Equations in One Variable

Complete each sentence.

38. If $3a = -11$, then $12a = $ __?__

39. If $2c = -5$, then $-8c = $ __?__

Find the solution set of each equation. Check your answer.

40. $|a| = 30$

41. $-12|x| = -144$

42. $\dfrac{|c|}{3} = 15$

43. $-6|2t| = -8$

44. True or false: Every equation in this lesson can be solved using the multiplication property for equations exclusively. Explain your answer.

Applications

Calculator Write each equation in calculation-ready form. Then solve the equation and state your answer to the nearest hundredth.

45. $\dfrac{m}{4.5} = 33$

46. $\dfrac{r}{29} = 1.1$

47. $\dfrac{1}{6}y = 0.89$

48. $17.25x = 50.59$

49. $81.23m = 62.4$

50. $\dfrac{x}{5.023} = -2.98$

WRITING IN ALGEBRA

Write one example of each of the expressions 1–5.

1. An algebraic expression involving subtraction of negative numbers.
2. An algebraic expression involving addition of negative numbers.
3. An algebraic expression involving subtraction of a negative number and multiplication of a positive number.
4. An algebraic expression containing parentheses and the operations of subtraction and multiplication.
5. An algebraic expression containing parentheses and the operations of addition and division.
6. In Exercises 4 and 5, explain what must be known before you can evaluate the expressions. Identify the order of the operations that would be used to evaluate each expression.
7. Write your own explanation of the procedures to be used in *evaluating* an algebraic expression.

3.3 Solving Equations: More Than One Property

Objective: To solve equations using more than one equation property

Louisa asked Jesse what time the math contest started on Saturday morning. In typical contest fashion, Jesse replied, "If you multiply the time by 3 and then add 6, the result is 33." Louisa made the following diagram:

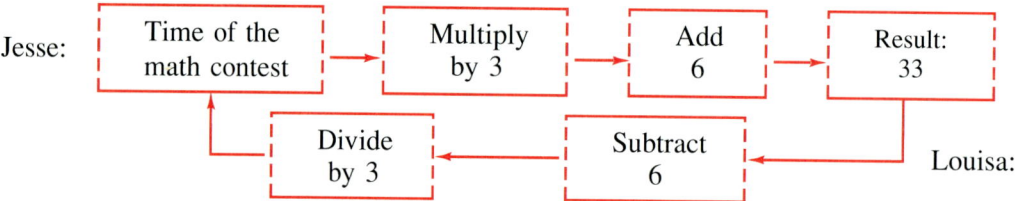

By "undoing" Jesse's steps, $(33 - 6 = 27;\ 27 \div 3 = 9)$, Louisa determined that the time of the math contest was 9:00 AM.

Capsule Review

To obtain n from the expression $4n - 20$, follow these steps:

(1) Add 20.
$4n - 20 + 20 = 4n + 0 = 4n$

(2) Divide by 4.
$4n \div 4 = 1n$, or n

Tell what operations to perform to get the variable n by itself.

1. $3n + 2$
2. $-4n + 24$
3. $\dfrac{n}{4} - 3$
4. $\dfrac{2}{3}n - 18$

To solve an equation when more than one equation property is required, first simplify the equation and then use the addition or subtraction property for equations. Then you are ready to use the multiplication or division property for equations.

EXAMPLE 1 Solve and check: $4n - 20 = 36$

$4n - 20 = 36$
$4n - 20 + 20 = 36 + 20$ First add 20 to each side of the equation.
$4n = 56$
$\dfrac{4n}{4} = \dfrac{56}{4}$ Then divide each side of the equation by 4.
$n = 14$

106 Chapter 3 Equations in One Variable

Check: $4n - 20 = 36$

$4 \cdot 14 - 20 \stackrel{?}{=} 36$ *Replace n with 14.*

$56 - 20 \stackrel{?}{=} 36$

$36 = 36$ ✔   *True*

The solution of the equation is 14.

In the next equation, $4(x - 3) = -24$, a number is decreased by 3 and the result is multiplied by 4. To solve equations of this type, the distributive property is used to simplify and then the inverse operations—addition and division—are used. It is important to simplify an equation first, then solve for the variable.

EXAMPLE 2   Solve: $4(x - 3) = -24$

$4(x - 3) = -24$

$4x - 12 = -24$ *Use distributive property.*

$4x - 12 + 12 = -24 + 12$ *Add 12 to each side of the equation.*

$4x = -12$

$\dfrac{4x}{4} = \dfrac{-12}{4}$ *Divide each side of the equation by 4.*

$x = -3$

The solution of the equation is -3. How do you check the solution?

EXAMPLE 3   Solve: $-\dfrac{8}{15} = \dfrac{6}{15} - \dfrac{2}{3}x$

$-\dfrac{8}{15} = \dfrac{6}{15} - \dfrac{2}{3}x$

$-\dfrac{8}{15} - \dfrac{6}{15} = \dfrac{6}{15} - \dfrac{2}{3}x - \dfrac{6}{15}$ *Subtract $\dfrac{6}{15}$ from each side of the equation.*

$-\dfrac{14}{15} = -\dfrac{2}{3}x$

$-\dfrac{3}{2}\left(-\dfrac{14}{15}\right) = -\dfrac{3}{2}\left(-\dfrac{2}{3}x\right)$ *Multiply each side of the equation by $-\dfrac{3}{2}$, the reciprocal of $-\dfrac{2}{3}$.*

$\dfrac{7}{5} = x$

The solution of the equation is $\dfrac{7}{5}$. Check the solution.

CLASS EXERCISES

Tell what must be done to each side of the first equation in order to get the second equation.

1. $-7b + 8 = -6; -7b = -14$
2. $2x - 15 = -9; 2x = 6$
3. $-19 + \frac{5}{4}y = 26; \frac{5}{4}y = 45$
4. $-17 = 11 + \frac{t}{7}; -28 = \frac{t}{7}$

Tell what must be done to each side of the equation to get x alone.

5. $2x - 3 = 8$
6. $3x + 7 = 17$
7. $11 - 4x = 1\frac{1}{2}$
8. $15 - 1.3x = 0$
9. $4x + 28 = 28$
10. $-4.7 + \frac{3}{4}x = 1.8$
11. $3x - 15 = 33$
12. $\frac{1}{2}x - 20 = 20$
13. $-24 = 2 - x$

Solve each equation by changing it to an equivalent equation as described. Check each answer to see if it makes the original equation true.

14. $7m + 5 = 26$ Subtract 5; divide by 7.
15. $-11h - 7 = -18$ Add 7; divide by -11.
16. $\frac{2}{3}x - 9 = 17$ Add 9; multiply by $\frac{3}{2}$.
17. $-10 = \frac{3}{8}y + 14$ Subtract 14; multiply by $\frac{8}{3}$.
18. $\frac{x}{5} + 15 = 0$ Subtract 15; multiply by 5.
19. $-8 - x = 11$ Add 8; divide by -1.

PRACTICE EXERCISES

Solve and check.

1. $3a - 1 = 7$
2. $2y - 18 = 44$
3. $3x - 1 = 8$
4. $20 - 5y = 45$
5. $25 - 3c = 36$
6. $14 - 2x = 18$
7. $2(x - 4) = 26$
8. $5(3 - x) = 40$
9. $\frac{1}{4}(x - 24) = 13$
10. $\frac{1}{3}(x + 27) = 4$
11. $-7(3 + 2x) = 84$
12. $6(5 - 3x) = 84$

Chapter 3 Equations in One Variable

13. $-\dfrac{3}{4} = \dfrac{12}{4} - \dfrac{1}{2}x$

14. $-\dfrac{5}{6} = \dfrac{5}{6} - \dfrac{2}{3}x$

15. $\dfrac{2}{9} = \dfrac{1}{3} - \dfrac{4}{9}x$

16. $\dfrac{1}{2} = \dfrac{5}{4} - \dfrac{3}{2}x$

17. $\dfrac{1}{5} = \dfrac{2}{3} + \dfrac{3}{5}x$

18. $\dfrac{1}{8} = \dfrac{3}{4} + \dfrac{1}{8}x$

19. $49w - 186 = 5351$

20. $44x - 728 = 1736$

21. $\dfrac{x}{3} + 3 = 42$

22. $\dfrac{f}{34} + 16 = 35$

23. $-\dfrac{7}{12} = \dfrac{5}{12} - \dfrac{2}{3}x$

24. $\dfrac{-3}{24} = \dfrac{5}{24} - \dfrac{4}{12}x$

Solve and check.

25. $2|x| - 7 = 1$

26. $3|n| + 6 = -3$

27. $5|t| - 10 = 0$

28. $9|x| - 7 = 7$

29. $-5|x| + 7 = 2$

30. $-7|r| - 8 = -1$

Complete each sentence.

31. If $4x + 2 = 14$, then $2x + 1 = $ __?__.

32. If $3x - 1 = 17$, then $6x - 2 = $ __?__.

33. If $x + 7 = 8$, then $7 - x = $ __?__.

34. If $2x + 5 = 21$, then $8x = $ __?__.

Applications

Calculator The calculation-ready form of $23x - 1.7 = 1.2$ is: $x = \dfrac{1.2 + 1.7}{23}$. Write each equation below in calculation-ready form. Then solve the equation and state your answer to the nearest hundredth.

35. $583r + 23.58 = 2.79$

36. $-51.5 = 29m - 4.06$

37. $\dfrac{x}{0.24} - 0.03 = -0.14$

38. $\dfrac{a}{2.5} + 11.9 = 0.02$

TEST YOURSELF

Solve each equation. Check your answer.

1. $y + 11 = 4$
2. $-7.6 = x - 1.4$
3. $-\dfrac{3}{4} + z = \dfrac{5}{2}$ 3.1

4. $9t = 30$
5. $24 = -\dfrac{2}{5}m$
6. $\dfrac{x}{7} = -3.15$ 3.2

7. $6k + 5 = 23$
8. $5 - \dfrac{j}{12} = 1$
9. $-7.13 = 0.15p - 7.13$ 3.3

3.4 Problem Solving Strategy: Make a Model

Many real-world problems can be solved by constructing and analyzing *mathematical models*. A **mathematical model** represents the known parts of a problem. Many different kinds of models may be used to solve problems in algebra. For example, a drawing, figure, graph, or table can be constructed as a model. Other models may include a formula, pattern, equation, or an inequality. This lesson presents writing an equation as a model.

EXAMPLE Janice bought a number of cassette tapes for $8 each. She then spent $3 more for a magazine. If she spent a total of $35, how many tapes did she buy?

Understand the Problem
Janice bought a number of cassette tapes at $8 each and spent $3 for a magazine. The total cost was $35.

You are asked to find how many cassette tapes Janice bought.

Plan Your Approach
Choose a strategy.
A successful strategy to solve a problem is to set up a mathematical model. An appropriate model for this problem is an equation.

Write a word equation.

Number of dollars spent for tapes $+$ Cost of magazine $=$ Total amount spent

Assign variables.
Let x represent the number of tapes bought.
Then $8x$ represents the total amount spent for the tapes.

Translate the word equation into an algebraic equation.

Number of dollars spent for tapes	plus	Cost of magazine	equals	Total amount spent
↓	↓	↓	↓	↓
$8x$	$+$	3	$=$	35

Complete the Work
Solve the equation.
$$8x + 3 = 35$$
$$8x + 3 - 3 = 35 - 3$$
$$\frac{8x}{8} = \frac{32}{8}$$
$$x = 4$$

Interpret the Results

State your conclusion.
Janice bought 4 cassette tapes.

Check your conclusion.
$$8x + 3 = 35$$
$$8(4) + 3 \stackrel{?}{=} 35$$
$$32 + 3 \stackrel{?}{=} 35$$
$$35 = 35 \checkmark$$

CLASS EXERCISES

Identify a mathematical model for each situation described. The model may be a geometric figure, a formula, table, or an algebraic expression.

1. The number of books in a box which holds 64 books if the box is n books short of being filled.

2. The area of a rectangular floor in order to calculate the number of floor tiles needed to cover it.

3. The shortest distance from one point on a map to another point.

4. The receipts from the sale of 95 tickets if each ticket costs d dollars.

5. The distance an automobile travels in a given time if its speed is 35 mi/h.

PRACTICE EXERCISES

Identify a mathematical model for each problem. Then use the model to solve the problem.

1. Jane bought a number of records for $7 each. She then spent $3 more for a poster. If she spent a total of $24, how many records did she buy?

2. Robert and Jeanine are having a dinner party. There are 24 guests to be expected. If there are 4 tables seating an equal number of guests, how many guests will each table accommodate?

3. John works for a shipping department where he packs an equal number of books into boxes. John has 40 books and 5 boxes. How many books did John pack into each box? How many books were left over?

4. At a graduation dinner, an equal number of people were seated at 8 tables. The host invited 35 people and 3 guests arrived late. A new table was set up for the late guests. How many people were seated at each of the other tables?

5. Lois worked part time after school and 4 h on Saturday. One Saturday, her boss paid her $25 for the day, which included a $5 tip. How much does Lois earn per hour?

6. Juan enjoys solving mathematics problems. A friend gave him this problem to solve: Seven times a number increased by five is 33. What is the number?

Write a word problem for which the equation or inequality is a mathematical model.

7. $\frac{1}{2}b = 8$ 8. $5y - 14 = 27$ 9. $a - 3 < 7$ 10. $\frac{m}{7} + 1 > 19$

11. Select a model and then describe the procedure you would follow to explain how our place-value numeration system works.

12. Select a model and show how you can measure exactly 4 quarts when you have only a 3-quart measure and a 5-quart measure to use.

Mixed Problem Solving Review

1. An investor in the stock market had the following gains and losses on the sale of 6 different stocks: +$400, −$700, +$1,000, +$200, −$500, +$100. What is the total gain or loss? What is the average gain or loss for the 6 stocks?

2. Write an equation for each word sentence. Then solve the equation, using the set of whole numbers from 0 through 10 as the replacement set.
   a. The product of a number and 2, increased by 5, is 11.
   b. A number less 8 is the first positive even number.
   c. The quotient of 20 and a number equals 4.
   d. The sum of a number and nine times the number equals 10.

PROJECT

A table is a useful problem-solving model. A table shows numerical information, called *data* (plural of datum). Tables are used to organize data and answer questions about the data.

Construct a table to record the amount of time you spend studying each of your school subjects each day for 30 consecutive days. Record the time to the nearest 5 minutes.

At the end of 30 days, find the total time for each subject and divide by 30. Your answer will be the average number of minutes you study each subject each day for 30 days.

By analyzing the data in your table, identify three other conclusions that can be made. What does this project tell you about your study habits?

3.5 Algebraic Proof

Objectives: To identify postulates, theorems, and proofs
To prove simple algebraic statements

Many properties about the sums and products of real numbers are stated earlier in this book. You have used them in computations and in solving equations. Some of these properties are also called *postulates*. **Postulates** are statements that are accepted as true without requiring a proof.

Capsule Review

Name the property of real numbers illustrated by each statement.

1. $5 + 9 = 9 + 5$
2. $6\left(\frac{2}{3} + 4\right) = 6 \cdot \frac{2}{3} + 6 \cdot 4$
3. $3 + (7 + x) = (3 + 7) + x$
4. $8 \cdot 1 = 8$
5. $10 + (4 + 0) = 10 + 4$
6. $7 \cdot 3 + 7 \cdot 2 = 7(3 + 2)$

A summary follows of some of the definitions and properties of real numbers. These definitions and properties will be useful in doing algebraic proofs.

Properties of Operations with Real Numbers

	Addition	Multiplication
Commutative	$a + b = b + a$	$a \cdot b = b \cdot a$
Associative	$(a + b) + c = a + (b + c)$	$(a \cdot b) \cdot c = a \cdot (b \cdot c)$
Inverse	$a + (-a) = 0$ and $-a + a = 0$	$a \cdot \frac{1}{a} = 1$ and $\frac{1}{a} \cdot a = 1, a \neq 0$
Identity	$a + 0 = a$ and $0 + a = a$	$a \cdot 1 = a$ and $1 \cdot a = a$

Distributive property	$a(b + c) = ab + ac$ and $(b + c)a = ba + ca$
Property of 0 for multiplication	$a \cdot 0 = 0$ and $0 \cdot a = 0$
Property of -1 for multiplication	$-1(a) = -a$ and $a(-1) = -a$
Definition of subtraction	$a - b = a + (-b)$
Definition of division	$a \div b = a \cdot \frac{1}{b}, b \neq 0$

The following properties of equality for all real numbers a, b, and c are also used in algebraic proofs.

Properties of Equality

Reflexive	$a = a$ (A number is equal to itself.)
Symmetric	If $a = b$, then $b = a$.
Transitive	If $a = b$ and $b = c$, then $a = c$.
Substitution	If $a = b$, then a may replace b or b may replace a in any statement.

A **theorem** is a general conclusion that is shown to be true by using postulates, definitions, given facts, and other proved theorems. The reasoning that takes you from the **hypothesis** (the given statement) to the **conclusion** (the final statement) is called a **direct proof**.

EXAMPLE 1 Prove: For all real numbers a, b, and c, if $a = b$, then $ac = bc$.

If $\underline{a = b}$ then $\underline{ac = bc}$.

hypothesis (given)   conclusion

Proof:

Statements	Reasons
1. $a = b$	1. Given
2. $ac = ac$	2. Reflexive property
3. $ac = bc$	3. Substitution property

Therefore, if $a = b$, then $ac = bc$.

Notice that the theorem proved in Example 1 is a formal statement of the Multiplication Property for Equations.

EXAMPLE 2 Prove: For all real numbers a and b, $-(a - b) = b - a$.

Proof:

Statements	Reasons
1. $-(a - b) = -[a + (-b)]$	1. Definition of subtraction
2. $ = -1[a + (-b)]$	2. Property of -1 for multiplication
3. $ = (-1)a + (-1)(-b)$	3. Distributive property
4. $ = (-a) + (b)$	4. Property of -1 for multiplication
5. $ = b + (-a)$	5. Commutative property for addition
6. $ = b - a$	6. Definition of subtraction

Therefore, for all real numbers a and b, $-(a - b) = b - a$.

CLASS EXERCISES

Write the missing statements or reasons for each step of the proof.

Prove: For all real numbers a and b, $a + b + (-b) = a$.

Proof:

Statements	Reasons
1. $(a + b) + (-b) = a + [b + (-b)]$	1. ?
2. $\quad = a + $?	2. ?
3. $\quad = a$	3. ?

PRACTICE EXERCISES

Write the missing statements or reasons in each proof.

1. Prove: For all real numbers a, b, and c, if $a = b$ and $c \neq 0$, then $\dfrac{a}{c} = \dfrac{b}{c}$.

   Proof:

Statements	Reasons
1. $a = b, c \neq 0$	1. ?
2. $\dfrac{a}{c} = $?	2. Reflexive property
3. $\dfrac{a}{c} = $?	3. ?

2. Prove: For all real numbers x and y, $-(x - y) + (y - x) = 2(y - x)$.

   Proof:

Statements	Reasons
1. $-(x - y) + (y - x)$ $= -[x + (-y)] + (y - x)$	1. Definition of subtraction
2. $= -1[x + (-y)] + (y - x)$	2. ?
3. $= (-1)(x) + (-1)(-y) + (y - x)$	3. ?
4. $= -x + y + (y - x)$	4. ?
5. $= y + y - x - x$	5. Commutative property for addition
6. $= 2y - 2x$	6. ?
7. $= 2(y - x)$	7. Distributive property

Prove. Assume that all variables represent real numbers.

3. If $a = b$ then $a - c = b - c$.

4. If $x = y$ then $z - x = z - y$.

5. $a + [b + (-a)] = b$

6. $x + [-(x + y)] = -y$

7. If $a = b$, then $a - d = -(d - b)$.

3.5 Algebraic Proof

Complete the proof.

8. Prove: For all real numbers a, $(a)(0) = 0$.

   Proof:

Statements	Reasons
1. a is a real number	1. ?
2. $a + 0 = a$	2. ?
3. $a(a + 0) = (a)(a)$	3. ?
4. $(a)(a) + (a)(0) = (a)(a)$	4. ?
5. $(a)(a) + (a)(0) - (a)(a) = 0$	5. ?
6. $(a)(a) - (a)(a) + (a)(0) = 0$	6. ?
7. $0 + (a)(0) = 0$	7. ?
8. $(a)(0) = 0$	8. ?

Applications

Language Arts The reflexive, symmetric, and transitive properties are true for the equality relationship. Which of these properties are true for the given relationship of the replacement set of all family members?

9. is a brother of

10. is younger than

11. is married to

12. is a cousin of

LOGICAL REASONING

Sprouts is a game for two people to play. It begins with a fixed number of sprouts (or dots) on paper. To play, each player must (1) draw a branch from one sprout to another or to the same sprout, and (2) make an additional sprout anywhere on the branch drawn. The object is to cut off sprouts with branches so no moves are possible. The rules of the game follow.

Rules: No sprout may have more than three branches. Branches may not intersect. The person who draws the last branch is the winner.

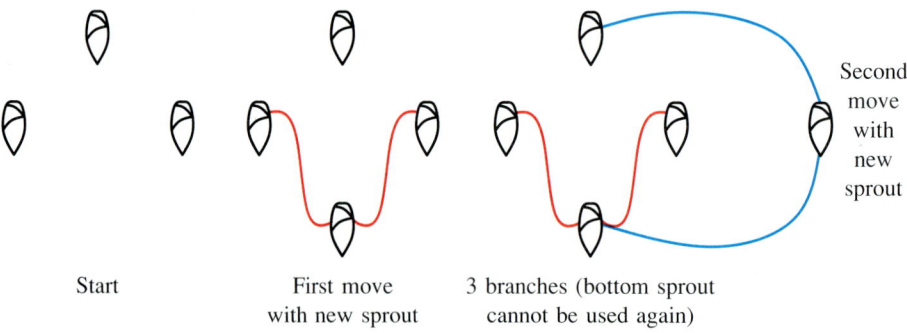

Start   First move with new sprout   3 branches (bottom sprout cannot be used again)   Second move with new sprout

What is the least number of moves needed to determine a winner?

3.6 Evaluating Formulas

Objectives: To solve for any variable in a formula when the values of the other variables are given
To use formulas to solve simple word problems

A **formula** is an equation that expresses a relationship among two or more quantities. The formula $P = 2l + 2w$ states a relationship involving the perimeter P, length l, and width w of a rectangle. If any two of the quantities are known, you can find the third.

You solve for variables in formulas in the same way as you do in other equations.

Capsule Review

EXAMPLES Solve: $27 = \frac{2}{3}r$

$$27 = \frac{2}{3}r$$
$$\frac{3}{2} \cdot 27 = \frac{3}{2} \cdot \frac{2}{3}r$$
$$40\frac{1}{2} = r$$

The solution is $40\frac{1}{2}$.

Solve: $3a + 6 = 27$

$$3a + 6 = 27$$
$$3a + 6 - 6 = 27 - 6$$
$$3a = 21$$
$$\frac{3a}{3} = \frac{21}{3}$$
$$a = 7$$

The solution is 7.

Solve each equation for x.

1. $-49 = 7x$
2. $\frac{1}{2}x = -35$
3. $\frac{x}{5} + 6 = 11$
4. $-4x - 9 = 31$
5. $13 - 13x = 0$
6. $\frac{1}{3}x - 9 = -6$
7. $16.2 = 2x + 6.2$
8. $0.7x - 5.6 = 0$

To solve a formula for one variable when values of other variables are given:

- Replace each variable by its given value.
- Solve the resulting equation, using the properties for equations.

In Example 1, a formula is used to find the length of a rectangle when its perimeter and width are known.

EXAMPLE 1 Given the formula $P = 2l + 2w$, find the length l of a rectangle when its perimeter P is 18.2 units and its width w is 3.6 units.

$$P = 2l + 2w$$
$$18.2 = 2l + 2(3.6)$$ *Replace P with 18.2 and w with 3.6.*
$$18.2 = 2l + 7.2$$ *Solve the equation for l.*
$$18.2 - 7.2 = 2l + 7.2 - 7.2$$
$$11 = 2l$$
$$\frac{11}{2} = \frac{2l}{2}$$
$$5.5 = l$$

The length of the rectangle is 5.5 units.

Example 2 uses the formula for the volume of a cylinder.

EXAMPLE 2 Evaluate $V = \pi r^2 h$ for h if $V = 9420$ ft^3, $\pi \approx 3.14$, and $r = 10$ ft.

$$V = \pi r^2 h$$
$$9420 = 3.14(10^2)h$$
$$9420 = 3.14(100)h$$
$$9420 = 314h$$
$$30 = h$$

The height of the cylinder is 30 ft.

CLASS EXERCISES

1. Area of triangle: $A = \frac{1}{2}bh$ Find the value of h if $A = 16$ and $b = 4$.

2. Amount in savings account: $A = p + prt$ Find the value of A when $p = \$800$, $r = 0.04$, and $t = \frac{1}{4}$.

3. Sum of n terms of arithmetic series: $S = \frac{n}{2}(a + l)$ Find the value of n if $S = 52$, $a = -4$, and $l = 12$.

4. Volume of rectangular prism: $V = lwh$ Find the value of l if $V = 3.6$, $w = 0.8$, and $h = 3$.

5. Temperature conversion: $F = \frac{9}{5}C + 32$ Find the value of C when $F = -31°$.

PRACTICE EXERCISES

1. Given the formula $P = 2l + 2w$, find the length l of a rectangle when its perimeter P is 15.5 units and its width w is 4.25 units.

2. Given the formula $P = 2l + 2w$, find the width w of a rectangle when its perimeter P is 17.6 units and its length l is 3.3 units.

3. Evaluate $V = \pi r^2 h$ for h if $V = 168$, $\pi \approx 3.14$, and $r = 3$.

4. Evaluate $V = \pi r^2 h$ for h if $V = 235.5$, $\pi \approx 3.14$, and $r = 5$.

5. Using the distance formula $d = rt$, find the value of r if $d = 28$ and $t = 4$.

6. Using the distance formula $d = rt$, find the value of t if $d = 45$ and $r = 5$.

7. Given the formula for the volume of a rectangular prism $V = lwh$. Find the value of h, if $V = 36$, $l = 4$, and $w = 3$.

8. Given the formula $V = lwh$, find the value of w if $V = 120$, $l = 5$, and $h = 3$.

In each formula, find the value of the indicated variable.

9. $P = 2l + 2w$; find w when $P = 240$ and $l = 10$.

10. $P = 2l + 2w$; find l when $P = 360$ and $w = 6$.

11. $A = \dfrac{b}{2} + i - 1$; find b when $A = 15$ and $i = 10$.

12. $V = \pi r^2 h$; find h when $V = 420$, $\pi \approx 3.14$, and $r = 3$.

Applications

13. **Travel** A train takes $3\frac{1}{2}$ h to go a distance of 334.25 km. Find its average speed.

14. **Travel** If a bicycle goes 219.8 cm in one turn of the wheels, what is the length of a radius of the wheels?

15. **Geometry** The floor of one of the classrooms in a new school has a perimeter of 34.6 m. Find the area if the length is 10.6 m.

WRITING IN ALGEBRA

1. Write an equation whose solution is found by using the addition and the division properties for equations.
2. Write an equation whose solution is found by using the subtraction and the multiplication properties for equations.
3. Write a short paragraph to explain how to check a solution for an equation.
4. Dan claims that to solve an equation like $3b - 2 = 19$, it is just as easy to use the division property before the addition property. Rosita insists that it is easier to use the addition property first. Show how each would solve the equation $3b - 2 = 19$. Which method is easier? Explain.

3.7 Problem Solving Strategy: Write an Equation

Equations serve as mathematical models for many situations in social sciences, physical sciences, and business. Before attempting to write an algebraic equation to solve a problem, it is helpful to write a *word equation,* as shown in Example 1.

EXAMPLE 1 Scientists use facts about how parts of the body relate to each other to study the human body. For example, it is a fact that the length of a woman's radius bone is about $\frac{1}{7}$ of her height. The radius bone connects the elbow and the wrist. If a scientist knows that the radius bone of a woman is 9 in. long, then about how tall is the woman?

Understand the Problem

Read the problem.

It is given that the length of a woman's radius bone is about $\frac{1}{7}$ of her height. It is also given that the radius bone of a woman is 9 in. long. You are asked to find the approximate height of the woman.

Plan Your Approach

Choose a strategy.

The word *is* implies that the length of the radius bone *equals* $\frac{1}{7}$ of a woman's height. Try the strategy of writing an equation.

Write a word equation.

The phrase "$\frac{1}{7}$ of her height" means "$\frac{1}{7}$ times her height," thus

$$\text{length of radius bone} = \frac{1}{7} \times \text{height}$$

Translate the word equation into an algebraic equation.
Let h represent the height of a woman whose radius bone measures 9 in.

$$9 = \frac{1}{7}h$$

Complete the Work

Solve the equation.

$$9 = \frac{1}{7}h$$

$$7 \cdot 9 = 7 \cdot \frac{1}{7}h \qquad \textit{Multiply each side of the equation by 7.}$$

$$63 = h$$

Chapter 3 Equations in One Variable

Interpret the Results

State your conclusion.
The woman is approximately 63 in., or 5 ft 3 in. tall.

Check your conclusion.
Is the length of the radius bone (9 in.) equal to $\frac{1}{7}$ of the woman's height (63 in.)?

$$9 = \frac{1}{7}h$$

$$9 \stackrel{?}{=} \frac{1}{7} \cdot 63 \qquad \text{Replace } h \text{ with 63.}$$

$$9 = 9 \checkmark \qquad \text{True}$$

Is 63 in. a reasonable height for a woman?

EXAMPLE 2 If twice the weight of a truck is increased by 1500 lb, it will weigh 9000 lb. Find the weight of the truck.

Understand the Problem
You are asked to find the actual weight of the truck after a certain number of pounds is added to twice the weight of the truck.

Plan Your Approach
Let w represent the weight of the truck.
Twice the weight increased by 1500 lb equals 9000 lb.

$$2w + 1500 = 9000$$

Complete the Work

$$2w + 1500 = 9000$$
$$2w + 1500 - 1500 = 9000 - 1500$$
$$2w = 7500$$
$$\frac{2w}{2} = \frac{7500}{2}$$
$$w = 3750$$

Interpret the Results
The weight of the truck is 3750 lb. Is 9000 lb equal to twice the weight of the truck increased by 1500 lb?

$$2w + 1500 = 9000$$
$$2(3750) + 1500 \stackrel{?}{=} 9000$$
$$7500 + 1500 \stackrel{?}{=} 9000$$
$$9000 = 9000 \checkmark$$

Problem Solving Reminders

- Write a word equation to express the relationships in a problem before you write an algebraic equation.
- Be sure that your solution to a problem *meets the conditions of the problem* and checks in the equation.

CLASS EXERCISES

Complete each of the following exercises.

1. A number reduced by 59 gives -85. If n represents the original number, __?__ represents the number reduced by 59.
   Complete the equation: __?__ $= -85$

2. Traveling for a number of hours at an average speed of 47 mi/h, Jon travels 564 mi. t represents the number of hours Jon travels.
   Complete the equation: $47t =$ __?__

3. The cost of an item is $1.59; n represents the number of these items purchased; $20.67 is the total cost. Complete the equation: __?__ $= 20.67$

4. The perimeter of a square is 148 cm. Let s represent the length in cm of one side of the square. Complete the equation: $4s =$ __?__

Solve each problem by writing and solving an equation.

5. The sum of a number and -32 is 79. Find the number.

6. Sue has $1809.46 in her savings account. This is $14.95 more than three times the amount Charles has in his account. How much money does Charles have in his savings account?

For Discussion

7. Name two types of open sentences.

8. What is a number called that makes an open sentence a true statement?

9. What word in a written sentence implies a relationship of equality?

10. What is a mathematical term for a relationship of equality?

11. Are equations and inequalities examples of mathematical models? Explain.

PRACTICE EXERCISES

Write an appropriate equation, but do not solve.

1. If $\frac{1}{3}$ of the money Jason earns each week consists of tips, how much money did he earn one week when he received $10 in tips?

2. Mary saves $\frac{1}{4}$ of her pay each week so she can purchase a bicycle. This week she saved $25. How much did she earn?

3. If twice the weight of an elephant is increased by 3 tons, it will weigh 7 tons. Find the weight of the elephant.

4. If three times the weight of a Shot Put is increased by 2 lb, it will weigh 18 lb. What is the weight of the Shot Put?

Solve each problem by writing and solving an equation.

5. Decreasing a number by 92 gives -28. What is the number?
6. Decreasing a number by 47 gives -15. What is the number?
7. The product of a number and 11 is -165. Find the number.
8. The product of a number and 15 is -240. Find the number.
9. A number increased by 7.13 is 2.09. Find the original number.
10. A number increased by 4.12 is 3.15. Find the original number.
11. Tom's age is $\frac{2}{5}$ of his mother's age. Tom is 16 years old. How old is his mother?
12. Dan's brother is $\frac{1}{2}$ his father's age. His brother is 22. How old is his father?
13. If 11 is added to 3 times a number, the result is 50. Find the number.
14. If 14 is added to 5 times a number, the result is 129. Find the number.
15. The perimeter of a rectangle is 54 cm. The width is 12 cm. What is the length?
16. What must the length of the two equal sides of an isosceles triangle be if the perimeter is 56 mm and the length of the third side is 18 mm?
17. Hester paid $245.97 for a VCR. This was $\frac{3}{5}$ of its original price. What is the original price?
18. Christy paid $25 for a sweater. This was $\frac{1}{2}$ of its regular price. What is the regular price?
19. At the local movie theater, children's tickets cost $2.50 each and adults' tickets cost $4.50 each. If 117 children's tickets were sold, and the total sales were $733.50, how many adults' tickets were sold?
20. A cheetah can sprint for short distances at a rate of 70 mi/h. At that rate, how many minutes should it take a cheetah to sprint 2.1 mi?
21. Mr. Sangupta invested $\frac{1}{2}$ of his money in land, $\frac{1}{10}$ in stock, and $\frac{1}{20}$ in machinery. The remainder, $35,000, is in a savings account. What is the total amount of money that Mr. Sangupta saved or invested?

Mixed Problem Solving Review

1. The weekly change in stock prices for the Pueschner Health Food Company was recorded as follows: Mon, $+3\frac{1}{8}$; Tue, $+2\frac{1}{4}$; Wed, -3; Thur, $-3\frac{3}{8}$; Fri, $+1\frac{1}{4}$. What was the net change for the week?

2. A square mat extends 2 in. beyond each side of a 12 in. × 12 in. photograph.
   a. Find the outer perimeter of the mat.
   b. Find the area of the mat surrounding the photograph.

3. Eduardo's age is a years and his mother's age is 1 year more than 3 times Eduardo's age. Write an algebraic expression to represent Eduardo's mother's age.

4. This is a diagram of the Smiths' new swimming pool. Each end is a semicircle. Mr. Smith needs to know its perimeter in order to purchase a special reflecting tape for the edge. If the dimension x is 22 ft and the dimension y is 28 ft, what is the pool's perimeter?

   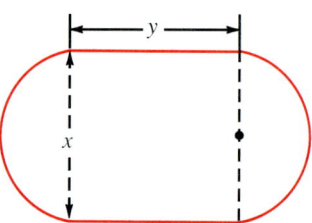

5. $r + s = 24$. If r and s are integers, what is the maximum value of rs?

PROJECT

Conduct a study of your classmates to determine if the following statement is true of your class.

> A girl's radius bone (bone that connects the elbow and the wrist) is approximately $\frac{1}{7}$ of her height.

- Talk to at least five female students who are willing to provide the necessary data.
- Record data for each volunteer—length of radius bone and height.
- Find the ratio (as a decimal to the nearest thousandth) of the length of the radius bone to height for each person.
- Find the average of the ratios you obtain.
- Compare the results that you obtain to the ratio $\frac{1}{7}$ (≈ 0.143) in Example 1. Using the average ratio you find, rewrite the equation in Example 1.
- Analyze your results. Would your results be different if the size of your sample of volunteers had been greater?

Conduct a study of your male classmates to determine if the height of an adult male is approximately four times the length of his tibia bone (the bone that goes from the ankle to the knee).

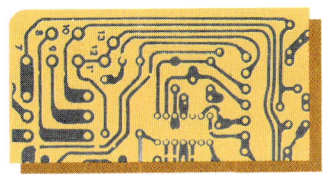

TECHNOLOGY: Using Spreadsheets

Most spreadsheets have special functions that enable you to do specific calculations. One very helpful function is the SUM function. This enables you to indicate what cells you would like the computer to add.

If the formula

    Function Name Range of Cells

$$@SUM(C6...C11)$$

appears in cell C13, it means that the computer will SUM the data in cells C6 through C11 and put the result in cell C13. The following spreadsheet uses the SUM function.

```
File: Grades                    REVIEW/ADD/CHANGE
======A========B========C========D========
 1|          GRADES—FALL SEMESTER
 2|
 3|                          Possible      Actual
 4|Description               Points        Points
 5|
 6|TEST 1                    100           85
 7|QUIZ 1                     10            9
 8|HOMEWORK P. 6               5            5
 9|HOMEWORK P. 10              5            5
10|QUIZ 2                     10            8
11|TEST 2                    100
12|                          _____
13|                          230           112
14|Score needed for an A      90
15|CURRENT GRADE IS          86.15384
16|
   _____
C15:
```

Solve.

1. What formula would the display area show for cell D13 if you wanted the sum?

2. If the formula in cell C15 is +D13/@SUM(C6...C10), what would it represent?

3. Use the spreadsheet to determine if it is possible for the student to get an A after taking TEST 2.

This spreadsheet is designed to compute a personal budget.

```
File: Budget                                    REVIEW/ADD/CHANGE
====== A======B======C======D======E======F======G======
 1|
 2|                    PERSONAL BUDGET
 3|                       SEPT.   OCT.    NOV.    DEC.
 4|AMOUNT FROM LAST MONTH                           .9
 5|ALLOWANCE                                        20
 6|                                               _____
 7|MONEY AVAILABLE THIS MONTH                      20.9
 8|
 9|EXPENSES:
10|   LUNCHES                                       10
11|   GAS                                            5
12|   TAPES                                          0
13|   DATES                                          0
14|   CLOTHES                                        0
15|   FAMILY                                        5.9
16|                                               _____
17|TOTAL MONEY SPENT                               20.9
18|
----------------------------------------------------------
G17:
```

Solve.

4. What formula should the display area show for cell G7?

5. What formula should be in the cell for G17?

6. During the summer Joe saved $800. During the school year he receives an allowance of $10.00 per week. His expenses for September, October, and November are as follows:

	Sept.	Oct.	Nov.
Lunches	$65	$60	$50
Gas	20	25	22
Tapes	7	0	18
Dates	0	20	30
Clothes	100	0	25
Family	10	12	0

Joe has set a spending limit of $150 for the holidays. Can he continue to spend at the above rates for the remainder of the school year?

7. Revise the spreadsheet to reflect your answer for Exercise 6.

Problem Solving: Mixed Types

Objective: To use equations to solve various types of word problems

The following exercises include different types of problems presented in this chapter. First, review the four-step approach to solving word problems.

Problem Solving Steps

Understand the Problem

Read the problem.
    What is given?
    What are you asked to find?
Identify important mathematical ideas.
Draw a diagram.

Plan Your Approach

Assign symbols. Label your drawing.

Write a word equation.

Translate the word equation into an algebraic equation.

Complete the Work

Solve the equation.
    Apply algebraic operations as needed.
    Keep an open mind.
    Change your approach if necessary.

Interpret the Results

State your conclusion.

Check your conclusion.

Keep in mind that some problems are easier to solve than others. All problems, however, can be solved by first carefully reading the information and understanding what you are being asked to find before proceeding.

CLASS EXERCISES

Solve each problem.

1. The sum of 8 times a number and 12 is -84. What is the number?

2. Two-thirds of a number decreased by 7 equals 27. Find the number.

3. The perimeter of a square is 196 m. How long is each side of the square?

4. In the freshman class, $\frac{8}{9}$ of the students are taking Algebra. Find the total number of students in the class if 160 are taking Algebra.

PRACTICE EXERCISES

Solve each problem.

1. Twelve increased by 3 times a number is 21. Find the number.

2. Fifteen increased by 3 times a number is 54. Find the number.

3. Five times a number, decreased by 1, is −26. Find the number.

4. Three times a number, decreased by 3, is −18. Find the number.

5. The product of a number and 0.5 divided by 3 equals 1.5. Find the number.

6. The product of a number and 2.4 divided by 6 equals 12. Find the number.

7. One-fifth of a number, decreased by 8, equals 7. Find the number.

8. One-fourth of a number, decreased by 5, equals 10. Find the number.

9. A sweater is on sale for $16. The sale price is $5.98 less than the original price. What was the original price of the sweater?

10. The cost of a shirt plus a sales tax of $0.87 equals $16.86. Find the cost of the shirt.

11. Eight less than twice Lisa's age is 36 yr. Find Lisa's age.

12. Ten less than three times Tony's age is 50. Find Tony's age.

13. If the quotient of a number and 5 is decreased by 15, the result is −100. Find the number.

14. If the quotient of a number and 7 is decreased by 14, the result is −150. Find the number.

15. If the difference of a certain number and 7 is multiplied by 6, the result is 4.2. Find the number.

16. If the difference of a certain number and 12 is multiplied by 8, the result is 9.2. Find the number.

17. The perimeter of a rectangle is 64 m. If the length of the rectangle is 14 m, find its width.

18. The perimeter of a rectangle is 96 m. If the width of the rectangle is 12 m, find its length.

19. Find the time it will take a cyclist to travel 40 mi at average speed of 12.5 mi/h.

20. Find the average speed of a runner who runs a distance of 2.5 mi in 20 min.

21. A pentagon, a five-sided figure, has two sides of equal measure and three sides measuring 12 in., 9 in., and 13 in. Find the lengths of the equal sides if the perimeter is 56 in.

22. A hexagon, a six-sided figure, has three equal sides and three sides measuring 15 cm, 20 cm, and 25 cm, respectively. Find the length of one of the equal sides if the perimeter is 150 cm.

23. A man invested one-half as much money in stock for National Computing, Inc. as he did in stock for Gateway Airlines. If his total investment was $4500, how much did he invest in stock for Gateway Airlines?

24. The distribution of grades in a class was 5 more B's than A's and twice as many C's as B's. If there are 35 students in the class, how many received each letter grade?

25. The formula $F = \frac{9}{5}C + 32$ can be used to convert a given Celsius temperature, C, to an equivalent Fahrenheit temperature, F. If the thermometer reads $-20°C$, find what the corresponding Fahrenheit reading would be.

26. The formula $C = \frac{5}{9}(F - 32)$ can be used to convert a given Fahrenheit temperature, F, to an equivalent Celsius temperature, C. If the thermometer reads $59°F$, find what the corresponding Celsius reading would be.

TEST YOURSELF

1. The quotient of a number and -6 is 36. What is the number?   3.4

2. Daphne paid $360 for a video cassette player. This was $30 less than 3 times the amount Barbara paid. How much did Barbara pay?

3. Prove: For all real numbers m, n, r ($n \neq 0$), if $m = r$, then $\frac{m}{n} = \frac{r}{n}$.   3.5

4. Evaluate $P = 4s$ for s if $P = \frac{1}{4}$.   3.6–3.7

5. Evaluate $A = \frac{1}{2}h(a + b)$ for h if $A = 40$, $a = 5.5$, and $b = 2.5$.

CHAPTER 3 SUMMARY AND REVIEW

Vocabulary

conclusion (114)
direct proof (114)
equivalent equations (96)
formula (117)
hypothesis (114)

inverse operations (96)
mathematical models (110)
postulate (113)
theorem (114)

Steps for Solving Equations   3.1–3.3

a. Use the Addition or Subtraction Property for Equations.

b. Use the Multiplication or Division Property for Equations.

c. Check your solution in the original equation.

Solve and check.

1. $x + 7 = 28$
2. $y - 11 = -38$
3. $-27 = a + 5$
4. $3.1 = -5.8 + b$
5. $9a = 45$
6. $\frac{1}{3}t = -15$
7. $-18 = -\frac{5}{2}x$
8. $-22y = -121$
9. $3a + 4 = 19$
10. $8 = 5x - 12$
11. $-5(y + 7) = 25$
12. $13a + 7 = 7$
13. $15.8 = 16 - 0.2d$
14. $4.9w - 18.6 = 535.1$
15. $0 = 1\frac{3}{5} - \frac{1}{5}y$
16. $\frac{1}{2} + \frac{3}{4}p = \frac{1}{4}$

Evaluating Formulas To use a formula as an equation, you must know the value of all variables but one.   3.6

17. $P = 2l + 2w$ If $P = 72$ and $l = 28$, find w.

18. $A = \frac{1}{2}h(a + b)$ If $A = 16$, $a = 3\frac{1}{4}$, and $b = 2$, find h.

19. $A = p + prt$ If $A = \$896$, $p = \$800$, and $t = 3$, find r.

20. $C = 2\pi r$ If $C = 16.956$ and $\pi \approx 3.14$, find r.

Chapter 3 Equations in One Variable

Algebraic Proof To prove statements about real numbers, use postulates, definitions, given facts, and other proved theorems.   3.5

21. State the Addition Property for Equations using real numbers x, y, and z.

22. Complete the unfinished statement and give the reason for each step of this proof.

    Prove: If $x = y$, then $x - z = y - z$.

    Proof:

Statements	Reasons
1. $x = y$	1. ?
2. $x - z = x - z$	2. ?
3. $x - z = $?	3. ?

Problem Solving Steps

Write an equation for each problem. Then solve the equation.   3.4, 3.7, 3.8

23. Three times the weight of a truck increased by 500 lb equals 12,200 lb. Find the weight of the truck.

24. The product of a number and $-\frac{3}{8}$ is -24. Find the number.

25. Alison has $809.45 in her savings account. This is $14.95 more than two times the amount Eric has in his account. How much money does Eric have in his account?

26. If Roberto's age is $\frac{3}{4}$ of his brother's age and Roberto is 45 years old, how old is his brother?

27. The perimeter of a rectangle is 54 cm. If the width is 12 cm, what is the length?

28. If Eli can average 37.5 mi/h, how long will it take him to drive 93.75 mi?

29. Sarat weighs twice as much as his sister Brenda. If Sarat weighs 210 lb, how much does Brenda weigh?

30. If Greg can walk 3.5 mi/h, how long will it take Greg to walk 5.25 mi?

CHAPTER 3 TEST

Solve and check.

1. $b - 2.7 = 2.5$

2. $7 = y + 3\frac{2}{5}$

3. $-\frac{1}{2}(y - 8) = 7$

4. $3t + 13 = -11$

5. $-\frac{5}{7}x = 35$

6. $-11 = a + 2$

7. $6d + \frac{3}{4} = -\frac{3}{4}$

8. $\frac{3}{4}(t + 8) = -6$

9. If $P = 2l + 2w$, $l = 15$, and $P = 80$, find w.

10. If $d = rt$, $d = 1140$, and $t = 4\frac{3}{4}$, find r.

11. Prove: For all real numbers x, y, and z, if $x = y$, then $xz = yz$.

12. Prove: For all real numbers a, b, c, and d, then $ab + cb + d = d + b(a + c)$.

Write an equation for each problem. Then solve the equation.

13. Anne Marie bought a number of shirts at $8 each. She also bought a pair of socks for $3. How many shirts did she buy if her total cost was $27.

14. If the product of a number and -3 is -87, find the number.

15. A bag of a dozen bagels cost 34¢ less than twelve individual bagels. If the price of the bag of bagels is $3.50, find the price of one individual bagel.

Challenge

Write an equation for each problem. Do not solve.

1. A 35 ft long wire is cut so that one piece is $9\frac{1}{2}$ ft longer than the other. What is the length of the shorter piece?

2. A computer system consisting of a monitor, a keyboard, and a computer with a hard disk costs $1800. The monitor costs $300 more than the keyboard, and the computer costs $600 more than the monitor. Find the cost of the keyboard.

PREPARING FOR STANDARDIZED TESTS

In each item you are to compare a quantity in Column 1 with a quantity in Column 2. Write the letter of the correct answer from these choices:

A. The quantity in Column 1 is greater than the quantity in Column 2.
B. The quantity in Column 2 is greater than the quantity in Column 1.
C. The quantity in Column 1 is equal to the quantity in Column 2.
D. The relationship cannot be determined from the given information.

Notes: Information centered over both columns refers to one or both of the quantities being compared. A symbol that appears in both columns has the same meaning in each column. All variables represent real numbers.

	Column 1	Column 2
1.	0.63	$\frac{17}{25}$
2.	$-\frac{2}{7}$	$-\frac{2}{5}$
3.	$\frac{0}{-1}$	$\frac{0}{1}$
4.	$10 + 4 \div 2$	$5 \cdot 6 \div 3$
5.	$\frac{1}{5}$ of $640	$\frac{1}{3}$ of $384
	$P = 2l + 2w$	
6.	$l = 10 \quad w = 5$	$l = 12 \quad w = 4$
7.	$-(a - b)$	$b - a$
8.	$\frac{ab}{3}$	$-\frac{3a}{b}$
9.	$2(2^2 + 3^2)$	$(2 + 3)^2$

	Column 1	Column 2
	$5x - 8 = 12$	
	$15 - 4y = 7$	
10.	x	y
11.	The sum of 2 times a number and 3	2 times the sum of a number and 3
	The product of	
12.	$\frac{55}{154}$ and $\frac{42}{70}$	$\frac{65}{78}$ and $\frac{54}{126}$

Use this problem to answer Exercises 13–15.

Ann's age is $\frac{2}{3}$ her mother's age.
Ann's age is $\frac{1}{2}$ her father's age.
Ann is 24.

	Column 1	Column 2
13.	father's age	mother's age
14.	30	mother's age
15.	father's age	48

MAINTAINING SKILLS

Simplify.

Example 1
$$8a - 3(a - 9) - 12$$
$$= 8a - 3a + 27 - 12 \quad \text{Distribute } -3.$$
$$= 5a + 15 \quad \text{Combine like terms.}$$

1. $7(b + 5) - 9b$
2. $6x + 5(x - 2) + 4$
3. $5y + 3 - (4y - 2)$
4. $-2(3c + 9) + 4c$
5. $m - 5 + (-5m + 9)$
6. $-(8 - 3n) + 6(2n - 12)$

Change each percent to a decimal. Change each decimal to a percent.

Example 2 $5\% \to 0.05 \to 0.05 \quad$ Divide by 100.

Example 3 $0.585 \to 058.5\% \to 58.5\% \quad$ Multiply by 100.

7. 20%
8. 42%
9. 6%
10. 12.5%
11. 10%
12. 30.3%
13. 0.75
14. 0.18
15. 0.07
16. 0.01
17. 0.385
18. 0.065

Change each fraction to a percent. Change each percent to a fraction.

Example 4 $\dfrac{5}{8} \quad 8\overline{)5.000}^{\,0.625} \quad 0.625 \longrightarrow 62.5\%$

Example 5 $41\dfrac{2}{3}\%$ means $\dfrac{41\frac{2}{3}}{100} = \dfrac{\frac{125}{3}}{100} = \dfrac{\cancel{125}^{\,5}}{3} \cdot \dfrac{1}{\cancel{100}_{\,4}} = \dfrac{5}{12}$

19. $\dfrac{1}{4}$
20. $\dfrac{2}{5}$
21. $\dfrac{7}{10}$
22. $\dfrac{1}{2}$
23. $\dfrac{3}{8}$
24. $\dfrac{2}{3}$
25. 35%
26. 80%
27. $37\dfrac{1}{2}\%$
28. 75%
29. $8\dfrac{1}{3}\%$
30. 36%

Use the formula given to solve each word problem.

31. Miriam lives 75 mi from Albany. How long would it take her to drive to Albany if she averages 50 mi/h? ($d = rt$)

32. The length of a rectangular garden is 12 m. The distance around the garden (P) is 42 m. What is the width? ($P = 2l + 2w$)

33. The Celsius (C) thermometer reads 35°. Find the temperature in degrees Fahrenheit. $\left(F = \dfrac{9}{5}C + 32\right)$

4 More Equations in One Variable

Manufacturers and designers of transportation systems must be able to solve equations involving variables. Such equations can aid in development of more efficient means of transportation.

4.1 Solving Equations: Combining Like Terms

Objective: To solve equations by removing parentheses and combining like terms

Some ninth graders are planning a trip to the Maritime Museum. The bus fare is $2 per student; admission to the museum is $3 per student.

If the variable n represents the number of students going on the trip, the entire cost can be thought of as $2n + 3n$.

Another way is to use the fact that the total cost per student is $5. Then $5n$ also expresses the entire cost of the trip. That is, $2n + 3n = 5n$.

Capsule Review

One of the key procedures needed in solving equations with the same variable is to combine like terms. Recall that $-5y$ and $3y$ are like terms but $3y$ and $3y^2$ are not like terms.

Combine like terms to simplify.

1. $7n + n - 2$
2. $5a^2 - b^2 + 8b^2$
3. $-5n - 6n - 1$
4. $-6 - 5 + 8x$
5. $3b + (-9b) - 8b^2$
6. $0 - 3w + 3$

EXAMPLE 1 It will cost each student $2 for bus fare and $3 for admission to the Maritime Museum. How many students plan to go on the trip if the total expense is $75?

Let n = number of students going on the trip.

$2n + 3n = 75$
$5n = 75$ *Combine like terms.*
$\dfrac{5n}{5} = \dfrac{75}{5}$ *Divide each side by 5.*
$n = 15$

So, 15 students plan to go on the trip.

Chapter 4 More Equations in One Variable

EXAMPLE 2 Solve and check: $28 - a + 4a = 7$

$28 - a + 4a = 7$
$28 + 3a = 7$ *Combine like terms: $-1a + 4a = 3a$*
$28 - 28 + 3a = 7 - 28$ *Subtract 28 from each side.*
$3a = -21$
$\dfrac{3a}{3} = \dfrac{-21}{3}$ *Divide each side by 3.*
$a = -7$

Check: $28 - a + 4a = 7$
$28 - (-7) + 4(-7) \stackrel{?}{=} 7$ *Replace a with -7.*
$28 + 7 - 28 \stackrel{?}{=} 7$
$7 = 7$ ✓   True

So, the solution is -7.

The commutative and associative properties for addition allow you to add and group terms in any order. In Example 3, notice how these properties are used to solve the equation $y + (y + 3) + (2y - 5) = 10$.

EXAMPLE 3 Solve and check: $y + (y + 3) + (2y - 5) = 10$

$y + (y + 3) + (2y - 5) = 10$
$(y + y + 2y) + [3 + (-5)] = 10$ *Group like terms.*
$4y\ \ +\ \ (-2) = 10$ *Combine like terms.*
$4y - 2 + 2 = 10 + 2$ *Add 2 to each side.*
$4y = 12$
$\dfrac{4y}{4} = \dfrac{12}{4}$ *Divide each side by 4.*
$y = 3$

Check: $y + (y + 3) + (2y - 5) = 10$
$3 + (3 + 3) + [(2 \cdot 3 - 5)] \stackrel{?}{=} 10$ *Replace y with 3.*
$3 +\ \ 6\ \ +\ \ 1\ \ \stackrel{?}{=} 10$
$10 = 10$ ✓   True

So, the solution is 3.

In the equation $4n - 7(n - 9) = 42$, parentheses are used to indicate multiplication. In such cases, use the distributive property to multiply before combining like terms and using the equation properties.

4.1 Solving Equations: Combining Like Terms

EXAMPLE 4 Solve: $4n - 7(n - 9) = 42$

$$4n - 7(n - 9) = 42$$
$$4n + (-7)(n - 9) = 42 \quad \text{Definition of subtraction}$$
$$4n + (-7)(n) + (-7)(-9) = 42 \quad \text{Distribute } -7.$$
$$4n - 7n + 63 = 42$$
$$-3n + 63 = 42 \quad \text{Combine like terms.}$$
$$-3n + 63 - 63 = 42 - 63 \quad \text{Subtract 63 from each side.}$$
$$-3n = -21 \quad \text{Divide by } -3.$$
$$n = 7$$

So, the solution is 7. The check is left for you.

CLASS EXERCISES

Use the distributive property to find each product.

1. $3(a + 2)$
2. $5(1 - 2b)$
3. $4(-3y - 8)$
4. $-3(-7 - 4x)$
5. $1.4(2m - 1)$
6. $-1(-1 + 3a)$

Solve and check.

7. $a + 3a = 8$
8. $7x + 2x - x = 30$
9. $5x - 3x - 6 = 6$
10. $x + x - 5 = 4$
11. $5 - t - t = -1$
12. $3(b + 5) = 21$
13. $4(y - 2) = 16$
14. $-3(-6 - 5m) = -12$
15. $(7x - 3) - (4x + 2) = 13$
16. $(5y + 4) + (3y - 9) = 11$

PRACTICE EXERCISES

Solve and check.

1. $6y + 2y = 16$
2. $4x - 2x = 18$
3. $4a + 3a - 7 = 21$
4. $7b - 2b = 30$
5. $3 - 6t - 5t = -19$
6. $5 - 8x - 2x = -25$
7. $-13 = 2x - x - 10$
8. $7x + 4 - 15x = 36$
9. $1 - 6y - 4y = 1$
10. $-5 = -6 - 3s + s + 1$
11. $c + (c + 2) + (c + 4) = 27$
12. $(2y + 5) - (y + 3) + (7y - 3y) = 17$
13. $x + (x + 3) + (x + 2) + (x + 7) = 40$
14. $(x + 7) + (x - 3) + (2x + 11) = 45$
15. $2(n - 3) = 12$
16. $3(b + 4) = 24$
17. $8(x - 1) = -24$
18. $3(5 - r) = 18$
19. $-2(3 - m) = 14$
20. $-5(3 - d) = 25$
21. $6(7 - 2t) = 30$
22. $4(5x - 5) = 0$
23. $-2(3x + 5) = 2$

Chapter 4 More Equations in One Variable

24. $4x - 7(x - 9) = 42$
25. $3n + 4(n - 9) = -78$
26. $-5(t - 7) + 5 = -5$
27. $2(7 - a) - 4 = 0$
28. $0.5d - 0.7d = 0.8$
29. $1.5c + 7.5c = 1.8$
30. $3z + z + z = 3.2$
31. $-4.5 = 5n - n - n$
32. $3(a - 5) = 4$
33. $-2(b + 4) = 11$
34. $x - 3(6 - x) = -14$
35. $6 - 22(p + 6) = 2$
36. $-4 - (7 + 3y) = 7$
37. $-9 - (8 - 5t) = 18$
38. $\frac{1}{3}(x - 9) = -6$
39. $\frac{2}{5}(y + 10) = 0$
40. $1 - \frac{3}{4}a - 9 = 2$
41. $\frac{1}{5}b - 6 + 11 = 1$
42. $\frac{2}{3} + 3(2 - m) = -2$
43. $-2(4 + n) - \frac{5}{8} = \frac{7}{8}$
44. $4.67y - 3.6y + 8.42 = 2$
45. $0.02 - 0.05x - 0.3x = -0.68$
46. $2[x + 3(x - 1)] = 18$
47. $4(2k - 7) + 3(k - 1) = 46$

Applications

Write an equation for each problem. Then solve the problem.

48. **Number Problem** Five times a number subtracted from 3 times the number equals 20. Find the number.

49. **Geometry** The length of a rectangular sign is 10 ft less than twice its width. Its perimeter is 118 ft. Find the length and width of the sign.

50. **Hobby** Jane, Jasmine, and Jocelyn have record collections. Jane has 3 times as many records as Jocelyn has. Jasmine has 2 more than twice the number Jocelyn has. Together the three girls have 56 records. How many records does each girl have?

WRITING IN ALGEBRA

Write an equation for each statement. By using operations and properties of numbers, think of a solution or solutions for each equation.

1. The sum of two numbers x and y is the same as their product.
2. A number n and its reciprocal are equal.
3. The difference of two numbers r and s is the same as their sum.
4. The quotient of two numbers c and d is the same as their product.
5. The sum of a number m and its reciprocal is 2.

4.2 Solving Equations: Variable on Both Sides

Objective: To solve equations that contain the variable on both sides

In the equations you have solved thus far, the terms containing variables were all on one side of the equation.

Capsule Review

To solve the equations below, you may have to use the distributive property to remove parentheses first, before combining like terms.

Solve and check.

1. $2x + 13 = 1$
2. $29 = 5 - 3y$
3. $4y - 2y = 16$
4. $-19 = 2a - a - 10$
5. $2(n - 3) = 14$
6. $5(4 - b) - 35 = 0$

In equations like $3x + 2 = -x - 2$, you need to place all the variable terms on one side of the equation.

EXAMPLE 1 Solve and check: $3x + 2 = -x - 2$

$$3x + 2 = -x - 2$$
$$3x + x + 2 = -x + x - 2 \quad \text{Add } x \text{ to each side.}$$
$$4x + 2 = -2 \quad \text{Combine like terms.}$$
$$4x + 2 - 2 = -2 - 2 \quad \text{Subtract 2 from each side.}$$
$$4x = -4$$
$$x = -1 \quad \text{Divide each side by 4.}$$

Check:
$$3x + 2 = -x - 2$$
$$3(-1) + 2 \stackrel{?}{=} -(-1) - 2 \quad \text{Replace } x \text{ with } -1.$$
$$-3 + 2 \stackrel{?}{=} 1 - 2$$
$$-1 = -1 \checkmark \quad \text{True}$$

So, the solution is -1.

In Example 1, you might choose to have the variable terms on the right. Then you would begin by subtracting $3x$ from each side of the equation. Solve $3x + 2 = -x - 2$ by first subtracting $3x$ from each side. Do you get -1 for the solution?

Some equations have more than one variable term on the same side. To solve these, the first step is to combine like terms that are on the same side.

EXAMPLE 2 Solve and check: $-36 + 2n = -3n + n - 5n$

$-36 + 2n = -3n + n - 5n$
$-36 + 2n = -7n$ *Combine like terms on the right side.*
$-36 + 2n - 2n = -7n - 2n$ *Subtract 2n from each side.*
$-36 = -9n$ *Combine like terms.*
$4 = n$ *Divide each side by -9.*

Check: $-36 + 2n = -3n + n - 5n$
$-36 + 2(4) = -3(4) + (4) - 5(4)$ *Replace n with 4.*
$-36 + 8 = -12 + 4 - 20$
$-28 = -28$ ✔   *True*

At this point, you should be familiar with the following steps for solving an equation.

Steps for Solving Equations

- Use the distributive property to remove parentheses on each side.
- Combine like terms on each side.
- Use the addition or subtraction property for equations so that
  (a) all variable terms are on one side of the equation, and
  (b) all numerical terms are on the other side.
- Use the multiplication or division property for equations.

EXAMPLE 3 Solve: $2y + 3(y - 9) = -(3y - 18) - y$

$2y + 3(y - 9) = -(3y - 18) - y$
$2y + 3(y - 9) = -1(3y - 18) - y$
$2y + 3y + 3(-9) = -1(3y) + (-1)(-18) - y$ *Distributive property*
$2y + 3y - 27 = -3y + 18 - y$ *Multiplication property for -1*
$5y - 27 = -4y + 18$ *Combine like terms.*
$5y + 4y - 27 = -4y + 4y + 18$ *Addition property for equations*
$9y - 27 = 18$
$9y - 27 + 27 = 18 + 27$ *Addition property for equations*
$9y = 45$
$y = 5$ *Division property for equations*

The check is left for you.

CLASS EXERCISES

In each case, tell what should be done to get an equivalent equation that has the variable terms on one side and the numerical terms on the other.

1. $7m = -30 + m$
2. $4x + 28 = 7 + x$
3. $6 - z = 2z$
4. $6n = 4(n + 5)$
5. $-7b = 3(4 - 3b)$
6. $2(y - 5) = 15y - 4y$

Solve and check each equation.

7. $3t - 4 = -3t - 4$
8. $8 - 5a = 3a$
9. $4(x + 7) = 7 + x$

For Discussion

10. To solve some equations in Exercises 1–6, you might prefer to get the variable terms on the right side rather than the left. For which equations? Why?
11. Which types of equations might not have a solution. Explain.

PRACTICE EXERCISES

Solve and check.

1. $2y + 5 = -y - 4$
2. $3m + 6 = -m - 6$
3. $3d - 8 = -6 + d$
4. $5a - 14 = -5 + 8a$
5. $4 - 7m = m + 4$
6. $4 - 9j = j + 4$
7. $6n = 4n + 20$
8. $11e = 9e + 14$
9. $5d - 4 = 2d + 6$
10. $2j - 5 = 8j + 7$
11. $-22 - 3d = -2d + d - 3d$
12. $-42 + 4c = -c + 3c - c$
13. $8y + 20 = y + 2y - 5y$
14. $4b - 10 = b + 3b - 2b$
15. $-7f + 2f = -36 + f + 3f$
16. $-5a + 7a = -18 + 3a + 5a$
17. $9y + 2 = 3(y + 4)$
18. $3x - 2 = 4(x - 2)$
19. $3y + 2(y - 5) = -(3y - 35) - y$
20. $4x + 3(x - 2) = -(5x - 20) - x$
21. $5y - 2(y + 5) = -(2y + 15) + y$
22. $2z - 3(z + 1) = -(5z + 3) + z$
23. $6m + 3(m + 2) = -(2m + 7) + m$
24. $9t + 5(t + 3) = -(t + 13) + t$
25. $5f - \frac{1}{2} = 4f + \frac{3}{4}$
26. $2g - \frac{5}{8} = g + \frac{5}{8}$
27. $7(2d - 1) + 5(2 - 3d) = 2d$
28. $3(6e - 2) + 4(1 - 5e) = e$
29. $5(3 - 4y) + 14y = 7(2 - 5y)$
30. $2(5 - 4x) + x = 3(3x - 11)$

31. $6(2n - 5) = -3(7 - 3n) + 2n$

32. $11(3 - 2q) = 3(9 - 7q) + q$

33. $5x - 1 = (4 - 3x)(-2) + 10$

34. $14t - 1 = (1 - 2t)(-3) + 12$

35. $-7y + 3 = -(3y + 5) - 2$

36. $-9m + 5 = -(2m + 8) - 6$

37. $-5(3x - 2) + 6(2 - 2x) = 3x$

38. $-6(6y - 4) + 5(3 - 7y) = -68y$

39. $-6t - [4 - (2 - 3t)] = 4(t + 1)$

40. $-8b - [2(11 - 2b) + 4] = 9b$

41. $4[5y - 4(y - 1)] = 3[4(y + 1)]$

42. $-2[5(2j - 6) - 7j] = 5[3(j - 2)]$

43. $\frac{1}{2}\left[\frac{2}{3}(4d - 1)\right] - (3d - 2) = \frac{5}{9}d$

44. $\left(4 - \frac{1}{2}c\right) - \frac{1}{3}\left[\frac{3}{4}(1 - 2c)\right] = \frac{1}{12}c$

45. $-3[4m + 2(m - 4)] = -5[2(m + 5)]$

46. $-2[3k + 3(k - 8)] = -2[3(k + 7)]$

Applications

47. **Geometry** The area A of a rectangle increased by 4 is equal to 36 decreased by 5 times the area. What is the area of the rectangle?

48. **Finance** Sally earns $25,000 per year as a computer programmer. Her mother's annual income is twice the amount of Sally's. Her brother's total income for 1 year is $\frac{1}{5}$ of Sally's income. What is her family's total income?

ALGEBRA IN TRANSPORTATION

Airplanes are one of the fastest means of transportation today. However, the velocities of different types of airplanes may vary greatly. Propeller planes may travel up to and over 250 mi/h, while a 747 can travel more than 600 mi/h. In 1947, the X-1 flew at the speed of sound, or Mach 1 (about 740 mi/h at sea level). Some years later, the X-15 flew at five times the speed of sound, or Mach 5. Presently, the Concorde provides passenger service at supersonic speeds for thousands of people.

Consider this problem: Two planes, one a propeller plane and one a supersonic transport (SST) travel the same distance. The propeller plane travels at a rate of 1300 mi/h less than that of the SST. If the SST makes the trip in 2 h and the propeller plane makes it in 15 h, what is the rate of the propeller plane? Use this equation: $2x = 15(x - 1300)$, where x is the rate of the SST.

4.3 Problem Solving: Consecutive Integers

Objective: To solve problems that involve consecutive integers

Consecutive integers are integers that differ by one. For example, 3, 4, and 5 are consecutive integers; $-3, -2,$ and -1 are also consecutive integers.

If x is an integer, a general representation of three consecutive integers is x, $x + 1$, $x + 2$.

EXAMPLE 1 Find three consecutive integers whose sum is 75.

■ **Understand the Problem**
You are told that the sum of three consecutive integers is 75. You are asked to find the integers.

■ **Plan Your Approach**
Let x = the first integer. Then $x + 1$ = the second integer and $x + 2$ = the third integer. The equation is:

$x + (x + 1) + (x + 2) = 75$

1st	x	?
2nd	$x+1$	?
3rd	$x+2$	?
Sum	$3x+3$	75

■ **Complete the Work**

Solve: $x + (x + 1) + (x + 2) = 75$
$(x + x + x) + (1 + 2) = 75$ *Group like terms.*
$3x + 3 = 75$
$3x + 3 - 3 = 75 - 3$ *Subtract 3 from each side.*
$3x = 72$
$x = 24$ *Divide each side by 3.*

Then $x + 1 = 25$ and $x + 2 = 26$.

■ **Interpret the Results**
Are 24, 25, and 26 three consecutive integers with a sum of 75?
$24 + 25 + 26 \stackrel{?}{=} 75$
$75 = 75$ ✔   True

Even integers are exactly divisible by 2. Beginning with an even integer and counting by twos gives consecutive even integers. For example, beginning with -8, four consecutive even integers are $-8, -6, -4,$ and -2. If n is an even integer, then four consecutive even integers are $n, n + 2, n + 4,$ and $n + 6$.

Chapter 4 More Equations in One Variable

EXAMPLE 2 The sum of four consecutive even integers is -148. Find the integers.

Understand the Problem This is an addition problem for which the sum is known; there are four consecutive addends, each divisible by 2.

Plan Your Approach Let $n =$ the first even integer.
Then $n + 2 =$ the second even integer,
$n + 4 =$ the third even integer,
and $n + 6 =$ the fourth even integer.

The equation is $n + (n + 2) + (n + 4) + (n + 6) = -148$.

Complete the Work
$4n + 12 = -148$
$4n = -160$ *Subtract 12 from each side.*
$n = -40$ *Divide each side by 4.*

Then $n + 2 = -40 + 2 = -38$,
$n + 4 = -40 + 4 = -36$,
and $n + 6 = -40 + 6 = -34$.

Interpret the Results The four consecutive even integers whose sum is -148 are -40, -38, -36, and -34. The check is left for you.

Odd integers are not exactly divisible by 2. Beginning with an odd integer and counting by twos gives consecutive odd integers. For example, three consecutive odd integers are 17, 19, and 21. If t is an odd integer, then four consecutive odd integers are t, $t + 2$, $t + 4$, and $t + 6$.

EXAMPLE 3 Find five consecutive odd integers if the sum of the first and the fifth is 1 less than 3 times the fourth.

Understand the Problem Let the five consecutive odd integers be t, $t + 2$, $t + 4$, $t + 6$, and $t + 8$.

Plan Your Approach The equation is $t + (t + 8) = 3(t + 6) - 1$.

Complete the Work
$2t + 8 = 3t + 18 - 1$
$2t + 8 = 3t + 17$
$8 = t + 17$ *Subtract 2t from each side.*
$-9 = t$ *Subtract 17 from each side.*

Interpret the Results If $t = -9$, counting by twos gives the five consecutive odd integers: -9, -7, -5, -3, and -1. Check the solution in Example 3.

4.3 Problem Solving: Consecutive Integers

CLASS EXERCISES

Complete the table where possible.

Begin with	Write the next three consecutive integers.	Write the next three consecutive even integers.	Write the next three consecutive odd integers.
1. 2	?	?	?
2. 0	?	?	?
3. −109	?	?	?

PRACTICE EXERCISES

Write an equation for each problem. Then solve the problem.

1. Find three consecutive integers whose sum is 99.
2. Find four consecutive integers whose sum is 26.
3. The sum of four consecutive even integers is −124. Find the number.
4. The sum of four consecutive even integers is −36. Find the number.
5. Find five consecutive even integers if the sum of the first and fifth is 2 less than 3 times the fourth.
6. Find four consecutive odd integers if the sum of the first and fourth is 3 less than 3 times the second.

Write an equation for each problem. Then solve the problem.

7. The sum of two consecutive integers is 105. Find the integers.
8. The sum of two consecutive integers is −35. Find the integers.
9. Find three consecutive integers whose sum is −354.
10. Find four consecutive integers whose sum is 50.
11. The sum of two consecutive even integers is −54. Find the integers.
12. The sum of three consecutive even integers is 312. Find the integers.
13. The sum of three consecutive odd integers is −45. Find the integers.
14. The sum of four consecutive even integers is 180. Find the integers.
15. Find four consecutive integers if the sum of the second and fourth is 48.
16. Find three consecutive integers. The sum of the first and third is −34.

17. Find two consecutive odd integers if twice the larger, increased by the smaller, equals 85.

18. Find three consecutive even integers if their sum, decreased by the third, equals -22.

19. The ages in years of three brothers are consecutive. The sum of their ages is 39 decreased by the age of the youngest. What are their ages?

20. The sum of three consecutive even integers is 50 more than the third integer. Find the integers.

21. The sum of three consecutive integers is equal to 9 less than 4 times the least of the integers. Find the three integers.

22. When the sum of four consecutive even integers is divided by 7, the result is 4. Find these integers.

23. Find four consecutive multiples of 5 whose sum is 90.

24. Find three consecutive multiples of 3 if the sum of the first and the third is 12.

25. Find three consecutive integers if twice the middle integer is equal to the sum of the first and the third.

26. Find two consecutive odd integers whose sum is 0.

27. Find four consecutive integers such that the sum of the two largest subtracted from three times the sum of the two smallest is 70.

28. Find four consecutive odd integers such that the sum of the two smallest added to four times the largest is 92.

MATH CLUB ACTIVITY

1. The sum of two consecutive integers is 2165. What is their difference?

2. Find four consecutive even integers if the sum of the first and the third is 8 less than the sum of the second and the fourth.

3. Find six consecutive multiples of 13 whose sum is 39.

4. Find three consecutive even integers if the first is $\frac{2}{5}$ of their sum.

5. A sequence of four integers begins with two consecutive even integers and contains the next two consecutive odd integers. Their sum is 18. What are the four integers?

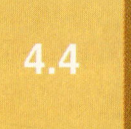

4.4 Shortcuts in Solving Equations

Objective: To use shortcuts to solve equations

Yasmin was asked to solve the equation $0.2x - 0.5x - 0.7 = 0.3 + 0.2x$. She knew that she could combine the two x-terms on the left side and then apply the equation properties. However, she noticed that there was a $0.2x$ term on each side of the equation. "Wouldn't it be simpler to subtract $0.2x$ from each side first?" she wondered. She tried it.

$$0.2x - 0.5x - 0.7 = 0.3 + 0.2x$$
$$0.2x - 0.2x - 0.5x - 0.7 = 0.3 + 0.2x - 0.2x \quad \text{Subtract } 0.2x \text{ from each side.}$$
$$-0.5x - 0.7 = 0.3 \quad \text{Add } 0.7 \text{ to each side.}$$
$$-0.5x = 1 \quad \text{Divide each side by } -0.5.$$
$$x = -2$$

Yasmin checked -2 in the original equation and found that it was correct.

Capsule Review

Multiplying a decimal by a power of 10 such as 10, 100, 1000 changes the place value of the digits. This can be thought of as "moving the decimal point."

EXAMPLE   **a.** $100(0.73) = 0\underset{\curvearrowright}{.}73.$ or 73   **b.** $1000(3.8) = 3\underset{\curvearrowright}{.}800.$ or 3800

                         2 places   3 places

Multiply.

1. $10(42.8)$ **2.** $8.75(100)$ **3.** $1000(-0.635)$ **4.** $7.001(10,000)$

State the power of 10 that gives the product at the right.

5. $(?)(0.041) = 41$ **6.** $(?)(32.08) = 3208$ **7.** $(?)(0.00032) = 32$

Equations with decimal coefficients can be simplified by multiplying each side of the equation by a power of 10. Consider the following:

$$0.11y = 1.5 + 0.1y$$

Since the greatest number of digits to the right of either decimal point is two, you can multiply each side of the equation by 100 to clear all decimals.

Chapter 4 More Equations in One Variable

EXAMPLE 1 Solve and check: $0.11y = 1.5 + 0.1y$

$$0.11y = 1.5 + 0.1y$$
$$(100)(0.11y) = 100(1.5 + 0.1y)$$ *Multiply each side by 100.*
$$11y = 100(1.5) + 100(0.1y)$$ *Distribute 100.*
$$11y = 150 + 10y$$
$$11y - 10y = 150 + 10y - 10y$$ *Subtract 10y from each side.*
$$y = 150$$

Check: $0.11y = 1.5 + 0.1y$
$$(0.11)150 \stackrel{?}{=} 1.5 + 0.1(150)$$ *Replace y with 150.*
$$16.5 \stackrel{?}{=} 1.5 + 15$$
$$16.5 = 16.5 \;\checkmark$$ *True*

So, 150 is the solution.

The equation $9a = 3(a - 2)$ can be simplified by distributing the 3, but note that both $9a$ and $3(a - 2)$ are divisible by 3.

EXAMPLE 2 Solve and check: $9a = 3(a - 2)$

$$9a = 3(a - 2)$$
$$\frac{9a}{3} = \frac{3(a - 2)}{3}$$ *Divide each side by 3.*

                               $\frac{3(a-2)}{3} = \frac{3}{3} \cdot \frac{a-2}{1} = a - 2$

$$3a = a - 2$$
$$3a - a = a - a - 2$$ *Subtract a from each side.*
$$2a = -2$$
$$a = -1$$

So, -1 is the solution. The check is left for you.

You need not always use shortcuts, but be aware that they frequently make solving equations much simpler.

CLASS EXERCISES

Identify an appropriate shortcut operation. Then give the resulting equation. Do not solve.

1. $4(3 - x) = 16$
2. $18 = -6(y + 5)$
3. $33 - 4a = 11a - 4a$
4. $-7(b + 4) + 5 = 5$
5. $0.3 - 0.1c = 1.5$
6. $2.1 + 0.3d = 0.33d$
7. $5(2m - 3) = 15m$
8. $5(p - 6) - 8 = 45 - 8$

For Discussion

9. If each side of the equation $0.4(0.2q + 5.1) = -7.2$ is multiplied by 10, what equation results?

10. In Exercise 8 above, two shortcut operations are possible. What are they? After the two shortcuts are used, what is the resulting equation?

PRACTICE EXERCISES

Solve and check. Use shortcuts where possible.

1. $0.2y - 0.4 = 1.5$
2. $0.8x + 0.3 = 59.7$
3. $1.80 - 0.02j = 1.32$
4. $3.21 - 4.27k = 7.48$
5. $23.4 - d = 0.25d$
6. $82.6 - e = 0.18e$
7. $6a = 2(a + 3)$
8. $15x = 5(x + 4)$
9. $2(5 - a) = 4$
10. $-18 = 9(t - 4)$
11. $(b + 7)(-12) = 36$
12. $(5 + c)(-13) = 65$
13. $5 + 7r = -25r + 7r$
14. $-8t - t = 24 - t$
15. $-8 + 5f = 3f - 11 + 5f$
16. $-13 + 7g = 4g - 23 + 7g$
17. $-7(2x - 7) = 21(3x - 5)$
18. $-8(4y - 6) = 16(2y - 5)$
19. $-35(14 - 2r) = -5(8 + r)$
20. $-36(7 - 3t) = -6(2 + 2t)$
21. $3(2y - 1) + y = y - 3$
22. $4(3x - 5) - x = -x + 16$
23. $962.5 = 805 + 1.5a$
24. $178.5 = 310 + 2.5b$
25. $0.045p = 0.06(2000 - p) + 27$
26. $0.075v = 0.03(50 - v) + 3$
27. $0.2q + 0.07(q + 4) = 0.01q$
28. $0.8r + 0.09(r + 3) = 0.08r$
29. $0.1a + 0.05(2 - a) = 0.08$
30. $0.7b + 0.15(5 - b) = 0.2$
31. $-4.5n + 0.02n = 19.2 - 4.5n$
32. $5.36m - 0.4m = 26.8 - 0.4m$
33. $8(3x - 4) + 16(x + 3) = -112$
34. $9(2y + 3) + 27(y + 2) = -36$
35. $2(3 - 4f) - 4(6 - f) = 4$
36. $10(2 - g) - 5(1 - 3g) = 30$
37. $4.85(2 - b) + 0.2b = 4.85(2 - b)$
38. $7.1(c - 11) + 3.4c = -7.1(11 - c)$
39. $6a - 3(2a - 3) + 9(3 - a) = 12$
40. $-5b - 15(1 - 2b) + 30(b - 2) = 60$
41. $0.1y - 0.05(y + 2) = 0.1y + 0.03(3 - 2y)$
42. $3.86x - 5.4(2x + 5) = 3.86x + 1.8(6 + 3x)$

Chapter 4 More Equations in One Variable

43. $0.31t - 0.1(0.02 - 0.2t) = 0.064$

44. $0.73w - 0.3(0.15 - 0.3w) = -5.785$

45. $-0.6x - 0.1[4 - (2 - 3x)] = 0.4(x + 1)$

46. $3.2y - 0.5[7 - (4 - 5y)] = 0.6(2y + 2)$

47. $7.03(2 - 5v) = 0.2[2v - (1 - v)] + 3.2v$

48. $-2.25(3 - 2z) = -0.5[-6z - (1 - 2z)] + z$

Applications

49. **Number Problem** Three times the sum of a number x and -5 equals -6 times the sum of x and -2. Find x.

50. **Number Problem** The product of -4 and a number y is equal to 20 times the sum of 7 and 3 times y. Find y.

51. **Geometry** If the perimeter of the rectangle shown is 84 m, find its length and width.

52. **Sales** Charles sold bottles of juice at a football game for $1.25 each. At the end of the game, he had $86.25. This included the $7.50 he had started with. How many bottles of juice did Charles sell?

TEST YOURSELF

1. Solve: $x + (x + 1) + (x + 2) = 45$ **4.1**

2. Solve: $-9 = 6m - (2 - m)$

3. The sum of a number and 2 more than twice the number is 8. Find the number.

4. Solve: $8y - 3 = -7 + 6y$ **4.2**

5. Solve: $2 - 3r = -3(1 + 2r)$

6. Solve: $-8j + 5 = 4(11 - 2j)$

7. If i is an even integer, write the next two consecutive even integers.   **4.3**

8. If $d - 1$ is an integer, write the next three consecutive integers.

9. Find four consecutive odd integers whose sum is 56.

10. Solve using a shortcut: $3t - 2t = 18 + 3t$ **4.4**

11. Solve using a shortcut: $0.01 + 0.2x = 7.0$

12. Solve using a shortcut: $-8k = 16(k - 1)$

4.5 Solving Equations: Percents

Objective: To use equations for solving simple percent problems

Number relationships are often expressed in terms of percent, as in these newspaper headlines. **Percent** (%) means per hundred or hundredth. For example, one way of thinking about an unemployment rate of 8% is to say 8 out of 100 people are out of work. Likewise, a 5% increase in gasoline prices translates into a 5-cent increase for every $1.00 spent. It is helpful to be able to use algebraic techniques to solve problems involving percent.

Capsule Review

EXAMPLE   12.5% = 0.12.5   Move the decimal point two places to the left.

Express each percent as a decimal.

1. 29%   2. 3.1%   3. 100%   4. 280%   5. 0.02%

EXAMPLE   0.82 = 0.82.%   Move the decimal point two places to the right.

Express each decimal as a percent.

6. 0.05   7. 0.13   8. 2.0   9. 0.4115   10. 15

Any situation that involves percent can be expressed in the general form:

$$a \text{ is } b\% \text{ of } c$$

Since there are three quantities involved here, there are three types of percent problems. If you know any two of the three quantities, the third can be found. These three types of percent problems are often referred to as the "three cases of percent."

CASE I   Find a percent of a number.

CASE II   Find what percent one number is of another.

CASE III Find a number when a percent of it is known.

152   Chapter 4 More Equations in One Variable

EXAMPLE 1 **What number is 25% of 700?**

Let n = the number.

n is 25% of 700.   **CASE I**

$n = 0.25(700)$ *25% = 0.25; "of" means times.*

$n = 175$

EXAMPLE 2 **1 is what percent of 3?**

Let r = the percent.

1 is what percent of 3?   **CASE II**

$1 = r(3)$

$\frac{1}{3} = r$ *Divide each side by 3.*

$r = 0.33\overline{3}$, or $33\frac{1}{3}\%$ *Change $\frac{1}{3}$ to a decimal and a percent.*

EXAMPLE 3 **33 is $37\frac{1}{2}\%$ of what number?**

Let t = the number.

33 is 37.5% of t.   **CASE III**

$33 = (0.375)t$

$33{,}000 = 375t$ *Multiply each side by 1000.*

$88 = t$ *Divide each side by 375.*

To solve percent problems, usually it is most efficient to change percents to decimals. However, it may be just as easy to work with fractions.

EXAMPLE 4 **$12\frac{1}{2}\%$ of what number is 30?**

Let m = the number.

$12\frac{1}{2}\%$ of m is 30.   **CASE III**

$0.125 \cdot m = 30$

$\frac{1}{8}m = 30$ *$0.125 = \frac{125}{1000} = \frac{1}{8}$*

$m = 240$ *Multiply each side by 8; $8 \cdot \frac{1}{8}m = m$.*

4.5 Solving Equations: Percents

In Example 5, this formula is used to compute annual simple interest.

$$\underbrace{\text{Annual Simple Interest}}_{i} = \underbrace{\text{Principal}}_{p} \times \underbrace{\text{Annual Interest Rate}}_{r} \times \underbrace{\text{Time (in years)}}_{t}$$

Explain how this relationship shows another case of "a is $b\%$ of c."

EXAMPLE 5 Dierdre earned $65.25 in simple interest for one year on an investment of $900. What annual rate of interest was paid?

Let r = the annual interest rate.

$65.25 is what percent of $900?   **CASE II**
$65.25 = 900r(1)$
$6525 = 90{,}000r$ *Multiply each side by 100.*
$0.0725 = r$ *Divide each side by 90,000.*

So, the annual rate of interest was 0.0725, or 7.25%.

CLASS EXERCISES

For each exercise, first give a word statement in the form a is $b\%$ of c. Then write it as an algebraic equation.

	a.	b.	c.
1.	?	25%	32
2.	150	?	200
3.	30	15%	?
4.	?	35%	400
5.	16	?	48
6.	8	4%	?

PRACTICE EXERCISES

1. What number is 2% of 49?
2. What number is 65% of 130?
3. Find 37.5% of 1000.
4. Find 6% of 248.
5. 5 is what percent of 25?
6. 35 is what percent of 35?
7. 6 is what percent of 9?
8. 30 is what percent of 80?
9. 0.75 is 5% of what number?
10. 8.9 is 100% of what number?
11. 80% of what number is 28?
12. $62\frac{1}{2}\%$ of 480 is what number?

13. $87\frac{1}{2}\%$ of $4000 is what amount?
14. 150% of what number is 60?
15. 0.8 is what percent of 32?
16. What percent of 75 is 2.4?
17. 36 is $66\frac{2}{3}\%$ of what number?
18. $33\frac{1}{3}\%$ of what number is 75?
19. What percent of 1.4 is 3.5?
20. 30 is what percent of 25?
21. $16\frac{2}{3}\%$ of what number is 1.2?
22. 12.5% of what number is 1.5?
23. What percent of 9 is 12?
24. 15 is what percent of 5?
25. What amount is $1\frac{1}{2}\%$ of $6000?
26. What amount is $8\frac{1}{4}\%$ of $720?
27. What amount is $\frac{1}{2}\%$ of $5000?
28. 10 is what percent of 4000?
29. $0.16a$ is what percent of $2a$?
30. What percent of $0.3b$ is $5b$?
31. $y\%$ of what number is y?
32. $x\%$ of what number is $3x$?

Applications

33. **Demography** Of 625 people surveyed, 44% agreed with the mayor. How many people agreed with the mayor?

34. **Finance** Ben's bank pays $6\frac{3}{4}\%$ annual simple interest on his account of $440. How much interest will the account earn in one year?

35. **Finance** Sara's investment of $1572 earned $129.69 in annual simple interest. What was the annual interest rate?

36. **Demography** The population of Andropolis is now 25,868. This is 145% of what it was ten years ago. What was the earlier population?

EXTRA

Draw a square like the one shown. Write the value of each variable in the appropriate space to form a magic square (the sum of the numbers in each row, each column, and each diagonal are all the same).

1. a is 5% of 80.
2. $4\frac{1}{2}\%$ of 200 is b.
3. 4 is 200% of c.
4. d is 15% of 20.
5. $\frac{1}{2}\%$ of 1000 is e.
6. $33\frac{1}{3}\%$ of 21 is f.
7. g is 12% of $66\frac{2}{3}$.
8. $\frac{1}{3}\%$ of 300 is h.
9. $90 is $i\%$ of $1500.

$a =$	$b =$	$c =$
$d =$	$e =$	$f =$
$g =$	$h =$	$i =$

4.6 Problem Solving: Percents

Objective: To solve percent of change problems

If you are thinking about taking a trip to the West Coast, this newspaper ad might be of interest to you.

The 30% mentioned is a *percent of change*. In this case, it is a **percent of increase**. The change will be 30% of the original price.

LAST WEEK

REDUCED FLIGHT RATES
ROUND TRIP TO WEST COAST $299
BEGINNING SUNDAY RATES WILL INCREASE BY 30%

EXAMPLE 1   If the price of a $299 round-trip plane ticket is increased by 30%, find the new price.

- **Understand the Problem**   Given: The original price is $299. The increase is 30%.
  Find: The new price.

- **Plan Your Approach**   What is 30% of the original price?
  Let x = the dollar change in price.

- **Complete the Work** $x = 0.30(299)$ *Write an equation.*
  $x = 89.7$ *The change in price is $89.70.*

- **Interpret the Results** $89.70 is 30% of the original price, $299.00.
  The new (higher) price is the sum of the original price and the increase: $299.00 + $89.70 = $388.70

  So, $388.70 is the new price.

Percent of change can involve a decrease as well as an increase. If you know an original value and a new lower value, you can find the **percent of decrease**.

EXAMPLE 2   The population of Warrensville has decreased from 696 to 580. Find the percent of decrease.

- **Understand the Problem**   Given: The change (the decrease): $696 - 580 = 116$
  Find: What percent the decrease is of the original.

Plan Your Approach Let r = percent of decrease.
116 is what percent of 696?

Complete the Work
$$116 = r(696)$$
$$\frac{116}{696} = r \quad \text{Divide each side by 696.}$$
$$r = \frac{1}{6} \qquad \frac{116}{696} = \frac{1}{6}$$
$$= 0.16\overline{6} \text{ or } 16\frac{2}{3}\%$$

Interpret the Results So, the population has decreased by $16\frac{2}{3}\%$, or approximately 16.7%. How can you check this answer?

CLASS EXERCISES

Copy and complete the table.

Original Price	New Price	Change (Inc/Dec)	% Change (Inc/Dec)
1. $36	$24	?	?
2. $72	?	$16 (inc)	?
3. ?	$64	$12 (inc)	?
4. $75	?	?	$33\frac{1}{3}\%$ (dec)
5. ?	?	$15 (dec)	10% (dec)

PRACTICE EXERCISES

Copy the table and find the missing entries in each row. Round amounts to the nearest cent, or to the nearest tenth of a percent.

Original Price	New Price	Change (Inc/Dec)	% Change (Inc/Dec)
1. $196	?	$49 (dec)	?
2. $144	?	?	$16\frac{2}{3}\%$ (inc)
3. ?	?	$29 (inc)	35% (inc)
4. ?	$1028	$64 (dec)	?
5. $883	$909	?	?
6. $684	?	$36 (inc)	?
7. $84	?	?	$6\frac{1}{4}\%$ (dec)
8. ?	$514	$18 (inc)	?

4.6 Problem Solving: Percents

9. When Marya began her new job in marketing, she was told that if her work was satisfactory she would receive a 15% increase after six months. If her starting salary was $19,500, what could she hope to earn after six months?

10. If oil prices decline by $12\frac{1}{2}\%$, what will be the price of oil now selling for $48 per barrel?

11. The number of subscribers to *ACE Magazine* fell from 810,153 to 708,417. What percent of decrease does this drop represent?

12. Jane bought shares of stock in MARTCO for $78.90 per share. Now each share is worth $84.95. By what percent has the value of MARTCO stock risen?

13. Over the last five years, the popoulation of Chowan increased by 1084 people. The earlier population was 9785. By what percent has Chowan's population risen?

14. Razi weighs 135 lb. In order to qualify for his wrestling division, he must lose 9 lb. What percent weight decrease does this loss represent?

15. In Dawn City, the number of registered voters fell by 115. Now there are 7826 registered voters. What percent of decrease is this change?

16. Sneakers at a local sports store cost $48.50. This is $6.75 more than they cost last year. By what percent has the price increased?

17. The price of a cheese stick at Dairy Heaven has gone up by $16\frac{2}{3}\%$, or 10¢. What was the old price and what is the new?

18. This year, 7 fewer students tried out for Lesterville High School's football team than tried out last year. This is a 14% decrease. How many students tried out last year? This year?

19. By buying last year's model, Mr. D'Andrea saved 20% on the sticker price of a new automobile. If the sticker price is $12,998, what did he actually pay?

20. At the Hairways Beauty Salon, women's haircuts cost $18.50. If the price increases by 15%, what will the new price be?

21. At the supermarket, Laura found that the price of tomatoes had risen from 89¢/lb to $1.19/lb. By what percent had the price increased?

22. Last year, the cost of Julio's dance lessons was $21.95 per week. If the price has increased by $8\frac{1}{3}\%$, what is the cost of the lessons this year?

23. In order to promote its new bagels, Bagel King plans to sell them initially at 25% less than the regular price of 88¢. What is the promotion price?

24. This year's assessment of the Mehtas' home is $189,000, an increase of $19,000 over last year's assessment. What percent increase does the new assessment represent?

25. The value of Ari's car, bought last year, has depreciated by $1850. If it is now worth $8990, by what percent did its value decrease?

26. This week at Record World, records by Girl Jill cost $1.10 less than usual. This is a 20% decrease. What is the current price?

27. At one service station the price of super unleaded gasoline is 20¢ per gal higher, or 26% higher than that of regular unleaded gasoline. What does super unleaded gasoline cost?

28. Pet World has reduced the price of angelfish by 30%. If they now cost $2.75 each, what did they cost before?

29. The number of students who signed up for a museum tour is 19% more than the number that went last year. If 88 plan to go this year, how many went last year?

30. Breathless Cosmetics now offers 24 shades of eye shadow, 20% more than last year. How many shades were offered last year?

31. After the price of T-shirt dresses was reduced by 39%, they sold for $13.98. What was the earlier price?

ALGEBRA IN ENGINEERING

Drag force is a vehicle's resistance to moving through air. When designing automobiles, engineers try to reduce drag force because 60% of an automobile's power may be needed to overcome it. Decreasing the surface area of the front of an auto and recessing bumpers and door handles helps to reduce drag force.

1. Rank these vehicles from 1 (best) to 5 (worst) at overcoming drag force.

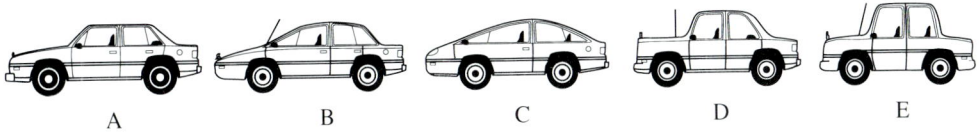

2. A 19% reduction in drag force will result in a 4.25% increase in gas mileage. If the previous mileage was 36.5 mi/gal, find the new mileage after the drag force has been reduced.

3. Luggage on top of a car increased the drag force by 19%, and caused a 4.5% decrease in gas mileage. Without luggage, mileage was 28.0 mi/gal. What was the gas mileage with luggage?

4. With a 15 mi/h tail wind, the gas mileage from Pittsburgh to Harrisburg was 43.0 mi/gal. With a head wind of 25 mi/h, the mileage was reduced by 9%. What was the mileage with the head wind?

APPLICATION:
Messenger Service

Bicycles are used throughout the world as a means of transportation. In large cities, messengers use bicycles because they can get in and out of traffic quickly. The messenger earns a percent of the amount of money the customer pays for the delivery. The percent increases each year he or she works for the service.

The table below gives the rates for a typical messenger service.

MESSENGER SERVICE

Location	Zone 1	Zone 2	Zone 3	Zone 4	Zone 5
Cost of Service	$6.00	$12.00	$18.00	$24.00	$30.00
Messengers' Earnings					
1st year	$3.00	$6.00	$ 9.00	$12.00	$15.00
2nd year	3.30	6.60	9.90	13.20	16.50
3rd year	3.60	7.20	10.80	14.40	18.00

Look for a pattern to find a formula for earnings in the first year.

Each amount in the first year is $\frac{1}{2}$ or 50% of the cost of the service.

Let c = cost of the service
e_1 = earnings the first year

Then e_1 = 50% of c
$e_1 = 0.5 \cdot c$

Solve.

1. Use guess and check to find a formula for the earnings in the second year.

2. Use guess and check to find a formula for the earnings in the third year.

3. The deliveries Jack made on Monday cost $228. This is his second year as a messenger. What were his earnings for the day?

4. The deliveries Ann makes on an average day cost $240. This is her third year as a messenger. How much does she earn in a 5-day week?

5. Sue is in her second year as a messenger. If she earned $90 in 1 day, what was the cost of the service to her customers for that day?

6. One week during Tom's third year as a messenger, he earned $132. What was the cost of the service to his customers for that week?

4.7 Problem Solving: Mixtures

Objective: To solve mixture problems involving percents

A type of problem that chemists and pharmacists often encounter is the need to change the concentration of solutions or other mixtures. In such problems, the amount of a particular ingredient in the solution or mixture is often expressed as a percent of the total.

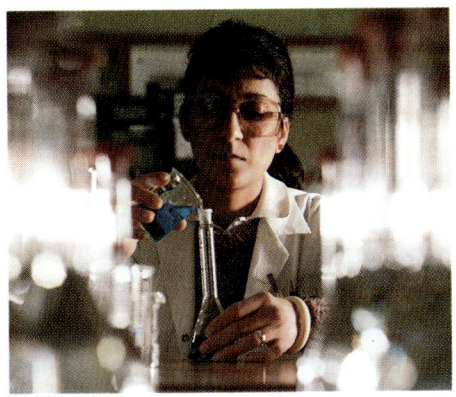

Suppose a solution of salt in water has a salt concentration of 15%. Then in 100 g of this solution there are 15 g of salt (and 85 g of water). If you have 20 g of this solution, it contains 0.15(20) or 3 g of salt.

EXAMPLE 1 A 50 mL solution of acid in water contains 25% acid. How much water would you add in order to make a 10% acid solution?

Understand the Problem
50 mL of 25% acid solution is to be changed to a 10% acid solution. Water is to be added. The number of milliliters of pure acid does not change.

Plan Your Approach
Let x = number of milliliters of water to be added.
Make a chart.

	Substance	Total No. mL	No. mL Pure Acid
Start with	25% acid solution	50	0.25(50)
Add	water	x	0
Finish with	10% acid solution	$x + 50$	$0.10(x + 50)$

Since the number of milliliters of pure acid stays the same, the first and third entries of the last column are equal.

$$0.25(50) = 0.10(x + 50)$$

Complete the Work

$$\begin{aligned}
0.25(50) &= 0.10(x + 50) \\
100(0.25)(50) &= 100(0.10)(x + 50) \quad \textit{Multiply each side by 100.}\\
25(50) &= 10(x + 50) \\
1250 &= 10x + 500 \\
750 &= 10x \\
75 &= x
\end{aligned}$$

■ **Interpret the Results** Check to see if 75 mL is a correct answer. Is the number of mL of pure acid, 0.25 · 50, equal to 10% of the total number of milliliters in the new solution, 0.10(75 mL + 50 mL) or 125 mL?

$$12.5 \stackrel{?}{=} 0.10(125)$$
$$12.5 = 12.5 \ ✔ \quad \text{True}$$

So, 75 mL of water must be added.

EXAMPLE 2 To increase the sugar concentration of a solution from 15% to 40%, how many ounces of pure sugar must be added to 80 oz of the 15% solution?

■ **Understand the Problem** An 80 oz solution which has a sugar concentration of 15% is to be changed to a solution with a 40% sugar concentration.

■ **Plan Your Approach** Let x = number of ounces of pure sugar to be added. Make a chart.

	Substance	Total No. Oz	No. Oz Pure Sugar
Start with	15% sugar solution	80	0.15(80)
Add	100% pure sugar	x	1.00(x), or x
Finish with	40% sugar solution	$80 + x$	$0.40(80 + x)$

In the last column, the number of ounces of pure sugar in the 15% solution plus the number of ounces of pure sugar added equals the number of ounces of pure sugar in the 40% solution.

■ **Complete the Work**

$$0.15(80) + x = 0.40(80 + x)$$
$$100(0.15)(80) + 100x = 100(0.40)(80 + x) \quad \text{Multiply each side by 100.}$$
$$15(80) + 100x = 40(80 + x)$$
$$1200 + 100x = 3200 + 40x$$
$$60x = 2000 \quad \text{Use the subtraction property.}$$
$$x = 33.\overline{3} \text{ or } 33\tfrac{1}{3}$$

■ **Interpret the Results** So, $33\tfrac{1}{3}$ oz of pure sugar must be added. How would you check this answer?

CLASS EXERCISES

Copy and complete the table. Do not simplify.

Substance	Total Amount of Solution	Amount of Substance in the Solution
1. 25% acid solution	150 mL	0.25(__?__) mL
2. 35% alcohol solution	250 oz	__?__ oz
3. pure acid	540 g	__?__ g
4. 15% salt solution	185 g	__?__ g

PRACTICE EXERCISES

1. Steven added 25 mL of water to 125 mL of a 20% salt solution in water. What is the salt concentration of the new solution?

2. Mary added 2 L of water to 8 L of a 48% nitric acid solution. What is the acid concentration of the new solution?

3. Jorge added 15 g of sugar to 210 g of a 5% sugar solution in water. What is the percent of sugar in his new solution?

4. Rita added 1 qt of alcohol to 4 qt of a solution that is 20% alcohol in water. What is the percent of alcohol in her new solution?

5. If you wish to increase the percent of acid in 50 mL of a 15% acid solution in water to 25% acid, how much pure acid must you add?

6. How many ounces of alcohol should be added to 125 oz of a 45% solution of sugar in alcohol in order to produce a 30% sugar solution?

7. In order to increase the concentration of 25 g of 5% salt solution in water to 15% salt, how much salt must be added?

8. How many gallons of milk that is $3\frac{1}{2}$% butterfat must be added to 80 gal of 1% butterfat milk to make milk that is 2% butterfat?

DID YOU KNOW?

Ancient Egyptians developed a base-ten system that used special symbols for powers of ten.

Our Symbols	1	10	100	1000	10,000	100,000
Egyptian Symbol	│	∩	ϡ	⚹	╒	∽

Our number 2345 can be written in expanded form as:

$$(2 \times 1000) + (3 \times 100) + (4 \times 10) + (5 \times 1)$$

In some Egyptian writings found on stone tablets, the order of their number symbols in comparison to ours was reversed. The Egyptians wrote their number for 2345 as:

ⅠⅠⅠⅠⅠ	∩∩∩∩	ϡϡϡ	⚹⚹
five units	four tens	three hundred	two thousands

1. Use Egyptian symbols to write five large numbers.

2. Conduct research to find out how the Egyptians performed the operations of addition and multiplication. Could they solve equations? Explain.

4.8 Literal Equations

Objective: To solve literal equations

A **literal equation** is one in which the constants or the coefficients of a variable are expressed by letters. Formulas are literal equations.

For the spring concert:
- a = number of adult tickets sold
- s = number of student tickets sold

The total revenue from ticket sales may be represented by:

$$t = 5a + 2s$$

The formula $t = 5a + 2s$ is a *literal equation*. You say that t *is solved in terms of a and s* because if you substitute values for a and s, the value of t can be found. The equation properties can be used to solve literal equations.

Capsule Review

EXAMPLE $2r - 7 = 3$
$2r - 7 + 7 = 3 + 7$ *Addition property for equations*
$2r = 3 + 7$

Name the equation property used to derive the second equation from the first.

1. $3 + y = 8$
   $y = 8 - 3$

2. $4a = 8 - 12$
   $a = 2 - 3$

3. $s - 9 = 6$
   $s = 15$

4. $4 = w - 3$
   $4 + 3 = w$

5. $3p = 16$
   $p = \frac{16}{3}$

6. $\frac{q}{4} = 17$
   $q = 68$

7. $3r + 11 = 6$
   $3r = -5$

8. $5y + 25 = 7$
   $5y = 7 - 25$

To solve a literal equation for one variable, the variable must be isolated on one side of the equation. In Example 1, the subtraction and multiplication properties are used.

EXAMPLE 1 Solve each equation for the underlined variable in terms of the other variable(s).

a. $\underline{x} + 11 = y$
$x + 11 - 11 = y - 11$
$x = y - 11$

b. $\frac{K}{m} = t$
$m \cdot \frac{K}{m} = m \cdot t$
$K = mt$

To solve some literal equations for a given variable, more than one of the properties may be needed.

EXAMPLE 2 Solve the equation $2x + 4a = 10$ for x.

$$2x + 4a = 10$$
$$2x + 4a - 4a = 10 - 4a \qquad \text{Subtract 4a from each side.}$$
$$2x = 10 - 4a$$
$$\frac{2x}{2} = \frac{10 - 4a}{2} \qquad \text{Divide each side by 2.}$$
$$x = \frac{10}{2} - \frac{4a}{2}$$
$$x = 5 - 2a$$

So, in terms of a, $x = 5 - 2a$.

Solving a literal equation may require combining like terms.

EXAMPLE 3 Solve the equation $ay - b = 3d + 7b$ for y, assuming a is not zero.

$$ay - b = 3d + 7b$$
$$ay - b + b = 3d + 7b + b \qquad \text{Add b to each side.}$$
$$ay = 3d + 8b \qquad \text{Combine like terms: 7b + b = 8b}$$
$$y = \frac{3d + 8b}{a} \qquad \text{Divide each side by a.}$$

Do you see why you needed to know that a is not zero?

In still other literal equations, it may be necessary to remove parentheses.

EXAMPLE 4 Solve $G = 3(t - s)$ for s. Then find the value of s when $G = 12$ and $t = -1$.

$$G = 3(t - s)$$
$$G = 3t - 3s \qquad \text{Distribute 3 on the right side.}$$
$$G - 3t = -3s \qquad \text{Subtract 3t from each side.}$$
$$\frac{G - 3t}{-3} = s \qquad \text{Divide each side by } -3.$$
$$s = \frac{G - 3t}{-3}$$
$$s = \frac{12 - 3(-1)}{-3} \qquad \text{Replace G with 12 and t with } -1.$$
$$s = \frac{12 + 3}{-3} = \frac{15}{-3} = -5$$

4.8 Literal Equations

CLASS EXERCISES

Solve each literal equation for the variable indicated.

1. $x - 7 = y$, for x
2. $pq = -12$, for q
3. $\frac{3}{4}r = st$, for r
4. $l = 2m + n$, for m
5. $3a = 2b + 2c$, for b
6. $3d + 14 = 7e - 4d$, for d

PRACTICE EXERCISES

Solve each literal equation for the underlined variable. Assume that all other literal coefficients are not zero.

1. $\underline{x} + 10 = y$
2. $\underline{a} + 13 = b$
3. $\underline{c} - 3 = d$
4. $y + \underline{z} = -8$
5. $2y = d\underline{a}$
6. $a\underline{x} = c$
7. $\frac{3}{4}\underline{n} = 2k$
8. $\frac{3}{2}\underline{l} = -4m$
9. $j = \frac{1}{2}kl\underline{m}$
10. $t = \frac{3}{4}\underline{p}sr$
11. $3x + 4\underline{y} = 12$
12. $\underline{j} + 3m = -2$
13. $\underline{y} - 4t = 0$
14. $6 = \underline{z} - 7p$
15. $-7 = 3\underline{a} + 11b$
16. $5\underline{p} - b = r$
17. $h = 2(l - 2\underline{m})$
18. $-3(2a - \underline{b}) = c$

Solve each literal equation for the underlined variable. Then find the value of the variable for the given values of the other variables.

19. $A = l\underline{w}$ Find w if $A = 63$, $l = 9$.
20. $V = lw\underline{h}$ Find h if $V = 64$, $l = 10$, and $w = 2$.
21. $p = 2a + \underline{b}$ Find b if $p = 74$ and $a = 27$.
22. $P = 2\underline{l} + 2w$ Find l if $P = 30$ and $w = 6$.
23. $V = \underline{b}h$ Find b if $V = 260$ and $h = \frac{3}{4}$.
24. $C = 2\pi\underline{r}$ Find r if $C = 81.64$ ($\pi \approx 3.14$).

Solve each literal equation for the underlined variable. Assume that all literal coefficients are not zero.

25. $3x - 4\underline{y} = 10x + 14z$
26. $5a + 9 = 2a - 3\underline{b}$
27. $0.02q + 2.1\underline{r} = 3.16 - 0.2r$
28. $3.5s - 0.12\underline{t} = 4.5 - 5.12t$
29. $-3(4t - \underline{w}) = 4t$
30. $-a = (\underline{j} - 3a)7$
31. $\frac{1}{3}(v - 2\underline{q}) = 2v + 3$
32. $\frac{1}{2}(3\underline{p} - t) = t - 8$

33. $\frac{3}{4}(l - \underline{h}) = l - 4j$ 34. $6 - 5m = \frac{3}{2}(\underline{n} + 2)$

35. $F = \frac{9}{5}\underline{C} + 32$ if $F = -4$ 36. $G = 3(t - \underline{s})$ if $G = 18$ and $t = 8$

37. $A = p + pr\underline{t}$ if $p = 800$, $A = 1000$, and $r = 0.05$

38. $V = \pi r^2 \underline{h}$ if $V = 6458.98$ and $r = 11$ ($\pi = 3.14$)

39. $L = a + (\underline{n} - 1)d$ 40. $A = \frac{1}{2}h(a + \underline{b})$ 41. $S = \frac{1}{2}n(a + \underline{l})$

42. $C = \frac{5}{9}(\underline{F} - 32)$ 43. $\frac{a + 2b}{3\underline{c}} = 4b$ 44. $\underline{x} - y = 5 - \frac{x}{2}$

Applications

Write a literal equation for each and solve for the desired variable. Then find the value of this variable for the given values of the other variables.

45. **Sports** In a Walk-a-thon, Terry walked 18 km in $3\frac{1}{2}$ h. What was her average speed?

46. **Geometry** The length of a rectangular room is 14 ft. If the perimeter is 42 ft, how wide is the room?

ALGEBRA IN AVIATION

In the formula $R = \frac{s^2}{A}$, R is the aspect ratio of a hang glider and indicates its ability to glide and soar. The area of the wing is A and the length of the wingspan is s.

If a hang glider has a wing area of 36 ft² and a wingspan of 10 ft, its aspect ratio is:

$$R = \frac{s^2}{A}$$

$$R = \frac{(10)^2}{36} = \frac{100}{36} = 2.78$$

1. Solve the aspect ratio formula for A.

Use the formula $R = \frac{s^2}{A}$ to find the values of the missing variables.

2. $R = 3$, $s = 9$ ft, $A = $ __?__   3. $s = 12.2$ ft, $A = 30.5$ ft², $R = $ __?__

4. $R = 4.50$, $A = 32$ ft², $s = $ __?__   Hint: $s^2 = s \cdot s$

4.9 Problem Solving Strategy: Make a Drawing or a Table

A very effective strategy to use in solving some problems is to make a simple drawing or sketch to illustrate the conditions of the problem. The use of a table can serve the same purpose, and drawings and tables are often used together.

Drawings are used to help solve problems involving motion. For example, if two cars leave a city traveling along the same road and travel the same distance, a drawing can show this situation.

The distance each car travels is the same and can be represented by d.

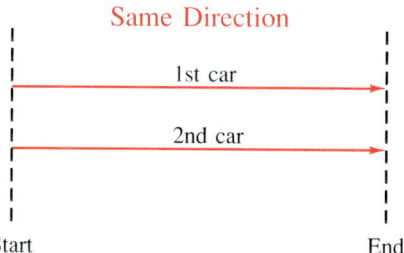

If two hikers leave from the same camp and walk in opposite directions, the drawing would look like this.

The distances may or may not be equal.

A car making a round trip to the beach can be shown in this way.

Since this is a round trip, the distances are equal.

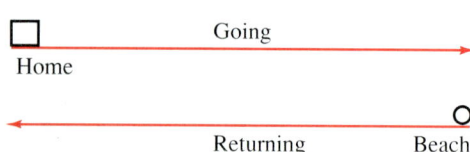

A table can also be used in solving motion problems. The formula for uniform motion is $d = rt$, where d is distance, r is rate or speed, and t is time.

Chapter 4 More Equations in One Variable

EXAMPLE Two buses leave from the same station at different times. Bus 2 leaves 2 h later than Bus 1. Bus 1 is traveling at 40 mi/h and Bus 2 is traveling at 60 mi/h. How long will it take Bus 2 to overtake Bus 1?

Understand the Problem

What are the given facts?
What are you asked to find?

You already know a great deal about the problem from the discussion above. You are asked to find how long it will take for Bus 2 to overtake Bus 1.

Plan Your Approach

Choose a strategy.

Use the first drawing on the previous page and a table.

Bus 1 travels $40t$ mi
Bus 2 travels $60(t - 2)$ mi

	Rate	× Time	= Distance
Bus 1	40	t	$40t$
Bus 2	60	$t - 2$	$60(t - 2)$

Since the distances the buses travel are equal, $40t = 60(t - 2)$.

Complete the Work

Solve the equation.

$40t = 60(t - 2)$
$40t = 60t - 120$
$0 = 20t - 120$
$120 = 20t$
$6 = t$ Bus 1's time: $t = 6$ h, Bus 2's time: $t - 2 = 6 - 2$, or 4 h

Interpret the Results

Conclusion: Bus 2 overtakes Bus 1 in 4 h.

Bus 1 travels 6 h at 40 mi/h for 240 mi.
Bus 2 travels 4 h at 60 mi/h for 240 mi.

CLASS EXERCISES

1. Make a drawing to show that one car has traveled twice as far as another.
2. Make a drawing to show that two planes have traveled the same distance in opposite directions.
3. Make a drawing to show that a train has traveled from Boston to New York City and back to Boston again.
4. If t represents the number of hours a car has traveled, how would you represent the time of a car that has traveled 1 h longer?
5. If t represents the number of hours an airplane has flown, how would you represent the time of a plane that has flown $\frac{1}{2}$ h less?
6. If $d = 45t$ and $d = 55(t + 1)$ and the distances are equal, write an equation involving t only.

PRACTICE EXERCISES

Use the table to write an equation. In each case, the distances are equal.

1.

	Rate	× Time	= Distance
Car 1	45	t	$45t$
Car 2	55	$t - 1$	$55(t - 1)$

2.

	Rate	× Time	= Distance
Train 1	65	$t + 1$	$65(t + 1)$
Train 2	70	t	$70t$

3.

	Rate	× Time	= Distance
Plane 1	550	$t + 2$	$550(t + 2)$
Plane 2	625	t	$625t$

Complete each table by writing an algebraic expression for each distance.

4.

	Rate	× Time	= Distance
1st Car	40	$t - 1$	?
2nd Car	30	t	?

5.

	Rate	× Time	= Distance
Truck 1	$r + 20$	3	?
Truck 2	r	4	?

6.

	Rate	× Time	= Distance
Going	20	t	?
Returning	35	$t - \frac{1}{2}$	?

Solve each equation.

7. $45t = 55(t - 1)$

8. $65(t + 1) = 70t$

9. $550(t + 2) = 625t$

10. $4(r - 10) = 3r$

11. $3(r + 30) = 4r$

12. $20x = 35\left(x - \frac{1}{2}\right)$

Organize the information in a table.

13. Two cars leave the same city. The first car is traveling at 40 mi/h. One hour later the second car leaves, traveling at 55 mi/h along the same road. In how many hours will the second car overtake the first car?

14. Two hikers start walking in opposite directions from a camp at the same time. The average rate of the northbound hiker is 3 mi/h faster than the rate of the southbound hiker. If they are 22 mi apart after 2 h, what is the rate of each hiker?

15. Gail and Bill drove to a beach at an average speed of 50 mi/h. They returned home over the same road at an average speed of 55/mi/h. The trip home took 30 min less time. How far is the beach from their home?

16. Write three equations, using your tables from Exercises 13, 14, and 15.

17. Solve the three equations you wrote in Exercise 16.

Solve the problem.

18. Two trains leave a city traveling at rates of 55 mi/h and 65 mi/h. The faster train leaves 30 min after the slower train. How long will it take for the faster train to pass the slower train?

Mixed Problem Solving Review

Write an equation for each sentence.

1. The product of a number squared and 4 is 64.
2. A number less 5 is twice 6.

Solve each problem.

3. Five times a number, decreased by 1, is −21. Find the number.
4. Ted bought a pair of sneakers at 25% off the price of $60. The sales tax was 5%. How much did he pay for the sneakers?

PROJECT

Working with a friend, review all the problem solving lessons in Chapters 1–4 and make a list of at least 10 problems that could have been solved by using a drawing, a table, or both. For example, on page 62, Exercise 33 could be solved using a table as follows:

Item	Quantity	Price	Total
Flour	3	$0.99	$2.97
Cereal	3	2.19	6.57
Juice	3	1.31	3.93

4.10 Problem Solving: Uniform Motion

Objective: To solve problems that involve uniform motion

An object that moves at a constant speed, or rate, is said to be in **uniform motion.** The formula $d = rt$ expresses the relationships among the distance d, the rate r, and the time t of a particular uniform motion.

Uniform motion problems may involve objects that move in the same direction, opposite directions, or round trips.

EXAMPLE 1 A car leaves a city traveling at a rate of 45 mi/h. One hour later, a second car leaves from the same place, along the same road at 54 mi/h. In how many hours will the second car overtake the first car?

Understand the Problem The two cars are traveling in the same direction. The first car, moving at 45 mi/h, travels 1 hour longer than the second car moving at 54 mi/h. Each car travels the same distance. You are asked to find the time of the second car.

Plan Your Approach Let $t =$ the time (h) of the second car. Draw a sketch and organize the information in a table. Fill in rate and time for each car. Multiply the expressions for rate and time to obtain an algebraic expression for the distance traveled by each car.

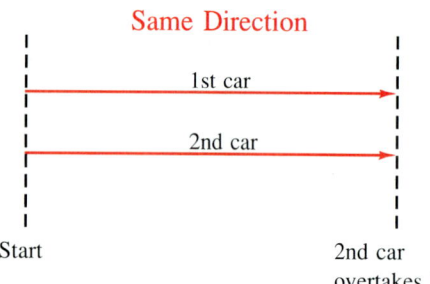

	Rate	× Time	= Distance
1st car	45	$t + 1$	$45(t + 1)$
2nd car	54	t	$54t$

Since each car travels the same distance,
$$45(t + 1) = 54t$$

Complete the Work
$$45(t + 1) = 54t$$
$45t + 45 = 54t$ Multiply by 45.
$$45 = 9t$$
$$5 = t$$

■ **Interpret the Results**   In 5 h, the second car travels 54 · 5, or 270 mi. In (5 + 1) h, the first car travels 45 · 6, or 270 mi. So, the second car overtakes the first in 5 h.

EXAMPLE 2   Two hikers walking in opposite directions start from camp at the same time. The average rate of the westbound hiker is 2 km/h more than the rate of the eastbound hiker. If they are 15 km apart after 1.5 h, what is the rate of each hiker?

■ **Understand the Problem**   Let r = the rate (km/h) of the eastbound hiker.

■ **Plan Your Approach**   Draw a sketch and make a table. The total of the two hikers' distances is 15 km.

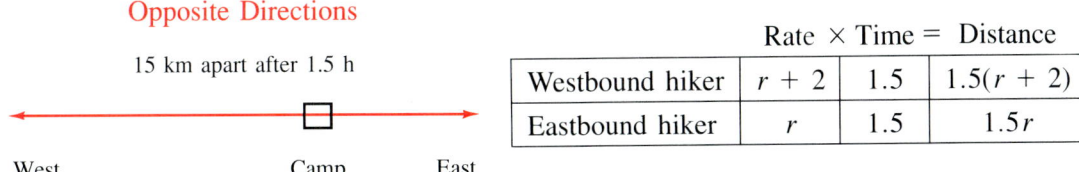

Opposite Directions

15 km apart after 1.5 h

West   Camp   East

	Rate	× Time	= Distance
Westbound hiker	$r + 2$	1.5	$1.5(r + 2)$
Eastbound hiker	r	1.5	$1.5r$

Since the total distance is 15 km, $1.5(r + 2) + 1.5r = 15$.

■ **Complete the Work**

$1.5(r + 2) + 1.5r = 15$

$15(r + 2) + 15r = 150$ *Multiply each side by 10.*

$15r + 30 + 15r = 150$

$30r + 30 = 150$ *Combine like terms.*

$30r = 120$

$r = 4$

$r + 2 = 6$

■ **Interpret the Results**   In 1.5 h the eastbound hiker walks 1.5(4), or 6 km. In 1.5 h the westbound hiker walks 1.5(6), or 9 km. 6 km + 9 km = 15 km.

So, the eastbound hiker walks at 4 km/h and the westbound hiker at 6 km/h.

In Example 2, why were the two distances added to get the equation?

EXAMPLE 3 A family drives to the beach at an average speed of 45 mi/h and returns home on the same road at an average speed of 54 mi/h. If the trip home takes 20 min less than the trip to the beach, how far is the beach from the family's home? (*Hint:* First find the time for each part of the trip.)

Understand the Problem Let x = the time (h) of the trip to the beach. The return trip takes 20 min less time. However, the rates 45 and 54 are miles per hour, so you must changes 20 min to $\frac{1}{3}$ h.

Plan Your Approach Draw a sketch and make a table.

Round Trip

	Rate	×	Time	=	Distance
Going	45		x		$45x$
Returning	54		$x - \frac{1}{3}$		$54\left(x - \frac{1}{3}\right)$

Since this is a round trip, the distances are equal.

$$54\left(x - \tfrac{1}{3}\right) = 45x$$

Complete the Work
$$54x - 18 = 45x$$
$$9x = 18$$
$$x = 2 \quad \text{Time to the beach is 2 h.}$$

Interpret the Results If it takes 2 h to travel to the beach, the distance is 2 · 45, or 90 mi. This can be checked by finding the distance home from the beach, which must be the same.

The return trip time was $\left(2 - \tfrac{1}{3}\right)$ h which is $1\tfrac{2}{3}$, or $\tfrac{5}{3}$ h.

$$54 \cdot \tfrac{5}{3} = 90$$

So, the distance from the family's home to the beach is 90 mi.

CLASS EXERCISES

For each problem, describe the motion as *same direction; opposite direction;* or *round trip.* Then draw a sketch, make a chart and write an equation for the problem.

1. The Smiths left for vacation traveling south at an average rate of 50 km/h. Three hours after the Smiths depart, the Duprees set out to overtake them. If the Duprees travel at a rate of 70 km/h, how long will it take them to overtake the Smiths? Let t = time (h) Duprees travel.

2. Two planes leave Newark Airport at the same time. The rate of speed of the northbound plane is 100 mi/h faster than that of the southbound plane. At the end of 5 h, the planes are 2000 mi apart. Find the rate of each. Let r = rate (mi/h) of northbound plane.

3. Andre traveled from his home to the office at an average rate of 25 mi/h. By traveling 5 mi/h faster, he took 30 min less to return home. How far is his office from home? Let t = time (h) going to office.

4. Two trains leave Union Station at noon. One train travels north 5 mi/h faster than the train traveling south. After $3\frac{1}{2}$ h the two trains are 399 mi apart. Find how far the northbound train is from the station. Let r = rate (mi/h) of the southbound train.

PRACTICE EXERCISES

Solve each problem.

1. A ship leaves a dock moving at 24 mi/h. Three hours later, a second ship sets out from the same dock at 32 mi/h. How long will it take for the second ship to overtake the first?

2. Starting one hour later, Jonathan sets out walking at 8 km/h to overtake his little sister who averages 5 km/h. How long will it take Jonathan?

3. Robinsport and Titusville are 324 mi apart on a railroad line. Trains leave each of these depots at the same time headed for the other depot. One travels at 43 mi/h, the other at 38 mi/h. How long will it take for the two trains to pass one another?

4. Two trains leave from a station at the same time. They travel in opposite directions, one at 62 km/h, the other at 48 km/h. How long will it take before they are 550 km apart?

5. Two cyclists simultaneously start in opposite directions down a straight road, one at 22 km/h and the other at 28 km/h. How long will they have ridden by the time they are 175 km apart?

6. Two planes leave simultaneously from an airport, one flying east and the other west. They are 870 mi apart after $\frac{3}{4}$ h. If the eastbound plane averages 120 mi/h more than the westbound plane, at what rate is each plane flying?

7. At the same time, two buses leave a depot and travel in opposite directions on a straight road. The first bus averages 5 mi/h more than the second. If they are 142.5 mi apart after $1\frac{1}{2}$ h, how fast is each bus going?

8. On a round trip over the same roads, Sophia averaged 24 mi/h going and 30 mi/h coming back. If the whole trip took $13\frac{1}{2}$ h, what is the distance one way? (*Hint:* First find the time for each part of the round trip.)

9. Enrique cycled to his grandmother's house at 15 km/h and returned home over the same route at 18 km/h. If the whole trip took 7 h and 20 min, how far does Enrique live from his grandmother's house? (*Hint:* First find the time for each part of the round trip.)

10. Two snails are 432 cm apart. If they travel toward each other at rates that differ by 2 cm/min and it takes them 27 min to meet, how fast is each snail going?

11. On his way to visit Soren, Peter averaged 7 mi/h on his skateboard and 4 mi/h walking. If Soren lives 6 mi away and the whole trip took 1 h, how many miles did Peter travel by skateboard? (*Hint:* First find the two times.)

12. After walking part of the way to a town 68 mi away at 4 mi/h, Erik was picked up by his friend Daniel who drove him the rest of the way. Daniel's car averaged 44 mi/h, and the whole trip took 2 h. How far did Daniel drive Erik? (*Hint:* First find the two times.)

13. Twenty minutes after Maria left for work on the bus, Juan, noticing that she had left her purse at home, set out by car to overtake the bus. Juan averaged 12 mi/h more than the bus and he overtook it in 40 min. On the average, how fast was each vehicle traveling?

14. On a recent 6-h bike trip of 70 mi, Lila spent 2 h on the first part of the trip where her average speed was half of her average speed on the second part. What was her speed on the first part?

15. On their 1200-km trip to Texas, the Wallaces first took a train and later a plane. The train, traveling at 48 km/h, took 2 h longer than the plane, traveling at 240 km/h. How long did the whole trip take?

16. At 7:30 PM, a freight train leaves a station traveling east. At 10:30 PM, a passenger train leaves the same station going west at 10 mi/h faster than the freight train. At 3:30 AM the next day, they are 635 mi apart. What is the rate of each train?

17. At 10:15 PM, a train left Petersburg, Virginia, traveling east at 40 mi/h. Two hours later, another train left the same station traveling west at 35 mi/h. At what time will the trains be 230 mi apart?

18. Two buses leave the same station at the same time and travel in opposite directions. After 8 h, they are 360 km apart. The speed of the faster bus is 3 km/h less than twice that of the slower bus. Find the rate of each bus.

19. At 6:00 AM, a freight train leaves a station at 40 mi/h, and at 7:30 AM a passenger train leaves at 64 mi/h, going in the same direction. At what time will the passenger train overtake the freight train?

20. Two cyclists are traveling toward one another, one at 25 km/h, the other at 30 km/h. If they are 68.75 km apart at 7:45 AM, at what time will they meet?

21. Benigno and Chris start at the same time from places 96 mi apart to cycle toward one another, Benigno at 16 mi/h and Chris at 12 mi/h. If Benigno rests for $\frac{1}{2}$ h on the way, after how many hours will they meet?

22. Rosa drives to Barnesville at 50 mi/h and returns by a road 5 mi longer at 40 mi/h. The return trip takes 15 min longer. How long is each road?

23. Tina can run 2 m/s faster than her sister can. If she runs one lap of a track in 100 s, which is 25 s less than her sister's time, what is the length of one lap?

24. Hendrik's average running speed is 375 m/min. This is 25 m/min faster than Esteban's running speed. If they want to finish at the same time, how much of a headstart, in time, should Esteban get for a 1500 m run?

TEST YOURSELF

1. Find 2.5% of 30.   **4.5**

2. 15% of what number is 120?

3. 84 is what percent of 28?

4. By how much will a $198 coat be reduced at a "30% off" sale?

5. A car originally cost $8400. It now costs $7980. What is the percent of decrease?   **4.6**

6. The price of oil, currently $28 per barrel, is expected to rise by 20%. What will the new price be?

7. To 25 mL of 10% acid solution, 15 mL of water is added. What percent acid is the new solution?   **4.7**

Solve each equation for the underlined variable.

8. $x + \underline{y} = 0$ 9. $5\underline{a} - 10b = 15$ 10. $M = -2(j - \underline{k})$ **4.8**

11. Two cyclists start from the same place at the same time. They travel in opposite directions. One averages 10 mi/h, and the other 12 mi/h. After how long will they be 66 mi apart?   **4.9, 4.10**

CHAPTER 4 SUMMARY AND REVIEW

Vocabulary

consecutive integers (144)
even integers (144)
literal equation (164)
odd integers (145)
percent (152)
percent of decrease (156)
percent of increase (156)
uniform motion (172)

Steps for Solving Equations   4.1, 4.2, 4.4

- Use the distributive property to remove parentheses on each side.
- Combine like terms on each side.
- Use the addition or subtraction property for equations so that
    (a) all variable terms are on one side of the equation, and
    (b) all numerical terms are on the other side.
- Use the multiplication or division property for equations.

Note: You may be able to use shortcuts.

Solve and check.

1. $27 - d + 3d = 9$
2. $5t - 4(2t - 3) = 0$
3. $c - 9 = 3\frac{3}{4} + 2c$
4. $-3(4 - 2r) + r = 2r$
5. $0.5t + 0.5(t + 6) = 0.18$
6. $-6(3p - 7) = 36$

Consecutive Integer Problems If x is an integer, then three consecutive   4.3
integers that follow x are $x + 1$, $x + 2$, and $x + 3$. If y is an even integer or
an odd integer, then each consecutive even or odd integer that follows y will
be 2 greater than the one before: $y + 2$, $y + 4$, $y + 6$, and so on.

7. z is an even integer. Write four consecutive even integers that follow $z + 6$.

8. Find three consecutive integers whose sum is -9.

Percent Problems The three cases of percent can be expressed in the   4.5
general form: a is $b\%$ of c. To solve any percent problem, try to identify the
values of a, b, and c.

9. What is 5% of 57?
10. 4.5 is what percent of 150?

11. For what investment, at an annual interest rate of 4.8%, would you receive an annual simple interest of $64.80?

12. If 4 machines in a 128-machine shipment are defective, what percent are defective?

13. Last year 90 students took auto mechanics. This year the enrollment increased by 20%. How many students now take the course?

14. A calculator that cost $60 in 1980 now sells for $10. What was the percent of decrease?

15. How many milliliters of water would you add to 75 mL of a 15% acid solution to make a 10% solution?

Literal Equations To solve a literal equation for one variable in terms of other variables, use the properties for equations to isolate the desired variable on one side of the equation. **4.8**

Solve each literal equation for the underlined variable.

16. $3 = \underline{y} - 12x$

17. $5(3m - \underline{n}) = p$

18. $-5d + 4b\underline{x} = 7d + 5b\underline{x}$

Solve each literal equation for the underlined variable. Then find the value of this variable in terms of the other variables.

19. $p = 2\underline{a} + b$ Find a if $p = 15$ and $b = 4$.

20. $C = 2\pi \underline{r}$ Find r if $C = 157$ and $\pi \approx 3.14$.

Uniform Motion Problems To solve problems that involve uniform motion, use the four problem-solving steps. Draw a sketch, and make a chart based on the formula $d = rt$. **4.9**

21. Two planes left Kennedy Airport in New York City at 1:00 PM. One flew north at 250 mi/h and the other flew south at 300 mi/h. How many miles apart were the planes at 4:00 PM?

22. A truck and a car travel the same route, the truck at 72 km/h, and the car at 80 km/h. If the car leaves 1 h after the truck, how long does it take the car to overtake the truck?

23. John takes a bus from home to school. The bus averages 32 km/h. He walks home at 8 km/h. The bus ride takes 15 min less time than the walk home. How far is John's home from school?

CHAPTER 4 TEST

Solve and check.

1. $-8t - 7t + 9 = -31$
2. $-5 - (6 + 3y) = 7$
3. $d - 2 = 1\frac{3}{4} - 2d$
4. $2(p + 1) - 3(4p - 2) = 6p$
5. $\frac{1}{3}(2y - 9) = -18$
6. $0.05x - 0.02(8000 - x) = 85$

Solve each literal equation for the underlined variable. Assume that the literal coefficients are not zero.

7. $\underline{x} + 10c = c + 8$
8. $A = \frac{1}{2}b\underline{h}$

Solve this literal equation for the underlined variable. Then find the value of this variable for the given values of the other variables.

9. $P = 2l + 2\underline{w}$ Then find the value of w if $P = 15.5$ and $l = 3.5$.

Write an equation and solve each problem.

10. What percent of 125 is 5?
11. 5.6 is 25% of what number?
12. What is $33\frac{1}{3}\%$ of $6.90?
13. If the state sales tax rate is 7.5%, how much tax will be paid on an item that costs $24.98?
14. Find three consecutive even integers whose sum is -42.
15. The population of Los Angeles increased from 2.5 million in 1960 to 3 million in 1980. What was the percent of increase?
16. To 150 g of a 5% sugar solution in water, 25 g of pure sugar was added. What was the percent of sugar in the new solution?
17. A car leaves Washington, D.C., for New York City at the same time another car leaves New York City for Washington, D.C. The car from Washington, D.C., travels at an average rate of 45 mi/h; the other car averages 55 mi/h. If the two cities are 240 mi apart, how soon will the cars meet and how far will each have traveled?

Challenge

An investment company invested one half its money at a 10% rate of simple interest, one third at 9%, and the remainder at 12%. If the company earned a total annual interest of $1110 from these investments, how much was invested at each rate?

PREPARING FOR STANDARDIZED TESTS

Select the best answer for each and write the appropriate letter.

1. Solve the equation
   $$2x + 5 = x + 9.$$
   A. 2 **B.** 4 **C.** 7 **D.** 8 **E.** 14

2. The equation representing the sentence "The sum of two consecutive integers is 329." would be:
   A. $x + 1 = 329$
   B. $x + 2 = 329$
   C. $x + x = 329$
   D. $x + (x + 1) = 329$
   E. $x + (x + 2) = 329$

3. Which of the following is *not* equivalent to 0.025?
   A. $\dfrac{25}{1000}$ **B.** $0.01(2.5)$
   C. $\dfrac{1}{40}$ **D.** 0.25% **E.** $\dfrac{5}{200}$

4. Solve the equation
   $$5x + 5 = 3x - 13$$
   A. -9 **B.** -1 **C.** 1 **D.** 4 **E.** 9

5. Find the value of the expression below when $x = -1$ and $y = 3$.
   $$3x(x + y) - 2xy^2$$
   A. 6 **B.** 12 **C.** 15 **D.** 18 **E.** 24

6. Solve for x in the equation:
   $$3x + 2b = 2x + a$$
   A. $a + 2b$ **B.** $a - 2b$
   C. $a + b$ **D.** $a - b$
   E. $2a - 2b$

7. A train and a car are traveling at different speeds. If the train travels 306 mi in 5 h, how many hours would it take the car to travel 153 mi?
   A. 2 **B.** $2\dfrac{1}{4}$ **C.** $2\dfrac{1}{2}$ **D.** 3
   E. It cannot be determined from the information given.

8. Which of the following is(are) true?
   I. If $x < y$ and $y < 4$, then $x < 4$.
   II. 1% of 160 is 16
   III. $\dfrac{3}{5} \times \dfrac{25}{6} \div \dfrac{5}{2} < \dfrac{3}{2}$

   A. I only **B.** II only
   C. III only **D.** II, III only
   E. I, III only

9. Mr. Miller bought a suit on sale at 30% off the original price of $150. If he also had to pay a sales tax of 6%, what was the total amount he paid?
   A. $154.23 **B.** $114
   C. $111.30 **D.** $106
   E. $98.70

10. In a factory, Pete and Al make the same machine part. On one day Al made 29 fewer than twice what Pete made and together they made 88 machine parts. How many did Pete make that day?
    A. 29 **B.** 38 **C.** 39 **D.** 48 **E.** 49

Preparing for Standardized Tests   **181**

CUMULATIVE REVIEW (CHAPTERS 1–4)

Perform the indicated operation.

1. $-\frac{3}{4} \cdot \frac{2}{3}$
2. $\frac{5}{8} + \left(-\frac{1}{4}\right)$
3. $-\frac{3}{10} \div \left(-\frac{2}{5}\right)$
4. $-\frac{2}{3} - \left(-\frac{5}{6}\right)$
5. $\frac{1}{2} \div \left(-\frac{1}{4}\right)$
6. $-\frac{3}{8} \cdot \left(-\frac{4}{5}\right)$
7. $-\frac{2}{3} + \frac{3}{4}$
8. $\frac{1}{3} \div \frac{3}{5}$
9. $-\frac{5}{6} + \left(-\frac{3}{8}\right)$
10. $-\frac{2}{3} \div \frac{4}{5}$
11. $-\frac{3}{4} - \frac{5}{12}$
12. $\frac{4}{9} \cdot -\frac{3}{8}$
13. $\frac{2}{5} - \left(-\frac{3}{10}\right)$
14. $-\frac{5}{6} + \frac{3}{4}$
15. $-\frac{4}{9} \div \left(-\frac{2}{3}\right)$
16. $\frac{5}{6} \cdot \frac{4}{15}$

Simplify.

17. $|8 - (-3)(-5)|$
18. $24 - 8 \cdot 2$
19. $-5[64 \div (8 - 12)^2]$
20. $7a^2 - 12a + 3a^2$
21. $5x - 3(x + 2)$
22. $-(-m^2 - 5)$
23. $(5h + 4) - (3h - 5)$
24. $5(a^2 + b) + 2(b - a^2)$
25. $7mn - 2(mn + 3) + 5$
26. $-3(x^2 - y^2) + 5(y^2 + x^2)$

Evaluate each expression if $x = 8$, $y = \frac{3}{4}$, and $z = -6$.

27. $xy + z$
28. $|z - 4y|$
29. $\frac{5x + 2}{y}$
30. $2(x^2 + 4z)$
31. $\frac{(x + z)^2}{y - \frac{1}{2}}$
32. $2x + 4(z + y)$
33. $(|x| - |z|)^3$
34. $x[2 + y(2 - z)]$
35. $x(y \div z)$

Answer each question or statement with an algebraic expression or equation.

Let m be Marc's age now.

36. What was Marc's age 6 years ago?
37. Nancy is 2 years younger than Marc. What is Nancy's age now?
38. What will Nancy's age be 4 years from now?
39. The sum of their ages now is 30.
40. 3 years ago the sum of their ages was 24.

Solve each equation and check your solution.

41. $12 = 3(c - 1)$
42. $20 = y + 20$
43. $81 = -9x$
44. $4p + 2 = 14$
45. $1 - 9d = 0$
46. $32 = -16 + 14n$
47. $g + 6 = -8$
48. $-\frac{1}{4} = \frac{t}{8}$
49. $2a + \frac{3}{5} = \frac{2}{5}$
50. $0.3r - 0.5 = 0.7$
51. $-40b = 40$
52. $5m + 30 - 3m = 42$
53. $4h - 3 = 11 - 3h$
54. $1 = 2\frac{1}{5} - \frac{2}{5}r$
55. $9t - 4(t + 3) = 8$
56. $8p - (2p - 5) = 23$

57. Given the formula $S = \frac{a}{1 - r}$, find the value of a if $S = 8$ and $r = -3$.

58. Evaluate $A = p + prt$ for t, if $A = \$840$, $p = \$600$, and $r = 0.08$.

59. In the formula $V = \pi r^2 h$, find the value of h if $V = 628$ and $r = 5$. Use 3.14 for π.

60. Find the value of E in the formula $I = \frac{E}{R + r}$ if $I = 80$, $R = 12$, and $r = 13$.

61. Given the formula $P = 2l + 2w$, find the value of w if $P = 200$ and $l = 75$.

Solve each problem by writing and solving an equation.

62. 9 less than 4 times a number is 55. Find the number.

63. The perimeter of a rectangle is 42 cm. Find the length if the width is 6 cm.

64. Three more than twice Frank's age is 35. How old is Frank?

65. The sum of two consecutive integers is 49. Find the integers.

66. Sue's allowance is 3 times as much as Gloria's allowance. If the sum of their allowances is $48, how much is Gloria's allowance?

For each exercise, translate the word equation into an algebraic equation and solve.

67. $a = ?$; $b = 15\%$; $c = 40$
68. $a = ?$; $b = 24\%$; $c = 68$
69. $a = 105$; $b = ?$; $c = 150$
70. $a = 125$; $b = ?$; $c = 2500$
71. $a = 48$; $b = 12\%$; $c = ?$
72. $a = 66$; $b = 15\%$; $c = ?$
73. $a = 91$; $b = 8\%$; $c = ?$
74. $a = 77$; $b = ?$; $c = 154$
75. $a = 4$; $b = 8\%$; $c = ?$
76. $a = 12$; $b = 12\%$; $c = ?$
77. What is 3% of 80?
78. What is 7% of 30?
79. What percent of 50 is 15?
80. What percent of 24 is 10?
81. 25% of $500 is what amount?
82. 45% of $650 is what amount?
83. 20% of what number is 5?
84. 80% of what number is 10?
85. 18 is what percent of 24?
86. 24 is what percent of 144?
87. Find $37\frac{1}{2}\%$ of 96.
88. Find $8\frac{1}{4}\%$ of 120.
89. 16 is 5% of what number?
90. 36 is 2% of what number?
91. 150% of what number is 30?
92. 250% of what number is 50?
93. What percent of 12 is 4?
94. What percent of 85 is 17?
95. What is $\frac{1}{2}\%$ of $200?
96. What is $\frac{1}{4}\%$ of $100?

Solve each literal equation for the underlined variable.

97. $x - \underline{y} = 9$
98. $3\underline{m} - 5 = n$
99. $\frac{1}{2}\underline{g} + 3 = h$
100. $a = 2(\underline{b} - 3c)$
101. $\frac{2}{3}\underline{t} = 8s$
102. $\frac{1}{2}\underline{x} + y = 1\frac{1}{2}$
103. $2d = 5d - \underline{c}$
104. $j = \frac{1}{3}\underline{k} + 5$
105. $0.2p - 0.3\underline{q} = 40$

Solve.

106. The price of an $85 jacket was decreased 25%. What is the new price?
107. The population of Haverstown has increased from 835 to 1002. Find the percent of increase.
108. If you have 200 mL of a 30% solution of acid in water, how much water would you add in order to make a 25% solution?
109. A car leaves a city traveling at a rate of 40 mi/h. Two hours later a second car leaves from the same place, along the same road traveling 50 mi/h. In how many hours will the second car overtake the first car?

5 Inequalities in One Variable

Knowing how to figure out distance, speed, and time is important in many sports activities and competitions. A good knowledge of mathematics can help produce solutions to otherwise complicated problems.

5.1 Graphing Equations and Inequalities

Objective: To graph equations and inequalities on a number line

The graph of a real number is a *point* on a number line. You can graph equations and inequalities by locating their solutions on a number line.

Capsule Review

The arrows on a number line indicate direction. Choose a starting point, the origin, and label it "0." Then mark off equal units of distance on both sides of the origin. The graph of point B corresponds to what number on the number line?

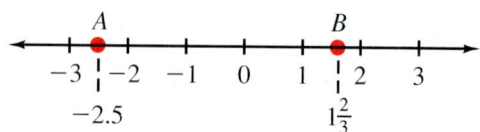

Point A is the graph of -2.5.

Graph these numbers on a number line.

1. 1.5 2. $-1\frac{2}{5}$ 3. $\frac{1}{4}$ 4. -1 5. -0.75

The solution set of $x = -2$ is $\{-2\}$. The graph of $x = -2$ is a point on the number line that corresponds to -2.

EXAMPLE 1 Solve $3a + 6 = a - 5$. Then draw its graph by locating the solution on a number line.

$3a + 6 = a - 5$
$2a + 6 = -5$ Subtract a from each side.
$2a = -11$ Subtract 6 from each side.
$a = -5\frac{1}{2}$ Divide each side by 2.

The solution is $-5\frac{1}{2}$.

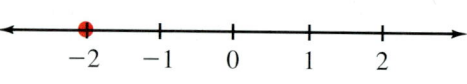

You have used the symbols $=$, $<$, $>$, and $\neq$ to compare numbers. The following property is basic to the comparison of real numbers.

Comparison Property

For all real numbers a and b, one and only one of the following is true:

$a = b$ $a < b$ $a > b$

---

186 Chapter 5 Inequalities in One Variable

Using the set of real numbers as the replacement set, there are many values for x that make $x < 1$ true.

Replace x with 0.	Replace x with $\frac{1}{2}$.	Replace x with -2.
$x < 1$	$x < 1$	$x < 1$
$0 < 1$ ✓ True	$\frac{1}{2} < 1$ ✓ True	$-2 < 1$ ✓ True

The solution set of $x < 1$ is {all real numbers less than 1}. To graph such a solution set, recall that every point on the number line corresponds to a real number.

EXAMPLE 2 Draw the graph of $x < 1$.

The graph of $x < 1$ includes all points on the number line to the left of the graph of 1. An "open circle" shows that 1 is not a solution.

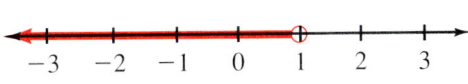

Sometimes comparisons are combined. For example:

Inequality	Read	Solution Set
$m \leq 5$	m is less than or equal to 5.	{5 and all real numbers less than 5}
$y \geq -\frac{1}{2}$	y is greater than or equal to $-\frac{1}{2}$.	$\{-\frac{1}{2}$ and all real numbers greater than $-\frac{1}{2}\}$

Unless stated otherwise, the replacement set for the variable in an inequality is the set of real numbers.

EXAMPLE 3 Draw the graph of $y \geq -\frac{1}{2}$.

A "closed circle" shows that $-\frac{1}{2}$ is a solution of $y \geq -\frac{1}{2}$.

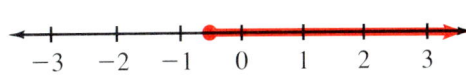

From the graph of $y \geq -\frac{1}{2}$ in Example 3, identify other numbers that are solutions and numbers that are not solutions of the inequality.

EXAMPLE 4 Draw the graph of $a \neq 1$.

All real numbers except 1 are solutions of $a \neq 1$.

5.1 Graphing Equations and Inequalities

If you are given the graph of an equation or an inequality, you can determine its solution set.

EXAMPLE 5 State the solution set of the inequality whose graph is given.

The solution set is {all real numbers greater than 0}.

CLASS EXERCISES

Classify each statement as *true* or *false*.

1. $-16 > -15$
2. $-1\frac{1}{3} \neq -1\frac{1}{3}$
3. $-1.5 < 0.5$
4. $-\frac{1}{6} \geq -\frac{1}{6}$

Graph each equation or inequality.

5. $x = -2.5$
6. $y < 2$
7. $c \geq -4$
8. $d \neq 0$

State the solution set of the equation or inequality whose graph is given.

9. 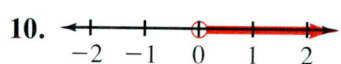 10. 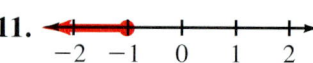 11.

PRACTICE EXERCISES

Solve each equation. Then graph the solution on a number line.

1. $4a + 2 = a - 7$
2. $6x + 3 = x - 12$
3. $-7m + 4 = -m - 2$
4. $-z + 8 = -7z + 2$
5. $3y - y = 12 - y$
6. $5n - n = 15 - n$

Graph each equation or inequality.

7. $a < 2$
8. $c > 5$
9. $z < -\frac{1}{2}$
10. $x < -\frac{3}{4}$
11. $t > 0$
12. $m > 3$
13. $a \geq -\frac{1}{4}$
14. $y \geq -\frac{1}{2}$
15. $x = 0.50$
16. $y = 0.25$
17. $x \neq 2$
18. $y \neq 3$

State the solution set of the equation or inequality whose graph is given.

19. 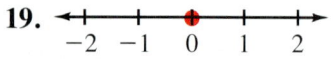 20. 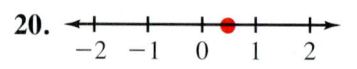 21.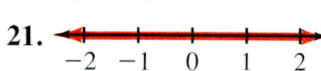

22. 23.  24.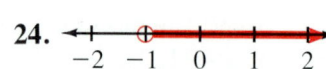

Solve each equation. Then graph the solution set.

25. $3x - 11 = -26$
26. $-5y + 21 = 36$
27. $7c - 9 = 6 + c$
28. $8r - 19 = -19 + 11r$
29. $6(2m + 3) = 4 - (m - 1)$
30. $15 + (3z - 6) = (9 + z)2$

Graph the solution set of each inequality. Let the replacement set for the variable be the set of negative integers.

31. $a \leq 5$
32. $b \geq -5$
33. $z \neq -4$
34. $v \neq -1$

The **Transitive Property of Order** for $<$ and $>$ is as follows:
For all real numbers x, y, and z,
(a) if $x < y$ and $y < z$, then $x < z$; and (b) if $x > y$ and $y > z$, then $x > z$.

35. Use number lines to draw graphs to illustrate the transitive property of order in (a) and (b) above.

For which real numbers a and b is the statement true?

36. If $a < b$, then $a^2 < b^2$.
37. If $a < b$, then $a^2 + b^2 > 2ab$.
38. If $a < b$, then $\frac{1}{a} > \frac{1}{b}$.
39. If $a < b$, then $a^2 = b^2$.

Applications

40. **Computer** Computers can graph inequalities involving decimal numbers quickly and very accurately. This program graphs solutions in the -3.25 to 3 range. By changing the range and the increment in 10–20 you can use the program for larger numbers and/or smaller increments. What would you change in the program to graph: $x > 2.25$; $x < 1.75$?

```
10 INPUT "GRAPH OF X < ";N: PRINT : PRINT
20 FOR X = - 3.25 TO 3 STEP .25
30 IF X < N THEN  PRINT "*";: GOTO 60
40 IF X = N THEN  PRINT "0";
50 PRINT "-";
60 NEXT X
70 PRINT : PRINT "-3..-2..-1...0...1..2...3"
80 END
```

LOGICAL REASONING

Lori has fifteen cubes that are identical except that one weighs less than each of the others. If she can compare these cubes on a balance scale, what is the fewest number of comparisons that will enable her to tell which cube weighs the least?

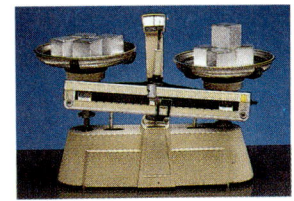

5.2 Solving Inequalities: Addition and Subtraction Properties

Objective: To solve inequalities using the addition and subtraction properties and to draw graphs of the solution sets

A solution of an inequality is any number from the replacement set that makes the inequality true. As with equations, some inequalities may be solved by using real number properties.

Capsule Review

The replacement set is the set of integers.

$2x < 9$	$x + 5 < 0$	$4 \geq x + 9$
Is 4 a solution?	Is -1 a solution?	Is -5 a solution?
$2 \cdot 4 \stackrel{?}{<} 9$	$-1 + 5 \stackrel{?}{<} 0$	$4 \stackrel{?}{\geq} -5 + 9$
$8 < 9$ ✔ True	$4 \not< 0$	$4 \geq 4$ ✔ True

Find the solution set of each inequality. Use $\{-2, -1, 0, 1, 2\}$ for the replacement set. If the solution set is the empty set, write ∅.

1. $3m > 0$ 2. $x + 2 < 0$ 3. $n + 1 \leq 0$ 4. $x - 4 \geq -4$

Consider the inequality $8 < 15$.

Add 5 to each side.   Is $13 < 20$?   Yes
Subtract 9 from each side.   Is $-1 < 6$?   Yes

Note that when the same real number is added to or subtracted from each side of $8 < 15$, the *order* (direction) of the inequality is unchanged.

Addition Property for Inequalities

For all real numbers a, b, and c: If $a > b$, then $a + c > b + c$
If $a < b$, then $a + c < b + c$

Subtraction Property for Inequalities

For all real numbers a, b, and c: If $a > b$, then $a - c > b - c$
If $a < b$, then $a - c < b - c$

To solve an inequality, you must express it as an equivalent inequality in which the variable term, with a coefficient of 1, is alone on one side.

EXAMPLE 1 Solve $n + 9 > 12$. Graph the solution set.

$n + 9 > 12$
$n + 9 - 9 > 12 - 9$ Subtract 9 from each side.
$n > 3$ $n > 3$ is equivalent to $n + 9 > 12$.

The solution set is {all real numbers greater than 3}.

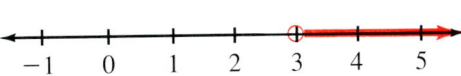

In Example 1, an open circle is drawn at 3 to show that 3 is *not* a solution. To check the solutions, replace the variable in the original inequality with the coordinates of some points on the graph and some points not on the graph.

Replace n with -1.	Replace n with 3.	Replace n with 10.
$n + 9 > 12$	$n + 9 > 12$	$n + 9 > 12$
$-1 + 9 \stackrel{?}{>} 12$	$3 + 9 \stackrel{?}{>} 12$	$10 + 9 \stackrel{?}{>} 12$
$8 \not> 12$	$12 \not> 12$	$19 > 12$ ✔ True
-1 is not a solution.	3 is not a solution.	10 is a solution.

EXAMPLE 2 Solve $1\frac{1}{2}y + 2 \leq \frac{1}{2}y + 2$. Graph the solution set.

$1\frac{1}{2}y + 2 \leq \frac{1}{2}y + 2$
$1\frac{1}{2}y - \frac{1}{2}y + 2 \leq \frac{1}{2}y - \frac{1}{2}y + 2$ Subtract $\frac{1}{2}y$ from each side.
$y + 2 \leq 2$
$y + 2 - 2 \leq 2 - 2$ Subtract 2 from each side.
$y \leq 0$ $y \leq 0$ is equivalent to $1\frac{1}{2}y + 2 \leq \frac{1}{2}y + 2$.

The solution set is {all real numbers less than or equal to 0}.

The closed circle at 0 shows that 0 *is* a solution.

EXAMPLE 3 Solve $5x - 4(x - 2) > 5$, using the set of integers as the replacement set. Graph the solution set.

$5x - 4(x - 2) > 5$
$5x - 4x + 8 > 5$ Use the distributive property.
$x + 8 > 5$ Combine like terms.
$x > -3$ Subtract 8 from each side.

The solution set is {all integers greater than -3}, or $\{-2, -1, 0, 1, \ldots\}$.

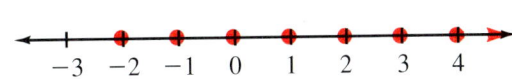

On the number line in Example 3, the arrow in the positive direction is shaded. This shows that there is no last positive integer in the solution set. How would the graph differ if the replacement set is the set of real numbers?

5.2 Solving Inequalities: Addition and Subtraction Properties

EXAMPLE 4 Solve $5x - 4x + 4 \geq 4$. The replacement set is the set of negative integers.

$$5x - 4x + 4 \geq 4$$
$$x + 4 \geq 4$$
$$x \geq 0$$

Using the replacement set of negative integers, there are no values of x that are greater than or equal to 0. The solution set is the empty set, $\emptyset$.

CLASS EXERCISES

Tell what must be done to the first inequality in order to get the second inequality.

1. $n - 4 \geq 6; n \geq 10$
2. $-2 < x - 4; 2 < x$
3. $z + 1 < 5.5; z < 4.5$
4. $p + 4 > 4; p > 0$
5. $4x + 8 \geq 3x + 6; x \geq -2$
6. $1\frac{1}{2}t - 1 < \frac{1}{2}t - 4; t < -3$
7. $5y - 4y + 3 \leq 0; y \leq -3$
8. $-\frac{1}{2}x + 8 + 1\frac{1}{2}x > 10; x > 2$

Solve each inequality. Graph the solution set.

9. $x + 4 > 0$
10. $y - 5 \leq -5$
11. $1\frac{1}{3}x - 3 \leq \frac{1}{3}x - 4$
12. $3\frac{1}{4}x - 6 > 2\frac{1}{4}x + 1$
13. $7(x - 1) - 6x < -8$
14. $3(2x + 3) \geq 5x + 5$

For Discussion

15. Find real numbers x, y, t and w where it is true that $x > y$ and $t > w$, but it is not true that $x - t > y - w$.

PRACTICE EXERCISES

Solve each inequality. Graph the solution set.

1. $m + 2 > 0$
2. $c + 5 > 4$
3. $3 + a < 4\frac{1}{2}$
4. $2 + x > 3\frac{1}{2}$
5. $x - 17 \geq -15$
6. $m - 25 \geq -35$
7. $3\frac{1}{2}y + 2 \leq 2\frac{1}{2}y + 2$
8. $4\frac{2}{3}z - 1 \leq 3\frac{2}{3}z - 1$
9. $1\frac{1}{2}x + 6 > 2\frac{1}{2}x - 6$
10. $3x - 2(x - 4) > 7$
11. $4a - 3(a - 2) > 10$
12. $6c - 5(c - 3) > -18$
13. $8z - 7z + 7 \geq 7$
14. $5x - 4x + 15 \geq 1$
15. $6x - 5x - 10 < 10$

Chapter 5 Inequalities in One Variable

Solve each inequality using the replacement set of negative integers.

16. $4z + 6 < 2z - 6$
17. $5x + 2 < x - 8$
18. $2t - 5 < t + 2$
19. $2x - 6 < -5 + x$
20. $3x + 6 < 2x + 6$
21. $6y + 5 < -5 + y$
22. $6x - 5(x - 1) > 2$
23. $6.5 < 2(t + 3) - t$
24. $7.2 < 3(x + 1) - x$

Solve each inequality. Graph the solution set.

25. $3t + 5.8 < 5.6 + 2t$
26. $2m - 0.5 > m + 3.5$
27. $5x + \frac{1}{4} > 4x + 10\frac{3}{4}$
28. $\frac{1}{2} + 3t + 1\frac{1}{2} \leq 8\frac{1}{2} + 2t$

Solve each inequality using the replacement set of positive integers.

29. $m - 5 > 1$
30. $-3 + 10x \geq 9x + 7$
31. $4z + 3 > z + 9$

32. Show by example that for all real numbers a, b, c, and d, if $a > b$ and $c < d$, then $a - c > b - d$.

33. Show by example that for all real numbers a, b, c, and d, if $a > b$ and $c > d$, then it is not always true that $a - c > b - d$.

34. True or *false*: $a + b \geq a - b$ for all real numbers a and b.

35. True or *false*: $x(x - y) < x(x + y)$ for all real numbers x and y, $y \neq 0$.

Applications

Write an inequality for the word sentence. Then solve the inequality.

36. **Number Problem** Seven more than some number y is greater than or equal to 0.

37. **Number Problem** A number k decreased by 6 is less than -6.

38. **Number Problem** The sum of x and $-\frac{1}{2}$ is greater than -5.

MATH CLUB ACTIVITY

1. Let the number of points that you have in a game be represented by the variable s. There are three other players in the game. The first of these has two times as many points as you do. The second has 7 more points than the first. The third has half as many points as the second. Write an algebraic expression to represent the number of points that the third player has.

2. During the first 6 plays of a football game, a team gained 6 yd, lost 3 yd, gained 8 yd, gained 9 yd, lost 7 yd, and then gained 2 yd. What is the total number of yd that the team lost and gained in all? How far from the starting point is the team now?

5.3 Solving Inequalities: Multiplication and Division Properties

Objective: To solve inequalities using the Multiplication and Division Properties and to draw graphs of the solution sets

Kato wants to improve his school's rushing record of 1200 yd during the football season of 15 games. How many yards must he average per game in order to achieve his goal?

Let g represent the average number of yards Kato must gain in each of the 15 games.

$$15g > 1200$$

How can you solve the inequality $15g > 1200$?

Capsule Review

Multiplying or dividing each side of an equation by the same nonzero number results in an equivalent equation.

$$14 = -2x$$
$$\frac{14}{-2} = \frac{-2x}{-2} \quad \text{Divide by } -2.$$
$$-7 = x$$

$$-\frac{1}{4}y = 6$$
$$(-4)\left(-\frac{1}{4}\right)y = (-4)(6) \quad \text{Multiply by } -4.$$
$$y = -24$$

Tell what must be done to each side of the equation so that the variable y has a coefficient of 1 and is alone on one side of the equation.

1. $3y = -12$
2. $25 = \dfrac{y}{-5}$
3. $-7y = 21$
4. $6y = 9$

If you multiply or divide each side of an inequality by a positive number, the resulting inequality is true.

When you multiply or divide each side of an inequality by a negative number, the *order* (direction) of the inequality is reversed. This leads to the following properties for inequalities.

> **Multiplication Property for Inequalities**
>
> For all real numbers a, b, and c: If $a > b$ and $c > 0$, then $ac > bc$.
> If $a > b$ and $c < 0$, then $ac < bc$.
>
> **Division Property for Inequalities**
>
> For all real numbers a, b, and c: If $a > b$ and $c > 0$, then $\dfrac{a}{c} > \dfrac{b}{c}$.
> If $a > b$ and $c < 0$, then $\dfrac{a}{c} < \dfrac{b}{c}$.

As a demonstration of the above properties, consider the inequality $15 > 10$.

Multiply by 2 $30 > 20$ ✔   True.
Multiply by -2 $-30 \not> -20$ Unless the inequality sign is reversed.
                  $-30 < -20$ ✔   True.

A similar demonstration can be shown using division.

The multiplication and division properties for inequalities are also true for "is less than" ($<$), "is less than or equal to" ($\leq$), and "is greater than or equal to" ($\geq$).

EXAMPLE 1   Solve $15g > 1200$. Graph the solution set.

$15g > 1200$

$\dfrac{15g}{15} > \dfrac{1200}{15}$ *Divide each side by 15.*

$g > 80$ *The direction of the inequality is unchanged.*

Kato must average more than 80 yd per game.

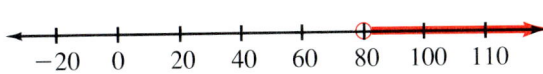

Substitute numbers such as 82, 90, and 100 for g in the inequality, $15g > 1200$ to check the solution set.

EXAMPLE 2   Solve $-7y > 21$. Graph the solution set.

$-7y > 21$

$\dfrac{-7y}{-7} < \dfrac{21}{-7}$ *Divide each side by -7. Reverse the order of the inequality: change $>$ to $<$.*

$y < -3$

The solution set is {all real numbers less than -3}.

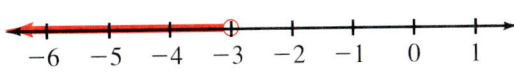

5.3 Solving Inequalities: Multiplication and Division Properties

In Example 3, each side of the inequality is multiplied by the reciprocal of the coefficient of the variable.

EXAMPLE 3 Solve $2 \leq 1\frac{2}{3}y - y$. Graph the solution set.

$$2 \leq 1\frac{2}{3}y - y$$

$$2 \leq \frac{2}{3}y \quad \text{Combine like terms.}$$

$$\frac{3}{2} \cdot 2 \leq \frac{3}{2} \cdot \frac{2}{3}y \quad \text{Multiply each side by } \frac{3}{2}.$$

$$3 \leq y$$

$3 \leq y$ is the same as $y \geq 3$.

The solution set is {all real numbers greater than or equal to 3}.

CLASS EXERCISES

Tell what must be done to each side of the first inequality in order to get the second inequality.

1. $4n \geq -24$; $n \geq -6$
2. $6 \leq -\frac{2}{3}y$; $-9 \geq y$
3. $-2t > 10$; $t < -5$
4. $\frac{3}{4}m < -\frac{27}{40}$; $m < -\frac{9}{10}$
5. $0.5y > -2.4$; $y > -4.8$
6. $35 < -25k$; $1.4 > k$

Solve each inequality. Graph the solution set.

7. $2x \geq 0$
8. $\frac{4}{5}y < -4$
9. $-3c > 9$
10. $-\frac{1}{2}d \leq 4$
11. $-\frac{2}{3}m \geq 12$
12. $-1.2t < -0.6$
13. $-4.6k > 1.38$
14. $1\frac{1}{5}y > -3\frac{1}{10}$

For Discussion

15. Why is zero (0) excluded in the statement of the Multiplication Property for Inequalities? *Hint:* What happens if you multiply both sides of $a > b$ by 0?

16. Why is zero (0) excluded in the statement of the Division Property for Inequalities?

17. Why is zero (0) *not* excluded in the statements of the Addition and Subtraction Properties for Inequalities?

PRACTICE EXERCISES

Solve and check each inequality. Draw a graph of the solution set.

1. $4x < 20$
2. $7y < 28$
3. $8a < 56$
4. $5z < 55$
5. $6b > 48$
6. $9x > 63$
7. $-7y > 42$
8. $-3t > 36$
9. $-4a < 28$
10. $-9c < 54$
11. $-2b > -64$
12. $-5z > -40$
13. $3x > -6$
14. $4a > -16$
15. $\frac{1}{2}y \leq 3$
16. $\frac{4}{5}z \leq 12$
17. $4 \leq 1\frac{4}{5}x - x$
18. $3 \leq 1\frac{1}{3}y - y$
19. $5 \geq 1\frac{1}{2}z - z$
20. $2 \geq 1\frac{1}{6}b - b$
21. $6p - 9p < -21$
22. $5x - 7x > 40$
23. $6y > 2y - 1$
24. $2c > 4c + 9$
25. $-\frac{5}{8}x \geq 25$
26. $-1.5t > -75$
27. $-11z \leq 1.21$
28. $-\frac{3}{5}x \geq 15$
29. $-0.5m \leq -1.5$
30. $-0.6t > 1.08$
31. $-1.3t > 0.6$
32. $-4h < 0.25$
33. $-\frac{1}{4}k > -2.4$
34. $-2\frac{1}{2}m \geq -1\frac{3}{4}$
35. $-15y - 7y \leq 44$
36. $-25 \geq 7w - 12w$

Applications

Write an inequality for the word sentence. Then solve the inequality.

37. **Number Problem** Three-fourths of a number y is greater than -18.

38. **Number Problem** $-2x$ is less than or equal to $\frac{1}{2}$.

EXTRA

A computer cannot directly solve an inequality. If you use an inequality such as $2x - 5 < 7$ in this program, you will get the correct solution. Note, however, the computer is really solving an equation in calculation ready form. It then decides whether it should print "X <" or "X >" as an answer by determining whether B and then A are positive or negative. Try using the program on some of the exercises.

```
10 INPUT "FOR AN INEQUALITY IN THE FORM
   AX + B < C, ENTER A, B, C  ";A,B,C
20 PRINT : LET X = (C - B) / A
30 PRINT "THE SOLUTION OF ";A;"X + ";B;
   " < ";C;" IS "
40 IF B < 0 THEN 50
50 IF A < 0 THEN 70
60 PRINT "X < ";X: PRINT : PRINT : GOTO 80
70 PRINT "X > ";X: PRINT : PRINT
80 INPUT "ENTER Y IF YOU WANT ANOTHER
   INEQUALITY.  ";N$: PRINT
90 IF N$ = "Y" THEN 10
100 END
```

5.4 Solving Inequalities: More Than One Property

Objective: To solve inequalities using more than one inequality property

To solve $9 + \frac{2}{3}x < 1$, you need to use more than one property of inequalities. The procedures are similar to those you used to solve equations.

Capsule Review

Solve: $-5z + 4 = -1$

$$-5z + 4 = -1$$
$$-5z + 4 - 4 = -1 - 4 \quad \text{Subtract 4 from each side.}$$
$$-5z = -5$$
$$\frac{-5z}{-5} = \frac{-5}{-5} \quad \text{Divide each side by } -5.$$
$$z = 1$$

The solution is 1.

Solve each equation.

1. $8x - 5 = 35$
2. $6 - 7y = -36$
3. $3(t - 4) + 14 = 16$

EXAMPLE 1 Solve: $9 + \frac{2}{3}x < 1$

$$9 + \frac{2}{3}x < 1$$
$$9 - 9 + \frac{2}{3}x < 1 - 9 \quad \text{Subtract 9 from each side.}$$
$$\frac{2}{3}x < -8$$
$$\frac{3}{2} \cdot \frac{2}{3}x < \frac{3}{2}(-8) \quad \text{Multiply each side by } \frac{3}{2}.$$
$$x < -12$$

The solution set is {all real numbers less than -12}.

The solution set in Example 1 can also be given using **set-builder** notation.
Read: $\{x: x < -12\}$, "the set of all real numbers x such that x is less than -12."

198 Chapter 5 Inequalities in One Variable

EXAMPLE 2 Solve $5p - 4 - 6p \geq -10$. State the solution set in set-builder notation.

$$5p - 4 - 6p \geq -10$$
$$-1p - 4 \geq -10 \quad \text{Combine like terms.}$$
$$-1p - 4 + 4 \geq -10 + 4 \quad \text{Add 4 to each side.}$$
$$-1p \geq -6$$
$$\frac{-1p}{-1} \leq \frac{-6}{-1} \quad \text{Divide each side by } -1. \text{ Reverse the order of the inequality.}$$
$$p \leq 6$$

The solution set is $\{p: p \leq 6\}$.

As you apply the properties to solve an inequality, the variable terms may "drop out." If the numerical inequality is false, the solution set is the empty set. If the resulting numerical inequality is true, the solution set is the set of all real numbers.

EXAMPLE 3 Solve $2(4 - a) - 2 \leq -2a + 6$. Graph the solution set.

$$2(4 - a) - 2 \leq -2a + 6$$
$$8 - 2a - 2 \leq -2a + 6 \quad \text{Use the distributive property.}$$
$$-2a + 6 \leq -2a + 6 \quad \text{Combine like terms.}$$
$$2a + (-2a) + 6 \leq 2a + (-2a) + 6 \quad \text{Add 2a to each side.}$$
$$6 \leq 6 \quad \text{✓ True}$$

The solution set is {all real numbers}. The graph is the entire number line.

EXAMPLE 4 Write and solve an inequality for the word sentence: Seven decreased by 2 times a number x is greater than or equal to 19.

Seven decreased by $2x$ is greater than or equal to 19.
$$7 - 2x \quad \geq \quad 19$$
$$7 - 7 - 2x \geq 19 - 7$$
$$-2x \geq 12$$
$$\frac{-2x}{-2} \leq \frac{12}{-2}$$
$$x \leq -6$$

The solution set is {all real numbers less than or equal to -6}, or $\{x: x \leq -6\}$.

CLASS EXERCISES

Tell what must be done to the first inequality to get the second inequality.

1. $2y - 10 > 18;\ y > 14$
2. $1 - \frac{x}{4} \geq -8;\ x \leq 36$
3. $\frac{2}{3}t + 15 \leq -1;\ t \leq -24$
4. $14 - 3(a + 4) < 0;\ a > \frac{2}{3}$

Solve each inequality. Then draw its graph by locating the solution on a number line.

5. $\frac{2}{3}t - 3 < 1$
6. $-4x + 24 > -4$
7. $-3 \leq \frac{a}{4} - 5$
8. $12 - 3d < 1\frac{1}{2}$
9. $6p - 7p + 1 \geq 5$
10. $2(x - 5) + 10 < -8$

PRACTICE EXERCISES

Solve each inequality. Graph the solution set.

1. $5 + \frac{1}{3}y > 4$
2. $5 + \frac{2}{3}d > 11$
3. $4x + 5 - 6x \leq 9$
4. $5y + 7 - 7y \leq 13$
5. $3(c + 4) \geq 15$
6. $2(y + 1) \geq 15$

Solve each inequality. State the solution set in set-builder notation.

7. $6 - \frac{2}{3}x < 4$
8. $2 - \frac{1}{4}c > 4$
9. $0.8b - 7 \leq 0.2$
10. $0.4p - 4 > 3.6$
11. $y + 2 - 5y \geq 22$
12. $n + 5 - 7n \geq 13$
13. $1 + a < 3 - 2a$
14. $1 + 4x < 5 - 6x$
15. $9d - 4 \geq 12 + 5d$
16. $6a - 5 > 11 - 2a$
17. $5y - 2(y - 15) < 10$
18. $4t - 2(t - 1) < 4$

Write an inequality for each word sentence and then find the solution set.

19. A number y increased by 4 is greater than -11.

20. The product of a number z and $-\frac{2}{3}$ is less than or equal to 16.

21. One decreased by 7 times a number k is less than 5 times the number k, increased by 3.

22. The product of 6 and the sum of d and 5 is greater than or equal to the sum of 3 and 6 times d.

Chapter 5 Inequalities in One Variable

Give the solution set for each inequality.

23. $2n - 3(n + 3) \leq 14$
24. $7x - (9x + 1) > -5$
25. $4(3m - 1) \geq 2(m + 3)$
26. $17 - (4y - 2) \geq 2(y + 3)$
27. $2(24 - w) - 11 \geq 17 - 2w$
28. $5d - \frac{1}{2}(3d + 8) \leq -4 + \frac{7}{2}d$
29. $\frac{1}{2} - t < -\frac{2}{3}$
30. $x + \frac{5}{12} \geq -\frac{3}{4}$
31. $2 - 3x \neq 4 + 5x$
32. $-1 + 2z \neq 5 - 2z$
33. $4 + 2(3 - 2w) \geq 4(3w - 5) - 6w$
34. $5(1 - y) - 2(y + 7) \leq -10y$
35. $\frac{1}{2} - \frac{1}{3}y < \frac{3}{8} + \frac{1}{6}y$
36. $\frac{1}{2}n - \frac{1}{8} > \frac{3}{4} + \frac{6}{5}n$
37. $6 - [2 - 5(7a + 9)] \geq a - (-a + 8)$
38. $y - [y - (y - 4)] \leq 3 - [4 - (2y + 8)3]$
39. $8m - (3 - m) \leq -\{-[-(3 - 9m)]\}$
40. $-2\{-2[-2(2 - z)]\} > -16 + 8z$

Applications

Write an inequality for the word sentence. Then solve the inequality.

41. **Number Problem** Two-thirds of a number y, increased by 14, is greater than or equal to 8.
42. **Number Problem** Six minus 3 times a number x is less than 30.

TEST YOURSELF

Graph each inequality.

1. $-2 < a$ 2. $c \geq 3$ 3. $b \neq 0$ **5.1**

State the solution set of the inequality whose graph is given.

4.    5.

Solve each inequality. Graph the solution set.

6. $0 > d - 5$ 7. $p + 5.7 < 5.9$ **5.2**
8. $-2.5y \leq 7.5$ 9. $\frac{2}{3}a < -8$ **5.3**
10. $3t - 5t > -2$ 11. $3 - 4(x - 2) \geq 5$ **5.4**
12. $-\frac{1}{3}y + 5 < -2$ 13. $2(a - 6) > 2a - 2$

5.5 Combined Inequalities

Objective: To solve combined inequalities and to draw graphs of the solution sets

In Olympic boxing competition, the mass of a welterweight must be at least 64 kg but no more than 67 kg. The mass of a welterweight can be expressed by a *combined inequality* with $>$, $<$, and $=$.

Capsule Review

Recall that $\leq$ means is less than or equal to, $\geq$ means is greater than or equal to, $<$ means is less than, and $>$ means is greater than.

Classify as *true* or *false*:
a. $13 > 5 \cdot 2 - (-3)$ b. $19 - 26 \leq 0 - 7$

a. $13 \stackrel{?}{>} 5 \cdot 2 - (-3)$ b. $19 - 26 \stackrel{?}{\leq} 0 - 7$
   $13 \not> 13$  $-7 \leq -7$ ✔ True

Classify each inequality as *true* or *false*.

1. $18 - 9 < 5$ 2. $6 \geq 4 + 2$ 3. $-5 > -8 + 2$

Let m represent the mass of a welterweight boxer described above. "At least 64 kg" means that the mass can be exactly 64 kg but it cannot be less than 64 kg. That is, $m \geq 64$.

Also, the mass must be "no more than 67 kg." The mass can be exactly 67 kg but it cannot be greater than that: $m \leq 67$. A **combined inequality**, called a *conjunction*, expresses the mass:

$$m \geq 64 \text{ and } m \leq 67$$

> A **conjunction** is a sentence formed by joining two sentences with the word *and*.

A conjunction is true only if *both* of its sentences are true. To solve a conjunction of two open sentences for a given variable, you find the values of the variable for which both sentences are true.

Chapter 5 Inequalities in One Variable

EXAMPLE 1 Graph the solution set for $m \geq 64$ and $m \leq 67$.

The solution set is {64, 67, and all real numbers between 64 and 67}.

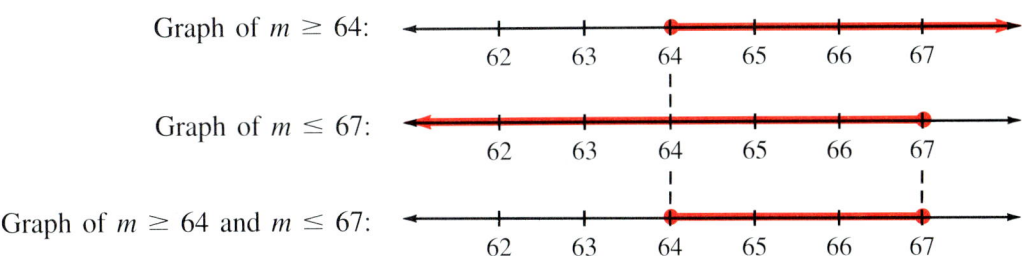

The graph of a combined inequality containing *and* is the **intersection** (overlap) of the graphs of the two inequalities. The conjunction $m \geq 64$ and $m \leq 67$ can be written in a more compact form as $64 \leq m \leq 67$.

EXAMPLE 2 Graph the solution set for $-1 < x < 2$.

Write the combined inequality with *and*: $x > -1$ and $x < 2$. Numbers greater than -1 and less than 2 make both inequalities true.

The solution set is {all real numbers between -1 and 2}.

In Example 2, notice that $-1 < x$ is the same as $x > -1$.

EXAMPLE 3 Graph: $4 < y$ and $y \leq 2$.

Can this be written as the combined inequality $4 < y \leq 2$?

There are no values of y that are both greater than 4 and less than or equal to 2. The graph of $y > 4$ and the graph of $y \leq 2$ have no points in common. They do not intersect.

The solution set is { }.

Sometimes a combined inequality contains the word *or*.

> A **disjunction** is a sentence formed by joining two sentences with the word *or*.

A disjunction is true if either sentence is true or if both sentences are true.

5.5 Combined Inequalities

EXAMPLE 4 Graph the solution set for $x \leq 4$ or $x > 7$.

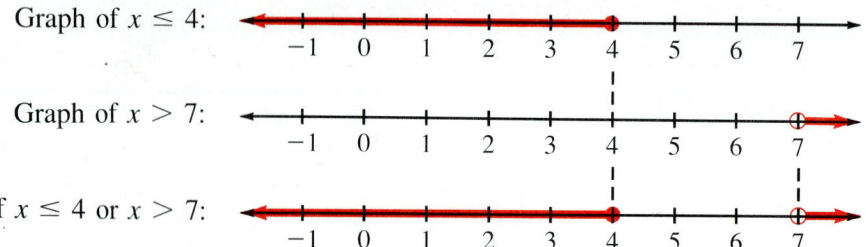

The solution set is {4, and all real numbers less than 4 or greater than 7}.

The graph of a combined inequality involving *or* is the **union** of the graphs of the two inequalities. The union contains both graphs.

EXAMPLE 5 Solve: $y < 2$ or $y \geq -2$

Numbers less than 2 are solutions of $y < 2$. Numbers greater than or equal to -2 are solutions of $y \geq -2$.

The solution set is {all real numbers}, or {y: y is a real number}. The graph of $y < 2$ or $y \geq -2$ is the entire number line.

To solve some combined inequalities, you may have to use the properties of inequalities.

EXAMPLE 6 Solve $3 < 5 - 2x < 7$. Graph the solution set.

First write the combined inequality with *and*. Then solve each inequality separately.

$$3 < 5 - 2x \quad \text{and} \quad 5 - 2x < 7$$
$$3 - 5 < 5 - 5 - 2x \quad \quad 5 - 5 - 2x < 7 - 5$$
$$\frac{-2}{-2} > \frac{-2x}{-2} \quad \quad \frac{-2x}{-2} > \frac{2}{-2}$$
$$1 > x \quad \text{and} \quad x > -1$$

The solution set is {all real numbers between -1 and 1}, or {x: $-1 < x < 1$}.

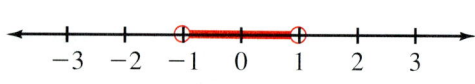

CLASS EXERCISES

Solve each inequality. Graph the solution set.

1. $-1 < 2 + 3x < 5$
2. $y - 1 < -2$ or $y - 1 \geq 2$

PRACTICE EXERCISES

Graph the solution set for each inequality.

1. $m \geq 34$ and $m \leq 35$
2. $x \geq 21$ and $x \leq 22$
3. $y < 1$ and $y > -1$
4. $t < 5$ and $t > -5$
5. $-2 < r < 3$
6. $-4 < z < 5$
7. $3 \leq w < 9$
8. $8 \leq t < 15$
9. $a \leq -4$ or $a > 4$
10. $t \leq 0$ or $t > 3$
11. $y < 5$ or $y \geq 5$
12. $m < 3$ or $m \geq 3$
13. $4 < 3 - 2x < 8$
14. $3 < 5 - 3x < 9$
15. $5 < 2z + 1 < 7$
16. $1 < 3a + 1 < 8$
17. $-1 \leq 2 - t < 3$
18. $-4 < 5 - p < 0$
19. $-3 \leq 8s + 5 \leq 21$
20. $-10 \leq 12r + 2 \leq 14$
21. $-4 \leq -4z - 8 \leq 8$
22. $-2 \leq -3m - 9 \leq 3$
23. $-6 < 2(w + 1) < 4$
24. $-6 < 3(x + 1) < 9$
25. $4a + 2a > 12$ and $3a - 9 < -2a$
26. $3(m - 3) > -3$ or $-8m > -24$
27. $4 - 2w < 3w + 1 < 7$
28. $z - 2 \geq 2 - 3z \geq 1 - 4z$
29. $5y - 2 < 3$ or $-2 < 2(y - 3) < 4$
30. $2 - x \leq \frac{2}{3} - 2$ or $\frac{2}{3} - 2 \geq -1 + x$

Applications

31. **Sports** A baseball measures between 23 cm and 23.5 cm in circumference.
32. **Sports** The mass of a baseball is between 142 g and 148.8 g.
33. **Sports** The length of a major league baseball bat may not be more than 107 cm.

ALGEBRA IN GEOMETRY

Graphs of inequalities may be lines, rays, segments, or points.

EXAMPLE 1 $x \leq -2$

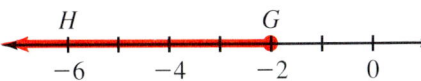

Ray GH starts at $G(-2)$ and goes forever in the direction of H.

EXAMPLE 2 $-1 \leq y \leq 3$

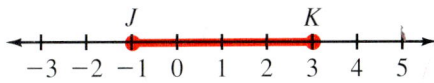

Segment JK has endpoints $J(-1)$ and $K(3)$.

Solve and graph each inequality. What part of a line does the graph represent?

1. $3r + 1 \leq -2$ or $\frac{1}{2}r + 4 \geq 6$
2. $1 - 2y \geq 0$ and $2y + 7 \geq 11$

5.6 Absolute Value Equations

Objective: To solve equations involving absolute value

The equation $|x| = 2$ is equivalent to a combined equation containing *or*, a disjunction. The equation has two solutions. Can you find them?

Capsule Review

The absolute value of a number is the distance of its graph from the origin on a number line. The absolute value of a number is always positive or zero.

Evaluate.

1. $|5|$
2. $|-4|$
3. $-|-6| - |-4|$
4. $-|6 + (-6)|$

In the equation $|x| = 2$, x is a number whose graph is 2 units from the origin on a number line. That is, $x = 2$ or $x = -2$.

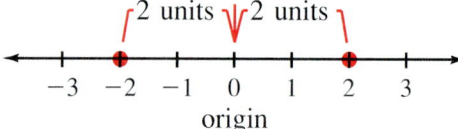

This suggests that to solve an equation that involves absolute value, first write it as a disjunction, then solve the two resulting equations.

EXAMPLE 1 Solve and check: $|2y + 1| = 6$

$$2y + 1 = 6 \quad \text{or} \quad 2y + 1 = -6$$
$$2y = 5 \qquad\qquad\quad 2y = -7$$
$$y = \frac{5}{2} \quad \text{or} \quad y = -\frac{7}{2}$$

Check: $|2y + 1| = 6$ **Check:** $|2y + 1| = 6$
$\left|2\left(\frac{5}{2}\right) + 1\right| \stackrel{?}{=} 6$ $\qquad \left|2\left(-\frac{7}{2}\right) + 1\right| \stackrel{?}{=} 6$
$|5 + 1| \stackrel{?}{=} 6$ $\qquad\qquad |-7 + 1| \stackrel{?}{=} 6$
$6 = 6 \;\checkmark\;$ True $\qquad\qquad 6 = 6 \;\checkmark\;$ True

The solution set is $\left\{\frac{5}{2}, -\frac{7}{2}\right\}$.

You can *count* to find the distance between points, $A(-5)$ and $B(3)$.

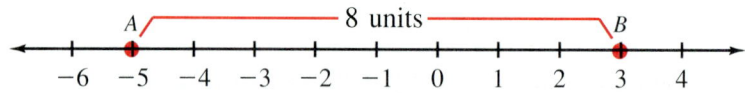

Chapter 5 Inequalities in One Variable

You can also *calculate* the distance between A and B, as follows:

$$|3 - (-5)| = |3 + 5| = 8 \text{ or } |-5 - 3| = |-8| = 8$$

Because of the absolute value signs, the coordinates -5 and 3 can be subtracted in either order. The equation $|3 - (-5)| = 8$ means that 3 is 8 units from -5. Similarly, the equation $|z - 5| = 6$ means that some number z is 6 units from 5. The equation can be solved by looking for appropriate values on the number line or by an algebraic method. Both methods are shown in Example 2.

EXAMPLE 2 Solve $|z - 5| = 6$ by two methods.

Solution by the number-line method:

$|z - 5| = 6$ Look for values of z that are 6 units from 5.

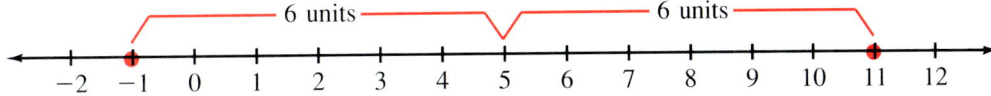

Since both 11 and -1 are 6 units from 5, $z = 11$ or $z = -1$.

Solution by the algebraic method:

$$\begin{array}{lll} z - 5 = 6 & \text{or} & z - 5 = -6 \\ z = 11 & \text{or} & z = -1 \end{array}$$

So, the solution set is $\{11, -1\}$.

In $|t + 6| = 4$, you can change $|t + 6|$ to $|t - (-6)|$ to represent the distance between two points as the difference between their coordinates.

EXAMPLE 3 Solve: $|t + 6| = 4$

Change $|t + 6| = 4$ to $|t - (-6)| = 4$.
Use the number line. -10 and -2
are each 4 units from -6.
The solution set is $\{-10, -2\}$.

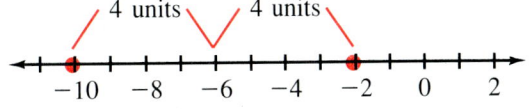

To solve an equation such as $|x| + 5 = 4 - 6$, you should isolate the term containing the absolute value.

EXAMPLE 4 Solve: $|x| + 5 = 4 - 6$

$$\begin{array}{ll} |x| + 5 = 4 - 6 & \\ |x| + 5 = -2 & \textit{Combine like terms.} \\ |x| + 5 - 5 = -2 - 5 & \textit{Subtract 5 from each side.} \\ |x| = -7 & \end{array}$$

What are the values of x that make this a true statement?

5.6 Absolute Value Equations

During a recent college bowl contest the following correct answer was given by one of the contestants:

The absolute value of a number represents the *distance* of its graph from the origin. This means that the absolute value of a number cannot be negative, so there are no values of x for which $|x| = -7$.

The solution set is the empty set, $\emptyset$.

CLASS EXERCISES

Translate each statement to an absolute value equation.

1. The distance from x to -9 is 5 units.
2. m is 6 units away from -2.

Change each absolute value equation to a disjunction. Then solve.

3. $|y + 4| = 11$
4. $|12 - 2t| = 9$
5. $|3z - 7| = 5$

PRACTICE EXERCISES

Solve and check. Use the number line method for Exercises 1–4 and the method of your choice for Exercises 5–38.

1. $|x + 1| = 4$
2. $|z + 7| = 8$
3. $|a + 3| = -2$
4. $|t + 2| = 0$
5. $|x + 6| = -4 + 7$
6. $|4m + 1| = 3$
7. $|2p + 7| = 9$
8. $|3x + 8| = 11$
9. $|6z + 4| = 10$
10. $|3x - 1| = 5$
11. $|5y - 2| = 13$
12. $|4m - 3| = 9$
13. $|7h - 4| = 17$
14. $\left|\frac{1}{3}a - 2\right| = 4$
15. $\left|\frac{3}{4}h - 5\right| = 1$
16. $|g| + 4 = 7 - 12$
17. $|m| - 4 = 8 - 10$
18. $|p| - 3 = 7 - 12$
19. $|2(1 - 2m)| = 3$
20. $|-3(2 - m)| = 6$
21. $|3b + 5 + 2b| = 10$
22. $|3y - 2(y - 2) - 1| = 7$
23. $|4x - 3(x - 1) - 2| = 9$
24. $|13 - (k + 2)| = 1$
25. $|y - (2y + 1)| = 3$
26. $|d + 3d| = 6$
27. $\left|\frac{c - 1}{2}\right| = 5$
28. $\left|\frac{2 - d}{3}\right| = 21$
29. $\left|\frac{4 - m}{3}\right| + 2 = 7$
30. $\left|\frac{p + 1}{5}\right| + 1 = 6$
31. $|4 - 2(n - 1)| = 3$
32. $|7x - (9x + 1)| = 5$
33. $|3c - 3| = 2c$
34. $|2t - 1| = 3t$
35. $|4 - a| = 2a$
36. $|2 - 3s| = s$
37. $|3y - 2| = 2y - 1$
38. $|2z - 1| = 3z + 4$

Applications

Translate each word statement into an absolute value equation. Then solve.

39. **Number Problem** The absolute value of the sum of 4 and some number is 16.

40. **Number Problem** If 5 is subtracted from the absolute value of a number, the result is 27.

41. **Number Problem** The absolute value of the difference of twice a number and the number is 33.

42. **Number Problem** The absolute value of the difference of four times a number and 4, multiplied by 5, is 20.

43. **Number Problem** If four is subtracted from the absolute value of the difference of twice a number and 2, multiplied by 2, the result is 24.

DID YOU KNOW?

Throughout history people have feared and been fascinated by comets. When Halley's comet appeared in 1910, people bought "comet pills" to protect themselves from "comet fever" and disaster.

Most comets are named after their discoverers. Japanese amateur astronomers Kaoru Ikeya and Tsutomu Seki discovered the comet called Ikeya-Seki. As Ikeya-Seki came close to our sun in 1965, it accelerated to more than 1,000,000 mi/h and broke into three parts.

There are about 100,000 comets in the solar system. Most of them orbit the sun over long periods of time. These long-period comets may take many thousands of years to complete one trip around the sun.

The most famous short-period comet is Halley's comet, named for the English astronomer Edmund Halley. Halley's comet completes one trip every 76 to 79 years. Its most recent visit to the earth was in 1986.

Write a combined inequality to express the date of the next visit of Halley's comet to the earth.

5.7 Absolute Value Inequalities

Objective: To solve and graph inequalities involving absolute value

An inequality involving absolute value can be written as a combined inequality containing either *and* or *or*.

Capsule Review

Conjunction: $r < 5$ and $r > -2$ can be written as $-2 < r < 5$.
Disjunction: $x > 0$ or $x = 0$ can be written as $x \geq 0$.

Write each combined inequality without *and* or without *or*.

1. $x > 0$ and $x < 5$
2. $y > \frac{1}{4}$ or $y = \frac{1}{4}$
3. $-3 < n$ and $n < 0$

The graph of $|x| = 3$ is found by locating points on a number line that are 3 units from the origin. The graph of $|x| > 3$, shown below, contains those points that are *more* than 3 units from the origin.

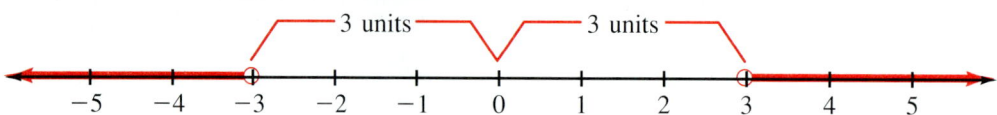

This suggests that $|x| > 3$ is equivalent to the disjunction $x < -3$ or $x > 3$.

EXAMPLE 1 Solve $|y - 2| > 4$. Graph the solution set.

Write $|y - 2| > 4$ as a disjunction.

$y - 2 > 4$ or $y - 2 < -4$
$y > 6$ or $y < -2$

The solution set is $\{y: y > 6 \text{ or } y < -2\}$.

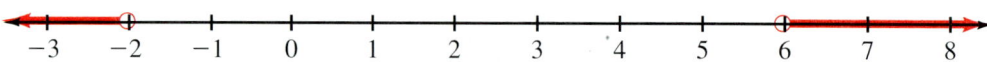

Since $|y - 2|$ represents the distance between y and 2 on a number line, $|y - 2| > 4$ means that y is more than 4 units from 2. This suggests that if the absolute value of an expression is *greater than* a given positive number, a disjunction results.

210 Chapter 5 Inequalities in One Variable

In $|w| < 3$, w represents numbers less than 3 units from the origin.

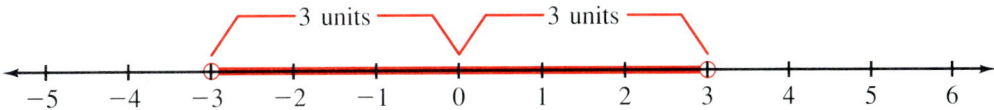

The graph shows that $w > -3$ and $w < 3$, which can be stated as $-3 < w < 3$. This suggests that if the absolute value of an expression is *less than* a given positive number, a conjunction results.

EXAMPLE 2 Solve $|z + 5| < 2$ by two methods. Graph the solution set.

Algebraic method:
$|z + 5| < 2$
$z + 5 < 2$ and $z + 5 > -2$ *Write the conjunction.*
$z < -3$ and $z > -7$

Number-line method:

$|z + 5| < 2$ must be written to show a *subtraction* of coordinates. $z + 5$ is equivalent to $z - (-5)$.

So, $|z + 5| < 2$ is equivalent to $|z - (-5)| < 2$. The values of z are less than 2 units from -5.

The solution set is $\{z: z > -7$ and $z < -3\}$, or $\{z: -7 < z < -3\}$.

EXAMPLE 3 Solve $|3t + 1| \leq 5$ by the algebraic method. Graph the solution set.

$|3t + 1| \leq 5$ means $3t + 1 \leq 5$ and $3t + 1 \geq -5$.

$3t + 1 \leq 5$ and $3t + 1 \geq -5$
$3t \leq 4$ $3t \geq -6$
$t \leq \dfrac{4}{3}$ $t \geq -2$

The solution set is $\left\{t: -2 \leq t \leq \dfrac{4}{3}\right\}$.

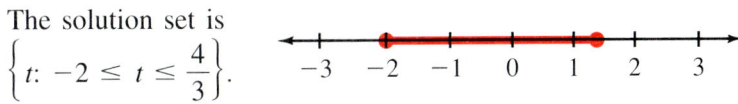

The inequality in Example 3 cannot be solved as easily by the number-line method. Why?

The expression within the absolute-value symbol may need to be simplified. Also, isolate the absolute-value expression on one side of the inequality.

EXAMPLE 4 Solve $8 - |-t + 2(t - 1)| \geq 6$. Graph the solution set.

$8 - |-t + 2(t - 1)| \geq 6$
$8 - |-t + 2t - 2| \geq 6$ Use the distributive property.
$8 - |t - 2| \geq 6$ Combine like terms.
$-|t - 2| \geq -2$ Subtract 8 from each side.
$|t - 2| \leq 2$ Multiply each side by -1.
             Change $\geq$ to $\leq$.

The values of t are 2 or less units from 2.

The solution set is $\{t: 0 \leq t \leq 4\}$.

CLASS EXERCISES

Pair each inequality in Exercises 1–4 with one of the statements in a–d.

1. $|x| < 2$ **a.** Values of x are more than 0 units from 2.
2. $|x| \geq 2$ **b.** Values of x are less than 2 units from the origin.
3. $|x - 2| > 0$ **c.** Values of x are more than 0 units from -2.
4. $|x + 2| > 0$ **d.** The distance from x to the origin is 2 or more units.

Pair each graph in Exercises 5–8 with one of the inequalities in a–d.

5. 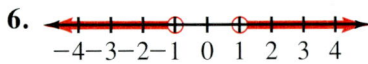   6.

7. 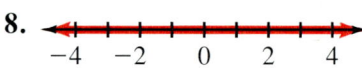   8.

a. $|y| > 1$
b. $|1 - y| \geq 0$
c. $|y| \leq 4$
d. $|y + 2| < 3$

PRACTICE EXERCISES

Solve and graph. Use the number line to solve Exercises 1–8.

1. $|y - 4| > 6$
2. $|x - 3| > 7$
3. $|z + 6| > 8$
4. $|a + 1| > 5$
5. $|c + 3| < 9$
6. $|b + 5| < 10$
7. $|t - 2| < 5$
8. $|w - 1| < 8$
9. $|x - 3| < 4$
10. $|a - 2| < 5$
11. $|2t - 3| < 0$
12. $|3t - 1| < 5$
13. $|2m - 9| > 1$
14. $|4q - 1| > 0$
15. $|2t + 3| \leq 6$
16. $|3x + 1| \leq 8$
17. $|5a - 2| \leq 9$
18. $|6b - 3| \leq 12$
19. $|a + 3| \geq 7$
20. $|b + 5| \geq 2$
21. $|4x - 8| \leq 2$
22. $|2q - 1| \leq 5$
23. $|2t| \geq 0$
24. $|3x| \geq 0$
25. $7 - |-z + 3(z - 1)| \geq 5$
26. $14 - |-t + 2(t - 3)| \geq 9$
27. $8 - |-x + 2(x - 2)| \geq 2$
28. $9 - |-m + 3(m - 4)| \geq 7$
29. $3 + |2 - w| < 2$
30. $|m - 4| - 4 > -6$
31. $\left|\dfrac{2q - 4}{3}\right| \leq 1$

Chapter 5 Inequalities in One Variable

32. $\left|\dfrac{2-3b}{2}\right| \leq 1$ 	33. $\left|\dfrac{2-2c}{3}\right| \geq 0$ 	34. $\left|\dfrac{3-4m}{4}\right| \geq 1$

Solve each inequality. Express your answer in set-builder notation.

35. $|5-(3d-4)| > 6$
36. $|2-(4-2f)| > 4$
37. $-6 + |2t + 5(1-t)| \leq -2$
38. $|3(1-v) - 6| + 5 \leq 7$
39. $\left|\dfrac{p+1}{2}\right| + 3 \geq 3$
40. $-2 + \left|\dfrac{1-2x}{3}\right| \geq -2$
41. $|3z - 5| + 8 > 7$
42. $-4 + |2 - 5m| \geq -6$
43. $|g - 1| < 3g$
44. $|4 - m| \geq 2m$
45. $|2 - y| > 2 - y$
46. $p + 6 \geq |p + 6|$
47. $|3k - 2| \leq 2k - 3$
48. $|2t - 1| > 3t + 5$

Applications

Write an absolute value inequality for each situation.

49. **Meteorology** A meteorologist reported that the overnight temperature in the tri-state area was within 5 degrees of 0.

50. **Sports** A poll shows that 85% of the sportscasters watched the Super Bowl. The poll has a plus or minus 3 percentage-point margin of error. (*Hint:* Let x = percentage who watched the Super Bowl.)

ALGEBRA IN MECHANICS

When you turn a faucet on and off, you expect water to flow and then to stop. A valve inside the faucet controls the flow of water. But what happens if the valve does not fit properly?

Manufactured parts may vary from the established standard by only a specified amount, called **tolerance**. For example, the standard diameter for a valve inside a faucet might be 2 cm, and a valve may not vary from this standard by more than 0.0001 cm; that is, the tolerance is ±0.0001.

The tolerance interval for any valve with a diameter x can be shown by writing $|x - 2| \leq 0.0001$. You can also use a combined inequality to show the interval: $1.9999 \leq x \leq 2.0001$.

Show the tolerance interval first using absolute value and then by using a combined inequality.

1. Standard: 6 mm
   Tolerance: ±0.001
2. Standard: 14 in.
   Tolerance: ±0.5
3. Standard: 2.75 in.
   Tolerance: ±0.001

5.8 Problem Solving Strategy: Write an Inequality

In many problem solving situations, an inequality can be written to solve the problem. The inequality is one type of *mathematical model* of a real-world situation.

Oftentimes situations requiring the use of inequalities occur in financial planning, personal budget development, and other money making decisions that you may encounter.

EXAMPLE 1 Rates at a car rental agency are $160 a week plus $0.10 a mile. If Jane rents a car for a week, how far can she drive if she wants to spend no more than $250?

▪ **Understand the Problem**
You are given the weekly rental ($160) and the mileage rate ($0.10 per mile). You must find the number of miles Jane can drive if she wants to spend *no more than* $250.

▪ **Plan Your Approach**
Let y = the number of miles Jane can drive.
Total amount to spend ≤ 250, or $160 + 0.10y \leq 250$

▪ **Complete the Work**
$160 + 0.10y \leq 250$
$0.10y \leq 90$ *Subtract 160 from each side.*
$y \leq 900$ *Divide each side by 0.10.*

▪ **Interpret the Results**
Jane can drive 900 or fewer miles. Test 900 mi in the original statement: If she drives 900 mi, Jane will pay $160 plus ($0.10)(900). Will that cost no more than $250?

$160 + (0.10)(900) \stackrel{?}{\leq} 250$
$160 + 90 \stackrel{?}{\leq} 250$
$250 \leq 250$ ✔ True

As shown in Example 1, *is less than, is less than or equal to, is greater than,* and *is greater than or equal to* are not the only phrases that are used in translating inequalities. For example, the symbols $\leq$ and $\geq$ are translated into the following equivalent phrases:

214   Chapter 5 Inequalities in One Variable

≤	≥
at most	at least
no more than	no less than

EXAMPLE 2 There are four math tests in the first marking period. Sam's scores on three tests are 85, 91, and 76. What is the lowest score he can get on the fourth test to have an average greater than 85 for the marking period?

Understand the Problem You are given three test scores: 85, 91, and 76. You must find the lowest possible fourth test score for the average to be greater than 85.

Plan Your Approach Let x = the fourth test score.

The average of the four scores is: $\dfrac{85 + 91 + 76 + x}{4}$.

Translate the information into an inequality.

The average is greater than 85.

$$\dfrac{85 + 91 + 76 + x}{4} > 85$$

Complete the Work
$85 + 91 + 76 + x > 340$ Multiply each side by 4.
$252 + x > 340$
$x > 88$ Subtract 252 from each side.

Interpret the Results The fourth test score must be greater than 88. Using whole numbers as the replacement set, the lowest possible fourth test score is 89.

CLASS EXERCISES

Write and solve an inequality that represents the given situation.

1. A freight elevator can safely hold no more than 2000 lb. What is the greatest number of 60-lb boxes the elevator can hold safely?

2. Sang-Ho received commissions of $150 and $225. How much must he receive from a third sale so his total commission will exceed $500?

PRACTICE EXERCISES

1. A company's policy is to spend no more than $350,000 a year for salaries. The president's salary is $110,000. The salaries of the eight other employees are equal. How much can be spent on each salary?

2. Sally is paid $250 a week plus a commission equal to 3% of the sales. What must her sales be if she is to have a weekly income of no less than $460?

3. Lisa's grades on four exams were 80, 92, 86, and 78. What is the lowest grade she can receive on the next exam to have an average greater than 85?

4. The amount of rainfall over a three-year period was 65 in., 72 in., and 59 in. How many inches of rain must fall during the fourth year for the average rainfall to be at least 68 in. for the four years?

5. Freddie works two jobs. He earns $6 per hour doing one job and $100 a week doing the other job. How many hours must he work per week on the first job so that this combined weekly income is no less than $244?

6. The length of a rectangle is 2 ft more than the width. Find the minimum dimensions if the perimeter is more than 16 ft and the length and width are integers.

7. The length of a rectangle is 4 m more than twice its width. Find the greatest possible value for the width if the perimeter is at most 38 m.

8. Manuel wishes to purchase 25¢ and 30¢ stamps. He decides to buy 10 fewer 30¢ stamps. How many of each can he buy for no more than $19.00?

9. A taxi charges 75¢ for the first quarter mile and 35¢ for each additional quarter mile or part. How far did Jim ride if he paid less than $7.40 for his taxi?

10. A washing machine technician charges $27 for the first half hour and $18 for each half hour or part thereafter. How long did the technician work if the bill did not exceed $100?

11. Mrs. Chavis will invest $10,000 in stock and bonds. She expects to receive $4\frac{1}{2}\%$ annual interest from the bonds and $2\frac{1}{2}\%$ annual interest from the stocks. She wants her annual income from the stocks and bonds to be at least $300. How much should Mrs. Chavis invest in each?

12. If $450 is invested annually, part at 5% and the remainder at $4\frac{1}{2}\%$, how much must be invested at $4\frac{1}{2}\%$ to receive at least $21.50 in interest?

Mixed Problem Solving Review

1. John bought some tapes on sale from a catalog for $5.69 each. The postage and handling charge was $2.50. If John paid a total of $70.78, how many tapes did he buy?

2. Two buses leave the terminal at the same time but go in opposite directions. One averages 10 mi/h more than the other. After 3 h, they are 270 mi apart. How fast is each bus traveling?

3. Find the four smallest consecutive odd integers whose sum is greater than 48.

PROJECT

A travel agent tries to get the best rate for clients when they rent a car. Use ads for different car rental companies and the spreadsheet to determine the cost of renting a car for 10 days.

```
======= A ===== B ===== C ===== D ===== E ===== F ===== G ===== H ===== I =====
 1 RENTAL
 2    WEEKLY RATE
 3    DAILY RATE
 4
 5 INSURANCE
 6    WEEKLY RATE
 7    DAILY RATE
 8
 9 FUEL CHARGES
10
11 TRIP MILEAGE
12 DAILY FREE MILES
13 ADDITIONAL MILES CHARGE
14
15 TOTAL COST
```

TEST YOURSELF

Solve each inequality. Graph the solution set. 5.5

1. $3p + 1 > -2$ or $p + 5 < 2$
2. $-3 < 2m + 1 < 9$
3. $1 - 2y > -9$ or $3y - 10 > 2$
4. $5 < -7k - 2$ and $\frac{3}{4}k + 6 < -3$

Solve each equation. 5.6

5. $|x - 3| = 4$
6. $|3t + 2| = 7$
7. $1 + \left|m + \frac{2}{3}\right| = 2$
8. $-5|6 - y| + 4 = -6$

Solve each inequality. Graph the solution set. 5.7

9. $|y| > \frac{1}{2}$
10. $|x - 4| < 2$
11. $|7 + f| > 13$
12. $|10 - 3g| + 4 > 3$

Write and solve an inequality that represents the given situation. 5.8

13. Find any number n such that 11 decreased by $\frac{1}{3}$ of n is more than 5.
14. Bill earns a 15% commission on his sales. If he hopes to earn at least $240 this week, what must his weekly sales be?
15. In the past five races, Margot's times for the 100 m have been 13.1, 12.8, 13.0, 13.3, and 12.7 s. What time must Margot run in her next race, to have an average time less than 12.8 s?

APPLICATION: Marathon Running

Did you know that the Olympic marathon race was originally run in the 1896 Olympics in Athens, Greece? This race commemorated an unknown messenger who ran 24 miles to announce a Greek victory. Preparing for a modern day marathon requires careful training and physical conditioning. In order to run a distance of 26 miles and 385 yards, marathon runners need to build up greater distances and faster times. Keeping careful track of their performances enables these long distance runners to structure a proper training schedule.

No official records are kept for marathon races due to varying course conditions and degrees of difficulty. The following table shows the men's and women's marathon gold medalists for the 1988 and 1984 Olympics.

Gelindo Bordin	Italy	1988	2 h 10 min 32 s
Rosa Mota	Portugal	1988	2 h 25 min 40 s
Carlos Lopes	Portugal	1984	2 h 9 min 55 s
Joan Benoit	U.S.A.	1984	2 h 24 min 52 s

EXAMPLE 1 While training for a marathon, Jane plans to run at least 40 mi each week. On Wednesday she is only able to run 4 mi. What is the least number of miles she must average daily by next Wednesday to meet her distance goal?

$$m = \text{number of miles run each day}$$
$$6m = \text{number of miles run for the 6 days}$$
$$6m + 4 \geq 40$$
$$6m \geq 36$$
$$m \geq 6$$

Therefore, Jane must average at least 6 mi each day for the next 6 days.

Chapter 5 Inequalities in One Variable

EXAMPLE 2 Dan wishes to run from 15 mi to 20 mi on weekdays. On Monday he ran 3 mi, on Tuesday $3\frac{1}{2}$ mi. Find the least and the greatest number of miles Dan needs to run over the next three days to maintain his training schedule.

$$m = \text{number miles run each day}$$
$$3m = \text{number of miles run for 3 days}$$
$$15 \leq 3m + 6\frac{1}{2} \leq 20$$

$15 \leq 3m + 6\frac{1}{2}$	and	$3m + 6\frac{1}{2} \leq 20$
$8\frac{1}{2} \leq 3m$		$3m \leq 13\frac{1}{2}$
$2\frac{5}{6} \leq m$		$m \leq 4\frac{1}{2}$

So, Dan will need to run from $2\frac{5}{6}$ mi to $4\frac{1}{2}$ mi each day, or writing this as a combined inequality: $2\frac{5}{6} \leq m \leq 4\frac{1}{2}$.

Solve.

1. During a week of training Carlos wishes to run at least 27 mi. Due to bad weather he is only able to run 3 mi the first day. Find the least number of miles he must run each day to meet his goal.

2. During the third week of Carlos' training he wishes to increase his total distance to 35 mi from 27 mi. How much more must he average each day compared to the previous week?

3. Christine ran for three consecutive days 10, 8.5, and 9 mi. What would be the fewest number of miles she should run over the next 3 days to reach her goal of 50 mi?

4. Katrina runs from 12 mi to 18 mi each week. Over a 2-day period she ran 2 mi and 3 mi, respectively. If she plans to run for only 4 more days, what is the least and greatest number of miles she must run to achieve her mileage goals?

5. Adrienne ran $2\frac{1}{2}$ mi Monday and $2\frac{1}{4}$ mi Wednesday, and she plans to run again on Friday and Sunday. What would be the least and greatest number of miles she would need to run to average from 10 mi to 12 mi for that week?

6. Using the table on page 218 find the average rate of each runner over the entire course of the marathon.

Applications: Marathon Running

CHAPTER 5 SUMMARY AND REVIEW

Vocabulary

combined inequality (202)
conjunction (202)
disjunction (203)
intersection (203)
set-builder notation (198)
union (204)

Graphing Simple Inequalities The graphs of $x > -2$ and $y \leq 3$ are shown.   5.1

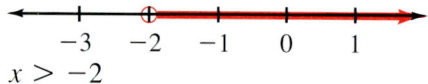

Note the meaning of the "open circle" and the "closed circle."

Graph each equation or inequality.

1. $z = -3$
2. $v < 1$
3. $x \geq -2$
4. $t \neq -5.5$

Properties for Inequalities   5.2–5.4

Addition Property   For all real numbers a, b, and c,
if $a > b$, then $a + c > b + c$.

Subtraction Property   For all real number a, b, and c,
if $a > b$, then $a - c > b - c$.

Multiplication Property   For all real numbers a, b, and c,
     (a) If $a > b$ and $c > 0$, then $ac > bc$.
     (b) If $a > b$ and $c < 0$, then $ac < bc$.

Division Property   For all real numbers a, b, and c,
     (a) If $a > b$ and $c > 0$, then $\dfrac{a}{c} > \dfrac{b}{c}$.
     (b) If $a > b$ and $c < 0$, then $\dfrac{a}{c} < \dfrac{b}{c}$.

These properties are also true for $a < b$, $a \leq b$, and $a \geq b$.

Solve each inequality. Graph the solution set.

5. $0 > x + 5$
6. $y - \frac{1}{2} < 1$
7. $-1.5a > 3$
8. $2(2 - x) + 3x \geq 0$
9. $g - 2(3g + 1) > 5(4 + g) + 8$

Combined Inequalities Combined inequalities containing "or" are disjunctions. Their graph is the *union* of the two graphs. Combined inequalities containing "and" are conjunctions. Their graph is the *intersection* of the two graphs.

5.5

Solve each inequality. Draw a graph of the solution set.

10. $3y + 7 < -8$ or $2(3 - y) < -4$
11. $-(5 - 2x) + 3 \leq 0$ and $2(7 - x) \leq x + 17$
12. $-3 < \frac{1}{2}(4 - 6v) < 6$

Absolute Value Equations An absolute value equation can be treated as a disjunction (or) and the two resulting equations then solved. An absolute value equation can also be considered as a statement about the distance between two points on a number line.

5.6

Solve and check each equation.

13. $|m - 2| = 7$
14. $|n + 3| = \frac{5}{3}$
15. $|6x - (8x + 1)| = 9$

Absolute Value Inequalities To solve an absolute value inequality such as $|2r + 3| > 7$, change to the equivalent disjunction $2r + 3 > 7$ or $2r + 3 < -7$. When the absolute value inequality involves $<$, such as $|j + 4| < 6$, change to the equivalent conjunction $j + 4 < 6$ *and* $j + 4 > -6$.

5.7–5.8

Solve each inequality. Draw a graph of the solution set.

16. $|g - 3| > 5$
17. $5 - |4 - y| \leq 3$
18. $\left|\frac{y - 8}{4}\right| \leq 4$

Write and solve an inequality that represents the given situation.

19. Half of a number x, increased by 3, is greater than 2. Find x.
20. The sum of two consecutive even integers is less than 28. What are the largest possible consecutive integers whose sum is less than 28?
21. Over a 5-week period, Wanda earned commissions of $17, $21, $19, $25, and $12. In order for her average for 6 weeks to be at least $20, how much must her commission for the sixth week be?
22. The length of a rectangle is 3 ft more than the width. If the perimeter is no more than 52 ft and the dimensions are integers, what are the largest possible dimensions?

Summary and Review

CHAPTER 5 TEST

Draw a graph of each equation or inequality.

1. $y = 2$
2. $x \leq 0$
3. $m > -4$

Solve each inequality. Draw a graph of the solution set.

4. $x - 5 > 8$
5. $1 - 3y > 7$
6. $18 - 3x > 0$ or $-16 > -8x$
7. $3y - 4 < 5$ and $2y > 1$

Solve and check each equation.

8. $|b| = 3$
9. $|y - 3| = 7$
10. $|x - (3x + 2)| = 3$

Solve each inequality. Draw a graph of the solution set.

11. $|a| > 4$
12. $|3 - 2x| \geq 5$
13. $|8 + a| < 2$

Write and solve an inequality that represents the given situation.

14. Amy plans to spend at most $200 at a department store. If she purchases a dress that costs $75 and a pair of shoes for $45, how much can she spend on other items in the store?

15. The sum of two consecutive odd integers is greater than -15. What are the smallest such integers?

16. If Seth increased the time he plans to spend cycling by 2 h, he would travel more than 3 times as far. If his average speed is 12 mi/h, how much time does he plan to spend cycling?

17. Oscar had a certain amount of money in his savings account. After he withdrew $125, there was less than $500 in the account. How much money was originally in the account?

18. If Lisa increased her rate by 15 mi/h, then in 2 h she would travel a greater distance than she does in 3 h at her present rate. What do you know about her present rate?

Challenge

Solve each inequality. Draw a graph of the solution set.

1. $|x - 1| < 0.5$ and $|x - 2| < 0.5$
2. $|x + 1| < 1$ or $|x + 2| > 1$

PREPARING FOR STANDARDIZED TESTS

Select the correct answer for each question.

1. Solve for x:
$$x + 3x + 4 = 2x - 6$$
   A. 10 B. 5 C. -1
   D. -5 E. -10

2. What is the sum of the following numbers?
$$8.03,\ 11.2,\ 7.863,\ 0.007$$
   A. 8.785 B. 26.09 C. 26.90
   D. 27.01 E. 27.10

3. Solve the inequality:
$$3x + 5 < x + 17$$
   A. $x < 6$
   B. $x > 6$
   C. $x = 6$
   D. $x < -6$
   E. $x > -6$

4. The winner of a mile race crossed the finish line in 4 min and 38.42 s, while the second runner took 4 min 39.8 s. By how many seconds did the winner beat the second place runner?
   A. 4.44 B. 1.48 C. 1.38
   D. 0.48 E. 0.38

5. Find the value of $|3a + 2b^2|$ when $a = -2$ and $b = -4$.
   A. -22 B. 16 C. 22
   D. 26 E. 38

6. Which sentence best describes the graph:

   A. $x \geq -2$ B. $x > -2$
   C. $x = -2$ D. $x \leq -2$
   E. $x < -2$

7. Find the value of the expression $m^2n - n^3p^2$ when $m = 3$, $n = -2$, and $p = -1$.
   A. 26 B. 18 C. 10
   D. -2 E. -10

8. Solve the equation:
$$3(x - 2) - 5(x + 4) = 4(2x + 11)$$
   A. -7 B. -6 C. -3
   D. 3 E. 6

9. A hardware store paid $299.95 for an order of 24 electric hedge clippers, and marked a price of $16.95 on each one. If all of the clippers were eventually sold, how much profit was made on the order?
   A. $16.85 B. $98.15
   C. $106.85 D. $108.15
   E. $109.65

10. Solve for b in the equation:
$$s = \tfrac{1}{2}(a + b + c)$$
    A. $b = \dfrac{s - a - c}{2}$
    B. $b = \dfrac{s - a + c}{2}$
    C. $b = s - a - c$
    D. $b = 2s - a + c$
    E. $b = 2s - a - c$

11. How much sales tax would be charged on the total of $3.84, $7.98, $14.95, and $9.95 in a city which has a sales tax of $5\tfrac{1}{2}\%$?
    A. $2.01 B. $2.02 C. $2.10
    D. $2.19 E. $2.20

Preparing for Standardized Tests

MAINTAINING SKILLS

Simplify.

Example 1 $(-3)^3$

$$(-3)^3 = (-3)(-3)(-3) = (9)(-3) = -27$$

Example 2 $(2^3)^2$

$$(2^3)^2 = (2^3)(2^3) = (2 \cdot 2 \cdot 2)(2 \cdot 2 \cdot 2) = 8 \cdot 8 = 64$$

1. 10^5
2. 2^3
3. $(-1)^4$
4. 5^2
5. $(-4)^3$
6. $-(-3)^4$
7. $(3^2)^3$
8. $(2^3)^3$
9. $[(-1)^3]^2$
10. $[(-2)^2]^2$
11. $(4^2)^2$
12. $[(-3)^2]^3$

Simplify.

Example 3 $\dfrac{18}{45}$

$$\dfrac{18}{45} = \dfrac{2 \cdot \overset{1}{\cancel{3}} \cdot \overset{1}{\cancel{3}}}{5 \cdot \underset{1}{\cancel{3}} \cdot \underset{1}{\cancel{3}}} = \dfrac{2}{5}$$

13. $\dfrac{7}{56}$
14. $\dfrac{9}{21}$
15. $\dfrac{15}{25}$
16. $\dfrac{12}{36}$
17. $\dfrac{20}{24}$
18. $\dfrac{14}{21}$

Simplify.

Example 4 $3a^2 + 5ab + 2a^2 - 7ab$ *Combine like terms.*

$$(3a^2 + 2a^2) + (5ab - 7ab) = 5a^2 + (-2ab) = 5a^2 - 2ab$$

19. $-2x^2 + 3xy - 3x^2 + 2xy$
20. $5m^2 - 2mn - 3m^2 + mn$
21. $2c^2 - (5cd + 3c^2) + c^2$
22. $7s^2 - 3(s^2 - st) + st$
23. $2p^2 + 4(q^2 - pq) + 2q^2$
24. $-3(2ab - b^2) - (-2ab)$

Solve.

25. Bob took out a $3,500 loan for 1 year at 12% interest. What was the total cost of the loan?

26. Kara bought a sweater for $24.98 and a pair of jeans for $25.89. How much did she spend if the sales tax is 8%?

27. Susan invested $1,800 for 3 years at 7% interest. How much simple interest did she earn?

6 Polynomials

Polynomial expressions are often used to represent very large or very small numbers. For example, the mass of the earth is approximately 5.9742×10^{24} kg, while the mass of a hydrogen atom is 1.6735×10^{-27} kg.

6.1 Monomials and Multiplying Monomials

Objective: To identify and to multiply monomials

Recall that in a^n where a is any real number and n is any positive integer, a is called the *base* and n is called the *exponent*. The exponent indicates how many times the base is used as a factor.

Capsule Review

Expressions with exponents can be simplified by first rewriting the base in factored form.

EXAMPLE Write in factored form. Then simplify: **a.** 3^3 **b.** $-(-2)^4$

a. $3^3 = 3 \cdot 3 \cdot 3 = 27$ **b.** $-(-2)^4 = -[(-2)(-2)(-2)(-2)] = -[16] = -16$

Write in factored form. Then simplify.

1. 4^2
2. 5^3
3. -2^6
4. $(-8)^2$
5. $-(-10)^3$

In the expression $-2y^4$, -2 is the coefficient (numerical factor) and y^4 is the variable factor. In factored form:

$$-2y^4 = -2 \cdot y \cdot y \cdot y \cdot y \qquad y \text{ is the base.}$$

In an expression such as $(-x)^5$, the exponent 5 applies to the entire quantity within the parentheses, as shown below.

$$(-x)^5 = (-x)(-x)(-x)(-x)(-x) \qquad -x \text{ is the base.}$$

EXAMPLE 1 Write in factored form: **a.** $2^3 x^2$ **b.** $-5a^3 b^4$

a. $2^3 x^2 = 2 \cdot 2 \cdot 2 \cdot x \cdot x$ **b.** $-5a^3 b^4 = -5 \cdot a \cdot a \cdot a \cdot b \cdot b \cdot b \cdot b$

The algebraic expression $-2y^4 + y^2 + \frac{1}{2}yx^3 - 7 + \frac{5}{y^2}$ consists of terms $-2y^4$, y^2, $\frac{1}{2}yx^3$, -7, and $\frac{5}{y^2}$. All these terms except $\frac{5}{y^2}$ are *monomials*.

> A **monomial** is a real number, a variable, or a product of a real number and one or more variables.

Chapter 6 Polynomials

$\dfrac{5}{y^2}$ is *not* a monomial because there is a variable in the denominator.

EXAMPLE 2 Which of these terms are monomials?

    a. $\dfrac{1}{2}x^2yz^2$ **b.** $\dfrac{-13}{2}$ **c.** $\sqrt{7}g^{23}$ **d.** $\dfrac{8a^2b^5}{c}$

    a. $\dfrac{1}{2}x^2yz^2$ Yes   **b.** $\dfrac{-13}{2}$ Yes

    c. $\sqrt{7}g^{23}$ Yes; $\sqrt{7}$ is a real number.

    d. $\dfrac{8a^2b^5}{c}$ No; a variable cannot be in the denominator.

To multiply two monomials with the same base, such as x^4 and x^3, add the exponents.

$$x^4 \cdot x^3 = \underbrace{(x \cdot x \cdot x \cdot x)}_{4 \text{ factors}} \cdot \underbrace{(x \cdot x \cdot x)}_{3 \text{ factors}}$$
$$= x \cdot x \cdot x \cdot x \cdot x \cdot x \cdot x = x^7$$

Property of Exponents for Multiplication

For all real numbers a and all positive integers m and n:

$$a^m \cdot a^n = a^{m+n}$$

In order to use the above property, the monomials must have the same base.

$$m^3 \cdot m^5 = m^{3+5} = m^8 \qquad\qquad 3 \cdot 3^6 = 3^1 \cdot 3^6 = 3^{1+6} = 3^7$$

When you multiply two monomials, use the commutative and associative properties along with the property of exponents for multiplication.

EXAMPLE 3 Simplify: $(2x^2y)(3x^5y^2)$

    $(2x^2y)(3x^5y^2)$
    $= (2 \cdot 3)(x^2 \cdot x^5)(y \cdot y^2)$ *Commutative and associative properties for multiplication*
    $= 6x^7y^3$ *Property of exponents for multiplication*

When some of the factors to be multiplied contain powers of 10, the product may be left in exponential form. Notice the form of the answer in Example 4.

EXAMPLE 4 Multiply and write in exponential form: $(3 \times 10^3)(7 \times 10^9)$

    $(3 \times 10^3)(7 \times 10^9)$
    $= (3 \times 7) \times (10^3 \times 10^9)$ *Commutative and associative properties for multiplication*
    $= 21 \times 10^{12}$ $10^3 \times 10^9 = 10^{3+9} = 10^{12}$

6.1 Monomials and Multiplying Monomials

CLASS EXERCISES

Write in factored form.

1. $7y^3$
2. $3^2 st^5$
3. $-4^2 x^2 z$

Which of these are monomials?

4. $2tu$
5. $\dfrac{3z^2 x}{y}$
6. $\dfrac{-2}{3}$
7. $5ab^3$

Simplify.

8. $a^4 \cdot a^3$
9. $3^3 \cdot 3^6$
10. $(2x^3 y)(2^3 xy)$

For Discussion

11. Explain how the associative property and the commutative property are used in multiplying $(a^2 b)(ab^4)$.

12. Determine whether the property of exponents for multiplication can be used to find the product of x^6 and y^5. Explain your answer. How can you represent the product of x^6 and y^5?

PRACTICE EXERCISES

Write in factored form.

1. $3^3 z^2$
2. $2^3 y^2$
3. $-7a^3 b^4$
4. $-5c^3 d^4$

Which of these are monomials?

5. $\dfrac{1}{2}a^2 bc^2$
6. $\dfrac{2}{3}m^2 np^2$
7. $\dfrac{7x^2 y^5}{z}$
8. $\dfrac{9c^2 d^5}{e}$
9. $\sqrt{5}h^{25}$
10. $\sqrt{3}l^{35}$
11. $\dfrac{-15}{2}$
12. $\dfrac{-17}{3}$

Simplify.

13. $(3x^2 y)(5x^5 y^2)$
14. $(7m^2 n)(2m^5 n^2)$
15. $(6r^3 s)(3r^4 s^5)$
16. $(8a^3 b)(2a^4 b^5)$
17. $(4g^3)(5g^5)$
18. $(7z^3)(3z^5)$

Multiply and write in exponential form.

19. $(3 \times 10^2)(5 \times 10^7)$
20. $(4 \times 10^5)(8 \times 10^8)$
21. $(2 \times 10^5)(1.5 \times 10^4)$
22. $(1.2 \times 10^4)(2 \times 10^6)$
23. $(4.2 \times 10^6)(0.5 \times 10^2)$
24. $(2.1 \times 10^3)(0.2 \times 10^3)$

Simplify. All variable exponents represent positive integers.

25. $k^m \cdot k^4$
26. $h \cdot h^b$
27. $b^{x+1} \cdot b^2$
28. $x^{a+4} \cdot x^3$
29. $2 \cdot 2^{m-4} \cdot 2^3$
30. $3 \cdot 3^{x-5} \cdot 3^2$

Use the power key on your calculator to do the evaluations if $a = -19$ and $b = 171$.

31. a^5
32. b^2
33. a^4
34. b^3

Simplify. All variable exponents represent positive integers.

35. $(2r^{3x})(r^{2x}d^x)$
36. $(e^{3y}f^x)(3e^yf^x)$
37. $(9z^{3a}y^{2m})(-3z^{2a}y^{4m})$

Evaluate each expression if $a = \frac{1}{2}$, $b = -3$, and $c = -1$.

38. $(6a^3)(2^{bc})$
39. $(5b^2)(3^{2a})$
40. $-5^{10a+b-c}$
41. $-3^{b+6a-3c}$
42. $-5^{6a}c^{-b}$
43. $\dfrac{-3b^{-c}}{a^5}$

Applications

Geometry Use the appropriate formula $A = lw$ for a rectangle to express the area.

44. The width of a rectangular mirror is $2a$. Its length is $4a$. Express the area in terms of a.

45. A rectangular picture has a length of x and a width of $\frac{1}{2}x$. Express the area in terms of x.

MATH CLUB ACTIVITY

In recent years, mathematicians have developed academic games that are both instructive and fun to play. Many cities have active math leagues in which competition is based on the ability to play a particular mathematical game.

Wff'n Proof is one such game used in many school competitions. Wff'n Proof is based on modern symbolic logic. It uses a form of reasoning called reverse Polish logic.

1. Find out what is meant by reverse Polish logic and give some examples.

2. Find out if your school has Wff'n Proof. If so, you may enjoy learning to play the game. If you belong to a Math Club, Wff'n Proof may provide an interesting stimulus to club activities.

3. Investigate other possible math club activities, such as written math test competitions sponsored by the Mathematics Association of America and the National Council of Teachers of Mathematics.

6.2 Dividing Monomials

Objectives: To divide monomials
To identify zero exponents

You can simplify a fraction by dividing out common factors that occur in the numerator and the denominator. The same principle can be used to divide monomials.

Capsule Review

To simplify a fraction, write the numerator and denominator in factored form. Then divide out common factors.

EXAMPLE $\dfrac{36}{48} = \dfrac{\cancel{2} \cdot \cancel{2} \cdot \cancel{3} \cdot 3}{\cancel{2} \cdot \cancel{2} \cdot 2 \cdot 2 \cdot \cancel{3}} = \dfrac{3}{2 \cdot 2} = \dfrac{3}{4}$

Simplify.

1. $\dfrac{33}{55}$ 2. $\dfrac{24}{18}$ 3. $\dfrac{30}{75}$ 4. $\dfrac{56}{42}$ 5. $\dfrac{108}{144}$ 6. $\dfrac{78}{52}$

To divide monomials, write each monomial in factored form. Then divide out common factors.

$$\dfrac{x^7}{x^5} = \dfrac{\overbrace{\cancel{x} \cdot \cancel{x} \cdot \cancel{x} \cdot \cancel{x} \cdot \cancel{x} \cdot x \cdot x}^{7 \text{ factors}}}{\underbrace{\cancel{x} \cdot \cancel{x} \cdot \cancel{x} \cdot \cancel{x} \cdot \cancel{x}}_{5 \text{ factors}}} = \dfrac{x^2}{1} = x^2$$

$$\dfrac{5^3}{5^6} = \dfrac{\cancel{5} \cdot \cancel{5} \cdot \cancel{5}}{\cancel{5} \cdot \cancel{5} \cdot \cancel{5} \cdot 5 \cdot 5 \cdot 5} = \dfrac{1}{5 \cdot 5 \cdot 5} = \dfrac{1}{5^3}$$

The examples above suggest the following property of exponents for division.

Property of Exponents for Division

For all real numbers a, $a \neq 0$, and for all positive integers m and n:

If $m > n$,
then $\dfrac{a^m}{a^n} = a^{m-n}$.

If $m < n$,
then $\dfrac{a^m}{a^n} = \dfrac{1}{a^{n-m}}$.

If $m = n$,
then $\dfrac{a^m}{a^n} = a^0 = 1$.

Chapter 6 Polynomials

EXAMPLE 1 Simplify: **a.** $\dfrac{y^6}{y^3}$ **b.** $\dfrac{y^2}{y^7}$ **c.** $\dfrac{y^{16}}{y^{16}}$ Assume that $y \neq 0$.

a. $\dfrac{y^6}{y^3} = y^{6-3}$ **b.** $\dfrac{y^2}{y^7} = \dfrac{1}{y^{7-2}}$ **c.** $\dfrac{y^{16}}{y^{16}} = y^{16-16}$

$\phantom{\dfrac{y^6}{y^3}} = y^3$ $\phantom{\dfrac{y^2}{y^7}} = \dfrac{1}{y^5}$ $\phantom{\dfrac{y^{16}}{y^{16}}} = y^0$

$ = 1$

As with multiplication of monomials, this property of exponents is used only when the bases are the same. The property does not apply to a quotient such as $\dfrac{x^4}{y^4}$, which cannot be simplified further.

EXAMPLE 2 Simplify: $\dfrac{4a^2 b^5}{2a^6 b^2}$ Assume that $a \neq 0$ and $b \neq 0$.

Rewrite $\dfrac{4a^2 b^5}{2a^6 b^2}$ using the rules for multiplication of fractions.

$\dfrac{4a^2 b^5}{2a^6 b^2} = \dfrac{4}{2} \cdot \dfrac{a^2}{a^6} \cdot \dfrac{b^5}{b^2}$ *Use the property of exponents for division.*

$\phantom{\dfrac{4a^2 b^5}{2a^6 b^2}} = 2 \cdot \dfrac{1}{a^4} \cdot b^3$ *Multiply the remaining factors.*

$\phantom{\dfrac{4a^2 b^5}{2a^6 b^2}} = \dfrac{2b^3}{a^4}$

You may want to use the commutative property for multiplication to align factors that have the same base.

EXAMPLE 3 Simplify: $\dfrac{12k^2 m^3 n}{-9 m^3 n^6 k^5}$ Assume that $k \neq 0$, $m \neq 0$, and $n \neq 0$.

$\dfrac{12k^2 m^3 n}{-9 m^3 n^6 k^5} = \dfrac{12 k^2 m^3 n}{-9 k^5 m^3 n^6}$ *Commutative property for multiplication*

$\phantom{\dfrac{12k^2 m^3 n}{-9 m^3 n^6 k^5}} = \dfrac{12}{-9} \cdot \dfrac{k^2}{k^5} \cdot \dfrac{m^3}{m^3} \cdot \dfrac{n^1}{n^6}$ *n means n^1.*

$\phantom{\dfrac{12k^2 m^3 n}{-9 m^3 n^6 k^5}} = \dfrac{4}{-3} \cdot \dfrac{1}{k^3} \cdot 1 \cdot \dfrac{1}{n^5}$

$\phantom{\dfrac{12k^2 m^3 n}{-9 m^3 n^6 k^5}} = \dfrac{4}{-3k^3 n^5}$, or $-\dfrac{4}{3k^3 n^5}$

In Example 4, there are powers of ten in both the numerator and the denominator. It is often convenient to keep the answer as a power of ten.

6.2 Dividing Monomials

EXAMPLE 4 Simplify: $\dfrac{3 \times 10^{12}}{2 \times 10^{6}}$

$$\dfrac{3 \times 10^{12}}{2 \times 10^{6}} = \dfrac{3}{2} \times \dfrac{10^{12}}{10^{6}} = 1.5 \times 10^{6}$$

The third part of the property of exponents for division permits you to simplify an expression with a zero exponent.

EXAMPLE 5 Simplify: a. 5^0 b. $(3x)^0$ c. $-x^0$ Assume that $x \ne 0$.

a. $5^0 = 1$ b. $(3x)^0 = 1$ c. $-x^0 = -1 \cdot x^0 = -1$

Use the power key on your calculator to verify that $5^0 = 1$.

CLASS EXERCISES

Simplify. Assume that no variable equals zero.

1. $\dfrac{x^6}{x^2}$
2. $\dfrac{a^3}{a^7}$
3. $\dfrac{m^8}{m^8}$
4. $\dfrac{3^3}{3^2}$
5. $\dfrac{9}{9^3}$
6. $\dfrac{10^2}{10}$
7. $\dfrac{-2x^2y}{3x^2y}$
8. $\dfrac{3^3 a^2 b^3}{3ab^2 c^4}$

9. In $\dfrac{12k^2 m^3 n}{-9k^5 m^3 n^6}$, why must it be assumed that $k \ne 0$, $m \ne 0$, and $n \ne 0$?

10. In the definition $a^0 = 1$, why must it be stated that $a \ne 0$? (*Hint:* Is the following statement true or false: $\dfrac{0^3}{0^3} = 0^{3-3} = 0^0 = 1$? Why?)

PRACTICE EXERCISES

Simplify. Assume that no variable equals zero.

1. $\dfrac{x^7}{x^4}$
2. $\dfrac{a^9}{a^2}$
3. $\dfrac{y}{y^9}$
4. $\dfrac{i}{i^8}$
5. $\dfrac{b^2}{b^2}$
6. $\dfrac{n^3}{n^3}$
7. $\dfrac{8^2}{8}$
8. $\dfrac{2^4}{2}$
9. $\dfrac{10^3}{10^7}$
10. $\dfrac{10^2}{10^5}$
11. $\dfrac{6a^2 b^5}{2a^6 b^2}$
12. $\dfrac{9x^2 y^5}{3x^6 y^2}$
13. $\dfrac{3ac^5}{7a^3}$
14. $\dfrac{5mn^7}{9m^4}$
15. $\dfrac{4th}{t^2}$
16. $\dfrac{8st}{s^2}$
17. $\dfrac{20 \times 10^7}{5 \times 10^4}$
18. $\dfrac{18 \times 10^6}{9 \times 10^5}$
19. $\dfrac{6 \times 10^4}{2 \times 10^3}$
20. $\dfrac{15 \times 10^7}{3 \times 10^2}$
21. 8^0
22. $(5x)^0$
23. $-t^0$
24. -10^0

25. $\dfrac{m^0}{2}$ 26. $-4b^0$ 27. $3(-x)^0$ 28. $-(-3x)^0$

29. $\dfrac{-3a^5b^7}{6a^4b^8}$ 30. $\dfrac{-2c^6d^4}{4c^2d^5}$ 31. $\dfrac{5x^2y^4}{-3x^3y^2}$ 32. $\dfrac{-4x^3y^5}{7x^4y}$

33. $\dfrac{7x^2y^3z^7}{3x^4z^4}$ 34. $\dfrac{9k^4m^5n}{-3k^3m}$ 35. $\dfrac{-4p^2q^3r^4}{2qrp^2}$ 36. $\dfrac{-6fde}{-4d^2ef^5}$

Simplify. Write each answer in exponential form.

37. $\dfrac{4.2 \times 10^{14}}{3 \times 10^5}$ 38. $\dfrac{1.6 \times 10^{11}}{0.8 \times 10^3}$ 39. $\dfrac{-1.2 \times 10^4}{0.2 \times 10^5}$ 40. $\dfrac{-4.6 \times 10^6}{2.3 \times 10^9}$

Simplify. Assume that no variable equals zero. Exponents are positive integers.

41. $\dfrac{x^a}{x^2}$ $(a > 2)$ 42. $\dfrac{z^c}{z^3}$ $(c < 3)$ 43. $\dfrac{y}{y^{2k}}$ 44. $\dfrac{r^{m-1}}{r}$ $(m > 2)$

Find the value of x in each equation.

45. $\dfrac{3^x}{3^9} = \dfrac{1}{3^3}$ 46. $\dfrac{r}{r^x} = \dfrac{1}{r^5}$ 47. $\dfrac{t^{5-x}}{t^x} = t^3$ 48. $\dfrac{2^{3+x}}{2} = 2^7$

Applications

Geometry The volume of a rectangular prism equals the product of the length, the width, and the height: $V = lwh$.

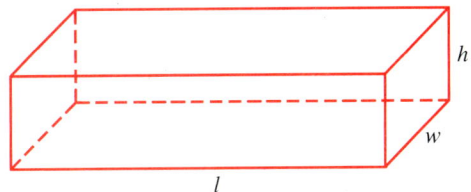

The volume and two dimensions of a rectangular prism are given. Find the third dimension.

49. volume $= x^6$, width $= x$, length $= x^3$ Find the height.

50. volume $= 36x^8$, height $= 2x^2$, length $= 9x^4$ Find the width.

DID YOU KNOW?

In 1939 the physicist J. Robert Oppenheimer demonstrated mathematically that a star could implode. This means that under certain conditions a star could suddenly collapse into a superdense body. The diameter of this body might be only 10^1 miles, and a cubic inch might weigh 10^9 tons.

Suppose a star implodes into a spherical shape with a diameter of 10^1 miles. If a cubic inch weighs 10^9 tons, find the approximate weight of this superdense body. (*Hint:* Use $V = \dfrac{4}{3}\pi r^3$ to find the volume of the sphere in cubic inches. Then find its weight. There are 5280 ft in one mile.)

6.3 Monomials and Exponents

Objectives: To raise a monomial to a power
To raise a quotient of monomials to a power

To simplify $(y^5)^3$, use the definition of exponents and the property of exponents for multiplication ($a^m \cdot a^n = a^{m+n}$).

$$(y^5)^3 = y^5 \cdot y^5 \cdot y^5 = y^{5+5+5} = y^{15}$$
So, $(y^5)^3 = y^{15}$

Capsule Review

Whenever you multiply expressions with the same base add the exponents.

EXAMPLE Simplify: **a.** $x^3 \cdot x^4$ **b.** $3m \cdot 3^2 m^2$

a. $x^3 \cdot x^4 = x^{3+4} = x^7$ **b.** $3m \cdot 3^2 m^2 = 3 \cdot 3^2 \cdot m \cdot m^2 = 3^3 m^3 = 27m^3$

Simplify.

1. -6^2 2. $(-1)^8$ 3. $ab \cdot a^2 b^3$ 4. $(-3a)(-3a)$

You can also simplify an expression such as $(y^5)^3$ by multiplying the exponents.

Property of Exponents

For all real numbers a and for all positive integers m and n:

$$(a^m)^n = a^{mn}$$

EXAMPLE 1 Simplify: $(x^6)^9$

$(x^6)^9 = x^{6 \cdot 9} = x^{54}$ *Multiply the exponents.*

Does $(x^6)^9$ mean the same as $x^6 \cdot x^9$? Explain.

EXAMPLE 2 Simplify: **a.** $-(2^2)^3$ **b.** $[(-2)^2]^3$ **c.** $(-2^2)^3$

a. $-(2^2)^3 = -2^6 = -64$ 2 is the base.
b. $[(-2)^2]^3 = (-2)^6 = 64$ -2 is the base.
c. $(-2^2)^3 = (-4)^3 = -64$ 2 is the base first; then -4 is the base.

Chapter 6 Polynomials

To raise the product uv to the fourth power, write:
$$(uv)^4 = uv \cdot uv \cdot uv \cdot uv = u^4v^4$$

> For all real numbers a and b and for all positive integers m:
> $$(ab)^m = a^m b^m$$

This property of exponents is often used together with other exponent properties.

EXAMPLE 3 Simplify: **a.** $(-7 \times 10^5)^2$ **b.** $(3x^2)^3 \cdot 2x^4$

a. $(-7 \times 10^5)^2 = (-7)^2 \times (10^5)^2$ Raise both factors, -7 and 10^5,
$= 49 \times 10^{10}$ to the 2nd power.

b. $(3x^2)^3 \cdot 2x^4 = 3^3(x^2)^3 \cdot 2x^4$ Raise 3 and x^2 to the 3rd power.
$= 27x^6 \cdot 2x^4$
$= 54x^{10}$ Multiply the two monomials, $27x^6$ and $2x^4$.

Now consider raising a quotient to a power:
$$\left(\frac{u}{v}\right)^5 = \frac{u}{v} \cdot \frac{u}{v} \cdot \frac{u}{v} \cdot \frac{u}{v} \cdot \frac{u}{v} = \frac{u \cdot u \cdot u \cdot u \cdot u}{v \cdot v \cdot v \cdot v \cdot v} = \frac{u^5}{v^5}$$

> For all real numbers a and b, $b \neq 0$, and for all positive integers m:
> $$\left(\frac{a}{b}\right)^m = \frac{a^m}{b^m}$$

EXAMPLE 4 Simplify: $\left(\frac{-2c}{d^2}\right)^4$; $d \neq 0$

$\left(\frac{-2c}{d^2}\right)^4 = \frac{(-2c)^4}{(d^2)^4}$ Raise the numerator and denominator to the 4th power.

$\phantom{\left(\frac{-2c}{d^2}\right)^4} = \frac{(-2)^4 c^4}{(d^2)^4} = \frac{16c^4}{d^8}$ Raise the two factors of the numerator to the 4th power.

CLASS EXERCISES

Simplify. Assume that no variable equals zero.

1. $(-d^2)^3$
2. $[(-d)^2]^3$
3. $-(3^2)^3$
4. $-(3^3)^2$
5. $(3x^2y)^2$
6. $(-4y^2z^5)^3$
7. $\left(\frac{-a}{b^2}\right)^5$
8. $\left(\frac{-m}{b^3}\right)^8$

PRACTICE EXERCISES

Simplify. Assume that no variable equals zero.

1. a. $(x^5)^6$ b. $(z^3)^9$
2. a. $(3^2)^3$ b. $(2^3)^4$
3. $-(10^2)^5$
4. $-(10^3)^7$
5. $[(-3)^2]^2$
6. $[(-2)^3]^3$
7. $\left(\dfrac{-a}{b^2}\right)^5$
8. $\left(\dfrac{-m}{b^3}\right)^8$
9. $(-2 \times 10^3)^4$
10. $(-5 \times 10^2)^3$
11. $(2x^2)^3 \cdot 3x^4$
12. $(3z^3)^2 \cdot 2z^5$
13. $\left(\dfrac{-3a}{b^2}\right)^3$
14. $\left(\dfrac{-2c}{7d}\right)^2$
15. $\left(\dfrac{4y}{5x}\right)^2$
16. $\left(\dfrac{7w}{12y}\right)^2$
17. $\left(\dfrac{-2k^4}{3j^3}\right)^3$
18. $\left(\dfrac{5x^3}{-2y^4}\right)^2$
19. $\left(\dfrac{2b}{18b^3}\right)^2$
20. $\left(\dfrac{15d^5}{3d^2}\right)^2$
21. $(2ab^3)^2$
22. $(3a^4b)^3$
23. $(-2m^3n)(5mn^2)^3$
24. $(-3yz^4)(9yz^3)^2$
25. $\dfrac{(3a^2)^3 a^2}{-9}$
26. $\left(\dfrac{4a^2b}{-3a^5b^2}\right)^2$
27. $\left(\dfrac{-xy}{2x^2y^4}\right)^5$
28. $\dfrac{(-2h^3)^5}{2h^4}$

Simplify. Assume that no variable equals zero. Exponents are positive integers.

29. $(-z^k)^4$
30. $(-y^3)^{2j}$
31. $(-2k^m)^5$
32. $(-3j)^t(-3j^2)^{t+1}$
33. $\left(\dfrac{a^{2m+3}}{a^{m+1}}\right)^2$
34. $\left(\dfrac{b^{2a+1}}{b^a}\right)^4$
35. $[(-2a^2)^3(3b)^2]^2$
36. $[(5c^3)^2(-2d^2)^3]^2$

Applications

Computer The squares of numbers present some very interesting number patterns. Use the program to print out the indicated squares. For $A = 1$ and $B = 20$, what pattern is true for the last digit of the square?

```
10 INPUT "ENTER THE SPAN, A TO B.  ";A,B
20 PRINT : PRINT : PRINT "NUMBER","SQUARE"
30 FOR X = A TO B
40 PRINT X,X * X
50 NEXT X
60 END
```

37. Use the pattern when $A = 46$ and $B = 60$ to write the squares of 46 to 60.

38. **Calculator** How would you solve Exercise 37 using a calculator?

LOGICAL REASONING

1. If 2^{27} is expanded (written without exponents), what will the last digit be?
2. Find a positive integer whose cube exceeds its square by 4624.

6.4 Scientific Notation

Objectives: To write numbers in scientific notation
To multiply and divide numbers written in scientific notation

Many numbers in science and engineering are very large or very small. Scientists use exponents to represent numbers:

$$\text{diameter of Earth} = 420{,}000{,}000 \text{ ft} = 4.2 \times 10^8 \text{ ft}$$
$$\text{diameter of a cell} = 0.000025 \text{ cm} = 2.5 \times 10^{-5} \text{ cm}$$

Capsule Review

All the properties of exponents apply to monomials with base 10.

EXAMPLES   **a.** $10^4 \times 10^2 = 10^6$ **b.** $\dfrac{10^3}{10^{10}} = \dfrac{1}{10^7}$

                c. $(1.5 \times 10^5)^2 = 1.5^2 \times 10^{10}$

Simplify.

1. $10^3 \times 10^5$ 2. $(13 \times 10^7)(3 \times 10^2)$ 3. $(4 \times 10^2)^3$ 4. $\dfrac{44 \times 10^8}{1.1 \times 10^4}$

Until now, you have used only nonnegative exponents. The following pattern suggests a definition for *negative exponents*.

$100 \div 10 = 10$	$10 \div 10 = 1$	$1 \div 10 = \dfrac{1}{10}$	$\dfrac{1}{10} \div 10 = \dfrac{1}{100}$
↑ ↑	↑ ↑	↑ ↑	↑ ↑
10^2 10^1	10^1 10^0	10^0 10^{-1}	10^{-1} 10^{-2}

Notice that $10^{-1} = \dfrac{1}{10^1}$, $10^{-2} = \dfrac{1}{10^2}$, and so on.

> For all real numbers a, $a \neq 0$, and for all positive integers n:
> $$a^{-n} = \dfrac{1}{a^n}$$

Using this definition, you need only one rule for dividing powers.

$$\dfrac{y^4}{y^6} = y^{4-6} = y^{-2}, \text{ or } \dfrac{1}{y^2}$$

Negative as well as positive exponents are used in *scientific notation*.

> A number is written in **scientific notation** when it is expressed in the form $n \times 10^m$, where n is a real number such that $1 \leq n < 10$ and m is an integer.

EXAMPLE 1 Is the number written in scientific notation? If not, explain.
  a. 3.459×10^{-3} b. 4×2^4 c. 18.7×10^3

  a. Yes   **b.** No; 2^4 is not a power of ten.   **c.** No; 18.7 is greater than ten.

EXAMPLE 2 Write 18,459 in scientific notation.

$$\begin{aligned}&18{,}459\\=\ &1.8459 \times 10^4\end{aligned}$$ Move the decimal point 4 places to the left. Then multiply by 10^4, or 10,000.

So, $18{,}459 = 1.8459 \times 10^4$. To check, multiply $1.8459 \times 10{,}000$.

EXAMPLE 3 Write 0.00987 in scientific notation.

$$\begin{aligned}&0.00987\\=\ &0.009.87 \times 10^{-3}\end{aligned}$$ Move the decimal point 3 places to the right. Then multiply by 10^{-3} or $\frac{1}{10^3}$.

So, $0.00987 = 9.87 \times 10^{-3}$. How would you check?

EXAMPLE 4 Simplify by mental computation: $(5.87 \times 10^{12})(1 \times 10^6)$. Give your answer in scientific notation.

$(5.87 \times 10^{12})(1 \times 10^6) = (5.87 \times 1)(10^{12} \times 10^6) = 5.87 \times 10^{18}$

EXAMPLE 5 Simplify by mental computation: $\dfrac{2.4 \times 10^8}{6 \times 10^4}$. Give your answer in scientific notation.

$\dfrac{2.4 \times 10^8}{6 \times 10^4} = \dfrac{2.4}{6} \times \dfrac{10^8}{10^4} = 0.4 \times 10^4 = 4 \times 10^{-1} \times 10^4 = 4 \times 10^3$

CLASS EXERCISES

Write each number in scientific notation.

1. 100,000   2. 0.00105   3. 72834.5   4. 6.2

Chapter 6 Polynomials

PRACTICE EXERCISES

Is the number written in scientific notation? If not, explain.

1. 27×10^{36}
2. 4×10^{-17}
3. 0.009×10^3
4. 0.7×10^{83}

Write each number in scientific notation.

5. 2,000,000
6. 8,000,000
7. 0.00765
8. 0.00438

Simplify. Write the answer in scientific notation.

9. $(3.98 \times 10^2)(1 \times 10^7)$
10. $(4.34 \times 10^5)(2 \times 10^8)$
11. $(5.23 \times 10^4)(4 \times 10^6)$
12. $(7.12 \times 10^3)(3 \times 10^2)$
13. $\dfrac{1.8 \times 10^{12}}{2 \times 10^5}$
14. $\dfrac{4.5 \times 10^{13}}{5 \times 10^6}$
15. $\dfrac{3.6 \times 10^{15}}{6 \times 10^5}$
16. $\dfrac{4.9 \times 10^9}{7 \times 10^3}$
17. $\dfrac{(2 \times 10^3)(4 \times 10^8)}{6 \times 10^2}$
18. $\dfrac{(1.5 \times 10^{11})(2 \times 10^{15})}{3 \times 10^9}$
19. $(-1.5 \times 10^3)^2(1.6 \times 10^2)^3$
20. $(1.2 \times 10^2)^3(-2.4 \times 10^3)^2$
21. $\left(\dfrac{6.4 \times 10^3}{8 \times 10}\right)^2(0.5 \times 10^2)$
22. $\left(\dfrac{4.8 \times 10^{12}}{1.6 \times 10^3}\right)^3(9 \times 10^3)$

Applications

23. **Computer** A computer can perform 8×10^8 arithmetic operations per second. How many operations is that per minute? per day? per year?

TEST YOURSELF

Is the expression a monomial? If not, tell why not.   **6.1**

1. $3x^2 + y^2$
2. $-\dfrac{3}{4}a^4$
3. m^2n
4. $\dfrac{-5r^2}{s^3}$

Simplify. Assume that no variable equals zero.   **6.2, 6.3**

5. $-4x(3x^3)$
6. $2^4 a^3 b^2 \cdot 2a^2 b$
7. -3^4
8. $\dfrac{a^{11}}{a^7}$
9. $\dfrac{3xy^4}{-1.5x^4y^5}$
10. $4s^0$

Simplify if necessary. Then write each answer in scientific notation.   **6.4**

11. 0.000312
12. 1,938,500
13. $\dfrac{23}{100,000}$

6.4 Scientific Notation

APPLICATION:
Fluid Motion

The mathematician John von Neumann played an important role in the development of the computer as we know it today. His special interest in fluid dynamics was a key factor in his work in developing a computer. Due to the complexity of the equations used to represent fluid motion, the calculations were extremely tedious and time consuming. This led to von Neumann's design and construction of a machine that served as a model for today's computers. He built the machine in the late 1940s.

Second-generation computers measured calculations in milliseconds.

$$1 \text{ millisecond} = \frac{1}{1000} \text{ second} = 10^{-3} \text{ of a second}$$

Third-generation computers measured calculations in nanoseconds. One nanosecond is the time that electricity takes to travel 1 foot.

$$1 \text{ nanosecond} = \frac{1}{1,000,000,000} \text{ second} = 10^{-9} \text{ of a second}$$

Today's fastest computers measure calculations in picoseconds.

$$1 \text{ picosecond} = \frac{1}{1,000,000,000,000} \text{ second} = 10^{-12} \text{ of a second}$$

Computers print these numbers in their own form of scientific notation called exponential notation. For example, 6.84273×10^7 would appear on the computer as 6.84273E+7 and 9.51×10^{-6} as 9.51E−6. The number after the E corresponds to the exponent when a number is in scientific notation.

This program allows you to utilize the speed of the computer to calculate about how many molecules of water flow over Niagara Falls in a given time.

```
10 LET D = 840000000000
20 LET M = 17000000000000000000000
30 PRINT "M = MOLECULES PER DROP    D = DROPS
      PER MINUTE": PRINT : PRINT
40 INPUT "HOW MANY MINUTES? ";X: PRINT : PRINT
50 PRINT "IN ";X;" MINUTES, ABOUT ";M * D * X;
   " MOLECULES OF WATER FLOW OVER NIAGARA
   FALLS." : END
```

Solve.

Run the program to determine the molecules that flow in:

1. 1 minute
2. 15 minutes
3. 30 minutes
4. 60 minutes

5. If the weight of water in a lake is 1.58×10^{18} tons, after how much time would the number of molecules flowing over the falls be approximately equal to the number of tons?

6.5 Polynomials

Objectives: To identify polynomials, binomials, and trinomials, and the degree of a monomial and polynomial
To simplify polynomials by combining like terms and writing them in ascending or descending order of exponents

Each of the following is an example of a monomial or term:

| y | variable | $5x^2$ | product of a constant and a variable |
| -16 | numeral or constant | $9ab^2c^3$ | product of a constant and several variables |

Capsule Review

You have been shown how to combine like terms using the distributive property:

a. $x + 3x$
$= 1 \cdot x + 3 \cdot x$
$= (1 + 3)x$
$= 4x$

b. $\frac{1}{2}mn^2 - 6mn^2 + mn = \left(\frac{1}{2} - 6\right)mn^2 + mn$
$= \left(\frac{1}{2} - \frac{12}{2}\right)mn^2 + mn = -\frac{11}{2}mn^2 + mn$

Simplify, if possible, by combining like terms.

1. $a + 4a$ **2.** $-2x^3 - \frac{7x^3}{3}$ **3.** $x^2y - 7x^2y^2$ **4.** $6s^3t^2 - 6s^3t^2$

The **degree of a monomial** is the sum of the exponents of its variables. If the monomial is a nonzero constant, the degree is 0. The constant 0 has no degree.

monomial	x	$4ab^2$	m^3n^5	$-\frac{7}{8}$	gh	0
degree	1	3	8	0	2	no degree

EXAMPLE 1 State the degree: **a.** $\frac{1}{2}x$ **b.** $8a^2b^5$ **c.** 6

Monomial	Degree	Reason
a. $\frac{1}{2}x$	1	$\frac{1}{2}x = \frac{1}{2}x^1$. Exponent is 1.
b. $8a^2b^5$	7	Sum of exponents (2 + 5) is 7.
c. 6	0	The degree of a nonzero constant is 0.

6.5 Polynomials   **241**

A **polynomial** is a monomial or a sum or difference of monomials. The following are examples of polynomials.

$-3x^2 \qquad \frac{7}{8}a + 2b^3 \qquad -y^4 + 5y - 11 \qquad 5x^3y + x^2 + 2y^4 - 19$

The monomials that make up a polynomial are called its terms. Polynomials of two or three terms are used so often they have special names.

monomial	one term	$0.006t$	mn	$-5x^3y^4z^2$
binomial	two terms	$3x + 2$	$3m + 5n$	$x + y$
trinomial	three terms	$3x^2 + 2x - 1$	$a^2 + 2ab + b^2$	$x + y + z$

The **degree of a polynomial** is the *highest* degree of any of its terms after it has been simplified.

The degree of $6x^3 + 4x^2 + 7$ is 3.

EXAMPLE 2 State the degree of $2x^2y^2 + \frac{2}{3}x^2y + 8xy^2 - 5$.

$$2x^2y^2 + \frac{2}{3}x^2y + 8xy^2 - 5$$
$$\downarrow \qquad \downarrow \qquad \downarrow \qquad \downarrow$$
$$4 \qquad 3 \qquad 3 \qquad 0$$

Add the exponents in each monomial. The highest degree is 4.

So, the degree of the polynomial $2x^2y^2 + \frac{2}{3}x^2y + 8xy^2 - 5$ is 4.

The polynomial $x^5 + 2x^3 - 7x^2 + 6x - 11$ is said to be arranged in **descending order,** since the degree of each term is lower than that of the preceding terms. In **ascending order,** the polynomial is written as $-11 + 6x - 7x^2 + 2x^3 + x^5$.

If a polynomial has more than one variable, it can be written in descending or ascending order of the exponents with respect to one variable. For example, $6ab^4 + 10a^2b^2 - 2a^3b + 3a^4$ is in descending order with respect to b.

To simplify a polynomial, combine like terms. Remember that like terms contain the same variables with the same exponents.

EXAMPLE 3 Simplify: $-2x^2y^2 - 3x^3y + 8x^2y^2 + 5xy^3$. Write the result in descending order of the exponents with respect to y.

$$-2x^2y^2 - 3x^3y + 8x^2y^2 + 5xy^3$$
$$= (-2x^2y^2 + 8x^2y^2) - 3x^3y + 5xy^3 \qquad \text{\textit{Group like terms.}}$$
$$= 6x^2y^2 - 3x^3y + 5xy^3 \qquad \text{\textit{Combine like terms.}}$$
$$= 5xy^3 + 6x^2y^2 - 3x^3y \qquad \text{\textit{Write in descending order.}}$$

EXAMPLE 4 Simplify: $6(x^3 - 3) - 5(3x^3 + 2x^2 + 1) + 25$. Write the result in ascending order of the exponents.

$$6(x^3 - 3) - 5(3x^3 + 2x^2 + 1) + 25$$
$$= 6x^3 + 6(-3) - 5(3x^3) - 5(2x^2) - 5(1) + 25 \quad \text{Distributive property}$$
$$= 6x^3 - 18 - 15x^3 - 10x^2 - 5 + 25$$
$$= -9x^3 - 10x^2 + 2 \quad \text{Combine like terms.}$$
$$= 2 - 10x^2 - 9x^3 \quad \text{Ascending order}$$

Note that the degree of the polynomial is 3.

CLASS EXERCISES

State the degree.

1. $3x$
2. $-12xy^3z^5$
3. $\frac{-7}{8}$
4. $wxyz$
5. $0.1h$
6. $\frac{3y}{4} - 7$
7. $2a^2 + \frac{4a}{5} - 8$
8. $4c^3 + 3c^2d^2 - 8cd + \frac{1}{2}d^3$
9. $3x^3 - 2x^2y^2 + 18x^2y + 3x^3$
10. $-2x^2y^3 - xy^4 + x^2y^2 + xy^4 + x^3y^2$

Simplify. Write the result in descending order of the exponents with respect to a.

11. $2a + 8a - a^2$
12. $2ab^2 + (-5ab^2)$
13. $\frac{1}{3}a^3c - 3a^3c + a$
14. $\frac{1}{2}a^4 - 2(a^5 - 2a^3 + 1) + a^2$

PRACTICE EXERCISES

State the degree.

1. $\frac{3}{4}x$
2. $\frac{7}{8}z$
3. $7m^2n^5$
4. $12r^3s^6$
5. 9
6. 18
7. $5x^2y^2 + \frac{1}{2}x^2y^2 + 6xy - 3$
8. $4r^2s^2 + \frac{5}{7}r^2s^2 + 10rs - 5$
9. $xy + 3x^2y^2 - \frac{4}{5}x^2y - 9x^3y^2$
10. $ab + 4a^2b^2 - \frac{1}{9}a^2b - 3a^3b^2$

Simplify. Write the result in descending order of the exponents with respect to x.

11. $-3x^2y^2 - 4x^3y + 9x^2y^2 + 2xy^3$
12. $-7x^2y^2 - 5x^3y + 10x^2y^2 + 9xy^3$
13. $5x^2y^2 + 2xy^3 - 4xy^3$
14. $-2x^3y + 6x^3y - 2xy^3$
15. $3(x^3 - 5) - 5(4x^3 + 5x^2 + 1) + 30$
16. $5(x^3 - 6) - 7(3x^3 + 3x^2 + 1) + 15$
17. $4x^2z + \frac{9}{8}xz^2 - 3xz^2$
18. $0.5xy^2 - 0.7x^2y + 0.4xy^2 + 0.7x^2y$
19. $-5x^2z + \frac{2}{5}x^2z + 2xz^2$
20. $0.2xy^2 - 0.6x^2y - 0.3xy^2 + 0.1xy^2$

Simplify. Write the result in descending order of the exponents of the underlined variable.

21. $\frac{1}{2}c\underline{d}e - \frac{1}{2}(c\underline{d}^2 + c\underline{d}e)$

22. $f\underline{g} + \frac{1}{3}(\underline{g}^2 - 3f\underline{g}) + \frac{1}{3}\underline{g}^2$

23. $4\underline{x}^2 + 3(\underline{x}^2 - 3) - 2(4 - 7\underline{x}^2)$

24. $a\underline{b}^2 - 4(a + \underline{b}) + 2(a\underline{b}^2 - \underline{b})$

25. $4(2x\underline{y}^2 + 3x^2\underline{y}) - 2(2\underline{y}x^2 + 5\underline{y}^2x)$

26. $6(x\underline{z}^3 + x^2\underline{z}) - 7(\underline{z}^2x - 2x^3\underline{z})$

Simplify. Write the result in descending order of the exponents with respect to x. Assume that a is a positive integer.

27. $x^a - x^{3a} + x^a + x^{2a}$

28. $x^{2a} + x^a - x^{4a} - x^{2a}$

29. $-2x^a + 3x^{2a} + x^a - 2x^{2a}$

30. $x^{2a} - 2x^a + 3x^a + x^a$

Simplify. Then evaluate if $x = -2$, $y = \frac{1}{2}$, and $a = 2$.

31. $3xy^a + 2x^ay - 3x^ay + y^a$

32. $x^ay^a - 2y^a + x^ay^a - 6y$

Applications

Geometry Find the perimeter of each figure.

33.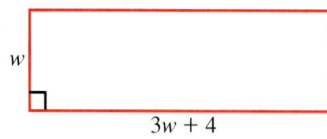

34.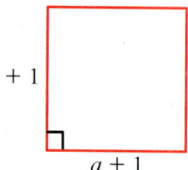

35.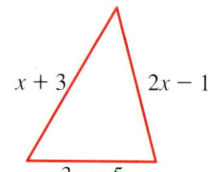

CRITICAL THINKING: Classifying

Your first task in classifying items is to decide on a characteristic or characteristics of interest. Then you must decide which item(s) contain the given characteristic.

Polynomials can be classified by these characteristics: (a) number of terms, (b) degree, and (c) number of variables.

$x^2 + y^2 \qquad 3a^2 + 2ab + b^2 \qquad 5xy^2 + 3 \qquad 0 \qquad y^3 - m^3 \qquad r^3$

Classify the polynomials above by listing the ones that belong to each of the following categories.

1. polynomials with two terms
2. second-degree polynomials
3. polynomials in two variables
4. third-degree binomials in two variables
5. List two other ways that polynomials may be classified.

6.6 Adding and Subtracting Polynomials

Objectives: To add polynomials
To subtract polynomials

You have used the distributive property to combine the like terms (monomials) of a polynomial. To *combine* like terms means to *add* or *subtract* like terms.

$$4x + 6y + 3x - 3y$$
$$= (4x + 3x) + (6y - 3y)$$
$$= (4 + 3)x + (6 - 3)y$$
$$= 7x + 3y$$

Capsule Review

When combining terms express the result in descending order of the exponents of one of the variables.

EXAMPLE
$$7m - 3m^2 - 2m^4 - 5 - 4m^4 + 3m^3 - 7m^2 + 1$$
$$= (-2m^4 - 4m^4) + 3m^3 + (-3m^2 - 7m^2) + 7m + (-5 + 1)$$
$$= -6m^4 + 3m^3 - 10m^2 + 7m - 4$$

Simplify.

1. $a^2 + 4a + 3a^2$
2. $2m - 18m$
3. $-2x^3 - 7x^2$
4. $m^3n - 7m^2n^2 + 5m^2n^2 - 3m$
5. $\frac{1}{2}s - \frac{1}{3}s^2 + s - s^2$

To add polynomials, add the like terms of the polynomials. The addition can be shown horizontally or vertically.

EXAMPLE 1 Add $3x^3 + x^2 - 2x - 5$ and $x^2 - 5x + 7$.

Add horizontally

$$(3x^3 + x^2 - 2x - 5) + (x^2 - 5x + 7)$$
$$= 3x^3 + (x^2 + x^2) + (-2x - 5x) + (-5 + 7) \quad \text{Group like terms.}$$
$$= 3x^3 + (1 + 1)x^2 + (-2 - 5)x + (-5 + 7) \quad \text{Use the distributive property.}$$
$$= 3x^3 + 2x^2 - 7x + 2$$

Add vertically

$$\begin{array}{r} 3x^3 + x^2 - 2x - 5 \\ x^2 - 5x + 7 \\ \hline 3x^3 + 2x^2 - 7x + 2 \end{array}$$

Align like terms vertically.
Add like terms in each column.

6.6 Adding and Subtracting Polynomials   **245**

To subtract a polynomial, add the opposite of each term of the polynomial being subtracted and then simplify.

EXAMPLE 2 Subtract $x^2 - 5x + 7$ from $3x^3 + x^2 - 2x - 5$.

$(3x^3 + x^2 - 2x - 5) - (x^2 - 5x + 7)$
$= 3x^3 + x^2 - 2x - 5 + (-x^2) + 5x + (-7)$ Add the opposite of each term.
$= 3x^3 + 3x - 12$ Combine like terms.

EXAMPLE 3 Subtract: $(5a^3 - a^2 - 6) - (4a^3 - a - 9)$

Align like terms. Add the opposite of each term subtracted.

$\begin{array}{l} 5a^3 - a^2 \phantom{{}-a} - 6 \\ 4a^3 \phantom{{}-a^2} - a - 9 \end{array}$ ← opposites → $\begin{array}{l} 5a^3 - a^2 \phantom{{}+a} - 6 \\ -4a^3 \phantom{{}-a^2} + a + 9 \\ \hline a^3 - a^2 + a + 3 \end{array}$

CLASS EXERCISES

Add.

1. $11r + 3s + 2t$
   $r + 2s - t$

2. $5a + 4b$
   $-3a + b$

3. $3d - 6e + 7f$
   $d + 4e - 4f$

Add or subtract, as indicated.

4. $(2x + 5) + (3x - 2)$
5. $(4c^2 + 3c) + (c^2 + 6c)$
6. $(3h + 2k + 4) + (7h - k + 3)$
7. $(9m^2 + 16m - 3) + (2m^2 + 3m - 1)$
8. $(4c^2 + 7c) - (3c^2 + 5c)$
9. $(3k - d) - (2k - 3d)$
10. $(-11a^2 + 2a - 1) - (7a^2 + 4a - 1)$
11. $(6r^3 - 3r + 7) - (6r^3 - 3r + 7)$

PRACTICE EXERCISES

Add.

1. $5x^3 + x^2 - 3x - 7$
   $x^2 - 6x + 9$

2. $8x^3 + x^2 - 5x - 9$
   $x^2 - 8x + 11$

3. $7r + 2s - 2t$
   $4s - 3t$

4. $8j - 3k + 6m$
   $-2j + 3m$

5. $2x^3 + 6x^2 + x$ and $3x^2 - 2x + 2$
6. $3x^3 + 7x^2 + 6x$ and $2x^3 + 3x$

7–12. For Exercises 1–6, subtract the second polynomial from the first polynomial.

Add or subtract, as indicated.

13. $(5x^3 + x^2 - 3x - 8) - (x^2 - 7x + 9)$
14. $(6x^3 + 3x^2 - x - 6) - (x^2 - 8x + 9)$
15. $(3m^3 - m^7 + 6) - (2m^3 - m^7 - 9)$
16. $(7n^3 - n^2 - 8) - (6n^3 - n^2 + 10)$
17. $(3c^3 - 3c) + (2c^3 - c)$
18. $(8x^2 + 3) + (7x^2 + 10)$
19. $(7a + 3b - c) + (6a - 2b + 3c)$
20. $(x^2 + 2x + 1) + (2x^2 - 3x + 4)$
21. $(9r + 3s + 8) + (3s + 2)$
22. $(4x + 10y + 7) + (x + 2y)$

Subtract the first polynomial from the second polynomial.

23. $(2j + k - 2m); (3j + 2k + 4m)$
24. $(4a - 3b + 2c); (6a + 5b - 3c)$
25. $(2x^2 - 2x + 5); (3x^2 + 3x + 5)$
26. $(5a^2 - 11a - 3); (7a^2 - 11a + 7)$
27. $(6x^3 + 4x^2 + 3x); (3x^4 - 7x^3 - 3x^2)$
28. $(4z^3 + 3z^2 + 4); (-7z^3 + 11z^2 + 12z)$

Add or subtract, as indicated.

29. $(11a^2 + 3a + 8) + (6a^2 - 2a + 4) - (3a^2 + 5)$
30. $(3y^2 - 7y + 9) + (2y^2 + 8y - 4) - (2y - 3)$
31. $(17x^3 + 3x^2 + 4x) - (7x^3 - 2x + 4) - (4x^2 - 3)$
32. $(4z^3 + 7z^2 + 8z) - (2z^3 - 11z + 3) - (2z^2 + 4)$

Applications

33. **Number Problem** If n is an integer, write an expression to represent the sum of n and the next three consecutive integers. Simplify.

34. **Geometry** The length of a rectangle is 4 cm greater than its width w. Write an expression to represent the perimeter. Simplify.

ALGEBRA IN HEALTH

A scientist observed that the growth of a bacteria over t hours could be approximated by the polynomial $0.0035t^3 - 0.15t^2 + 1.5t + 10$. Use this program to print out a chart for times of 0 h to 40 h.

```
10 PRINT "TIME","BACTERIA GROWTH"
20 FOR T = 0 TO 40
30 PRINT T,0.0035 * T ^ 3 - 0.15 *
    T ^ 2 + 1.5 * T + 10
40 NEXT T
```

At what times does the bacteria start to decay? To grow?

6.7 Multiplying a Polynomial by a Monomial

Objectives: To multiply a polynomial by a monomial
To simplify algebraic expressions that involve multiplication of a polynomial by a monomial

The distributive property provides you with a method for multiplying a polynomial by any monomial. The monomial may be a constant or it may have variable factors. The polynomial may have any number of terms.

Capsule Review

EXAMPLE $3(m^3 - 7m^2 + m - 1) = 3(m^3) + 3(-7m^2) + 3(m) + 3(-1)$
$= 3m^3 - 21m^2 + 3m - 3$

Simplify. Use the distributive property.

1. $3(m^2 - 7m + 3)$
2. $-3(4n^2 + 5p^2)$
3. $\frac{1}{2}(a + 2b)$
4. $0.2(x^2 + 4x - 6)$
5. $2(-6x^3 + x^2 - 0.3x + 0.1)$
6. $-\frac{2}{3}(12y^2 - 9y + 18)$

To multiply a polynomial by a monomial, you will often use the property of exponents for multiplication: $a^m \cdot a^n = a^{m+n}$.

EXAMPLE 1 Simplify: $-3x^2(x^3 - 5x^2 + 7x - 1)$

Horizontal format

$-3x^2(x^3 - 5x^2 + 7x - 1)$
$= (-3x^2)(x^3) + (-3x^2)(-5x^2) + (-3x^2)(7x) + (-3x^2)(-1)$
$= -3x^5 + 15x^4 - 21x^3 + 3x^2$

Vertical format

$$\begin{array}{r} x^3 - 5x^2 + 7x - 1 \\ -3x^2 \\ \hline -3x^5 + 15x^4 - 21x^3 + 3x^2 \end{array}$$

Multiply each term in the top row by $-3x^2$.

In Example 2, the parentheses are removed by multiplying each term of each polynomial by a monomial factor. Then the like terms are combined.

EXAMPLE 2 Simplify: $-5xy^2(3xy - 2x^2) + 4x^2y(y^2 - 2xy + 7)$

$$-5xy^2(3xy - 2x^2) + 4x^2y(y^2 - 2xy + 7)$$
$$= [(-5xy^2)(3xy) + (-5xy^2)(-2x^2)] + [(4x^2y)(y^2) + (4x^2y)(-2xy) + (4x^2y)(7)]$$
$$= -15x^2y^3 + 10x^3y^2 + 4x^2y^3 - 8x^3y^2 + 28x^2y$$
$$= (-15x^2y^3 + 4x^2y^3) + (10x^3y^2 - 8x^3y^2) + 28x^2y \quad \text{Group like terms.}$$
$$= -11x^2y^3 + 2x^3y^2 + 28x^2y \quad \text{Combine like terms.}$$

CLASS EXERCISES

Multiply.

1. $3x - 2y$
   $\underline{4y}$

2. $9r + 3s + 2t$
   $\underline{2rst}$

3. $-3ab(2a - 4b)$
4. $-3d^2f(2d - 4e + 6f)$
5. $2x(3x - 2)$
6. $-3c(2c + 6)$
7. $(3c - 5)(-5c)$
8. $(-2k - 3)(-3k)$
9. $-11a^2b^3(7a^2 + 3b + c)$
10. $-8rs^4(-6r^3 - 2s^2 + rs)$

PRACTICE EXERCISES

Multiply and simplify, where possible.

1. $-5x^2(x^3 - 6x^2 + 8x - 5)$
2. $-2x^2(x^3 - 7x^2 + 9x - 8)$
3. $2p^3(-3p^4 + 2p^2 - 5)$
4. $6t^3(-t^3 - 7t - 4)$
5. $3p^2q(4p + 2q - r)$
6. $2a^2b(3a + 4b - 5c)$
7. $-2x^2(-x^3 + x - 3)$
8. $-10y^3(y^4 - y^2 + 2)$
9. $5mn^2(3m^3 - 2n^2 - mn + p)$
10. $7xy^2(2y^3 - 3y^2 - xy + z)$
11. $7hk(7h - k + 8)$
12. $5mn(2m + 3n - 5)$
13. $-2ab^2(3ab - 2a^2) + 5a^2b(b^2 - 3ab + 8)$
14. $-4xy^2(5xy - 3x^2) + 3x^2y(y^2 - 5xy + 9)$
15. $2y(-y + 1) + 5(y - 2)$
16. $3x(2 - 3x) + 4(3x - 1)$
17. $2m(1 - 3m^2) + 3m^2(4m - 2)$
18. $5g^2(3 - 2g) + g(g + 3g^2)$
19. $(-3ab^2 + 2a^2b)(-2ab)$
20. $(-6d^2e^2 + 7de)(-d^2e^2)$
21. $5rs^4(3r^3 - 2s^2r + s)$
22. $6a^2b(7a^2 + 3ab - b)$
23. $3abc^3(-4a^2c + 2b^2c^3 - ab)$
24. $8x^2yz(3x^2y + 7xz^2 - yz)$
25. $-7m(m^2 - m) + 2m^2(n + m)$
26. $-9t^2(r^3 - s) + 3t(tr^3 + 5ts)$

6.7 Multiplying a Polynomial by a Monomial

Simplify. All variable exponents represent nonnegative integers.

27. $4a^x(6a^x - 2a^2 + 4)$

28. $2y^b(7y^b + 3y^3 - 8)$

Applications

29. **Number Problem** If j is an odd integer, write an expression to represent the product of j and the next consecutive odd integer. Simplify.

30. **Number Problem** If e is an even integer, write an expression to represent the product of e and the *preceding* consecutive even integer. Simplify.

31. **Geometry** The length of a rectangle is 2 cm greater than its width w. Write an expression to represent the area. Simplify.

32. **Geometry** The length of a rectangle is 5 in. greater than twice its width w. Write an expression to represent the area. Simplify.

33. **Geometry** The width of a rectangle is 10 cm less than three times its length l. Write an expression to represent the area.

EXTRA

The 4-digit base 10 numeral 3134_{10} can be represented as a polynomial in x with $x = 10$, as shown below.

$$3134_{10} = 3 \cdot 10^3 + 1 \cdot 10^2 + 3 \cdot 10 + 4$$

The polynomial is $3x^3 + x^2 + 3x + 4$.

The 4-digit base 5 numeral 3134_5 can also be represented as a polynomial in x with $x = 10$.

$$\begin{aligned} 3134_5 &= 3 \cdot 5^3 + 1 \cdot 5^2 + 3 \cdot 5 + 4 \\ &= 3 \cdot 125 + 1 \cdot 25 + 3 \cdot 5 + 4 \\ &= 375 + 25 + 15 + 4 \\ &= 419 \end{aligned}$$

The polynomial is $4x^2 + x + 9$.

Express these numerals as polynomials in x with $x = 10$.

1. 301_{10}
2. 1092_{10}
3. 4123_5
4. 100101_2
5. 5482_9
6. 11111_2

6.8 Multiplying Polynomials

Objectives: To multiply two binomials
To multiply any two polynomials using both vertical and horizontal formats

The distributive property can be used to multiply two binomials such as $(x + y)(a + b)$.

Capsule Review

EXAMPLE $3a(a + 2) + 4a(2a - 5) = 3a^2 + 6a + 8a^2 - 20a = 11a^2 - 14a$

Simplify by using the distributive property and combining like terms.

1. $5z(z + 7) - 2(z + 7)$
2. $6y(y^2 - y) + 4(y + y^2)$
3. $2ab(a - b) + b(ab + a)$
4. $(2x - 3)4x + (2x - 3)9$

To multiply two binomials such as $(x + y)(a + b)$, think of $(x + y)$ as one factor. Distribute $(x + y)$ over both terms of $(a + b)$. Then use the distributive property a second time.

$$(x + y)(a + b) = (x + y)a + (x + y)b$$
$$= xa + ya + xb + yb$$

The four terms of the product are xa, ya, xb, and yb. Each term is the product of one term from the first factor, $(x + y)$, and one term from the second factor, $(a + b)$. Try to relate these terms to the steps in the **FOIL method** for multiplying two binomials, shown below.

FOIL Method for Multiplying Two Binomials

To multiply two binomials, find:

F   The product of the two **FIRST** terms

O   The product of the two **OUTSIDE** terms

I   The product of the two **INSIDE** terms

L   The product of the two **LAST** terms

$(x + y)(a + b)$

$xa + xb + ya + yb$
 F   O   I   L

If the outer and inner products (O and I) are like terms, they can be combined.

EXAMPLE 1 Use the FOIL method to multiply $(x + 3)(x + 7)$.

$$
\begin{array}{l}
(x + 3)(x + 7) = \overset{F}{x \cdot x} + \overset{O}{x \cdot 7} + \overset{I}{3 \cdot x} + \overset{L}{3 \cdot 7} \\
 = x^2 + 7x + 3x + 21 \\
 = x^2 + 10x + 21
\end{array}
$$

Multiplying polynomials that involve subtraction may be easier if you think of subtracting as adding the opposite.

EXAMPLE 2 Multiply: $(3a + b)(a - 2b)$

$$
\begin{aligned}
&(3a + b)(a - 2b) \\
&= (3a + b)[a + (-2b)] \quad \text{$a - 2b$ means $a + (-2b)$.} \\
&= \overset{F}{3a \cdot a} + \overset{O}{3a \cdot (-2b)} + \overset{I}{b \cdot a} + \overset{L}{b \cdot (-2b)} \\
&= 3a^2 + (-6ab) + ab + (-2b^2) \\
&= 3a^2 - 5ab - 2b^2
\end{aligned}
$$

When one or both of the polynomials have more than two terms, arrange the terms in descending order with respect to the degree of a variable. Then multiply.

Compare the horizontal and the vertical formats in the next example.

EXAMPLE 3 Multiply: $(-2y + y^2 - 3y^3)(4y - 5)$

$$
\begin{aligned}
&(-2y + y^2 - 3y^3)(4y - 5) \\
&= (-3y^3 + y^2 - 2y)(4y - 5) \quad \text{Arrange in descending order.} \\
&= [-3y^3 + y^2 + (-2y)][4y + (-5)] \quad \text{Rewrite the subtractions as additions.}
\end{aligned}
$$

Horizontal format

$$
\begin{aligned}
&[-3y^3 + y^2 + (-2y)][4y + (-5)] \\
&= [-3y^3 + y^2 + (-2y)] \cdot 4y + [-3y^3 + y^2 + (-2y)] \cdot (-5) \quad \text{Distributive property} \\
&= [-3y^3 \cdot 4y + y^2 \cdot 4y + (-2y) \cdot 4y] + [-3y^3 \cdot (-5) + y^2 \cdot (-5) + (-2y) \cdot (-5)] \\
&= [-12y^4 + 4y^3 - 8y^2] + [15y^3 - 5y^2 + 10y] \\
&= -12y^4 + 19y^3 - 13y^2 + 10y \quad \text{Combine like terms.} \\
& \quad \text{Arrange in descending order.}
\end{aligned}
$$

Vertical format

$$
\begin{array}{r}
-3y^3 + y^2 + (-2y) \\
4y + (-5) \\
\hline
15y^3 - 5y^2 + 10y \\
-12y^4 + 4y^3 - 8y^2 \\
\hline
-12y^4 + 19y^3 - 13y^2 + 10y
\end{array}
$$

Multiply by -5.
Multiply by $4y$. Align like terms.
Add.

In Example 3, does one format seem more convenient? Does one format make you more aware of the sign of each term?

CLASS EXERCISES

Multiply.

1. $(x + 1)(x - 5)$
2. $(y + 2)(y - 4)$
3. $(c + d)(e + f)$
4. $(a - b)(c - d)$
5. $(3c + 4)(2c - 5)$
6. $(g - 5)(3g + 7)$
7. $(x + 3)(-2x + x^2 - 3)$
8. $(z + 2)(-7z + z^2 - 6)$
9. $(2h - 3)^2$ means $(2h - 3)(2h - 3)$. Write $(2h - 3)^2$ as a polynomial with three terms.
10. Write $(3x - 2)(x + 3)$ as a polynomial with three terms.
11. Write $(2x - 4)(2x + 4)$ as a polynomial with two terms.

For Discussion

12. Use the vertical format to multiply $(x + y)(a + b)$. Are your results equivalent to those of the FOIL method? Explain.
13. Without changing the order of terms in the first factor, use FOIL to multiply $(1 - 2y)(3y + 4)$. Why would it have been better to rewrite the first factor in descending order of the exponents before multiplying?
14. When multiplying two binomials, when will the product be a trinomial?

PRACTICE EXERCISES

Multiply.

1. $(a + 5)(a - 8)$
2. $(x + 5)(x - 6)$
3. $(h + 1)(h - 7)$
4. $(m + 4)(m - 2)$
5. $(a + 5)(3a - 4)$
6. $(p + 6)(4p - 3)$
7. $(2m - 9)(6 + 5m)$
8. $(3n - 2)(5 + 6n)$
9. $(7r - 3)(r - 5)$
10. $(3j - 7)(j - 2)$
11. $(5x - 8)(2x - 3)$
12. $(2y - 5)(7y - 3)$
13. $(4y + z)(y - 2z)$
14. $(5b + d)(b - 2d)$
15. $(2j - 3k)(4j + k)$
16. $(p - q)(2p + 3q)$
17. $(3b + 7)(3 - 2b)$
18. $(4a + 5)(2 - 4a)$

6.8 Multiplying Polynomials

19. $(-2y + y^2 - 4y^3)(6y - 3)$
20. $(-5x + x^2 - 3x^3)(4x - 6)$
21. $(4z + z^2 - 5z^3)(2z + 3)$
22. $(8m + 3m^2 - 2m^3)(m + 5)$
23. $(x^2 + 3x - 5)(x + 2)$
24. $(y^2 + 11y - 12)(y + 5)$
25. $(z - 3)(z^2 - 5z - 12)$
26. $(w - 4)(w^2 - 5w + 1)$
27. $(c^2 + 2c + 1)(c - 5)$
28. $(d^2 - 2d + 3)(d + 2)$
29. $(3 + k)(7k - k^2 + 8)$
30. $(5 + n)(3n + 2 - n^2)$
31. $(y - 3)(-2y + 1 + y^2)$
32. $(x + 2)(-1 + x^2 + 3x)$
33. $(3x - 4)^2$
34. $(2y - 7)^2$
35. $(2a - 3b)^2$
36. $(3p - 4q)^2$
37. $(2a^2 - a + 5)(3a + 2)$
38. $(4b^2 + b - 2)(5b + 3)$
39. $(2c + 1)(-c^2 + 5c - 2)$
40. $(3d + 2)(-d^2 - 3d + 1)$
41. $(5g^2 + 3 - 2g)(2g - 1)$
42. $(4 + 2h^2 - 5h)(3h - 1)$
43. $(x - 5)(3x - x^2 + x^3)$
44. $(y - 3)(2y^2 - y^3 + y)$
45. $(2y^3 + 2y - y^2)(2y - 3)$
46. $(-2x^2 + 3x^3 - x)(3x - 4)$
47. $(t^2 + 2t)(5t^2 - 6t - 2)$
48. $(2v^2 + v)(3v^2 - 2v - 3)$
49. $(a + b)(a^2 - ab + b^2)$
50. $(c - d)(c^2 + cd + d^2)$
51. $(y + 3)(y^3 - 2y^2 + 3)$
52. $-4(x - 3)(x + 2)$
53. $(x + 3)(x + 3)(x + 3)$
54. $(x - 2)(x - 2)(x - 2)$
55. $(4b - 1)^3$
56. $(3d - 1)^3$

Hint: In Exercises 57–60, some powers of the variable do not appear in the polynomials. In $x^3 - x + 2$, there is no x^2 term. In $3x^2 - 1$, there is no x-term. When you multiply such polynomials vertically, be sure that like terms are aligned in the columns.

57. $(x^3 - x + 2)(3x^2 - 1)$
58. $(5y^3 - y)(y^4 + y^2 - 1)$
59. $(2a^3 - a^2 + 3)(5a^2 - 2a + 1)$
60. $(3b^2 + b - 2)(4b^3 - 2b^2 - 6)$

Applications

61. **Number Problem** If *n* is an integer, write an expression to represent the product of the next two consecutive integers. Simplify.

62. **Geometry** The width of a rectangle is $(x + 2)$ cm. Its length is 3 cm greater than its width. Write an expression to represent the area. Simplify.

63. **Geometry** The length of a rectangle is 5 in. greater than twice its width, $(x + 1)$ in. Write an expression to represent the area. Simplify.

64. Geometry One rectangle is $(2x + 1)$ units wide and $(x + 4)$ units long. A second rectangle has a width of $(2x - 9)$ units and length of $(x + 9)$ units. Which has the greater area? How much greater?

65. Number Problem If e is an even integer, write an expression to represent the product of the next two consecutive even integers. Simplify.

ALGEBRA IN SPACE TECHNOLOGY

Radio signals travel so rapidly, approximately 1.86×10^5 mi/s, that on Earth the signals seem to be received the instant they are sent. You have to wait, however, before hearing any message from space. For example, sunspot activity causes radio signals to be emitted. How many seconds does it take one such signal to reach Earth when Earth is at its nearest point to the Sun?

Distance from Sun (mi)

Planet	Maximum	Minimum
Mercury	4.34×10^7	2.86×10^7
Venus	6.77×10^7	6.68×10^7
Earth	9.46×10^7	9.14×10^7
Mars	1.55×10^8	1.29×10^8
Jupiter	5.07×10^8	4.61×10^8
Saturn	9.37×10^8	8.38×10^8
Uranus	1.86×10^9	1.67×10^9
Neptune	2.82×10^9	2.76×10^9
Pluto	4.55×10^9	2.76×10^9

Distance = rate × time, or $d = rt$.
Solving for t, you have $t = \dfrac{d}{r}$.

$d = 9.14 \times 10^7$ mi and $r = 1.86 \times 10^5$ mi/s

$$t = \frac{9.14 \times 10^7}{1.86 \times 10^5}$$
$$= 4.91 \times 10^2, \text{ or } 491 \text{ seconds}$$

Find the number of seconds it takes for a radio signal to travel from the Sun to each of the planets when the planet is (a) at its maximum distance from the Sun, and (b) at its minimum distance from the Sun. Use a calculator.

6.9 Multiplying Polynomials: Special Cases

Objectives: To find the square of a binomial
To find the product of the sum and the difference of two terms

To square a number or an algebraic expression means to use it as a factor two times. Suppose the number is positive or the algebraic expression represents a positive quantity. Then the number or expression could represent the length of a side of a square, as shown below.

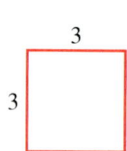

Area = 3 · 3, or 3^2

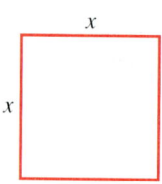

Area = $x · x$, or x^2

Area = $(x + y)(x + y)$, or $(x + y)^2$

Capsule Review

EXAMPLE a. $(b^3)^2 = b^6$ b. $(-3x^2y)^2 = (-3)^2(x^2)^2y^2$
$= 9x^4y^2$

Find the square of each number or expression.

1. -12
2. $3r$
3. $-5m$
4. $7p^2$
5. $\dfrac{a}{2}$
6. $\dfrac{2x^4}{3}$
7. $\dfrac{1}{3}s^4t$
8. $6st^2u$
9. $0.2xy$
10. $-1.2gh^5$
11. $\dfrac{2}{5}d^2e^2$
12. $0.1a^5b^3c^2$

Notice what happens when you square a binomial that is a sum of two terms.

$(x + y)^2 = (x + y)(x + y) = x · x + x · y + y · x + y · y$ FOIL
$= x^2 + xy + yx + y^2$
$= x^2 + 2xy + y^2$

There is a change when the binomial is a difference of two terms. Use FOIL.

$(x - y)^2 = (x - y)(x - y) = x · x + x · (-y) + (-y) · x + (-y) · (-y)$
$= x^2 + (-xy) + (-yx) + y^2$
$= x^2 - 2xy + y^2$

> **Square of a Binomial: $(m + n)^2$ or $(m - n)^2$**
>
> - Square the first term: m^2
> - Double the product of the two terms: $+2mn$ for $(m + n)^2$
> $-2mn$ for $(m - n)^2$
> - Square the last term: n^2
> - Write the sum of the three terms: $(m + n)^2 = m^2 + 2mn + n^2$
> $(m - n)^2 = m^2 - 2mn + n^2$

EXAMPLE 1 Simplify: $(2a + 3b)^2$

$$(2a + 3b)^2 = (2a)^2 + 2(2a \cdot 3b) + (3b)^2$$
$$= 4a^2 + 12ab + 9b^2$$

EXAMPLE 2 Simplify: $(3m^2 - 11)^2$

$$(3m^2 - 11)^2 = (3m^2)^2 + 2[3m^2 \cdot (-11)] + (-11)^2$$
$$= 9m^4 - 66m^2 + 121$$

If you forget the rule for squaring a binomial, you can rewrite the square as a product of two binomials and apply the FOIL method.

Example 3 shows how squaring a binomial can apply to *mental arithmetic*.

EXAMPLE 3 Simplify: 39^2

$$39^2 = (40 - 1)^2$$
$$= 40^2 - 2(40 \cdot 1) + 1^2$$
$$= 1600 - 80 + 1 = 1521$$

Another special case of binomial multiplication is a product of the form $(x + y)(x - y)$.

$$(x + y)(x - y) = x \cdot x + x \cdot (-y) + yx + y \cdot (-y)$$
$$= x^2 + (-xy) + yx + (-y^2)$$
$$= x^2 - y^2$$

> **Product of the Sum and Difference of the Same Two Terms: $(m + n)(m - n)$**
>
> - Square the first term: m^2
> - Square the last term: n^2
> - Write the difference of the two squares: $(m + n)(m - n) = m^2 - n^2$

EXAMPLE 4 Simplify: **a.** $(z + 3)(z - 3)$ **b.** $(2r^3 + 5)(2r^3 - 5)$

a. $(z + 3)(z - 3) = (z)^2 - (3)^2$
$= z^2 - 9$

b. $(2r^3 + 5)(2r^3 - 5) = (2r^3)^2 - (5)^2$
$= 4r^6 - 25$

CLASS EXERCISES

Simplify.

1. $(c + d)^2$
2. $(r + 3s)^2$
3. $(m - n)^2$
4. $(2r - s)^2$
5. $(u + v)(u - v)$
6. $(7m + 1)(7m - 1)$
7. $(62)^2$
8. 89^2
9. $(30 + 3)^2$
10. $(70 - 1)(70 + 1)$
11. $(3x^3 - 4y^2)(3x^3 + 4y^2)$
12. $(5x^3 - 2y^2)(5x^3 + 2y^2)$
13. $(3m^2 + 16m)(3m^2 - 16m)$
14. $(10x^4 - 15x^2)(10x^4 + 15x^2)$
15. $(4y^5 + 3y^4)(4y^5 - 3y^4)$
16. $(16a^2b^2 - 4ab)(16a^2b^2 + 4ab)$

For Discussion

17. How would you use the procedure shown in Example 3, on the previous page, to simplify 102^2?
18. Would it be useful to use $(47 - 2)^2$ to simplify 45^2? Explain.
19. True or false? The square of a binomial is always a trinomial.

PRACTICE EXERCISES

Simplify.

1. $(3x + 2y)^2$
2. $(5a + 3b)^2$
3. $(8m + 2n)^2$
4. $(4j + 6k)^2$
5. $(4m^2 - 6)^2$
6. $(6r^2 - 4)^2$
7. $(7s^2 - 3)^2$
8. $(3j^2 - 7)^2$
9. $(24)^2$
10. $(45)^2$
11. $(12)^2$
12. $(52)^2$
13. $(x + 4)(x - 4)$
14. $(h + 9)(h - 9)$
15. $(y + 5)(y - 5)$
16. $(m + 2)(m - 2)$
17. $(4x^3 - 3)(4x^3 + 3)$
18. $(5b^3 - 8)(5b^3 + 8)$
19. $(2t - u)^2$
20. $(3j - h)^2$
21. $(d^2 + e^2)^2$
22. $(f^2 + g^2)^2$
23. $(h^2 - j^2)(h^2 + j^2)$
24. $(k^2 - m^2)(k^2 + m^2)$
25. $(1 - 12g^3h^2)^2$
26. $(2 - 6x^3y^2)^2$
27. $(5ab^3 + 6c^2d^4)^2$
28. $(3ef^3 + 4g^2h^4)^2$
29. $(61x^3y^2z)^2$
30. $(82a^4b^0c^5)^2$

Chapter 6 Polynomials

Simplify and express as a polynomial. All variable exponents represent nonnegative integers.

31. $(4 - a^x)(4 + a^x)$
32. $(y^b + 2)(y^b - 2)$
33. $(x^{a+1} + x)^2$
34. $(a^{x+2} + 2a)^2$
35. $(3^{2y+1} - 2)^2$
36. $(5^{1-3x} - 4)^2$

Applications

Geometry Find the area.

37. square with side $a + 1$

38. square with side $b + 3$

39. rectangle with width $2x + 3$ and length $2x - 3$

40. rectangle with width $4m + 3$ and length $4m - 3$

TEST YOURSELF

Simplify. Write the result in descending order of the exponents with respect to x.   6.5

1. $4x^3 + x - x^3 - 5x$
2. $-5x^2y + x^3y^3 + 9x + 3x^2y$

Add or subtract, as indicated.   6.6

3. $(5y^2 + 6) + (-3y^2 - 10)$
4. $(a^3 + 5a^2 - 6) - (-4a^2 - a + 9)$

Multiply. The multiplication may be done in horizontal or in vertical form.   6.7

5. $(-2y)(-3y^2 + y)$
6. $(4x^3 + 2x^2 - 1)(-3x^4)$
7. $(11a^2b)(5a^3b^2 - 3a^2b^3 + 7ab^4)$
8. $-3t(1 - 2t^2) + 8(5t - 4)$

Simplify.   6.8, 6.9

9. $(3r + 7)(r + 11)$
10. $(1 - 5j)(3 + 4j)$
11. $(8n - 15)(3n + 5)$
12. $(z - 6)(-2z + 3z^2 + 9)$
13. $(9y + 3)^2$
14. $(2x + 3)(2x - 3)$
15. $(5a - 9b)^2$
16. $(b^4 - 11a)(b^4 + 11a)$

6.10 Problem Solving Strategy: Look for a Pattern

The numbers 1, 1, 2, 3, 5, 8, 13, 21, . . . are arranged in a definite pattern. Each number is the sum of the two preceding numbers. When numbers are arranged in a pattern, they form a **sequence**. Finding a pattern is a problem solving strategy. If a pattern exists and is identified, then the pattern can be used to solve the problem.

EXAMPLE 1 Write an algebraic expression that represents the sum of the first n odd positive integers.

Understand the Problem

Read the problem.

What are the given facts?
The odd positive integers 1, 3, 5, 7, 9, ...

What are you asked to find?
An algebraic expression that represents the sum of any number n of odd positive integers.

Plan Your Approach

Choose a strategy.

Positive Odd Integers	Sums
1	1 = 1
1, 3	1 + 3 = 4
1, 3, 5	1 + 3 + 5 = 9
1, 3, 5, 7	1 + 3 + 5 + 7 = 16
1, 3, 5, 7, 9	1 + 3 + 5 + 7 + 9 = 25
1, 3, 5, 7, 9, 11	1 + 3 + 5 + 7 + 9 + 11 = 36
1, 3, 5, 7, 9, 11, 13	1 + 3 + 5 + 7 + 9 + 11 + 13 = 49

Complete the Work

Find the pattern.

Number of Odd Integers	Sums
1 (1)	$1 = 1^2$
2 (1, 3)	$4 = 2^2$
3 (1, 3, 5)	$9 = 3^2$
4 (1, 3, 5, 7)	$16 = 4^2$
5 (1, 3, 5, 7, 9)	$25 = 5^2$
6 (1, 3, 5, 7, 9, 11)	$36 = 6^2$
7 (1, 3, 5, 7, 9, 11, 13)	$49 = 7^2$

Interpret the Results	**State your conclusion.** A pattern does exist. For any number, n, of positive odd integers, the pattern suggests that the sum is n^2.

Check your conclusion.
Suppose there are eight positive odd integers. According to the pattern, the sum of the first eight positive odd integers should be 8^2, or 64.
$$1 + 3 + 5 + 7 + 9 + 11 + 13 + 15 = 64 \checkmark$$

Recall the sequence; 1, 1, 2, 3, 5, 8, 13, 21, . . . which was presented at the beginning of the lesson. This sequence is known as the **Fibonacci Sequence.**

The following numbers follow the same pattern as the Fibonacci sequence. They are called a *Fibonacci-like* sequence.

5, 9, 14, 23, 37, 60, 97, 157, 254, 411, 665, 1076, 1741, 2817, 4558

EXAMPLE 2 Given fifteen consecutive numbers of any Fibonacci-like sequence, show that the sum of the first thirteen of them is equal to the fifteenth number minus the second number.

Understand the Problem
You are asked to show that the sum of the first thirteen numbers of the given sequence is equal to the fifteenth number minus the second number.

Plan Your Approach
In this problem, the pattern is given: any number in the sequence is the sum of the two preceding numbers.

Let a represent the first number in the sequence.

Let b represent the second number in the sequence.

Write the first fifteen numbers in the sequence using a and b.

a
b
$a + b$
$a + 2b$
$2a + 3b$
$3a + 5b$
$5a + 8b$
$8a + 13b$
$13a + 21b$
$21a + 34b$
$34a + 55b$
$55a + 89b$
$89a + 144b$
$144a + 233b$
$233a + 377b$

Complete the Work

Sum of First Thirteen Numbers

$233a + 376b$

Difference of Fifteenth and Second Numbers

$(233a + 377b) - b$

Interpret the Results
Is $233a + 376b$ equal to $(233a + 377b) - b$?
$(233a + 376b) \stackrel{?}{=} (233a + 377b) - b$
$233a + 376b = 233a + 376b \checkmark$

CLASS EXERCISES

Look for a pattern to find the next three numbers in each sequence.

1. 1, 1, 2, 3, 5, 8, 13, 21, _?_, _?_, _?_
2. 2, 4, 6, 8, 10, 12, 14, _?_, _?_, _?_
3. 3, 7, 10, 17, 27, 44, 71, _?_, _?_, _?_
4. 1, 2, 3, 5, 8, 13, 21, 34, _?_, _?_, _?_

PRACTICE EXERCISES

Look for a pattern to find the next three numbers in each sequence.

1. 10, 100, 1000, 10,000, _?_, _?_, _?_
2. 2, 6, 18, 54, 162, _?_, _?_, _?_
3. 0, 3, 8, 15, 24, _?_, _?_, _?_
4. Find the sum of these thirteen Fibonacci-like numbers: 5, 9, 14, 23, 37, 60, 97, 157, 254, 411, 665, 1076, 1741.
5. Find the fifteenth number in the sequence given in Exercise 4.
6. In Exercise 4, is the fifteenth number in the sequence minus the second number equal to the sum of the first thirteen numbers?
7. Write a Fibonacci-like sequence that starts with the numbers 6, 12.
8. Find the sum of the first thirteen numbers in your sequence for Exercise 7. Is the fifteenth number minus the second number equal to the sum of the first thirteen numbers?
9. Use a pattern to find an algebraic expression that can be used to find the sum of the first n positive integers. *Hint:* Use the product of n and $n + 1$.

   $1 + 2 = 3$
   $1 + 2 + 3 = 6$
   $1 + 2 + 3 + 4 = 10$
   $1 + 2 + 3 + 4 + 5 = 15$
   $1 + 2 + 3 + 4 + 5 + 6 = 21 \ldots$ and so on.

10. Starting with two 2's, construct a sequence using the same rule used in the Fibonacci sequence.
11. In Exercise 10, how does the eighth number in the sequence compare with the eighth number in the Fibonacci sequence?
12. Use a pattern and algebraic notation to show that the sum of the first ten consecutive numbers in a Fibonacci-like sequence is 11 times the seventh number in the sequence.

13. Use your calculator to determine which ten consecutive terms of the Fibonacci sequence have a sum of 45,991.

14. The figure below is called Pascal's triangle. The first seven rows and diagonals are shown. Look for patterns to find numbers in the eighth row.

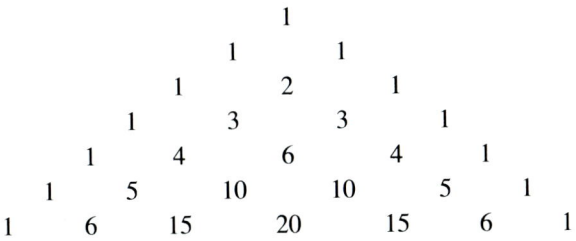

15. Can you discover the Fibonacci sequence using Pascal's triangle? It is hard to spot, but it is there. *Hint:* Draw a series of parallel sloping diagonals so that the first just starts at the top 1 and so on.

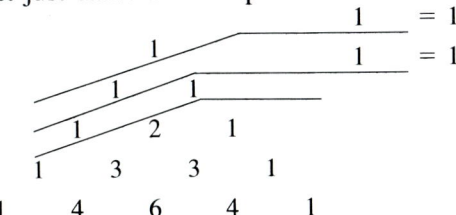

Mixed Problem Solving Review

1. Two jet planes leave Chicago at 2 PM. One travels north at 850 km/h and the other south at 750 km/h. At what time will they be 4000 km apart?

2. There are three tests given during the school term. Lauren received grades of 83 and 74 on the first two tests. What grade must Lauren get on the last test in order to have an average for the term of no less than 80?

PROJECT

A **prime number** is divisible only by itself and 1. For example, the first ten prime numbers are 2, 3, 5, 7, 11, 13, 17, 19, 23, and 29. Observe a pattern in the following numbers and write a *conjecture* about them. A **conjecture** is a guess that something is true. Try a few more examples of your own. Do you think your conjecture is true?

$$4 = 2 + 2 \qquad 16 = 3 + 13 \qquad 6 = 3 + 3 \qquad 18 = 7 + 11$$
$$8 = 3 + 5 \qquad 20 = 7 + 13 \qquad 10 = 5 + 5 \qquad 22 = 11 + 11$$
$$12 = 5 + 7 \qquad 24 = 5 + 19 \qquad 14 = 7 + 7 \qquad 26 = 3 + 23$$

CHAPTER 6 SUMMARY AND REVIEW

Vocabulary

ascending order of a polynomial (242)
binomial (242)
degree of a monomial (241)
degree of a polynomial (242)
descending order of a polynomial (242)
Fibonacci Sequence (261)

FOIL method (251)
monomial (226)
polynomial (242)
scientific notation (238)
sequence (260)
trinomial (242)

Properties of Exponents For all real numbers a and b, $a \neq 0$, and all positive integers m and n: **6.1–6.3**

$a^m \cdot a^n = a^{m+n}$ $(a^m)^n = a^{mn}$ $(ab)^m = a^m b^m$

$\dfrac{a^m}{a^n} = a^{m-n}$, if $m > n$ $\dfrac{a^m}{a^n} = \dfrac{1}{a^{n-m}}$, if $m < n$ $\dfrac{a^m}{a^n} = a^0 = 1$, if $m = n$

Simplify.

1. $a^3 \cdot a^4$
2. $3 \cdot 3^2$
3. $(-3ac^2)(5a^3c^3)$

4. Multiply and write in exponential form: $(-3 \times 10^2)(2.5 \times 10^5)$

Simplify. Assume that no variable equals zero.

5. $\dfrac{x^5}{x^3}$
6. $\dfrac{y^2}{y^8}$
7. $\dfrac{x^{12}}{x^{12}}$
8. $\dfrac{12x^3y^4}{-9x^2y^9}$

Simplify. Write each answer in exponential form.

9. $\dfrac{3.9 \times 10^5}{1.3 \times 10}$
10. $\dfrac{7.5 \times 10^5}{2.5 \times 10^6}$

Simplify. Assume that no variable equals zero.

11. y^0
12. $(-3)^0$
13. $\left(\dfrac{5a}{c^4}\right)^2$
14. $\dfrac{-3x^2yz^7}{9yx^3z^4}$

Scientific Notation A number is in scientific notation when it is expressed in the form $n \times 10^m$ where n is a real number such that $1 \leq n < 10$ and m is an integer. **6.4**

Simplify. Write the answer in scientific notation.

15. 289,000
16. 0.041
17. $(3.2 \times 10^3)(5 \times 10^2)$
18. $(4.2 \times 10^4)(3 \times 10^2)$
19. $\dfrac{8.1 \times 10^8}{9 \times 10^4}$
20. $\dfrac{2.4 \times 10^9}{8 \times 10^6}$

Chapter 6 Polynomials

State the degree of each monomial term and the degree of the polynomial. 6.5

21. $3x$　　　　　　**22.** $-8a^3b^5 + 2a^2b + 4$

Simplify. Write the result in descending order of the exponent with respect to x.

23. $5 + x^2 + 3x^3 - 2x^2$　　**24.** $-2x^3y^2 + 3x^4y - xy - x^3y^2 + y^3 - 4y^3$

Adding and Subtracting Polynomials　To find the sum of polynomials, add their like terms. To subtract one polynomial from another, add the opposite of each term of the polynomial you are subtracting.　6.6

Add or subtract, as indicated.

25. $\quad 3x - 2y$　　　　**26.** $\quad 10r - 3s + 2t$　　**27.** $(2y + 3) - (3y - 2)$
$\quad\underline{+ 6x + 4y}$　　　　　　$\quad\underline{-\ \ \ r + 2s - t}$

28. $(-9a^3 + 6a^2 - 3) - (5a^2 + a - 3)$

Multiply. Simplify if possible.　6.7

29. $3x(x^2 + 2y - 1)$　　**30.** $-2a^5(a^2 - 3a + 5)$　　**31.** $(-7xy + y^2)5x^2y$

The FOIL Method of Multiplying Two Binomials　6.8

$$(x + y)(a + b) = xa + xb + ya + yb$$
$$\qquad\qquad\qquad\quad\ \ \text{F}\ \ \ \ \text{O}\ \ \ \ \text{I}\ \ \ \ \text{L}$$

Multiply.

32. $(2a + 4)(a + 7)$　　**33.** $(b - 3)(3b - 8)$　　**34.** $(c + 4d)(5c - 2d)$

Special Cases of Polynomial Multiplication　　$(m + n)^2 = (m + n)(m + n) = m^2 + 2mn + n^2$
$(m - n)^2 = (m - n)(m - n) = m^2 - 2mn + n^2$
$(m + n)(m - n) = m^2 - n^2$　6.9

Simplify.

35. $(d - 9)^2$　　**36.** $(3z^2 + 7)^2$　　**37.** $(5x - 4)(5x + 4)$

Look for a pattern to find the next three numbers in each sequence.　6.10

38. $2, 8, 32, 128,$ ____, ____, ____

39. $0, 4, 9, 15, 22,$ ____, ____, ____

CHAPTER 6 TEST

Simplify. Assume that no variable equals zero.

1. $3^3 \cdot 3^2$
2. $\dfrac{c}{c^7}$
3. $(8m)^0$
4. $-9b^0$
5. $(2^3)^2$
6. $(-4xy^2)(-6x^3y^5)$
7. $\dfrac{14c^4d^3e^2}{7c^5de^2}$
8. $\dfrac{(-3a)^3}{2}$
9. $\left(\dfrac{5b^2c^3}{10b^4}\right)^4$

Rewrite in scientific notation any number that is not already in scientific notation. Simplify first if necessary.

10. 9.9×10^{-5}
11. 5100
12. $\dfrac{125 \times 10^5}{25 \times 10^2}$
13. 1.034
14. $(2.4 \times 10)(5 \times 10^3)$
15. $(0.2 \times 10^2)^3$

State the degree of each monomial and the degree of the polynomial.

16. $-\dfrac{5}{8}$
17. $4t^3 + 7t^2v^2 - tv^3 - 6v^5$

Simplify. Write the result in descending order of the exponent of y.

18. $-3y^2 + 2y - 9 + y^2 - y^3$
19. $5x^2y - 3(xy^2 + y^3 - 2x^2y) + x^3$

Add or subtract, as indicated.

20. $x - 9y$
    $+ 3x + 6y$

21. $4a - 3b + 2c$
    $-a - 7b + 5c$

22. $(2x^2 - x + 7) + (x^2 + 2x - 3)$
23. $(2c^4 + c^2 - c) - (c^2 + 2c)$

Simplify.

24. $2x(x - 3y + z)$
25. $(2b^2 + 3b - 1)(-3b^3)$
26. $(k + 3)(2k + 5)$
27. $(4y - 3)^2$
28. $(5r^2 - 1)(5r^2 + 1)$
29. $(7c^2 - 2c + 3)(c^2 + 6)$

Challenge

Simplify: $(3x^3 - 2y^2)^3$

PREPARING FOR STANDARDIZED TESTS

Select the best choice for each question.

1. Which fraction is exactly halfway between $\frac{1}{6}$ and $\frac{3}{2}$?
   A. $\frac{4}{5}$ B. $\frac{5}{6}$ C. $\frac{4}{7}$
   D. $\frac{5}{7}$ E. $\frac{5}{9}$

2. When $n = -3$ and $s = -1$, which of the following would be largest?
   A. $13s^2$ B. $5ns$
   C. $n^2 + s^2$ D. $2n^2$
   E. $-4(n + s)$

3. Find the value of $\frac{2}{3}$ of the quantity (60% of 150).
   A. 60 B. 75 C. 90
   D. 120 E. 190

4. When $3xy - 2xz + 5yz$ is subtracted from $5xy + 3yz$, the result is:
   A. $-2xy - 2xz + 2yz$
   B. $2xy - 2xz - 2yz$
   C. $2xy + 2xz - 2yz$
   D. $2xy - 2xz + 2yz$
   E. $2xy + 2xz + 2yz$

5. The number 7,130,000, when written in scientific notation, becomes:
   A. 7.13×10^6
   B. 71.3×10^6
   C. 7.13×10^5
   D. 71.3×10^5
   E. 7.13×10^4

6. When a $20 bill is used to pay for items costing $4.50, $7.95, $1.98, 79¢, and 5¢, what is the change?
   A. $4.18 B. $4.73 C. $5.18
   D. $5.73 E. $5.63

7. The solution set for $2(x + 3) > x + 4$ is:
   A. $\{x|x > 10\}$
   B. $\{x|x > 2\}$
   C. $\{x|x > 0\}$
   D. $\{x|x > -2\}$
   E. $\{x|x > -10\}$

8. What percent of 480 is 2.4?
   A. 50% B. 5% C. 1%
   D. $\frac{1}{2}$% E. $\frac{1}{4}$%

9. Solve the equation: $2(x - 1) = 3(x + 2) - 8$
   A. 4 B. 2 C. 0
   D. -2 E. -4

10. $(2x + 5y)(x - 3y)$ equals:
    A. $2x^2 + xy + 15y^2$
    B. $2x^2 - xy + 15y^2$
    C. $2x^2 - 15y^2$
    D. $2x^2 + xy - 15y^2$
    E. $2x^2 - xy - 15y^2$

11. $(-2a^2b^3c)(7ab^2c^4) =$
    A. $-14a^2b^5c^4$
    B. $-14a^3b^5c^5$
    C. $-14a^2b^6c^4$
    D. $-14a^3b^5c^4$
    E. None of the above

12. One day Mr. Prompt left home at 7:47 AM and drove to his office. If he arrived there at 8:39 AM, how long did it take for the drive to work?
    A. 52 min
    B. 58 min
    C. 62 min
    D. 68 min
    E. 1 h 32 min

Preparing for Standardized Tests

CUMULATIVE REVIEW (CHAPTERS 1–6)

Perform the indicated operation.

1. $0.1 \times (-5)$
2. $8.6 - (0.6)$
3. $-2.1 \div (-0.3)$
4. $-5.2 + 3.1$
5. $(2x - 3) - (7x - 4)$
6. $(8h^2 - 6) + (7 - 5h^2)$
7. $(9m^2 + 8) - (m + 6)$
8. $3y(y^2 - 2y + 4)$
9. $-2b^2(-2b + 3a - 1)$
10. $(g + 2)(g - 3)$
11. $(2n - 1)(3n - 4)$
12. $(p + 1)(p^2 + 3p - 2)$

Simplify. Assume that no variable equals zero.

13. $3 + 2 \cdot 6$
14. $[2 - (-2)^3]^2$
15. $|6 - 9| + 5$
16. $3a^2 + 5a - 7a^2$
17. $-(3xy) - (xy)$
18. $d^5 \cdot d^3$
19. $(3n^2)(2n^3)$
20. $(s^2)(st^2)$
21. $(2p^2)^3$
22. $\dfrac{b^2}{b^3}$
23. $\dfrac{3m^4}{12m^2}$
24. $\dfrac{(-9k^5)^2}{18k^3}$

Solve each equation or inequality.

25. $-5 = r + 9$
26. $c - 8 > 1$
27. $-3m = 24$
28. $-2p < 12$
29. $\dfrac{x}{2} + 1 = 11$
30. $|3a - 5| = 7$
31. $3(t - 2) = 9$
32. $5d + 7 \geq 2$
33. $0.5n - 1 = 4.5$
34. $3y + 8 = 7y - 4$
35. $b - (2b - 3) + (5b - 2) = 17$
36. $f \geq 1$ and $f < 3$
37. $z < -2$ or $z \geq 0$

Solve.

38. Find three consecutive integers. The sum of the second number and the third number is 23.

39. Two cyclists simultaneously start in opposite directions down a straight road. The average rate of one cyclist is 5 km/h more than the average rate of the other cyclist. If they are 90 km apart after 2 h, what is the rate of each cyclist?

40. The length of a rectangle is 2 more than 3 times its width. Find the least possible value for the width if the perimeter is at least 24 m.

7 Factoring Polynomials

A gardener must visualize how a garden will be landscaped before it is tilled and planted. Very often mathematics can be helpful in determining the structure of the topsoil and the sizes of different plots.

7.1 Factors and Exponents

Objectives: To identify numbers as prime or composite
To identify the positive integral factors of a number and to find its prime factorization

Steven is designing a rectangular vegetable garden that has an area of 20 yd². If the dimensions in yards are to be whole numbers, how many choices are there for the length and the width? Steven set up the table below.

$l \cdot w = A$
$l \cdot w = 20$

l	1	2	4	5	10	20
w	20	10	5	4	2	1

When two or more numbers are multiplied, each number is called a *factor*. That is, 2 and 10 are factors of 20 because $2 \cdot 10 = 20$. Also, 20 is *divisible by* 2 and 10, since the remainder is 0 when 20 is divided by 2 or by 10 in both cases. Notice that 1, 2, 4, 5, 10, and 20 are all factors of 20. Unless stated otherwise, the term *factors* will mean positive integers in this lesson.

Capsule Review

An exponent tells how many times the base is used as a factor.

EXAMPLES $2 \cdot 2 \cdot 2 \cdot 3 \cdot 3 = 2^3 \cdot 3^2$ $3 \cdot 5 \cdot 5 \cdot 7 \cdot 5 \cdot 3 = 3^2 \cdot 5^3 \cdot 7$

Write each as the product of its factors by using exponents.

1. $2 \cdot 2 \cdot 2 \cdot 2$
2. $5 \cdot 5 \cdot 5$
3. $2 \cdot 3 \cdot 5 \cdot 5$
4. $3 \cdot 5 \cdot 7 \cdot 7 \cdot 7$
5. $1 \cdot 1 \cdot 1 \cdot 1 \cdot 1$
6. $3 \cdot 3 \cdot 3 \cdot 3 \cdot 3$
7. $5 \cdot 3 \cdot 5$
8. $2 \cdot 7 \cdot 2 \cdot 7$
9. $10 \cdot 10 \cdot 10 \cdot 11 \cdot 11$

Integers greater than 1 are classified as *prime* or *composite*, depending on their factors. The integer 1 is neither prime nor composite.

A **prime number** is a positive integer with exactly two positive integral factors, itself and 1.

A **composite number** is a positive integer that has more than two positive integral factors.

Chapter 7 Factoring Polynomials

The prime and composite numbers that are less than 20 are given below.

Prime: 2, 3, 5, 7, 11, 13, 17, 19
Composite: 4, 6, 8, 9, 10, 12, 14, 15, 16, 18

Every composite number can be written as the product of prime numbers. This is called its **prime factorization.** The prime factorization of 30 is $2 \cdot 3 \cdot 5$, since the factors 2, 3, and 5 are all prime numbers.

If a prime number appears as a factor more than once, you may use an exponent to express the power of the factor.

EXAMPLE 1 Find the prime factorization of 198.

$198 = 2 \cdot 99$ *Divide by 2, the first prime number.*
$= 2 \cdot 3 \cdot 33$ *99 is not divisible by 2, so try division by 3.*
$= 2 \cdot 3 \cdot 3 \cdot 11$ *33 is divisible by 3.*
$= 2 \cdot 3^2 \cdot 11$ *11 is a prime number.*

So, the prime factorization of 198 is $2 \cdot 3^2 \cdot 11$.

A prime factorization is not complete until all the factors are prime.

EXAMPLE 2 Find the prime factorization of 4675.

$4675 = 5 \cdot 935$ *4675 is not divisible by 2 or 3. Divide by 5.*
$= 5 \cdot 5 \cdot 187$ *Divide by 5 again.*
$= 5 \cdot 5 \cdot 11 \cdot 17$ *187 is not divisible by 7. Divide by 11.*
$= 5^2 \cdot 11 \cdot 17$ *17 is a prime number.*

So, the prime factorization of 4675 is $5^2 \cdot 11 \cdot 17$.

CLASS EXERCISES

List all positive integers that are factors of the number.

1. 15   **2.** 6   **3.** 16   **4.** 24

State whether the number is *prime* or *composite*.

5. 23   **6.** 21   **7.** 1   **8.** 51

Find the prime factorization of the number.

9. 14   **10.** 18   **11.** 13   **12.** 60

For Discussion

13. Is there more than one prime factorization of any composite number? (The order of the factors does not count.) Explain.

14. Explain why 1 is neither prime nor composite.

PRACTICE EXERCISES

Find the prime factorization of the number.

1. 26	**2.** 33	**3.** 52	**4.** 70
5. 154	**6.** 135	**7.** 195	**8.** 315
9. 105	**10.** 165	**11.** 143	**12.** 273
13. 3575	**14.** 1925	**15.** 9625	**16.** 7425
17. 2205	**18.** 1155	**19.** 1760	**20.** 1716
21. 242	**22.** 338	**23.** 288	**24.** 455
25. 101	**26.** 107	**27.** 1000	**28.** 1875

Find the next prime number after the given prime number.

29. 53   **30.** 71   **31.** 107   **32.** 127

If x is a prime number, must the expression represent a composite number? Justify your answer.

33. $3x$ **34.** $x + x$ **35.** x^2 **36.** $x + 4$

Applications

Computer Suppose you have a given length of fencing. You need to determine what length and width dimensions will enable you to enclose the greatest area. This program considers the length of the fencing as the perimeter and then checks each combination of length and width dimensions.

```
10 INPUT "ENTER THE FENCING LENGTH.
   ";P: PRINT
20 PRINT "LENGTH","WIDTH","AREA"
30 FOR W = 1 TO P / 2 - 1
40 LET L = (P - 2 * W) / 2
50 LET A = L * W: PRINT L,W,A
60 NEXT W
70 END
```

Run the program for the following lengths.

37. 12   **38.** 79   **39.** 250   **40.** 86.4

41. Run the program for a variety of fencing lengths. Does any pattern emerge for the maximum dimensions?

BIOGRAPHY: Hypatia

Hypatia was born in Alexandria, Egypt, in 370 A.D. Her father, Theon, was a mathematician and philosopher, which probably stimulated her interest in these fields. Historians today identify Hypatia as the first notable mathematician who was a woman.

Hypatia devoted most of her scholarly attention to mathematics and astronomy, and wrote several commentaries on the works of scholars such as Diophantus and Ptolemy.

The philosophy of Hypatia was more scholarly and scientific and less mystical and pagan than that of the Athenian school then in existence. Because her school symbolized learning and science, however, it was identified with paganism by early religious fanatics. Around 412 A.D., a group of these fanatics brutally murdered Hypatia. Her death signaled the beginning of the decline of Alexandria as an important center of learning.

1. Diophantine problems (or equations) are a particular type of problem named after Diophantus, a contemporary of Hypatia. Diophantine problems are indeterminate algebraic problems with restriction of the solutions to integers. One possible integral solution of $x^2 + y^2 = z^2$ is $8^2 + 6^2 = 10^2$, with $x = 8$, $y = 6$, and $z = 10$. Find three other integral solutions.

2. A student's transcript shows x 5-hour courses, and y 3-hour courses. The total number of course hours is 64. Study the graph and find the number of each type of course taken. Possible solutions are $x = 2$ and $y = 18$, or $x = 5$ and $y = 13$. How many other integer solutions are there? How would you check your answers?

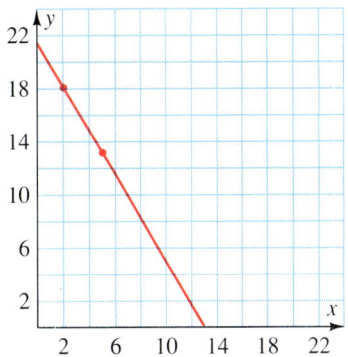

3. Research the life of one or more contemporaries of Hypatia who were associated with the remarkable scholarship in Egypt and Greece at that time.

7.1 Factors and Exponents

7.2 Monomial Factors of Polynomials

Objectives: To find the greatest common factor (GCF) of two or more integers
To factor the greatest common monomial factor from a polynomial

Jenny raises golden retriever puppies. She wants to place a rectangular pen for the dogs along a wall. The area of the pen should be 112 yd². If the length of the pen is 16 yd, what is the width?

Area = length × width
$112 = 16 \cdot \underline{?}$ *Missing factor*
$\frac{112}{16} = 7$ *Divide 112 by 16 to find the missing factor.*

So, $112 = 16 \cdot 7$, and the width of the pen is 7 yd.

Capsule Review

EXAMPLES $12a^7 = 4a^3 \cdot \underline{?}$ $18a^5b^4 = -3ab \cdot \underline{?}$

$\frac{12a^7}{4a^3} = 3a^{7-3}$, or $3a^4$ $\frac{18a^5b^4}{-3ab} = -6a^{5-1}b^{4-1}$, or $-6a^4b^3$

So, the missing factor is $3a^4$:   So, the missing factor is $-6a^4b^3$:
$12a^7 = 4a^3 \cdot 3a^4$ $18a^5b^4 = -3ab(-6a^4b^3)$

Divide to find the missing factor.

1. $x^5 = x^2 \cdot \underline{?}$
2. $4y^4 = y \cdot \underline{?}$
3. $-9m^3 = 3m \cdot \underline{?}$
4. $a^4b^3 = b \cdot \underline{?}$
5. $20x^5y^4 = -4xy^3 \cdot \underline{?}$
6. $56rs = 7r \cdot \underline{?}$
7. $32m^3n^3 = 4mn^2 \cdot \underline{?}$
8. $-15x^3 = 3x \cdot \underline{?}$
9. $-52mn^3 = -m \cdot \underline{?}$

A **common factor** of two or more integers is a factor of each integer. For example, 5 is a common factor of 15 and 25. The **greatest common factor (GCF)** of two or more integers is the greatest integer that is a factor of each.

EXAMPLE 1 Find the GCF of 45 and 63.

First, find the prime factorization of each number.
$45 = 3 \cdot 3 \cdot 5$, or $3^2 \cdot 5$
$63 = 3 \cdot 3 \cdot 7$, or $3^2 \cdot 7$ *3^2 appears in both factorizations.*

So, the greatest common factor of 45 and 63 is 3^2, or 9.

The concept of greatest common factor applies to monomials and polynomials. Write the monomials $21x^2$ and $-28xy^3$ in factored form:

$$21x^2 = 3 \cdot 7 \cdot x \cdot x$$
$$-28xy^3 = -1 \cdot 2 \cdot 2 \cdot 7 \cdot x \cdot y \cdot y \cdot y$$

The greatest common factor of these two monomials is $7x$, the product of their common factors. The GCF of two or more monomials is the product of the common factors of both the coefficients and all the common variables. For example,

$$4x^3y = 2 \cdot 2 \cdot x \cdot x \cdot x \cdot y$$
$$12x^2y^3 = 2 \cdot 2 \cdot 3 \cdot x \cdot x \cdot y \cdot y \cdot y$$

To find the GCF multiply the common factors: $2 \cdot 2 \cdot x \cdot x \cdot y = 4x^2y$. The monomials contain powers of x and powers of y. x^2 and y are the greatest powers of the variables that occur in both monomials.

To factor a polynomial means to write the polynomial as a product of other polynomials. First find the greatest common factor of its terms. Then use the distributive property to write the polynomial in factored form.

Polynomial	**Find the GCF of the terms**	**Use the distributive property**
$21x^2 - 28xy^3$	$\longrightarrow 7x(3x) - 7x(4y^3)$	$\longrightarrow 7x(3x - 4y^3)$

In this example, $7x$ is the GCF of the terms $21x^2$ and $-28xy^3$. It is a *common monomial factor* of the terms. A *binomial factor* of $21x^2 - 28xy^3$ is $(3x - 4y^3)$.

EXAMPLE 2 Factor: $3x^3y - 15x^2y^4$

$3x^3y - 15x^2y^4$ The GCF of the terms is $3x^2y$.
$3x^2y(x) - 3x^2y(5y^3)$ Factor out the GCF from each term.
$3x^2y(x - 5y^3)$ Write the polynomial in factored form.

Check by multiplying $3x^2y$ and $x - 5y^3$. The product should be the original polynomial.

EXAMPLE 3 Factor: $8m^4n^2 + 18m^3n^2 - 6m^2n$

$8m^4n^2 + 18m^3n^2 - 6m^2n$ The GCF is $2m^2n$.
$2m^2n(4m^2n) + 2m^2n(9mn) - 2m^2n(3)$ Factor out the GCF from each term.
$2m^2n(4m^2n + 9mn - 3)$ Factored form of the polynomial

Check by multiplying: Is $2m^2n(4m^2n + 9mn - 3) = 8m^4n^2 + 18m^3n^2 - 6m^2n$?

When a measure is given as a polynomial, consider whether it can be expressed in factored form.

EXAMPLE 4 Find the area of the shaded region.

7.2 Monomial Factors of Polynomials

A side of the square measures $2r$. Why?

$$\begin{aligned}
\text{Area}_{\text{shaded}} &= A_{\text{square}} + A_{\text{triangle}} - A_{\text{circle}} \\
&= (2r)^2 + \frac{1}{2}(2r \cdot r) - (\pi r^2) \\
&= 4r^2 + r^2 - \pi r^2 \\
&= 5r^2 - \pi r^2 \\
&= r^2(5 - \pi) \quad \text{Factored form of the polynomial}
\end{aligned}$$

The area of the shaded region is $r^2(5 - \pi)$.

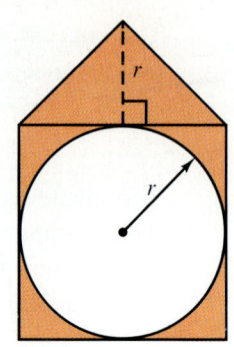

CLASS EXERCISES

Find the GCF of each group of numbers.

1. 14, 24
2. 15, 19
3. 24, 36, 42
4. 7, 35, 105

Factor.

5. $3p + 6$
6. $10z + 15$
7. $8k - 4$
8. $27r + 9t$
9. $2x + 7x^2$
10. $14y - y^4$
11. $8a^2 + 12a$
12. $12 - 12a$

For each of the following, first find the GCF, then factor.

13. $3pq^2 + 18p^2q$
14. $11j^4k^5 - 13j^3k^2$

PRACTICE EXERCISES

Find the GCF of each group of numbers.

1. 35, 49
2. 39, 52
3. 144, 126
4. 154, 198
5. 25, 75, 100
6. 32, 40, 56
7. 112, 224, 104
8. 174, 216, 162

Factor.

9. $2x^3y - 12x^2y^4$
10. $5a^2b - 15a^3b^3$
11. $7cd^3 + 14c^3d^5$
12. $6mn^4 + 18m^5n^2$
13. $13x^2y^3 + 26x^2y^2$
14. $33w^3y^2 + 11w^2y^2$
15. $12m^4n^5 - 18mn^3$
16. $7a^2b^3 - 9a^4b^3$
17. $4j^3k - 6jk^2 + 8jk$
18. $5m^2n - 15mn^2 + 10mn$
19. $4a^2b + 8a^2b^2 + 12ab$
20. $9cd^2 + 6c^2d + 3cd$
21. $12x^3y^4 + 36x^4y^4 - 60x^3y^5$
22. $10x^4y^4 + 15x^4y^7 - 25x^5y^6$
23. $24m^3n^2 + 21m^2n^3 - 39m^2n^4$
24. $49jk^2 - 21j^2k^3 + 84j^3k^4$
25. $13x^5y^4 - 11x^3y^4 + 17x^4y^3$
26. $23s^3t^4 - 19s^2t^5 + 29s^4t^3$
27. $42l^3m^2 - 36l^2m^5 - 54l^4m$
28. $72c^2d^5 - 64c^3d - 28c^2d^4$
29. $33x^5y^7 - 99x^6y^9 - 66x^5y^8$
30. $76a^4b^5 - 38a^5b^5 - 114ab$
31. $89r^2s^9 + 113r^3s^7 + 73r^4s^3$
32. $71x^8y^9 + 119x^7y^6 + 97x^4y^8$

Chapter 7 Factoring Polynomials

33. $-15a^3b^4 - 35a^4b^5 - 55a^2b^4$
34. $-22w^3z^4 - 28w^2z^5 - 34wz^4$
35. $-115x^4y^4 - 225x^5y^5 - 285x^6y^6$
36. $-145m^4n^7 - 280m^5n^6 - 310m^6n^9$
37. $77c^4d^6e^3 + 28c^5d^2e^4$
38. $56j^3k^2l^4 + 72j^2kl^3$
39. $54x^2y^5z^3 - 36x^3y^2z^2$
40. $42s^2t^3u^4 + 35s^3t^2u^3$

Applications

Geometry Find the area of the shaded region. Give the answer in factored form.

41.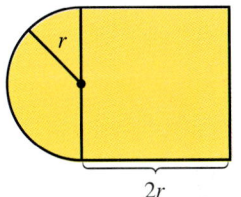

42.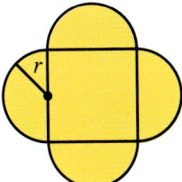

43.

EXTRA

For some numbers, it is difficult to determine if there is a GCF greater than 1, and if so, what it is. You can use Euclid's Algorithm to find the GCF. This program will do it for you or you can use a calculator and follow the steps in the example below.

```
10 INPUT "ENTER TWO NUMBERS.  ";X,Y
20 IF X > Y THEN GOTO 40
30 R = Y - INT (Y / X) * X:A = X:Y = R: GOTO 50
40 R = X - INT (X / Y) * Y:A = Y:Y = R
50 IF R = 0 THEN PRINT "GCF = ";A: GOTO 70
60 GOTO 20
70 END
```

EXAMPLE Factor $368x + 805$ by finding the GCF of 368 and 805.

Solution
1. Divide the larger number (805) by the smaller number (368).
2. Divide the previous divisor (368) by the remainder (69).
3. Repeat step 2 until the remainder is 0.
4. When the remainder is 0, the GCF is the divisor in that step.

$$368\overline{)805} \quad 2 \text{ R}69$$
$$69\overline{)368} \quad 5 \text{ R}23$$
$$23\overline{)69} \quad 3 \text{ R}0$$
GCF = 23

Answer $368x + 805 = 23(16x + 35)$

Use Euclid's Algorithm to factor each expression.

1. $765x + 187$
2. $136x + 255$
3. $460x + 161$
4. $1276x + 435$
5. $264x + 605$
6. $286x + 385$
7. $736x + 483$
8. $351x + 455$

7.3 Factoring $x^2 + bx + c$, $c > 0$

Objective: To factor a trinomial of the form $x^2 + bx + c$, $c > 0$

Factoring a trinomial of the form $x^2 + bx + c$ means to express the trinomial as the product of two binomials of the form $(x + r)(x + s)$.

Capsule Review

$$\begin{array}{c} \text{F} \quad \text{O} \quad \text{I} \quad \text{L} \end{array}$$

EXAMPLES $(x + 2)(x + 7) = x^2 + 7x + 2x + 14 = x^2 + 9x + 14$
$(1 - y)(12 - y) = 12 - y - 12y + y^2 = 12 - 13y + y^2$

Multiply.

1. $(x + 1)(x + 2)$
2. $(t - 2)(t - 5)$
3. $(y + 3)(y + 4)$
4. $(m - 10)(m - 50)$
5. $(100 + z)(8 + z)$
6. $(11 - a)(10 - a)$

Factoring a trinomial involves recognizing patterns, estimating, looking for clues, and multiplying to check.

Consider how FOIL can be used in factoring trinomials into two binomial factors $(x + r)(x + s)$ in which the coefficient of the first term is 1.

EXAMPLE 1   Factor: $x^2 + 7x + 10$

The product of the second terms, r and s of the binomials, must be 10, the last term of the trinomial. The sum of the outer and inner products must be $7x$. The table below shows how to test pairs of factors of 10.

Factors of 10	Possible Binomial Factors	Sum of Outer and Inner Products	
1, 10	$(x + 1)(x + 10)$	$10x + 1x = 11x$	
$-1, -10$	$(x - 1)(x - 10)$	$-10x + (-1x) = -11x$	
2, 5	$(x + 2)(x + 5)$	$5x + 2x = 7x$	←Correct middle term
$-2, -5$	$(x - 2)(x - 5)$	$-5x + (-2x) = -7x$	

So, $x^2 + 7x + 10 = (x + 2)(x + 5)$.
Check by multiplying: $(x + 2)(x + 5) = x^2 + 5x + 2x + 10 = x^2 + 7x + 10$

Example 1 suggests a *factoring clue:* When the first term of a trinomial has a coefficient of 1 and the middle and the last terms are positive, you need to test only the positive factors of the last term.

EXAMPLE 2 Factor: $m^2 - 5m + 6$

Factors of 6	Possible Binomial Factors	Sum of Outer and Inner Products
1, 6	$(m + 1)(m + 6)$	$6m + 1m = 7m$
$-1, -6$	$(m - 1)(m - 6)$	$-6m + (-1m) = -7m$
2, 3	$(m + 2)(m + 3)$	$3m + 2m = 5m$
$-2, -3$	$(m - 2)(m - 3)$	$-3m + (-2m) = -5m$ ←*Correct middle term*

So, $m^2 - 5m + 6 = (m - 2)(m - 3)$. Check by multiplying.

Example 2 leads to another *factoring clue:* When the first term of a trinomial has a coefficient of 1 and the middle term is negative and the last term is positive, you need to test only the negative factors of the last term.

EXAMPLE 3 Factor: $12 - 8a + a^2$

$$12 - 8a + a^2$$
$$= a^2 - 8a + 12$$ *Write the terms in descending order.*
$$= (a - \underline{?})(a - \underline{?})$$ *Both second terms are negative.*

Factors of 12	Possible Binomial Factors	Sum of Outer and Inner Products
$-1, -12$	$(a - 1)(a - 12)$	$-12a + (-1a) = -13a$
$-4, -3$	$(a - 4)(a - 3)$	$-3a + (-4a) = -7a$
$-6, -2$	$(a - 6)(a - 2)$	$-2a + (-6a) = -8a$ ←*Correct middle term*

So, $12 - 8a + a^2 = (a - 6)(a - 2)$. Check by multiplying.

If you so desire, exercises of this type can also be factored without rearranging the terms of the polynomial: $12 - 8a + a^2 = (6 - a)(2 - a)$.

You may use the factoring process to factor a trinomial that contains more than one variable.

EXAMPLE 4 Factor: $a^2 + 7ab + 12b^2$

$a^2 + 7ab + 12b^2 = (a + \underline{?})(a + \underline{?})$ Both factors of $12b^2$ must be positive.

Factors of $12b^2$	Possible Binomial Factors	Sum of Outer and Inner Products
$1b, 12b$	$(a + 1b)(a + 12b)$	$12ab + 1ab = 13ab$
$2b, 6b$	$(a + 2b)(a + 6b)$	$6ab + 2ab = 8ab$
$3b, 4b$	$(a + 3b)(a + 4b)$	$4ab + 3ab = 7ab$ ←*Correct middle term*

So, $a^2 + 7ab + 12b^2 = (a + 3b)(a + 4b)$. Check by multiplying.

7.3 Factoring $x^2 + bx + c, c > 0$

CLASS EXERCISES

Find the missing factors. Check by multiplying.

1. $x^2 + 5x + 6 = (x + 2)(x + \underline{?})$
2. $b^2 - 13b + 12 = (b - 1)(b - \underline{?})$
3. $u^2 - 8u + 7 = (u - \underline{?})(u - \underline{?})$
4. $z^2 - 18z + 17 = (z - \underline{?})(z - \underline{?})$

Factor. Check by multiplying.

5. $x^2 + 4x + 3$
6. $m^2 + 3m + 2$
7. $y^2 - 6y + 5$
8. $p^2 - 2p + 1$
9. $7 + 8m + m^2$
10. $11 - 12d + d^2$

PRACTICE EXERCISES

Factor. Check by multiplying.

1. $x^2 + 10x + 9$
2. $y^2 + 14y + 13$
3. $a^2 + 12a + 11$
4. $r^2 + 6r + 8$
5. $z^2 + 30z + 29$
6. $t^2 + 16t + 39$
7. $y^2 + 14y + 33$
8. $x^2 + 20x + 91$
9. $s^2 - 7s + 6$
10. $k^2 - 5k + 4$
11. $m^2 - 42m + 41$
12. $b^2 - 20b + 19$
13. $y^2 - 13y + 22$
14. $n^2 - 9n + 14$
15. $z^2 - 20z + 51$
16. $g^2 - 12g + 35$
17. $65 - 18b + b^2$
18. $26 - 15n + n^2$
19. $36 - 15z + z^2$
20. $54 - 21a + a^2$
21. $48 + 19y + y^2$
22. $12 + 7d + d^2$
23. $72 - 17f + f^2$
24. $54 - 15r + r^2$
25. $a^2 + 12ab + 27b^2$
26. $n^2 + 12np + 35p^2$
27. $x^2 - 8xy + 7y^2$
28. $s^2 - 14st + 13t^2$
29. $r^2 + 5rt + 6t^2$
30. $z^2 + 8xz + 12x^2$
31. $m^2 - 26mn + 25n^2$
32. $r^2 + 33rs + 32s^2$
33. $s^2 + 11st + 18t^2$
34. $j^2 - 7jk + 12k^2$
35. $y^2 - 10yz + 16z^2$
36. $m^2 + 9mn + 20n^2$
37. $x^2 - 27xy + 50y^2$
38. $s^2 - 16st + 48t^2$
39. $r^2 + 20rt + 64t^2$
40. $a^2 + 32ab + 60b^2$
41. $m^2 - 27mn + 72n^2$
42. $n^2 - 29nq + 100q^2$
43. $m^2 - 6mn - 16n^2$
44. $s^2 - 9st + 18t^2$
45. $c^4 + 8c^2 + 12$
46. $u^4 - 16u^2 + 28$
47. $y^4 - 11y^2 + 24$
48. $n^4 - 18n^2 + 32$
49. $(a + 1)^2 + 8(a + 1) + 7$
50. $(x + 2)^2 + 4(x + 2) + 3$
51. $(x - 1)^2 + 2(x - 1) + 1$
52. $(a + 2)^2 + 2(a + 2) + 1$
53. $36 - 15(z + 1) + (z + 1)^2$
54. $44 - 15(x + 2) + (x + 2)^2$
55. $a^{2x} + 3a^x + 2$
56. $b^{2y} + 5b^y + 6$
57. $x^{4n} - 7x^{2n} + 12$
58. $y^{4m} - 13y^{2m} + 12$

Applications

Geometry The area of each rectangle is given. Find the binomials to represent the length and the width.

59.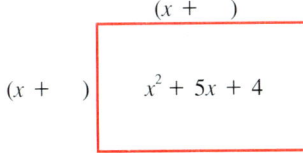

60.

61. Area = $m^2 + 4m + 3$

62. Area = $n^2 + 10n + 16$

63. Area = $x^2 + 3xa + 2a^2$

64. Area = $b^2 + 4bc + 3c^2$

DID YOU KNOW?

The distributive property can be illustrated geometrically. Think of a rectangle divided into two smaller rectangles as illustrated. The area of the largest rectangle is $a(b + c)$. The area of the right rectangle is ab. The area of the left rectangle is ac. The sum of the areas of the two smaller rectangles is equal to the area of the large rectangle. Therefore,

$$a(b + c) = ab + ac$$

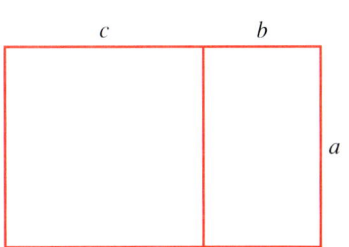

Can you find a way to illustrate geometrically the multiplication of two binomials: $(a + b)(c + d)$?

7.4 Factoring $x^2 + bx + c$, $c < 0$

Objective: To factor a trinomial of the form $x^2 + bx + c$, $c < 0$

Susan made up this riddle for Rebecca: "I'm thinking of two numbers. The product of the numbers is -24, and the sum of the numbers is -2." Rebecca tried a few combinations in her head before coming up with the answer: "The numbers must be -6 and $+4$." Is any other combination possible?

Capsule Review

EXAMPLES $(x - 3)(x + 1) = x^2 + 1x - 3x - 3 = x^2 - 2x - 3$
$(x + 6)(x - 2) = x^2 - 2x + 6x - 12 = x^2 + 4x - 12$

Multiply.

1. $(x + 4)(x - 2)$
2. $(x - 5)(x + 7)$
3. $(x - 4)(x + 5)$
4. $(x - 7)(x + 6)$
5. $(y - 9)(y + 11)$
6. $(s + 3)(s - 12)$

Many factoring problems are like the riddle described above. You look for two numbers whose sum and product you know. Consider how to factor trinomials in which the coefficient of the first term is 1 and the sign of the last term is negative.

EXAMPLE 1 Factor: $x^2 + 3x - 10$

Factors of -10	Possible Binomial Factors	Sum of Outer and Inner Products	
$-1, 10$	$(x - 1)(x + 10)$	$10x + (-1x) = 9x$	
$1, -10$	$(x + 1)(x - 10)$	$-10x + 1x = -9x$	
$2, -5$	$(x + 2)(x - 5)$	$-5x + 2x = -3x$	
$-2, 5$	$(x - 2)(x + 5)$	$5x + (-2x) = 3x$	←Correct middle term

So, $x^2 + 3x - 10 = (x - 2)(x + 5)$. Check by multiplying.

Notice that the placement of signs in the binomial factors is important. In Example 1, the factors are incorrect if we change the signs:

$(x + 2)(x - 5) = x^2 - 5x + 2x - 10 = x^2 - 3x - 10$
                                              ↑
                                     incorrect sign

Chapter 7 Factoring Polynomials

EXAMPLE 2 Factor: $a^2 - 5a - 24$

Factors of -24 (not all listed)	Possible Binomial Factors	Sum of Outer and Inner Products
8, -3	$(a + 8)(a - 3)$	$-3a + 8a = 5a$
-8, 3	$(a - 8)(a + 3)$	$3a + (-8a) = -5a$ ←Correct middle term

So, $a^2 - 5a - 24 = (a - 8)(a + 3)$. Check by multiplying.

EXAMPLE 3 Factor: $m^2 - 5mn - 14n^2$

Factors of $-14n^2$	Possible Binomial Factors	Sum of Outer and Inner Products
$-1n$, $14n$	$(m - 1n)(m + 14n)$	$14mn + (-1mn) = 13mn$
$1n$, $-14n$	$(m + 1n)(m - 14n)$	$-14mn + 1mn = -13mn$
$-2n$, $7n$	$(m - 2n)(m + 7n)$	$7mn + (-2mn) = 5mn$
$2n$, $-7n$	$(m + 2n)(m - 7n)$	$-7mn + 2mn = -5mn$ ←Correct middle term

So, $m^2 - 5mn - 14n^2 = (m + 2n)(m - 7n)$. Check by multiplying.

Examples 1, 2, and 3 lead to a *factoring clue:* When the first term of a trinomial has a coefficient of 1 and the last term is negative, one factor of the last term is positive and the other is negative.

CLASS EXERCISES

Find the missing signs.

1. $x^2 + 2x - 3 = (x \underline{?} 3)(x \underline{?} 1)$
2. $b^2 - 11b - 12 = (b \underline{?} 1)(b \underline{?} 12)$
3. $m^2 - 6m - 7 = (m \underline{?} 1)(m \underline{?} 7)$
4. $z^2 + z - 6 = (z \underline{?} 2)(z \underline{?} 3)$

Find the missing factors.

5. $x^2 + x - 6 = (x + \underline{?})(x - \underline{?})$
6. $b^2 - 15b - 16 = (b + \underline{?})(b - \underline{?})$
7. $u^2 - 8u - 9 = (u - \underline{?})(u + \underline{?})$
8. $z^2 + 3z - 18 = (z - \underline{?})(z + \underline{?})$

Factor. Check by multiplying.

9. $x^2 + 4x - 5$
10. $m^2 - 3m - 10$
11. $y^2 - 6y - 16$

For Discussion

12. The coefficient of the third term of the trinomials in this lesson is negative. What clue does this give about the signs of the second terms in the binomial factors?

PRACTICE EXERCISES

Factor. Check by multiplying.

1. $x^2 + 4x - 5$
2. $a^2 + 6a - 7$
3. $m^2 - 9m - 10$
4. $z^2 - 23z - 24$
5. $x^2 + 2x - 8$
6. $a^2 + 7a - 18$
7. $m^2 - 4m - 12$
8. $z^2 - 8z - 20$
9. $x^2 + 2x - 3$
10. $y^2 + 12y - 13$
11. $a^2 + 5a - 6$
12. $z^2 + 8z - 9$
13. $m^2 - m - 12$
14. $b^2 - b - 56$
15. $k^2 - 13k - 30$
16. $s^2 - 2s - 24$
17. $x^2 - xy - 6y^2$
18. $m^2 - 2mn - 15n^2$
19. $r^2 - 4rs - 21s^2$
20. $n^2 - 11np - 42p^2$
21. $x^2 - 6xy - 40y^2$
22. $a^2 - 6ab - 27b^2$
23. $k^2 - 5kj - 36j^2$
24. $w^2 - 8wz - 48z^2$
25. $x^2 - xy - 2y^2$
26. $s^2 - 4st - 5t^2$
27. $r^2 + 5rt - 14t^2$
28. $a^2 - 6ab - 27b^2$
29. $y^2 - 14yz - 32z^2$
30. $n^2 - 2np - 48p^2$
31. $x^2 - 8xy - 20y^2$
32. $s^2 - st - 42t^2$
33. $r^2 + 3rt - 54t^2$
34. $a^2 - 9ab - 52b^2$
35. $y^2 - 15yz - 100z^2$
36. $n^2 - 11np - 80p^2$
37. $y^4 + 23y^2 - 50$
38. $z^6 - 10z^3 - 75$
39. $(a - 1)^2 - 4(a - 1) - 32$
40. $(b + 2)^2 + 15(b + 2) - 54$

Applications

41. **Gardening** A surveyor's map shows a plan for a rectangular rose garden whose area is $a^2 + 25ab - 350b^2$. Find an algebraic expression for the length and the width. If $a = 200$ ft and $b = 10$ ft, find the actual dimensions of the garden.

ALGEBRA IN RECREATION

A large target for skydivers is marked out on a field. The target is made up of concentric circles with radii of lengths 2 m, 4 m, 6 m, 8 m, and 10 m. The skydivers score points by landing within the target—the closer to the center, the greater the number of points.

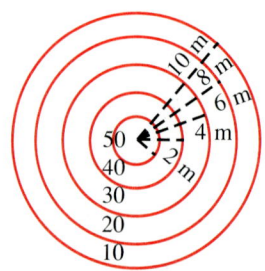

Find the area of the bull's-eye or ring that gives each number of points.

1. 50
2. 40
3. 30
4. 20
5. 10

284 Chapter 7 Factoring Polynomials

7.5 Factoring $ax^2 + bx + c$

Objective: To factor a trinomial of the form $ax^2 + bx + c$

When the coefficient of the first term of a trinomial is *not* 1, the number of possible binomial factors increases. A partial list of possible factors of $2y^2 - 11y + 12$ is given below.

$(2y - 4)(y - 3)$ $(y - 6)(2y - 2)$ $(2y - 3)(y - 4)$ $(2y - 12)(y - 1)$

What are some of the other possible binomial factors? Which are correct? You can check your answer by multiplication.

Capsule Review

EXAMPLES

$(2x + 1)(x - 2)$
$= 2x^2 - 4x + x - 2$
$= 2x^2 - 3x - 2$

$(3a - 4)(2a - 3)$
$= 6a^2 - 9a - 8a + 12$
$= 6a^2 - 17a + 12$

Multiply.

1. $(2m + 1)(m + 1)$
2. $(t + 3)(3t + 1)$
3. $(3y + 2)(y - 2)$
4. $(5y - 2)(y + 4)$
5. $(3h - 1)(h - 7)$
6. $(2x + 3)(3x + 1)$
7. $(3a + b)(a - b)$
8. $(2s - t)(3s + 2t)$
9. $(4u - 3v)(3u - 4v)$

The technique for factoring trinomials like $2x^2 - 3x - 5$ is much the same as for trinomials in which the coefficient of x^2 is 1. However, there is no straightforward way in which to find the factors other than by simply listing all the possibilities. You must test various factors of *both* the first term and the last term of the trinomial, then look for combinations that will make the middle term work out.

EXAMPLE 1 Factor: $2x^2 - 3x - 5$

Factors of $2x^2$	Factors of -5	Possible Binomial Factors	Sum of Outer and Inner Products
$2x, x$	$-1, 5$	$(2x - 1)(x + 5)$	$10x + (-1x) = 9x$
		$(x - 1)(2x + 5)$	$5x + (-2x) = 3x$
	$1, -5$	$(2x + 1)(x - 5)$	$-10x + 1x = -9x$
		$(x + 1)(2x - 5)$	$-5x + 2x = -3x$ ←Correct middle term

So, $2x^2 - 3x - 5 = (x + 1)(2x - 5)$. Check by multiplying.

EXAMPLE 2 Factor: $6x^2 + 19x + 3$

There are two possible ways to factor $6x^2$.

Factors of $6x^2$	Factors of 3	Possible Binomial Factors	Sum of Outer and Inner Products
$3x, 2x$	3, 1	$(3x + 3)(2x + 1)$	$3x + 6x = 9x$
		$(3x + 1)(2x + 3)$	$9x + 2x = 11x$
$6x, x$	3, 1	$(6x + 3)(x + 1)$	$6x + 3x = 9x$
		$(6x + 1)(x + 3)$	$18x + 1x = 19x$ ←Correct middle term

So, $6x^2 + 19x + 3 = (6x + 1)(x + 3)$.

Again, you should check all factorizations by multiplying the resulting factors. The check for this example is left for you.

To factor a given trinomial, there may be many binomial factors to check. Practice and experience will help you make better estimates. It is not efficient to list and check every possible combination.

EXAMPLE 3 Factor: $6x^2 - x - 12$

There are two possible ways to factor $6x^2$.

Factors of $6x^2$	Factors of -12	Possible Binomial Factors	Sum of Outer and Inner Products
$3x, 2x$	12, -1	$(3x + 12)(2x - 1)$	$-3x + 24x = 21x$
	$-12, 1$	$(3x - 12)(2x + 1)$	$3x + (-24x) = -21x$
	4, -3	$(3x + 4)(2x - 3)$	$-9x + 8x = -1x$ ←Correct middle term

Note that there are many other possible pairs of binomial factors, but you do not need to list all of them. Once you find the correct pair of factors, you can stop.

So, $6x^2 - x - 12 = (3x + 4)(2x - 3)$. Check by multiplying.

Not all trinomials have binomial factors in which the coefficients are integers. For example:

$$3x^2 - 2x + 6 \neq (3x - 2)(x - 3)$$
$$\neq (3x - 3)(x - 2)$$
$$\neq (3x - 1)(x - 6)$$
$$\neq (3x - 6)(x - 1)$$

When a polynomial has no polynomial factors with integral coefficients except itself and 1, it is a **prime polynomial** *with respect to the integers.* $3x^2 - 2x + 6$ is a prime polynomial.

EXAMPLE 4 Factor: $4x^2 + 59x - 15$

There are two possible ways to factor $4x^2$.

In the product of the binomial factors $(2x + \underline{?})(2x - \underline{?})$, the coefficient of the middle term is an even number since 2 is a factor of both the inner and outer products. In this case, you can immediately eliminate all such combinations, since the coefficient of the middle term, $59x$, is an odd number.

Factors of $4x^2$	Factors of -15	Possible Binomial Factors	Sum of Outer and Inner Products	
$4x, x$	$3, -5$	$(4x + 3)(x - 5)$	$-20x + 3x = -17x$	
	$-3, 5$	$(4x - 3)(x + 5)$	$20x + (-3x) = 17x$	
	$1, -15$	$(4x + 1)(x - 15)$	$-60x + 1x = -59x$	
	$-1, 15$	$(4x - 1)(x + 15)$	$60x + (-1x) = 59x$	←Correct middle term

So, $4x^2 + 59x - 15 = (4x - 1)(x + 15)$. Check by multiplying.

CLASS EXERCISES

Find the missing signs or factors.

1. $2x^2 + x - 3 = (2x \underline{?} 3)(x \underline{?} 1)$
2. $2x^2 + 3x - 5 = (2x + \underline{?})(x - \underline{?})$
3. $3b^2 - 20b - 7 = (3b \underline{?} 1)(b \underline{?} 7)$
4. $5b^2 - 34b - 7 = (5b + \underline{?})(b - \underline{?})$

Factor. Check by multiplying.

5. $3x^2 + 20x - 7$
6. $2m^2 - 3m - 14$
7. $6y^2 + 25y + 21$
8. $4x^2 - 4x - 3$

PRACTICE EXERCISES

Factor. Check by multiplying.

1. $3x^2 - 22x - 16$
2. $2m^2 - 11m - 21$
3. $5z^2 - 13z - 6$
4. $7r^2 - 23r - 20$
5. $2m^2 + 7m + 5$
6. $3s^2 + 17s + 20$
7. $15r^2 + 44r + 21$
8. $14k^2 + 29k + 12$
9. $81l^2 + 72l + 15$
10. $9q^2 + 27q + 20$
11. $6z^2 - z - 5$
12. $12c^2 - 7c - 10$
13. $4m^2 - 16m - 9$
14. $16x^2 - 8x - 15$
15. $2x^2 + 9x - 11$
16. $5a^2 + 2a - 7$
17. $14d^2 + 11d - 15$
18. $10k^2 + 3k - 4$
19. $24x^2 - 47x + 20$
20. $34d^2 - 41d + 15$
21. $20z^2 + 49z + 30$
22. $22n^2 + 47n + 6$
23. $26c^2 + 29c - 15$
24. $18n^2 + 11n - 24$
25. $3x^2 - xy - 2y^2$
26. $7s^2 - 20st - 3t^2$
27. $2r^2 + 15rt + 7t^2$
28. $4y^2 + 8yz + 3z^2$
29. $3a^2 - 16ab + 5b^2$
30. $7s^2 - 19st + 10t^2$

31. $6n^2 - np - 2p^2$ **32.** $8n^2 - 9np - 14p^2$ **33.** $5x^2 + 18xy - 8y^2$
34. $5r^2 + 29rt - 5t^2$ **35.** $3a^2 + 15ab - 7b^2$ **36.** $10c^2 + 99cd - 10d^2$
37. $12x^2 + 35xy + 18y^2$ **38.** $24g^2 + 50gh + 21h^2$ **39.** $14k^2 - 83km + 33m^2$
40. $20r^2 - 53rt - 21t^2$ **41.** $27a^2 - 12ab - 32b^2$ **42.** $36m^2 + 12mn - 35n^2$
43. $4(x + 2)^2 + 11(x + 2) + 6$ **44.** $6(y + 5)^2 + 11(y + 5) - 10$
45. $12(a - 3)^2 - 19(a - 3) - 21$ **46.** $15(b - 2)^2 - 37(b - 2) + 18$
47. $6x^{2k} + 25x^k + 14$ **48.** $15c^{4r} + 14c^{2r} - 16$
49. $10x^{4k+6} - 7x^{2k+3} - 12$ **50.** $20d^{2r+16} - 23d^{r+8} + 6$

Applications

Geometry Using the given area, find the dimensions of each figure if the length of each side is expressed as a binomial.

51.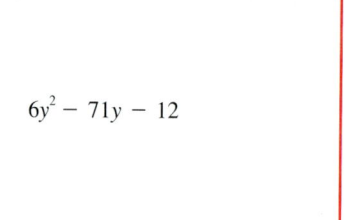
$6y^2 - 71y - 12$

52.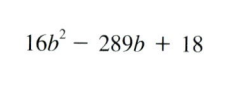
$16b^2 - 289b + 18$

53.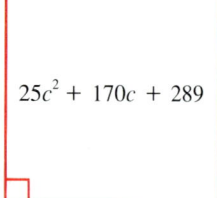
$25c^2 + 170c + 289$

54.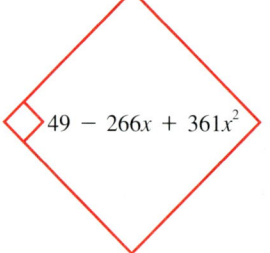
$49 - 266x + 361x^2$

TEST YOURSELF

Find the prime factorization of the number.   **7.1**

1. 110   **2.** 105   **3.** 180   **4.** 250   **5.** 147

Factor. Check by multiplying.   **7.2–7.6**

6. $9a + 18$ **7.** $42x - 14x^3$ **8.** $51p^2q + 3pq$
9. $a^2 + 12a + 35$ **10.** $t^2 + 20t + 36$ **11.** $m^2 - 16m + 60$
12. $x^2 + 5x - 50$ **13.** $y^2 - 11y - 60$ **14.** $m^2 - mn - 6n^2$
15. $2x^2 + 25x + 12$ **16.** $8y^2 - 2y - 3$ **17.** $3x^2 + 4x - 4$

7.6 Factoring: Special Cases

Objectives: To factor perfect square trinomials
To factor a difference of two squares

When you square a binomial, the product is called a **perfect square trinomial**.

$$(x + 4)^2 = (x + 4)(x + 4) = x^2 + 8x + 16$$

- square of the first term
- twice the product of the two terms
- square of the last term

$x^2 + 8x + 16$ is a perfect square trinomial.

Capsule Review

Squaring a binomial results in a trinomial.

EXAMPLE $(x + 3)^2 = (x + 3)(x + 3) = x^2 + (2)(3)x + (3)^2$
$= x^2 + 6x + 9$

Multiply.

1. $(x + 2)(x + 2)$ **2.** $(m - 5)(m - 5)$ **3.** $(2 - 3x)^2$

EXAMPLE 1 Is the polynomial a perfect square trinomial?
 a. $x^2 - 8x + 16$ **b.** $36m^4 + 36m^2n + 9n^2$

a. Yes; x^2 and 16 are the squares of x and 4, respectively, and $8x$ is twice the product of x and 4:

$$x^2 - 8x + 16 = (x)^2 - 2(4 \cdot x) + (4)^2$$

b. Yes; $36m^4$ and $9n^2$ are the squares of $6m^2$ and $3n$, and $36m^2n$ is twice the product of $6m^2$ and $3n$:

$$36m^4 + 36m^2n + 9n^2 = (6m^2)^2 + 2(6m^2 \cdot 3n) + (3n)^2$$

After you identify a polynomial as a perfect square trinomial, its factors are relatively easy to find, as shown in Example 2.

EXAMPLE 2 Factor: $64m^{12} + 16m^6 + 1$

$64m^{12} + 16m^6 + 1$
$= (8m^6)^2 + 2(8m^6 \cdot 1) + 1^2$ *The polynomial is a perfect square trinomial.*
$= (8m^6 + 1)(8m^6 + 1)$, or $(8m^6 + 1)^2$

7.6 Factoring: Special Cases

When you multiply binomials that are the sum and difference of the same two numbers, the product is a **difference of two squares**.

$$(x + 7)(x - 7) = (x)^2 - (7)^2 = x^2 - 49$$

square of the first term — square of the last term

$x^2 - 49$ is a difference of two squares.

EXAMPLE 3 Is the binomial a difference of two squares?
   a. $25x^2 - 1$ b. $x^4y - 9$ c. $144m^6n^2 - 625$

a. Yes; $25x^2 - 1 = (5x)^2 - (1)^2$.   b. No; the exponent of y is 1, and y^1 is not a square.

c. Yes; $144m^6n^2 - 625 = (12m^3n)^2 - (25)^2$.

After you identify a polynomial as a difference of two squares, its factors are relatively easy to find.

EXAMPLE 4 Factor: $m^{10}n^8 - 49$

$m^{10}n^8 - 49 = (m^5n^4)^2 - (7)^2$ *The polynomial is a difference of two squares.*
$= (m^5n^4 - 7)(m^5n^4 + 7)$

CLASS EXERCISES

Factor the polynomial.

1. $4x^2 - 9y^2$
2. $100a^2 - 16b^4z^8$
3. $16r^2s^2 - 81t^{10}$
4. $x^2 + 2xy + y^2$
5. $25m^2 + 20m + 4$
6. $81m^4 + 72m^2n + 16n^2$

PRACTICE EXERCISES

Is the polynomial a perfect square trinomial?

1. $x^2 + 2x + 1$
2. $y^2 + 8y + 16$
3. $3a^2 + 18ab + 6b^2$
4. $25z^2 + 5xz + x^2$
5. $36m^2 - 36m + 9$
6. $16b^2 - 24b + 9$

Factor the polynomial.

7. $x^2 - 12x + 36$
8. $x^2 + 14x + 49$
9. $4d^2 + 36d + 81$
10. $9h^2 + 24h + 16$
11. $k^2 + 20k + 100$
12. $s^2 + 26s + 169$
13. $81y^2 - 36y + 4$
14. $16n^2 - 56n + 49$
15. $25t^2 + 10t + 1$
16. $49r^2 + 14r + 1$
17. $100k^{10} + 20k^5 + 1$
18. $36y^{10} + 12y^5 + 1$

19. $25y^8 + 10y^4 + 1$
20. $9x^{12} + 6x^6 + 1$
21. $a^{10}b^4 - 16$
22. $m^{16}n^8 - 25$
23. $x^{18}y^{10} - 36$
24. $c^{12}d^6 - 64$

Is the binomial a difference of two squares?

25. $x^2 - 9$
26. $a^2 - 121$
27. $4a^2 - 100$
28. $16z^2 - 49$
29. $49k^3 - 25$
30. $81z^2 - 16z$
31. $225a^2 - 100$
32. $144m^2 - 81$

Factor the polynomial.

33. $144x^4y^2 - 625$
34. $121m^6n^2 - 81$
35. $a^6b^2c^4 - d^2$
36. $e^6f^4g^6 - h^2$
37. $4p^4q^4r^8 - 81$
38. $9x^8y^4z^2 - 64$
39. $49r^2s^2 + 14rs + 1$
40. $25a^2b^2 + 20ab + 4$
41. $36x^6y^4 - 36x^3y^2 + 9$
42. $64c^6d^8 - 32c^3d^4 + 4$
43. $e^{64}f^{100} - g^{144}h^{36}$
44. $25x^{50} - 49y^{200}$
45. $(a + 1)^2 - a^2$
46. $(x - 2)^2 - x^2$
47. $(x - 1)^2 + 2(x - 1) + 1$
48. $(a + 2)^2 + 2(a + 2) + 1$
49. $(a^2 + 2ab + b^2) - c^2$
50. $m^2 - (n^2 + 2np + p^2)$
51. $(x + 1)^2 + 2(x + 1)(2x + 3) + (2x + 3)^2$

Applications

Geometry Find the area of the shaded region. Factor, if possible.

52.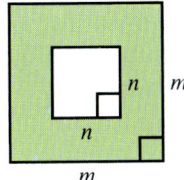

53.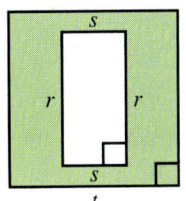

54. **Calculator** Factor $x^2 - 187x - 4368$.

LOGICAL REASONING

Find the fallacy in the following argument.
    Suppose that for all real numbers, $x = y$.
    Since $x^2 = y^2$, then $x^2 - y^2 = 0$.
    Multiply each side of the equation by 2. $2(x^2 - y^2) = 0$.
    Factor the difference of two squares. $2(x + y)(x - y) = 0$.
    Divide each side of the equation by $(x - y)$. $2(x + y) = 0$.
    Divide each side of the equation by $(x + y)$, and $2 = 0$.

7.7 Factoring by Grouping

Objectives: To factor a polynomial by removing a common binomial factor
To factor a polynomial by grouping the terms

Some polynomials may contain common binomial factors. Sometimes these binomial factors are opposites, or additive inverses.

Capsule Review

The inverse of a is $-a$. To find the inverse of a polynomial, use this property.

EXAMPLES

a. $x - y, y - x$
$-1(x - y) = -x + y$
$= y - x$
So, $x - y$ and $y - x$ are inverses.

b. $2x + 1, 2x - 1$
$-1(2x + 1) = -2x - 1$
$-2x - 1 \neq 2x - 1$
So, $2x + 1$ and $2x - 1$ are *not* inverses.

Are these polynomials inverses of each other?

1. $1 - m, m - 1$
2. $3t - 4, 4 - 3t$
3. $3y - 2, -2 - 3y$
4. $5y - 2, 5y + 2$
5. $4h^2 + 5, 5 - 4h^2$
6. $-2x - 3, 2x + 3$

In the polynomial $6(a + b) + 3(a + b)$, the binomial $(a + b)$ is common to both terms. The distributive property can be used to factor out $(a + b)$.

$$6(a + b) + 3(a + b) = (6 + 3)(a + b)$$
$$= 9(a + b)$$

EXAMPLE 1 Factor: $8(x + y) - 5(x + y) + (x + y)$

$8(x + y) - 5(x + y) + (x + y) = [8 - 5 + 1](x + y)$
$= 4(x + y)$

Another method of using the distributive property to factor is as follows:

$8x + 8y - 5x - 5y + x + y$ *Multiply.*
$= 8x - 5x + x + 8y - 5y + y$ *Group like terms.*
$= 4x + 4y = 4(x + y)$ *Combine and use distributive property.*

If opposite polynomials (additive inverses) are present, rewrite one of them as the product of -1 and its additive inverse, then examine the polynomial for common factors. In Example 2, $b - a$ is rewritten as $-1(a - b)$.

Chapter 7 Factoring Polynomials

EXAMPLE 2 Factor: $16(a - b) - 19(b - a)$

$$16(a - b) - 19(b - a) = 16(a - b) - 19[-1(a - b)]$$
$$= 16(a - b) + 19(a - b)$$
$$= (16 + 19)(a - b)$$
$$= 35(a - b)$$

When a polynomial has four terms, it may help to group pairs of terms. This method of factoring is called *factoring by grouping*.

EXAMPLE 3 Factor: $16a - 16b + 19b - 19a$

$$16a - 16b + 19b - 19a = [16a - 16b] + [19b - 19a]$$
$$= [16(a - b)] + [19(b - a)]$$
$$= [16(a - b)] + [-19(a - b)]$$
$$= [16 + (-19)](a - b)$$
$$= (16 - 19)(a - b)$$
$$= -3(a - b)$$

Quite often a polynomial can be factored by employing more than one of the techniques already discussed. At times the basic approach to be used can be recognized after rearranging and regrouping of terms. Sometimes you must factor more than once.

EXAMPLE 4 Factor: $4h - 4k - h^3 + h^2k$

$$4h - 4k - h^3 + h^2k$$
$$= 4(h - k) - h^2(h - k)$$ *Factor out $-h^2$ from the last two terms.*
$$= (h - k)(4 - h^2)$$ *$4 - h^2$ is a difference of two squares.*
$$= (h - k)(2 - h)(2 + h)$$

In the next example, the first two terms are a difference of two squares. However, this arrangement does not result in a complete factoring.

EXAMPLE 5 Factor: $25y^2 - 4x^2 + 12x - 9$

$$25y^2 - 4x^2 + 12x - 9 = (25y^2 - 4x^2) + (12x - 9)$$
$$= (5y - 2x)(5y + 2x) + 3(4x - 3)$$

There is no common factor. Look for a different grouping of the terms.

$$25y^2 - 4x^2 + 12x - 9$$
$$= [25y^2] - [4x^2 - 12x + 9]$$ *$4x^2 - 12x + 9$ is a perfect square trinomial.*
$$= (5y)^2 - (2x - 3)^2$$ *$(5y)^2 - (2x - 3)^2$ is a difference of two squares.*
$$= [5y - (2x - 3)][5y + (2x - 3)]$$
$$= (5y - 2x + 3)(5y + 2x - 3)$$

7.7 Factoring by Grouping

CLASS EXERCISES

Is there a common factor for the polynomial? If so, what is it?

1. $2x^2 + 4x$
2. $15x^2 + 12x + 6x^3$
3. $8(j + k) + 3(j + k)$
4. $2a(c - d) + 3(-c - d)$
5. $3(a - b) + c(a + b)$
6. $2k(m - n) + 4(m - n)$

Factor.

7. $8(x + y) - 3(x + y)$
8. $7(x - 3) + 2(3 - x)$
9. $11(j - 2) + 17(2 - j)$
10. $ab + b^2 + ca + cb$
11. $[r - rt] + [st - s]$
12. $[9 + 9c] - [d^2 + cd^2]$

For Discussion

13. How can factoring be used to simplify: $(4^2)(5)^2(3^2) - (16)(25)(5^2)$?

PRACTICE EXERCISES

Factor.

1. $7(a + 3) - 5(a + 3) + 3(a + 3)$
2. $13(g + 2) - 2(g + 2) + 8(g + 2)$
3. $5(x - 3) + 2(x - 3) - 3(x - 3)$
4. $7(r - 4) + 5(r - 4) - 4(r - 4)$
5. $3x(x - 2) + 2x(x - 2) - 4(x - 2)$
6. $2m(n - 2p) + (4mn - 8mp) + (-2n + 4p)$
7. $11(j - 2) - 6(2 - j) + 2(j - 2)$
8. $9(x - 5) - 4(5 - x) + 5(x - 5)$
9. $9a(b - c) + 2(c - b) + 3a(b - c)$
10. $4x(y - z) + 5(z - y) + 7x(y - z)$
11. $4(3 - j) + 2j(j - 3) + 8(j - 3)$
12. $5(2 - m) + 3m(m - 2) + 7(m - 2)$
13. $10a - 10b + 12b - 12a$
14. $21x - 21y + 15x - 15y$
15. $13r - 13s - 8s + 8r$
16. $11m - 11n + 10n - 10m$
17. $6h - 6k - h^4 + h^3k$
18. $5m - 5n - m^5 + m^4n$
19. $rs + rt - 3s - 3t$
20. $mn + np + 2m + 2p$
21. $3rt + st + 3rw + sw$
22. $2ab - b + 14a - 7$
23. $ac - 2a + 3bc - 6b$
24. $4wx - 6wy + 6xz - 9yz$
25. $36y^2 - 9x^2 - 24x - 16$
26. $49w^2 - 16x^2 - 24x - 9$
27. $x^2 + zy + xy + zx$
28. $3ab + 12c + 9a + 4bc$
29. $15ab - 9bc + 20ac - 12c^2$
30. $6mn + 12mp - 5np - 10p^2$
31. $6rt - 5s + 2t - 15rs$
32. $2pq - 5qr - 4p + 10r$

33. $r^2 - (4s + t)^2$

34. $m^2 - (2n - 3p)^2$

35. $a^2 + 6a + 9 - c^2$

36. $x^2 - 4x + 4 - y^2$

37. $25y^2 - 4x^2 + 12x - 9$

38. $121m^2 - 4n^2 - 52n - 169$

39. $x^{2r+1} + xy + 3x^{2r}z^{3r} + 3z^{3r}y$

40. $2a^{k+3} + 2a^2b^k + a^{k+1}c^{2k+3} + b^kc^{2k+3}$

41. $m^rp + np + m^rn^{2r} + n^{2r+1}$

42. $2a^{n+2} + 2a^2b^2 + a^nc^{2n} + b^2c^{2n}$

43. $x^{2a} + 2x^ay^b + y^{2b} - 1$

44. $m^{2x} - 2m^xn^y + n^{2y} - 4$

Applications

Geometry Write a polynomial in factored form for each.

45. A square is circumscribed about a circle. The area of the circle is $49y^2$ and the area of the square is $9x^2 - 6xy + y^2$. Write a polynomial in factored form to represent the difference of the two areas.

46. A square is enclosed in a circle. The area of the square is $(4r^2 - 32r + 64)$ and the area of the circle is $484r^2$. Write a polynomial in factored form to represent the difference of the two areas.

HISTORICAL NOTE

Throughout the history of mathematics, many of the very best scholars have been fascinated by the study of prime numbers. Sometime between 300 and 200 B.C., Euclid developed his proof that there is an infinite number of primes, and Eratosthenes developed his "sieve" for determining the primes less than a given number. An unwieldy formula was derived from the Sieve of Eratosthenes for determining the number of primes less than a given number when the primes less than the square root of the given number are known. In 1870, Ernst Meissel improved on the formula derived from the work of Eratosthenes by which he was able to show that the number of primes below 10^8 is 5,761,455. Twenty-three years later, the Danish mathematician Bertelsen proclaimed that the number of primes less than 10^9 is 50,847,478.

Mathematicians also have long been challenged by finding some method to test whether a large number is prime. In 1876, the French mathematician Anatole Lucas verified the 39-digit number $2^{127} - 1$ as a prime, and for more than 75 years this was the greatest verified prime. Since the advent of modern computer techniques, greater primes have been verified.

Research: Find out what the largest verified prime number is today.

7.8 Factoring Completely

Objective: To factor a polynomial completely

Factoring is an important skill in algebra. The techniques for factoring polynomials that have been developed in previous lessons will enable you to factor a polynomial completely. A polynomial is **factored completely** when it is written as a product of prime polynomials. Recall that a prime polynomial is one that cannot be factored.

Capsule Review

When you factor a polynomial of any number of terms, the first step is to factor out the greatest common monomial factor, if any.

EXAMPLE $4x^2y^3 - 6x^2y^2 + 2xy^2 = 2xy^2(2xy - 3x + 1)$

Factor out the greatest common factor (GCF).

1. $5x + 10y$
2. $2h^4 + 4h^3$
3. $9c^2 - 3c$
4. $6ab^2 + 8a^3b^2$
5. $2m^3n - 8m^5n^2$
6. $5s^2t - 15st^4$
7. $9x^2y^3 - 12xy^4 + 8xy^2$
8. $15a^4b^4 - 10a^4b^2 - 5b^4$

After you have factored out the greatest common monomial factor, if any, factor the remaining polynomial as before.

> To factor a polynomial completely:
> 1. Factor out the greatest monomial factor (GCF).
> 2. If the polynomial has two or three terms, look for:
>   - a perfect square trinomial,
>   - a difference of two squares, or
>   - a pair of binomial factors.
> 3. If there are four or more terms, group terms, if possible, in ways that can be factored. Then factor out any common polynomials.
> 4. Check that each factor is prime.
> 5. Check your answer by multiplying all the factors.

EXAMPLE 1 **Factor:** $5a^3 - 20ab^2$

$5a^3 - 20ab^2$
$= 5a(a^2 - 4b^2)$ 5a is the GCF.
$= 5a(a - 2b)(a + 2b)$ $a^2 - 4b^2$ is a difference of two squares: $(a)^2 - (2b)^2$.

Check: $5a(a - 2b)(a + 2b) = 5a(a^2 + 2ab - 2ab - 4b^2)$
$ = 5a(a^2 - 4b^2) = 5a^3 - 20ab^2$

EXAMPLE 2 **Factor:** $32x^3 - 16x^2 + 2x$

$32x^3 - 16x^2 + 2x$
$= 2x(16x^2 - 8x + 1)$ Factor out 2x.
$= 2x(4x - 1)(4x - 1)$ $16x^2 - 8x + 1$ is a perfect square trinomial.
$= 2x(4x - 1)^2$

Check: $2x(4x - 1)(4x - 1) = 2x(16x^2 - 8x + 1)$
$ = 32x^3 - 16x^2 + 2x$

EXAMPLE 3 **Factor:** $18x^4y + 15x^3y - 12x^2y$

$18x^4y + 15x^3y - 12x^2y = 3x^2y(6x^2 + 5x - 4)$
$ = 3x^2y(3x + 4)(2x - 1)$

Check: $3x^2y(3x + 4)(2x - 1) = 3x^2y(6x^2 + 5x - 4)$
$ = 18x^4y + 15x^3y - 12x^2y$

Sometimes a pattern for factoring is not immediately obvious. After you have factored out the GCF, the pattern for further factoring may be easier to see.

EXAMPLE 4 **Factor:** $3x^9 - 48x$

$3x^9 - 48x$
$= 3x(x^8 - 16)$ Factor out 3x.
$= 3x(x^4 + 4)(x^4 - 4)$ $x^8 - 16$ is a difference of squares:
$= 3x(x^4 + 4)(x^2 - 2)(x^2 + 2)$ $(x^4)^2 - (4)^2$.

Why is $x^2 + 2$ not factored? Why is $x^4 + 4$ not factored?

CLASS EXERCISES

Find the GCF.

1. $2x^2 + 4x + 6$
2. $2r^3 + 4r^2$
3. $3xy^2 + 9x^3y$
4. $12a^2b^3 + 14ab^4$
5. $9r^2s^3 + 12rs^4 + 8r^3s^2$
6. $128x^3yz^3 + 16xy^5z^4 + 48x^4y^3z$

Factor.

7. $3x^2 + 21x^4$
8. $3b^2 + 6b + 3$
9. $4z^2 - 4$
10. $2w^2 + 8w - 24$
11. $t^3 + 7t^2 + 12t$
12. $4h^3 + 12h^2 + 8h$

PRACTICE EXERCISES

Factor completely.

1. $5x^2 - 20$
2. $6a^2 - 6$
3. $3k^2 - 147$
4. $5m^2 - 125$
5. $4y^3 - 36yz^2$
6. $27p^3 - 108pq^2$
7. $6x^3 - 24xy^2$
8. $3r^3 - 48rs^2$
9. $2x^2 + 6x - 20$
10. $3x^2 - 6x - 24$
11. $6k^2 + 12k + 6$
12. $8c^2 - 24c + 16$
13. $10k^2 + 35k + 15$
14. $12r^2 - 45r - 12$
15. $-4x^2 - 4x + 24$
16. $-3d^2 - 6d + 24$
17. $-10m^2 + 40m + 210$
18. $-6x^2 + 36x + 96$
19. $75x^3 - 30x^2 + 3x$
20. $48y^3 - 24y^2 + 3y$
21. $12x^3 + 24x^2 + 12x$
22. $27m^3 + 36m^2 + 12m$
23. $8x^4y + 4x^3y - 12x^2y$
24. $24m^4n - 12m^3n - 12m^2n$
25. $12r^4s^2 + 6r^3s^2 - 6r^2s^2$
26. $45j^4k^2 + 45j^3k^2 - 20j^2k^2$
27. $18x^5y^3 - 15x^4y^4 - 18x^3y^5$
28. $32x^5y^2 - 48x^4y^3 - 32x^3y^4$
29. $16x^6y^4 - 48x^5y^5 + 36x^4y^6$
30. $18a^5b^3 - 60a^4b^4 + 50a^3b^5$
31. $-12m^7p^2 - 60m^6p^3 - 75m^5p^4$
32. $-36d^5f^5 - 96d^4f^6 - 64d^3f^7$
33. $2x^9 - 50x$
34. $3r^9 - 27r$
35. $2x^7 - 32x$
36. $3x^7 - 75x$
37. $3m^5 - 60m^3 + 192m$
38. $2m^5 - 68m^3 + 450m$
39. $18r^6 - 170r^4 + 72r^2$
40. $8r^6 - 50r^4 + 72r^2$
41. $192x^6y^5 - 144x^5y^5 + 27x^4y^5$
42. $75x^6y^5 - 60x^5y^5 + 12x^4y^5$
43. $6(a - 1)^2 - 15(a - 1) - 9$
44. $4(b + 2)^2 + 20(b + 2) - 24$
45. $5(a + 1)^2 - 5(a - 1) - 20$
46. $3(a + 2)^2 - 3(a - 1) - 27$
47. $a^4 + 4a^3 - 5a^2 - 36a - 36$
48. $a^4 + 4a^3 - 12a^2 - 64a - 64$
49. $3a^{17} + 30a^9 + 75a$
50. $4a^{19} + 24a^{10} + 36a$
51. $x^{3k+21} + 2x^{2k+14} + x^{k+7}$
52. $3a^{9z-27} - 13a^{6z-18} - 10a^{3z-9}$

Applications

Geometry The volume (*lwh*) of each rectangular prism below is expressed as a polynomial. Factor each polynomial into three factors corresponding to the length, width, and height of the prism.

53.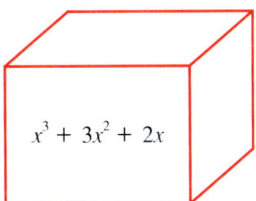
$x^3 + 3x^2 + 2x$

54.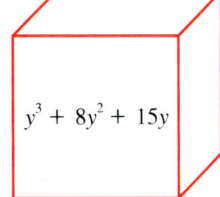
$y^3 + 8y^2 + 15y$

55.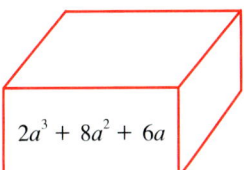
$2a^3 + 8a^2 + 6a$

56.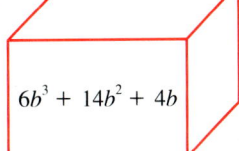
$6b^3 + 14b^2 + 4b$

ALGEBRA IN GEOMETRY

Applying what you know about factoring can often make computations easier to perform. Suppose, for example, that you must find the area of the shaded region in the figure below for different values of *x* and *y*. One way is to find the area of each square and then find the difference.

$$x^2 - y^2$$

Another way is to use the factored form.

$$(x - y)(x + y)$$

Suppose $x = 16$ and $y = 14$. Which of these two ways is easier to use?

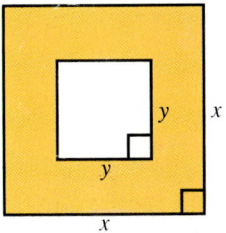

For each figure below, write and factor an expression that makes the computation for finding the area of the shaded region easy to perform.

1.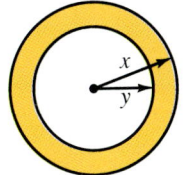

2.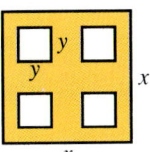

3.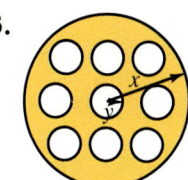

7.8 Factoring Completely **299**

7.9 Solving Polynomial Equations by Factoring

Objective: To solve polynomial equations by factoring

One important use of factoring is in finding the solutions of *polynomial equations*. A **polynomial equation** is an equation whose sides are both polynomials. Some polynomial equations are given special names.

For all real numbers a, b, and c and $a \neq 0$:

$$ax + b = 0 \text{ is called a } linear\ equation$$
$$ax^2 + bx + c = 0 \text{ is called a } quadratic\ equation$$

Capsule Review

A number that gives a true sentence is a solution of the equation. The set of all solutions is the solution set of the equation. You can check solutions by substituting in the original equation.

EXAMPLE $x^2 = x + 6$ Replacement set: $\{0, 3\}$

Try 0. $0^2 \stackrel{?}{=} 0 + 6$
$0 \neq 6$

Try 3. $3^2 \stackrel{?}{=} 3 + 6$
$9 = 9$ ✓ True

So, 3 is a solution of $x^2 = x + 6$ and 0 is *not* a solution.

Test the numbers in the replacement set to see which are solutions.

1. $3x^2 - 12 = 0$ $\{4, -2\}$
2. $y^2 - 7y = -12$ $\{3, 4\}$
3. $9m^2 = 3m$ $\{0, -\frac{1}{3}\}$
4. $a^2 = \frac{1}{4}$ $\{\frac{1}{2}, -\frac{1}{2}\}$

Some polynomial equations can be solved by first factoring and then applying the *zero-product property*. This property states that if the product of two factors is zero, then one or both of the factors must be zero.

Zero-Product Property

For all real numbers a and b, if $ab = 0$, then $a = 0$ or $b = 0$ or both a and $b = 0$.

The zero-product property is true for any number of factors.

EXAMPLE 1 Solve: $(2x - 1)(x + 6) = 0$

By the zero-product property, at least one of the factors must equal zero.
$(2x - 1)(x + 6) = 0$

$2x - 1 = 0$	$x + 6 = 0$	Set each factor equal to 0.
$2x = 1$	$x = -6$	Solve for x.
$x = \frac{1}{2}$		

Check: Replace x with $\frac{1}{2}$.
$(2x - 1)(x + 6) = 0$
$\left(2 \cdot \frac{1}{2} - 1\right)\left(\frac{1}{2} + 6\right) \stackrel{?}{=} 0$
$(0)\left(6\frac{1}{2}\right) \stackrel{?}{=} 0$
$0 = 0$ ✓

Replace x with -6.
$(2x - 1)(x + 6) = 0$
$[2(-6) - 1](-6 + 6) \stackrel{?}{=} 0$
$(-12 - 1)(0) \stackrel{?}{=} 0$
$(-13)(0) \stackrel{?}{=} 0$
$0 = 0$ ✓

So, the solutions are $\frac{1}{2}$ and -6.

EXAMPLE 2 Solve: $a^2 - 7a = 0$

$a^2 - 7a = 0$
$a(a - 7) = 0$ Factor.

$a = 0$	$a - 7 = 0$	Set each factor equal to 0.
	$a = 7$	

The solutions are 0 and 7. The check is left for you.

Always begin with the equation in the form $ax^2 + bx + c = 0$.

EXAMPLE 3 Solve: $x^2 - x = 6$

$x^2 - x = 6$
$x^2 - x - 6 = 6 - 6$ Subtract 6 from each side.
$x^2 - x - 6 = 0$
$(x - 3)(x + 2) = 0$ Factor.

$x - 3 = 0$	$x + 2 = 0$	Set each factor equal to 0.
$x = 3$	$x = -2$	

Check the solutions by substituting in the original equation.

Check:
$x^2 - x = 6$
$3^2 - 3 \stackrel{?}{=} 6$
$6 = 6$ ✓   Replace x with 3.

$x^2 - x = 6$
$(-2)^2 - (-2) \stackrel{?}{=} 6$
$6 = 6$ ✓   Replace x with -2.

The solutions are 3 and -2.

A polynomial equation may have more than one *solution*, or *root*. Sometimes, however, the roots may be the same number.

EXAMPLE 4 Solve: $25x^2 - 10x + 1 = 0$

$$25x^2 - 10x + 1 = 0$$
$$(5x - 1)(5x - 1) = 0$$
$$5x = 1 \quad | \quad 5x = 1$$
$$x = \frac{1}{5} \quad | \quad x = \frac{1}{5}$$

The solution is $\frac{1}{5}$. Check by substituting $\frac{1}{5}$ for x in the original equation.

CLASS EXERCISES

Solve. Check your solutions.

1. $2r(r - 3) = 0$
2. $(x - 4)(3x + 2) = 0$
3. $3m^2 - m = 0$
4. $2z^2 + 3z = 0$
5. $y^2 + 7y + 6 = 0$
6. $2x^2 + 13x - 7 = 0$
7. $2m^2 + m - 1 = 0$
8. $c^2 + 8c + 16 = 0$
9. $x^2 - 4x = 12$
10. $3w^2 + 5w = 12$

PRACTICE EXERCISES

Solve. Check your solutions.

1. $(3x - 2)(x + 4) = 0$
2. $(4z - 3)(z + 5) = 0$
3. $(2x - 6)(x + 6) = 0$
4. $(5x - 8)(x + 9) = 0$
5. $(5x + 3)(x - 2) = 0$
6. $(7x + 5)(x - 3) = 0$
7. $m^2 + 4m = 0$
8. $n^2 + 5n = 0$
9. $2w^2 - 8w = 0$
10. $3z^2 - 6z = 0$
11. $5x^2 + 15x = 0$
12. $4r^2 + 16r = 0$
13. $x^2 + 2x = 8$
14. $z^2 - 4z = 21$
15. $3m^2 + 4m = -1$
16. $2z^2 + 4z = 16$
17. $4m^2 - 11m = 3$
18. $5m^2 - 7m = 6$
19. $r^2 - r - 90 = 0$
20. $t^2 - 3t - 18 = 0$
21. $x^2 - 8x + 15 = 0$
22. $3k^2 + 17k + 10 = 0$
23. $3m^2 - 7m - 20 = 0$
24. $2n^2 + 13n - 24 = 0$

Chapter 7 Factoring Polynomials

25. $3r^2 + 4r = 15$
26. $2x^2 - 11x = 21$
27. $6y^2 - 10 = -11y$
28. $15s^2 - 28 = s$
29. $14a^2 = 29a + 15$
30. $6b^2 = -b + 35$
31. $24d^2 + 18d = -4d + 2$
32. $2t^2 + 12t = 18t + 108$
33. $n^3 + 4n^2 - 21n = 0$
34. $m^3 - m^2 - 20m = 0$
35. $6x^3 - 7x^2 - 20x = 0$
36. $10n^3 - 29n^2 - 21n = 0$
37. $16j^3 + 44j^2 = 126j$
38. $15k^3 - 114k^2 = 189k$
39. $30x^3 = 21x^2 + 135x$
40. $12b^3 = 86b^2 + 80b$
41. $72a^3 - 132a^2 = -108a^2 + 198a$
42. $120y^3 + 72y^2 = 140y^2 + 84y$
43. $(t - 2)^2 + 7(t - 2) + 12 = 0$
44. $(y + 4)^2 + 3(y + 4) - 10 = 0$
45. $(x + 3)^3 + 2(x + 3)^2 - 8(x + 3) = 0$
46. $(z - 3)^3 + 9(z - 3)^2 + 14(z - 3) = 0$
47. $a^5 - 10a^3 + 9a = 0$
48. $2b^5 - 100b^3 + 98b = 0$

Applications

Number Problems For each problem, let n represent the missing number. Write a quadratic equation to solve the problem.

49. A number squared added to five times the number equals 24.
50. Three times a number, subtracted from the number squared, equals 18.
51. Twice the square of a number equals the difference of the number and 10.

TEST YOURSELF

Factor. If a polynomial cannot be factored write *prime*.   7.6–7.8

1. $2y^2 - 98$
2. $5y + 25y^3$
3. $4a^3 - 12a^2 + 8a$
4. $7(m + n) - w(m - n)$
5. $x(1 - y) - 2x(y - 1)$
6. $9y^2 - 4x^2 + 12x - 9$
7. $16x^2 + 64$

Solve. Check your answers.   7.9

8. $m(m - 4) = 0$
9. $a^2 - 9a - 10 = 0$
10. $24b^2 - 32b + 12 = 9b$
11. $2x^2 - 5x - 12 = 0$

7.10 Problem Solving Strategy: Use Polynomial Equations

Many practical problems are solved by setting up and solving a polynomial equation. The polynomial equations in this lesson can be solved by factoring.

EXAMPLE 1   A gardener is planning to make a rectangular garden with an area of 80 ft². She has 12 yd of fencing to put around the perimeter of the garden. What should the dimensions of the garden be?

Understand the Problem

What are the given facts?
The problem involves a rectangular garden with area 80 ft² and perimeter 12 yd.

What are you asked to find?
You are asked to find the length and width of the rectangular garden.

Plan Your Approach

Choose a strategy.
1. Draw a diagram.
2. Units of measure must be the same. Change 12 yd to 36 ft.
$36 = 2l + 2w$
$18 = l + w$
$18 - w = l$

Complete the Work

Write a word equation.
Area of a rectangle equals length times width.

$$A = (18 - w)w$$

$$80 = (18 - w)w \quad \text{Replace A with 80.}$$
$$80 = 18w - w^2$$
$$w^2 - 18w + 80 = 0 \quad \text{Solve the equation.}$$
$$(w - 10)(w - 8) = 0$$

$w - 10 = 0 \qquad\qquad w - 8 = 0$ Set each factor equal to 0.
$w = 10 \qquad\qquad\qquad w = 8$

$l = 18 - 10 = 8 \qquad l = 18 - 8 = 10$

Interpret the Results

State your answer.
The dimensions of the garden are 8 ft and 10 ft.

304   Chapter 7 Factoring Polynomials

Check the conclusion: A rectangle 8 ft by 10 ft has an area of 8 × 10, or 80 ft^2. The perimeter is 2(10) + 2(8), or 36 ft, which is also 12 yd.

EXAMPLE 2 Find three consecutive even integers such that the second times the third is 12 times the first.

Understand the Problem

What are the given facts?
The second number times the third is 12 times first.
Let n = first even integer.
Then $n + 2$ = second even integer and
$n + 4$ = third even integer.

Plan Your Approach

Write an equation.
$$(n + 2) \cdot (n + 4) = 12 \cdot (n)$$

Complete the Work

$(n + 2)(n + 4) = 12n$
$n^2 + 6n + 8 = 12n$ *Multiply $(n + 2)(n + 4)$.*
$n^2 - 6n + 8 = 0$ *Subtract $12n$ from each side.*
$(n - 4)(n - 2) = 0$

$n - 4 = 0 \quad | \quad n - 2 = 0$ *Set each factor equal to 0.*
$n = 4 \quad\quad | \quad n = 2$

$n + 2 = 6 \quad | \quad n + 2 = 4$
$n + 4 = 8 \quad | \quad n + 4 = 6$

Interpret the Results

There are two sets of answers.
The three integers are 4, 6, and 8 or 2, 4, and 6.
Check: $6 \times 8 \stackrel{?}{=} 12 \times 4$ $4 \times 6 \stackrel{?}{=} 12 \times 2$
             $48 = 48$ ✓ $24 = 24$ ✓

CLASS EXERCISES

The length of a rectangular piece of cloth is 3 in. more than its width. If the area of the cloth is 130 in.^{2}, find the width.

1. What information is given? What are you asked to find?
2. What mathematical idea is important in this problem?
3. Draw a diagram to represent the piece of cloth. Label the dimensions in terms of one variable.
4. What equation could you use to solve this problem?
5. Solve the equation and check the answer(s).
6. If the solution set includes a negative number, what does this mean?

PRACTICE EXERCISES

Solve. Check your solutions.

1. The area of a rectangular garden is 140 ft². If its length is 4 ft more than its width, find the dimensions.

2. The length of a rectangular rug is 3 ft more than the width. If the area of the rug is 180 ft², find the dimensions.

3. The perimeter of a rectangular piece of tin is 40 cm. If the area is 84 cm², what are its dimensions?

4. A rectangular patio has an area of 96 m² and its perimeter is 44 m. Find the dimensions of the patio.

5. Find two consecutive even integers whose product is 48.

6. Find two consecutive even integers whose product is 120.

7. The product of two consecutive odd integers is 63. Find the numbers.

8. The product of two consecutive odd integers is 143. Find the numbers.

9. Find the least of three consecutive even integers if the first times the third is 4 greater than seven times the second integer.

10. Find the greatest of three consecutive odd integers if the product of the second and third is 8 more than 13 times the first.

11. Last year Rosa's garden measured 10 m by 25 m. This year she increased the length and decreased the width by the same amount. If her garden has an area of 216 m², find its dimensions.

12. Iris has a square garden and wants to make it larger. If she extended one side 5 ft and the adjacent side 2 ft, the resulting garden would be rectangular with an area of 130 ft². What would its dimensions be?

13. When Seth draped a rectangular tablecloth over a square table, 5 in. of cloth hung over each of two opposite sides and 6 in. hung over the other two sides. Find the dimensions of the table if the area of the tablecloth is 1680 in.².

14. A rectangular tablecloth has an area of 1085 in.². When it is draped over a square table, it hangs 5 in. over two opposite sides and 3 in. over the other two sides. Find the dimensions of the table.

15. A rectangular photograph is mounted in a frame which is 1 in. wide. If the photograph is twice as long as it is wide and the area of the photograph and frame together is 60 in.², what are the dimensions of the photograph?

16. A rectangular swimming pool, 10 ft by 25 ft, is surrounded by a deck that is the same width all around the pool. If the pool and deck together have an area of 594 ft², how wide is the deck?

Mixed Problem Solving Review

1. The length of a rectangle is 6 times its width. The perimeter of the rectangle is 70 cm. Find the length and width.

2. Ms. Wayne, a chemist, needs to make a 12% alcohol solution. How many milliliters of an 8% alcohol solution must be added to 10 mL of a 20% alcohol solution to get a 12% solution?

3. Lauren and Erik are 228 km apart. To meet, Lauren drives 64 km/h and Erik drives 56 km/h. Lauren is delayed by traffic for 45 minutes. How soon will they meet?

4. David is 10 kg heavier than Joanne. Together they weigh less than 100 kg. How much do they each weigh?

PROJECT

Work together in small groups. Let your group be a club that studies, builds, and launches model rockets. It is important to know the formula $h = vt - 5t^2$. It tells you the height h in meters of a rocket t seconds after launch at an initial velocity of v meters per second.

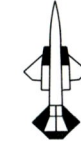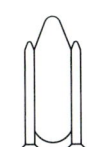

EXAMPLE Suppose you launch a rocket with an initial velocity of 30 m/s. How high will the rocket be 4 s after launch?

Write the formula. $h = vt - 5t^2$
$h = 30(4) - 5(4^2) = 120 - 80 = 40$

So, the rocket will be 40 m high.

1. You launch your rocket with an initial velocity of 45 m/s. How high will it be after 6 s?

2. What initial velocity is necessary for a rocket if you want it to have an altitude of 50 m, 10 s after launch?

3. If you launch a rocket at an initial velocity of 50 m/s, how many seconds after launch will it strike the ground?

4. Suppose the record flight in your state is 32 seconds. Your club wants to beat this record. This means you must design and build a rocket that can be launched at a velocity that will keep it in the air for longer than 32 seconds. What velocity is that?

APPLICATION: Agriculture

Did you know that $\frac{1}{5}$ of the total number of farms in the United States account for 80% of the produce sold in this country? Farmland covers about $\frac{1}{2}$ of the area of the United States, that is more than one billion acres. There are about 2,300,000 farms with an average size of about 440 acres. This intensive agriculture makes the best use of the land.

Most farmers grow their produce in fields and store it in a building or bin. Rectangular and round bins are commonly used on farms for storing small grains. To determine their storage needs farmers take into consideration the capacity of a bin. The capacity of a bin is represented in units, usually **bushels.** A bushel is a unit of dry measure for grain, fruit, and so on, equal to 4 pecks or 32 qt.

To compute the number of bushels of grain in a bin use the following formulas, in which the dimensions are in feet.

Rectangular Bins
Length × width × height × 0.8 = $lwh(0.8)$ = number of bushels

Round Bins
Diameter squared × height × 0.625 = $d^2(h)(0.625)$ = number of bushels

EXAMPLE 1 A farmer has a rectangular bin with the following dimensions: $l = 20$ ft, $w = 18$ ft and $h = 12$ ft. How much grain can be stored in two bins?

$lwh(0.8)$ = number of bushels
$(20)(18)(12)(0.8) = 3456$ bushels
$3456 \times 2 = 6912$ bushels

The farmer can store 6912 bushels in two bins. Farmers must also use other formulas in their work. Often they must estimate the amount of fencing needed to enclose their land.

EXAMPLE 2 A farmer has 2 mi of fencing to enclose 100 acres for a wheat field. Estimate the dimensions of the field.

$$\text{Hint: } 1 \text{ mi} = 5280 \text{ ft} \qquad 1 \text{ acre} = 43{,}530 \text{ ft}^2$$

Estimate: 2 mi is about 10,000 ft
$2l + 2w = 10{,}000$
$l = 5000 - w$

100 acres is about 4,000,000 ft²
$l \cdot w = 4{,}000{,}000$
$(5000 - w)w = 4{,}000{,}000$
$w^2 - 5000w + 4{,}000{,}000 = 0$
$(w - 4000)(w - 1000) = 0$
$l = 4000 \qquad w = 1000$

The wheat field is about 4000 ft by 1000 ft.

A farmer can intensify his efforts by using the same amount of fencing to enclose a greater area. With the same 2 mi of fencing, find the approximate dimensions of the field with the greatest area.

Explain the problem solving strategy you would use to solve this problem.

You could use the computer program on page 272 to solve the problem.

How are the problem solving strategy and the computer program similar? How do they differ?

Solve. Use the computer program on page 272 for Exercises 4 and 6, if you wish.

1. What is the capacity in bushels of a rectangular bin that has 5040 ft³ of space?

2. What is the capacity in bushels of a round bin that has a diameter of 16 ft and a height of 12 ft?

3. The length of a rectangular bin is twice its width. The height of the bin is 14 ft and the area of the floor of the bin is 450 ft². Find the length, width, and the capacity of bushels of the bin.

4. A farmer has 2 mi of fencing to enclose 150 acres for a corn field. Estimate the dimensions of the field.

5. The length of a corn field is 4 times its width. The perimeter is 3 mi. Estimate how many plants can be grown in the field if you grow 28,000 plants per acre.

6. A corn field is bounded on one side by a stone wall 1 mi long. Estimate how many acres a farmer can enclose with 2 mi of fencing.

Application: Agriculture

CHAPTER 7 SUMMARY AND REVIEW

Vocabulary

common factor (274)
composite number (270)
difference of two squares (290)
factor completely (296)
greatest common factor (GCF) (274)
perfect square trinomial (289)
polynomial equation (300)
prime number (270)
prime factorization (270)
quadratic equation (300)
zero-product property (300)

Factors and Powers   A prime number has exactly two factors—itself and 1. A composite number can be written as a product of prime numbers.   7.1

State whether the number is *prime* or *composite*.

1. 23
2. 33
3. 45
4. 89

Find the prime factorization.

5. 48
6. 300
7. 120
8. 64

Factoring a Polynomial   The greatest common factor (GCF) of two or more integers is the greatest integer that is a factor of each.   7.2

Find the GCF of each group of numbers.

9. 35 and 200
10. $12x^2$ and $16x^3y$

Factor.

11. $3ab^2 - 9a^2b^2$
12. $6m^3n^2 + 12m^2n^2 - 18m^2n$

Factoring a Trinomial $x^2 + bx + c$ Factoring a trinomial means to express the trinomial as the product of two binomials of the form $(x + r)(x + s)$.   7.3, 7.4

Factor. Check by multiplying.

13. $x^2 + 5x + 6$
14. $y^2 - 14y + 33$
15. $a^2 + 12ab + 11b^2$
16. $c^2 + c - 56$
17. $y^2 - 10y - 24$
18. $x^2 + 2xy - 15y^2$

Factoring a Trinomial $ax^2 + bx + c$ To factor trinomials when the coefficient of the first term is not 1, you must test various factors of both the first term of the trinomial and the last term.   7.5

Factor. Check by multiplying.

19. $2m^2 + 15m + 7$ **20.** $3x^2 + 5x - 12$ **21.** $6x^2 - x - 15$

Special Cases of Polynomial Factoring 7.6

$a^2 + 2ab + b^2 = (a + b)(a + b) = (a + b)^2$
$a^2 - 2ab + b^2 = (a - b)(a - b) = (a - b)^2$
$a^2 - b^2 = (a + b)(a - b)$

Factor.

22. $w^2 - 49$ **23.** $100m^2 - 16n^2$ **24.** $y^2 + 2y + 1$

Factoring a Polynomial Completely To factor a polynomial 7.7, 7.8
completely, look for:

- greatest common monomial factor, GCF
- a perfect square trinomial
- a difference of two squares
- a pair of binomial factors
- a grouping of terms

Check that each factor is prime.

Check your answer by multiplying all the factors.

Factor.

25. $7m^2 - 28$ **26.** $x^3y - 9xy$ **27.** $6a^3 + 12a^2 + 9a$
28. $3w(z + 3) - 2(z + 3)$ **29.** $9(x - y) + 4(y - x)$
30. $cx - cd + dy - xy$ **31.** $36x^3 - 24x^2 + 4x$
32. $75x^2 - 3$ **33.** $18x^2 - 45$

Solving Polynomial Equations Factoring can be used to find solutions 7.9
of polynomial equations.

Solve. Check your solutions.

34. $m^2 + 6m = 0$ **35.** $r^2 - 2r - 80 = 0$ **36.** $t^2 - 3t = 18$

Using Polynomial Equations Practical problems can be solved by 7.10
setting up and solving a polynomial equation.

37. Find two consecutive even integers whose product is 728.

38. The length of a rectangular rug is 3 ft more than the width. If the area of the rug is 270 ft^2, find its dimensions.

CHAPTER 7 TEST

State whether the number is *prime* or *composite*.

1. 25
2. 29
3. 17
4. 63

Find the prime factorization.

5. 24
6. 200
7. 34
8. 75

Factor. Check by multiplying. Write *prime* if the polynomial is prime.

9. $9a + 6b$
10. $25 - 5y^2$
11. $8a^2b^3 - 6ab^2$
12. $a^2 - 16$
13. $x^2 - 25y^2$
14. $y^2 + 4y + 4$
15. $x^2 + 6x + 5$
16. $n^2 - 16n + 55$
17. $p^2 + p + 20$
18. $x^2 + 4xy - 21y^2$
19. $2m^2 + 7m + 3$
20. $6x^3 - 2x^2 - 20x$
21. $2d(f + 2) - (f + 2)$
22. $5(a - b) - 2(b - a)$
23. $dy - de + ez - yz$
24. $16x^2 - 8xy + y^2 - 9$

Solve. Check your solutions.

25. $(x - 2)(3x + 1) = 0$
26. $b^2 + 5b = 0$
27. $s^2 - 6s + 8 = 0$
28. $z^2 - 14z = -49$

29. The sum of a negative number and its square is 72. Find the number.

30. Find two consecutive odd integers whose product is 255.

31. The length of a rectangular garden is three times it width. Increasing its width by 1 ft and its length by 3 ft results in a garden with an area of 75 ft². Find the new dimensions.

Challenge

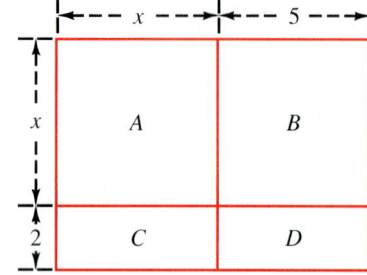

a. Find the area of the smaller rectangles A, B, C, and D.

b. Write the total area of the large rectangle as the sum of the areas of the smaller rectangles.

c. Simplify the expression in (b).

d. Write the total area in factored form.

PREPARING FOR STANDARDIZED TESTS

Directions: In each item you are to compare a quantity in column 1 with a quantity in column 2. Write the letter of the correct answer from these choices:

A. The quantity in column 1 is greater than the quantity in column 2.
B. The quantity in column 2 is greater than the quantity in column 1.
C. The quantity in column 1 is equal to the quantity in column 2.
D. The relationship cannot be determined from the given information.

Notes: Information centered over both columns refers to one or both of the quantities being compared. A symbol that appears in both columns has the same meaning in each column. All variables represent real numbers.

Column 1	Column 2
1. $\frac{3}{2} + \frac{4}{5}$	$\frac{2}{3} + \frac{5}{4}$
2. $\frac{1}{3}$ of 276	$\frac{3}{5}$ of 150

$12 < n$ and $n < 42$

3. Average of 12, n, and 42	27

$a = -2, b = 3$

4. $a^2b - ab$	$1 - ab^2$
5. 0.0064	$\frac{2}{625}$

$n > 5$
$n < t$

6. 5	t
7. 30% profit on $650	25% profit on $765
8. 3.4×10^3	3400
9. $|3(2-6)+4|$	$|6-4(-2)|$

Column 1	Column 2

$x < 0$

10. x	$\frac{1}{x}$
11. $(a+b)(a-b)$	$a^2 - b^2$

Use this circle graph to answer questions 12–14.

History test grades given to 35 9th graders

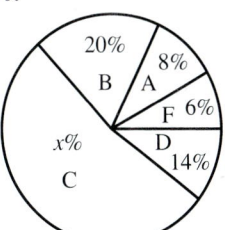

(*Note:* All numbers rounded to the nearest whole number and figure not drawn to scale.)

12. x	48
13. Total number of A's given.	4
14. Total receiving a C or better	25

Preparing for Standardized Tests **313**

MAINTAINING SKILLS

Multiply.

Example 1 $\dfrac{3}{4} \times \dfrac{4}{15} \qquad \dfrac{\overset{1}{\cancel{3}}}{\underset{1}{\cancel{4}}} \times \dfrac{\overset{1}{\cancel{4}}}{\underset{5}{\cancel{15}}} = \dfrac{1 \times 1}{1 \times 5} = \dfrac{1}{5}$

Divide.

Example 2 $\dfrac{2}{3} \div \dfrac{8}{9} \qquad \dfrac{2}{3} \div \dfrac{8}{9} = \dfrac{\overset{1}{\cancel{2}}}{\underset{1}{\cancel{3}}} \times \dfrac{\overset{3}{\cancel{9}}}{\underset{4}{\cancel{8}}} = \dfrac{1 \times 3}{1 \times 4} = \dfrac{3}{4}$

Perform the indicated operation.

1. $\dfrac{3}{4} \times \dfrac{2}{3}$
2. $\dfrac{2}{7} \times \dfrac{5}{10}$
3. $\dfrac{5}{8} \times \dfrac{4}{25}$
4. $\dfrac{3}{4} \times \dfrac{3}{7}$
5. $\dfrac{4}{5} \div \dfrac{14}{15}$
6. $\dfrac{3}{8} \div \dfrac{9}{10}$
7. $\dfrac{5}{7} \div \dfrac{2}{5}$
8. $\dfrac{5}{6} \div \dfrac{5}{8}$

Rewrite each pair of fractions using the least common denominator (LCD).

Example 3 $\dfrac{5}{12}$ and $\dfrac{7}{8} \qquad \dfrac{5}{12} = \dfrac{5}{12} \cdot \dfrac{2}{2} = \dfrac{10}{24}$ and $\dfrac{7}{8} = \dfrac{7}{8} \cdot \dfrac{3}{3} = \dfrac{21}{24}$

9. $\dfrac{5}{9}$ and $\dfrac{2}{3}$
10. $\dfrac{6}{11}$ and $\dfrac{1}{3}$
11. $\dfrac{5}{18}$ and $\dfrac{7}{12}$
12. $\dfrac{3}{8}$ and $\dfrac{5}{6}$

Perform the indicated operation.

Example 4 $\dfrac{5}{12} + \dfrac{7}{8} \qquad \dfrac{5}{12} + \dfrac{7}{8} = \dfrac{5}{12} \cdot \dfrac{2}{2} + \dfrac{7}{8} \cdot \dfrac{3}{3} = \dfrac{10}{24} + \dfrac{21}{24} = \dfrac{31}{24}$

13. $\dfrac{2}{3} + \dfrac{3}{8}$
14. $\dfrac{7}{12} - \dfrac{3}{18}$
15. $\dfrac{3}{10} + \dfrac{5}{12}$
16. $\dfrac{3}{4} - \dfrac{7}{12}$

Find the area.

Example 5

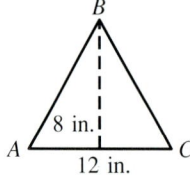

Area $= \dfrac{1}{2}$ base $\times$ height

$= \dfrac{1}{2}(8)(12) = (4)(12) = 48$

Area $= 48$ in.^{2}

17. rectangle: 4 in. wide, 10 in. long
18. triangle: 9 in. base, 5 in. height

8 Rational Expressions

Food packaging and labels contain nutritional information that is important to a healthy diet. Calculating the calories per serving and the recommended daily percentages of nutrients involves the use of rational expressions.

8.1 Simplifying Rational Expressions

Objectives: To identify values for variables that make a rational expression undefined
To simplify rational expressions

A *rational number* is a number that can be expressed in the form $\frac{m}{n}$, where m and n are integers, $n \neq 0$. These numbers are examples of rational numbers:

$$5 \qquad \frac{2}{3} \qquad 6\frac{1}{2} \qquad -\frac{3}{8}$$

Recall that division by 0 is undefined. For example, $\frac{1}{z}$ is undefined when $z = 0$. Identifying values for variables that make a rational expression undefined may require solving a quadratic equation.

Capsule Review

To solve a quadratic equation, factor and apply the *zero product property*.

EXAMPLE Solve: $x^2 - 8x = 0$

$$x^2 - 8x = x(x - 8) = 0$$
$$x = 0 \quad | \quad x - 8 = 0$$
$$ x = 8$$

Solve.

1. $a^2 + 3a = 0$
2. $2b^2 - 3b = 0$
3. $5c^2 - c = 0$
4. $x^2 - 7x + 6 = 0$
5. $2y^2 + y - 6 = 0$
6. $3z^2 + 7z - 20 = 0$

Rational numbers and *rational expressions* are similar. A rational expression is usually expressed as the quotient of two polynomials. Examples of rational expressions are:

$$za + 1 \qquad \frac{1}{z} \qquad \frac{x}{z^2 - 8x} \qquad \frac{a + 1}{4a^2 - 1}$$

> A **rational expression** is an expression that can be written in the form $\frac{p}{q}$ where p and q are polynomials, $q \neq 0$.

Chapter 8 Rational Expressions

Recall that rational numbers cannot have denominators equal to zero. In a rational expression it is important to determine the values of the variable for which the denominator $q = 0$. For these values the rational expression is undefined.

EXAMPLE 1 State the values for which $\dfrac{a + 1}{4a^2 - 1}$ is undefined.

The rational expression is undefined if the denominator is equal to 0. Set the denominator equal to 0 and solve by factoring as shown in this example.

$$\text{Let } 4a^2 - 1 = 0$$
$$(2a - 1)(2a + 1) = 0$$
$$2a - 1 = 0 \quad | \quad 2a + 1 = 0$$
$$a = \tfrac{1}{2} \quad | \quad a = -\tfrac{1}{2}$$

The expression is undefined when $a = \tfrac{1}{2}$ or when $a = -\tfrac{1}{2}$. How do you know? One way to check is to store $\tfrac{1}{2}$, or 0.5, into the memory of your calculator and use it to evaluate the expression. What happens? Why? How would you check the value when $a = -\tfrac{1}{2}$?

A rational expression is in **simplest form** when the numerator and the denominator have no common factors other than 1.

EXAMPLE 2 Simplify: $\dfrac{12x}{42x^2}$

$\dfrac{12x}{42x^2} = \dfrac{\cancel{12}^{2}\cancel{x}}{\cancel{42}_{7}\cancel{x^2}_{x}}$ *Divide 12 and 42 by the common factor, 6.*
Divide x and x^2 by the common factor, x.

$= \dfrac{2}{7x} \quad x \neq 0$ *The numerator and denominator have no common factors other than 1. The expression is in simplest form. x cannot equal zero, since the expression would be undefined.*

EXAMPLE 3 Simplify $\dfrac{5 + x}{2x^2 + 9x - 5}$, and state any restrictions on the variable.

$\dfrac{5 + x}{2x^2 + 9x - 5} = \dfrac{\cancel{5 + x}^{1}}{(2x - 1)\cancel{(x + 5)}_{1}}$ *Factor the denominator.*
Divide by the GCF, x + 5.

$= \dfrac{1}{(2x - 1)1} = \dfrac{1}{2x - 1}$

So, $\dfrac{5 + x}{2x^2 + 9x - 5}$ in simplest form is $\dfrac{1}{2x - 1}$. The factored expression $\dfrac{5 + x}{(2x - 1)(x + 5)}$ shows restrictions on the variable: $x \neq \tfrac{1}{2}$ and $x \neq -5$.

8.1 Simplifying Rational Expressions   **317**

It is important to notice that when two binomials are *opposites*, the property of -1 can be used.

EXAMPLE 4 Simplify: $\dfrac{c^2 + 2c - 8}{4 - c^2}$

$\dfrac{c^2 + 2c - 8}{4 - c^2} = \dfrac{(c - 2)(c + 4)}{(2 - c)(2 + c)}$ Factor. Look for opposites.
$c - 2$ and $2 - c$ are opposites.
$2 - c = -1(-2 + c)$, or $-1(c - 2)$

$= \dfrac{(c + 4)}{-1(c + 2)}$ Write $2 - c$ as $-1(c - 2)$.
Divide by the GCF.

$= \dfrac{c + 4}{-1(c + 2)} = -\dfrac{c + 4}{c + 2}$

So, $\dfrac{c^2 + 2c - 8}{4 - c^2} = -\dfrac{c + 4}{c + 2}$; $c \neq -2$ and $c \neq 2$.

CLASS EXERCISES

State the value of the variable for which the expression is undefined.

1. $\dfrac{4}{d - 2}$
2. $\dfrac{x - 7}{2x + 1}$
3. $\dfrac{5m}{(m - 5)(3m - 2)}$
4. $\dfrac{11}{y^2 - y - 12}$

Simplify, and state any restrictions on the variable.

5. $\dfrac{15b^3}{25b^2}$
6. $\dfrac{18m^3n^2}{6m^2n^3}$
7. $\dfrac{4d - 20}{12}$
8. $\dfrac{3x + 12}{4 + x}$
9. $\dfrac{12c + 18}{9c + 6}$
10. $\dfrac{3 - z}{(z + 5)(z - 3)}$
11. $\dfrac{3a - 9}{a^2 - a - 6}$
12. $\dfrac{16 - y^2}{y^2 - 7y + 12}$

PRACTICE EXERCISES

State the value of the variable for which the expression is undefined.

1. $\dfrac{11}{m + 4}$
2. $\dfrac{5}{a + 7}$
3. $\dfrac{3x}{10x + 4}$
4. $\dfrac{7n}{8n - 6}$
5. $\dfrac{x + 1}{9x^2 - 1}$
6. $\dfrac{m + 4}{16m^2 - 1}$
7. $\dfrac{b - 4}{b^2 - 9}$
8. $\dfrac{y - 3}{y^2 - 16}$

Simplify, and state any restrictions on the variable.

9. $\dfrac{3c}{12c^2}$
10. $\dfrac{4x}{28x^2}$
11. $\dfrac{6a + 9}{12}$
12. $\dfrac{24z + 18}{36}$
13. $\dfrac{7a - 14}{a - 2}$
14. $\dfrac{5r - 15}{r - 3}$
15. $\dfrac{5 - 2m}{6m - 15}$
16. $\dfrac{24 - 2p}{4p - 48}$

Chapter 8 Rational Expressions

17. $\dfrac{3+x}{2x^2+5x-3}$ 18. $\dfrac{2+x}{3x^2+2x-8}$ 19. $\dfrac{x+2}{5x^2+7x-6}$

20. $\dfrac{k^2+4k-12}{4-k^2}$ 21. $\dfrac{c^2+3c-18}{9-c^2}$ 22. $\dfrac{a^2+2a-8}{16-a^2}$

23. $\dfrac{y^2+5y+6}{y^2-4}$ 24. $\dfrac{h^2+h-12}{h^2-9}$ 25. $\dfrac{m^2+m-20}{m^2-25}$

Simplify. State the values for which the expression is undefined.

26. $\dfrac{18xy^3}{24x^3y^2}$ 27. $\dfrac{25a^2c^5}{15a^4c^3}$ 28. $\dfrac{32a^3}{16a^2-8a}$

29. $\dfrac{12m^2+8m^2n}{28m^5n^3}$ 30. $\dfrac{4r^2-4r-15}{2r^2-13r+20}$ 31. $\dfrac{7z^2+23z+6}{z^2+2z-3}$

32. $\dfrac{2r^2+9r-5}{r^2+10r+25}$ 33. $\dfrac{4c^2+12c+9}{2c^2-11c-21}$ 34. $\dfrac{4-s-3s^2}{6s^2-s-2}$

35. $\dfrac{4a^2+8a-5}{15-a-2a^2}$ 36. $\dfrac{16+16g+3g^2}{g^2-3g-28}$ 37. $\dfrac{9+11x+2x^2}{x^2-10x-11}$

38. $\dfrac{5z^2+6z-8}{3z^2+5z-2}$ 39. $\dfrac{6r^2-13r+5}{6r^2-r-1}$ 40. $\dfrac{10+c-3c^2}{5c^2-6c-8}$

41. $\dfrac{a^2-5ab+6b^2}{a^2+2ab-8b^2}$ 42. $\dfrac{c^2-6cd+8d^2}{c^2+4cd-12d^2}$ 43. $\dfrac{x^2-y^2}{x^2+4xy+3y^2}$

44. $\dfrac{m^2-n^2}{m^2+11mn+10n^2}$ 45. $\dfrac{9r^2-16s^2}{6r^2-11rs+4s^2}$ 46. $\dfrac{36g^2-49h^2}{18g^2+9gh-14h^2}$

Applications

Baking The cooking time for baked goods can be expressed as the ratio of the surface area and the volume, $\dfrac{S}{V}$. For each item below, express the ratio $\dfrac{S}{V}$ in simplest form.

47. Loaf (cube)
    $S = 6s^2$
    $V = s^3$
    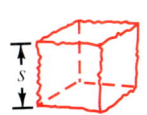

48. Hard roll (half-sphere)
    $S = 2\pi r^2$
    $V = \tfrac{2}{3}\pi r^3$

EXTRA

For all real numbers a, b, and x, tell whether each statement is *always true*, *sometimes true*, or *never true*.

1. $\dfrac{x}{x} = 1$ 2. $\dfrac{ab^2}{b^2} = ab^3$ 3. $\dfrac{x^2+6x-5}{2x+2} = \dfrac{x+5}{2}$

8.2 Multiplying Rational Expressions

Objective: To multiply rational expressions

When you multiply rational numbers, you look for common factors to simplify the product or to simplify the multiplication.

$$\frac{2}{3} \cdot \frac{9}{10} = \frac{2 \cdot 9}{3 \cdot 10} = \frac{\overset{3}{\cancel{18}}}{\underset{5}{\cancel{30}}} = \frac{3}{5} \quad \text{or} \quad \frac{2}{3} \cdot \frac{9}{10} = \frac{\overset{1}{\cancel{2}}}{\underset{1}{\cancel{3}}} \cdot \frac{\overset{3}{\cancel{9}}}{\underset{5}{\cancel{10}}} = \frac{3}{5}$$

This same skill may also be used when you multiply rational expressions.

Capsule Review

EXAMPLE Simplify: $\dfrac{15a^2b}{3a^3}$

$$\frac{15a^2b}{3a^3} = \frac{\overset{5}{\cancel{15}}a^2b}{\underset{a}{\cancel{3a^3}}} = \frac{5b}{a} \qquad \textit{Divide by common factors.}$$

Simplify.

1. $\dfrac{10m^2}{2m^3}$ 2. $\dfrac{12x^2y^3}{4x^3y^2}$ 3. $\dfrac{3m+6}{2m+4}$ 4. $\dfrac{2x^2+x-6}{4-x^2}$

The rule for multiplying rational expressions is similar to the rule for multiplying rational numbers.

> **Multiplication of Rational Expressions**
>
> If $\dfrac{P}{Q}$ and $\dfrac{R}{S}$ are rational expressions with $Q \neq 0$ and $S \neq 0$,
>
> then $\dfrac{P}{Q} \cdot \dfrac{R}{S} = \dfrac{P \cdot R}{Q \cdot S}.$

From now on, to make your work simpler, you will not be asked to state the restrictions upon the variables. You may assume that all denominators do not equal zero.

EXAMPLE 1 Multiply: $\dfrac{2}{g} \cdot \dfrac{5}{3gh}$

$$\frac{2}{g} \cdot \frac{5}{3gh} = \frac{2 \cdot 5}{g \cdot 3gh} = \frac{10}{3g^2h} \qquad \text{Can the product be simplified? Why?}$$

320 Chapter 8 Rational Expressions

EXAMPLE 2 Multiply: $\dfrac{4b}{3a^3} \cdot \dfrac{15ab^2}{2b}$

$\dfrac{4b}{3a^3} \cdot \dfrac{15ab^2}{2b} = \dfrac{\cancel{4b}^{\,2}}{\cancel{3a^3}_{\,a^2}} \cdot \dfrac{\cancel{15}^{\,5}ab^2}{\cancel{2b}_{\,1}}$ *Divide by common factors.*

$\dfrac{2}{a^2} \cdot \dfrac{5b^2}{1} = \dfrac{10b^2}{a^2}$ *Multiply.*

When multiplying rational expressions that have polynomials in the numerator and in the denominator, factor all the polynomials and then look for common factors.

EXAMPLE 3 Multiply: $\dfrac{m-1}{2m+4} \cdot \dfrac{3m+6}{4m-1}$

$\dfrac{m-1}{2m+4} \cdot \dfrac{3m+6}{4m-1} = \dfrac{m-1}{2\cancel{(m+2)}} \cdot \dfrac{3\cancel{(m+2)}}{4m-1}$ *Factor, and then divide by common factors.*

$= \dfrac{3(m-1)}{2(4m-1)} = \dfrac{3m-3}{8m-2}$ *Multiply the remaining factors.*

EXAMPLE 4 Multiply: $\dfrac{2x^2 - 5x - 12}{6x} \cdot \dfrac{-3x - 12}{x^2 - 16}$

$\dfrac{2x^2 - 5x - 12}{6x} \cdot \dfrac{-3x - 12}{x^2 - 16} = \dfrac{(2x+3)\cancel{(x-4)}}{\cancel{6x}_{\,2}} \cdot \dfrac{\cancel{-3}^{\,-1}\cancel{(x+4)}}{\cancel{(x+4)}\cancel{(x-4)}}$ *Factor. Divide.*

$= \dfrac{-1(2x+3)}{2x} = -\dfrac{2x+3}{2x}$ *Multiply.*

EXAMPLE 5 Multiply: $\dfrac{2x^2 - 2x - 4}{4 - x^2} \cdot \dfrac{2x^2 + x - 6}{4x^2 - 2x - 6}$

$\dfrac{2x^2 - 2x - 4}{4 - x^2} \cdot \dfrac{2x^2 + x - 6}{4x^2 - 2x - 6}$

$= \dfrac{2(x+1)(x-2)}{(2-x)(2+x)} \cdot \dfrac{(2x-3)(x+2)}{2(x+1)(2x-3)}$ *Factor.*

$= \dfrac{\cancel{2}\cancel{(x+1)}\cancel{(x-2)}}{-1\cancel{(x-2)}\cancel{(2+x)}} \cdot \dfrac{\cancel{(2x-3)}\cancel{(x+2)}}{\cancel{2}\cancel{(x+1)}\cancel{(2x-3)}}$ *Rewrite opposites. Divide.*

$= \dfrac{1}{-1} = -1$ *Multiply.*

CLASS EXERCISES

Multiply. Express the product in simplest form.

1. $\dfrac{1}{2} \cdot \dfrac{5}{9}$
2. $\dfrac{4}{7} \cdot \dfrac{21}{6}$
3. $\dfrac{m}{6} \cdot \dfrac{m}{2}$
4. $\dfrac{a}{b} \cdot \dfrac{b}{-a}$
5. $\dfrac{uv}{u^2} \cdot \dfrac{uv^2}{v}$
6. $\dfrac{3b^2}{a} \cdot \dfrac{2a^2}{b}$
7. $\dfrac{5ab^2}{c^2} \cdot \dfrac{3ac^3}{10b^3}$
8. $\dfrac{xy + y}{xy} \cdot \dfrac{x}{y}$
9. $\dfrac{s^2 + 3s}{6s + 12} \cdot \dfrac{2s + 4}{s + 3}$
10. $\dfrac{t^2 + 5t + 6}{t - 3} \cdot \dfrac{t^2 - 2t - 3}{t^2 + 3t + 2}$

PRACTICE EXERCISES

Multiply. Express the product in simplest form.

1. $\dfrac{3}{x} \cdot \dfrac{2}{5xy}$
2. $\dfrac{4}{a} \cdot \dfrac{3}{7ab}$
3. $\dfrac{3}{xy} \cdot \dfrac{4}{5x}$
4. $\dfrac{7}{2ab} \cdot \dfrac{3}{2b}$
5. $\dfrac{3x}{2y^3} \cdot \dfrac{10yx^2}{3x}$
6. $\dfrac{2b}{5a^3} \cdot \dfrac{20ab^2}{4b}$
7. $\dfrac{3rs}{2t^2} \cdot \dfrac{4rt}{5s^2}$
8. $\dfrac{10xy}{3z^2} \cdot \dfrac{9yz}{5x^2}$
9. $\dfrac{12pq}{7r^3} \cdot \dfrac{49qr}{14p^2}$
10. $\dfrac{24ab}{3c^3} \cdot \dfrac{5bc}{15a^2}$
11. $\dfrac{28m^3n^2}{3p} \cdot \dfrac{18n^2p^2}{7mn}$
12. $\dfrac{m - 2}{3m + 9} \cdot \dfrac{2m + 6}{2m - 4}$
13. $\dfrac{x - 5}{4x + 6} \cdot \dfrac{6x + 9}{3x - 15}$
14. $\dfrac{n + 3}{2n - 8} \cdot \dfrac{6n - 24}{2n + 1}$
15. $\dfrac{2c + 4}{6c - 8} \cdot \dfrac{c - 5}{c + 2}$
16. $\dfrac{r + 4}{2r + 18} \cdot \dfrac{3r + 27}{r - 7}$
17. $\dfrac{b + 2}{3b + 36} \cdot \dfrac{b + 12}{2b - 3}$
18. $\dfrac{2x^2 - 2x - 4}{8x} \cdot \dfrac{-8x - 16}{x^2 - 4}$
19. $\dfrac{3r^2 - 10r - 8}{2r} \cdot \dfrac{-2r - 8}{r^2 - 16}$
20. $\dfrac{x^2 - 4x - 5}{25 - x^2} \cdot \dfrac{x^2 + 2x - 15}{x^2 - 2x - 3}$
21. $\dfrac{2z^2 + z - 6}{4 - z^2} \cdot \dfrac{z^2 + 3z - 10}{2z^2 + 7z - 15}$
22. $\dfrac{32a^5b^3}{7c^2d^3} \cdot \dfrac{28c^3d^2}{24a^2b^5}$
23. $\dfrac{45q^2r^3}{8s^3t^5} \cdot \dfrac{24s^5t^2}{18q^4r}$
24. $\dfrac{7t^2 - 28t}{2t^2 - 5t - 12} \cdot \dfrac{6t^2 - t - 15}{49t^3}$
25. $\dfrac{18i^2 - 6i}{3i^4} \cdot \dfrac{10i^2 - 7i - 12}{15i^2 + 7i - 4}$
26. $\dfrac{2\theta^2 + 7\theta + 3}{3\theta^2 + 14\theta + 15} \cdot \dfrac{6\theta^2 + 19\theta + 15}{4\theta^2 + 8\theta + 3}$
27. $\dfrac{15n^2 + 16n + 4}{20n^2 + 43n + 14} \cdot \dfrac{8n^2 + 50n + 63}{6n^2 + 31n + 18}$
28. $\dfrac{9a^2 + 43a - 10}{27a^2 + 12a - 4} \cdot \dfrac{6a^2 - 11a - 10}{5a^2 + 29a + 20}$
29. $\dfrac{10l^2 - 67l - 21}{20l^2 - 84l - 27} \cdot \dfrac{3l^2 + 8l - 35}{3l^2 - 28l + 49}$
30. $\dfrac{10a^2 - 7a - 12}{9 - 4a^2} \cdot \dfrac{2a^2 - a - 6}{7a^2 - 10a - 8}$
31. $\dfrac{7l^2 + 33l - 10}{15 - 17l - 4l^2} \cdot \dfrac{28l^2 - 9l - 9}{49l^2 + 7l - 6}$

Chapter 8 Rational Expressions

32. $\dfrac{6w^2 - 19w - 7}{2w^2 + 3w - 35} \cdot \dfrac{15 - 37w - 8w^2}{24w^2 - w - 3}$

33. $\dfrac{2a^2 + 5a + 2}{10 + 29a - 21a^2} \cdot \dfrac{3a^2 + 7a - 20}{a^2 + 6a + 8}$

34. $\dfrac{4y^2 + 14y + 6}{18y^2 + 69y + 21} \cdot \dfrac{12y^2 + 102y + 210}{y^2 + 4y + 3}$

35. $\dfrac{4s^2 - 2s - 2}{3s^2 + s - 10} \cdot \dfrac{3s^2 - 17s + 20}{8s^2 + 28s + 12}$

36. $\dfrac{f^2 - r^2}{f^2 + fr - 2r^2} \cdot \dfrac{f^2 + 3fr + 2r^2}{f^2 + 2fr + r^2}$

37. $\dfrac{a^2 + 5ae + 6e^2}{2a^2 + 7ae + 3e^2} \cdot \dfrac{2a^2 + 11ae + 5e^2}{a^2 + 7ae + 10e^2}$

38. $\dfrac{2c^2 + cd - 3d^2}{2c^2 - 9cd - 35d^2} \cdot \dfrac{2c^2 + 7cd + 5d^2}{2c^2 + 5cd + 3d^2}$

39. $\dfrac{6t^2 + 5tu + u^2}{2t^2 + 3tu + u^2} \cdot \dfrac{7t^2 + 4tu - 3u^2}{6t^2 + 29tu + 9u^2}$

40. $\dfrac{2\theta^2 + 9\theta c + 7c^2}{3\theta^2 + 2\theta c - c^2} \cdot \dfrac{3\theta^2 + 13\theta c + 4c^2}{2\theta^2 + 13\theta c + 21c^2}$

41. $\dfrac{3r^2 - 4re - 4e^2}{4r^2 - 4re - 3e^2} \cdot \dfrac{2r^2 + re - 6e^2}{6r^2 + re - 2e^2}$

Applications

Calculator Can the following expressions be entered directly into your calculator? Write the input sequence for each using $a = -2$, $b = 3$, and $c = 8$ and evaluate.

42. $5a^3 + c$

43. $\dfrac{-6ab^2c}{bc}$

44. $\dfrac{a^4 + 2b}{a + b}$

45. $\dfrac{(a - b)^3}{(a + b)^2}$

46. $\dfrac{2(b + c)}{2c}$

47. $\dfrac{3b + 8c}{4bc}$

Geometry Find the volume of each rectangular solid with the given dimensions. ($V = lwh$)

48. length: $\dfrac{x - 2}{x^2 + 2x - 35}$; width: $\dfrac{3x + 2}{4}$; height: $\dfrac{x - 5}{3x + 2}$

49. length: $\dfrac{3x^2 + 8x - 3}{2x^2 + 5x - 3}$; width: $\dfrac{4x - 6}{3x - 1}$; height: $\dfrac{2x - 1}{2x^2 + 7x - 15}$

WRITING IN ALGEBRA

Use the expression $x^2 + 5x + 6$ as part of your answer for each of the following.

1. Write an equation that shows the factored form of the expression.
2. Write a rational expression that can be simplified.
3. Write a rational expression that cannot be simplified.
4. Write a rational expression that is undefined when $x = -3$.

8.3 Dividing Rational Expressions

Objective: To divide rational expressions

Multiplication and division are opposite (inverse) operations. Recall that the rules for dividing involve *reciprocals* or *multiplicative inverses*. Reciprocals are two numbers whose product is 1. For example, $\frac{3}{4} \cdot \frac{4}{3} = 1$. Thus $\frac{3}{4}$ and $\frac{4}{3}$ are reciprocals. Are $2\frac{1}{2}$ and $\frac{2}{5}$ reciprocals? Why?

Capsule Review

To divide by a fraction, multiply by the reciprocal of the fraction.

EXAMPLE Divide: $\frac{5}{8} \div \frac{3}{4}$

$$\frac{5}{8} \div \frac{3}{4} = \frac{5}{8} \cdot \frac{4}{3} = \frac{5}{\underset{2}{\cancel{8}}} \cdot \frac{\overset{1}{\cancel{4}}}{3} = \frac{5}{6}$$

(reciprocals)

Divide. Simplify if possible.

1. $\frac{2}{3} \div \frac{4}{5}$ 2. $6 \div \frac{3}{8}$ 3. $\frac{3}{4} \div 2$ 4. $3\frac{1}{2} \div 4$

The rule for dividing rational expressions is similar to the rule for dividing fractions.

Division of Rational Expressions

If $\frac{P}{Q}$ and $\frac{R}{S}$ are rational expressions and $Q \neq 0$, $R \neq 0$, and $S \neq 0$,

then $\frac{P}{Q} \div \frac{R}{S} = \frac{P}{Q} \cdot \frac{S}{R} = \frac{P \cdot S}{Q \cdot R}$

EXAMPLE 1 Divide: $\frac{3}{a} \div \frac{6b}{5a}$

$\frac{3}{a} \div \frac{6b}{5a} = \frac{3}{a} \cdot \frac{5a}{6b}$ Multiply by $\frac{5a}{6b}$, the reciprocal of $\frac{6b}{5a}$.

$= \frac{\cancel{3}}{\cancel{a}} \cdot \frac{5\cancel{a}}{\underset{2}{\cancel{6}}b} = \frac{5}{2b}$

324 Chapter 8 Rational Expressions

EXAMPLE 2 **Divide:** $\dfrac{x^2 + x - 2}{3x^2 + 9x + 6} \div (x - 1)$

$\dfrac{x^2 + x - 2}{3x^2 + 9x + 6} \div (x - 1)$

$= \dfrac{x^2 + x - 2}{3x^2 + 9x + 6} \cdot \dfrac{1}{(x - 1)}$ Multiply by the reciprocal of $\dfrac{(x - 1)}{1}$.

$= \dfrac{(x - 1)(x + 2)}{3(x + 2)(x + 1)} \cdot \dfrac{1}{(x - 1)}$ Factor.

$= \dfrac{\cancel{(x - 1)}\cancel{(x + 2)}}{3\cancel{(x + 2)}(x + 1)} \cdot \dfrac{1}{\cancel{(x - 1)}}$ Divide by common factors.

$= \dfrac{1}{3(x + 1)}$ The answer may be left in factored form.

Rational expressions involving both multiplication and division should be simplified by performing operations from left to right unless parentheses indicate a different order.

EXAMPLE 3 **Simplify:** $\dfrac{2m^2 - 5m - 3}{9 - m^2} \div \dfrac{4m + 2}{2m^2 + 2m - 12} \cdot \dfrac{2}{m - 2}$

$\dfrac{2m^2 - 5m - 3}{9 - m^2} \div \dfrac{4m + 2}{2m^2 + 2m - 12} \cdot \dfrac{2}{m - 2}$

$= \dfrac{2m^2 - 5m - 3}{9 - m^2} \cdot \dfrac{2m^2 + 2m - 12}{4m + 2} \cdot \dfrac{2}{m - 2}$

$= \dfrac{\cancel{(2m + 1)}\cancel{(m - 3)}}{-1\cancel{(m - 3)}\cancel{(m + 3)}} \cdot \dfrac{2\cancel{(m + 3)}\cancel{(m - 2)}}{2\cancel{(2m + 1)}} \cdot \dfrac{2}{\cancel{(m - 2)}}$

$= \dfrac{2}{-1} = -2$

CLASS EXERCISES

Simplify.

1. $\dfrac{3}{4} \div \dfrac{9}{8}$

2. $\dfrac{3}{7}m \div \dfrac{2m^3}{21m}$

3. $\dfrac{d}{e} \div \dfrac{e^2}{d^3}$

4. $\dfrac{a^2b}{2a} \div ab^2$

5. $\dfrac{a - 2}{ab} \div \dfrac{a - 2}{a}$

6. $\dfrac{x - 3}{6} \div \dfrac{3 - x}{2}$

7. $\dfrac{y + 3}{y + 2} \div (y + 2)$

8. $\dfrac{x^2 + 3x + 2}{x^2 - 4x + 3} \div \dfrac{x + 2}{x - 3}$

9. $\dfrac{p}{q} \div \dfrac{r}{q} \cdot \dfrac{p}{q}$

PRACTICE EXERCISES

Divide.

1. $\dfrac{15}{a} \div \dfrac{3b}{2a}$

2. $\dfrac{9}{x} \div \dfrac{3y}{4x}$

3. $\dfrac{6}{2c} \div \dfrac{3d}{4c}$

4. $\dfrac{r^2}{7s^2} \div \dfrac{3r}{28s}$

5. $\dfrac{d^2}{3e^2} \div \dfrac{4d}{2e}$

6. $\dfrac{15a^2}{4b^2} \div \dfrac{5a}{2b}$

7. $\dfrac{3z - 51}{2z + 5} \div (z - 17)$

8. $\dfrac{11k + 121}{7k - 15} \div (k + 11)$

9. $\dfrac{x^2 + 10x - 11}{x^2 + 12x + 11} \div (x - 1)$

10. $\dfrac{z^2 + 2z - 15}{z^2 + 9z + 20} \div (z - 3)$

11. $\dfrac{3r - 21}{5r + 15} \div \dfrac{3r + 6}{7r + 21}$

12. $\dfrac{5a + 10}{2a - 20} \div \dfrac{7a + 14}{14a - 20}$

13. $\dfrac{x^2 + 7x + 10}{x^2 - 36} \div \dfrac{x + 5}{x - 6}$

14. $\dfrac{a^2 - 9}{a^2 - 2a - 24} \div \dfrac{a - 3}{a - 6}$

15. $\dfrac{2a^2}{3b} \div \dfrac{a}{b} \div \dfrac{5b^2}{2a}$

16. $\dfrac{r^4 s}{3} \div \dfrac{r^3}{s^2} \div \dfrac{3s}{4}$

17. $\dfrac{b - a}{2b + a} \cdot \dfrac{a + 2b}{b + a} \div \dfrac{b - a}{b + a}$

18. $\dfrac{2r + s}{3s - 1} \div \dfrac{s - 3}{1 - 3s} \cdot \dfrac{s - 3}{s + 2r}$

19. $\dfrac{18a^3 c^2}{25b^2} \div \dfrac{12a^2 c}{5b}$

20. $\dfrac{24x^5 y^3}{18z^2} \div \dfrac{15x^2 y}{12z}$

21. $\dfrac{5x^2}{y^2 - 36} \div \dfrac{25xy - 25x}{y^2 - 7y + 6}$

22. $\dfrac{4a^3}{b^2 - 4} \div \dfrac{6ab - 18a}{b^2 - b - 6}$

23. $(16 - c^2) \div \dfrac{2c^2 - c - 36}{12c - 54}$

24. $(27 - 3r^2) \div \dfrac{5r^2 - 9r - 18}{45r + 54}$

25. $\dfrac{10h^2 + 21h - 10}{12h^2 - 7h - 12} \div \dfrac{2h^2 + 9h + 10}{4h^2 + 11h + 6}$

26. $\dfrac{6n^2 + 7n - 24}{2n^2 + 3n - 9} \div \dfrac{-9n^2 + 64}{8n^2 + 21n - 9}$

27. $\dfrac{8t^2 + 2t - 15}{6t^2 - 2t - 4} \div \dfrac{15 - 22t + 8t^2}{6t^2 - 5t - 6}$

28. $\dfrac{3a^2 + 7a - 6}{4a^2 + 8a - 5} \div \dfrac{6 - 7a - 3a^2}{2a^2 + a - 1}$

29. $\dfrac{5x^2 + 10x - 15}{5 - 6x + x^2} \div \dfrac{2x^2 + 7x + 3}{4x^2 - 8x - 5}$

30. $\dfrac{12x^2 + 19x + 5}{2x^2 - 7x + 3} \div \dfrac{10x + 5}{3x - 2} \cdot \dfrac{2x^2 - 5x - 3}{12x^2 - 5x - 3}$

31. $\dfrac{14y^2 + 13y + 3}{30y^2 - 27y - 21} \div \dfrac{6y^2 + 11y - 10}{25y^2 - 50y + 21} \cdot \dfrac{2y + 5}{5y - 3}$

32. $\dfrac{x^2 + 6x + 8}{x^2 + x - 2} \div \dfrac{x + 4}{2x + 4} \div \dfrac{x + 3}{x - 1}$

33. $\dfrac{2y^2 - 5y - 3}{4y^2 - 12y - 7} \div \dfrac{4y + 5}{2y - 7} \div \dfrac{y - 3}{3y - 1}$

34. $\dfrac{3a^2}{b^2 - 16} \div \dfrac{3ab + 6a}{b^2 + 6b + 8}$

35. $\dfrac{5x^2}{y^2 - 25} \div \dfrac{5xy - 25x}{y^2 - 10y + 25}$

36. $\dfrac{2a^2 - ab - 6b^2}{2b^2 + 9ab - 5a^2} \div \dfrac{2a^2 + 7ab + 6b^2}{a^2 - 4b^2}$

37. $\dfrac{4g^2 - 8gh - 12h^2}{8h^2 + 10gh + 2g^2} \div \dfrac{12h^2 + 18gh + 6g^2}{g^2 - 16h^2}$

38. $\dfrac{m^2 + 5mn - 6n^2}{6m^2 - mn} \div \dfrac{m^2 - n^2}{2m^2 - 5mn - 3n^2} \div \dfrac{m^2 + 3mn - 18n^2}{3m^2 + mn - 2n^2}$

39. $\dfrac{3c^2 - 7cd - 6d^2}{c^2 + 6cd + 9d^2} \div \dfrac{3c^2 - 17cd + 24d^2}{c^2 - 9d^2} \div \dfrac{6c^2 + 7cd - 3d^2}{3c^2 + cd - 24d^2}$

Applications

Solve.

40. **Physics** June conducted an experiment to determine the stress on a copper bar. She used the formula stress $= \dfrac{\text{force}}{\text{area}}$. The force she applied was $4x - 6$ lb. She found the area of the bar to be $2x^2 + 5x - 12$ ft^2. Calculate the stress.

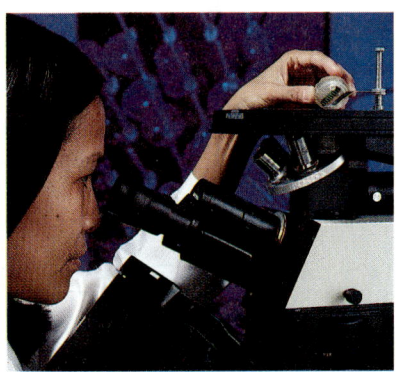

41. **Manufacturing** The efficiency of a machine is defined as the work output divided by the work input. Crystal had to calculate the efficiency of an industrial lathe. She found the work output to be $3x^2 + 5x - 2$ J (joules) and the work input to be $x + 2$ J. What is the efficiency?

42. **Transportation** A train travels $x^2 - 9$ mi in $x + 3$ h. Find the rate. Use the formula distance $=$ rate $\times$ time.

LOGICAL REASONING

Identifying and ordering a series of steps is helpful in understanding a computational skill. A series of steps is needed to divide rational expressions.

1. What are the steps?
2. If you change the order of any of the steps, would it make a difference? Why?
3. Which steps can be interchanged?

8.4 Least Common Denominator (LCD)

Objectives: To find the LCD of two or more rational expressions
To express a fraction as an equivalent fraction with a given denominator

Fractions that have the same denominator can easily be added or subtracted. Fractions that have different denominators must first be expressed as equivalent fractions with a common denominator before they can be added or subtracted. The **least common denominator (LCD)** is the smallest possible common denominator of the fractions.

$$\frac{1}{2a} + \frac{1}{3a} \qquad \frac{6}{12a} + \frac{4}{12a} \qquad \frac{3}{6a} + \frac{2}{6a}$$

common denominator   least common denominator

You can use prime factorization to find the LCD of fractions or rational expressions.

Capsule Review

EXAMPLE Find the prime factorization of $3575m^3n$.

$3575 = $ ___?___ ← Not divisible by 2 or 3, try 5.
$ = 5 \times 715$ ← Try 5 again.
$ = 5 \times 5 \times 143$ ← 143 is not divisible by 7; try 11.
$ = 5 \times 5 \times 11 \times 13$ ← 13 is a prime number.

Prime factorization of $3575m^3n$: $5 \cdot 5 \cdot 11 \cdot 13 \cdot m \cdot m \cdot m \cdot n$

Find the prime factorization of the following.

1. 36
2. $12g^2h$
3. $140x^3$
4. $825abc$
5. $23m^3np^2$
6. $1875e^2f$
7. $1000x^2y^3z$
8. $100,000n^5$

Sometimes you can find the LCD by looking for common factors.

EXAMPLE 1 Find the least common denominator (LCD): $\frac{1}{12ab}$ and $\frac{2b}{3a}$

Factor: $12ab = 3a \cdot 4b$ *3a is a factor of 12ab.*

So, $12ab$ is the LCD of $\frac{1}{12ab}$ and $\frac{2b}{3a}$.

Chapter 8 Rational Expressions

Sometimes it is difficult to recognize common factors.

> To find the least common denominator (LCD):
>
> - Factor each denominator into primes.
> - Write the greatest power of each prime factor that appears in any denominator. Use exponents.
> - Multiply these factors.

EXAMPLE 2 Find the LCD: $\dfrac{5}{12a^3b}$ and $\dfrac{-3}{10ab^2}$

Factor denominators: $12a^3b = 2 \cdot 2 \cdot 3 \cdot a \cdot a \cdot a \cdot b$
$10ab^2 = 2 \cdot 5 \cdot a \cdot b \cdot b$
Greatest Powers: $2^2, 3, 5, a^3, b^2$
Multiply: $2^2 \cdot 3 \cdot 5 \cdot a^3 \cdot b^2 = 60a^3b^2$

The LCD of $\dfrac{5}{12a^3b}$ and $\dfrac{-3}{10ab^2}$ is $60a^3b^2$.

To obtain equivalent expressions that have a given LCD, compare each denominator with the LCD to find the missing factors. The LCD from Example 2 is used in Example 3.

EXAMPLE 3 Change $\dfrac{5}{12a^3b}$ to an equivalent expression with denominator $60a^3b^2$.

$\dfrac{5}{12a^3b} = \dfrac{?}{60a^3b^2}$ *Look for missing factors.*
$12a^3b \cdot 5b = 60a^3b^2$

$\dfrac{5}{12a^3b} \cdot \dfrac{5b}{5b} = \dfrac{25b}{60a^3b^2}$ *Multiply numerator and denominator by 5b.*
$\dfrac{5b}{5b} = 1$; *the expressions are equivalent.*

How would you change $\dfrac{-3}{10ab^2}$ to an equivalent expression that has the denominator $60a^3b^2$?

You can use prime factorization to find the LCD when the denominators are polynomials.

EXAMPLE 4 Find the LCD: $\dfrac{3}{5n+5}$ and $\dfrac{n+2}{n^2+5n+4}$

Factor denominators: $5n + 5 = 5(n+1)$
$n^2 + 5n + 4 = (n+1)(n+4)$
Greatest powers: $5, (n+1), (n+4)$

The LCD is $5(n+1)(n+4)$. You may leave the LCD in factored form.

EXAMPLE 5 Change $\dfrac{3}{5n+5}$ to an equivalent expression that has the denominator $5(n+1)(n+4)$.

$$\dfrac{3}{5n+5} = \dfrac{3}{5(n+1)}$$ Factor the denominator. Notice that $(n+4)$ is missing.

$$= \dfrac{3}{5(n+1)} \cdot \dfrac{n+4}{n+4}$$ Multiply by 1 in the form $\dfrac{n+4}{n+4}$.

$$= \dfrac{3(n+4)}{5(n+1)(n+4)}$$

How would you change $\dfrac{n+2}{n^2+5n+4}$ to an equivalent expression that has the denominator $5(n+1)(n+4)$?

Be on the alert for binomial factors that are the same or that are opposites.

EXAMPLE 6 Change $\dfrac{2a}{3-a}$ to an equivalent expression that has the denominator a^2-9.

$$a^2 - 9 = (a+3)(a-3)$$ Factor.

$$\dfrac{2a}{3-a} = \dfrac{2a}{-1(a-3)} = \dfrac{-2a}{a-3}$$ Look for common factors. The opposite of $3-a$ is a factor of a^2-9.

$$\dfrac{-2a}{a-3} \cdot \dfrac{a+3}{a+3} = \dfrac{-2a(a+3)}{(a-3)(a+3)}$$ Multiply by the missing factor, $a+3$.

CLASS EXERCISES

Find the LCD.

1. $\dfrac{3}{8}; \dfrac{9}{24}$
2. $\dfrac{5}{12}; \dfrac{7}{18}$
3. $\dfrac{1}{5x}; \dfrac{3}{20x^2}$
4. $\dfrac{7b}{3a^2}; \dfrac{2a}{5b^2}$
5. $\dfrac{5}{18m^2n}; \dfrac{7}{24n^2}$
6. $\dfrac{3}{2a}; \dfrac{5}{2a+4}$
7. $\dfrac{1}{x-3}; \dfrac{1}{x+5}$
8. $\dfrac{1}{3m+9}; \dfrac{1}{m^2+4m+3}$

Write equivalent expressions. Use the least common denominator.

9. $\dfrac{4}{3xy}; \dfrac{7}{12x^2}$
10. $\dfrac{3}{a^2-16}; \dfrac{5}{12+a-a^2}$

PRACTICE EXERCISES

Find the LCD.

1. $\dfrac{1}{5ab}; \dfrac{3b}{2a}$
2. $\dfrac{1}{10rs}; \dfrac{2s}{12r}$
3. $\dfrac{1}{6x}; \dfrac{3}{14y}$
4. $\dfrac{7}{6c}; \dfrac{c^2}{9d}$

5. $\dfrac{11}{14a^3b}; \dfrac{-15}{21ab^2}$
6. $\dfrac{5}{24s^3t}; \dfrac{-13}{6st^2}$
7. $\dfrac{-4}{18xy^3}; \dfrac{2}{6x^2y}$
8. $\dfrac{-7}{13cd^3}; \dfrac{5}{3c^2d}$

9. $\dfrac{2}{3n+3}; \dfrac{n+4}{n^2+6n+5}$
10. $\dfrac{7}{6x+6}; \dfrac{x+3}{x^2+9x+8}$
11. $\dfrac{4}{3r-27}; \dfrac{5}{4r+10}$

12. $\dfrac{15}{8y-36}; \dfrac{17}{5y+15}$
13. $\dfrac{3}{t^2-36}; \dfrac{7}{t-6}$
14. $\dfrac{7}{x^2-4}; \dfrac{3}{x-2}$

Find the LCD and write the equivalent expressions with the LCD as the denominator.

15. $\dfrac{5}{12mn^2}; \dfrac{3m}{4n}$
16. $\dfrac{1}{45ab^2}; \dfrac{5a}{18b}$
17. $\dfrac{5}{7y+7}; \dfrac{3}{y^2+4y+3}$

18. $\dfrac{5}{2h+2}; \dfrac{2}{h^2+4h+3}$
19. $\dfrac{3y}{y^2-9}; \dfrac{2}{y-3}$
20. $\dfrac{3a}{a^2-25}; \dfrac{1}{5-a}$

Find the LCD.

21. $\dfrac{1}{x^2+5x+6}; \dfrac{x}{x^2+7x+10}$
22. $\dfrac{3r}{r^2-r-12}; \dfrac{5}{r^2-4r-21}$

23. $\dfrac{6}{12c^2+13c-35}; \dfrac{10c}{3c^2-11c-42}$
24. $\dfrac{13}{6h^2-17h+12}; \dfrac{6h}{8h^2+2h-21}$

25. $\dfrac{1}{16x^2-43x-15}; \dfrac{3}{2x^2+7x+3}$
26. $\dfrac{17x}{12x^2-7x-12}; \dfrac{3}{6x^2-5x-6}$

Write equivalent expressions having the same denominator. Use the LCD.

27. $\dfrac{3}{2a^2}; \dfrac{9}{3b}; \dfrac{a+b}{6ab^2}$
28. $\dfrac{m}{3n^2}; \dfrac{n+1}{15m^2n}; \dfrac{3}{4mn}$

29. $\dfrac{8n}{5-3n}; \dfrac{5n^2}{9n^2-25}$
30. $\dfrac{5r}{3-2r}; \dfrac{3r^2}{4r^2-9}$

31. $\dfrac{3d}{2d^2-4d}; \dfrac{d+3}{d^2-d-2}$
32. $\dfrac{5x+1}{x^2+3x-4}; \dfrac{x}{3x^2+12x}$

33. $\dfrac{c}{3c+3}; \dfrac{c^2}{c^2+3c+2}; \dfrac{3c}{c+2}$
34. $\dfrac{y+1}{5y+15}; \dfrac{4}{y^2+5y+6}; \dfrac{2y}{y+2}$

Find the LCD.

35. $\dfrac{1}{15x^2-xy-28y^2}; \dfrac{x+11}{12x^2+7xy-12y^2}$
36. $\dfrac{6}{8x^2-14xy-15y^2}; \dfrac{y-4}{4x^2-25y^2}$

8.4 Least Common Denominator (LCD)

37. $\dfrac{1}{8a^4 - 8b^4}$; $\dfrac{b-7}{(3a-3b)^2}$

38. $\dfrac{7}{3a^4 + 6a^2b^2 + 3b^4}$; $\dfrac{3}{2a^4 - 2b^4}$

39. $\dfrac{1}{15m^2 - 2m - 8}$; $\dfrac{3}{3m^2 - 10m - 8}$; $\dfrac{5}{6m - 24}$

40. $\dfrac{7}{2t^2 - 2t - 12}$; $\dfrac{13}{4t^2 + 10t + 4}$; $\dfrac{17}{6t + 3}$

Applications

41. Photography When a camera is in focus it satisfies the equation $\dfrac{1}{p} + \dfrac{1}{q} = \dfrac{1}{f}$. Find the LCD if $p = x^2$, $q = 2x$ and $f = 3x^3$.

42. Construction The forces acting on a concrete slab can be found using the rational expression $1 - \dfrac{4c}{\pi l} + \dfrac{c^3}{3l^3}$. Write the expression using the least common denominator.

LOGICAL REASONING

A statement written in if-then form is called a conditional.

EXAMPLES   **a.** If it rains on Monday, then Carlos will stay home.
            ↑   ↑
      hypothesis   conclusion

When the hypothesis is true, ⟶ It rains on Monday.
  the conclusion is true. ⟶ Carlos stays home.
When the hypothesis is not true, ⟶ It does not rain on Monday.
  the conclusion may or may not be true. ⟶ Carlos may or may not stay home.

b. If a and b are both even numbers, then $a + b$ is even.

  When the hypothesis is true, ⟶ a and b both even
    the conclusion is true. ⟶ $a + b$ even
  When the hypothesis is not true, ⟶ a and b not both even
    the conclusion may or may not be true. ⟶ $a + b$ even or odd

The sentences in the exercise are true. What can you conclude?

1. If x is divisible by 10, then x is divisible by 2. x is not divisible by 10.

2. If a polygon has 4 sides, then it has 4 angles. $ABCD$ is a polygon with 4 sides.

3. All integers are rational numbers. x is an integer.

4. All factors of x are factors of y. n is not a factor of x.

332   Chapter 8 Rational Expressions

8.5 Adding and Subtracting Rational Expressions

Objective: To add and subtract rational expressions

Rational expressions are added or subtracted in the same way as fractions. As with fractions, you may need to find the LCD of the rational expressions before you add or subtract.

Capsule Review

EXAMPLE Find the LCD: $\dfrac{3}{5x^2y}$; $\dfrac{x}{5x^2 + 10x}$

Factor the denominators. $5x^2y = 5 \cdot x \cdot x \cdot y$
$5x^2 + 10x = 5 \cdot x \cdot (x + 2)$

Multiply the greatest powers of the prime factors.

$$5x^2y(x + 2)$$

Find the LCD.

1. $\dfrac{2}{x}$; $\dfrac{3}{6x}$ 2. $\dfrac{2}{ab^3}$; $\dfrac{6}{a^2b}$ 3. $\dfrac{-1}{2xy}$; $\dfrac{x}{2x^2 + 6x}$ 4. $\dfrac{3}{d-3}$; $\dfrac{d}{d^2 - 9}$

The method for adding or subtracting rational expressions may be generalized as follows.

> **Addition and Subtraction of Rational Expressions**
>
> If $\dfrac{P}{Q}$ and $\dfrac{R}{Q}$ are rational expressions with $Q \neq 0$, then
> $$\dfrac{P}{Q} + \dfrac{R}{Q} = \dfrac{P + R}{Q} \quad \text{and} \quad \dfrac{P}{Q} - \dfrac{R}{Q} = \dfrac{P - R}{Q}.$$

EXAMPLE 1 Add: $\dfrac{4}{3m} + \dfrac{2}{3m}$

$$\dfrac{4}{3m} + \dfrac{2}{3m} = \dfrac{4 + 2}{3m} = \dfrac{\cancel{6}^{\,2}}{\cancel{3}m} = \dfrac{2}{m}$$

8.5 Adding and Subtracting Rational Expressions   **333**

When subtracting it is often helpful to use parentheses when you rewrite polynomial numerators to clearly indicate the expression(s) being subtracted.

EXAMPLE 2 Subtract: $\dfrac{5n}{n+2} - \dfrac{n-8}{n+2}$

$$\dfrac{5n}{n+2} - \dfrac{n-8}{n+2} = \dfrac{5n - (n-8)}{n+2} \quad \longleftarrow \text{Parentheses are necessary.}$$

$$= \dfrac{5n - n + 8}{n+2}$$

$$= \dfrac{4n + 8}{n+2} = \dfrac{4}{n+2} = 4$$

To add or subtract rational expressions with unlike denominators, first write them as equivalent expressions with a least common denominator.

EXAMPLE 3 Combine: $\dfrac{3}{4a} + \dfrac{5}{6a^2} - \dfrac{1}{a}$

$\dfrac{3}{4a} + \dfrac{5}{6a^2} - \dfrac{1}{a}$ *The LCD is $2^2 \cdot 3 \cdot a^2$, or $12a^2$.*

$= \dfrac{3}{4a} \cdot \dfrac{3a}{3a} + \dfrac{5}{6a^2} \cdot \dfrac{2}{2} - \dfrac{1}{a} \cdot \dfrac{12a}{12a}$ *Write equivalent expressions that have the LCD.*

$= \dfrac{9a}{12a^2} + \dfrac{10}{12a^2} - \dfrac{12a}{12a^2}$

$= \dfrac{9a + 10 - 12a}{12a^2} = \dfrac{10 - 3a}{12a^2}$ *Combine like terms in the numerator and simplify.*

It is important to notice binomial factors that are the same or opposites.

EXAMPLE 4 Add: $\dfrac{2y+1}{9-y^2} + \dfrac{2}{y-3} - \dfrac{1}{y+3}$

$\dfrac{2y+1}{9-y^2} + \dfrac{2}{y-3} - \dfrac{1}{y+3} = \dfrac{2y+1}{(3-y)(3+y)} + \dfrac{2}{y-3} - \dfrac{1}{y+3}$

$= \dfrac{2y+1}{-(y-3)(y+3)} + \dfrac{2}{y-3} - \dfrac{1}{y+3}$

$= \dfrac{-(2y+1)}{(y-3)(y+3)} + \dfrac{2}{y-3} \cdot \dfrac{y+3}{y+3} - \dfrac{1}{y+3} \cdot \dfrac{y-3}{y-3}$

$= \dfrac{-(2y+1) + 2(y+3) - 1(y-3)}{(y-3)(y+3)}$

$= \dfrac{-2y - 1 + 2y + 6 - y + 3}{(y-3)(y+3)}$

$= \dfrac{8 - y}{(y-3)(y+3)}$

CLASS EXERCISES

Combine. Simplify if possible.

1. $\dfrac{9}{11} - \dfrac{3}{11}$
2. $\dfrac{x}{4} + \dfrac{3x}{4}$
3. $\dfrac{3}{2ab} - \dfrac{5}{2ab}$
4. $\dfrac{m}{m-3} + \dfrac{2}{m-3}$
5. $\dfrac{2c}{3d^2} + \dfrac{3}{2cd}$
6. $\dfrac{3r}{r^2-9} - \dfrac{5r}{r+3}$

PRACTICE EXERCISES

Combine. Simplify if possible.

1. $\dfrac{5}{2m} + \dfrac{3}{2m}$
2. $\dfrac{5}{6n} + \dfrac{7}{6n}$
3. $\dfrac{5}{4x} - \dfrac{3}{4x}$
4. $\dfrac{6}{5z} - \dfrac{1}{5z}$
5. $\dfrac{3z}{z+2} - \dfrac{z-4}{z+2}$
6. $\dfrac{5c}{c+7} - \dfrac{c-28}{c+7}$
7. $\dfrac{2}{a-2} - \dfrac{a}{a-2}$
8. $\dfrac{3}{b-3} - \dfrac{b}{b-3}$
9. $\dfrac{3}{2a} + \dfrac{2}{4a^2} - \dfrac{1}{a}$
10. $\dfrac{3}{3x} + \dfrac{6}{9x^2} - \dfrac{1}{x}$
11. $\dfrac{1}{3m^2} - \dfrac{4}{6m} - \dfrac{1}{m}$
12. $\dfrac{8}{16r^2} - \dfrac{3}{8r} - \dfrac{1}{r}$
13. $\dfrac{4}{2a+8} - \dfrac{a}{5a+20}$
14. $\dfrac{a}{a+3} - \dfrac{3}{a+5}$
15. $\dfrac{5}{l+4} + \dfrac{3}{l-1}$
16. $\dfrac{i}{7i+14} + \dfrac{6}{3i+6}$
17. $\dfrac{5x+1}{25-x^2} + \dfrac{5}{x-5}$
18. $\dfrac{3z+2}{16-z^2} + \dfrac{3}{z-4}$
19. $\dfrac{4r+1}{9-r^2} + \dfrac{4}{r-3}$
20. $\dfrac{y+3}{4-y^2} + \dfrac{1}{y-2}$
21. $\dfrac{2}{y-1} + \dfrac{6y-2}{y^2+2y-3}$
22. $\dfrac{5}{d-3} - \dfrac{d-4}{d^2-d-6}$
23. $\dfrac{2a^2-a}{2a^2+a-3} - \dfrac{6}{2a^2+a-3}$
24. $\dfrac{l^2}{2l^2-l-6} + \dfrac{l-6}{2l^2-l-6}$
25. $\dfrac{g^3+3g}{g^2+2g} + \dfrac{g^2-g}{g^2+2g} - \dfrac{4}{g^2+2g}$
26. $\dfrac{e^3+e^2}{2e^2+6e} + \dfrac{e^3+2e^2}{2e^2+6e} - \dfrac{e^2+12}{2e^2+6e}$
27. $\dfrac{b-2}{2b} + \dfrac{b+3}{3b} - \dfrac{b-2}{6b^2}$
28. $\dfrac{r+3}{4r} - \dfrac{r+2}{3r^2} + \dfrac{r-4}{12r^2}$
29. $\dfrac{u-5}{2u+4} + \dfrac{u+3}{3u-6}$
30. $\dfrac{n+5}{4n+16} + \dfrac{n-1}{3n-9}$
31. $\dfrac{3a-2}{a^2-a-12} + \dfrac{a+3}{a-4}$
32. $\dfrac{n^2+1}{n^2-2n-15} + \dfrac{n-1}{n+3}$

8.5 Adding and Subtracting Rational Expressions

33. $\dfrac{7d-2}{d^2+2d-8} - \dfrac{4}{d+4} - \dfrac{d}{d-2}$

34. $\dfrac{i-24}{i^2-3i-18} - \dfrac{3}{i+3} + \dfrac{i}{i-6}$

35. $\dfrac{t}{8t^2+10t-3} - \dfrac{3}{4t^2+19t-5}$

36. $\dfrac{a}{6a^2-7a-20} - \dfrac{2}{3a^2-2a-8}$

37. $\dfrac{n+2}{n^2+n-6} + \dfrac{n-5}{3n^2+13n+12}$

38. $\dfrac{2t+1}{3t^2-31t-22} + \dfrac{t-7}{12t^2-19t-18}$

39. $\dfrac{6}{k^2+2k-15} + \dfrac{3}{k^2-3k} + \dfrac{5}{k^2-9}$

40. $\dfrac{b}{2b^2+b-6} - \dfrac{2}{2b^2-3b} + \dfrac{b+1}{b^2-4}$

Applications

41. Physics Kinetic energy (E_K) is the energy of motion. The work of an external force is equal to the change in kinetic energy:

$$w = E_{K_2} - E_{K_1}$$

The data from an experiment on kinetic energy is at the right. Find the work for each time the experiment was run.

Work and Kinetic Energy

	Work	E_{K_1}	E_{K_2}
a.	?	$\dfrac{2}{a}$	$\dfrac{2}{a+4}$
b.	?	$\dfrac{3}{h+1}$	$\dfrac{3}{h-1}$
c.	?	$\dfrac{5}{d-3}$	$\dfrac{2d}{d^2-9}$

ALGEBRA IN PHYSICS

An electric circuit is the path followed by an electric current. A simple circuit consists of a switch, a battery, and a light bulb. The amount of current (I) is related to the voltage (V) in the battery and the resistance (R) in the light bulb. What happens if more light bulbs are added to the circuit? How does the resistance change? Light bulbs can be added in series or parallel.

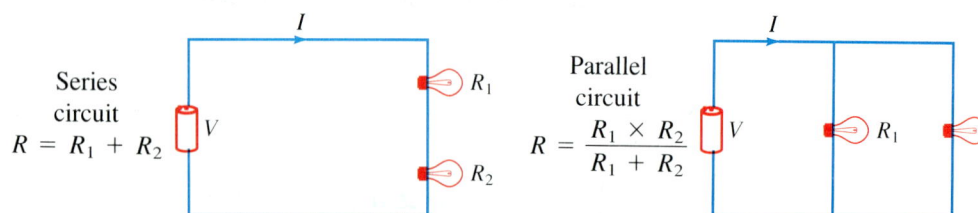

Series circuit
$R = R_1 + R_2$

Parallel circuit
$R = \dfrac{R_1 \times R_2}{R_1 + R_2}$

Solve for R if $R_1 = 0.5$ ohms and $R_2 = 3.0$ ohms.

1. Find the total resistance if two light bulbs are placed in series.
2. Find the total resistance if they are parallel.
3. Which circuit provides the least amount of total resistance?

Chapter 8 Rational Expressions

8.6 Mixed Expressions and Complex Rational Expressions

Objectives: To add or subtract a polynomial and a rational expression
To simplify complex rational expressions

Recall that a mixed number such as $2\frac{3}{4}$ is the sum of an integer and a fraction $\left(2 + \frac{3}{4}\right)$. You can express a mixed number as a fraction.

Capsule Review

EXAMPLE Express $2\frac{3}{4}$ as a fraction.

$$2\frac{3}{4} = 2 + \frac{3}{4}$$
$$= \frac{2}{1} \cdot \frac{4}{4} + \frac{3}{4} \quad \text{Express the whole number 2 as a}$$
$$\quad\quad\quad\quad\quad\quad\quad \text{fraction with the LCD 4. Combine.}$$
$$= \frac{8}{4} + \frac{3}{4} = \frac{11}{4}$$

Express each mixed number as a fraction.

1. $2\frac{2}{3}$
2. $3\frac{1}{4}$
3. $1\frac{3}{8}$
4. $11\frac{14}{15}$
5. $2\frac{1}{100}$
6. $99\frac{99}{100}$

A **mixed expression** is the sum or difference of a polynomial and a rational expression. Examples are: $2 + \frac{1}{x}$ and $m + 2 - \frac{5}{m-1}$. The procedure for combining a polynomial and a rational expression is similar to expressing a mixed number as a fraction.

EXAMPLE 1 Combine: $2 + \frac{1}{x}$

$$2 + \frac{1}{x} = \frac{2}{1} \cdot \frac{x}{x} + \frac{1}{x} \quad \text{Express 2 as a fraction with a denominator of } x.$$
$$= \frac{2x}{x} + \frac{1}{x} = \frac{2x + 1}{x}$$

A **complex rational expression** is a rational expression whose numerator or denominator contains one or more rational expressions.

$$\dfrac{\dfrac{3}{x}}{\dfrac{5}{x^2}}; \quad \dfrac{\dfrac{1}{m} + \dfrac{1}{n}}{\dfrac{2}{3m} + \dfrac{2}{3n}}; \quad \dfrac{a - \dfrac{2}{2a-3}}{2a - 1 - \dfrac{8}{2a-3}}$$

One way to simplify $\dfrac{\frac{3}{x}}{\frac{5}{x^2}}$ is to rewrite it as $\dfrac{3}{x} \div \dfrac{5}{x^2}$ and divide. Another way to simplify a rational expression is to multiply the numerator and denominator by the LCD.

EXAMPLE 2 Simplify: $\dfrac{\frac{1}{m} + \frac{1}{n}}{\frac{2}{3m} + \frac{2}{3n}}$

$\dfrac{\frac{1}{m} + \frac{1}{n}}{\frac{2}{3m} + \frac{2}{3n}} = \dfrac{\left(\frac{1}{m} + \frac{1}{n}\right)}{\left(\frac{2}{3m} + \frac{2}{3n}\right)} \cdot \dfrac{3mn}{3mn}$ *Multiply numerator and denominator by $3mn$, the LCD of $\frac{1}{m}, \frac{1}{n}, \frac{2}{3m}, \frac{2}{3n}$.*

$= \dfrac{\frac{1}{m} \cdot 3mn + \frac{1}{n} \cdot 3mn}{\frac{2}{3m} \cdot 3mn + \frac{2}{3n} \cdot 3mn} = \dfrac{3n + 3m}{2n + 2m} = \dfrac{3(n + m)}{2(n + m)} = \dfrac{3}{2}$

EXAMPLE 3 Simplify: $\dfrac{a - \frac{2}{2a - 3}}{2a - 1 - \frac{8}{2a - 3}}$

$\dfrac{a - \frac{2}{2a - 3}}{2a - 1 - \frac{8}{2a - 3}} = \dfrac{\left[a - \frac{2}{(2a - 3)}\right]}{\left[(2a - 1) - \frac{8}{(2a - 3)}\right]} \cdot \dfrac{2a - 3}{2a - 3}$

$= \dfrac{a(2a - 3) - \frac{2}{(2a-3)} \cdot (2a-3)}{(2a - 1)(2a - 3) - \frac{8}{(2a-3)} \cdot (2a-3)}$

$= \dfrac{2a^2 - 3a - 2}{4a^2 - 8a + 3 - 8}$

$= \dfrac{2a^2 - 3a - 2}{4a^2 - 8a - 5} = \dfrac{(2a + 1)(a - 2)}{(2a + 1)(2a - 5)} = \dfrac{a - 2}{2a - 5}$

CLASS EXERCISES

Combine.

1. $4 + \dfrac{1}{3}$
2. $a + \dfrac{b}{c}$
3. $x + 3 - \dfrac{4}{x - 2}$
4. $\dfrac{t}{t + 1} + 5t$

Simplify.

5. $\dfrac{\frac{2}{3}}{\frac{5}{6}}$

6. $\dfrac{\frac{3s}{8}}{\frac{s}{4}}$

7. $\dfrac{\frac{2n}{m^2} - \frac{1}{m}}{1 + \frac{2n}{m^2}}$

8. $\dfrac{h + \frac{3}{2h+5}}{h - \frac{4h+3}{2h+5}}$

PRACTICE EXERCISES

Combine. Simplify if possible.

1. $2 + \dfrac{3}{x}$

2. $4 - \dfrac{8}{a}$

3. $3 - \dfrac{10}{x}$

4. $4 - \dfrac{5}{b}$

5. $2z - \dfrac{z+1}{z}$

6. $3c - \dfrac{c+1}{c}$

7. $\dfrac{2y+3}{4y} + y - 3$

8. $\dfrac{3r+5}{3r} + 2 - r$

9. $d + \dfrac{d-3}{2d+1}$

10. $c + \dfrac{c-2}{3c-1}$

11. $\dfrac{2u-1}{u+2} + u$

12. $\dfrac{3v-1}{v+3} - v$

13. $\dfrac{\frac{1}{x} + \frac{1}{z}}{\frac{1}{2x} + \frac{1}{2z}}$

14. $\dfrac{\frac{1}{r} + \frac{1}{s}}{\frac{3}{5r} + \frac{3}{5s}}$

15. $\dfrac{\frac{1}{u} - \frac{1}{v}}{\frac{5}{2u} - \frac{5}{2v}}$

16. $\dfrac{\frac{1}{a} - \frac{1}{b}}{\frac{3}{4a} - \frac{3}{4b}}$

17. $\dfrac{\frac{m}{n} + \frac{2}{n^2}}{2 - \frac{m}{n^2}}$

18. $\dfrac{\frac{5}{v^2} + \frac{u}{v}}{\frac{u}{v} - 3}$

19. $\dfrac{1 + \frac{4}{a}}{1 - \frac{16}{a^2}}$

20. $\dfrac{1 + \frac{5}{b}}{1 - \frac{25}{b^2}}$

21. $\dfrac{3 - \frac{12}{d+4}}{2 - \frac{8}{d+4}}$

22. $\dfrac{5 - \frac{25}{c+5}}{3 - \frac{15}{c+5}}$

23. $\dfrac{u - \frac{4}{u+3}}{1 + \frac{1}{u+3}}$

24. $\dfrac{v - \frac{3}{v+2}}{1 + \frac{1}{v+2}}$

25. $\dfrac{x - \frac{3}{3x-4}}{3x - 1 - \frac{9}{3x-4}}$

26. $\dfrac{b - \frac{6}{2b-4}}{2b - 1 - \frac{10}{2b-4}}$

27. $\dfrac{z - \frac{20z+10}{3z+7}}{z - \frac{7z+5}{3z+7}}$

28. $\dfrac{a - \frac{36a+6}{5a+11}}{a - \frac{3a-3}{5a+11}}$

29. $x + 1 - \dfrac{4}{x+1}$

30. $y + 3 - \dfrac{1}{y+3}$

31. $a - 2 + \dfrac{3}{a+1}$

32. $b - 4 - \dfrac{2}{b+3}$

33. $\dfrac{m-3}{2m+5} + m - 4$

34. $\dfrac{n+2}{5n-3} + n - 6$

8.6 Mixed Expressions and Complex Rational Expressions

35. $\dfrac{1 + \dfrac{1}{c} - \dfrac{6}{c^2}}{\dfrac{1}{c} - \dfrac{2}{c^2}}$

36. $\dfrac{1 - \dfrac{2}{z} - \dfrac{8}{z^2}}{\dfrac{1}{z} + \dfrac{2}{z^2}}$

37. $\dfrac{a - 1 - \dfrac{3}{a+1}}{a + 5 + \dfrac{3}{a+1}}$

38. $\dfrac{b - 2 - \dfrac{25}{b-2}}{b + 2 - \dfrac{5}{b-2}}$

39. $\dfrac{\dfrac{2n+6}{n+3} - \dfrac{n-2}{n+2}}{\dfrac{n^2-9}{n^2+5n+6}}$

40. $\dfrac{\dfrac{2t+2}{t} - \dfrac{t-5}{t-3}}{\dfrac{t^2-4}{t^2-3t}}$

41. $\dfrac{\dfrac{u-1}{u+1} - \dfrac{u+1}{u-1}}{\dfrac{u-1}{u+1} + \dfrac{u+1}{u-1}}$

42. $\dfrac{\dfrac{v+2}{v-2} + \dfrac{v-2}{v+2}}{\dfrac{v+2}{v-2} - \dfrac{v-2}{v+2}}$

43. $\dfrac{\dfrac{x+2}{x} - \dfrac{3}{x+1}}{\dfrac{x-1}{x^2+x} + \dfrac{3}{x+1}}$

44. $\dfrac{\dfrac{r+3}{r^2+4r} + \dfrac{r-1}{r+4}}{\dfrac{r-2}{r+4} - \dfrac{3}{r}}$

45. $\dfrac{\dfrac{2}{a^2-4} + \dfrac{2}{a^2+a-2}}{\dfrac{6}{a^2-1} - \dfrac{3}{a^2-a-2}}$

Applications

46. **Statistics** Find the mean for the two rational expressions $\dfrac{2}{3x}$ and $\dfrac{1}{x^2}$.

47. **Electricity** The total resistance in a parallel circuit is $R = \dfrac{R_1 \times R_2}{R_1 + R_2}$. Simplify the expression for R if $R_1 = \dfrac{3}{a}$ and $R_2 = \dfrac{a+3}{a^3}$.

TEST YOURSELF

Simplify. 8.1–8.3

1. $\dfrac{m^2 + 2m - 24}{6m} \cdot \dfrac{2m + 6}{m^2 - m - 12}$

2. $\dfrac{y - x}{2y + 3x} \cdot \dfrac{3x + 2y}{x + y} \div \dfrac{x - y}{x + y}$

Write equivalent expressions having the same denominator. Use the LCD. 8.4

3. $\dfrac{3}{10ab^2}; \dfrac{2a}{5b}$

4. $\dfrac{4}{p - 2}; \dfrac{p}{p^2 - 5p + 6}$

Combine. Simplify if possible. 8.5–8.6

5. $\dfrac{m - 1}{3m^2} + \dfrac{m + 3}{2m} - \dfrac{m - 2}{6m^2}$

6. $\dfrac{5x}{x + 4} - \dfrac{(x - 3)}{(x + 4)}$

7. $y - \dfrac{1}{3}$

8. $\dfrac{\dfrac{r+1}{r^2-4}}{\dfrac{r^2-1}{r+2}}$

APPLICATION: Interest

Did you know that you can use a formula to find the monthly payment required when you borrow to make a purchase? You will need a calculator to use the formula.

Suppose you borrow $100,000 to buy a health food franchise. The annual interest rate is 18%. The loan is to be repaid over a 10-year period. What will be your monthly payment?

$$m = \frac{A\left(\frac{r}{12}\right)\left(1 + \frac{r}{12}\right)^n}{\left(1 + \frac{r}{12}\right)^n - 1}$$

m = monthly payment
A = amount borrowed
r = annual rate of interest
n = number of months of the loan

Calculation-Ready Form

$$m = \frac{100{,}000\left(\frac{0.18}{12}\right)\left(1 + \frac{0.18}{12}\right)^{120}}{\left(1 + \frac{0.18}{12}\right)^{120} - 1}$$

$= 1801.85$

The monthly payment is $1801.85.

Solve these problems.

1. What is the monthly payment on a loan of $1500 at 12% interest for 18 months?

2. What is the monthly payment on a loan of $3000 at 9.5% interest for 24 months?

3. Sue has $1000 for the down payment on a used car. She has budgeted a monthly car payment of $150, and would like to pay off a car loan in two years. Which of these cars should she consider?

Car	Price	Annual Interest
A	$4200	8%
B	3800	9.5%
C	5200	6%

4. Harvey borrows $40,000 to buy a bookstore. The annual rate of interest is 18%, and he will repay the loan in 36 monthly payments. Find his monthly payment and how much interest he will pay in all.

5. The Jacksons are buying a health food franchise. The purchase price is $145,000. They will make a down payment of 20%, and will finance the remaining amount over a period of 20 years at an annual rate of 11.4% interest. What will their monthly payments be?

8.7 Dividing Polynomials

Objective: To divide polynomials

Division of one polynomial by another when there are no common factors is similar to long division in arithmetic.

Capsule Review

EXAMPLE Divide and check: $13,680 \div 325$

$$
\begin{array}{r}
42 \leftarrow \text{quotient} \\
325\overline{)13,680} \\
-13\ 00 \\
\hline
680 \\
-650 \\
\hline
30 \leftarrow \text{remainder}
\end{array}
$$

Check:

Quotient $\times$ divisor $+$ remainder $=$ dividend
$42 \quad \times \quad 325 \quad + \quad 30 \quad = \quad 13,680$

Divide and check.

1. $19,663 \div 612$
2. $13,820 \div 488$
3. $38,040 \div 421$
4. $97,979 \div 126$
5. $50,505 \div 500$
6. $37,925 \div 350$

To divide a polynomial by a monomial, divide *each* term of the polynomial by the monomial. Check the results.

EXAMPLE 1 Divide and check: $(9w^4 + 3w^3 - w^2) \div 3w^3$

$$\frac{9w^4 + 3w^3 - w^2}{3w^3} = \frac{9w^4}{3w^3} + \frac{3w^3}{3w^3} - \frac{w^2}{3w^3}$$ *Divide each term by the monomial.*

$$= 3w + 1 - \frac{1}{3w}$$ *Simplify.*

Check: $\left(3w + 1 - \dfrac{1}{3w}\right)3w^3 = 9w^4 + 3w^3 - w^2$ ✓

When you divide one polynomial by another, factor if possible and then divide by any common factors.

EXAMPLE 2 Divide: $(a^2 + 3a - 4) \div (a + 4)$

$$\frac{a^2 + 3a - 4}{a + 4} = \frac{(a - 1)\cancel{(a + 4)}}{\cancel{(a + 4)}} = a - 1$$

Chapter 8 Rational Expressions

If the polynomials cannot be factored or if there are no common factors, follow a procedure similar to long division.

EXAMPLE 3 **Divide and check:** $(2y^2 + 3y - 11) \div (y - 3)$

$$\begin{array}{r} 2y + 9 \\ y - 3 \overline{\smash{)}2y^2 + 3y - 11} \\ \underline{-(2y^2 - 6y)} \downarrow \\ 9y - 11 \\ \underline{-(9y - 27)} \\ 16 \end{array}$$

Divide.   Think: $2y^2 \div y = 2y$
Multiply. $(y - 3)2y = 2y^2 - 6y$
Subtract. $2y^2 + 3y - (2y^2 - 6y) = 9y$

Divide.   Think: $9y \div y = 9$
Multiply. $(y - 3)9 = 9y - 27$
Subtract. $9y - 11 - (9y - 27) = 16$

So, $\dfrac{2y^2 + 3y - 11}{y - 3} = 2y + 9 + \dfrac{16}{y - 3}$ Write the remainder as a fraction.

Check: Quotient × divisor + remainder = dividend
$(2y + 9) \cdot (y - 3) + 16 = (2y^2 + 3y - 27) + 16$
$= 2y^2 + 3y - 11$ ✓

When you divide polynomials, write the terms of the dividend in descending order of the exponents of a variable. If a term is missing, insert a zero or a zero coefficient, as a placeholder.

EXAMPLE 4 **Divide:** $(7b + 4b^3) \div (2b - 1)$

$$\begin{array}{r} 2b^2 + b + 4 \\ 2b - 1 \overline{\smash{)}4b^3 + 0b^2 + 7b + 0} \\ \underline{-(4b^3 - 2b^2)} \\ 2b^2 + 7b \\ \underline{-(2b^2 - b)} \\ 8b + 0 \\ \underline{-(8b - 4)} \\ 4 \end{array}$$

Terms in descending order, with zeros as place holders.

So, $(7b + 4b^3) \div (2b - 1) = 2b^2 + b + 4 + \dfrac{4}{2b - 1}$

Check: Does $(2b^2 + b + 4)(2b - 1) + 4 = 4b^3 + 7b$?

CLASS EXERCISES

Divide and check.

1. $(8q^2 - 32q^3) \div 2q$
2. $(14t^4 - 28t^3 + 35t^2 - 7t) \div 7t^2$
3. $(n^2 - 5n + 4) \div (n - 4)$
4. $(6s^2 - 7s + 5) \div (2s - 3)$
5. $(2x^3 - 3x^2 - 10x + 3) \div (x - 3)$
6. $(3a^3 - 16) \div (a - 2)$

CE EXERCISES

and check.

1. $(6x^4 + 3x^3 - x^2) \div 3x^3$
2. $(10z^4 + 5z^3 - z^2) \div 5z^3$
3. $(x^3 - 18x^2 + 3x - 7) \div x$
4. $(c^3 + 11c^2 - 15c + 8) \div c$
5. $(3y^4 - 15y^3 + 21y^2 + 6y - 9) \div 3y$
6. $(4m^4 + 6m^3 - 20m^2 + 8m - 12) \div 2m$
7. $(n^2 + n - 30) \div (n + 6)$
8. $(a^2 - 2a - 24) \div (a + 4)$
9. $(k^2 + 11k + 18) \div (k + 6)$
10. $(r^2 + 3r + 5) \div (r + 2)$
11. $(3j^2 + 7j - 5) \div (j - 3)$
12. $(s^2 + 2s - 24) \div (s - 8)$
13. $(10m^2 + 9m - 36) \div (2m - 3)$
14. $(15h^2 - 62h + 40) \div (5h - 4)$
15. $(d^2 - 7d + 4) \div (d - 3)$
16. $(2m^2 - 3m + 5) \div (m - 2)$
17. $(8b + 2b^3) \div (b - 1)$
18. $(9a + 3a^3) \div (a - 1)$
19. $(3x^3 - 3) \div (x + 1)$
20. $(6 + 3q^3) \div (q - 2)$
21. $(18s^2 - 51s) \div 3s$
22. $(108a^3 - 72a^4) \div 9a^3$
23. $(18p^5q^4 - 54p^3q^3 + 27p^2q^5 - 9pq^2) \div 9pq^2$
24. $(21a^3b^2 + 56a^5b - 28a^2b + 63a^5b^2) \div 7a^2b$
25. $(p^3 + 3p^2 - 5p - 4) \div (p + 2)$
26. $(a^3 + 2a^2 + 4a - 5) \div (a + 3)$
27. $(6h^3 + 5h^2 - 9h - 7) \div (2h + 1)$
28. $(9t^3 - 3t^2 - 9t + 5) \div (3t + 2)$
29. $(x^3 - 3x + 5) \div (x - 4)$
30. $(y^3 + 4y - 7) \div (y - 3)$
31. $(9h^3 - 4h + 2) \div (3h - 1)$
32. $(4j^3 - 5j - 3) \div (2j + 3)$
33. $(27t^3 - 343) \div (3t - 7)$
34. $(64h^3 - 125) \div (4h - 5)$
35. $(15x^2 + 7xy - 2y^2) \div (5x - y)$
36. $(10m^2 - 11mn + 3n^2) \div (2m - n)$
37. $(6p^2 + 10pq + 6q^2) \div (3p + 2q)$
38. $(12c^2 + 25cd + 12d^2) \div (3c + 4d)$

Applications

39. **Finance** The amount (A) in a savings account is $A = p + prt$. Solve for p if $A = x^2 - 1$, $r = \dfrac{1}{x}$, and $t = x^2$. Find the principal (p) if $x = \$101$.

ALGEBRA IN HEALTH

An almanac provides a table of the Recommended Daily Allowances from infancy to 51+ years. The nutritive value of various foods is also given. Use this information to plan three nutritious meals for each member of your family.

Chapter 8 Rational Expressions

8.8 Ratios and Proportions

Objectives: To express ratios in simplest form
To solve equations involving proportions

One bag of fruit labeled A has 3 lb of apples and 4 lb of oranges. Another bag labeled B has 6 lb of apples and 8 lb of oranges. In bag A, you can say that the assortment has a 3 to 4 mixture of apples to oranges. This relationship is called a *ratio*.

A **ratio** is the comparison of two quantities by division.

What is the ratio of apples to oranges in bag B? Are the ratios found for each bag equal? How do you know?

A statement that ratios are equal is called a **proportion.** The proportion of apples to oranges can be written in several ways.

$$3 \text{ is to } 4 \text{ as } 6 \text{ is to } 8 \qquad 3:4 = 6:8 \qquad \frac{3}{4} = \frac{6}{8}$$

Capsule Review

To simplify ratios you will need to express fractions in simplest form, and to solve proportions, you will need to use the properties of equations.

Divide and express as a fraction in simplest form.

1. $\frac{225}{400}$
2. $3 \div 126$
3. $1\frac{1}{4} \div 2$
4. $3\frac{1}{2} \div 7$

Solve.

5. $6d = -20$
6. $12y = 72$
7. $\frac{3}{4}a = 3$
8. $\frac{-x}{3} = \frac{1}{4}$

A ratio can be expressed in a variety of ways as shown in the example below.

EXAMPLE 1 Express in simplest form:

a. $2.40 per $6
b. $1\frac{1}{2}$ out of 5
c. 2.4 cm to 40 mm

$\dfrac{2.40}{600} = \dfrac{240}{600}$ 　　 b. $1\dfrac{1}{2} \div 5 = \dfrac{3}{2} \cdot \dfrac{1}{5}$ 　　 c. $\dfrac{2.4 \text{ cm}}{40 \text{ mm}} = \dfrac{24 \text{ mm}}{40 \text{ mm}}$

$= \dfrac{4}{10}, \text{ or } \dfrac{2}{5}$ 　　　　　　　$= \dfrac{3}{10}$ 　　　　　　　$= \dfrac{24}{40}, \text{ or } \dfrac{3}{5}$

Notice that in Example 1c, 2.4 cm was changed to 24 mm. When simplifying ratios with different units within the same kind of measure, begin by expressing the measures using the same units.

In the proportion $3:4 = 6:8$ in the opening problem, 4 and 6 are called the **means**; and 3 and 8 are called the **extremes**.

In general, $\dfrac{a}{b} = \dfrac{c}{d}$ or $a:b = c:d$ 　　 b and c are the means and a and d are the extremes

---

Property of Proportions

In a proportion, the product of the means equals the product of the extremes. For all real numbers a, b, c, and d; $b \neq 0$ and $d \neq 0$:

If $\dfrac{a}{b} = \dfrac{c}{d}$, then $a \cdot d = b \cdot c$.

If $\dfrac{a}{b} = \dfrac{c}{d}$, how would you show that $a \cdot d = b \cdot c$?

This property is useful when you check to see whether two ratios are equal.

EXAMPLE 2 　True or false? 　a. $15:24 = 100:160$ 　　b. $\dfrac{3 \text{ lb}}{7 \text{ in.}^2} = \dfrac{10 \text{ lb}}{24 \text{ in.}^2}$

a. 　　$15:24 \stackrel{?}{=} 100:160$ 　　　b. 　$\dfrac{3 \text{ lb}}{7 \text{ in.}^2} \stackrel{?}{=} \dfrac{10 \text{ lb}}{24 \text{ in.}^2}$

　　　　$\dfrac{15}{24} \diagup\!\!\!\!\diagdown \dfrac{100}{160}$ 　　　　　　　$\dfrac{3 \text{ lb}}{7 \text{ in.}^2} \diagup\!\!\!\!\diagdown \dfrac{10 \text{ lb}}{24 \text{ in.}^2}$

　　　$24 \times 100 \stackrel{?}{=} 15 \times 160$ 　　　　　$7 \times 10 \stackrel{?}{=} 3 \times 24$
　　　　　$2400 = 2400$ ✓ 　　　　　　　　　$70 \neq 72$
　　　　　　true

The property of proportions is used to solve for an unknown quantity in a proportion.

EXAMPLE 3 Solve: **a.** $\frac{3}{7} = \frac{10}{x}$ **b.** $\frac{2s - 21}{3} = \frac{s}{5}$

a. $\frac{3}{7} = \frac{10}{x}$ **b.** $\frac{2s - 21}{3} = \frac{s}{5}$

$3x = 7(10)$ $3s = 5(2s - 21)$
$x = \frac{70}{3}$ $3s = 10s - 105$
           $-7s = -105$
           $s = 15$

Many real-world problems are solved by using proportions.

EXAMPLE 4 The scale on a map states: 1.5 cm represents 24 km. What distance does 4.25 cm on the map represent?

$\frac{1.5 \text{ cm}}{24 \text{ km}} = \frac{4.25 \text{ cm}}{x}$ ⟶ map distance
⟶ real distance in km

Write the ratios in the same order.

$(1.5 \text{ cm})(x \text{ km}) = (4.25 \text{ cm})(24 \text{ km})$

$x = \frac{(4.25 \text{ cm})(24 \text{ km})}{1.5 \text{ cm}}$

Units common to both numerator and denominator can be divided out.

$x = \frac{102 \text{ km}}{1.5} = 68 \text{ km}$

So, on the map 4.25 cm represents 68 km.

CLASS EXERCISES

Express as a ratio in simplest form.

1. 0.2 to 3
2. $2\frac{1}{2}$ out of 3
3. $4a^2 : 2ab$
4. 1.5 cm to 45 mm

True or false?

5. $\frac{2}{5} = \frac{6}{15}$
6. $\frac{2}{7} = \frac{9}{31}$
7. $\frac{1\frac{1}{2}}{1} = \frac{3}{2}$
8. $2:9 = 3:18$

Solve.

9. $\frac{8}{x} = \frac{2}{5}$
10. $\frac{3n}{2} = \frac{-9}{10}$
11. $\frac{a+1}{9} = \frac{2}{3}$
12. $8 = \frac{40}{m-3}$

13. If $\frac{1}{2}$ in. represents 20 mi, what distance does 4 in. represent?

8.8 Ratios and Proportions

PRACTICE EXERCISES

Express as a ratio, in simplest form.

1. 6 to 8
2. 48 to 32
3. $\frac{1}{2}$ to 3
4. 4 to $2\frac{2}{3}$
5. 0.5 to 2.5
6. 7.5 to 6
7. 2 to $4x$
8. $3y$ to y^2
9. 5 ft to 2 yd
10. 2.4 cm to 36 mm
11. $3.20 to $0.80
12. 0.72 to 0.6

True or False?

13. $\frac{5}{12} = \frac{3}{7}$
14. $\frac{4}{9} = \frac{7}{16}$
15. $\frac{6}{15} = \frac{16}{40}$
16. $\frac{9}{30} = \frac{16}{55}$

17. If 3 cans of Miller's Chunky Vegetable Soup cost $0.93, then 5 cans cost $1.55.

18. A supermarket sells 5 lemons at a cost of $0.95, then 8 lemons will cost $1.52.

Solve.

19. $\dfrac{5}{6} = \dfrac{30}{s}$
20. $\dfrac{28}{r} = \dfrac{4}{7}$
21. $\dfrac{s+4}{12} = \dfrac{7}{4}$
22. $\dfrac{9}{5} = \dfrac{18}{t+2}$
23. $\dfrac{8}{a} = \dfrac{16}{3}$
24. $\dfrac{5}{24} = \dfrac{x}{12}$
25. $\dfrac{18p}{54} = \dfrac{12}{9}$
26. $\dfrac{12c}{28} = \dfrac{15}{7}$
27. $\dfrac{5}{3n+5} = \dfrac{5}{5n-2}$
28. $\dfrac{5}{3+b} = \dfrac{3}{7b+1}$
29. $\dfrac{4}{x+3} = \dfrac{2}{2x+1}$
30. $\dfrac{4}{y+4} = \dfrac{2}{3y+2}$
31. $\dfrac{s-2}{5} = \dfrac{2s+3}{3}$
32. $\dfrac{t+3}{2} = \dfrac{2t-5}{6}$
33. $\dfrac{2}{2m+3} = \dfrac{3m-2}{4}$
34. $\dfrac{a}{a+3} = \dfrac{4}{5a}$
35. $\dfrac{q}{3} = \dfrac{4}{q+4}$
36. $\dfrac{5}{p+2} = \dfrac{p}{7}$
37. $\dfrac{w-3}{3} = \dfrac{3}{w+5}$
38. $\dfrac{11}{u-2} = \dfrac{u+7}{2}$

Applications

39. Entertainment A party punch contained 2 qt of pineapple juice, 3 pt of orange juice, $2\frac{1}{2}$ c of cranberry juice, and $1\frac{3}{4}$ c of ice. Find the following ratios.
   a. pineapple juice to ice
   b. cranberry juice to orange juice
   c. orange juice to pineapple juice

40. Cartography According to a map's scale, 1.5 cm represents 6 km. What distance does 3.25 cm represent?

41. Cartography If 2.5 cm represents 10 km, what distance does 4.25 cm represent?

42. Travel A car traveled 66 mi in $1\frac{1}{2}$ h. At the same speed, how many miles will the car travel in 2 h?

43. Travel If a cyclist traveled 45 mi in $2\frac{1}{2}$ h, how long would it take her to travel 153 mi?

BIOGRAPHY

Jean-Victor Poncelet (1788–1867), an officer in the French army, was captured during Napoleon's Russian campaign. While he was in prison, Poncelet created the area of mathematics known as *projective geometry*.

Investigation

One of the important ideas in projective geometry is cross-ratio.

- Draw four lines, *a*, *b*, *c*, and *d* through a point *O*.
- Draw a line *l* that intersects these lines at points *A*, *B*, *C*, and *D*.
- Consider the ratios $\frac{AC}{AD}$ and $\frac{BC}{BD}$.
  Form the cross-ratio: $\frac{(AC)(BD)}{(AD)(BC)}$
  An extraordinary fact is that this ratio has the same numerical value for all positions of the line *l*. Make several drawings, and measure to verify this fact.

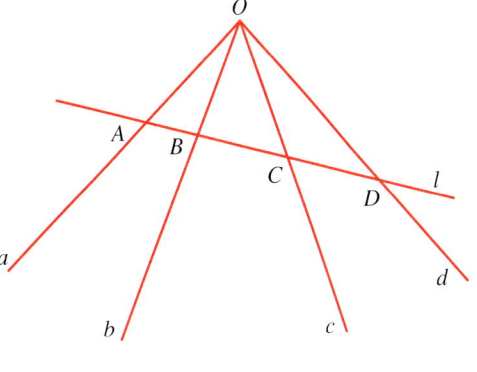

8.8 Ratios and Proportions

Solving Rational Equations

Objective: To solve rational equations, including some with extraneous solutions

In an earlier lesson you used the least common denominator (LCD) to help you add and subtract rational expressions. In this lesson the LCD will help you simplify and then solve equations.

Capsule Review

EXAMPLE Find the LCD: $\dfrac{5}{6xy}, \dfrac{x}{2x-4}$, and $\dfrac{10x}{4-2x}$

Factor the denominators:
$6xy = 2 \cdot 3 \cdot x \cdot y$
$2x - 4 = 2(x - 2)$
$4 - 2x = (-1)(2)(x - 2)$

The LCD is the product of the prime factors, each raised to the greatest power that appears in any one denominator. The negative one from the last denominator need not be included.

$2 \cdot 3xy(x - 2)$, or $6xy(x - 2)$

Find the LCD.

1. $\dfrac{2}{x}; \dfrac{3}{5x}; \dfrac{x}{5}$

2. $\dfrac{1}{ab}; \dfrac{2}{ab^3}; \dfrac{7}{a^2b}$

3. $\dfrac{9}{11m}; \dfrac{13}{44m^2}; \dfrac{m}{2}$

4. $\dfrac{4}{y}; \dfrac{3}{2+y}; \dfrac{x}{2y^2+4y}$

5. $\dfrac{s}{4}; \dfrac{3}{r+s}; \dfrac{r}{2r+2s}$

6. $\dfrac{d}{3-d}; \dfrac{3}{d-3}; \dfrac{d}{d^2-9}$

An equation that contains rational expressions is easier to solve if you first eliminate all denominators other than 1.

To solve an equation that contains rational expressions:
- Multiply each term of the equation by the least common denominator.
- Solve the resulting equation.

When checking a solution, you need to replace each variable in the original equation, not in the resulting equation, to make sure that the denominator does not equal zero.

EXAMPLE 1 Solve and check: $2m - \dfrac{4}{5} = \dfrac{-2m}{15}$

$2m - \dfrac{4}{5} = \dfrac{-2m}{15}$ *The LCD is 15.*

$(15)2m - (15)\dfrac{4}{5} = (15)\dfrac{-2m}{15}$ *Multiply each term by the LCD.*

$30m - 12 = -2m$ *Solve the transformed equation.*
$32m = 12$
$m = \dfrac{12}{32}$, or $\dfrac{3}{8}$

Check: $2\left(\dfrac{3}{8}\right) - \dfrac{4}{5} \stackrel{?}{=} \dfrac{-2}{15}\left(\dfrac{3}{8}\right)$ *In the original equation, replace m with $\dfrac{3}{8}$.*

$\dfrac{3}{4} - \dfrac{4}{5} \stackrel{?}{=} -\dfrac{1}{20}$

$-\dfrac{1}{20} = -\dfrac{1}{20}$ ✓

Since division by zero is undefined, -4 cannot be used as a solution in the next example. This is called an *extraneous solution*. An **extraneous solution** is an apparent solution which does not satisfy the original equation.

EXAMPLE 2 Solve and check: $\dfrac{a}{a+4} = 3 - \dfrac{4}{a+4}$

$\dfrac{a}{a+4} = 3 - \dfrac{4}{a+4}$ *The LCD is $(a+4)$.*

$\dfrac{a}{a+4}(a+4) = 3(a+4) - \dfrac{4}{a+4}(a+4)$ *Multiply by the LCD.*

$a = 3(a+4) - 4$ *Solve.*
$a = 3a + 12 - 4$
$-2a = 8$
$a = -4$

Check: $\dfrac{-4}{-4+4} \stackrel{?}{=} 3 - \dfrac{4}{-4+4}$ Replace a with -4 in the original equation. The denominator is 0, and so -4 is an extraneous solution. The equation is said to have no solution.

EXAMPLE 3 Solve and check: $\dfrac{d}{d+2} - \dfrac{2}{2-d} = \dfrac{d+6}{d^2-4}$

$\dfrac{d}{d+2} - \dfrac{2}{2-d} = \dfrac{d+6}{d^2-4}$

$\dfrac{d}{d+2} + \dfrac{2}{d-2} = \dfrac{d+6}{(d+2)(d-2)}$ *Factor. Rewrite $\dfrac{-2}{2-d}$ as $\dfrac{2}{d-2}$.*

8.9 Solving Rational Equations

Multiply by the LCD: $(d+2)(d-2)$

$$\frac{d}{d+2}(d+2)(d-2) + \frac{2}{d-2}(d+2)(d-2) = \frac{d+6}{(d+2)(d-2)}(d+2)(d-2)$$

$$d(d-2) + 2(d+2) = d+6 \quad \text{Solve.}$$
$$d^2 - 2d + 2d + 4 = d+6$$
$$d^2 - d - 2 = 0$$
$$(d+1)(d-2) = 0 \quad \text{Factor.}$$
$$d+1 = 0 \quad | \quad d-2 = 0 \quad \text{Zero Product Rule}$$
$$d = -1 \quad | \quad d = 2$$

Check: In the original equation, replace d with -1 and then with 2. Which of these is an extraneous solution?

CLASS EXERCISES

Solve and check. If the equation has no solution, write "no solution."

1. $\dfrac{y}{3} + \dfrac{2}{3} = 1$
2. $\dfrac{1}{3} + \dfrac{5z}{6} = 2$
3. $\dfrac{3}{a} - \dfrac{5}{a} = 2$
4. $\dfrac{2}{3} - \dfrac{5}{m} = \dfrac{1}{3m}$
5. $x + \dfrac{2}{3} = \dfrac{5x}{6}$
6. $5 + \dfrac{2}{p} = \dfrac{17}{p}$
7. $\dfrac{b}{b+3} = 2 - \dfrac{3}{b+3}$
8. $\dfrac{5}{f} + \dfrac{3}{f+1} = \dfrac{7}{f}$
9. $\dfrac{1}{g-3} + 1 = \dfrac{3}{g^2 - 3g}$

PRACTICE EXERCISES

Solve and check. If the equation has no solution, write "no solution."

1. $3m - \dfrac{3}{4} = \dfrac{2m}{3}$
2. $2y - \dfrac{3}{4} = \dfrac{3y}{8}$
3. $5x - \dfrac{2}{3} = \dfrac{-5x}{6}$
4. $4t - \dfrac{3}{5} = \dfrac{-2t}{3}$
5. $\dfrac{5e}{3} - \dfrac{7e}{6} = 2$
6. $\dfrac{7f}{3} - \dfrac{8f}{15} = 9$
7. $\dfrac{b}{b+3} = 5 - \dfrac{3}{b+3}$
8. $\dfrac{x}{x+5} = 3 - \dfrac{5}{x+5}$
9. $\dfrac{2t}{t-4} = 5 - \dfrac{1}{t-4}$
10. $\dfrac{3s}{s-5} = 7 - \dfrac{1}{s-5}$
11. $\dfrac{2}{i} - \dfrac{8}{i} = -15$
12. $\dfrac{7}{j} - \dfrac{9}{j} = -14$
13. $2 - \dfrac{8}{m} = 6$
14. $1 - \dfrac{9}{k} = 4$
15. $\dfrac{5}{2s} + \dfrac{3}{4} = \dfrac{9}{4s}$
16. $\dfrac{2}{3t} + \dfrac{1}{2} = \dfrac{3}{4t}$
17. $\dfrac{u+1}{u} + \dfrac{1}{2u} = 4$
18. $\dfrac{v+2}{v} + \dfrac{4}{3v} = 11$
19. $\dfrac{4w+5}{w-4} = \dfrac{5w}{w-4}$
20. $\dfrac{2x+4}{x-3} = \dfrac{3x}{x-3}$

Chapter 8 Rational Expressions

21. $\dfrac{y}{y-3} = \dfrac{3}{y-3} - 1$ 22. $\dfrac{z}{z+2} = 3 - \dfrac{2}{z+2}$

23. $\dfrac{c}{c+6} = \dfrac{1}{c+2}$ 24. $\dfrac{h}{h+5} = \dfrac{2}{h+5}$ 25. $\dfrac{3}{e-1} = \dfrac{2e}{e+4}$

26. $\dfrac{4}{c-4} = \dfrac{3c}{c+3}$ 27. $\dfrac{4}{k} + \dfrac{2}{3} = 10k$ 28. $5f = \dfrac{7}{2} + \dfrac{6}{f}$

29. $\dfrac{2}{e-2} = 2 - \dfrac{4}{e}$ 30. $6 - \dfrac{2}{r} = \dfrac{-5}{r-3}$ 31. $\dfrac{2e}{e-4} - 2 = \dfrac{4}{e+5}$

32. $\dfrac{r+1}{r-1} = \dfrac{r}{3} + \dfrac{2}{r-1}$ 33. $\dfrac{2}{a+2} = \dfrac{a}{a-2} + \dfrac{13}{4-a^2}$

34. $\dfrac{e+1}{e+2} = \dfrac{-1}{e-3} + \dfrac{e-1}{e^2-e-6}$ 35. $\dfrac{f-2}{f-4} = \dfrac{1}{f+2} + \dfrac{f+3}{f^2-2f-8}$

36. $\dfrac{u}{2u-2} + \dfrac{u+1}{2u+1} = \dfrac{-3u}{2u^2-u-1}$ 37. $\dfrac{s}{3s+2} + \dfrac{s+3}{2s-4} = \dfrac{-2s}{3s^2-4s-4}$

Applications

38. **Geometry** The perimeter of a triangle is 24 cm. Side c of the triangle is 2 cm longer than side a. Side b is $\tfrac{3}{5}$ as long as side c. What are the lengths of the three sides of the triangle?

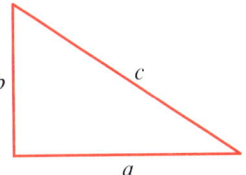

39. **Accounting** The librarian reports that $940 of the total book allowance for the three classes in the school has been spent. Class A spent $\tfrac{1}{3}$ of the total allowance; Class B spent $\tfrac{1}{4}$; Class C spent $\tfrac{1}{5}$. How much of the book allowance is left?

CRITICAL THINKING: Generalizing

Shown below are three different rectangles that have one thing in common: The number of units in the perimeter is equal to the number of units in the area. Give the dimensions for another rectangle in which this is true.

 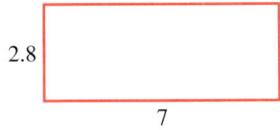 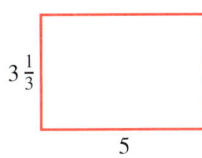

State a generalization to give the dimensions for an infinite number of rectangles that satisfy the given conditions. There is a limit for one of the dimensions. What is it?

8.10 Problem Solving: Work and Motion

Objective: To solve work and motion problems involving rational equations

Work, like motion, is related to rate and time.

[work RATE] · [TIME worked] = [part of JOB DONE]

Manuel can word process a monthly report in 4 h. What is his rate per hour?

$$r \cdot 4 = 1 \text{ monthly report}$$
$$r = \tfrac{1}{4} \text{ of the monthly report per hour}$$

What part of the report can be typed in 3 h?

$$\tfrac{1}{4} \cdot 3 = \tfrac{3}{4} \text{ of the report}$$

What part of the report can be typed in 15 min? 15 min = $\tfrac{1}{4}$ h

$$\tfrac{1}{4} \cdot \tfrac{1}{4} = \tfrac{1}{16} \text{ of the report}$$

The solution to all work problems is based on the amount of work done per unit of time. To solve problems where two or more people work together at different rates, find that part of the job each person completes and set up an appropriate equation.

EXAMPLE 1 Eric can wash the family van in 1 h. Stephanie can wash it in 45 min. How long will it take them if they work together?

Understand the Problem
What are the given facts?
The job can be completed by Eric in 1 h and by Stephanie in 45 min.

What are you asked to find?
The time needed to complete the job if they work together.

Plan Your Approach
Measurement units must be the same. Change 1 h to 60 min.
Find each person's rate of work.
Let n = the number of minutes working together.

	Work rate	· time worked	= part of job done
Eric	$\tfrac{1}{60}$	n	$\tfrac{n}{60}$
Stephanie	$\tfrac{1}{45}$	n	$\tfrac{n}{45}$

> The sum of the parts of the job done must be equal to 1, the complete job.

Write an equation.

$$\underset{\text{of job done.}}{\text{Eric's part}} + \underset{\text{of job done.}}{\text{Stephanie's part}} = \text{complete job}$$

$$\frac{n}{60} + \frac{n}{45} = 1$$

■ **Complete the Work**

$\frac{n}{60} + \frac{n}{45} = 1$ *Multiply by the LCD, 180.*

$3n + 4n = 180$

$7n = 180$

$n = \frac{180}{7}$, or $25\frac{5}{7}$ min

■ **Interpret the Results**

State your conclusion.

In $25\frac{5}{7}$ min the job is completed.

Check your conclusion.

Notice that $25\frac{5}{7}$ min is less time than it would take either to do the job alone.

Eric did $\frac{180}{7} \cdot \frac{1}{60}$, or $\frac{3}{7}$ of the job.

Stephanie did $\frac{180}{7} \cdot \frac{1}{45}$, or $\frac{4}{7}$.

$\frac{3}{7} + \frac{4}{7} = 1$ complete job

EXAMPLE 2 Phillip left his home at 2:15 PM and had driven 60 mi when he ran out of gas. He had to walk 2 mi to a gas station, where he arrived at 4:15 PM. If he drives 10 times faster than he walks, how fast does he walk?

■ **Understand the Problem**

Let r = rate walking and $10r$ = rate driving

■ **Plan Your Approach**

$\frac{d}{r} = t$

	Distance	Rate	Time
Driving	60	$10r$	$\frac{60}{10r}$
Walking	2	r	$\frac{2}{r}$

8.10 Problem Solving: Work and Motion

	Time driving + time walking = 2 h (2:15 to 4:15)
	$$\frac{60}{10r} + \frac{2}{r} = 2$$
Complete the Work	$$(10r) \cdot \frac{60}{10r} + (10r) \cdot \frac{2}{r} = (10r)2$$
	$$60 + 20 = 20r$$
	$$r = 4 \text{ mi/h}$$
	Phillip walked at the rate of 4 mi/h.
Interpret the Results	**Check:** Driving time is $\frac{60}{40}$, or $1\frac{1}{2}$ h.
	Walking time is $\frac{2}{4}$, or $\frac{1}{2}$ h.
	Total time is $1\frac{1}{2} + \frac{1}{2} = 2$ h. ✓

CLASS EXERCISES

Barry can paint a room in 12 h, Carrie can paint it in 10 h, and Harry takes only 9 h. If Barry and Carrie work together for 2 h, how long will it take Harry to finish the job?

1. What information is given? What are you asked to find?

2. What is each person's rate?

3. After 2 h, what part of the job has Barry done? Has Carrie done? Have they done together?

4. What equation could you use to solve this problem? Solve it.

PRACTICE EXERCISES

Solve. Check your answers.

1. Marian can weed a garden in 3 h, and Robin can do it in 4 h. How long will it take them if they work together?

2. David can unload a delivery truck in 20 min, and Allie can do it in 35 min. If they work together, how long will it take?

3. Art can paint a set of kitchen cabinets in 5 h. His mother can do it in 4 h. How long will it take them if they work together?

4. Peggy can gather a bushel of apples in 45 min. Peter can gather a bushel in 75 min. How long will it take Peggy and Peter to gather a bushel of apples if they work together?

5. It took Rhoda the same time to drive 275 mi as it took Van to drive 240 mi. If Rhoda's rate was 7 mi/h faster than Van's rate, how fast did each person drive?

6. On Saturday, Earl rode his bike for 3 h longer than Alice. Earl traveled 135 km and Alice traveled 90 km. If they both averaged the same rate of speed, how long did Earl ride?

7. It took Maggie a total of 4 h to drive 75 mi to the airport and then fly 2100 mi to a city in Mexico. If the plane rate is 12 times faster than her car, how fast did she drive?

8. To get to his grandmother's house, Fred must drive 135 mi on a freeway and then 45 mi along a country road. If the trip takes 5 h, and he can travel three times faster on the freeway than on the country road, how fast does he travel on the freeway?

9. Two pipes fill a storage tank in 9 hours. If the larger pipe fills the tank three times as fast as the smaller one, how long would it take the larger pipe to fill the tank alone?

10. A bathtub can be filled in 20 min with both faucets running. If the cold water faucet runs twice as fast as the hot water, how long would it take the cold water faucet to fill the tub by itself?

11. Sumi can wash the windows of an office building in $\frac{3}{4}$ the time it takes her apprentice. One day they worked on the building together for 2 h 16 min, and then Sumi continued alone. It took her 4 h 32 min more to complete the job. How long would it take her apprentice to wash all the windows?

12. Tim can trim 10 trees in $\frac{2}{3}$ the time it takes Tom. They trim trees together for 1 h 11 min. Then Tom continues alone until a total of 10 trees are trimmed (it took him 35 min 30 s). Working alone, how long would it take Tim to trim 10 trees?

MATH CLUB ACTIVITY

This problem is similar to one used in a high school mathematics examination sponsored by the Mathematical Association of America. Find the solution.

> A town's population increased by 1500 people, and then the new population decreased by 6%. The town now had 510 more people than it did before the 1500 increase. What was the original population?

8.11 Problem Solving Strategy: Solve a Simpler Problem

Sometimes a difficult problem can be solved by breaking it up into simpler parts, and then putting these parts back together to represent the original problem.

EXAMPLE Sawyer Automotive Co. often buys large shipments of surplus tires for a given amount of money. Adding $15 to their cost per tire, they then sell as many of each shipment as they can. They need a formula for their profit on each shipment in terms of the total shipment cost t, the original number of tires in the shipment, n, and the number of tires unsold, u.

Understand the Problem

What are the given facts?
Sawyer bought n tires for t dollars. They sold all but u tires. Each tire sold was sold for $15 more than it cost.

What are you asked to find?
Find a formula which gives the profit on each shipment.

Plan Your Approach

Choose a strategy.
Write a word equation which expresses the profit in terms of total sales and total cost. Total cost is t. Consider the parts which make up total sales.

Complete the Work

Write a word equation.
Profit = total sales − total cost
= total sales − t

Total sales = (number sold)(price per tire)
= $(n - u)$(price per tire)

Price per tire = original cost per tire + 15

Original cost per tire = $\dfrac{\text{total cost}}{\text{original number of tires}} = \dfrac{t}{n}$

Putting this all together from the opposite direction:

Original cost per tire = $\dfrac{t}{n}$

Price per tire = $\dfrac{t}{n} + 15$

Total sales = $(n - u)\left(\dfrac{t}{n} + 15\right)$

Chapter 8 Rational Expressions

$$\text{Profit} = (n - u)\left(\frac{t}{n} + 15\right) - t$$

Simplify the rational expression.

$$(n - u)\left(\frac{t}{n} + 15\right) - t = n\left(\frac{t}{n} + 15\right) - u\left(\frac{t}{n} + 15\right) - t$$

$$= t + 15n - \frac{ut}{n} - 15u - t$$

$$= 15n - 15u - \frac{ut}{n}$$

Interpret the Results

State your answer.
For a shipment costing t dollars, the profit, p, for n tires, of which u are unsold, is given by:

$$p = 15n - 15u - \frac{ut}{n}$$

Check your answer.
Choose convenient numbers for t, n, and u. Find the profit, p, first without the formula, then with the formula.

Assume there was a shipment of 100 tires with a total cost of $5000, with 10 tires unsold.

Without the formula:

Each tire cost $\frac{5000}{100}$ or $50.
The selling price was 50 + 15 or $65 per tire.
100 − 10 or 90 tires were sold.
So, total sales were 90(65) or $5850.
Profit was 5850 − 5000 or $850.

With the formula:
$t = 5000$, $n = 100$, and $u = 10$. Find p.

$$p = 15n - 15u - \frac{ut}{n}$$

$$p = 15(100) - 15(10) - \frac{10(5000)}{100} \qquad \textit{Calculation-ready form}$$

$$p = 850$$

The two methods confirm one another.

CLASS EXERCISES

Solve.

1. In the Example, if the company wishes to raise its profit by adding $18 to the original price per tire, what will be its new formula?

2. If the company always succeeds in selling the whole shipment of tires when they add $15 to the cost of each tire, how should its profit formula be changed?

PRACTICE EXERCISES

Solve.

1. What profit will be made if all but 14 of a shipment of 650 tires, purchased for $12,225, are sold?

2. What profit will be made if all but 23 of a shipment of 545 tires, purchased for $11,500, are sold?

3. If the shipments purchased by the company always consist of 500 tires, change the profit formula. Use the new formula to determine the profit if such a shipment costs $11,000 and 45 tires are not sold?

4. If the cost of the shipments purchased by the company is always $10,000, regardless of the number of tires obtained, change the profit formula to correspond to this fact. Use the new formula to determine what profit will be made if such a shipment contains 475 tires and 45 tires are not sold.

5. If Andres works r h/wk at the deli for $5.50 per h and s h/wk at the service station for $4.50 per h, find a formula for t, his total weekly income in dollars before deductions.

6. Unleaded premium gasoline costs c cents/gal at the full service pump and z cents/gal at the self service pump. Write a formula for finding the amount saved, s, using the self service pump, rather than the full service pump, to fill a 12.5 gal gasoline tank.

7. If Camilo averages u mi/h over the first 10 mi of his commute to work and v mi/h over the next 7 mi, write a formula for his commuting time, t.

8. If Juana can do a job alone in j hours and Kate can do it alone in k hours, write a formula for the time, t, it takes them working together.

9. Sue bought n dress patterns for $200. She sold all but 3 of them for $2 more per pattern than she paid for them. In terms of n, write a formula for r, the amount she received for the patterns sold.

10. For $125 Zhian bought h hot dogs to sell at the football game. He sold all but 11 of them for $0.25 more per hot dog than he paid. In terms of h, write a formula for a, the amount of his sales.

11. If the company in the Example on page 358 is concerned not with the total cost of a tire shipment, t, but with its cost per tire, c, revise the profit formula to use the variable c instead of t.

Chapter 8 Rational Expressions

12. If the company in the Example on page 358 later finds that the unsold tires may be disposed of by selling them for $5 more than the original cost per tire, revise the profit formula to take account of this.

Mixed Problem Solving Review

1. An express train traveling at 0.75 km/min passes a subway platform 5 min after a local train traveling at 0.50 km/min. In how many minutes should the express overtake the local?

2. If 25 oz of a 15% alcohol solution in water is to have the alcohol concentration doubled, how much alcohol must be added?

3. A floor 36 ft × 24 ft is partially covered by a rug so that a uniform border of bare floor is left. If the area of the rug is 540 ft², how wide is the border?

PROJECT

Form a team with two classmates to solve the following problem: Suppose that your team has some sort of small business. Considering your costs, mark-up, sales, and at least two other factors, devise a profit formula for your business. Show that your formula will give reasonable results.

TEST YOURSELF

Divide these polynomials.   8.7

1. $(18y^3 - 3y^2 + 15y) \div 3y^3$

2. $(3m^3 - 5m^2 - 9) \div (m + 3)$

Express as a ratio in simplest form.   8.8–8.9

3. 7.2 mm to 18 cm

4. 3 gal to 4 pt

5. If 2.5 cm on a map represents 100 m, what distance is represented by 32.5 cm?

Solve these equations.   8.10–8.11

6. $\dfrac{5}{2y} - \dfrac{12}{y} = -19$

7. $\dfrac{3z}{z+4} - 2 = \dfrac{3}{z-5}$

8. If Said can clean a room in 2 h while his little sister requires 3 h, how long should it take them working together?

9. If you buy posters for a total of $25 and sell each of them for $1.50 more per poster than the cost, write a profit formula.

CHAPTER 8 SUMMARY AND REVIEW

Vocabulary

complex rational expression (337)
extraneous solution (351)
extremes (346)
least common denominator (328)
means (346)
mixed expression (337)
proportion (345)
ratio (345)
rational expression (316)
simplest form (317)

Simplifying Rational Expressions To simplify a rational expression, factor the numerator and denominator. Divide both by the common factors. Restrictions on the denominator include values that make the denominator equal to zero.

8.1

Simplify and state the restrictions, if any.

1. $\dfrac{3b^3}{9b^2}$
2. $\dfrac{4}{2x+6}$
3. $\dfrac{m-1}{(m-1)(m+2)}$
4. $\dfrac{4-y}{y^2-16}$

Multiplying and Dividing Rational Expressions To multiply rational expressions, factor the numerators and the denominators, divide common factors, then multiply remaining factors. To divide rational expressions, multiply by the reciprocal of the divisor.

8.2–8.3

Multiply or divide. Simplify if possible.

5. $\dfrac{32x^2y}{25z^2} \cdot \dfrac{20z}{8xy}$
6. $\dfrac{2w}{3} \div \dfrac{4w}{9}$
7. $\dfrac{2+a}{1+a} \cdot \dfrac{a^2-1}{a^2-4}$
8. $\dfrac{2r^2+9r-5}{r-3} \div \dfrac{r^2+2r-15}{r}$

Adding and Subtracting Rational Expressions To add or subtract rational expressions, you need common denominators. Write the rational expressions with a common denominator and combine.

8.4–8.5

Find the LCD. Write equivalent expressions, use the LCD.

9. $\dfrac{1}{8m^3n}; \dfrac{5}{12mn^2}$
10. $\dfrac{5}{c+1}; \dfrac{3}{c-2}$
11. $\dfrac{2x}{x^2-9}; \dfrac{3}{x+3}$

Add or subtract.

12. $\dfrac{1}{8m^3n} + \dfrac{5}{12mn^3}$
13. $\dfrac{5}{c+1} - \dfrac{3}{c-2}$
14. $\dfrac{3}{x+3} - \dfrac{2x}{x^2-9}$

362 Chapter 8 Rational Expressions

Simplifying a Complex Rational Expression To simplify a complex rational expression, multiply the numerator and denominator by the LCD of all the rational expressions in the complex rational expression. 8.6

Simplify.

15. $u + 2 - \dfrac{1}{u+2}$

16. $\dfrac{1 - \dfrac{1}{r}}{\dfrac{1}{r^2}}$

17. $\dfrac{\dfrac{y}{x^2} - \dfrac{1}{y}}{\dfrac{1}{xy} + \dfrac{1}{x^2}}$

Dividing Polynomials To divide polynomials, divide by common factors. If the polynomial cannot be factored, divide as in long division. 8.7

Divide.

18. $\dfrac{12z^2 + 42z}{6z}$

19. $\dfrac{2r^3 + 19r^2 + 40r - 25}{r + 5}$

Solving a Proportion To solve for an unknown quantity in a proportion, use the property of proportions. In a proportion, the product of the means equals the product of the extremes. 8.8

Solve.

20. $\dfrac{10}{y} = \dfrac{15}{3}$

21. $\dfrac{5}{12} = \dfrac{n}{6}$

22. $\dfrac{7}{9} = \dfrac{28}{x+3}$

Solving Rational Equations To solve a rational equation, multiply each term of the equation by the least common denominator, then solve. Check for any extraneous solutions. 8.9

Solve and check. If the equation has no solution, write *no solution*.

23. $\dfrac{d}{5} + \dfrac{3d}{2} = 17$

24. $\dfrac{5}{x+3} - \dfrac{2}{x} = \dfrac{9}{2}$

Solving Work and Motion Problems To solve work and motion problems involving different rates, use rational equations. 8.10

Solve.

25. James can wash a car in 45 min, and Bertha can do it in 30 min. How long will it take them if they work together?

26. It took a plane the same time to fly 1125 mi as it took a car to go 125 mi. If the plane's rate is 400 mi/h faster than the car's rate, how fast did the car go?

CHAPTER 8 TEST

Simplify.

1. $\dfrac{9m - 15}{3}$

2. $\dfrac{3x^2 + 8x - 3}{x^2 - x - 12}$

3. $\dfrac{25ab}{3c} \cdot \dfrac{12}{5a}$

4. $\dfrac{2d}{5} \div \dfrac{4d}{15}$

5. $\dfrac{5}{x + 1} \cdot \dfrac{2x + 2}{6}$

6. $\dfrac{a^2 + a - 2}{a^2 - 9} \div \dfrac{a + 2}{a + 3}$

7. $\dfrac{4}{y - 4} - \dfrac{y}{y - 4}$

8. $\dfrac{5}{6k} + \dfrac{3}{4k^2} - \dfrac{2}{3k}$

9. $\dfrac{1}{a + 1} - \dfrac{2}{a - 3}$

10. $\dfrac{3}{u} - 6$

11. $g + 5 - \dfrac{1}{g + 5}$

12. $\dfrac{\dfrac{1}{x^2} - 4}{\dfrac{1}{x}}$

Divide and check.

13. $\dfrac{14p^2 + 21p}{7p}$

14. $\dfrac{c^2 - c - 4}{c - 2}$

15. $\dfrac{3x^3 + 2x^2 - 6x + 4}{x + 2}$

Express as a ratio or rate in simplest form.

16. 8 out of 20

17. 3 pt to 2 qt

18. 1.5 cm to 30 mm

Solve.

19. $\dfrac{b}{5} + \dfrac{5b}{3} = 2$

20. $\dfrac{x + 1}{x} - 7 = \dfrac{9}{2x}$

21. $\dfrac{1}{r} + \dfrac{2}{r - 1} = -2$

22. Sadie can make a pizza in 40 min, and Alvin can make one in 30 min. How long will it take them if they work together?

Challenge

1. Solve for r: $\dfrac{-2}{r - 2} - \dfrac{r}{r + 2} = \dfrac{r + 6}{4 - r^2}$

2. A tub can be filled in 15 min by a hot-water faucet and in 12 min by the cold-water faucet. The drain can empty the tub in 20 min. How long will it take the tub to fill if both faucets and the drain are open?

PREPARING FOR STANDARDIZED TESTS

Select the best choice for each question.

1. $\frac{5}{6} + \frac{2}{5} - \frac{7}{10} = \underline{\ ?\ }$
   A. $\frac{1}{2}$ B. $\frac{8}{15}$ C. $\frac{17}{30}$ D. $\frac{19}{30}$ E. $\frac{13}{15}$

2. $\left(\frac{6a^2b^3}{2ab}\right)^2$ equals
   A. $3a^2b^2$ B. $9a^2b^2$ C. $3a^2b^4$
   D. $9a^2b^4$ E. $9a^3b^5$

3. $\frac{8 \times 10^8}{4 \times 10^6} = \underline{\ ?\ }$
   A. 200 B. 20 C. 2
   D. 0.2 E. 0.02

4. Solve for x if $\frac{16}{7} = \frac{x}{28}$.
   A. 64 B. 32 C. 28 D. 14 E. 4

5. A $450 stereo is on sale at a $33\frac{1}{3}\%$ discount. If there is a 6% sales tax, what is the total cost of the stereo while it is on sale?
   A. $480 B. $327 C. $318
   D. $301.80 E. $300.18

6. $\frac{4}{3y} + \frac{1}{2x}$ equals
   A. $\frac{5}{6xy}$ B. $\frac{8x + 3y}{5xy}$
   C. $\frac{8x + 3y}{6xy}$ D. $\frac{5}{3y + 2x}$
   E. $\frac{4x + y}{5xy}$

7. If light travels 1.86×10^5 mi in one second, how many miles does it travel in two seconds?
   A. 3.72×10^5 B. 3.72×10^7
   C. 3.72×10^{10} D. 1.86×10^7
   E. 1.86×10^{10}

8. Solve the equation: $3(2x - 1) + 4(x + 1) = 2(3x + 4) - 9$
   A. 2 B. $\frac{1}{2}$ C. $-\frac{23}{4}$ D. -2 E. $-\frac{1}{2}$

Use this bar graph to answer questions 9–11.

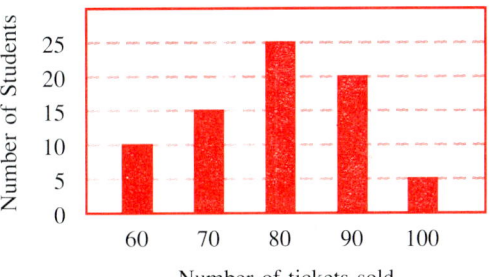

9. How many students sold 70 tickets?
   A. 5 B. 10 C. 15 D. 20 E. 25

10. How many students sold 80 or more tickets?
    A. 40 B. 50 C. 60 D. 65
    E. It cannot be determined from the information given.

11. What was the total number of tickets sold?
    A. 75 B. 400 C. 4050
    D. 5450 E. 5950

12. If x is subtracted from y and this difference is divided by the sum of x and $2y$, the result is:
    A. $\frac{x - y}{x + 2y}$ B. $\frac{2y - x}{x + y}$ C. $\frac{y - x}{x + 2y}$
    D. $\frac{x + 2y}{y - x}$ E. $\frac{y + x}{2y - x}$

Preparing for Standardized Tests

CUMULATIVE REVIEW (CHAPTERS 1–8)

For Exercises 1–8 refer to the number line below.

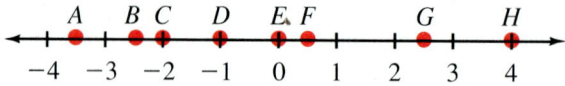

Give the coordinate of each point.

1. D
2. E
3. C
4. F
5. G
6. H
7. A
8. B

State the solution set of the equation or inequality whose graph is given.

9.
10.
11.
12.
13.
14.

Solve each equation or inequality.

15. $5x + 8 = -7$
16. $\dfrac{a}{3} - 6 = 2$
17. $t - (-3) \geq 9$
18. $-8m < 32$
19. $|y| + 7 = 3$
20. $3b - 4 = b + 1$
21. $\dfrac{2n}{3} - 5 = \dfrac{1}{3}$
22. $2c - 3 > 20$
23. $5 - \dfrac{4}{r} = 3$
24. $z^2 + 2z = 0$
25. $-2 \leq 3p + 4 \leq 13$
26. $|4d - 7| = 29$
27. $8n - 2(3n - 1) = 26$
28. $g^2 + 2g - 24 = 0$
29. $2.7 - 0.2k = 0.7k$
30. $2y + 3 > 3$ or $3y - 4 \leq -7$
31. $\dfrac{5}{m + 3} = \dfrac{3}{m - 1}$
32. $\dfrac{1}{p + 1} - \dfrac{1}{p - 2} = 0$

Write each number in scientific notation.

33. 6,000,000
34. 450
35. 12,800
36. 790,000,000

Evaluate each expression if $x = \dfrac{1}{2}$, $y = \dfrac{2}{3}$ and $z = -6$.

37. $y \cdot z - 1$
38. $\dfrac{2}{x} - z$
39. $2(x - 1) + z^2$

40. $y\left(\dfrac{x+3}{z-1}\right)$

41. x^2yz

42. $\dfrac{y-z}{y+1} - x^2$

43. $[z(x+y) + 5]^3$

44. $\left(\dfrac{x}{y} + \dfrac{y}{x}\right)\dfrac{z}{5}$

45. $|(2+z)^2 - 20|$

46. $x^2 - y^2$

47. $(z)(x^2)(y^2)$

48. $\dfrac{|x-y||x+y|}{z}$

Simplify. Assume that no denominator equals zero.

49. $4 + 12 \div (-3 + 1)$

50. $[(-2)^3 + 3^2]^5$

51. $|2-7| - (-3)^2$

52. $(3m^2n)(2m^3n^2)$

53. $(-2b^3)^2$

54. $(2p+3)^2$

55. $\dfrac{3x^2y}{9y^2x}$

56. $\dfrac{3c+9}{15}$

57. $\dfrac{a-b}{b-a}$

58. $\dfrac{5g^2 - 5}{g+1}$

59. $\dfrac{z+2}{z^2-4}$

60. $\dfrac{d^2-9}{3-d}$

61. $5a^2 + 3a - 2(a^2 - a)$

62. $2t(3t-4) - (t-1)^2$

63. $\dfrac{r^2-25}{r+5} + 2(r-5)$

64. $\dfrac{f-2}{f^2-4} + \dfrac{f+1}{f+2}$

65. $\dfrac{h^2-9}{h+1} \cdot \dfrac{h^2+2h+1}{h-3}$

66. $\dfrac{h^2+2h-15}{h^2+h-12} \cdot \dfrac{h^2+2h-8}{h^2+3h-10}$

67. $\dfrac{k^2-25}{k^2+4k+4} \div \dfrac{k^2+10k+25}{k^2-4}$

68. $\dfrac{m^2-3m-10}{m^2-25} \div \dfrac{m^2-m-6}{m^2+2m-15}$

Write the missing reasons in the proof.

69. **Prove:** For all real numbers x, y, and z, if $x = y$, then $x + z = z + y$.
Proof:

Statement	Reason
$x = y$	Given
$x + z = x + z$	a. ?
$x + z = y + z$	b. ?
$x + z = z + y$	c. ?

Solve.

70. If you have 70 mL of a 35% solution of acid in water, how much water would you add in order to make a 25% acid solution?

71. How many ounces of pure salt must be added to 40 oz of a 20% solution of salt in water to produce a 50% salt solution?

72. Adding 30 mL of water to 70 mL of a 20% sugar solution in water will result in a new solution of what concentration?

Factor completely.

73. $4a - 8b$
74. $12x + 15y$
75. $6c^2 - 9c$
76. $10m^2 + 5m$
77. $21p^3 + 14p^2 - 28p$
78. $k^2 - 9$
79. $d^2 + 8d + 16$
80. $5t^2 - 20$
81. $g^2 + 2g - 15$
82. $9a^2 - 25$
83. $r^2 - 10r + 24$
84. $6y^2 + 5y - 6$

Find the LCD of each group of fractions.

85. $\dfrac{3}{2a^3}; \dfrac{2}{3a^2}$
86. $\dfrac{1}{a^2b}; \dfrac{5}{ab^2}$
87. $\dfrac{1}{3x}; \dfrac{x}{x+1}$

Perform the indicated operation.

88. $\left(-\dfrac{3}{2}\right) + \dfrac{3}{4} - \left(-\dfrac{1}{2}\right)$
89. $\dfrac{2}{3}\left[\dfrac{1}{2} + \left(-\dfrac{5}{6}\right)\right]$
90. $\dfrac{3}{5}\left(-\dfrac{1}{3} \div \dfrac{1}{6}\right)$

91. $\dfrac{5}{4b} + \dfrac{3}{2b^2}$
92. $\dfrac{t}{t^2 - 9} - \dfrac{3}{3 - t}$
93. $5h + \dfrac{h+3}{h+2}$

94. $\dfrac{m-2}{m+2} - \dfrac{2m^2 + 6}{m^2 - 4} + \dfrac{-m+2}{2-m}$
95. $\dfrac{(a-2)^2 - (a-2)}{a^2 - 5a + 6}$

96. $(2m + 5)(5m - 2)$
97. $4x^2y(x - 2y^2 + 1)$
98. $3d(d + 2)^2$

99. $(3a^2 + 5a - 3) - (2a^2 - 2a + 1)$
100. $(4p^5 + 6p^3 - 2p^2) \div 2p^2$

Solve.

101. The price of a $65 jacket is decreased by 15%. Find the new price.

102. Rates at a car rental agency are $128 a week plus $0.12 a mile. Frank rents a care for a week. How far can he drive if he wants to spend no more than $200?

103. Find three consecutive odd integers such that the first times the third is 1 more than 4 times the second.

104. If a map scale is $\frac{1}{4}$ inch = 15 miles, what distance does $2\frac{1}{2}$ inches represent?

105. Allison can cut her parents lawn in 2 h, while her brother Jason does it in 2.5 h. Working together how long will it take them?

106. A train left San Mateo Station and traveled east at 75 mi/h. A second train left the same station 1.5 h later and traveled east at 85 mi/h. How many hours will it take for the second train to catch the first one.

107. The Ward family drove to the beach in $3\frac{1}{2}$ h at an average speed of 50 mi/h. The return trip took $4\frac{1}{5}$ h. What was the average speed for the trip home?

9 Linear Equations

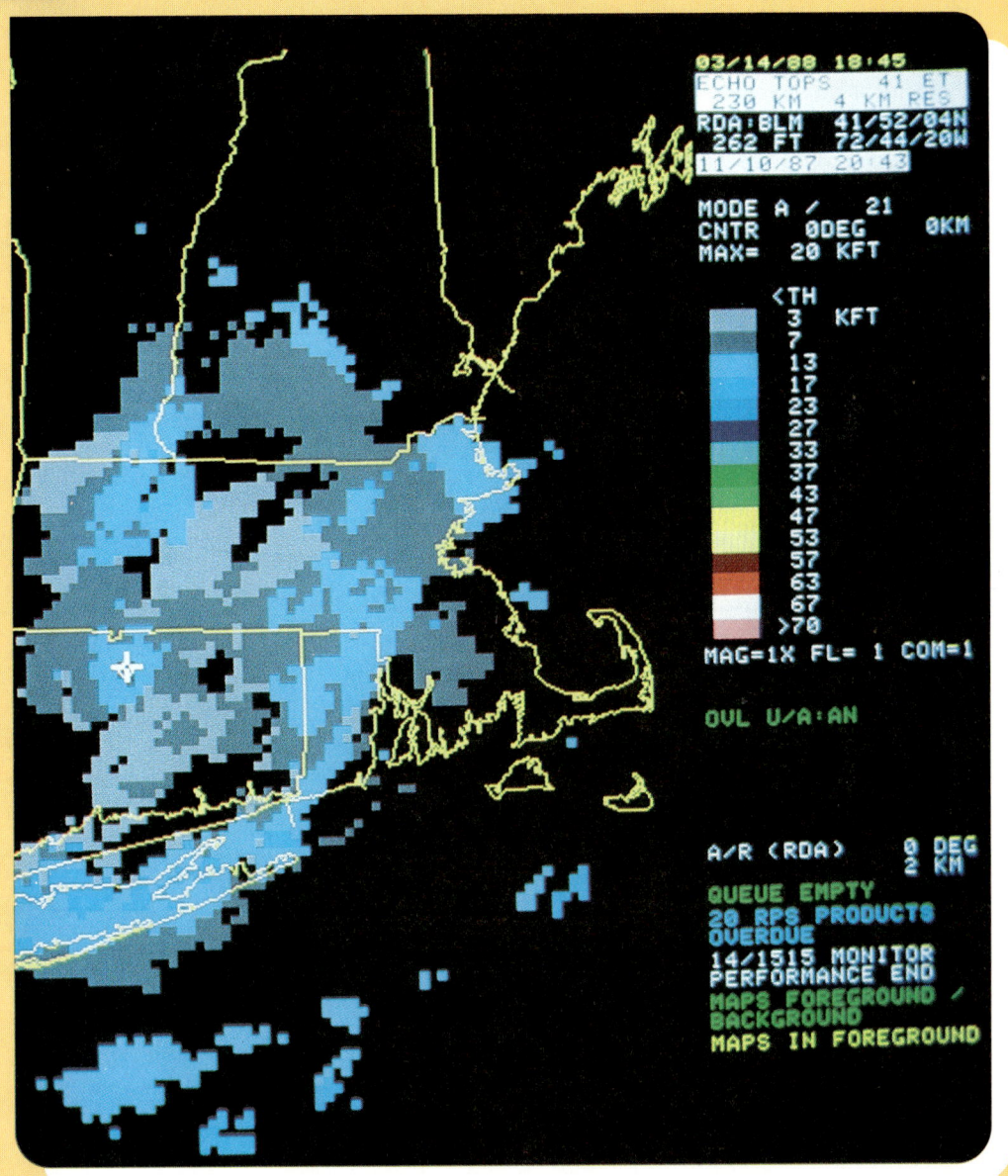

Physical scientists use graphs to represent numerical data. These graphs provide an easily interpreted method for analyzing trends, patterns, or direction of the numerical data.

9.1 Graphing Ordered Pairs

Objectives: To graph ordered pairs of numbers
To identify a solution of an equation in two variables

Many maps use a grid system to help you locate a street or town. You are directed to a particular section of the map by finding the intersection of a vertical column with a horizontal row. On the map at the right, Eaton is in the intersection of column C and row 2.

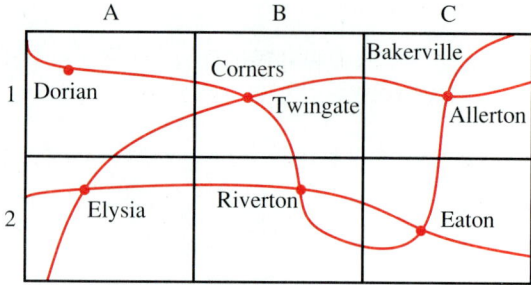

Capsule Review

The graph of a real number is a point on a number line. The number is called the *coordinate* of the point.

Use the coordinates to locate these points on a number line.

1. $A(15)$
2. $B(-2)$
3. $C(0)$
4. $D\left(-1\frac{1}{4}\right)$

A **coordinate system** in a plane is determined when two number lines are drawn in the plane so that they intersect at right angles. The zero point of each line is common to both lines and is called the **origin.** The horizontal number line is called the **x-axis.** The vertical number line is called the **y-axis.** The negative directions are to the left of the origin on the x-axis and down from the origin on the y-axis.

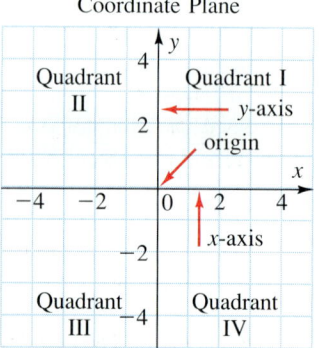

The axes separate the coordinate plane into four **quadrants,** as shown at the right. Points on the axes are not in any quadrant.

Each point in a coordinate plane is assigned a unique **ordered pair** of real numbers, (x, y). The two numbers paired with a given point are called the *coordinates* of that point.

$$P(5, -3)$$

x-coordinate   y-coordinate

Read: Point P, with coordinates 5 and −3.

370   Chapter 9 Linear Equations

The *x*- and *y*-coordinates are known, respectively, as the **abscissa** and the **ordinate**.

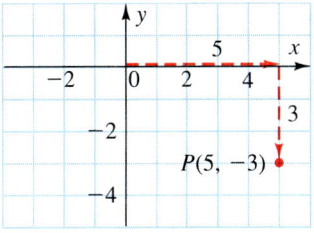

To locate or *plot* a point in a coordinate plane, always start at the origin. For $P(5, -3)$ move 5 units to the right from the origin and then move down 3 units. A point at that location is the *graph* of the ordered pair $(5, -3)$.

The order is important: $(5, -3)$ and $(-3, 5)$ are two different ordered pairs. How would you graph the point $Q(-3, 5)$?

EXAMPLE 1 Graph the points $A(4, 1)$, $B(-1, 3)$, $C(-2, 0)$, and $D(-2, -4)$ in a coordinate plane.

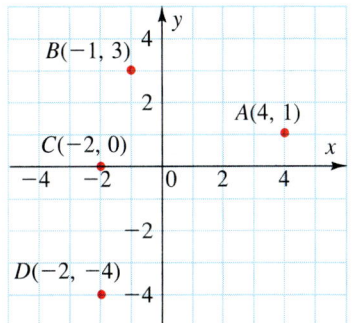

$A(4, 1)$ — right 4, up 1

$B(-1, 3)$ — left 1, up 3

$C(-2, 0)$ — left 2, stay on axis

$D(-2, -4)$ — left 2, down 4

Any ordered pair of real numbers names a unique point in the plane. Conversely, every point in the plane has a unique pair of coordinates. The coordinates of the origin are $(0, 0)$.

A solution of an equation containing two variables consists of two real numbers, one for each variable. For example:

$$x - y = 10 \qquad \text{A solution is } x = 12 \text{ and } y = 2.$$

You can write this solution as the ordered pair $(12, 2)$. Of course, there are other ordered pairs of numbers whose difference is 10: $(10, 0)$, $(15, 5)$, $(5, -5)$, and so on.

You can use substitution to decide whether a given ordered pair is a solution of an equation. The first number in the ordered pair is the value for *x*, and the second number is the value for *y*.

EXAMPLE 2 Determine whether $(-2, -4)$, $(4, 16)$, and $(-3, 9)$ are solutions of the equation $2x - y = 0$.

$(-2, -4)$
$2x - y = 0$
$2(-2) - (-4) \stackrel{?}{=} 0$
$-4 + 4 \stackrel{?}{=} 0$
$0 = 0$ ✓

$(4, 16)$
$2x - y = 0$
$2(4) - 16 \stackrel{?}{=} 0$
$8 - 16 \stackrel{?}{=} 0$
$-8 \neq 0$

$(-3, 9)$
$2x - y = 0$
$2(-3) - 9 \stackrel{?}{=} 0$
$-6 - 9 \stackrel{?}{=} 0$
$-15 \neq 0$

$(-2, -4)$ is a solution. $(4, 16)$ and $(-3, 9)$ are not solutions.

9.1 Graphing Ordered Pairs

EXAMPLE 3 The ordered pairs (0, 6), (4, 9), and (−4, 3) are solutions of the equation $3x = 4y - 24$. Graph the corresponding points in a coordinate plane and then connect them.

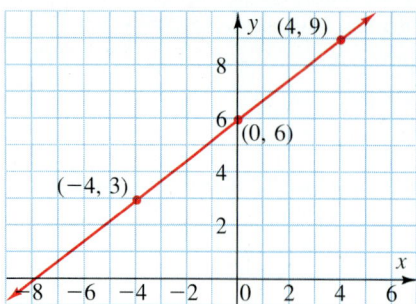

These points appear to lie on a line. (−8, 0) is another solution of the equation. Does the point that corresponds to this ordered pair lie on the same line as the other points?

CLASS EXERCISES

Graph the points in a coordinate plane.

1. $T(4, 1)$
2. $R(-3, -6)$
3. $U(-2, 5)$
4. $V(2, -5)$

State the coordinates and the quadrant (or axis) for each point.

5. D
6. A
7. K
8. J
9. C
10. F
11. H
12. I

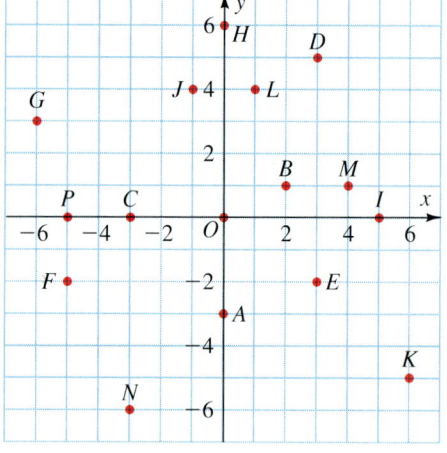

Name the point that is the graph of each ordered pair.

13. (2, 1)
14. (3, −2)
15. (0, 0)
16. (−6, 3)
17. (−5, 0)
18. (1, 4)
19. (4, 1)
20. (−3, −6)

State whether (0, 1) is a solution for each equation.

21. $x + 2y = 2$
22. $3x - y = -1$
23. $x = y + 1$
24. $-x + y = -1$
25. $4x + y = 3$
26. $xy = 0$

PRACTICE EXERCISES

Graph the points in a coordinate plane.

1. $A(1, 2)$
2. $B(3, 4)$
3. $C(0, 4)$
4. $D(3, 0)$
5. $E(5, 9)$
6. $F(3, 2)$
7. $G(-1, 5)$
8. $H(-3, 1)$

9. $I(-11, 10)$ 10. $J(-7, 4)$ 11. $K(-4, 2)$ 12. $L(-6, -6)$
13. $M(-3, -5)$ 14. $S(-1, -7)$ 15. $P\left(-1, 1\tfrac{1}{2}\right)$ 16. $Q\left(-4, \tfrac{1}{2}\right)$

Determine which of the ordered pairs listed are solutions of the equation.

17. $3x - y = -2$
    $(-3, -7), (3, 5)$

18. $5x - y = -7$
    $(2, 5), (0, 7)$

19. $x + 3y = 6$
    $(3, 1), (4, 2)$

20. $x + 5y = 1$
    $\left(0, \tfrac{1}{5}\right), (3, 2)$

Graph the ordered pairs in a coordinate plane, and then connect them.

21. $(0, 1), (2, 4), (-1, 6)$
22. $(1, 0), (3, 2), (-2, 3)$
23. $(2, 5), (-3, 2), (0, 5)$
24. $(1, 6), (-5, 0), (1, 4)$
25. $(-3, -2), (-5, 0), (6, -1)$
26. $(-4, -3), (-2, 1), (4, -3)$
27. $(-5, 0), (0, -5), (5, 0)$
28. $(-8, -1), (4, 0), (3, 5)$

Identify the axis on which each point lies.

29. $P(x, y)$, if $x = 0$ and y is any real number.
30. $T(x, y)$, if x is any real number and $y = 0$.

Give two equations for which the ordered pair is a solution.

31. $(1, -3)$ 32. $\left(\tfrac{1}{2}, -2\right)$ 33. $(-16, -3)$ 34. $(7, 49)$

The coordinates of three vertices of a rectangle are given. Graph the three vertices and determine the coordinates of the fourth vertex.

35. $(-1, -4), (4, -4), (4, 4)$ 36. $(1, 3), (1, -2), (-2, -2)$

Complete each statement.

37. A line through points that have the same ordinate is ___?___ to the x-axis and ___?___ to the y-axis.

38. A line through points that have the same abscissa is ___?___ to the x-axis and ___?___ to the y-axis.

39. A line through points where $x = -2$ and a line through points where $y = 4$ have ___?___ point(s) in common. Give the coordinates of the point(s), if any.

Graph the six ordered pairs in the same coordinate plane. Make a statement about the relationship between x and y. Then write an equation in terms of x and y that states the relationship.

40. $(-3, -3), (-2, -2), (-1, -1), (0, 0), (1, 1), (2, 2)$

41. $(-2, -1), (-1, 0), (0, 1), (1, 2), (2, 3), (3, 4)$

Applications

Draw and label the axes. Then graph the data to answer the question.

42. Chemistry A chemistry student measured the temperature c (in degrees Celsius) of a chemical reaction after m minutes and recorded the ordered pairs (c, m): $(2.5, 7.5), (5, 10), (10, 15), (20, 25)$. Find the temperature after 25 min. At what time will the temperature be 100°C?

43. Transportation In the equation $d = 24t$, d represents the number of miles traveled by a truck moving at an average speed of 24 mi/h and t represents the time in hours.
  a. For this situation, what do the pairs (t, d) represent?
  b. Are the following pairs (t, d) members of the solution set of this equation: $(2.5, 60)$? $(1.75, 42)$? $(0.5, 14)$? $(3.5, 84)$? $(0, 0)$?

44. Economics A computer programmer's salary s is represented by the equation $s = 50h + 100$, where h represents the number of hours needed to complete a program.
  a. Which of the following pairs of numbers (h, s) are members of the solution set for the equation? $(0.5, 125)$? $(1.25, 162)$? $(2.5, 225)$? $(3.0, 300)$? $(5.5, 375)$?
  b. Use the graph to determine the minimum amount the programmer earns.

MATH CLUB ACTIVITY

The coordinates of point A are $(-1, -1)$, and the coordinates of point B are $(2, 1)$. Graph these ordered pairs in a coordinate plane. An electric current can flow along the grid lines, but it can flow only to the right or up. How many different paths could the current follow to flow from A to B?

The coordinates of point C are $(0, 0)$, and the coordinates of point D are $(3, 2)$. Without counting them, tell how many different paths the current could follow to flow from C to D.

Give the coordinates of two other points E and F, where the number of paths that the current could follow to flow from point E to point F is the same as the number that it could follow to flow from C to D.

9.2 Graphs of Linear Equations

Objectives: To graph linear equations by making tables of ordered pairs
To graph linear equations from their x- and y-intercepts

An equation such as $3x = 4y - 24$ is called a *linear equation*. A **linear equation** is an equation whose graph is a line.

Capsule Review

Each set of ordered pairs are solutions of the given equation. Graph the ordered pairs in a coordinate plane and then connect them.

1. $\{(0, -2), (1, 1), (-1, -5)\}$; $y = 3x - 2$
2. $\{(-1, -1), (0, 0), (3, 3)\}$; $y - x = 0$
3. $\{(0, -3), (1, -1), (2, 1)\}$; $y - 2x = -3$
4. $\{(2, 0), (-2, -2), (4, 1)\}$; $y - \frac{1}{2}x = -1$

In many situations, it is important to write linear equations in *standard form*.

> Linear equations (with the variables x and y) can be written in the form $Ax + By = C$, where A, B, and C are real numbers, and A and B are not both equal to 0. If A, B, and C are integers, the equation is said to be in **standard form.**

$3x - y = 2$ and $-x + y = 0$ are examples of linear equations in standard form. $y = 2x - 3$ and $y - \frac{1}{2}x = -1$ are examples of linear equations that are not in standard form. How would you write the last two equations in standard form?

To graph a linear equation, you may follow these steps:

1. Solve the equation for y.
2. Choose several values for x. Substitute each x-value in the equation and find the corresponding y-value.
3. Record the ordered-pair solutions in a table.
4. Graph the ordered pairs. Draw a line through the points.

EXAMPLE 1 Make a solution table and graph: $3x + 4y = 12$.

1. Solve for y. $3x + 4y = 12$

    $4y = -3x + 12$ Subtract $3x$ from each side.

    $y = -\frac{3}{4}x + 3$ Divide each side by 4.

2. Choose the x-values. Solve to find corresponding y-values. This computer program may be used to generate the table of values.

```
10  PRINT "X","Y"
20  FOR X = - 4 TO 4 STEP 4
30  Y = - 3 / 4 * X + 3
40  PRINT X,Y
50  NEXT X
60  END
```

x	y
-4	$-\frac{3}{4}(-4) + 3 = 6$
0	$-\frac{3}{4}(0) + 3 = 3$
4	$-\frac{3}{4}(4) + 3 = 0$

3. Graph the ordered pairs from the above table. This computer program may be used to generate the graph.

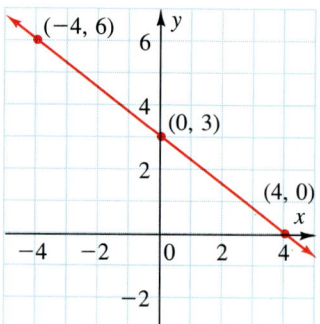

```
10  FOR Y = 6.5 TO 0 STEP - .5
20  FOR X = - 4 TO 4 STEP .25
30  IF Y = - 3 / 4 * X + 3 THEN
      PRINT "+(";X;",";Y;")"; :X = 4: GOTO 60
40  IF X = 0 THEN PRINT "1";: GOTO 60
50  PRINT " ";
60  NEXT X: PRINT
70  NEXT Y
80  FOR X = 1 TO 8: PRINT "*---";:
      NEXT X: PRINT "*"
90  END
```

In Example 1, why do you think multiples of 4 were chosen for values of x? Do you think the coordinates of every point on the graph satisfy the equation $3x + 4y = 12$?

At the right, the graph of $3y = x - 6$ intersects the x-axis at point $P(6, 0)$. 6 is called the **x-intercept** of the graph.

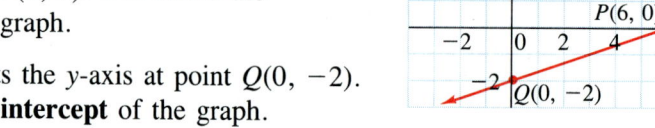

The graph intersects the y-axis at point $Q(0, -2)$. -2 is called the **y-intercept** of the graph.

The x-intercept and y-intercept can be used to graph a linear equation. As shown in Example 2, sometimes it is helpful first to write the equation in standard form.

Chapter 9 Linear Equations

EXAMPLE 2 Use the intercepts to graph $x = \frac{1}{3}y + 1$.

Write the equation in standard form: $3x - y = 3$.
Substitute 0 for y to find the x-intercept. $3x - 0 = 3;\ 3x = 3;\ x = 1$
Substitute 0 for x to find the y-intercept. $3(0) - y = 3;\ -y = 3;\ y = -3$
Plot the intercept points and draw the graph. Plot $A(1, 0)$ and $B(0, -3)$.

The computer program will find the intercepts to be used in graphing.

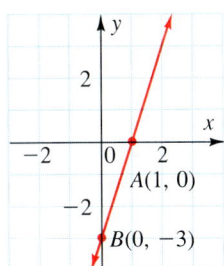

```
10 Y = 0:X = 1 / 3 * Y + 1
20 PRINT "WHEN Y = 0, X = ";X;" SO THE
   X-INTERCEPT IS A(";X;",0)"
30 X = 0:Y = 3 * X - 3
40 PRINT : PRINT "WHEN X = 0, Y = ";Y;" SO
   THE Y-INTERCEPT IS B(0,";Y;")"
50 END
```

Any third solution such as (2, 3) can be used as a check.

When you graph an equation such as $x = -4$ in a coordinate plane, notice that the equation places no restrictions on y. Set-builder notation may help you to see which ordered pairs are solutions of $x = -4$.

$\{(x, y): x = -4\}$ *Read:* The set of all ordered pairs of real numbers x and y, such that $x = -4$.

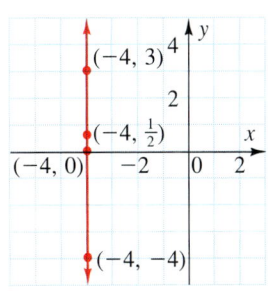

In all ordered pair solutions of $x = -4$, the x-coordinate is -4. The y-coordinate can be any real number.

Graphing the ordered pairs in which the abscissa is always -4 yields a graph that is parallel to the y-axis and 4 units to the left of the y-axis.

EXAMPLE 3 Graph $y = 3$ in a coordinate plane.

The solution set of $y = 3$ is $\{(x, y): y = 3\}$. Some solutions are: $(-3, 3)$, $(0, 3)$, $\left(1\frac{1}{2}, 3\right)$, and $(3, 3)$.
The graph is parallel to the x-axis and 3 units above the x-axis.

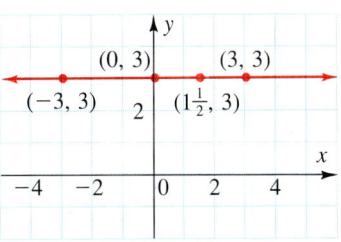

How do the solutions for the graph of $x = -4$ compare with the solutions for the graph of $y = 3$?

9.2 Graphs of Linear Equations   **377**

EXAMPLE 4 The graph of $y = -2x + 1$ is shown at the right. Find two solutions from the graph. Check these solutions.

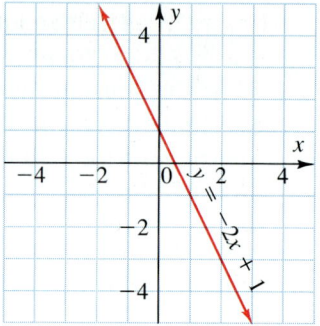

$(0, 1)$ and $(1, -1)$ appear to be solutions. Check by substituting these values in the equation.

$(0, 1)$
$y = -2x + 1$
$1 \stackrel{?}{=} -2(0) + 1$
$1 \stackrel{?}{=} 0 + 1$
$1 = 1$ ✓

$(1, -1)$
$y = -2x + 1$
$-1 \stackrel{?}{=} -2(1) + 1$
$-1 \stackrel{?}{=} -2 + 1$
$-1 = -1$ ✓

CLASS EXERCISES

Find the *x*- and *y*-intercepts of the graph of each equation.

1. $y - 3x = 4$
2. $x + y = 8$
3. $y = -7$
4. $x = 3$

For Discussion

5. What is the fewest number of points needed to determine a line? Why is it a good idea to locate three points when you graph an equation?

6. The standard form of a linear equation is $Ax + By = C$, where A, B, and C are integers and A and B are not both 0. What type of graph results if $A = 0$? If $B = 0$?

PRACTICE EXERCISES

Make a solution table. Then graph the equation.

1. $2x + 3y = 12$
2. $4x + 5y = 20$
3. $x - 2y = 4$
4. $x - 3y = 15$

Use the *x*- and *y*-intercepts to graph the equation.

5. $x = \frac{1}{2}y + 3$
6. $x = \frac{2}{3}y + 5$
7. $2x + y = 8$
8. $4x + y = 16$
9. $y = \frac{2}{5}x + 2$
10. $y = \frac{1}{2}x + 2$
11. $y = \frac{1}{4}x - 1$
12. $y = \frac{3}{4}x - 3$

Graph each equation in a coordinate plane. Give three solutions of each.

13. $y = 2$
14. $y = 0$
15. $x = 1$
16. $x = 5$
17. $2x + y = 4$
18. $3x + y = 9$
19. $y = -\frac{1}{3}x$
20. $y = -\frac{1}{4}x$

Draw the graph to determine whether each ordered pair is a solution of $y = \frac{2}{3}x - 1$. Check your answer.

21. (0, 1) 22. (3, 1) 23. (−3, −3) 24. (1, 0)

Graph each equation.

25. $x = -\frac{5}{2}$ 26. $y + 6 = 0$ 27. $y = \frac{3}{5}x - \frac{2}{5}$ 28. $y = -\frac{1}{4}x + \frac{3}{4}$
29. $2x - y = 9$ 30. $4x - 6y = 6$ 31. $3y + 2x = 5$ 32. $4y - x = 10$

Draw the graph to determine the missing coordinate of each ordered pair solution of $y = x + 3$.

33. (0, y) 34. (−2, y) 35. (x, 0) 36. (x, −1)

Determine the missing coordinate of the given solution.

37. $y = 3x$; $(2a^2, \underline{})$
38. $y = 2x + 1$; $(b^2, \underline{})$
39. $y = \frac{1}{2}x - 3$; $(\underline{}, 4d^4)$
40. $y = \frac{1}{4}x + 1$; $(\underline{}, 8b^3)$

Applications

Finance In the equation $w = 0.25s + 75$, w (dollars) is the total weekly salary in a sales job in which there is a base salary of $75 and a 25% commission on sales s (dollars). You may wish to use a computer to help you in solving Exercises 41–42.

41. Using w as the vertical axis and s as the horizontal axis, graph this linear equation from a table of values.

42. What is the w-intercept, and what does it represent in this situation?

DID YOU KNOW?

Why does it take longer to fly from New York to Miami than from Miami to New York?

Did you know that the **Coriolis** force plays an important role in flying?

The Coriolis force is caused by the earth's rotation, which affects the way airplanes travel.

Therefore, a plane never flys on a straight line, since it has to continuously make corrections in order to reach its destinations.

Research how these corrections are made.

9.3 Problem Solving Strategy: Estimate from Graphs

Graphs in the coordinate plane are often used in mathematics, science, and business to show relationships between two variables and to make predictions.

EXAMPLE Mrs. Clark earns $100 a week plus 3% commission on her car sales during that week. Her income can be expressed by $I = 0.03s + 100$, where I is her income and s is the amount of her sales. From a graph, estimate Mrs. Clark's income when her weekly sales are:
a. $10,000   b. $22,554

Understand the Problem

What are the given facts?
You are given: $I = 0.03s + 100$. You are asked to graph the equation and estimate the income for sales of $10,000 and $22,554.

Plan Your Approach

Choose a strategy.
1. Income I depends on sales s. Label the horizontal axis s and the vertical axis I. Titles may be useful.
2. Look at the data; choose appropriate scales for the axes.
3. Graph: $I = 0.03s + 100$.

Complete the Work

Solve.
Units of $5000 are convenient for the horizontal axis. You can make a table of some values for (s, I). Units of $100 are convenient for the vertical axis.

s	I
0	100
5000	250
20,000	700

Interpret the Results

Conclusion.
Estimates from the graph:
An income of $400 corresponds to $10,000 in sales. An income of about $775 corresponds to sales of $22,554.

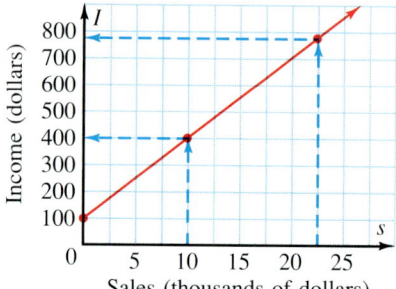

Check your conclusions.

(10,000; 400)
$I = 0.03s + 100$
$400 \stackrel{?}{=} 0.03(10,000) + 100$
$400 = 400$ ✓

(22,554; 775)
$I = 0.03s + 100$
$775 \stackrel{?}{=} 0.03(22,554) + 100$
$775 \approx 776.62$ *Very close estimate*

CLASS EXERCISES

The graph shows how the distance a package travels is related to the amount charged by Expert Courier.

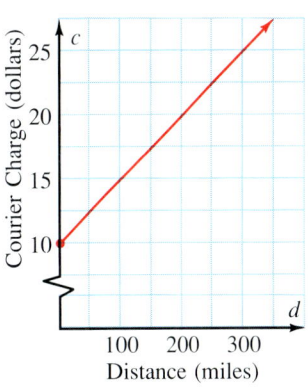

1. Estimate the cost of sending a package 250 mi.
2. Estimate how far you can send a package for $20.

Assume that the graph continues in a straight line indefinitely.

3. Estimate the cost of sending a package 600 mi.

Below is a rate schedule Jill developed for shoveling snow. Plot the points for the data listed in the table, then connect them.

x-axis	Time worked (hours)	1	2	3	4
y-axis	Charge (dollars)	10	15	20	25

4. How much does Jill charge for $2\frac{1}{2}$ h? for 6 h?

For Discussion

5. The graphs of linear relationships involve two variables. How do you decide which variable to represent along the horizontal axis and which along the vertical axis? Explain.

6. Why should you examine the data before choosing scales for the axes of a graph? Can the two axes have different scales? Explain.

PRACTICE EXERCISES

The fee for renting a word processor is $25 plus $15 for each day you keep the machine. The total fee can be expressed by $F = 25 + 15d$, where F is the total fee and d is the number of days the machine is rented. Use this information for Exercises 1–4.

1. Complete the table.

Number of days (d)	1	2	3
Rental fee (F)	$40	?	?

2. Graph the solutions given in the table. Label the horizontal axis d, and the vertical axis F. Connect the points with a line.

3. Use the graph to estimate the rental fee for 6 days.

4. Jose can spend no more than $200 in rental fees. For how many days can he rent a word processor?

9.3 Problem Solving Strategy: Estimate from Graphs **381**

Sally went shopping for school clothes in Blue City. The city sales tax is 2%. Her total cost can be expressed by $C = x + 0.02x$, where C is her total cost including tax and x is the amount of her purchase in dollars. Use this information for Exercises 5–8.

5. Complete the table.

Amount of purchase (x)	$50	$100	$200
Amount of total cost (C)	$51	?	?

6. Graph the solutions in the table. Label the horizontal axis x, and the vertical axis C. Connect the points with a line.

7. Use the graph to estimate the total cost of a purchase of $300.

8. Use the graph to estimate the total cost of a purchase of $450.

A company claims that a certain fertilizer multiplies the yield of cucumbers per plant by 3. The expected number of cucumbers per plant, with fertilizer, can be expressed by $C = 3p$, where C is the actual number of cucumbers per plant, with fertilizer, and p is the average number of cucumbers per plant, without fertilizer. Use this information for Exercises 9–12.

9. Complete the table.

Variety	A	B	C
Average number of cucumbers per plant, without fertilizer	4	10	12
Expected number of cucumbers per plant, with fertilizer	12	?	?

10. Graph the solutions given in the table.

11. An actual test produced this data:

Variety	A	B	C
Average number of cucumbers per plant, without fertilizer	4	10	12
Actual number of cucumbers per plant, with fertilizer	15	28	35

On your graph for Exercise 10, graph these points in a different color.

12. Does the company's claim seem justified?

The power p in watts delivered to a certain element with current i in amperes is given by the equation $p = 100i - 50i^2$. Use this for Exercises 13–14.

13. Use 0 amps, 0.5 amp, 1 amp, 1.5 amps, and 2 amps to graph the given equation.

14. Estimate the power that corresponds to 0.25 amp and to 1.75 amps.

For a science project, a stone is thrown upward from the ground at a rate of 48 ft per second. Its distance d (in feet) above the ground is given by the equation $d = -16t^2 + 48t$, where t is time in seconds.

15. Use $t = 0$, $t = 0.5$, $t = 1$, $t = 1.5$, $t = 2$, and $t = 2.5$ to graph the equation.

16. Use the graph to estimate when the stone will hit the ground.

Mixed Problem Solving Review

1. The length of a rectangle is $\frac{2}{w+2}$ in., and the area is $\frac{4}{w^2-4}$ in.^{2}. Find the width of the rectangle.

2. A jet plane and a prop plane leave the same airport at the same time and travel in opposite directions. The jet travels at 960 km/h and the prop plane travels at 560 km/h. In how many hours will they be 4560 km apart?

3. The lengths of two adjacent sides of a parallelogram are consecutive even integers. The perimeter of the parallelogram is 108 m. What are the dimensions of the parallelogram?

4. The area of a rectangle is 176 cm^2. Find the dimensions of the rectangle if the length exceeds the width by 5 cm.

5. Erik traveled 1120 km in 2 days. At this rate, how far would he travel in 13 days?

PROJECT

Long before units of measure were standardized, people used parts of their bodies as units. For example, a **foot** was the distance from the end of a person's heel to the tip of the toe, and a **cubit** was the distance from the end of the middle finger to the elbow.

1. Measure and record the *foot* and the *cubit* for 20 or more different people. Use the centimeter as the unit of measure and record each pair of measures as an ordered pair: (*foot, cubit*).

2. On a piece of graph paper, plot all the points for the ordered pairs that result from your measurements. (*Note:* When data is presented this way, the result is called a **scatter diagram**, or a **scatter plot**.)

3. If the points appear to cluster around a line, then a relationship may exist between the two units of measure. If they do not obviously cluster around a line, then a relationship may not exist between the two units. Look at your graph. Does it show a relationship or no relationship?

9.4 Slope

Objectives: To find the slope of a line from its graph or from the coordinates of two points of the line
To draw a line with a given slope through a given point

Which ramp is easier to climb? Why? To describe the steepness of an incline, you can say that it rises a certain vertical distance for a given horizontal distance. Ramp A rises 6 ft for every 100 ft of horizontal run, while Ramp B rises 15 ft for every 100 ft of horizontal run.

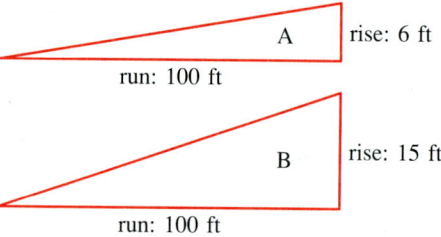

The relationship of the rise to the run can be expressed as a *ratio*.

$$\frac{\text{rise}}{\text{run}} = \frac{6}{100} = \frac{3}{50} \qquad \frac{\text{rise}}{\text{run}} = \frac{15}{100} = \frac{3}{20}$$

Capsule Review

You can use division to simplify a ratio.

EXAMPLE $\frac{-12}{60} = \frac{-12 \div 12}{60 \div 12} = \frac{-1}{5}$, or $-\frac{1}{5}$

Simplify.

1. $\frac{24}{32}$ 2. $\frac{51}{17}$ 3. $\frac{-18}{54}$ 4. $\frac{16}{-6}$ 5. $\frac{-11}{-11}$

You can use the coordinates of two points of a line to determine the slope, or steepness, of the line. The **slope** of a line is the ratio of the change in y to the change in x between any two points on the line. This is often thought of as the ratio of the rise, or vertical change, to the run, or horizontal change. This is illustrated with line PQ (written $\overleftrightarrow{PQ}$).

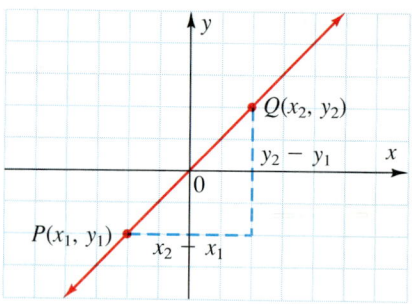

For any two points of a line, $P(x_1, y_1)$ and $Q(x_2, y_2)$, the slope m is given by the formula:

$$m = \frac{\text{rise}}{\text{run}} = \frac{\text{vertical change}}{\text{horizontal change}} = \frac{y_2 - y_1}{x_2 - x_1}, \text{ where } x_2 - x_1 \neq 0.$$

Note: It does not matter which point you call P or Q so long as you stick to your choice once it is made.

EXAMPLE 1 Find the slope of a line that contains the points $A(2, 1)$ and $B(4, 2)$.

Substitute in the formula $m = \dfrac{y_2 - y_1}{x_2 - x_1}$.

$$\begin{array}{cc} A(2,1) & B(4,2) \\ \uparrow \uparrow & \uparrow \uparrow \\ x_1 \; y_1 & x_2 \; y_2 \end{array}$$

$$m = \frac{y_2 - y_1}{x_2 - x_1} = \frac{2 - 1}{4 - 2} = \frac{1}{2}$$

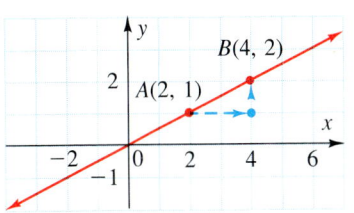

The slope is positive. The line slants up 1 unit for every 2 unit change to the right. Note that either point can be assigned the coordinates (x_1, y_1) or (x_2, y_2). The slope of the line will be the same. How can you show this?

EXAMPLE 2 Find the slope of a line that contains the points $C(-3, 1)$ and $D(2, -2)$.

$$\begin{array}{cc} C(-3,1) & D(2,-2) \\ \uparrow \uparrow & \uparrow \uparrow \\ x_1 \; y_1 & x_2 \; y_2 \end{array}$$

$$m = \frac{y_2 - y_1}{x_2 - x_1} = \frac{-2 - 1}{2 - (-3)}$$

$$= \frac{-3}{5} = -\frac{3}{5}$$

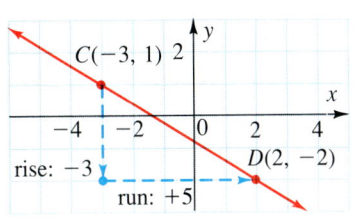

In Example 2, the slope is negative. The line slants down 3 units for every 5-unit change to the right.

Lines that slant up to the right have positive slopes. Lines that slant down to the right have negative slopes. In Example 3, the slopes of a horizontal line and a vertical line are considered.

EXAMPLE 3 Find the slope of:
a. a horizontal line that contains the points $M(0, 2)$ and $N(5, 2)$
b. a vertical line that contains the points $P(-3, 4)$ and $R(-3, 2)$

a.

$$m = \frac{y_2 - y_1}{x_2 - x_1}$$

$$= \frac{2 - 2}{5 - 0}$$

$$= \frac{0}{5}, \text{ or } 0$$

9.4 Slope **385**

b.

$$m = \frac{y_2 - y_1}{x_2 - x_1}$$

$$= \frac{4 - 2}{-3 - (-3)}$$

$$= \frac{2}{0}, \text{ or undefined}$$

If a line is horizontal, then its slope is 0. If a line is vertical, then its slope is undefined, or the line has no slope.

If you are given the coordinates of one point of a line and its slope, you can draw the line. The procedure is shown in Example 4.

EXAMPLE 4 Through point $R(1, 2)$, draw a line with slope of $-\frac{3}{2}$.

1. Plot $R(1, 2)$.
2. From R, go down 3 units (for the rise, -3), since $-\frac{3}{2} = \frac{-3}{2}$. Go to the right 2 units (for the run, 2). Mark point, S.
3. Draw a line through R and S.

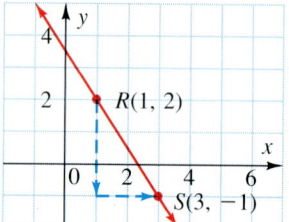

Collinear points are points that lie on the same line. If you know the coordinates of any collinear points, you can find the slope of the line and determine if any other points are also collinear. It is not necessary to draw the graph.

EXAMPLE 5 Determine if points $A(-4, -1)$, $B(-2, 1)$, and $C(2, 5)$ are collinear.

If the points *are* collinear, then the slopes of $\overleftrightarrow{AB}$ and $\overleftrightarrow{BC}$ will be the same.

slope of $\overleftrightarrow{AB}$

$$m = \frac{1 - (-1)}{-2 - (-4)} = \frac{2}{2}, \text{ or } 1$$

slope of $\overleftrightarrow{BC}$

$$m = \frac{5 - 1}{2 - (-2)} = \frac{4}{4}, \text{ or } 1$$

Since the slope of $\overleftrightarrow{AB}$ equals the slope of $\overleftrightarrow{BC}$, the points are collinear. Check by graphing the three points.

CLASS EXERCISES

Refer to the graph at the right. Classify the slope of the line as positive, negative, zero, or no slope.

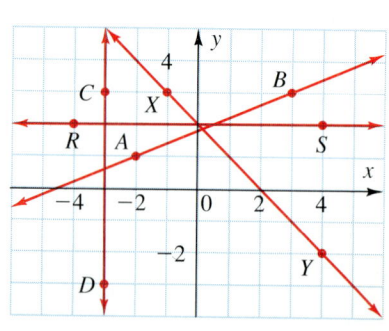

1. $\overleftrightarrow{AB}$
2. $\overleftrightarrow{CD}$
3. $\overleftrightarrow{XY}$
4. $\overleftrightarrow{RS}$

Find the slope of a line that contains the given points.

5. A(6, 7); B(1, 3)
6. R(−1, 3); S(2, 4)
7. C(−2, 3); D(2, 1)
8. E(1, 3); F(5, −3)
9. P(−5, 2); Q(3, 2)
10. M(−1, 2); N(−1, 4)

PRACTICE EXERCISES

Find the slope of a line that contains the given points.

1. A(3, 2), B(5, 6)
2. R(3, 8), T(1, 4)
3. C(2, 9), D(5, 14)
4. E(4, 7), F(7, 11)
5. M(−4, 4), N(2, −5)
6. J(−3, 1), K(3, −4)
7. S(−2, 1), T(3, −5)
8. P(−5, 2), Q(2, −4)
9. A(9, −2), B(3, 4)
10. R(6, −3), S(1, 2)
11. W(7, −1), X(2, 3)
12. G(5, −2), H(4, 3)

Find the slope of a line that contains the following points. State whether it is a horizontal or vertical line.

13. M(1, 3), N(2, 3)
14. X(0, 5), Y(3, 5)
15. P(−5, 2), R(−5, 1)
16. J(−3, 0), K(−3, 2)
17. A(4, 7), B(3, 7)
18. B(2, 1), C(0, 1)

Through the given point, draw a line with the given slope.

19. R(3, 4) slope $\frac{1}{2}$
20. P(2, 5) slope $-\frac{4}{3}$
21. T(−1, 5) slope $-\frac{2}{3}$
22. U(−2, 3) slope $-\frac{1}{3}$

The coordinates of three points are given. Use slope to determine if the points are collinear.

23. A(−1, −5), B(1, −2), C(5, 4)
24. X(−2, −6), Y(0, −2), Z(6, 10)
25. D(−3, 4), E(0, 2), F(−3, 0)
26. G(3, 3), H(1, −1), I(0, 0)
27. J(−2, 1), K(−2, 4), L(2, 4)
28. M(−2, 1), N(0, 4), P(2, 7)
29. R(1, −2), S(−1, −5), T(5, 4)
30. S(−3, 4), M(0, 2), C(−3, 0)

Find the slope, if it exists, of each side of the given figure.

31.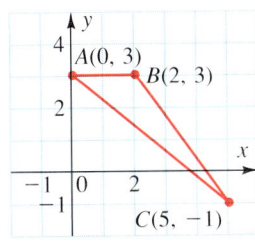

32.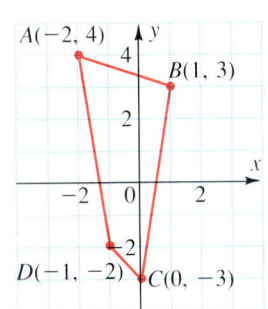

9.4 Slope **387**

Through the given point, draw a line with the given slope.

33. $P(6, -1)$
    slope: -1
34. $R\left(0, \frac{1}{2}\right)$
    slope: 6
35. $S\left(\frac{1}{2}, -\frac{3}{2}\right)$
    slope: 0
36. $T\left(1\frac{1}{3}, -2\frac{2}{3}\right)$
    slope: -3

Find the value of the missing coordinate using the given slope.

37. $A(4, y)$ and $B(8, 12)$; slope: $-\frac{1}{2}$
38. $C(x, -3)$ and $D(2, 1)$; slope: $\frac{1}{4}$
39. $E(x, 0)$ and $F(3, 4)$; slope: 2
40. $G(-1, y)$ and $H(1, -1)$; slope: 3

Find the slope of a line that contains the given points.

41. $A(a, 3a)$ and $B(b, 3b)$
42. $C(c, c + 4)$ and $D(d, d + 4)$

Applications

43. **Geometry** The vertices of a parallelogram $RSTU$ are $R(-2, 4)$, $S(2, 6)$, $T(7, 2)$, and $U(x, 0)$. Find the slope of each side of the parallelogram.

44. **Science** If 35 cm^3 of gold weigh 392 g and 21 cm^3 of gold weigh 235.2 g, draw a graph with cubic centimeters of gold as the horizontal axis and grams of gold as the vertical axis. Assuming that weight and volume of gold are related through a linear equation, find the slope. What does the slope represent in this problem?

TEST YOURSELF

Is the given ordered pair a solution of the equation? 9.1

1. $2x + 3y = -2$; $(5, -4)$
2. $y - x = 2$; $\left(\frac{1}{2}, \frac{3}{2}\right)$

Graph each equation. 9.2

3. $5y - 2x = 10$
4. $y - 3x = 9$
5. $y = -x + 1$
6. $y = \frac{1}{4}x + 2$

The amount of dog food ordered per week by the Poodle Kennel depends on how many dogs are registered. The table shows data for three weeks. 9.3

7. Graph the ordered pairs and connect the points.

No. of Dogs	10	12	16
Food (pounds)	36	42	54

8. From the graph, estimate the amount of food to order for 6 dogs.
9. How many pounds of food should be ordered for 20 dogs?
10. If 60 pounds were ordered, how many dogs were registered that week?

Find the slope of a line that contains the given points. 9.4

11. $A(5, 3)$, $B(3, 2)$
12. $C(-2, -7)$, $D(4, -3)$

APPLICATION:
Radio Waves

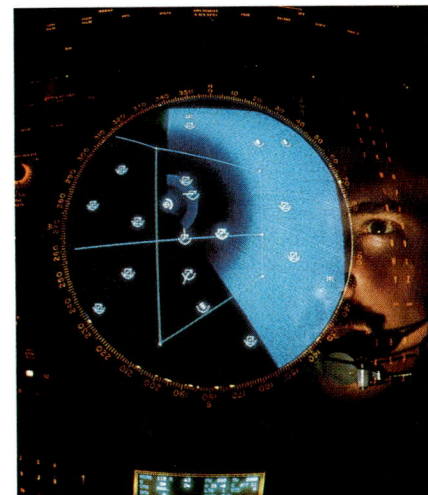

Did you know that radar is used to detect and locate fixed or moving objects? It can determine the direction, distance, height, and speed of objects that are much too far away for us to see. It can locate objects as small as insects or as large as mountains. The word radar comes from *ra*dio *d*etection *a*nd *r*anging.

Almost every radar set works by sending radio waves toward an object and receiving waves that are reflected from the object. The time it takes for the reflected waves to return indicates how far away the object is. The direction from which the reflected waves return tells the location of the object. The ability of radar to do so many tasks makes it useful for a wide variety of purposes.

Radio waves as transmitted by radar have definite frequency. They travel at the rate of 186,000 mi/s. The frequencies of such waves are measured in units called megahertz (MHz). One megahertz equals 1 million hertz (cycles per second).

EXAMPLE How far away from each other are two boats, if a signal is sent out by one ship and returned by the other in $\frac{1}{100,000}$ s?
Hint: Use the distance formula.

Let $r = 186,000$ and $t = \frac{1}{100,000}$ and solve for d.

$$d = r \times t$$
$$= 186,000 \times \frac{1}{100,000}$$
$$= \frac{186,000 \text{ mi/s}}{100,000 \text{ s}}$$
$$= 1.86 \text{ mi}$$

The signal traveled 1.86 mi altogether. So the distance between the ships is one-half of the distance or 0.93 mi.

Modern radar has provided useful images from space. The Apollo 17 flight in December 1972 gathered data with radar to develop maps of the moon. In 1978, the satellite Seasat carried the first space-based imaging radar. It operated for three months. The second imaging radar was aboard the United States space shuttle Columbia during its second flight in November 1981.

Radar is influenced by the temperature, air pressure, and the water vapor content of the atmosphere. The program below gives you the radar distance and the geometric distance when you input the optical horizon distance.

```
10 INPUT "ENTER OPTICAL HORIZON DISTANCE:  ";D
20 DR = 1.07 * D:DG = DR / 1.15
30 PRINT : PRINT "RADAR DISTANCE IS ";DR
40 PRINT : PRINT "GEOMETRIC DISTANCE IS ";DG
50 END
```

Solve. Use the distance formula and the above program.

1. How far away are two boats if a signal is sent out and returns in $\frac{1}{10,000}$ of a second?

2. How far must a signal travel from one ship to another to verify that they are 3.5 mi apart?

3. Find the distance between two airplanes if a signal is sent out from one airplane and returned in 0.0002 s.

4. An air traffic control tower receives a signal from an airplane 500 mi away. How many seconds does it take for the signal to reach the station?

5. The moon is approximately 240,000 mi from the earth. How long does it take for a radar signal to reach the moon and return?

6. A satellite orbiting the earth sends a radar signal to a tracking station and receives a reply in 0.003 s. How far is the satellite from the earth?

7. A satellite orbiting the earth at a distance of 150 mi sends a signal to a tracking station which relays the signal up to second satellite. The total time needed is 0.0014 s. How far above the earth is the second satellite?

8. Calculate the radar distance and geometric distance for an optical horizon distance of:
   a. 15 mi on flat land.   **b.** 8 mi on open sea.   **c.** 100 mi on a mountain top.

9. Calculate the radar distance and geometric distance for the following optical horizon distances:
   a. Los Angeles, CA 15 mi   **b.** Ely, NV 45 mi   **c.** Jackson, MS 7 mi

9.5 Slope-Intercept Form of a Linear Equation

Objectives: To use the slope-intercept form to graph a linear equation
To use slope to determine if two lines are parallel

When a linear equation such as $3x + y - 2 = 0$ is solved for y, an equivalent form of a linear equation is obtained.

Capsule Review

You can solve an equation by addition, subtraction, multiplication, or division.

EXAMPLE Solve for y:
$2x + 3y = 3$
$3y = -2x + 3$ *Subtract 2x from both sides.*
$y = \dfrac{-2x + 3}{3} = -\dfrac{2}{3}x + 1$ *Divide each side by 3.*

Solve each equation for y.

1. $y - 2x = 8$
2. $y + 2x = -5$
3. $3y - x = 9$
4. $2y + 3x = 4$
5. $2y + x + 5 = 0$
6. $3y + 2x - 7 = 0$

Look at the graphs of $y = -3x$ and $y = -3x + 2$.

$y = -3x$

x	y
0	0
1	-3
-1	3

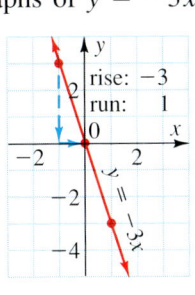

$y = -3x + 2$

x	y
0	2
1	-1
2	-4

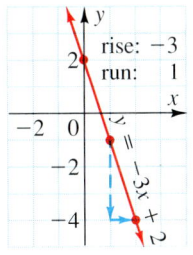

The graphs of both $y = -3x$ and $y = -3x + 2$ are lines with slope -3. The lines intersect the y-axis at different points. The y-intercept of $y = -3x$ is 0. The y-intercept of $y = -3x + 2$ is 2. This suggests the following property.

The equation $y = mx + b$ is called the **slope-intercept form** of a linear equation. The graph of the equation is a line with slope m and y-intercept b.

EXAMPLE 1 Find the slope and the *y*-intercept of the line for the equation $4x + 3y = 12$.

$4x + 3y = 12$ *Solve for y.*
$3y = -4x + 12$ *Subtract 4x from each side.*
$y = -\frac{4}{3}x + 4$ *Divide each side by 3.*
     ↑   ↑   *The equation is in slope-intercept form.*
     m   b

The slope is $-\frac{4}{3}$; the *y*-intercept is 4.

EXAMPLE 2 Write an equation of the line with slope -2 and *y*-intercept 0.

$m = -2$ and $b = 0$
$y = mx + b$
$y = -2x + 0$ *Substitute -2 for m and 0 for b.*
$y = -2x$

Although a linear equation can be graphed using a table of values, a more efficient method uses only the slope and *y*-intercept.

EXAMPLE 3 Graph: $x - 2y = 4$

Write the equation in the form $y = mx + b$.

$x - 2y = 4$
$-2y = -x + 4$
$y = \frac{1}{2}x - 2$ $m = \frac{1}{2}; b = -2$

1. The *y*-intercept is -2. Plot the point whose coordinates are $(0, -2)$.

2. From that point, use the slope, $\frac{1}{2}$, to locate another point. Go up 1 unit and 2 units to the right.

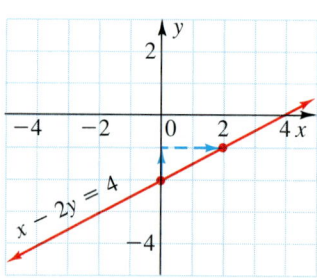

$$\text{slope} = \frac{1}{2} \begin{array}{l} \leftarrow \text{up 1} \\ \leftarrow \text{right 2} \end{array}$$

3. Draw a line through the two points.

4. Use a third point to check. Is $(-4, -4)$ a solution of $y = \frac{1}{2}x - 2$? Is the point for $(-4, -4)$ on the line?

Look at the graphs of $y = x + 1$ and $y = x - 1$.

The lines seem to be *parallel*. Note that the slopes are the same, but the y-intercepts are different.

$y = x + 1$ $y = x - 1$
$m = 1, b = 1$ $m = 1, b = -1$

Graph $y = x + 4$. What do you notice about this line and the other two lines?

This suggests the following property.

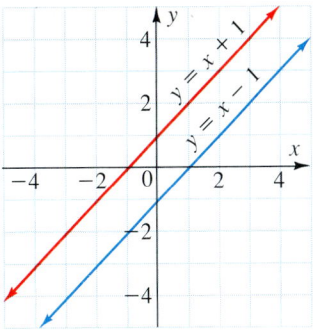

> Two nonvertical lines have the same slope if and only if they are **parallel**.

Note that vertical lines are parallel but have slopes that are undefined.

EXAMPLE 4 Determine whether the graphs of $y = -2x + 3$ and $4x + 2y = 5$ are parallel lines.

Find the slope of each line.

$y = -2x + 3$ $4x + 2y = 5$
slope $= -2$ $2y = -4x + 5$
                     $y = -2x + \frac{5}{2}$
                     slope $= -2$

Since the lines have the same slope, they are parallel.

CLASS EXERCISES

Find the slope and the y-intercept of the line for the given equation.

1. $y = -\frac{2}{3}x$
2. $x + 5y = 10$
3. $8x - y = 2$
4. $x = 3y + 7$

Graph each equation using the slope and the y-intercept.

5. $y = -\frac{4}{3}x + 2$
6. $y - 2x = 0$
7. $2x + 5y = 10$
8. $y - 4 = 0$

For Discussion

9. You have learned to draw a line with a given slope through a given point. How does this relate to using the slope and y-intercept to graph a linear equation?

10. Describe the relationship of the graphs of three linear equations if b is the same for all three but m is different.

Are the graphs of the two equations parallel lines? Explain.

11. $y = 2x + 5$
    $-2x + y = 9$

12. $x + 3y = 7$
    $y = \frac{1}{3}x - 4$

13. $x - y = 3$
    $y - x = 2$

14. $y = 8$
    $y + 2 = 3$

PRACTICE EXERCISES

Find the slope and the y-intercept of the line for the given equation.

1. $3x + 2y = 6$
2. $5x + 3y = 15$
3. $x - 3y = 9$
4. $x - 4y = 20$
5. $y = -2x$
6. $y = -5x$
7. $y = 3x + 6$
8. $y = 4x + 8$

Write an equation of the line with the given slope and y-intercept.

9. $m = -2$
   $b = 0$

10. $m = -5$
    $b = 2$

11. $m = 1$
    $b = 3$

12. $m = 4$
    $b = 2$

13. $m = \frac{2}{3}$
    $b = 4$

14. $m = \frac{3}{4}$
    $b = 3$

15. $m = 0$
    $b = -1$

16. $m = 0$
    $b = -3$

Graph each equation.

17. $x - 3y = 9$
18. $x - 6y = 18$
19. $x + 2y = 8$
20. $x + 4y = 20$
21. $3x - 2y = 4$
22. $4x - y = 2$
23. $3x = 4y$
24. $2x = 5y$

Determine whether the graphs of the two equations are parallel lines.

25. $y = -3x + 2$
    $3x + 2y = 7$

26. $x - y = 2$
    $4x = 4y + 6$

27. $2x - 3y = 9$
    $-4x + 6y = 12$

28. $2x = 4y - 7$
    $2y = 6x + 1$

Write the equation in slope-intercept form. Then state the slope and the y-intercept.

29. $3y = 6x$
30. $2y = -x + 7$
31. $3x - 5y = 15$
32. $2x - 3y = 7$
33. $\frac{x}{2} = \frac{y}{5}$
34. $\frac{y}{2} - \frac{x}{3} = \frac{1}{4}$
35. $2x + 3y = 4y$
36. $x + 4 = y - 1$

Graph each equation using the slope and the y-intercept.

37. $5x + 2y + 10 = 0$
38. $2y - 3x - 2 = 0$
39. $4x = 2y + 3$
40. $3x = 4y + 1$
41. $\frac{x}{2} + 6y = 15$
42. $2x - 3y = 2$

Find the value of *a* so that the graph of the equation has the given slope.

43. $y = 2ax + 4$; $m = -1$
44. $3y = 2ax - 2$; $m = -2$

45. $ax + 2y = 3$; $m = 4$ **46.** $y - 3ax = 5$; $m = \frac{1}{3}$

Two lines are perpendicular if the product of their slopes is -1. That is, $m_1 m_2 = -1$, where m_1 is the slope of one line and m_2 is the slope of the other line.

For each pair of equations, tell if the lines are parallel or perpendicular.

47. $3x - 4y = 12$; $-6x + 8y = -4$ **48.** $2x + 3y = 8$; $4y - 3 = 6x$

49. $ax + by = c$; $bx - ay = d$ **50.** $ax - by = c$; $-ax + by = d$

51. Can two parallel lines have the same y-intercept? Explain.

52. Can two perpendicular lines have the same y-intercept? Explain.

Applications

Computer The following program prints a graph that enables you to estimate the temperature in either Fahrenheit or Celsius when the temperature is given in the other scale.

```
10 PRINT "FAHRENHEIT"
20 FOR C = 100 TO 0 STEP  - 5
30 F = (9 / 5) * C + 32:J = C / 5 + 2
40 IF F < 100 THEN J = J + 1
50 PRINT F; SPC( J);"* (";C;",";F;")"
60 NEXT C
70 PRINT   SPC( 5);"+5";
80 FOR I = 20 TO 100 STEP 20: PRINT "--";
90 PRINT : PRINT ; TAB( 10);"CELSIUS"
100 END
```

Use the graph to answer the following.

53. Write the equation of the line.

54. What is the temperature Celsius if the temperature Fahrenheit is
  a. 75   **b.** 100   **c.** 207

55. What is the temperature Fahrenheit if the temperature Celsius is
  a. 75   **b.** 100   **c.** 207

EXTRA

The tip of the minute hand on Big Ben, the clock at the Houses of Parliament in London, England, travels 146 mi each year. Determine the number of times a year that the minute and the hour hands have the same slope when pointing in opposite directions.

9.5 Slope-Intercept Form of a Linear Equation

9.6 Equation of a Line

Objectives: To determine an equation of a line, given the slope of the line and the coordinates of one point
To determine an equation of a line, given the coordinates of two points of the line

You have been working with equations and graphs that show how two variables are related. If you have only a graph, you may be able to read enough information from the graph to write the related equation.

Capsule Review

Determine the following from the graph at the right. Use $m = \dfrac{y_2 - y_1}{x_2 - x_1}$.

1. Slope, using A and B
2. Slope, using B and C
3. y-intercept
4. x-intercept

When you can determine the y-intercept and the slope of a line from its graph, you can write an equation of the line.

EXAMPLE 1 Write an equation of the given line.

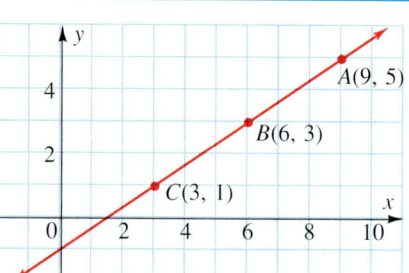

1. The line intersects the y-axis at the point whose coordinates are $(0, -2)$. The y-intercept is -2.

2. To find the slope, use two points whose coordinates you can determine, such as $(2, -1)$ and $(-2, -3)$.

$$m = \dfrac{y_2 - y_1}{x_2 - x_1} \quad \underset{\underset{x_2\ y_2}{\uparrow\ \uparrow}}{(2, -1)} \quad \underset{\underset{x_1\ y_1}{\uparrow\ \uparrow}}{(-2, -3)}$$

$$= \dfrac{-1 - (-3)}{2 - (-2)} = \dfrac{2}{4} = \dfrac{1}{2} \quad \text{The slope is } \dfrac{1}{2}.$$

3. Write the equation in slope-intercept form.

$$y = mx + b \quad m = \dfrac{1}{2};\ b = -2$$

$$y = \dfrac{1}{2}x + (-2),\ \text{or}\ y = \dfrac{1}{2}x - 2$$

Chapter 9 Linear Equations

The linear equation in Example 1, $y = \frac{1}{2}x - 2$, can be placed in standard form. Recall that $Ax + By = C$ is the standard form of a linear equation if A, B, and C are integers and A and B are not both 0.

$2y = x - 4$ *Multiply each side by 2. Subtract x from each side.*
$-x + 2y = -4$ ←——— *Standard form of the equation*

You can also write an equation of a line when you know only the slope and the coordinates of any point of the line.

EXAMPLE 2   Write an equation of a line in standard form that has a slope of $\frac{2}{3}$ and contains the point $P(3, 6)$.

1. Write the equation in slope-intercept form.

$y = \frac{2}{3}x + b$ $m = \frac{2}{3}$

$6 = \frac{2}{3}(3) + b$ *Substitute 6 for y and 3 for x.*

$6 = 2 + b$
$4 = b$ *Solve for b.*

$y = \frac{2}{3}x + 4$ *Slope-intercept form*

2. Change the equation to standard form.

$y = \frac{2}{3}x + 4$

$3y = 2x + 12$
$-2x + 3y = 12$

So, the standard form of the equation is $-2x + 3y = 12$.

An equation of a line can be determined when the coordinates of two points of the line are given.

EXAMPLE 3   Write an equation of a line that contains points $A(-3, -2)$ and $B(5, 2)$.

1. Find the slope of the line through the two points.

$m = \frac{y_2 - y_1}{x_2 - x_1}$ $A(-3, -2)$ $B(5, 2)$
 ↑   ↑   ↑   ↑
 x_1 y_1 x_2 y_2

$= \frac{2 - (-2)}{5 - (-3)} = \frac{4}{8}$, or $\frac{1}{2}$

The slope is $\frac{1}{2}$.

2. Write in slope-intercept form.

$y = \frac{1}{2}x + b$ *Substitute the coordinates of either point for x and y and solve for b.*

$2 = \frac{1}{2}(5) + b$

$-\frac{1}{2} = b$

$y = \frac{1}{2}x - \frac{1}{2}$ *Substitute $\frac{1}{2}$ for m and $-\frac{1}{2}$ for b.*

So the slope-intercept form of the equation is $y = \frac{1}{2}x - \frac{1}{2}$.

EXAMPLE 4 Write an equation of a line that is parallel to the graph of $y = 2x - 1$ and contains $P(1, -2)$.

If lines are parallel, their slopes are equal. The slope of the graph of $y = 2x - 1$ is 2. The slopes of the two lines must be equal.

$y = mx + b$ Slope-intercept form
$y = 2x + b$ It is known that $m = 2$.
$-2 = 2(1) + b$ Substitute the coordinates $(1, -2)$ for x and y.
$-4 = b$ Solve for b.

So, the equation is $y = 2x - 4$.

CLASS EXERCISES

Refer to the graph at the right. Give an equation in slope-intercept form for each.

1. $\overleftrightarrow{AB}$ 2. $\overleftrightarrow{CD}$

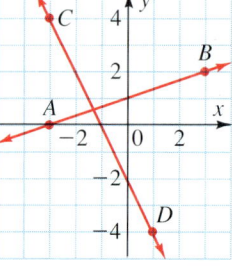

Give each linear equation in standard form.

3. $y = 3x - 1$ 4. $y = \frac{1}{4}x + \frac{3}{8}$

5. $x = \frac{2}{3}y$ 6. $\frac{x}{4} + \frac{y}{3} = 8$

Write an equation of a line in standard form given the slope and a point.

7. $m = 3$ 8. $m = 1$ 9. $m = \frac{3}{4}$ 10. $m = -\frac{1}{2}$
$P(-2, -4)$ $R(-4, -1)$ $T(1, 1)$ $V(0, 3)$

11. Write an equation in standard form of a line through $P(-3, 8)$ and $Q(-1, 12)$.

12. Write an equation in standard form of a line that is parallel to the graph of $y - 2x = 4$ and contains the point $P(2, -2)$.

PRACTICE EXERCISES

Use two points whose coordinates you can determine from the graph to find the slope. Then write an equation in slope-intercept form.

1. The line intersects the y-axis at the point $(0, -3)$. The y-intercept is -3.

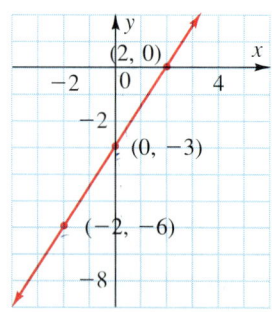

Chapter 9 Linear Equations

2. The line intersects the y-axis at the point (0, 4). The y-intercept is 4.

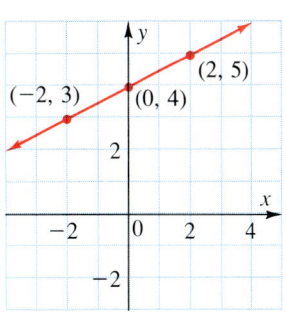

Write an equation of a line in standard form given the slope and a point.

3. $m = \frac{1}{2}$
   $R(0, 0)$

4. $m = \frac{3}{2}$
   $C(2, 6)$

5. $m = -2$
   $B(-1, 1)$

6. $m = -3$
   $A(-1, 5)$

7. $m = -\frac{5}{4}$
   $P(4, -3)$

8. $m = -\frac{1}{5}$
   $Q(6, 1)$

9. $m = \frac{7}{3}$
   $S\left(0, -\frac{1}{6}\right)$

10. $m = \frac{5}{2}$
    $T\left(3, -\frac{1}{4}\right)$

Write an equation in slope-intercept form that contains the given points.

11. $A(-1, -7)$
    $B(2, 8)$

12. $U(-2, -9)$
    $V(1, 8)$

13. $R(5, 6)$
    $S(6, 9)$

14. $X(3, 1)$
    $Y(4, 8)$

15. $T(1, 5)$
    $R(-1, 11)$

16. $C(2, 0)$
    $D(-2, 8)$

17. $L(3, 3)$
    $M(-6, 9)$

18. $E(7, 2)$
    $F(-2, 2)$

Write an equation of a line in standard form with the given characteristics.

19. The line is parallel to the graph of $y = 3x - 2$ and contains $P(2, -3)$.

20. The line is parallel to the graph of $y = 3x - 5$ and contains $S(1, -4)$.

21. The line is parallel to the graph of $2x + 5y = 3$ and the x-intercept is -2.

22. The y-intercept of the line is $-\frac{1}{2}$, and it is parallel to the graph of $-3x + y - 4 = 0$.

Write an equation of a line in standard form that contains the given points.

23. $A(-2, -12)$
    $C(5, 2)$

24. $M(-6, 8)$
    $N(-9, 16)$

25. $G(2, -7)$
    $H(-1, -10)$

26. $J(5, -1)$
    $K(-1, 8)$

27. $L\left(\frac{1}{2}, 0\right)$
    $M\left(3, 2\frac{1}{2}\right)$

28. $N\left(\frac{1}{3}, 1\right)$
    $P\left(2\frac{2}{3}, 5\right)$

29. $Q\left(-1\frac{1}{2}, 2\right)$
    $R\left(1\frac{1}{2}, 4\right)$

30. $S\left(-1, 1\frac{1}{2}\right)$
    $T(0, 3)$

Write an equation of a line in standard form with the given characteristics.

31. The line has a y-intercept of 3 and an x-intercept of -5.

32. The line passes through $B(-3, 7)$ and is parallel to the line through $C(1, 4)$ and $D(2, 6)$.

33. The line passes through $N(2, -5)$ and is perpendicular to the graph of $y + 3x - 8 = 0$.

34. The line passes through $P(c, 4c)$ and is perpendicular to the graph of $3y + 4x = 6$.

Write an equation of a line in slope-intercept form given the slope and a point.

35. $m = 2g$; $T(2, 7)$
36. $m = -5$; $U(0, -3k)$
37. $m = \frac{1}{n}$; $V(-2, 1)$

Applications

38. **Chemistry** The solubility s, in $\frac{g}{100}$ cm^3 of water, of sodium chlorate is 100 at 20°C and 170 at 70°C. Assuming that s and the temperature t are related through a linear equation, determine this equation. (*Hint: s* will take the place of y, and t will take the place of x in the equation $y = mx + b$.)

CRITICAL THINKING: Predicting Consequences

Predicting results or consequences is a critical thinking skill that helps you draw conclusions about the probable effects of some event. Sometimes called "making an educated guess," it involves linking what you *know* with what you *think* will happen. Using information, data, or facts that are readily available, you can predict a logical result if the trend suggested by the information is maintained.

The slope of a line is an indicator of its relative position. If you know the slope of a line, you know something about the line. In this lesson you are given two indicators that can be used to predict whether a line rises or falls moving from left to right. What are they?

You can predict how steep a line is by knowing its slope. As the slope of a line gets closer to zero, what happens to the line? As the slope increases from 0, what happens to the line? Make a general statement that helps you predict the steepness of a line.

9.7 Linear Inequalities in Two Variables

Objective: To graph linear inequalities in two variables

If peanuts cost $4 a pound and cashews cost $8 a pound, how would you draw a graph to show the different combinations of nuts you could buy for $24 or less?

To draw such a graph, you would need to graph a *linear inequality* in two variables. A **linear inequality** relates a linear expression with inequality signs.

Capsule Review

You can write an equation for the graph of a line when given certain characteristics of the line.

EXAMPLE What is the equation of this line?

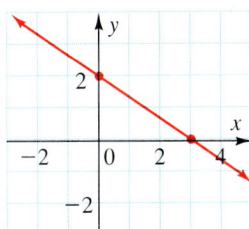

- Find the y-intercept, b. $b = 2$
- Find the slope, m. $m = -\frac{2}{3}$
- Write the equation. $y = mx + b$

$$y = -\frac{2}{3}x + 2$$
$$3y = -2x + 6$$
$$2x + 3y = 6$$

Write the equation of the line drawn.

1.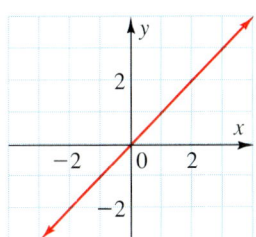

2.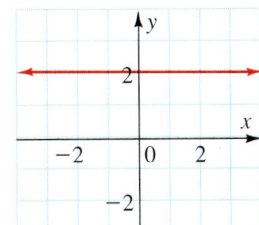

3.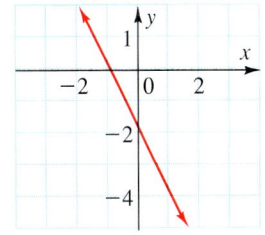

The graph of a linear equation separates the coordinate plane into three sets of points: points on the line, points above the line, and points below the line.

These regions are called **half-planes** and the line is called the **boundary** of each half-plane.

The graph of an inequality related to a linear equation includes all points (x, y) on one side of the line and sometimes the line as well.

EXAMPLE 1 State which of the given points belong to the graph of the inequality.
  a. $y > x + 2$; $(-5, -1)$, $(0, 0)$ b. $x < 3$; $(-2, 4)$, $(3, 2)$

a. Try $(-5, -1)$. Try $(0, 0)$. b. Try $(-2, 4)$. Try $(3, 2)$.
  $y > x + 2$ $y > x + 2$ $x < 3$ $x < 3$
  $-1 \overset{?}{>} -5 + 2$ $0 \overset{?}{>} 0 + 2$ $-2 < 3$ $3 \not< 3$
  $-1 > -3$ $0 \not> 2$ Yes ✓
  Yes ✓

The graphs of inequalities are shown by shading. If the inequality uses $<$ or $>$, the boundary line is not part of the graph and is drawn as a dashed line. This is called an **open half-plane.** If the inequality uses $\le$ or $\ge$, the boundary line is part of the graph and is drawn as a solid line. This is called a **closed half-plane.**

To graph a linear inequality in the two variables x and y when the coefficient of y is not zero:

- Write the given inequality as an equivalent inequality that has y alone on one side.
- Graph the related linear equation. Use a dashed line if the inequality involves $<$ or $>$. Use a solid line if the inequality involves $\le$ or $\ge$.
- Test a point in each half-plane to see which satisfies the inequality.
- Shade the appropriate half-plane.

EXAMPLE 2 Write the inequality whose graph is shown. Tell whether it is an open or closed half-plane.

a. 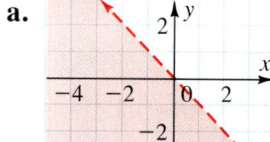   b. 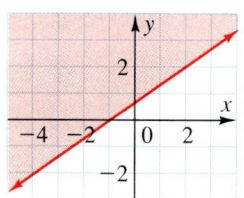   c.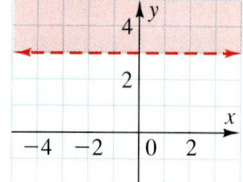

$y < -x$; open half-plane $y \ge \frac{2}{3}x + \frac{2}{3}$; closed half-plane $y > 3$; open half-plane

EXAMPLE 3 Graph $y < 2x + 1$.

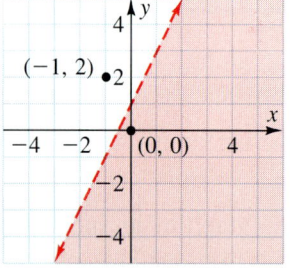

- Draw the graph of $y = 2x + 1$ as a dashed line.
- Test a point in each half-plane.
  Try $(0, 0)$.   Try $(-1, 2)$.
  $0 \overset{?}{<} 2(0) + 1$ $2 \overset{?}{<} 2(-1) + 1$
  $0 < 1$ ✓ Yes $2 \not< -1$
- Shade the half-plane containing the point $(0, 0)$.

EXAMPLE 4 Graph $x \geq 2$.

- Draw the graph of $x = 2$ as a solid line.
- Test a point in each half-plane.
  Try $(0, 0)$.   Try $(3, 1)$.
  $0 \not\geq 2$ $3 \geq 2$ ✔ Yes
- Shade the half-plane containing the point $(3, 1)$.

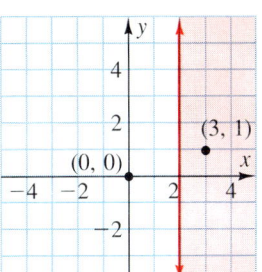

CLASS EXERCISES

State whether the given point belongs to the graph of the inequality.

1. $y > x + 1$; $(1, -1)$
2. $y \leq 2x + 4$; $(3, 2)$
3. $y < 2x$; $(-3, -6)$
4. $y > 5$; $(6, -1)$
5. $x \geq -1$; $(-3, 0)$
6. $x \leq 0$; $(-1, 4)$

Write the inequality whose graph is shown. Tell whether it is an open or closed half-plane.

7.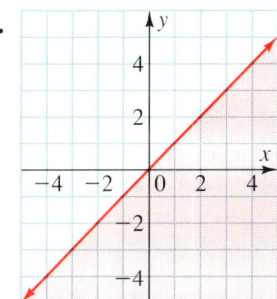
8.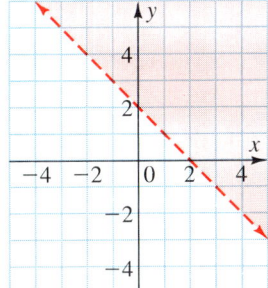
9.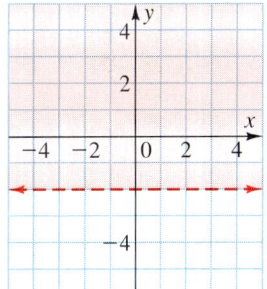

PRACTICE EXERCISES

State which of the given points belong to the graph of the inequality.

1. $y > x + 5$; $(3, 1)$, $(-6, 0)$
2. $y > x + 1$; $(3, 1)$, $(0, 2)$
3. $y + 2 > x + 1$; $(0, 0)$, $(3, 5)$
4. $y + 3 \geq x + 4$; $(1, 2)$, $(3, 0)$

Write the inequality whose graph is shown. Tell whether it is an open or closed half-plane.

5.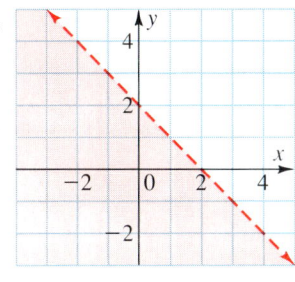
6.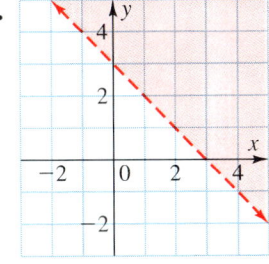
7.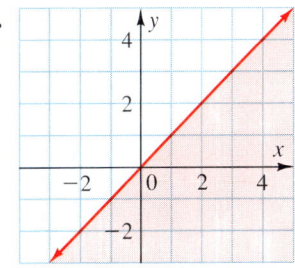

9.7 Linear Inequalities in Two Variables   **403**

8. 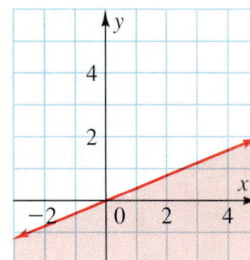 9. 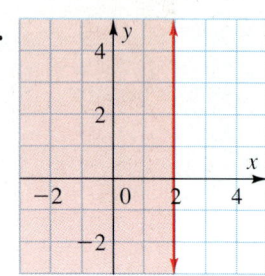 10.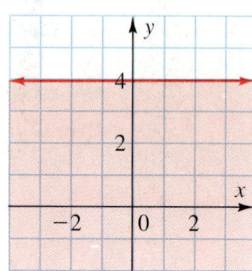

Graph the inequality and test a point in each half-plane. Shade the half-plane containing the point.

11. $y < 3x + 2$
12. $y < 4x + 1$
13. $y > x + 3$
14. $y > 2x + 1$
15. $x \geq 3$
16. $x \geq 5$
17. $y > 3x + 4$
18. $y > 2x + 5$
19. $y \geq 1$
20. $y \geq -4$
21. $y < 3x - 1$
22. $y < 2x - 3$
23. $y - 1 \leq x + 3$
24. $y - 4 \leq x + 5$
25. $y + 3 < x - 3$
26. $y + 2 < x - 1$
27. $y + 6 > 2x - 1$
28. $y + 3 > 3x - 4$
29. $y - 5 > 5x - 10$
30. $y - 3 > 4x - 9$
31. $y + 8 > x + 9$

Write the inequality whose graph is shown.

32. 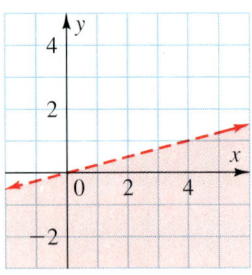 33. 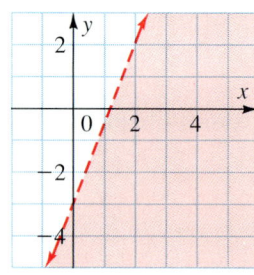 34.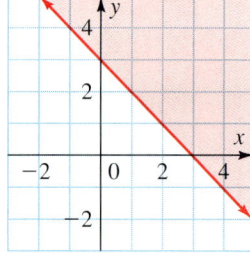

35. 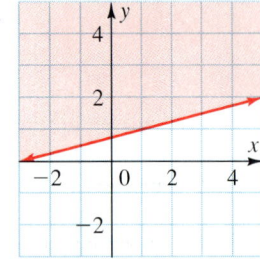 36. 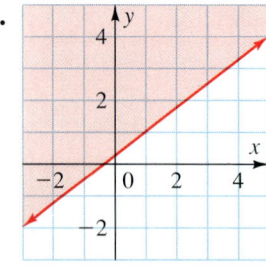 37.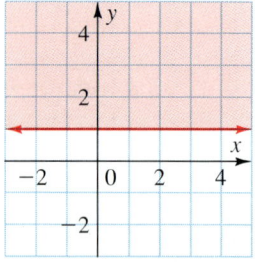

Graph the inequality and tell whether it is an open or closed half-plane.

38. $2y + 3 < \frac{1}{2}x + 5$ **39.** $3y + 4 < \frac{1}{3}x + 7$ **40.** $2y - 3 \geq \frac{1}{4}x + 4$

41. $4y - 1 \geq \frac{1}{8}x + 6$ **42.** $\frac{3}{5}y + 1 < \frac{1}{3}x + 2$ **43.** $\frac{2}{3}y + 2 < \frac{2}{5}x + 1$

Applications

Computer Computer programs are used in business to make decisions. A club that sells cookies or doughnuts decides to use the computer to help them decide what to sell. Use this program to help you decide.

```
10 PRINT   TAB( 15);"PROFIT MADE": PRINT
20 PRINT "DOZEN","COOKIES","DONUTS"
30 FOR I = 1 TO 12
40 C = 2.5 * I - 10
50 D = 4 * I - 25
60 PRINT I,C,D
70 NEXT I
80 END
```

44. Which product would you sell if the club members only volunteered to bring
   a. 2 dozen   **b.** 8 dozen   **c.** 10 dozen   **d.** 12 dozen

45. Use the program to graph the two profit lines. Would you still answer Exercise 44a–d the same way? Why?

46. **Nutrition** Suppose that a person working at 21°C uses about 3000 calories per day, and for each 1°C increase in temperature this person uses 30 fewer calories. Find a linear equation to represent this data. Draw a graph to determine the theoretical temperature at which a person at work would need 0 calories.

TEST YOURSELF

Find the slope and the y-intercept of the line for the given equation. Then graph the equation.   9.5

1. $4x + y = 2$ **2.** $3x - 2y = 6$

Write an equation in standard form of a line for each situation.   9.6

3. Slope m of line is -2, and the line contains $P(3, 2)$.

4. Line contains the points $A(2, -1)$ and $B(3, -4)$.

State which of the given points belong to the graph of the inequality. Graph each inequality.   9.7

5. $y > 2x + 3$; $(0, 3)$, $(1, 7)$ **6.** $y < 3x - 2$; $(-1, -7)$, $(-2, 0)$

7. $y \leq x + 5$; $(0, 6)$, $(1, 3)$ **8.** $y \geq \frac{1}{2}x - 4$; $(0, 4)$, $(2, -5)$

CHAPTER 9   SUMMARY AND REVIEW

Vocabulary

abscissa (371)
boundary (401)
closed half-plane (402)
collinear points (386)
coordinate system (370)
half-plane (401)
linear equation (375)
linear inequality (401)
open half-plane (402)
ordinate (371)
origin (370)
parallel (393)
quadrant (370)
slope (384)
slope-intercept form (391)
standard form (375)
x-axis (370)
x-intercept (376)
y-axis (370)
y-intercept (376)

Graphing Ordered Pairs   To graph an ordered pair of numbers in the coordinate plane, move left or right from the origin for the abscissa, or x-coordinate, and then up or down for the ordinate, or y-coordinate.

9.1

Graph the points in a coordinate plane.

1. $A(-2, -4)$
2. $B(1, -3)$
3. $C(0, 2)$
4. $D(-3, 2)$

Finding Solutions   To determine if an ordered pair (x, y) is a solution of an equation in two variables, substitute the x- and y-coordinates in the equation to see if they produce a true sentence.

Determine which of the ordered pairs listed are solutions of the equation.

5. $2x + y = 7$; $(2, 3)$, $(4, 2)$
6. $-x - y = -3$; $(-3, 0)$, $(3, 0)$

Graphing Linear Equations   Linear equations are equations that can be written in the form $Ax + By = C$ where A, B, and C are real numbers and A and B are not both 0. If A, B, and C are integers, the equation is said to be in **standard form**. You can use several methods to graph a linear equation.

9.2

1. From a table of x- and y-values, graph at least three ordered pairs.
2. Plot points for the x- and y-intercepts. To find the x-intercept, let $y = 0$ and solve for x. To find the y-intercept, let $x = 0$ and solve for y.

Make a solution table. Then graph the equation.

7. $y = -\frac{1}{2}x + 1$
8. $2x - y = 4$
9. $3x + 2y = 6$
10. $x = -6$

406   Chapter 9 Linear Equations

Estimating from Graphs When the data in a problem can be graphed, the graph can be used to estimate solutions to the problem.

9.3

The Fahrenheit temperature is approximately equal to the number of times a cricket chirps in 15 seconds, t, plus 39. Use this for Exercises 11–12.

11. Make a table of ordered pairs for $F = t + 39$ and graph the function.

12. If the outdoor temperature is 95°F, how many cricket chirps would you expect to hear in 15 s?

Slope of a Line The formula is $m = \dfrac{y_2 - y_1}{x_2 - x_1}$, where $P(x_1, y_1)$ and $Q(x_2, y_2)$ are any two points of a line and $x_2 - x_1 \neq 0$.

9.4

Find the slope of a line that contains the given points.

13. $A(-1, 5); B(1, 6)$
14. $C(-2, -4); D(1, -2)$

Through the given point, draw a line with the given slope.

15. $m = \frac{1}{4}; C(2, -2)$
16. $m = -\frac{2}{3}; D(0, 1)$

Slope-Intercept Form of a Linear Equation The slope-intercept form of an equation is $y = mx + b$, where m is the slope and b is the y-intercept.

9.5

Find the slope and the y-intercept. Then graph the equation.

17. $y = \frac{1}{2}x - 4$
18. $y = -\frac{3}{4}x + 3$
19. $x + 2y = 6$
20. $3x - 5y = 15$

Determine whether the graphs of the two equations are parallel lines.

21. $5x - y = 1; 5x + y = 7$
22. $2x + 3y = 2; 4x + 6y = -4$

Equation of a Line An equation can be written using the slope of the line and the coordinates of one point, or the coordinates of two points.

9.6

Write an equation in standard form.

23. The line has a slope of $\frac{2}{3}$ and contains the point $A(0, -4)$.

24. The line contains the points $P(-1, 4)$ and $Q(-4, -2)$.

Linear Inequalities in Two Variables The solution set of a linear inequality in two variables is an open or closed half-plane.

9.7

Graph each inequality.

25. $y > x + 1$
26. $x \geq 3$
27. $y + 2x \leq 2$

Summary and Review

CHAPTER 9 TEST

Graph the ordered pairs in a coordinate plane.

1. $(2, -1)$
2. $(1, -2)$
3. $(5, -6)$
4. $(-1, -4)$
5. $(0, 2)$
6. $(3, 0)$
7. $(-3, 0)$
8. $(-5, 4)$

Determine which ordered pairs listed are solutions of the equation.

9. $x - 3y = 6$
   $(3, -1), (6, -1)$

10. $y = \frac{1}{2}x - 4$
    $(-2, -5), (1, -3)$

Make a solution table. Then graph the equation.

11. $y = 2x - 1$
12. $y = -\frac{1}{3}x$

Use the slope and the intercepts to graph.

13. $3x - y = 9$
14. $2y = 4x - 8$

15. The formula used to convert temperature from a Fahrenheit (F) scale to a Celsius (C) scale is $C = \frac{5}{9}(F - 32)$. Make a graph of the given equation using points for 32°F, −4°F, and 5°F. Use the graph to estimate the Celsius temperature corresponding to 50°F.

16. Find the slope of a line that contains the points $C(8, -3)$ and $D(4, 1)$.

17. Through the point $Q(0, 3)$, draw the graph of a line having a slope of 5.

18. Write $3x + 2y = 1$ in slope-intercept form. Then graph the equation.

19. Determine if the graphs of $y = 4x - 1$ and $16x - 4y = 12$ are parallel.

20. Write an equation in standard form of a line that has slope $\frac{1}{4}$ and contains the point $C(-1, -2)$.

21. Write an equation in standard form of a line that contains $A(-5, 6)$ and $B(1, 2)$.

Graph each inequality.

22. $y < x - 1$
23. $y - \frac{1}{2}x \geq 1$

Challenge

Write an equation in standard form of a line that contains the point $M(a, 3a)$ and is perpendicular to the graph of $4y + 3x = 6$.

PREPARING FOR STANDARDIZED TESTS

Select the best choice for each question.

1. An equation of a line through $P(3, 1)$ with slope of 2 is:
   A. $y = 2x - 5$ B. $y = 2x + 5$
   C. $y = x + 2$ D. $y = 2x + 1$
   E. $y = x - 2$

2. A driver gets an average of 28 mi/gal of gas with her car. How many gallons would be used on a trip of 518 mi?
   A. 18 gal B. 18.5 gal C. 19 gal
   D. 19.5 gal E. 20 gal

3. Find the value of $|3x - 2y + 4|$ when $x = -5$ and $y = 3$.
   A. -25 B. 5 C. 13 D. 17 E. 25

4. $\dfrac{1}{x} + \dfrac{3}{2x} - \dfrac{4}{3x} =$
   A. 0   B. $\dfrac{7}{6}$ C. $\dfrac{7}{6x}$
   D. $\dfrac{7}{6x^2}$ E. $\dfrac{7}{6x^3}$

5. Find the slope of a line which passes through the points $P(-3, 2)$ and $Q(1, 5)$.
   A. $-\dfrac{3}{4}$ B. $\dfrac{3}{4}$
   C. $-\dfrac{9}{2}$ D. $\dfrac{4}{3}$
   E. $-\dfrac{4}{3}$

6. Find the product of $(2x - 3)(3x + 2)$
   A. $6x^2 - 6$
   B. $6x^2 + 4x - 6$
   C. $6x^2 + 5x - 6$
   D. $6x^2 - 4x - 6$
   E. $6x^2 - 5x - 6$

7. An artist bought 4 tubes of paint for $11.95 a tube, and 3 brushes, one for $8.50 and the other 2 for $6.75 each. What was the total amount spent for these supplies?
   A. $68.50 B. $68.80 C. $69.50
   D. $69.80 E. $70.80

8. Solve for x when $\dfrac{3}{x} + \dfrac{2}{x} = \dfrac{5}{2}$.
   A. $\dfrac{1}{2}$ B. 1 C. $1\dfrac{1}{2}$ D. 2 E. $2\dfrac{1}{2}$

9. Find the value of $\left(\dfrac{3x^2 y}{2ab}\right)\left(\dfrac{4ab^2}{6xy^2}\right)$ when $a = 1$, $b = -2$, $x = 3$, and $y = 2$.
   A. -3 B. -1 C. 1 D. 2 E. 3

10. Which of the following represents a point on a line for the equation $2x - 3y = 8$?
    I. $A(7, 2)$
    II. $B(3, -1)$
    III. $C(-2, -4)$
    A. I only   B. II only
    C. III only   D. I, II only
    E. I, III only

11. A homeowners' loan of $4000 for a new roof requires that $4500 be paid back by the end of one year. Find the annual rate of simple interest on the loan.
    A. 1.11% B. 1.25% C. 10%
    D. 11.1% E. 12.5%

12. If x is 30% of y and y is 20% of z, then x is what percent of z?
    A. 5 B. 6 C. 10 D. 50 E. 60

MAINTAINING SKILLS

Write the coordinates for each point.

Example 1   Point A See graph.
The coordinates of A are $(-2, 3)$.

1. C
2. B
3. E
4. D
5. G
6. F

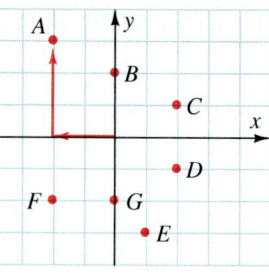

Graph each ordered pair.

7. $(3, 2)$
8. $(-1, 4)$
9. $(2, 0)$
10. $(-3, -2)$
11. $(3, -3)$
12. $(0, -2)$

Draw a graph of each equation.

Example 2 $y = 2x + 1$
$y = 2(-1) + 1$
$= -2 + 1$
$= -1$

x	y
-1	-1
0	1
1	3

13. $y = -2$
14. $y = 3x - 2$
15. $y = 2x$
16. $-2x + 3y = 6$

Evaluate each expression for the given value of the variable.

Example 3 $x^2 - 1,\ x = -4$
$x^2 - 1 = (-4)^2 - 1 = 16 - 1 = 15$

17. $3x - 2,\ x = -1$
18. $2x + 5,\ x = 2$
19. $-3x + 3,\ x = -3$
20. $x^2 + 3,\ x = 4$
21. $-2x^2 - 3,\ x = 2$
22. $3x^2 - 2,\ x = -3$
23. $\frac{1}{2}x^2 - 2,\ x = 3$
24. $|x|,\ x = -2$
25. $|x| - 1,\ x = -1$

Solve.

26. At Oradell High School one hundred twenty-five students study French. Seventy-five students study Spanish. Write in simplest form the ratio of the number of students who study Spanish to the number of students who study French.

27. During an election 7 out of 10 students voted for Hilary for class president. If 42 students voted for Hilary, how many students voted in the election?

28. The ratio of the number of boys to girls in the Senior High is 6 to 7. There are 350 girls in the Senior High. How many boys are there?

10 Relations, Functions, and Variation

Money earned at an hourly wage is directly related to the number of hours spent on the job. As the number of hours increases, the amount of take-home pay also increases.

10.1 Relations

Objectives: To identify relations from tables, graphs, and diagrams
To determine the domain and range of a relation

Ernie makes extra money as a word processor. It took him 3 hours to input a 40-page report, 8 hours to input a 100-page manuscript, and $\frac{1}{2}$ hour to input a 5-page letter. The relation between the number of pages p he types and the time t it takes him to do so can be shown as $\{(40, 3), (100, 8), (5, \frac{1}{2})\}$. This set of ordered pairs can be represented by a *table*, by a *mapping*, or by a *graph*.

p	t
5	$\frac{1}{2}$
40	3
100	8

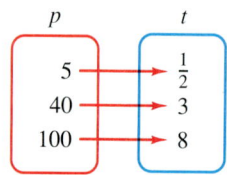

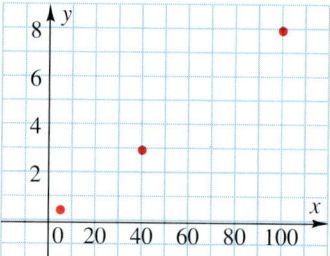

Capsule Review

The points on a graph can be shown in a table or as a set of ordered pairs.

EXAMPLE Write the points shown on the graph in a table and as a set of ordered pairs.

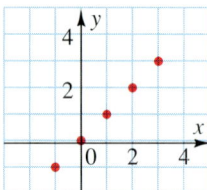

x	y
-1	-1
0	0
1	1
2	2
3	3

$\{(-1, -1); (0, 0); (1, 1); (2, 2); (3, 3)\}$

1. Determine y for the points shown on the graph.

x	-3	-1	0	2	5
y					

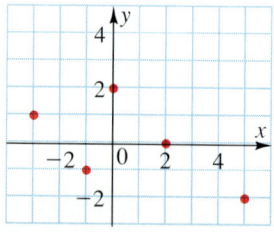

2. Write the points as a set of ordered pairs.

412 Chapter 10 Relations, Functions, and Variation

> A **relation** is a set of one or more ordered pairs. It can be described in a set of ordered pairs, a graph, a mapping, or an open sentence in two variables.

The **domain** of a relation is the set of all its first elements or *x-coordinates*.

The **range** of a relation is the set of all its second elements or *y-coordinates*.

EXAMPLE 1 State the relation specified in the table below as a set of ordered pairs. State the domain and the range of this relation.

x	y
-1	0
0	1.5
1	3

Ordered pairs: $\{(-1, 0), (0, 1.5), (1, 3)\}$
Domain: $\{-1, 0, 1\}$
Range: $\{0, 1.5, 3\}$

When indicating the domain or range of a relation in set notation, use each element once.

EXAMPLE 2 State the relation specified in the mapping below as a set of ordered pairs. State the domain and the range of the relation.

Ordered pairs: $\left\{\left(0, \frac{1}{2}\right), (1, 3), (-2, 0), (-2, 3)\right\}$
Domain: $\{-2, 0, 1\}$
Range: $\left\{0, \frac{1}{2}, 3\right\}$

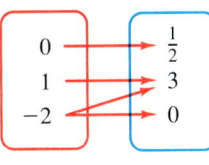

EXAMPLE 3 The graph shows the closing price of MAX stock over a period of one week. State the relation as a set of ordered pairs. State the domain and the range of the relation.

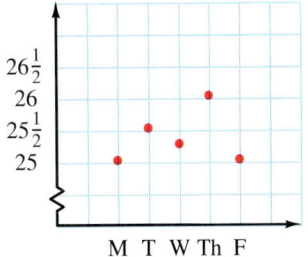

Ordered pairs: $\left\{(\text{Mon.}, 25), \left(\text{Tue.}, 25\frac{1}{2}\right), \left(\text{Wed.}, 25\frac{1}{4}\right), (\text{Thur.}, 26), (\text{Fri.}, 25)\right\}$
Domain: $\{\text{Mon., Tue., Wed., Thur., Fri.}\}$
Range: $\left\{25, 25\frac{1}{4}, 25\frac{1}{2}, 26\right\}$

Sometimes a relation is defined by a given domain and an equation. When variables other than x and y are used, assume that the first letter in alphabetic order represents the domain.

EXAMPLE 4 Given the domain $\{-2, 0, 1\}$ of the relation $2x + y = -1$, determine the range. Graph your results.

Rewrite $2x + y = -1$ in terms of y.

$$y = -2x - 1$$

Let $x =$ an element in the domain. Solve for y.

x	y
-2	3
0	-1
1	-3

The range is $\{3, -1, -3\}$.

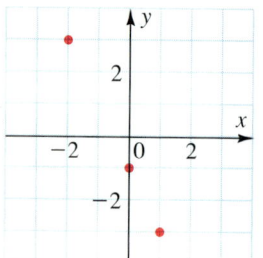

CLASS EXERCISES

1. Is the set $\{1, 2, 3, 4, 5\}$ a relation? Explain.
2. A grocer sells 5 lb of potatoes for $1.80. Show the relation.

Determine the domain and the range of the following relations.

3. (4, 1), (4, 2)

4.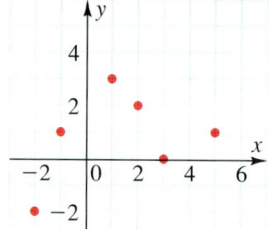

5.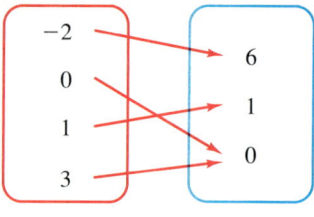

6.
x	y
Sept.	$140
Oct.	$255
Nov.	$235
Dec.	$180

PRACTICE EXERCISES

State the relation as a set of ordered pairs. Determine the domain and the range of the relation.

1.
x	y
-3	5
-1	4
1	4
3	-3

2.
x	y
2	-1
5	3
7	-1
9	-7

3.
x	y
$\frac{1}{3}$	-1
0	0
$\frac{1}{3}$	2
$\frac{1}{2}$	$2\frac{1}{2}$

4.
x	y
2	-1
$2\frac{1}{2}$	0
3	-1
$3\frac{1}{2}$	$-3\frac{1}{2}$

414   Chapter 10 Relations, Functions, and Variation

5. 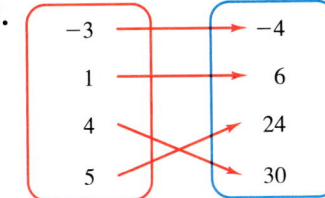   6.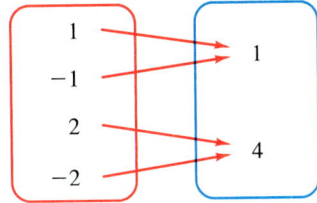

7. 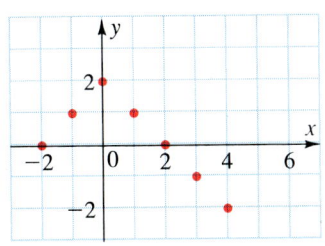   8.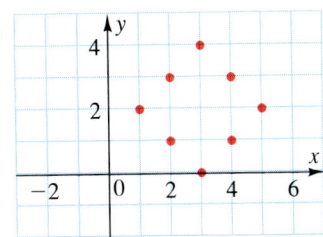

Given the domain $\{-1, 0, 2\}$, determine the range for each relation and graph the results.

9. $y = 3x + 5$
10. $y = 2x - 3$
11. $y = 9 - 2x$
12. $x - 2y = 14$
13. $y = \dfrac{x}{2} + 1$
14. $y = \dfrac{x}{2} - 2$

State the relation as a set of ordered pairs. State the domain and the range. Graph the relation.

15.
Year	1890	1910	1930	1950	1970
Price of milk	$0.07	$0.08	$0.14	$0.21	$0.33

16.
Time	10 AM	11 AM	Noon	1 PM	2 PM
Stock price	$22\tfrac{1}{2}$	23	$23\tfrac{1}{4}$	$22\tfrac{5}{8}$	23

17. 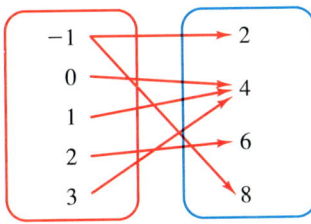   18.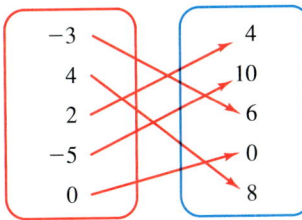

Given the domain $\left\{-1\tfrac{1}{4}, \tfrac{1}{2}, 2\right\}$, determine the range for each relation and graph the results.

19. $8x + \tfrac{1}{2}y = 8$
20. $4x - \tfrac{1}{2}y = -5$
21. $4x - y = 2$
22. $16x - 2y = -4$

10.1 Relations **415**

Applications

State the domain and the range of the following relations. Describe your results as a set of ordered pairs.

23. **Finance** In the first year, Zelda made $5000 selling encyclopedias. She increased her earnings by $200 each year for the next four years.

24. **Finance** John started out at 9 AM to go door-to-door selling brushes. He sold 6 brushes an hour until 3 PM.

25. **Meteorology** Sun Valley was covered with 3 ft of snow when it started to snow again. The snow fell at a rate of 1 ft/h for 5 h.

ALGEBRA IN TAXATION

Did you know that relationships play a very important role in our tax structure? There are three types of tax structures in our society:

Proportional tax — A tax in which the percentage of tax paid is the same for all incomes. An example of this is a constant or "flat rate" state income tax.

Progressive tax — A tax in which the percentage of tax paid increases as total income increases. An example of this is the federal income tax.

Regressive tax — A tax in which the percentage of tax paid decreases as total income increases. An example of this is social security tax. The amount of social security tax paid in 1989 was 7.51% of the first $48,000 earned.

1. Dan had an income of $11,000 in a state where the income tax is 2.5%. What was the amount Dan paid in taxes?

2. Mildred is single and earns $35,000 a year. The federal government has three tax rates shown at the right. Find the amount Mildred paid in income tax.

Income is between	Then the tax rate is
$0–$15,000	15%
$15,001–$40,000	28%
above $40,001	30%

3. Marcia earned $20,000 and Alex earned $50,000. Who paid the higher percent of total income to social security?

10.2 Functions and Function Notation

Objectives: To identify functions
To understand standard function notation

Rose earns $5.00 per hour in a flower shop. The relation between her salary s and the number of hours h she works is represented in this graph. Her salary depends on how many hours she works. For 7 hours she earns $35. For $1\frac{1}{2}$ hours she earns only $7.50. The equation $s = 5h$ represents this relation.

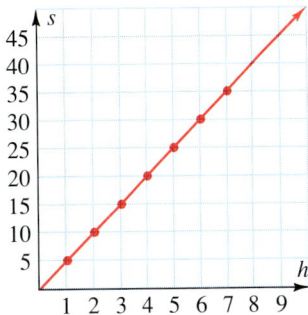

Capsule Review

Recall that relations are sets of ordered pairs and can be indicated by mappings. Give the relations indicated as sets of ordered pairs.

1.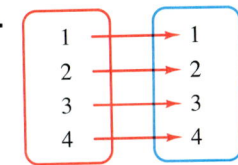
2.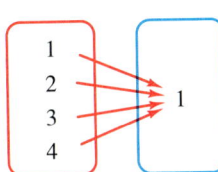
3.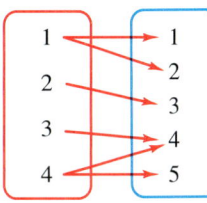

A **function** is a relation in which each element of the domain is paired with *exactly* one element of the range.

EXAMPLE 1 Tell whether each relation specifies a function. Explain.

a. $(0, -1), (2, -2), (1, -1), (1, 5), (-3, 4)$

b.
Time	Temp.
1 PM	15°C
2 PM	16°C
3 PM	17°C
4 PM	19°C
5 PM	17°C

c.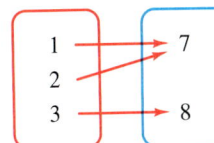

10.2 Functions and Function Notation   **417**

a. No; the element 1 is paired with both −1 and 5.

b. Yes; each element of time corresponds with exactly one temperature.

c. Yes; each element of the domain is paired with exactly one element of the range.

Another method for determining whether a relation is a function involves drawing a graph of the ordered pairs and then drawing vertical lines through the graph. This is called the **vertical line test.**

Vertical Line Test

If a vertical line intersects a graph in more than one point, then the graph is not the graph of a function.

EXAMPLE 2 Graph each of the following relations. Use the vertical line test to determine whether each relation is a function.

a. (1, 4), (0, 1), (2, 1), (3, 3), (5, 0)   **b.** (1, 3), (3, 5), (3, 2), (3, 0)

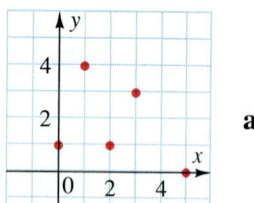   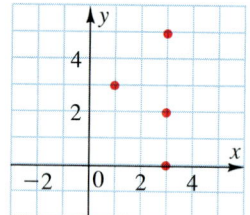

a. Yes.   **b.** No.

For the relation to be a function, any vertical line can only pass through a single point. Example (b) is not a function while Example (a) is a function.

Functions can be specified using different types of **function notation.**

Arrow notation, $F: x \rightarrow 2x + 1$ Read: The function F that pairs x with $2x + 1$.

f of x notation: $f(x) = 2x + 1$ Read: f of x equals $2x + 1$.
Equation in two variables: $y = 2x + 1$ Explain why y is called a function of x.

All the ordered pairs of a function are in the form (x, y) or $(x, f(x))$. Note that $f(x)$ does not mean f times x; rather, it refers to the range of the function whose domain is represented by x. Letters other than f, such as F, g, G, h, and so on, are often used for *function notation*.

EXAMPLE 3 Evaluate, given $f(x) = x - 3$ and $g(x) = x^2$.
a. $f(0)$ **b.** $g(2)$ **c.** $f(0) - g(2)$

a. $f(0) = 0 - 3 = -3$ **b.** $g(2) = 2^2 = 4$ **c.** $f(0) - g(2) = -3 - 4 = -7$

EXAMPLE 4 Evaluate, given the domain $\{-2, 0, 1\}$.

a. $y = 4x - 1$ b. $g(x) = \frac{1}{3}x^2 - 2$

a. $y = 4x - 1$

x	$4x - 1$	y
-2	$4(-2) - 1$	-9
0	$4(0) - 1$	-1
1	$4(1) - 1$	3

b. $g(x) = \frac{1}{3}x^2 - 2$

$g(-2) = \frac{1}{3}(-2)^2 - 2 = \frac{1}{3}(4) - 2 = -\frac{2}{3}$

$g(0) = \frac{1}{3}(0)^2 - 2 = -2$

$g(1) = \frac{1}{3}(1)^2 - 2 = \frac{1}{3}(1) - 2 = -\frac{5}{3}$

Some functions have a point where the graph for the function changes direction. Some of these points can be determined by examining the coefficients and constants in the function.

EXAMPLE 5 Compare the graphs of $y = |2x|$, $y = |2x + 3|$, and $y = |2x + 3| - 1$ using a graphing calculator.

The graph of $y = |2x|$ changes direction at $(0, 0)$. The graph of $y = |2x + 3|$ is three units to the left of the graph of $y = |2x|$ and changes direction at $\left(-\frac{3}{2}, 0\right)$. The graph of $y = |2x + 3| - 1$ is one unit below the graph of $y = |2x + 3|$ and changes direction at $\left(-\frac{3}{2}, -1\right)$.

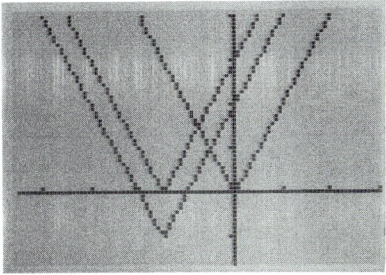

CLASS EXERCISES

For Discussion

Is each relation a function? Explain.

1. (x, y) where x is a person and y is the type of car that person owns.
2. (x, y) where x is each person in your class and y is that person's birthdate.
3. (x, y) where x is a person and y is that person's social security number.

PRACTICE EXERCISES

Tell whether each relation specifies a function.

1. $\{(-3, -2), (-2, -1), (-1, 0), (0, 1), (1, 2)\}$
2. $\{(5, 0), (0, 5), (5, 1), (1, 5), (5, 2), (2, 5)\}$
3. $\{(-1, -3), (4, 0), (5, 2), (2, 3), (-2, -3), (1, 1)\}$
4. $\{(2.5, -1.5), (-0.5, 3), (-1.5, -5), (2.5, 2)\}$

5. 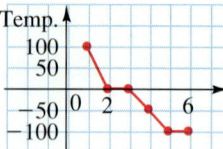 **6.** 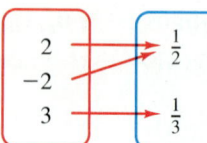 **7.** 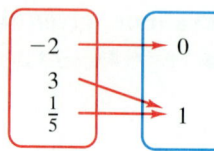 **8.**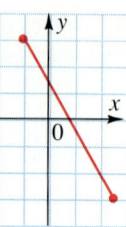

9. 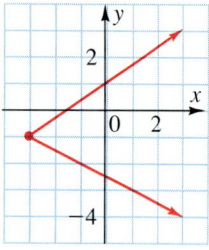 **10.** 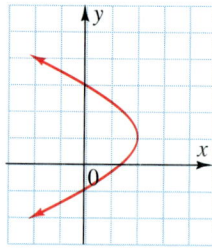 **11.**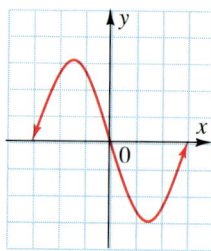

Evaluate, given $f(x) = x + 2$.

12. $f(2)$ **13.** $f(3)$ **14.** $f(-1)$ **15.** $f(-4)$

Graph each relation given the domain: $\{-1, 0, 3, 6\}$. State the range and tell whether each is a function.

16. $y = x + 4$
17. $3x - y = 5$
18. $f(x) = |x| + 1$
19. $f(x) = -|x - 1|$
20. $f: x \rightarrow x - 25$, domain: {all real numbers}. State the range and whether the relation is a function.
21. $f(x) = |x| - 1$, domain: {all real numbers}. State the range and whether the relation is a function.

Evaluate given that $f(x) = x^2 + 1$ and $g(x) = 5x - 3$.

22. $f(-2)$ **23.** $g(1)$ **24.** $f(-2) + g(1)$
25. $f(2) \times g(0)$ **26.** $f(4) \times g(1)$ **27.** $f(-2) - g(1)$

Graph each absolute value function using a graphing calculator or a computer program to find the point where the graph changes direction.

28. $f(x) = |x + 4|$ **29.** $g(x) = |2x + 4|$ **30.** $y = |2x + 4| - 5$

Determine the relation by evaluating for the conditions. Graph the relation and tell whether or not it is a function.

31. $3a + 2b = 1$; Domain: $\{a: -5 < \text{integers} < 5\}$

32. $7j + 3k = 9$; Domain: $\{j: \text{counting numbers} < 6\}$

33. $y = |x| + 2$; Range: $\{\text{integers} < 4\}$

34. $y = |x| - 3$; Range: $\{\text{whole numbers} < 5\}$

Applications

35. Finance Graph the closing prices of AEM stock over the period of one week: (Mon., 38), $\left(\text{Tue.}, 38\frac{1}{8}\right)$, $\left(\text{Wed.}, 35\frac{1}{4}\right)$, (Thur., 37), $\left(\text{Fri.}, 37\frac{3}{8}\right)$. Is this relation a function?

36. Meteorology Graph the relation between month m and average temperature t as described in this chart. Is this relation a function?

Average Temperatures for Anchorage, Alaska

m	Jan.	Feb.	Mar.	Apr.	May	Jun.	Jul.
t	13°F	18°F	24°F	35°F	46°F	54°F	58°F

READING IN ALGEBRA

Perhaps you read entire lines in one or two glances as a way of reading quickly. Since many sentences in a math textbook are concerned with definitions, examples, and explanations, you may miss important ideas when you "scan" a page.

Below are several statements from this book. Examine each statement and decide whether it is exactly as given in the book. If not, restate it so that it duplicates the original.

1. (page 412) The relationship between the pages he types and the time can be represented as a set of pairs, a table, or a graph.

2. (page 417) A function is a special relation where different ordered pairs have different first coordinates.

3. (page 345) A proportion is a statement with equal ratios.

4. (page 385) Recall that a line with a positive slope slants up. Lines with negative slopes slant down.

10.3 Linear, Constant, and Composite Functions

Objectives: To identify linear and constant functions from a graph
To evaluate composite functions

Linear equations such as $y = 3x + 4$ and $y = 2$ define functions. Recall that the graph of a linear equation is a straight line.

Capsule Review

The graph of the equation $y = mx + b$ where m and b are real numbers is a line whose slope is m and whose y-intercept is b.

EXAMPLE Use the slope and the y-intercept to graph the equation $y = -x + 3$.

$y = -x + 3$, so the slope m is -1; the y-intercept b is 3.

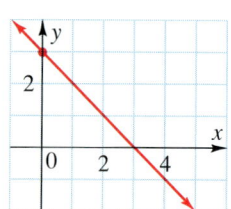

Use the slope and the y-intercept to graph the equation.

1. $x + 3y = 9$ **2.** $-5y - x = 10$ **3.** $2x - 3y = 8$

A **linear function** is a special kind of function with a domain equal to the set of real numbers, whose graph is a straight line. A linear function is a function that can be defined by an equation in slope-intercept form.

$$y = mx + b \quad m = \text{slope}; \quad b = y\text{-intercept}$$

EXAMPLE 1 Graph each relation. Is the relation a function? A linear function?

a. $x + y = -2$

x	0	1	2
y	-2	-3	-4

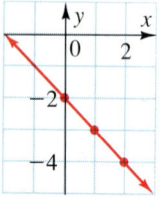

b. $y = |x| + 1$

x	2	1	0	-1	-2
y	3	2	1	2	3

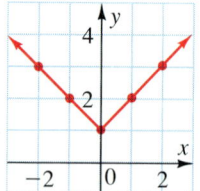

a. Yes, because the vertical line test holds. Yes, because it can be written in slope-intercept form $y = -x - 2$, where m is -1 and b is -2.

b. Yes, because the vertical line test holds. No, because the equation cannot be written in slope-intercept form (the graph is not a straight line).

Chapter 10 Relations, Functions, and Variation

A **constant function** is a linear function whose range contains only one element. The graph of a constant function is a horizontal line.

EXAMPLE 2 Graph each relation. Is the relation a function? A linear function? A constant function?

a. $g(x) = x - 2$ **b.** $x = -2$ **c.** $f(x) = 3$

a. Label the vertical axis $g(x)$.

x	$g(x)$
0	-2
1	-1
2	0

Linear function; not a constant function

b. All x values equal -2.

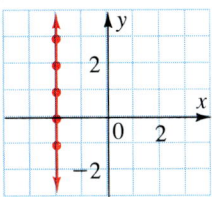

Not a function

c. All $f(x)$ values equal 3.

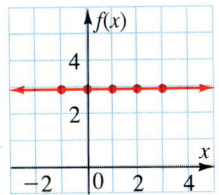

Linear function; constant function

A **composite function** combines two or more functions. To evaluate a composite function, evaluate the inner function first, then evaluate the outer function.

EXAMPLE 3 Evaluate $f(g(2))$, given $f(x) = 3x + 2$ and $g(x) = \frac{1}{2}x^2 - 1$.

First, evaluate the innermost function.

$g(2) = \frac{1}{2}(2)^2 - 1$
$= \frac{1}{2}(4) - 1$
$= 2 - 1 = 1$

Now, since $g(2)$ is 1,
$f(g(2)) = f(1)$
$= 3(1) + 2$
$= 3 + 2$
$= 5$

CLASS EXERCISES

Graph each relation. Is the relation a function? A linear function? A constant function?

1. $f(x) = 2x - 1$ **2.** $m(x) = 4$ **3.** $x = 5$

Evaluate, given that $f(x) = 2x + 2$ and $g(x) = 6x + 5$.

4. $f(g(2))$ **5.** $f(g(7))$ **6.** $g(f(-1))$ **7.** $g(f(-4))$

PRACTICE EXERCISES

Graph each relation. Is the relation a function? A linear function? A constant function?

1. $f(x) = 3x - 2$ **2.** $G: x \to x + 7$ **3.** $f(x) = |x| + 1$ **4.** $h(x) = 2|x|$

10.3 Linear, Constant, and Composite Functions **423**

5. $f(x) = -7$ 6. $g(x) = 5$ 7. $x = 5$ 8. $x = -4$

9. $f(x) = -5$ 10. $H: x \to 2$ 11. $f(x) = -\frac{x}{2} + 3$ 12. $f(x) = \frac{x}{3} + 6$

Evaluate, given $f(x) = 2x + 1$ and $g(x) = 4x - 5$.

13. $f(-1)$ 14. $g(-1)$ 15. $f(-2)$ 16. $g(-2)$
17. $f(g(1))$ 18. $f(g(2))$ 19. $g(f(3))$ 20. $g(f(4))$

Graph each relation. Is the relation a function? A linear function? A constant function?

21. $f(x) = 2x - 9$ 22. $f(x) = 3x - 7$ 23. $h: x \to -|x| - 3$

24. $h: x \to \left|\frac{x}{2}\right| + 11$ 25. $g(x) = -4|x| + 2$ 26. $g(x) = -\frac{3}{2}|x| - 3$

Evaluate, given $f(x) = 5x - 3$ and $g(x) = x^2 - 3x + 1$.

27. $f(g(1))$ 28. $f(g(2))$ 29. $g(f(0))$ 30. $g(f(3))$
31. $f(g(3))$ 32. $f(g(-1))$ 33. $g(f(1))$ 34. $g(f(2))$

35. Evaluate $f(t(15))$, given $f(x) = |-x| - 2$ and $t(x) = \frac{1}{x}$.

36. Evaluate $g(r(-2))$, given $g(x) = |x| + x$ and $r(x) = -1.1$.

Applications

37. **Consumer** The cost of a long distance phone call of t minutes is found by using the function $f(t) = 0.75t + 1.45$. Find the cost of a 15-min phone call.

TEST YOURSELF

Given the domain $\{-3, 0, 4\}$, state the range and graph the results. 10.1

1. $y = 2x + 3$ 2. $y = 3x - 4$ 3. $y = \frac{1}{4}x - 1$ 4. $y = \frac{1}{3}x + 1$

Graph each relation. Is the relation a function? A linear function? A constant function? 10.2

5. $\{(0, 1), (1, 3), (-1, 4), (2, 5)\}$ 6. $\{(0, -2), (1, -1), (0, -3), (4, 2)\}$
7. $y = -2x + 4$ 8. $y = -4$ 9. $y = |x| + 5$ 10. $x = -5$

Evaluate, given $f(x) = 3x + 2$ and $g(x) = 4x - 1$. 10.3

11. $f(-3)$ 12. $g(4)$ 13. $f(g(2))$ 14. $g(f(3))$

Chapter 10 Relations, Functions, and Variation

APPLICATION: Cost Analysis

Did you know that cost analysis plays an integral part in mathematical modeling with respect to business applications? In businesses where various products are produced, a manufacturer must take fixed and variable costs into consideration. Fixed costs are costs which remain constant, while variable costs can change at any time during production.

The following example illustrates the above.

EXAMPLE A computer manufacturer needs to purchase microchips. The supplier charges a fixed cost of $50.00 on the first 100 chips, and $0.35 as a variable cost for each chip purchased over this amount.

How much would the computer manufacturer have to pay the supplier in order to purchase 2010 chips?

$c(x)$ = cost function
m = variable cost of $0.35 per chip after the first 100
b = fixed cost of $50.00 on the first 100 chips
x = amount of chips over 100 purchased
$c(x) = mx + b$

Thus the equation would be:

$$c(x) = (0.35)(1910) + 50$$
$$c(x) = 668.50 + 50$$
$$c(x) = 718.50$$

The total cost to the computer maker would be $718.50 for the purchase of 2010 microchips.

Solve.

1. The fixed cost of purchasing the first 20 televisions is $480. The variable cost per television is $12. Find the cost function $c(x)$. If 186 televisions are to be purchased, what is the total cost?

2. Determine the total cost of producing 190 chairs if the fixed cost of rent and taxes is $6600 and the variable cost of materials and shipping expenses is $20 per chair.

Application: Cost Analysis **425**

10.4 Direct Variation

Objectives: To identify direct variations
To solve word problems involving direct variation

Karla earns money by waxing automobiles. She charges $15 per car. Karla waxed 2 cars on Friday, 6 cars on Saturday, and 3 cars on Sunday. How much did she earn each day?

The amount Karla earns depends on the number of cars she waxes. This is called a *direct variation*. The direct variation can be written as $y = 15x$. The constant 15 is called the **constant of variation** or the **constant of proportionality**.

Let $x =$ number of cars waxed		Let $y =$ dollars earned
Friday	2	$30
Saturday	6	$90
Sunday	3	$45

Capsule Review

A linear equation shows how the values of *x* and *y* depend on each other.

EXAMPLE Show how *y* depends on *x* in the equation $3x - y = 2$.

$$3x - y = 2$$
$$-y = -3x + 2 \quad \text{Solve for y.}$$
$$y = 3x - 2$$

y is always 2 less than 3 times the value of *x*.

Show how *y* depends on *x* in the following equations.

1. $-x + y = 4$
2. $\frac{1}{2}x - \frac{1}{2}y = 1$
3. $3x = 4y - 3$

A **direct variation** is a function in the form: $y = kx$, $k \neq 0$, where it is said that *y* varies directly as *x*, or *y* is directly proportional to *x*.

Direct variation is a special case of the linear function $y = mx + b$, where $m = k$ and $b = 0$.

Chapter 10 Relations, Functions, and Variation

The graph of a direct variation is a straight line with slope of k passing through the origin.

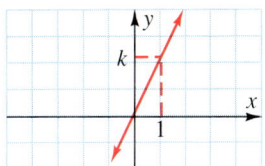

EXAMPLE 1 For each equation, tell whether y varies directly as x. Explain.

a. $2x + y = 0$
   $y = -2x$

b. $x = y + 1$
   $x - 1 = y$

a. Yes; $y = -2x$ is in the form $y = kx$ with $k = -2$. (y is always negative two times x.)

b. No; $y = x - 1$ is not in the form $y = kx$.

EXAMPLE 2 For the following table tell whether y varies directly as x. If so, name the constant of variation and the equation that shows the relationship.

x	y
3	3.6
6.1	7.32
11	13.2

Substitute each x and y in $\frac{y}{x} = k$.

$\frac{3.6}{3} = 1.2$ $\frac{7.32}{6.1} = 1.2$ $\frac{13.2}{11} = 1.2$

y varies directly as x. $k = 1.2$ and $y = 1.2x$.

In general, given two ordered pairs (x_1, y_1) and (x_2, y_2) of the same direct variation, the constant of variation is the same for each equation: $y_1 = kx_1$ and $y_2 = kx_2$. From these equations, you can write $\frac{y_1}{x_1} = k$ and $\frac{y_2}{x_2} = k$. Then $\frac{y_1}{x_1} = \frac{y_2}{x_2}$ because each ratio equals k.

Recall that an equation which states that two ratios are equal is a proportion. To solve a proportion use the property of proportions: if $\frac{a}{b} = \frac{c}{d}$, then $ad = bc$.

EXAMPLE 3 The amount of rice used in a casserole recipe is directly proportional to the number of people served. If $1\frac{1}{4}$ cups of uncooked rice serve 6 people, how many cups of rice would be needed to serve a group of 40 people?

Let r = number of cups of rice needed and p = number of people to be served. Then $r = kp$.

$\frac{r_1}{p_1} = \frac{r_2}{p_2}$ $\frac{r_1}{p_1} = k$ and $\frac{r_2}{p_2} = k$ $6r = 50$ Property of Proportions

$\frac{1\frac{1}{4}}{6} = \frac{r}{40}$ Substitute. $r = 8\frac{1}{3}$

$8\frac{1}{3}$ cups of rice are needed.

10.4 Direct Variation 427

EXAMPLE 4 y varies directly as x. If $y = 5$ when $x = 7.5$, find x when $y = 9$.

$\dfrac{y_1}{x_1} = \dfrac{y_2}{x_2}$ $\dfrac{y_1}{x_1} = k$ and $\dfrac{y_2}{x_2} = k$ $5x = 67.5$ *Property of Proportions.*
  $x = 13.5$

$\dfrac{5}{7.5} = \dfrac{9}{x}$ *Substitute.*

CLASS EXERCISES

Find the constant of variation k if y varies directly as x.

1. $y = 5.1$, $x = 3$
2. $y = 3.2$, $x = 0.8$

For Discussion

Does y vary directly with x?

3. $y = \dfrac{1}{2}x$
4. $y = 2x + 3$
5. $xy = 3$

PRACTICE EXERCISES

For each equation tell whether y varies directly as x.

1. $y = \dfrac{4}{5}x$
2. $y = \dfrac{7}{9}x + 1$
3. $y = 3x - 2$
4. $y = 5x$

For each table tell whether y varies directly as x. If so, give the constant of variation and the equation that shows the relationship.

5.
x	1	3	5	9
y	4	12	20	36

6.
x	8	18	20	21
y	4	9	10	11

7.
x	0.6	4.5	6.3	15.6
y	0.2	1.5	2.1	5.2

8.
x	1.6	2.2	2.4	2.8
y	1.2	1.75	1.8	2.1

Solve.

9. y varies directly with x. $y = 5$ when $x = 25$. Find y when $x = 55$.
10. y varies directly with x. $y = 3$ when $x = 2$. Find y when $x = 34$.
11. B varies directly with A. $B = 4$ when $A = 150$. Find B when $A = 200$.
12. N varies directly with M. $N = 84$ when $M = 12$. Find N when $M = 14$.
13. y varies directly with x. $y = 5$ when $x = 6$. Find y when $x = 42$.
14. y varies directly with x. $y = 35$ when $x = 56$. Find y when $x = 8$.
15. E varies directly with M. $E = 110$ when $M = 17.6$. Find M when $E = 80$.
16. C varies directly as W. $C = 3.78$ when $W = 3$. Find C when $W = 4.5$.
17. W is directly proportional to L. $W = 9$ when $L = 12.5$. Find W when $L = 16.5$.
18. F is directly proportional to T. $F = 2.25$ when $T = 12$. Find F when $T = 50$.

19. l is directly proportional to w.
    $l = 3.9$ when $w = 3$. Find l when $w = 40$.

20. d is directly proportional to t.
    $d = 36$ when $t = 0.9$. Find d when $t = 11$.

21. A varies directly as l. $A = 5.4$ when $l = 1.8$. Find A when $l = 3$.

22. C varies directly as r. $C = 1.9$ when $r = 0.213$. Find C when $r = 10$.

23. y is directly proportional to x.
    $y = 2.56$ and $x = 3.2$. Find y when $x = 25$.

24. y is directly proportional to x.
    $y = 7.29$ when $x = 0.9$. Find y when $x = 1.2$.

25. If y varies directly as the square of x, and $y = 1.2$ when $x = 2$, what does y equal when $x = 5$?

26. If y varies directly as the square of x, and $y = 40.5$ when $x = 9$, what does y equal when $x = 12$?

27. Does the area of a circle vary directly as the radius?

28. Does a square's perimeter vary directly as the length of its side?

Applications

29. **Finance** The amount of interest earned on a savings account is directly proportional to the amount of money in the account. If $5000 earns $350 interest, how much interest is earned on $8000?

ALGEBRA IN ACCOUNTING

The following function uses the *straight line method* to determine the annual rate of depreciation.

$$D = \frac{C - S}{N}$$

where C = original cost N = estimated life
       S = scrap value D = the amount to be depreciated each year

EXAMPLE A car is purchased at a cost of $9600. It is estimated that it will be used for 20 years, after which it will be worth $600 for scrap. Find the annual depreciation and the annual rate of depreciation of the car.

$$D = \frac{9600 - 600}{20} = \$450 \text{ annual depreciation}$$

$450 \div 9600 = 4.69\%$ annual rate of depreciation

Use the above function to solve.

A mainframe computer is purchased at a cost of $110,000. It is estimated it will be used for 5 years, after which it will have a scrap value of $10,000. Find the annual depreciation and the annual rate of depreciation.

10.5 Inverse Variation

Objectives: To identify inverse variations
To solve problems involving inverse variations

Ron is a SCUBA diver and he knows that as he dives the amount of pressure p increases, while the volume v of a confined gas will decrease. A relation of this type is called an *inverse variation*. This relation between pressure and volume is known as *Boyle's Law*.

Capsule Review

Determining k identifies the equation for finding any pair (x, y) related in direct proportion by that constant.

EXAMPLE For $y = kx$, find k if $x = 3$ and $y = 7.5$. Then use k to state the equation.

$y = kx$
$7.5 = k(3)$ *Substitute the elements of the known pair.*
$2.5 = k$ *The equation is $y = 2.5x$.*

Evaluate for k, given $y = kx$. Then use k to state the equation.

1. $x = 2$ and $y = 8$ **2.** $y = -3$ and $x = \frac{2}{3}$ **3.** $x = 7$ and $y = 28$

An **inverse variation** is a function in the form: $xy = k$ or $y = \frac{k}{x}$, $k \neq 0$, where it is said that y varies inversely (or indirectly) as x, or y is inversely proportional to x. The *constant of variation* is k.

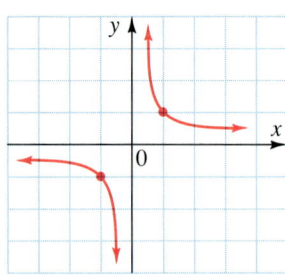

Inverse variation is not a linear function. The graph of an inverse variation is a hyperbola described by the function $xy = k$, $k \neq 0$.

EXAMPLE 1 From the table below tell whether y varies inversely as x. If so, name the constant of variation and the equation that shows the relationship.

x	y
5	4.8
6.4	3.75
3	8
20	1.2

Substitute each (x, y) in $xy = k$.

$5(4.8) = 24$ $6.4(3.75) = 24$ $3(8) = 24$ $20(1.2) = 24$

y varies inversely as x. $k = 24$; $xy = 24$

430 Chapter 10 Relations, Functions, and Variation

Suppose (x_1, y_1) and (x_2, y_2) are ordered pairs of the same inverse variation. The constant of variation would be the same for each: $x_1 \cdot y_1 = k$ and $x_2 \cdot y_2 = k$.

Therefore, $x_1 \cdot y_1 = x_2 \cdot y_2$ or $\dfrac{x_1}{x_2} = \dfrac{y_2}{y_1}$.

EXAMPLE 2 y varies indirectly as x. If $y = 3$ when $x = 10.5$, find x when $y = 9$.

$$
\begin{aligned}
x_1 \cdot y_1 &= x_2 \cdot y_2 & x_1 \cdot y_1 &= k \text{ and } x_2 \cdot y_2 = k \\
10.5(3) &= x(9) & &\text{Substitute.} \\
31.5 &= 9x \\
3.5 &= x
\end{aligned}
$$

EXAMPLE 3 At a depth of 33 ft in salt water, the pressure is measured to be 2 atmospheres (atm). The volume of air in a container is 10 ft³. Find the volume of air in the container at a depth where the pressure is 1.25 atm.

Let x = pressure at 33 ft and y = volume of air in the container; then $xy = k$.

$$
\begin{aligned}
x_1 \cdot y_1 &= x_2 \cdot y_2 & 20 &= 1.25y \\
(2)(10) &= (1.25)(y) & 16 &= y \qquad \text{The volume of air is 16 ft}^3.
\end{aligned}
$$

Another example of an inverse variation is the principle of the lever. A lever is a bar that pivots about a fixed point called the fulcrum. If masses m_1 and m_2 are placed at distances d_1 and d_2 from the fulcrum and the lever is balanced, then $m_1 d_1 = m_2 d_2$.

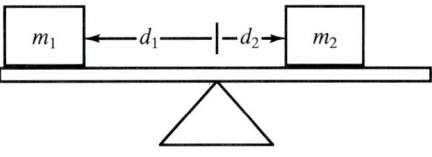

EXAMPLE 4 Eric and Stephanie were playing on a seesaw. Eric, whose mass is 32 kg, sat 1.5 m from the fulcrum. What must be Stephanie's mass if the seesaw balanced when she sat 2 m from the fulcrum?

Let m = mass in kg and d = distance in m; then $md = k$.

$$
\begin{aligned}
m_1 \cdot d_1 &= m_2 \cdot d_2 \\
32(1.5) &= m_2 \cdot 2 & &\text{Substitution.} \\
48 &= m_2 \cdot 2 \\
24 &= m_2 & &\text{Stephanie's mass is 24 kg.}
\end{aligned}
$$

CLASS EXERCISES

For each table tell whether y varies inversely as x. If so, name the constant of variation and the equation that shows the relationship.

1.
x	6	8	1.2	48
y	4	3	20	0.5

2.
x	3	4.5	5	1.2
y	15	10	9	38

Find the constant of variation k if y varies indirectly as x.

3. $y = 8$, $x = 4$
4. $y = 3.3$, $x = 3$
5. $y = 8.1$, $x = 0.7$

6. y varies inversely as x. If $y = 6$ when $x = 0.8$, find y when $x = 3$.

7. In a rectangle with a constant area, the width w is inversely proportional to the length l. If $w = 9$ when $l = 32$, find w when $l = 2.4$.

PRACTICE EXERCISES

For each table tell whether y varies inversely as x. If so, name the constant of variation and the equation that shows the relationship.

1. x	y	2. x	y	3. x	y	4. x	y
2	6	9	4	−0.8	1.8	−1.6	0.3
−3	−4	−2	−18	1.2	−1.2	0.4	7.2
12	1	3	12	−3.6	0.4	2	−0.24

Solve.

5. y varies inversely as x. If $y = 12$ when $x = 6$, find y when $x = 9$.

6. y varies indirectly as x. If $y = 8$ when $x = 18$, find y when $x = 12$.

7. l is inversely proportional to w. If $l = 1.6$ when $w = 30$, find l when $w = 12$.

8. s is inversely proportional to t. If $s = 25$ when $t = 8$, find s when $t = 0.5$.

9. w varies indirectly as l. If $w = 3.2$ when $l = 6$, find w when $l = 0.8$.

10. C varies inversely as k. If $C = 24$ when $k = 1.6$, find C when $k = 0.12$.

11. y is inversely proportional to x. If $y = 4.2$ and $x = 3.9$, find y when $x = 2.6$.

12. y varies inversely as x. If $y = 2.25$ when $x = 3.6$, find y when $x = 0.3$.

13. j is inversely proportional to k. If $j = 9$ when $k = 2.5$, find j when $k = 4.25$.

14. h varies inversely as l. If $h = \frac{1}{4}$ when $l = \frac{1}{6}$, find h when $l = \frac{2}{3}$.

15. m varies indirectly as the square root of t. If $t = 144$ when $m = 6$, find t when $m = 132$.

16. l varies indirectly with z^2. If $l = 9$ when $z = 2$, find z when $l = 90$.

17. n varies inversely with the cube of q. If $n = 9.3$ when $q = 4.2$, find n when $q = 3.2$.

18. h varies indirectly as the square of i. If $h = 4$ when $i = 1.6$, find h when $i = 6.2$.

19. The volume of air in a container is 15 in.^{3} at 1.5 atm. Find the volume at 2.25 atm.

20. The volume of air in a container is 12 m^3 at 1.75 atm. Find the pressure if the volume is 4 m^3.

21. How much mass must be placed 2.5 m away from the fulcrum in order to balance a 35-kg object that is 3 m from the fulcrum?

22. If a 60-kg object is 3 m from the fulcrum and balances a 45-kg object, how far is the 45-kg object from the fulcrum?

23. Two camels traveled the same distance. Do their rates of speed vary indirectly as the number of hours traveled? Explain.

24. Two factory workers had the same gross salary per week. Do their rates of pay vary inversely as the number of hours they work? Give an example to explain your answer.

For Exercises 25 and 26, the rate r varies inversely as time t.

25. A distance was traveled in 5 h at 42 mi/h. The return trip was done in 6 h. What was the average speed on the return trip?

26. A trip was made in a light plane traveling at 160 mi/h and took 3 h 30 min. The average speed on the return trip was 200 mi/h. How long did the return trip take?

Applications

27. **Geometry** If the area of a triangle stays constant, does the base of the triangle vary indirectly as the height?

28. **Geometry** Does the circumference of a circle vary inversely as its radius?

EXTRA

The cost of owning a car depends upon the number of miles it is driven. D.D. kept records and found that it cost $362.00 to drive 400 mi a month and $397.00 to drive 650 mi a month. Use this information to answer these questions.

1. Assume that a linear equation is a reasonable mathematical model of the cost-distance function. Write the equation.

2. Predict the cost of driving D.D.'s car 2500 mi per month.

3. What meaning does slope have in the case of D.D.'s car?

4. According to your equation, how much will it cost to drive 0 mi per month? Is that amount reasonable?

10.6 Problem Solving Strategy: Use an Appropriate Formula

A formula is an equation that expresses a relationship among different variables. Mathematical formulas are used to solve problems in science, engineering, economics, business, and the social sciences. Formulas are also used in everyday life to solve consumer problems.

People working in different occupations have reference books containing formulas needed to solve specific problems. For example, an accountant uses a formula to calculate how much business equipment depreciates each year. Depreciation is the amount of decrease in the value of the equipment over a period of time.

EXAMPLE 1 James is an accountant. He uses the formula given below to find the linear depreciation for business equipment owned by a client. Tax rules allow the property to be depreciated over a period of 10 years. If the equipment had an original cost of $10,000, find its value at the end of 3 years.

$$V = C\left(1 - \frac{n}{N}\right)$$

where V = value at the end of n years
C = original cost
N = total number of years the equipment can be depreciated

Understand the Problem

What are the given facts?

The equipment can be depreciated for 10 years.
The original cost was $10,000.

What are you asked to find?

You are asked to find the value of the equipment at the end of 3 years.

Plan Your Approach

Choose a strategy.

The problem involves linear depreciation. The strategy requires applying the formula for calculating linear depreciation.

Substitute the given numbers into the formula to solve the problem.

$$V = 10,000\left(1 - \frac{3}{10}\right)$$

Chapter 10 Relations, Functions, and Variation

■ **Complete the Work** Solve the equation.
$$V = 10{,}000\left(1 - \tfrac{3}{10}\right)$$
$$= 10{,}000\left(\tfrac{7}{10}\right) = 7000$$

■ **Interpret the Results** State your answer.
The value of the equipment at the end of 3 years is $7000.

CLASS EXERCISES

Write a formula that can be used to solve a problem in each case.

1. area of a rectangle
2. area of a triangle
3. equation of a line
4. If Ralph bought 18 pens for $0.86 each, how much did he spend in all?
5. Use the formula $I = prt$. Determine the interest if $p = \$5000$, $r = 0.0825$, and $t = 4$.
6. Use the linear depreciation formula given in the example to find the value V if $C = \$5000$, $N = 10$ yr, and $n = 3$.

PRACTICE EXERCISES

Select a formula that can be used to solve a problem in each case.

1. area of a square
2. circumference of a circle
3. linear depreciation
4. simple interest
5. volume of a sphere
6. volume of a cube

$V = \tfrac{4}{3}\pi r^3$
$V = s^3$
$C = 2\pi r$
$I = prt$
$A = s^2$
$V = C\left(1 - \tfrac{n}{N}\right)$

Solve each problem.

7. Ernie bought 36 ballpoint pens for $1.12 each. How much did he spend in all?
8. A weather report gave the temperature as 20°C. Use the formula $F = \tfrac{9}{5}C + 32$ to find the equivalent temperature in degrees Fahrenheit.
9. Annie spent $169 for 130 tennis balls. How much did she spend for each tennis ball?
10. Robert and Mary bought 5 gallons of milk for $9.65. How much did they spend for each gallon of milk?

Using the formula $s = \frac{1}{2}gt^2$, solve for g.

11. $s = 9$, $t = 1$ **12.** $s = 36$, $t = 2$ **13.** $s = 64$, $t = 4$ **14.** $s = 81$, $t = 3$

Use the linear depreciation formula, $V = C\left(1 - \frac{n}{N}\right)$, to find the value V for each set of conditions.

	C	N	n		C	N	n
15.	$5000	5 yr	1	**16.**	$20,000	4 yr	4
17.	$3000	3 yr	3	**18.**	$15,000	10 yr	8
19.	$50,000	20 yr	15	**20.**	$60,000	25 yr	17

Use the formula, $V = \frac{1}{3}\pi r^2 h$ to solve for the following unknowns. Assume $\pi \approx 3.14$. This formula can be used to find the volume of a cone.

21. Find h when $V = 64$ and $r = 3$. **22.** Find r when $h = 6$ and $V = 64$.

23. Find V when $r = \frac{1}{4}$ and $h = \frac{1}{3}$. **24.** Find r when $h = 0.03$ and $V = 0.07$.

25. The ratio of the speed of an airplane to the speed of sound is called a Mach number. Find the Mach number (to one decimal place) for an airplane designed to fly at a maximum speed of 1800 mi/h. Use 740 mi/h for the speed of sound.

26. If A represents the speed of an aircraft and s represents the speed of sound, write a formula for finding the Mach number M of the aircraft.

Mixed Problem Solving Review

1. William commutes 648 mi a week to work by car. If the car averages 18 mi/gal, how much does he spend on gasoline over a 7-week period if gasoline sells for $0.96 per gallon?

2. It takes Carlos 6 h to paint a room and 7 h for Jerome to paint the same room. How long would it take them to do the job together?

3. The length of a rectangle is 3 in. more than 6 times the width. If the perimeter is 48 in., find the length and the width of the rectangle.

PROJECT

Find out the interest rates offered by 5 banks in your neighborhood. Determine the variables involved in obtaining a $5000 loan.

TEST YOURSELF

For each equation, tell whether y varies directly as x. Explain. 10.4

1. $3x + y = 0$
2. $x = y + 2$
3. $\frac{1}{2}x + y = 0$
4. $y = \frac{4}{7}x + 3$

Solve.

5. y varies directly with x. If $y = 3$ when $x = 2$, find y when $x = 4$.
6. y varies directly with x. If $y = 5$ when $x = 4$, find y when $x = 8$.

For each table tell whether y varies inversely as x. If so, name the constant of variation and the equation that shows the relationship. 10.5

7.
x	y
4	3.5
5.6	2.5
20	0.7
2	7

8.
x	y
6	2.5
1.2	12.5
30	0.5
3	5

9.
x	y
−2.3	0.3
0.5	4.3
2	−3.2
4	2.7

10.
x	y
3.2	0.5
1.6	2
0.3	5
4	2

Solve. 10.5–10.6

11. y varies indirectly as x. If $y = 5$ when $x = 4$, find x when $y = 2$.
12. y varies indirectly as x. If $y = 8$ when $x = 10$, find x when $y = 5$.
13. Robert bought 5 albums for $8.98 each. How much did he spend in all?
14. Dan spent $6.45 on five notebooks for school. How much did he spend for each notebook?

Given the linear depreciation formula, $V = C\left(1 - \frac{n}{N}\right)$, find the value of V, for each set of conditions.

	C	N	n
15.	4,000	4 yr	1
16.	12,000	10 yr	8

CHAPTER 10 SUMMARY AND REVIEW

Vocabulary

composite function (423)
constant function (423)
constant of proportionality (426)
direct variation (426)
domain (413)
function (417)
function notation (420)
inverse variation (430)
linear function (422)
range (413)
relation (413)
vertical line test (418)

A **relation** is any set of ordered pairs.
The **domain** of a relation is the set of all the first elements or *x*-coordinates.
The **range** of a relation is the set of all the second elements or *y*-coordinates.
A **function** is a relation in which each element of the domain is paired with exactly one element of the range. Function notation includes, as examples:
$y = 4x$, $f(x) = 4x$, and $H: x \rightarrow 4x$.

State the relation as a set of ordered pairs. Also state the domain and the range and tell whether the relation is a function. 10.1

1.
x	2	5	2	−2
y	−2	−5	0	1

2. (mapping diagram: 3 → −1, 3 → 1, −2 → 0, −2 → −1)

Vertical Line Test A relation is not a function if any vertical line can pass through the graph in more than one point.

Linear Function A set of ordered pairs whose graph is a nonvertical straight line can be defined by an equation in the slope-intercept form.

$$y = mx + b \qquad m = \text{slope};\ b = y\text{-intercept}$$

Graph each relation and tell whether or not it is a function. If so, is it a linear function? 10.2–10.3

3. $2x + y = 4$; domain $\{-1, 0, 4\}$
4. $y = |x|$; domain {all real numbers}
5. $f(x) = 2x + 2$; domain {all real numbers}
6. $x = 2$

7. Graph the relation between month m and average temperature t as described in this chart. Is this relation a function? Is it a linear function?

Average Temperature for Eureka, California

Month (m)	Jan.	Feb.	Mar.	Apr.	May	Jun.	Jul.
Temperature (t)	47°F	49°F	48°F	49°F	52°F	55°F	56°F

8. Evaluate $f(-2) + g(0)$, given $f(x) = x^2$ and $g(x) = 2x - 1$.

Direct Variation A *direct variation* is a function in the form
$$y = kx, \; k \neq 0$$
where y varies directly as x, or y is directly proportional to x. The constant of variation is k.

Inverse Variation An *inverse variation* is a function in the form
$$xy = k, \; k \neq 0$$
where y varies inversely (or indirectly) as x, or y is inversely proportional to x. The constant of variation is k.

State whether y varies directly or inversely as x. Also determine the equation which shows the relationship and name the constant of variation.   10.4–10.6

9. $\dfrac{y}{4} = x$

10.
x	36	6	8	90
y	0.5	3	2.25	0.2

11. If y varies directly with x, and $y = 5$ when $x = 25$, find y when $x = 55$.

12. If r is inversely proportional to t, and $r = 46$ when $t = 1.5$, find r when $t = 2$.

13. The amount of interest owed each month on a fixed mortgage varies directly as the current balance. If the balance for one month is $65,000 and the interest owed is $650, what is the interest owed five years later when the balance is $48,000?

14. Charles' Law says that the volume V of a gas varies directly as its temperature t at a constant pressure. If 20 L of neon is kept at a constant pressure while its temperature changes from $360°K$ to $513°K$, determine its new volume.

15. The length of a rectangle of constant area is inversely proportional to the width. If the length of the rectangle is 6 m and the width is 4 m, what would the length be if the width is 3 m?

CHAPTER 10 TEST

State each relation as a set of ordered pairs. Name the domain and range of each relation, and tell whether the relation is a function.

1.
x	2	5	0	−2
y	−5	−14	1	7

2.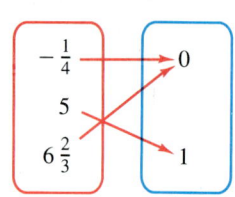

Graph each relation and tell whether or not it is a function and, if so, if it is a linear function.

3. $y - 3x = 2$;
Domain: $\{\ldots -1, 0, 1 \ldots\}$

4. $f(x) = 3x - 1$
Domain: {all real numbers}

5. $y = -1$
Domain: {all real numbers}

6. Graph the closing prices of a stock over a period of one week. Is this relation a function?

Day	Mon.	Tue.	Wed.	Thu.	Fri.
Closing price	22	$23\frac{1}{8}$	$23\frac{1}{4}$	23	$23\frac{1}{4}$

7. Evaluate $f(0) + g(-2)$, given $f(x) = x^2 + 1$ and $g(x) = 3x$.

Tell whether y varies directly or inversely as x. Name the equation which shows the relationship and give the constant of variation.

8. $\frac{y}{x} = -2$

9.
x	18	3	4	45
y	0.5	3	2.25	0.2

10. If y varies inversely as x, and $y = 6$ when $x = 9$, find y when $x = 2.7$.

11. If the cost of kumquats c varies directly as their weight w, and $c = 3.78$ when $w = 3$, find c when $w = 4.5$.

12. The amount of flour used in baking bread is directly proportional to the number of loaves of bread. If 6 cups of flour makes 2 loaves of bread, how many cups of flour would be used to make 5 loaves of bread?

Challenge

1. Graph $y = |x|$; domain $\{-3 \leq x \leq 3\}$

2. The distance a spring stretches is directly proportional to the mass hooked onto the bottom of it. If a spring stretches 1.1 cm when holding a 3-kg weight, how far will it stretch when a 7-kg weight is hooked onto its end?

PREPARING FOR STANDARDIZED TESTS

Select the best choice for each question.

1. The equation of the line through $(4, -1)$ and $(0, 7)$ is:
   A. $y = 7x + 2$
   B. $y = -7x + 2$
   C. $y = 2x + 7$
   D. $y = -2x + 7$
   E. $y = \frac{-3}{2}x + 7$

2. When $\frac{2}{3}$ of 690 is subtracted from $\frac{3}{5}$ of 865, the result is:
   A. 59   B. 69   C. 89
   D. 159   E. 169

3. A customer used a $100 bill to pay for purchases of $29.95, $14.50, $8.90, $2.75, and 79¢. What did the customer receive in change?
   A. $42.11
   B. $42.91
   C. $43.01
   D. $43.11
   E. $43.91

4. If x varies directly as y, and if $x = 12$ when $y = 33$, then what is the value of x when y is 198?
   A. 544.5
   B. 177
   C. 90
   D. 76
   E. 72

5. The equation for the graph of the vertical line through the point $(-1, 3)$ is:
   A. $x = -1$
   B. $y = -1$
   C. $x = 3$
   D. $y = 3$
   E. $y = x + 4$

6. Find the value of $f(-2)$ when $f(x) = 2x^2 + 4x + 7$.
   A. -9 B. -7 C. 7
   D. 9   E. 23

7. Martha took 1 h and 15 min to ride her bike the 15 mi to her cousin's house. She later returned by the same route but took 1 h and 45 min for the return trip. What was her average rate for the round trip?
   A. 8.5 mi/h
   B. 9 mi/h
   C. 9.5 mi/h
   D. 10 mi/h
   E. 10.5 mi/h

8. If the function $y = 5x - 2$ has a domain restricted to $x \geq 1$, then it has a range of:
   A. $y \leq 3$
   B. $y \geq 3$
   C. $y \leq 1$
   D. $y \geq 1$
   E. $y = 3$

9. Of two cereals, type A contains 160 mg of sodium per oz while type B contains 110 mg per oz. Which of the following would be true about the number of mg of sodium in a $1\frac{1}{2}$ oz serving of type A as compared to a 2 oz serving of type B?
   A. they have the same number
   B. serving A has 50 mg more
   C. serving B has 50 mg more
   D. serving A has 20 mg more
   E. serving B has 20 mg more

CUMULATIVE REVIEW   CHAPTERS 1–10

State the value of the variable for which the expression is undefined.

1. $\dfrac{15}{x+3}$
2. $\dfrac{4}{5y+3}$
3. $\dfrac{m+2}{m^2+8m+12}$
4. $\dfrac{s-4}{s^2-16}$

Simplify.

5. $\dfrac{15x^2y^4}{5xy^2}$
6. $\dfrac{24r^3s^5}{3r^2s^2}$
7. $\dfrac{14m^2+2mn}{14m^2n}$
8. $\dfrac{16x^3+4xy}{8x^3y}$

9. 4 to 8
10. 0.3 : 2.1
11. 3 ft to 5 yd
12. $4.40 to $0.22

13. $5 + 7x - \dfrac{3}{x}$
14. $\dfrac{\dfrac{3n}{n^2} - \dfrac{1}{n}}{1 + \dfrac{3m}{n^2}}$
15. $\dfrac{\dfrac{4x}{x^2} - \dfrac{1}{x}}{1 + \dfrac{4y}{x^2}}$

Factor completely.

16. $4x^4 - 16x^2$
17. $4x^2 - 4x - 15$
18. $16x^3y^2 - 8x^2y^2 + 24x^2y^3$

Write an equation of a line in standard form given the following slope and/or points on the line.

19. $m = 3$, $(0, -3)$
20. $m = -\dfrac{1}{2}$, $(6, 0)$
21. $(6, -9)$, $(4, -10)$
22. $(11, -13)$, $(-9, 7)$

Graph each equation or inequality.

23. $y = 2x + 3$
24. $y \geq -2x + 5$
25. $y \leq 4x - 4$

Find the range of each function.

26. $f(t) = \dfrac{1}{1+3t}$ $D = \{-1, 0, 1\}$
27. $g: x \to x^2 - 16$ $D = \left\{-4, \dfrac{1}{2}, 3\right\}$

Solve.

28. $|2m - 3| = 17$
29. $|8 - 3t| < 6$
30. $|4a - 7| \leq 1$

31. Two trains are traveling east from the same city. Train 2 traveling at 120 mi/h leaves 1 hr after train 1. Train 1 travels at a speed of 110 mi/h. How long will it take train 2 to overtake train 1?

32. Find four consecutive even integers whose sum is 52.

33. The perimeter of a rectangle is 110 cm. If the length of the rectangle is 5 cm more than the width, what are the dimensions of the rectangle?

11 Systems of Linear Equations

Large corporations as well as small companies need managers with good skills in business mathematics. Projecting the maximum profit and minimum cost involves an understanding of linear systems of equations.

11.1 Graphing Systems of Linear Equations

Objective: To use graphs to solve systems of linear equations

The graph of a linear equation in two variables is a line in the coordinate plane. The coordinates of any point on the line are said to belong to the solution set of the equation.

Capsule Review

The ability to locate and plot key points is an essential skill in graphing equations.

EXAMPLE Graph $y - 2x = 2$

x	y
0	2
-1	0
1	4

One way to graph $y - 2x = 2$ is to use the x- and y-intercepts. That is, plot the points with coordinates $(0, 2)$ and $(-1, 0)$. As a check, plot a third ordered pair that satisfies the equation.

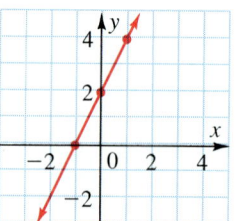

Graph the linear equations.

1. $y = -x + 4$ 2. $y = -2x + 1$ 3. $x - y = -3$ 4. $2y + 5x = 0$

Two (or more) linear equations using the same variables form a **system of linear equations**. Such equations are commonly known as **simultaneous equations**. Graphing is one method of solving such a system. Solving a system of equations means to determine the point of intersection of the lines.

EXAMPLE 1 Solve by graphing: $\begin{cases} x - y = 2 \\ x + y = 8 \end{cases}$

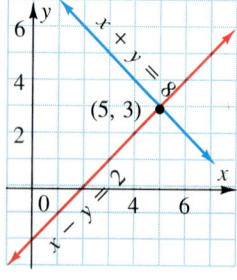

From the graph the point of intersection of the system of linear equations has the coordinates $(5, 3)$. This point is the only solution of the system.

$$x - y = 2 \qquad x + y = 8$$
$$5 - 3 \stackrel{?}{=} 2 \qquad 5 + 3 \stackrel{?}{=} 8$$
$$2 = 2 \checkmark \qquad 8 = 8 \checkmark$$

Systems of equations with at least one solution are called **consistent**. If the graph of each equation is different, the system is known as an **independent system**. So, the system in the above example is considered to be both an *independent* and *consistent system*.

444   Chapter 11 Systems of Linear Equations

Some systems of linear equations have more than one solution.

EXAMPLE 2 Solve by graphing: $\begin{cases} x - y = 3 \\ 4x - 4y = 12 \end{cases}$

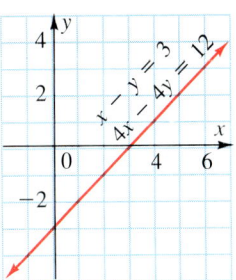

The graphs are the same line, that is, they coincide. The coordinates of any point on the line satisfy both equations, so this system *has infinitely many solutions*. This type of system is known as a **dependent** and **consistent system**. By dividing each side of the equation, $4x - 4y = 12$ by 4, you can show that the equations are equivalent.

You can state the solution as the set of all points on the line $x - y = 3$, or in symbols write:

$$\{(x, y): x - y = 3\}$$

Some systems have no solutions, as shown in Example 3.

EXAMPLE 3 Solve by graphing: $\begin{cases} x + 2y = 6 \\ x = 10 - 2y \end{cases}$

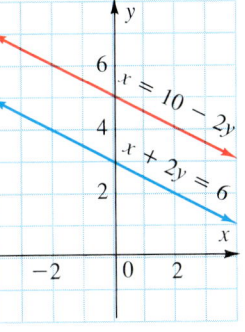

The lines do not intersect. They are parallel lines. Since the lines have no points in common, the system of equations has no solution and therefore can be classified as an **independent** and **inconsistent system**.

What can you tell about a system of equations without graphing? Consider the equations in Examples 1, 2, and 3 written in slope-intercept form, $y = mx + b$.

Example 1
$\begin{cases} y = x - 2 \\ y = -x + 8 \end{cases}$

Example 2
$\begin{cases} y = x - 3 \\ y = x - 3 \end{cases}$

Example 3
$\begin{cases} y = -\frac{1}{2}x + 3 \\ y = -\frac{1}{2}x + 5 \end{cases}$

The table below describes the nature of the graphs of these equations.

Slope *m*	*y*-intercept *b*	Graph of the system	Number of solutions
Different	Different or Same	Intersecting lines	One
Same	Same	Same line	Infinitely many
Same	Different	Parallel lines	None

Using this table, what generalizations can be made about lines that intersect? that coincide? that are parallel? In what ways does this relate to the solution sets of their equations?

EXAMPLE 4 Without graphing, describe the graph and determine the number of solutions: $\begin{cases} 4x - y = 2 \\ x - \frac{1}{2}y = 1 \end{cases}$

Write each equation in $y = mx + b$ form. Compare the slopes and the y-intercepts.

$4x - y = 2$ $y = 4x - 2$ $m_1 = 4; b_1 = -2$

$x - \frac{1}{2}y = 1$ $y = 2x - 2$ $m_2 = 2; b_2 = -2$

The slopes are different. The y-intercepts are the same. The graphs are intersecting lines with one solution, $(0, -2)$. You can use a computer or graphing calculator to check your conclusion. Graph $4x - y = 2$ and then graph $x - \frac{1}{2}y = 1$ on the same grid. The lines intersect and there is one solution.

CLASS EXERCISES

Using the graphs given, find the solution of each system. Then check the solution in each equation of the system.

1. $\begin{cases} y = x + 2 \\ y = 3x - 2 \end{cases}$
2. $\begin{cases} y = 3x - 2 \\ x - y = 4 \end{cases}$

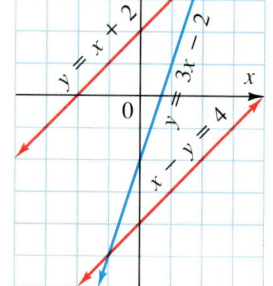

Solve by graphing.

3. $\begin{cases} 2x + y = 4 \\ 2y = 8 - 4x \end{cases}$
4. $\begin{cases} x + y = 9 \\ 3x + 3y = 18 \end{cases}$

Without graphing, describe the graphs of the system, and determine the number of solutions.

5. $\begin{cases} y = 3x + 6 \\ y = 3x \end{cases}$
6. $\begin{cases} 2y = -x + 2 \\ -2y = x - 2 \end{cases}$
7. $\begin{cases} x - y = 4 \\ 2x + y = 1 \end{cases}$

8. $\begin{cases} y = \frac{1}{2}x + 4 \\ 2y = x - 6 \end{cases}$
9. $\begin{cases} 4y = -3x + 5 \\ 2y = -x - 4 \end{cases}$
10. $\begin{cases} 5y = 15x + 10 \\ y = 3x + 2 \end{cases}$

For Discussion

11. What kinds of systems would be difficult to solve by graphing?

PRACTICE EXERCISES

Solve by graphing.

1. $\begin{cases} x - y = 6 \\ 2x + y = 3 \end{cases}$
2. $\begin{cases} x + 2y = 8 \\ x - y = -4 \end{cases}$
3. $\begin{cases} x + 2y = -4 \\ 2x + 4y = -8 \end{cases}$

4. $\begin{cases} x - y = 5 \\ 3x - 3y = 15 \end{cases}$

5. $\begin{cases} 2x - y = 6 \\ 2x - y = 1 \end{cases}$

6. $\begin{cases} y = 3x - 1 \\ y = 3x + 2 \end{cases}$

7. $\begin{cases} y - 2x = 4 \\ y + \frac{1}{3}x = -3 \end{cases}$

8. $\begin{cases} y - 2x = -6 \\ y - \frac{2}{3}x = 2 \end{cases}$

9. $\begin{cases} y = -\frac{1}{2}x + 2 \\ y = -\frac{1}{2}x - 4 \end{cases}$

10. $\begin{cases} y = \frac{3}{4}x - 3 \\ y = \frac{3}{4}x + 1 \end{cases}$

11. $\begin{cases} y = -2x \\ x = -2y + 6 \end{cases}$

12. $\begin{cases} y = 3x - 1 \\ x = \frac{1}{4}y \end{cases}$

For each graph state whether the system shown is independent and consistent, independent and inconsistent, or dependent and consistent.

13.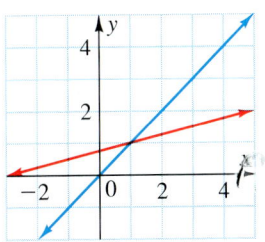

14.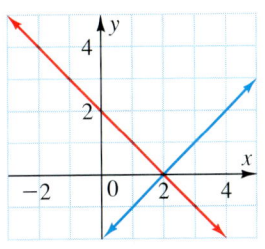

15.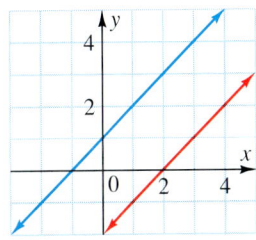

16.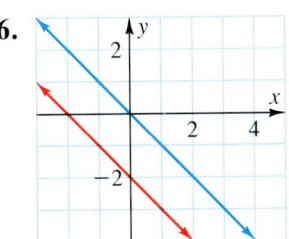

17.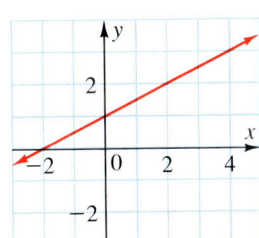

18.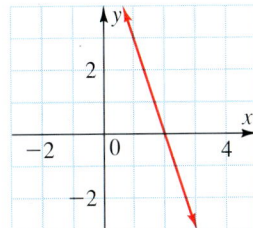

19.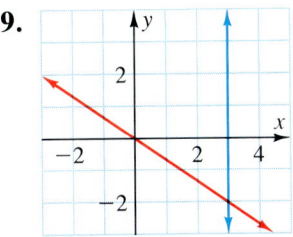

20.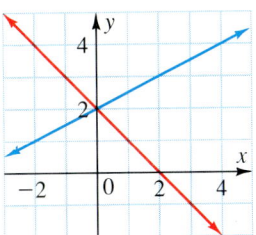

21.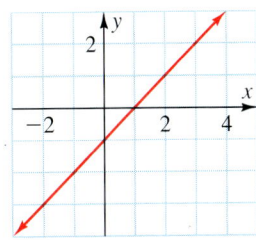

22.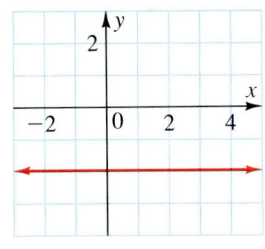

23.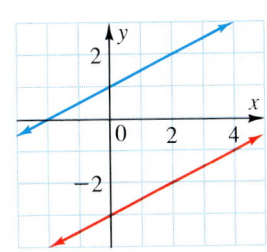

24.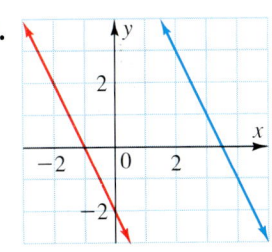

Without graphing, describe the graph, and tell the number of solutions.

25. $\begin{cases} 2x - y = 3 \\ x - \frac{1}{3}y = 1 \end{cases}$

26. $\begin{cases} 5x - y = 6 \\ 2x - \frac{1}{2}y = 3 \end{cases}$

27. $\begin{cases} 3x + y = -2 \\ 2 - y = 3x \end{cases}$

28. $\begin{cases} 3x - y = 15 \\ x - \frac{1}{3}y = 5 \end{cases}$

29. $\begin{cases} 4y + x = 8 \\ y + \frac{1}{4}x = 3 \end{cases}$

30. $\begin{cases} 2x + y = 3 \\ 2x + y = 1 \end{cases}$

Solve by graphing.

31. $\begin{cases} 5x - 4y = 20 \\ y = \frac{3}{2}x - 5 \end{cases}$

32. $\begin{cases} 7x = 12 - 2y \\ \frac{5}{3}x + y = 6 \end{cases}$

33. $\begin{cases} x = 2y \\ x - 2y = 4 \end{cases}$

34. $\begin{cases} 2x = 4 - y \\ 4x = -2y + 6 \end{cases}$

35. $\begin{cases} 3x - 2y = -4 \\ 2x + 6y = 1 \end{cases}$

36. $\begin{cases} 4y = 3x + 8 \\ 2y = 3x + 1 \end{cases}$

Without graphing, describe the graph, and tell the number of solutions.

37. $\begin{cases} \frac{1}{2}x + y = -1 \\ 1 - y = \frac{1}{2}x \end{cases}$

38. $\begin{cases} \frac{2}{3}x - y = \frac{1}{3} \\ 2x - 3y = 1 \end{cases}$

39. $\begin{cases} 4y = 3x - 8 \\ y - 2 = \frac{3}{4}x \end{cases}$

40. $\begin{cases} y + 2x = -3 \\ \frac{1}{3}y = \frac{2}{3}x + 1 \end{cases}$

41. $\begin{cases} 4x - 3y = -1 \\ 3x - 4y = 1 \end{cases}$

42. $\begin{cases} x = \frac{3}{4}y + \frac{1}{2} \\ y = \frac{4}{3}x - 2 \end{cases}$

Solve by graphing.

43. $\begin{cases} y + x = 0 \\ x - 2y = 6 \\ y = \frac{3}{2}x - 5 \end{cases}$

44. $\begin{cases} y = |x| \\ x + y - 2 = 0 \end{cases}$

45. $\begin{cases} y = |x + 2| \\ y = 2x + 3 \end{cases}$

$y = m_1 x + b_1$ and $y = m_2 x + b_2$ form a linear system. Write the general relationship between the slopes and between the y-intercepts for the systems below.

46. Dependent   47. Inconsistent   48. Independent   49. Consistent

Applications

Solve.

50. **Calculator** In Exercises 37–42, use a graphing calculator or a computer graphing program to check if you have predicted the correct number of solutions.

51. **Business** At the ABC Printing Company, it takes 18 min to produce 8 pages of letters and 5 pages of charts. It takes 15 min to print 5 pages of each. Find the time required to print 1 page of charts by graphing the following system, $8x + 5y = 18$ and $5x + 5y = 15$, and finding the value of y that satisfies each equation.

52. **Business** A contracting company rents a generator and a heavy-duty saw for 6 h at a total cost of $60. For another job the company rents the generator for 4 h and the saw for 8 h for a total cost of $56. Find the hourly rates g (for the generator) and s (for the saw) by graphing the system of equations $6g + 6s = 60$ and $4g + 8s = 56$.

53. **Economics** Joanne plans to invest $5000 in two accounts, a certificate of deposit (CD) and a mutual fund. The CD pays 7% interest and the mutual fund pays 8.5%. Joanne wishes to receive $380 a year in interest income. Find how much she must invest in a CD, x, and a mutual fund, y, by graphing the system of equations: $x + y = 5000$ and $0.07x + 0.085y = 380$.

54. **Geometry** A trapezoid is formed by lines with the equations $x + y = 5$ and $x = 3$ and the x- and y-axis. Find the area of that figure.

CAREER: Air Traffic Controller

Air traffic controllers work in the control tower, the nerve center of the airport. They use sophisticated technology to keep track of all the incoming and outgoing aircraft. Air controllers must interpret mathematical data as it relates to aircraft positions and speeds.

1. An airplane leaves Atlanta and flies about 800 mi to Dallas in 3 h 15 min. At the same time, an airplane leaves Dallas and flies to Atlanta in 2 h. How long after takeoff will the planes pass each other? How far from Atlanta will they be?

2. One plane flies 1200 mi from Boston to Miami in 3 h. At the same time, another plane flies from Miami to Boston in 4 h 30 min. Use a graph to find about how long after takeoff these planes will pass each other and about how far they will be from Boston at that time.

3. An airplane flew roundtrip from Denver to Chicago, a total distance of 1840 mi. On the return flight strong headwinds decreased the plane's speed to 75% of its speed during the first flight. The total time required for both flights was 4 h. Find the speed of the plane for both trips.

APPLICATION: Break-Even Point

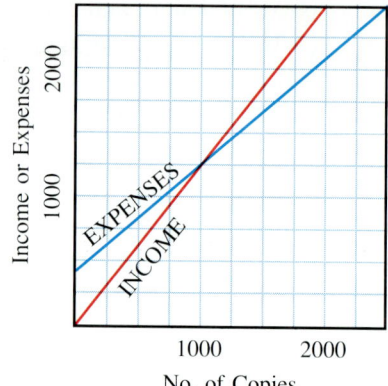

Did you know that graphing systems of linear equations can be used to determine profit and loss? The Franklins are thinking of publishing a newsletter for coin collectors. They determine that the expenses for research and writing will be $400. They then add the expenses for printing and mailing. The formula they devise for the total expenses y of printing and mailing x copies is $y = 400 + 0.85x$.

They plan to sell the newsletter for $1.25 per copy. Thus, their formula for the total income is $y = 1.25x$.

The Franklins graph the two equations. The point at which their income begins to exceed their expenses is the break-even point. What is the x-coordinate of the break-even point, and what does it represent? What is the y-coordinate of the break-even point, and what does it represent?

EXERCISES

Use the graph above for Exercises 1 and 2.

1. What is the loss if only 500 copies are sold during the first month?
2. What is the profit if 1500 copies are sold during the second month?

After the first year, the Franklins find that their expenses have increased. A better formula for their expenses now is $y = 600 + 0.9x$. The Franklins increase the price of the newsletter to $1.50, which gives the formula $y = 1.5x$ for their income.

3. Make a graph to show the new break-even point.
4. How many copies must be sold before there is a profit?
5. If one month after their price increase the Franklins sell only 300 copies of their newsletter, what will be their loss for the month?
6. In a few months, if the Franklins are selling an average of 2000 copies a month, what will be their average profit per month?

11.2 The Substitution Method

Objective: To solve a system of linear equations in two variables by the substitution method

If the solution of a system of equations involves fractions, it may be difficult to accurately read these values from a graph. Algebraic methods are used to provide more accurate solutions to linear systems than the graphing method.

Capsule Review

Isolating the variable is a key part of solving equations using certain algebraic methods.

EXAMPLE Solve for x: $y - x = 9$
$-x = 9 - y$
$x = y - 9$

Solve the equation for the indicated variable.

1. $x - y = 3$; y
2. $x - 2y = 4$; y
3. $2x - 3y = 5$; x
4. $3y - x = 9$; x
5. $\dfrac{x}{2} = \dfrac{y}{4}$; y
6. $\dfrac{1}{2}x = 4y$; x

The **substitution method** is one technique that can be used to solve a system of linear equations. The key to this method of solving a system of equations is the elimination of one of the variables. In this method the variable selected is eliminated by substitution.

The Substitution Method

- In either equation, solve for one variable in terms of the other.
- Substitute for that variable in the other equation. Solve.
- Substitute the result from Step 2 in either equation. Solve for the other variable.
- Check the solution in both original equations.

EXAMPLE 1 Solve: $\begin{cases} x - y = 3 \\ 2x - 3y = 5.5 \end{cases}$

In the first equation, if $x - y = 3$, then $x = y + 3$. Why?

$$\begin{aligned} 2x - 3y &= 5.5 & &\text{\textit{Restate the second equation of the system.}} \\ 2(y + 3) - 3y &= 5.5 & &\text{\textit{Substitute } y + 3 \text{ for } x.} \\ 2y + 6 - 3y &= 5.5 & &\text{\textit{Solve for y.}} \\ -y &= -0.5 & &\text{\textit{Multiply each side by } -1.} \\ y &= 0.5 \end{aligned}$$

Replace y with 0.5 in either equation.

$$\begin{aligned} x - y &= 3 & &\text{\textit{Solve for x.}} \\ x - 0.5 &= 3 \\ x &= 3.5 \end{aligned}$$

Check the solution (3.5, 0.5) in both equations. Why is it easier to use the first equation in solving for x?

Look at the linear equations in Example 2. How would you begin? Why?

EXAMPLE 2 Solve: $\begin{cases} 2a - b = -4 \\ 4a = 7 + 2b \end{cases}$

In the first equation, if $2a - b = -4$, then $b = 2a + 4$.

$$\begin{aligned} 4a &= 7 + 2b & &\text{\textit{Restate the second equation of the system.}} \\ 4a &= 7 + 2(2a + 4) & &\text{\textit{Substitute } 2a + 4 \text{ for } b.} \\ 4a &= 7 + 4a + 8 & &\text{\textit{Solve for a.}} \\ 4a &= 4a + 15 \\ 0 &\neq 15 & &\text{There is no solution.} \end{aligned}$$

Some systems lend themselves more to the substitution method than others. This method is best used when the coefficient of one of the variables is 1 or -1.

CLASS EXERCISES

Solve by substitution. Check the results in the original equations.

1. $\begin{cases} y = 2x \\ 7x - y = 35 \end{cases}$
2. $\begin{cases} a + 3b = 5 \\ 2a - 4b = -5 \end{cases}$
3. $\begin{cases} 3m + 6n = -15 \\ 2m - 3n = 4 \end{cases}$
4. $\begin{cases} 5g - 2f = -8 \\ 3f + 5g = -3 \end{cases}$

For Discussion

5. What happens when you try to solve these systems? Give the solution sets of each system.

   a. $\begin{cases} 3x - y = -2 \\ y = 3x + 2 \end{cases}$
   b. $\begin{cases} 3u - v = 0 \\ 6u - 2v = 5 \end{cases}$

PRACTICE EXERCISES

Solve by the substitution method. Check your results.

1. $\begin{cases} x - y = 0 \\ x + y = 2 \end{cases}$
2. $\begin{cases} x + 3y = -4 \\ y + x = 0 \end{cases}$
3. $\begin{cases} 3u - v = 17 \\ v + 2u = 8 \end{cases}$
4. $\begin{cases} 5 = p + 5q \\ p + q = -3 \end{cases}$
5. $\begin{cases} 3x + 2y = 9 \\ x + y = 3 \end{cases}$
6. $\begin{cases} x + 3y = -4 \\ 2y + 3x = 3 \end{cases}$
7. $\begin{cases} 2y - 3x = 4 \\ x = -2 \end{cases}$
8. $\begin{cases} y = 3 \\ 2x = 5y + 7 \end{cases}$
9. $\begin{cases} 2d = 5e \\ e - 3d = -13 \end{cases}$
10. $\begin{cases} 5g = 3f + 6 \\ g - 2f = 46 \end{cases}$
11. $\begin{cases} k = 10 - 2j \\ k = 4j + 36 \end{cases}$
12. $\begin{cases} n = 39 - 3m \\ n = 2m - 61 \end{cases}$
13. $\begin{cases} 25r - 10q = 100 \\ 5q + 15r = 60 \end{cases}$
14. $\begin{cases} 10m - 5n = -20 \\ 10n - 40m = 80 \end{cases}$
15. $\begin{cases} 3c + 4d = 6 \\ d - 6c = 6 \end{cases}$
16. $\begin{cases} 12e + 5f = 5 \\ f + 18e = 14 \end{cases}$
17. $\begin{cases} 4y - x = 2 \\ -2x + 12y = 17 \end{cases}$
18. $\begin{cases} 18x = y - 10 \\ 18x = 30y + 19 \end{cases}$
19. $\begin{cases} 3y + x = -1 \\ x = -3y \end{cases}$
20. $\begin{cases} y - x = 4 \\ x + 3 = y \end{cases}$
21. $\begin{cases} 5b - 2 = 2a \\ 3a + 6 = 25b \end{cases}$
22. $\begin{cases} 4d - c = -3 \\ 2c - 6 = 8d \end{cases}$
23. $\begin{cases} 4s - 3t = 8 \\ 2s + t = -1 \end{cases}$
24. $\begin{cases} j + 6k = 0 \\ 4j - 3k = 9 \end{cases}$
25. $\begin{cases} 5e - 7f = 1 \\ 4e - 2f = 16 \end{cases}$
26. $\begin{cases} 13h - 10g = 45 \\ 6g - 3h = -3 \end{cases}$
27. $\begin{cases} 20x - 15y = 17 \\ x = y + 1 \end{cases}$
28. $\begin{cases} 8x - 1 = 4y \\ 3x = y + 1 \end{cases}$
29. $\begin{cases} 2x - 3y = 19 \\ 5y - 2x = -37 \end{cases}$
30. $\begin{cases} 3y + 5x = -3 \\ 2x - 3y = -30 \end{cases}$
31. $\begin{cases} 15n = -270m \\ 15m + 2n = -7 \end{cases}$
32. $\begin{cases} -5r + 14t = 13 \\ -9r = -72t \end{cases}$
33. $\begin{cases} 2x - y = 1 \\ 2x - 5y = -1 \end{cases}$
34. $\begin{cases} 12y - 3x = 11 \\ x - 2y = -2 \end{cases}$
35. $\begin{cases} 4v + 3u = -6 \\ 5v - 6u = -27 \end{cases}$
36. $\begin{cases} 2s - 5t = 6 \\ 4s + 3t = -1 \end{cases}$
37. $\begin{cases} 5x - 7y = -21 \\ 14y - 5x = 22 \end{cases}$
38. $\begin{cases} 5x - 2y = 4 \\ 8y + 15x = 2\frac{2}{3} \end{cases}$
39. $\begin{cases} 2a + 3b = 5a - 6 \\ 4a - 2b - 8 = 2b \end{cases}$
40. $\begin{cases} 5d + 4c + 7 = 3d - 1 \\ -7c + 2d - 12 = 6 - 8c \end{cases}$
41. $\begin{cases} -4a + 3b - 5 = 2a - 5 \\ -8b + 2a + 3 = 10 - 6b \end{cases}$
42. $\begin{cases} \frac{1}{4}(3x - 4) = -\frac{1}{2}(2y - x) \\ -(y - 3x) = 2x - y \end{cases}$
43. $\begin{cases} 3(y - 2x) = 5(y - 2x) - 8 \\ -\frac{1}{3}(x - 5y) = (y - x) - 4 \end{cases}$
44. $\begin{cases} \frac{1}{3}(r - 1) = \frac{1}{2}(3r + s) \\ \frac{r - s}{4} = 1 \end{cases}$
45. $\begin{cases} \frac{2}{3}(2x + 1) - \frac{3}{2}(3y - 1) = -1 \\ \frac{y - 3x}{2} = -1 \end{cases}$

11.2 The Substitution Method

46. $\begin{cases} \dfrac{j+4k}{3} = -2 \\ -\dfrac{1}{4}(j+2) = \dfrac{1}{2}(j+2k) \end{cases}$

47. $\begin{cases} \dfrac{-2y-x}{3} = 1 \\ \dfrac{3}{4}(1-5x) - \dfrac{1}{3}(2y-1) = 4 \end{cases}$

The equations of the following systems are not linear. However, if c is substituted for $\dfrac{1}{x}$ and d for $\dfrac{1}{y}$, the equations become linear in c and d. In each, substitute c for $\dfrac{1}{x}$ and d for $\dfrac{1}{y}$. Then solve.

48. $\begin{cases} \dfrac{1}{x} + \dfrac{1}{y} = \dfrac{1}{3} \\ \dfrac{2}{x} - \dfrac{3}{y} = \dfrac{1}{4} \end{cases}$

49. $\begin{cases} \dfrac{1}{x} - \dfrac{1}{y} = -\dfrac{5}{6} \\ \dfrac{3}{x} + \dfrac{2}{y} = -\dfrac{5}{6} \end{cases}$

Applications

50. **Business** An accounting student took a test of verbal skills and a test of mathematical skills. Her total score was 1250, ... er math score was 200 more than the verbal. Find the two scores ... ving the system:

$$\begin{cases} x + y = 1250 \\ x = y + 200 \end{cases}$$

51. **Calculator** Graph Exercises 25–41 using a graphing calculator or using a computer graphing program. Check the answers you got using the substitution method with the solutions that the graph indicates.

WRITING IN ALGEBRA

Write an algebraic expression for each of the following.

1. After buying a newspaper, you have x dimes and y nickels. Write their total value in cents. (*Hint:* If x represents the number of dimes, then $10x$ represents the total value.)

2. You have x dimes and twice as many quarters. Write the total value of your quarters in cents.

3. If t represents the digit in the tens place of a two-digit number and u represents the digit in the ones place, write the number.

4. If h represents the hundreds digit in a three-digit number, t the tens digit, and 5 the ones digit, write the number.

5. You sell t tickets at $5 each. Your friend sells 10 more such tickets. Write the total number of dollars collected by your friend.

11.3 Problem Solving Strategy: Write a Linear System to Solve a Problem

When a problem requires that you find two numbers, you may find it easier to solve the problem by using a system of equations. To solve, assign different variables to the two unknown numbers, write two equations, and then solve the system.

EXAMPLE Arkville paid a contractor to have a rectangular grassplot fenced in. One side of the plot ran along a roadway, and the special fencing for this side cost $10/ft. Fencing for the other three sides cost $5/ft. The contractor claimed that the perimeter of the plot was 1740 ft, and his bill amounted to $11,325. If his calculations were correct, what are the dimensions of the plot?

Understand the Problem

What is given?
The perimeter of a rectangular plot is 1740 ft. Fencing along one side cost $10/ft and along the other sides $5/ft. The total cost of fencing was $11,325.

You are asked to find the length and width of the plot.

Plan Your Approach

Choose a strategy.
Draw a diagram.

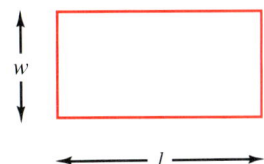

Let l = the length (in ft)
w = the width (in ft)

Write a system of equations.

$$\begin{cases} 2l + 2w = 1740 \\ 10l + 5l + 5w + 5w = 11{,}325 \end{cases}$$

The perimeter is 1740 ft.
The terms on the left represent the cost of fencing all four sides.

11.3 Problem Solving Strategy: Write a Linear System to Solve a Problem

Complete the Work

Solve the system of equations.
To simplify the system, divide each side of the first equation by 2. Combine like terms in the second equation, and then divide each side by 5.

$$\begin{cases} 2l + 2w = 1740 \\ 10l + 5l + 5w + 5w = 11{,}325 \end{cases} \rightarrow 15l + 10w = 11{,}325 \rightarrow \begin{cases} l + w = 870 \\ 3l + 2w = 2265 \end{cases}$$

The simplified system to be solved is:

$l + w = 870$ Solve the first equation for l.
$l = 870 - w$
$3l + 2w = 2265$ Substitute $870 - w$ for l in the second equation.

$3(870 - w) + 2w = 2265$ Solve for w.
$2610 - 3w + 2w = 2265$
$w = 345$

$l + w = 870$ Substitute 345 for w in the first equation.
$l + 345 = 870$ Solve for l.
$l = 525$

Interpret the Results

State your conclusion.
The width is 345 ft, and the length is 525 ft.
Check your conclusion.
Do these dimensions give a perimeter of 1740 ft?

$$2(525) + 2(345) \stackrel{?}{=} 1740$$
$$1740 = 1740 \checkmark$$

Do these dimensions give a total fencing cost of $11,325?

$$10(525) + 5(525) + 5(345) + 5(345) \stackrel{?}{=} 11{,}325$$
$$11{,}325 = 11{,}325 \checkmark$$

CLASS EXERCISES

Write a system of linear equations for the problem. Do not solve.

1. One positive number equals 3 times another number. The difference between the numbers is 10. Find the numbers.

2. Twenty-six students entered their projects in the science fair. The number of girls was 4 fewer than twice the number of boys. How many boys entered their science projects in the fair?

3. The perimeter of an isosceles triangle is 48 in. The length of the base is 1 in. less than half the length of one of the two equal sides. Find the lengths of the sides of the triangle.

PRACTICE EXERCISES

Solve the problem by writing and solving a system of equations.

1. The length of a rectangle is 2 ft less than three times the width. If the perimeter is 68 ft, what are the dimensions of the rectangle?

2. The perimeter of a rectangle is 10 m. Twice the width is equal to one-half the length. Find the length and the width.

3. The sum of two numbers is 48. If the smaller number is subtracted from the larger, the difference is 12. Find the numbers.

4. Find two numbers whose sum is −64, if twice the first is 1 more than the second.

5. Said's father has 13 more than 5 times as many coins in his antique coin collection as Said has. The total number of coins in both collections is 247. How many coins does each one have?

6. Bella and Irina together sold 137 tickets for a benefit concert. If Irina sold 10 fewer than twice as many as Bella, how many tickets did each girl sell?

7. For a mathematics project, 27 students were divided into two groups. One group had 3 more than twice the number of students in the other. How many students were in each group?

8. A 50-ft rope is cut into two pieces. The length of one piece is 9 times the length of the other. What is the length of the longer piece?

9. Find two positive numbers if one is $\frac{2}{3}$ the other, and their difference is 5.

10. One number is 4 more than half of another. The difference between the two numbers is 2. Find these numbers.

11. Last year Zachary received $469.75 interest from two investments. The interest rates were 7.5% on one account and 8% on the other. If the total amount invested was $6000, how much was invested at each rate?

12. The federal tax on a $12,000 salary was 8 times the state tax. If the combined taxes were $2700, find the state's share of taxes.

13. Leila left home at noon, traveling at 24 mi/h. An hour later her brother Josh, driving 36 mi/h, set out to overtake her. How long did it take him? (*Hint:* Let l = Leila's time in h, let j = Josh's time in h.)

14. It took Henri 50 h of cycling to finish his bike tour. On the return trip over the same route, he increased his cycling rate by 1 mi/h and took only 45 h. What was his rate on the return trip?

15. If the length of one pair of opposite sides of a rectangle were doubled and the other pair of opposite sides were each decreased by 14 m, the new rectangle would be a square with a perimeter of 84 m. Find the rectangle's original dimensions.

16. If the length of one pair of opposite sides of a square were tripled and the other pair were each increased by 7 cm, the new figure would be a rectangle with a perimeter of 96 cm. Find the dimensions of the original square.

17. Alice and Jeanine live 18 blocks apart in opposite directions from the Coles Sports Center. If Alice lives 3 more blocks than $\frac{1}{4}$ as far from the center as Jeanine does, how far does Alice live from the sports center?

Mixed Problem Solving Review

1. If you add 2.5 L of water to 10 L of a 10% solution of salt in water, what will be the percent of salt in the new solution?

2. On a round trip of 70 mi each way Diana averaged 7 mi/h more on the return trip, which took 30 min less than the trip going of $2\frac{1}{2}$ h. Find her two speeds.

3. Ian has one job at which he must work a 20 h/wk at $4.50/h. He may also work at a second job which pays $6.25/h. If he must earn at least $160 this week, how long must he work at the second job?

PROJECT

In this lesson some of the problems that you solved by writing a system of equations were the same types you solved using a single variable and one equation. Which method do you prefer? Why? Solve the following word problems with a partner. One student should use one equation to solve the problem. The other student should use a system of equations. Take turns using the two methods. Discuss your preferences.

1. The denominator of a fraction is 12 more than its numerator. If 5 is subtracted from both numerator and denominator, the new fraction reduces to $\frac{2}{3}$. Find the original fraction.

2. One number is 30% more than the other. If their sum is 161, find the two numbers.

3. Working together, two girls can finish a job in 4 days. If one girl works alone for 8 days, the other girl can then finish the job by herself in 2 days. How long would it take each girl working alone?

11.4 The Addition/Subtraction Method

Objective: To solve systems of linear equations by the addition/subtraction method

The substitution method is one algebraic method for solving simultaneous equations. Another is the **addition/subtraction method,** in which a new equation is formed by adding or subtracting the equations in the system. The new equation contains just one of the original variables and can be solved for that variable. Before using this technique, the equations of the system should be in standard form: $Ax + By = C$, where A, B, and C are integers and both A and B do not equal zero.

Capsule Review

EXAMPLE   Rewrite $4 - x = 3y$ in standard form.

$$-x - 3y = -4 \quad \text{Subtract 4 and 3y from each side.}$$
$$\text{or} \quad x + 3y = 4 \quad \text{Multiply each side by } -1.$$

Rewrite the following equations in standard form.

1. $5x + 4 = -9y$ **2.** $y = -3x - 5$ **3.** $14 - x = 0$

The Addition/Subtraction Method

- Add or subtract the given equations to eliminate one variable.
- Solve the resulting equation for the remaining variable.
- Replace the value of the known variable in one of the original equations to find the value of the unknown variable.
- Check the solution in both original equations.

EXAMPLE 1   Solve by the addition method: $\begin{cases} x + y = 6 \\ x - y = 2 \end{cases}$

$$\begin{cases} x + y = 6 \\ x - y = 2 \end{cases}$$
$$\underline{}$$
$$2x = 8 \quad \text{Add the equations to eliminate the y-terms.}$$
$$x = 4 \quad \text{Solve for x.}$$

$$x + y = 6 \quad \text{Replace x with 4 in either original equation.}$$
$$4 + y = 6 \quad \text{Solve for y.}$$
$$y = 2$$

So, the solution is $(4, 2)$.

Check the solution (4, 2) in both equations:

$$x + y = 6 \qquad x - y = 2$$
$$4 + 2 \stackrel{?}{=} 6 \qquad 4 - 2 \stackrel{?}{=} 2$$
$$6 = 6 \checkmark \qquad 2 = 2 \checkmark$$

In the next example, one equation is rewritten in standard form then one of the variables of the system is eliminated by subtraction.

EXAMPLE 2 Solve: $\begin{cases} 8m + 12n = 50 \\ 12n = 3m - 27 \end{cases}$

$$\begin{cases} 8m + 12n = 50 \\ -3m + 12n = -27 \end{cases}$$ *Express the second equation in standard form.*

$\overline{}$
$11m = 77$ *Subtract the equations to eliminate the n-terms.*
$m = 7$ *Solve for m.*

$12n = 3m - 27$ *Replace m with 7 in either original equation.*
$12n = 3(7) - 27$
$12n = 21 - 27$ *Solve for n.*
$12n = -6$
$n = -\frac{1}{2}$

The check of the solution $\left(7, -\frac{1}{2}\right)$ is left for you.

Notice that in Example 1 the coefficients of *y* are opposites, and in Example 2 the coefficients of *n* are the same. The addition/subtraction method can be used whenever the two equations have the same or opposite coefficients for one of the variables. This method works best when the equations are in standard form.

CLASS EXERCISES

Tell whether addition or subtraction is appropriate. Give the resulting one-variable equation. Then solve the system.

1. $\begin{cases} x + y = 4 \\ x - y = -10 \end{cases}$
2. $\begin{cases} x + y = 4 \\ 2x + y = 5 \end{cases}$
3. $\begin{cases} -7a + b = 1 \\ 3 - 3b = 7a \end{cases}$

4. $\begin{cases} 2p - q = 1 \\ 2p - q = -3 \end{cases}$
5. $\begin{cases} 3m - 4n = 1 \\ 3m - 2n = -1 \end{cases}$
6. $\begin{cases} 9s + 4t = -13 \\ -9s = \frac{1}{2}t - 1 \end{cases}$

7. Use the substitution method for Class Exercise 1, and show that it gives the same result as the addition method.

For Discussion

8. Explain why either the addition or subtraction method can be used to solve the system $-3x + y = -3$ and $3x + y = 9$.
9. The subtraction method produces no solution in Class Exercise 4. Why?

PRACTICE EXERCISES

Solve by the addition/subtraction method.

1. $\begin{cases} x - y = -8 \\ x + y = 12 \end{cases}$
2. $\begin{cases} x + y = 10 \\ x - y = -2 \end{cases}$
3. $\begin{cases} a + b = 0 \\ a - b = -6 \end{cases}$
4. $\begin{cases} a - b = -14 \\ a + b = -4 \end{cases}$
5. $\begin{cases} p + q = -2 \\ q = p + 10 \end{cases}$
6. $\begin{cases} q = 16 + p \\ -p - q = 0 \end{cases}$
7. $\begin{cases} 3a - b = 21 \\ 2a + b = 4 \end{cases}$
8. $\begin{cases} 7s + t = 22 \\ 5s - t = 14 \end{cases}$
9. $\begin{cases} m - 4n = 6 \\ m - 2n = 18 \end{cases}$
10. $\begin{cases} 7g + h = 42 \\ -3g + h = -8 \end{cases}$
11. $\begin{cases} 3j - k = -10 \\ -5j - k = 14 \end{cases}$
12. $\begin{cases} -v + 5w = 12 \\ -v - 3w = -4 \end{cases}$
13. $\begin{cases} 3y = -5 - 2x \\ 7x - 3y = 23 \end{cases}$
14. $\begin{cases} -2m - n = 5 \\ -2m - 3n = -7 \end{cases}$
15. $\begin{cases} 2a - b = 1 \\ b = 2a - 1 \end{cases}$
16. $\begin{cases} 3x + 8y = -7 \\ -8y = 7 + 3x \end{cases}$
17. $\begin{cases} 3x - y = 8 \\ \frac{1}{3}y - 3x = \frac{2}{3} \end{cases}$
18. $\begin{cases} 2p - 3q = 11 \\ \frac{1}{3}q - 2p = 1 \end{cases}$
19. $\begin{cases} 2c + 5d = 44 \\ -6c + 5d = 8 \end{cases}$
20. $\begin{cases} 4a - 7b = 3 \\ 16a - 7b = 12 \end{cases}$
21. $\begin{cases} 6a - 5b = 6 \\ -5b + 7a = 7 \end{cases}$
22. $\begin{cases} 4g - 2h = -14 \\ 5h = 4g + 32 \end{cases}$
23. $\begin{cases} 2m - 5n = 17 \\ 6m = 5n + 1 \end{cases}$
24. $\begin{cases} -7a + 4b = 13 \\ 2b = 3 + 7a \end{cases}$
25. $\begin{cases} -7c + 2d = 31 \\ -17c = 17 + 2d \end{cases}$
26. $\begin{cases} y = -3x + 1 \\ 5x - 2 = -y \end{cases}$
27. $\begin{cases} 2x = y + 1 \\ 5y - 2x = 1 \end{cases}$
28. $\begin{cases} 3d = 3 - 4c \\ c = 1 - 3d \end{cases}$
29. $\begin{cases} 4e = 5 + 2f \\ 6f = 9 - 4e \end{cases}$
30. $\begin{cases} 0.4a - 0.2b = -1.4 \\ 0.4a - 0.5b = -3.2 \end{cases}$
31. $\begin{cases} -0.7p + 0.2q = 3.1 \\ -1.7p - 0.2q = 1.7 \end{cases}$
32. $\begin{cases} 2v + 3w = 6 \\ 2v - 27w = 18 \end{cases}$
33. $\begin{cases} 12j + 5k = 76 \\ 4j = 52 - 5k \end{cases}$
34. $\begin{cases} 0.12x - 1.2y = 3.024 \\ 1.34x - 1.2y = 6.928 \end{cases}$
35. $\begin{cases} 2.3x - 0.45y = 7.99 \\ -2.3x + 1.6y = -4.31 \end{cases}$
36. $\begin{cases} \frac{3}{2}x + \frac{5}{4}y = \frac{18}{13} \\ -\frac{3}{2}x + \frac{3}{4}y = \frac{8}{13} \end{cases}$
37. $\begin{cases} 6r + \frac{3}{7}s = \frac{19}{3} \\ r + \frac{3}{7}s = \frac{4}{3} \end{cases}$

38. $\begin{cases} \dfrac{m+n}{2} = \dfrac{3n-2}{2} \\ \dfrac{3n+8}{2} = -\left(\dfrac{m+n}{2}\right) \end{cases}$

39. $\begin{cases} \dfrac{f+3}{2} + g = 2 \\ g - \dfrac{f+3}{2} = -5 \end{cases}$

40. $\begin{cases} \dfrac{p+2}{3} = q + 3 \\ 4q = \dfrac{p+2}{3} \end{cases}$

41. $\begin{cases} \dfrac{2j+3}{9} = k + 6 \\ -3k = \dfrac{2j+3}{9} \end{cases}$

In Exercises 42–45, substitute c for $\dfrac{1}{x}$, d for $\dfrac{1}{y}$ to solve for x and y.

42. $\begin{cases} \dfrac{6}{x} + \dfrac{4}{y} = 16 \\ \dfrac{3}{x} - \dfrac{4}{y} = 2 \end{cases}$

43. $\begin{cases} \dfrac{8}{x} - \dfrac{2}{y} = -2 \\ \dfrac{8}{x} + \dfrac{7}{y} = 25 \end{cases}$

44. $\begin{cases} \dfrac{1}{x} + \dfrac{3}{y} = 1 \\ -\dfrac{4}{x} + \dfrac{3}{y} = -3 \end{cases}$

45. $\begin{cases} -\dfrac{5}{x} + \dfrac{2}{y} = 3 \\ \dfrac{5}{x} - \dfrac{3}{y} = -4 \end{cases}$

Applications

Use a system of equations to solve.

46. **Agriculture** The Allens grow only soybeans and corn on their 240-acre farm. This year they plan to plant 80 more acres of soybeans than of corn. How many acres will they plant of each crop?

47. **Agriculture** The Bensons are planting an apple orchard on their farm. They decide to grow 85 acres of wheat along with some acres of apples. If they had doubled the size of the orchard there would have been only 60 acres of wheat. How large is the orchard they planted?

48. **Science** Two batteries produce a total voltage of 4.5 v. The difference in their voltage is 1.5 v. Determine the voltages of the two batteries.

49. **Engineering** When analyzing a certain electric circuit, an electric engineer found these equations, $3i_1 + 4i_2 = 3$ and $3i_1 - 5i_2 = -6$. Solve for the electric currents i_1 and i_2 (in amperes).

EXTRA

The graphs of these equations form a triangle. Find the vertices of the triangle. Then find the area of the triangle.

$x + y = 6$
$2x + y = 0$
$y = 0$

11.5 The Multiplication-Addition/Subtraction Method

Objective: To solve appropriate systems of linear equations by using the multiplication-addition/subtraction method

Sometimes the substitution method and the addition/subtraction method are not the most effective ways of solving simultaneous equations. Consider a system such as:

$$\begin{cases} 3x - 2y = -12 \\ 5x + 4y = 2 \end{cases}$$

Adding or subtracting will not result in an equation in just one variable, since neither the x nor the y terms are the same or opposites. If necessary, the substitution method can always be used. What is the difficulty?

Note that the addition method could be used to solve the system if both sides of the first equation were multiplied by 2. The resulting equation is an **equivalent system of equations.**

$$\begin{cases} 6x - 4y = -24 \\ 5x + 4y = 2 \end{cases}$$

Equivalent systems may look different but must have the same solution. Compare this second system of equations with the one above. Are the two systems equivalent?

Capsule Review

In using multiplication with the addition/subtraction method, it is important to remember to multiply both sides of the equation using the multiplication property of equality.

EXAMPLE Multiply: $x - 2y = 3$ by 4
$4(x - 2y) = 4(3)$
$4x - 8y = 12$

Multiply each side of the equation by the given integer.

1. $2m + 5n = 1$ by 3
2. $5p - q = 2$ by -1
3. $-3s + 6t = -8$ by -2
4. $-4x - 7y = -2$ by -5

The main idea in solving a system of equations by the **multiplication-addition/subtraction method** is to use the multiplication property of equality on one or both of the original equations.

EXAMPLE 1 **Solve:** $\begin{cases} 3a - 4b = 18 \\ a + 2b = -4 \end{cases}$

The b-terms may be eliminated. If the second equation is multiplied by 2, the coefficients of the b-terms will each be 4.

$\begin{cases} 3a - 4b = 18 \\ a + 2b = -4 \end{cases} \longrightarrow \begin{array}{l} 3a - 4b = 18 \\ 2(a + 2b) = 2(-4) \end{array} \longrightarrow \begin{array}{l} 3a - 4b = 18 \\ 2a + 4b = -8 \end{array}$

$\begin{array}{l} 3a - 4b = 18 \\ \underline{2a + 4b = -8} \\ 5a = 10 \\ a = 2 \end{array}$ Add to eliminate the b-terms.
Solve for a.

$\begin{array}{l} a + 2b = -4 \\ 2 + 2b = -4 \\ 2b = -6 \\ b = -3 \end{array}$ Replace a with 2 in either equation.
Solve for b.

The solution is $(2, -3)$. How would you check the solution?

EXAMPLE 2 **Solve:** $\begin{cases} 4x + 15y = 7 \\ 9y = -6x + 21 \end{cases}$

$\begin{cases} 4x + 15y = 7 \\ 6x + 9y = 21 \end{cases}$ *Write the second equation in standard form.*

The x-terms may be eliminated. If the first equation is multiplied by 3 and the second equation by 2, the coefficients of the x-terms will each be 12.

$\begin{cases} 4x + 15y = 7 \\ 6x + 9y = 21 \end{cases} \longrightarrow \begin{array}{l} 3(4x + 15y) = 3(7) \\ 2(6x + 9y) = 2(21) \end{array} \longrightarrow \begin{array}{l} 12x + 45y = 21 \\ 12x + 18y = 42 \end{array}$

$\begin{array}{l} 12x + 45y = 21 \\ \underline{12x + 18y = 42} \\ 27y = -21 \\ y = \frac{-21}{27} = -\frac{7}{9} \end{array}$ Subtract to eliminate the x-terms.
Solve for y.

$\begin{array}{l} 9y = -6x + 21 \\ 9\left(-\frac{7}{9}\right) = -6x + 21 \\ -7 = -6x + 21 \\ -28 = -6x \\ \frac{-28}{-6} = \frac{14}{3} = x \end{array}$ Replace y with $-\frac{7}{9}$ in either original equation.
Solve for x.

The check of the solution $\left(\frac{14}{3}, -\frac{7}{9}\right)$ is left for you.

CLASS EXERCISES

For each system, write an equivalent system in which addition or subtraction of the two equations will eliminate (a) the first variable and (b) the second variable. Then solve.

1. $\begin{cases} a + 2b = 4 \\ a + 3b = -2 \end{cases}$
2. $\begin{cases} 2x - y = 4 \\ x + 3y = 16 \end{cases}$
3. $\begin{cases} 2a - 3b = -4 \\ -6a + 9b = -15 \end{cases}$
4. $\begin{cases} 3s - 7t = 13 \\ 6s + 5t = 7 \end{cases}$
5. $\begin{cases} 8x + 3y = 13 \\ 3x + 2y = 11 \end{cases}$
6. $\begin{cases} 4c - 15d = -13 \\ 6c + 10d = 13 \end{cases}$

7. Explain how multiplication can be used with the addition/subtraction method to solve the system:
$$\begin{cases} 7x + 9y = 3 \\ 5x + 4y = 1 \end{cases}$$

PRACTICE EXERCISES

Solve by the multiplication with addition or subtraction method.

1. $\begin{cases} 3a - 4b = 1 \\ 12a - b = -11 \end{cases}$
2. $\begin{cases} -5c + 3d = -16 \\ -10c + d = -22 \end{cases}$
3. $\begin{cases} 8u - 4v = 16 \\ 4u + 5v = 22 \end{cases}$
4. $\begin{cases} -5m + 3n = -31 \\ -2m - 5n = 0 \end{cases}$
5. $\begin{cases} -3e + 4f = -6 \\ 5e - 6f = 8 \end{cases}$
6. $\begin{cases} 5k + 3l = 17 \\ 2k - 9l = 17 \end{cases}$
7. $\begin{cases} 8r - 5s = -11 \\ 3s = 4r - 11 \end{cases}$
8. $\begin{cases} 9p + 4q = -17 \\ 12q = -3 - 3p \end{cases}$
9. $\begin{cases} 3x - 2y = 6 \\ 5x + 7y = 41 \end{cases}$
10. $\begin{cases} 5m - 2n = 8 \\ 3m - 5n = 1 \end{cases}$
11. $\begin{cases} 7a - 3b = -9 \\ 9b - 4a = -24 \end{cases}$
12. $\begin{cases} -3x + 4y = -6 \\ -6y + 5x = 8 \end{cases}$
13. $\begin{cases} 8c - 3d = 5 \\ 16c + 9d = 5 \end{cases}$
14. $\begin{cases} 3s - 8t = 4 \\ 9s - 4t = 5 \end{cases}$
15. $\begin{cases} 3x = y + 1 \\ -9x + 2y = -5 \end{cases}$
16. $\begin{cases} 2a + 5b = 6 \\ 3a = 10b + 2 \end{cases}$
17. $\begin{cases} u = 4v \\ 3u + 2v = 7 \end{cases}$
18. $\begin{cases} p = 2q \\ 2p + 6q = 5 \end{cases}$
19. $\begin{cases} 6m + 12n = 7 \\ 8m - 15n = -1 \end{cases}$
20. $\begin{cases} 10r - 9s = 18 \\ 6s + 2r = 1 \end{cases}$
21. $\begin{cases} 2(p + 3) = 3 - q \\ 3(p - 1) = q - 4 \end{cases}$
22. $\begin{cases} 4(x + 3) = 3y + 7 \\ 2(y - 5) = x + 5 \end{cases}$
23. $\begin{cases} c - 2d = 500 \\ 0.03c + 0.02d = 51 \end{cases}$
24. $\begin{cases} 3e - 2f = -80 \\ 0.05e - 0.03f = 2 \end{cases}$
25. $\begin{cases} \dfrac{3m + 2n}{4} = -5m + n \\ 2n - 23m = 5 \end{cases}$
26. $\begin{cases} \dfrac{7u - 4v}{2} = 5v \\ 5 + v = 7u \end{cases}$
27. $\begin{cases} \dfrac{2a + b}{5} = \dfrac{4a - 2b}{3} \\ \dfrac{a + 2}{3} = \dfrac{1 - b}{9} \end{cases}$

11.5 The Multiplication-Addition/Subtraction Method

28. $\begin{cases} \dfrac{1}{x} + \dfrac{1}{y} = 5 \\ \dfrac{2}{x} + \dfrac{3}{y} = 13 \end{cases}$
29. $\begin{cases} \dfrac{1}{x} - \dfrac{2}{y} = 8 \\ \dfrac{3}{x} + \dfrac{1}{y} = 10 \end{cases}$
30. $\begin{cases} \dfrac{1}{x} + \dfrac{3}{y} = 10 \\ \dfrac{2}{x} + \dfrac{1}{y} = 6 \end{cases}$

Solve the following systems in terms of coefficients *a*, *b*, and *c*.

31. $\begin{cases} ax + y = c \\ x + by = c \end{cases}$
32. $\begin{cases} ax + y = c \\ ax + by = 0 \end{cases}$
33. $\begin{cases} ax + y = c \\ ax + y = 0 \end{cases}$

34. $(-1, 1)$ and $(3, -15)$ are two solutions of the nonlinear equation $y = ax^2 + b$. Find *a* and *b*.

Applications

35. **Business** A department store is having a sale on women's shoes. The shoes are selling for $20 and $25 a pair. At the end of the day the total receipts for the sale of 55 pairs of shoes were $1250. If *x* and *y* represent the numbers of pairs of $20 and $25 shoes respectively, find *x* by solving the system: $20x + 25y = 1250$ and $x + y = 55$.

36. **Food Preparation** The cafeteria usually makes 112 cups of orange-pineapple fruit punch. One Friday the amount of orange juice was tripled, and the pineapple juice was doubled to make a total of 274 c of punch. How many cups of the two juices were used on that Friday?

TEST YOURSELF

Describe the graph and tell the number of solutions. Verify by graphing. 11.1

1. $\begin{cases} x + y = 6 \\ 2x - y = 3 \end{cases}$
2. $\begin{cases} x + y = 6 \\ \frac{1}{2}x + \frac{1}{2}y = 3 \end{cases}$
3. $\begin{cases} 3x + y = 1 \\ x + \frac{1}{2}y = -1 \end{cases}$

Solve by an appropriate method of your choice: substitution, addition, subtraction, or multiplication-addition/subtraction. 11.2–11.5

4. $\begin{cases} 2x + y = -10 \\ x + 3y = 0 \end{cases}$
5. $\begin{cases} 3x + 2y = 10 \\ x - 2y = 14 \end{cases}$
6. $\begin{cases} 4x - 3y = 15 \\ x + y = 9 \end{cases}$

7. $\begin{cases} \frac{1}{3}x = -\frac{1}{3}y + 4 \\ \frac{8}{3} = x + \frac{1}{3}y \end{cases}$
8. $\begin{cases} -\frac{2}{3}y + 2x = -1 \\ \frac{1}{3}y + x = \frac{1}{2} \end{cases}$

9. A marketing firm budgets $33,000 each year for advertising. It spends twice as much money on advertisements in newspapers as in magazines. How much does it spend on advertising in newspapers? in magazines?

11.6 Problem Solving: Digit Problems

Objective: To use systems of equations to solve digit problems

Digit problems are problems involving the digits of a number. The number 25 consists of the digit 2 in the tens place and the digit 5 in the ones place. What happens when the digits of a number are reversed?

Number	Expanded Notation	Reversed Digits	Expanded Notation
25	$2 \cdot 10 + 5 \cdot 1$	52	$5 \cdot 10 + 2 \cdot 1$
48	$4 \cdot 10 + 8 \cdot 1$	84	$8 \cdot 10 + 4 \cdot 1$
tu	$10t + u$	ut	$10u + t$

In general, if $10t + u$ is the original number, then $10u + t$ is the number with the digits reversed.

EXAMPLE The sum of the digits of a two-digit number is 8. If the digits are reversed, the new number is 18 more than the original number. Find the original number.

Understand the Problem

Find a two-digit number that satisfies these conditions.
1. The sum of the digits is 8.
2. When the digits are reversed, the new number is 18 more than the original number.

Plan Your Approach

Write a system of equations.

$$\begin{cases} t + u = 8 \\ 10u + t = (10t + u) + 18 \end{cases}$$

The sum of the digits is 8.
The number with the digits reversed is 18 more than the original number.

Complete the Work

Solve the system by the easiest method.

$$\begin{cases} t + u = 8 \\ 10u + t = 10t + u + 18 \longrightarrow -9t + 9u = 18 \end{cases} \longrightarrow \begin{cases} t + u = 8 \\ -t + u = 2 \end{cases}$$

$$2u = 10$$
$$u = 5$$

Now find the value of t.
$$t + u = 8$$
$$t + 5 = 8$$
$$t = 3$$

Interpret the Results   The number $(10t + u)$ is 35.
Check the number in the word problem.
Find the sum of the digits. $3 + 5 = 8$
Reverse the digits. $53 \stackrel{?}{=} 35 + 18$
    $53 = 53$ ✔

CLASS EXERCISES

Write an equation for each word sentence.

1. The sum of the digits is 9.
2. The units digit is 4 more than the tens digit.
3. The tens digit is half the units digit.
4. The number with its digits reversed is 9 more than the original number.

Solve the system of equations to find the number from the Class Exercises above.

5. Exercises 1 and 3
6. Exercises 3 and 4
7. Exercises 1 and 4
8. Exercises 2 and 3

PRACTICE EXERCISES

Find the original two-digit number described by the problem.

1. The sum of the digits is 6. The number is 6 times its units digit.
2. The sum of the digits is 6. Reversing the digits gives a new number that is 18 less than the original.
3. The units digit is 3 times the tens digit. When the digits are reversed, the new number is 54 more than the original.
4. The units digit is 3 times the tens digit. The number is 12 more than the tens digit.
5. The units digit exceeds the tens digit by 6. The number is 10 more than 9 times the tens digit.
6. The sum of the digits is 12. The number is 1 more than 8 times its units digit.
7. The sum of the digits is 10. The original number decreased by 54 equals the number formed when the digits are reversed.
8. The sum of the digits is 12. The original number increased by 36 equals the number formed when the digits are reversed.

9. Twice the tens digit of a number increased by the units digit is 22. If the digits are reversed, the new number is 45 less than the original.

10. Three times the units digit is 2 less than the tens digit. If the digits are reversed, the new number is 54 less than the original.

11. The sum of the digits is 11. If the digits are reversed, the original number is 7 more than twice the new number.

12. A number is 9 more than the sum of its digits. If the digits are reversed, the new number is 3 less than 4 times the original.

13. A number is 7 times the sum of its digits. If the digits are reversed, the new number is 30 less than twice the original.

14. The hundreds digit of a three-digit number equals the sum of the tens and units digits. The units digit is twice the tens digit. The difference between the number and the new number with reversed digits is 297.

15. Write a single equation in digits t and u that has just one solution.

16. Write a two-digit integer problem that has no solution.

ALGEBRA IN BOOKKEEPING

When two digits in a number are turned around, it is called a **transposition.** In bookkeeping, the bookkeeper often has to check for **transposition errors.** To see if there is an error the bookkeeper subtracts the figure in question from the figure he/she obtained. If the difference is divisible by 9, there is usually a transposition error. This computer program checks for transposition errors.

```
10 INPUT "ENTER A THREE DIGIT
            NUMBER: ";N
20 NH = INT (N / 100):NT = INT
       ((N - NH * 100) / 10)
30 ND = (N - NH * 100 - NT * 10)
40 Y1 = ND * 100 + NT * 10 + NH
50 Y2 = NH * 100 + ND * 10 + NT
60 Y3 = NT * 100 + NH * 10 + ND
70 PRINT "NUMBER"; TAB( 15);"DIFFERENCE";
       TAB( 30);"DIFF/9"
80 PRINT N: PRINT Y1,N - Y1,(N - Y1) / 9
90 PRINT Y2,N - Y2,(N - Y2) / 9
100 PRINT Y3,N - Y3,(N - Y3) / 9
110 END
```

Exercises

1. Is it possible to use the first transposed number printed by the program to prove that the difference between any three-digit number and the number with the digits reversed is divisible by 99?

2. Run the program several times, and then explain what Y1 in line 40, Y2 in line 50, and Y3 in line 60 mean.

11.7 Problem Solving: Age Problems

Objective: To use systems of equations to solve age problems

Kim's father is now $2\frac{1}{2}$ times as old as Kim. Ten years ago he was 4 times as old as Kim.

This type of problem can readily be solved by setting up a system of equations to represent the two statements. Think of algebraic expressions to represent the two ages.

EXAMPLE Reread the problem above about Kim and her father. Find their present ages.

Understand the Problem Two facts about the relationship between the ages of Kim and her father are given.

Now: Kim's father's age is $2\frac{1}{2}$ times Kim's age.

10 yr ago: Kim's father's age was 4 times Kim's age.

Plan Your Approach **Represent the variables.**
Let k = Kim's age now.
Let f = Kim's father's age now.
Make a table to show their ages in terms of the variables. Then translate the two age relationships into a system of equations:

	Age now	Age 10 yr ago
Kim	k	$k - 10$
Father	f	$f - 10$

$$\begin{cases} f = 2\frac{1}{2}k \\ f - 10 = 4(k - 10) \end{cases}$$

Complete the Work

$$\begin{cases} f = 2\frac{1}{2}k \longrightarrow f = \frac{5}{2}k \\ f - 10 = 4(k - 10) \longrightarrow f - 10 = 4k - 40 \end{cases}$$

Substitute $f = \frac{5}{2}k$ into the second equation. Solve for k.

$\frac{5}{2}k - 10 = 4k - 40$

$5k - 20 = 8k - 80$ *Multiply each side by 2.*

$-3k = -60$ *Subtract 8k and add 20 to each side.*

$k = 20$ *Divide each side by -3.*

If $k = 20$, then $f = \frac{5}{2}k = \frac{5}{2}(20) = 50$.

> **Interpret the Results**
> Conclusion: Kim is 20 years old; her father is 50 years old.
> Check: $50 = 2\frac{1}{2}(20)$. Ten years ago Kim was 10, her father 40, so Kim's father was then 4 times as old as Kim.

CLASS EXERCISES

1. The headings for each column in the table are given. Complete the table.

Ages of Two Students

	Age 8 years ago	Age 5 years ago	Age 2 years ago	Present age	Age in 1 year	Age in 4 years	Age in 7 years
Anne	?	?	$a - 2$	a	?	$a + 4$	?
Bill	$b - 8$	?	$b - 2$	?	?	$b + 4$	$b + 7$

Use entries from the table above to write an equation for the statements.

2. Anne is 3 years younger than Bill.

3. Eight years ago Bill was twice as old as Anne.

4. In 4 years Anne will be $\frac{5}{6}$ as old as Bill.

5. Seven years from now Bill's age will be $\frac{7}{6}$ of Anne's age.

6. The sum of their present ages is 25.

Solve systems consisting of the equations from the Class Exercises above.

7. Exercises 2 and 6   8. Exercises 2 and 3   9. Exercises 4 and 5

PRACTICE EXERCISES

Write and solve a system of linear equations to find the present ages.

1. Cordell is twice as old as Beth. Eight years ago he was 3 times as old.

2. Dawn is 3 years older than Lois. Four years ago Dawn was twice as old as Lois.

3. Mario's age is 2 years more than twice Nadia's age. In 9 years her age will be $\frac{2}{3}$ of his.

4. Tanya's younger brother's age is $\frac{3}{4}$ of hers. In 7 years the sum of their ages will be 28.

5. Two years ago Adam was 6 times as old as his son. In 3 years he will be $3\frac{1}{2}$ times as old.

6. Six years ago Bob was 4 times as old as Peri. In 9 years he will be $\frac{3}{2}$ as old.

7. Sarat is 4 years younger than his brother Akhil. Fourteen years ago Akhil was twice as old as Sarat.

8. In 2 years Peter will be twice as old as his sister Eve. The sum of their present ages is 26.

9. Lauren is 18 years younger than Joyce. Twelve years ago Joyce was 4 times as old as Lauren was.

10. In 3 years Rufus will be $\frac{2}{3}$ the age of Jason. The sum of their present ages is 44.

11. Six years ago the sum of Harvey's and Carol's ages was 71. Harvey is 27 years older than Carol.

12. 21 years ago Alvin was 11 times older than Tomio. The sum of their present ages is 54.

13. Mary is 9 years older than Seth. In a year she will be 3 times as old as he is.

14. Roz is twice as old as Gerry. In 5 years the sum of their ages will be 28.

15. Susan is 3 times as old as Randi. One-fourth the sum of their ages is 8.

16. The sum of the ages of Ira and Sofia is 26. The difference between 4 times Sofia's age and twice Ira's age is 8.

17. In 1945, Liz was twice as old as Bill. That same year the difference in their ages was 18 years. In what year was each born?

18. In 1987, Joy was 3 times as old as Ali. If Ali was born 8 years after Joy, what age will each be in 1999?

19. Six years ago Bill was 17 years younger than twice Ann's age. Now their combined age is 100 years. Find their present ages.

20. Patty was 3 times as old as Tom in 1982. In 1986, the difference in their ages was 4 years. Find their ages now.

21. The sum of Spero's and Chris's ages is 50 years. In $2\frac{1}{2}$ years Spero will be $\frac{5}{6}$ of Chris's age. Find their present ages.

22. Presently, Susie is $\frac{3}{4}$ of Weiva's age. In 4 years she will be $\frac{4}{5}$ of Weiva's age. Find their present ages.

23. Armando is $1\frac{1}{3}$ times as old as Zelda. Five years ago his age was $1\frac{1}{2}$ hers. Find Armando's age now.

24. Nina's age is $\frac{2}{3}$ of Liu's. Four years ago Liu's age was $\frac{5}{3}$ of Nina's. What is Nina's age now?

Solve by using a system of three equations in three variables.

25. The sum of all Emma's and Vangie's books equals 45. One fourth of Kim's books are equal to twelve less than Emma's books. Kim's books are equal to $\frac{8}{5}$ of Emma's. Find the present number of books for each person.

26. Diana had three classes this semester, history, mathematics, and English. Her total grade points for all three classes was 255. Her history grade was 10 more points than her English grade. Her mathematics grade was 95 points less than twice her history grade. What grade did she receive in each class? What was her grade-point average?

27. Thirteen years ago Mary's age was $\frac{5}{17}$ of Grace's age. Two years from now Mary will be $\frac{1}{2}$ of Jim's age. Together all their ages add up to 118. Find their ages five years from now.

EXTRA

A **palindrome** sentence reads the same backward as forward. Solve the systems of equations to decode the palindromes. Each solution corresponds to a point on the graph. The letter for the point replaces the number of the system in the coded palindrome.

1. $x + y - 6 = 0$
   $x - 4 = y$

2. $x + y = 2$
   $2y - 10 = x$

3. $2x + 3y - 8 = 0$
   $3x + y = 5$

4. $2x + 3y + 5 = 0$
   $5x + 3y = 1$

5. $4x + 3y = -2$
   $8x = 2y + 12$

6. $2y = 5x - 11$
   $3x + 5y = 19$

7. $2x + 5y - 18 = 0$
   $y - 5x + 18 = 0$

8. $3x + y = 10$
   $y + 2x = 7$

9. $4x = 2y + 20$
   $x + 5y = -17$

10. $4x = 7 - 3y$
    $4y = 12 - 4x$

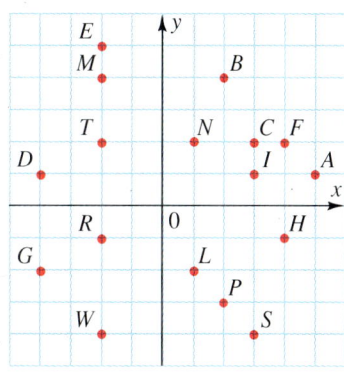

Palindrome 1: 2–1 8–9 1–9 9–10–5–7–5–10–9–9 1–9 8 1–2

Palindrome 2: 1 2–1–3, 1 4–5–1–3, 1 6–1–3–1–5, 4–1–3–1–2–1

11.8 Problem Solving: Money and Mixture Problems

Objective: To use systems of equations to solve money and mixture problems

Randy has 83 coins worth $10.70. Some are dimes and the rest are quarters. By trial and error, it is possible to find how many of each coin he has, although it would probably take a long time. Using algebraic equations might prove easier.

EXAMPLE 1 Use a system of equations to find how many of each kind of coin is in the mixture described above.

■ **Understand the Problem**

There are 83 coins in all, consisting only of dimes and quarters. The value of the coins, in dollars, is $10.70. (In cents, the value of the coins is 1070¢.)

■ **Plan Your Approach**

Let d = the number of dimes.
Let q = the number of quarters.
Organize the information in terms of the variables.

No. of Coins	Value of Each (¢)	Total Value
d	10	$10d$
q	25	$25q$
83		1070

Use the first and third columns:
$$\begin{cases} d + q = 83 \\ 10d + 25q = 1070 \end{cases}$$

Use the multiplication-addition/subtraction method to solve the system.

■ **Complete the Work**

$d + q = 83 \rightarrow \begin{cases} -10d - 10q = -830 & \text{Multiply each side by } -10. \\ \underline{10d + 25q = 1070} \end{cases}$

$15q = 240$ *Add to eliminate the d-terms.*
$q = 16$

If $q = 16$ and $d + q = 83$, then $d = 83 - 16$, or 67.

■ **Interpret the Results**

Conclusion: There are 16 quarters and 67 dimes.
Check: 16 quarters and 67 dimes make a total of 83 coins.
Value in cents: $16(25) + 67(10) = 400 + 670 = 1070$
So, there are 67 dimes and 16 quarters.

EXAMPLE 2 A chemist has two alcohol-in-water solutions: a 20% alcohol solution and a 50% alcohol solution. He needs 12 L of a solution that is 45% alcohol. How many liters of the two starting solutions should he mix?

■ Understand the Problem

How many liters of the 20% alcohol solution and of the 50% alcohol solution must be mixed to produce 12 L of a 45% solution?

■ Plan Your Approach

Let x = number of liters of 20% alcohol solution.
Let y = number of liters of 50% alcohol solution.

	No. of liters of solution	No. of liters of alcohol
20% solution	x	$0.20x$
50% solution	y	$0.50y$
45% solution	12	$0.45(12)$

The amount of pure alcohol in the first two solutions, $0.20x + 0.50y$, must equal the alcohol in the new solution, $0.45(12)$. Write two equations; solve by substitution.

■ Complete the Work

From the table: $\begin{cases} x + y = 12 \longrightarrow x = 12 - y \\ 0.20x + 0.50y = 0.45(12) \rightarrow 20x + 50y = 45(12) \end{cases}$

$20x + 50y = 45(12)$ Solve by substitution.
$20(12 - y) + 50y = 540$ Substitute $12 - y$ for x.
$2(12 - y) + 5y = 54$ Divide each side by 10.
$24 - 2y + 5y = 54$ Solve for y.
$y = 10$

If $x + y = 12$ and $y = 10$, then $x = 12 - 10$, or 2.

■ Interpret the Results

Conclusion: 2 L of 20% solution and 10 L of 50% solution are required to make 12 L of 45% solution.

Check: $0.20(2) + 0.50(10) \stackrel{?}{=} 0.45(12)$
$0.4 \quad + \quad 5 \quad \stackrel{?}{=} 0.45(12)$
$5.4 = 5.4$ ✔

CLASS EXERCISES

Write an algebraic expression for each word sentence. Use the following information.

h = number of lb of hazelnuts at $4.98/lb, and c = number of lb of cashews at $6.50/lb.

1. What is the total number of pounds of nuts?

2. What is the value of the hazelnuts in dollars? in cents?

11.8 Problem Solving: Money and Mixture Problems

3. What is the value of the cashews in dollars? in cents?
4. What is the total value of the nuts?
5. Write a system of equations for this problem. A 15-lb mixture of hazelnuts and cashews is worth $5.79/lb. How many pounds of each kind does the mixture contain?
6. What is the advantage of expressing the values of the nuts in cents?

x = amount of money (dollars) invested at 7.5% annual interest, and
y = amount of money (dollars) invested at 8% annual interest.

7. What yearly interest (dollars) is earned on the 7.5% investment?
8. What yearly interest (dollars) is earned on both investments?
9. What is the total amount of money invested?
10. Give a system of equations for this problem: part of $17,000 was invested at 7.5% annual interest and the remainder at 8.0%. If the combined annual interest was $1347.50, how much was invested at each rate?

PRACTICE EXERCISES

Write a system of equations, then solve.

1. George saves nickels and dimes for tolls. If he has 28 coins worth $2.60, how many are nickels and how many are dimes?
2. Lorena has 26 coins in nickels and quarters, which are worth a total of $3.10. How many of each coin does she have?
3. Movie tickets cost $6 for adults and $2 for children under twelve. If 175 tickets were sold, with cash receipts of $750, how many children's tickets were sold?
4. To raise funds, a school sells two kinds of raffle tickets, some for $6, others for $1.50. Sales for both amounted to $822. If 371 tickets were sold, how many were $1.50 tickets?
5. The health food store wishes to blend peanuts that cost $1.20/lb with raisins that cost $2.10/lb to make 50 lb of a mixture that costs $1.47/lb. How many pounds of peanuts and of raisins are needed?
6. A 100 kg mixture of $0.69/kg pinto beans and $0.89/kg kidney beans is valued at $81. How many kilograms of each does it contain?
7. Part of an investment of $32,000 earns 7.5% annual interest; the rest earns 9%. If the annual interest from both is $2670, how much is invested at the higher rate?

8. The total annual interest from two bank accounts is $481. One account earns 5.9% annual interest and the other earns 6.75%. If the two accounts contain a total of $7600, how much is in the account at the lower interest?

9. A bank teller has a total of 124 bills in fives and tens. The total value of the money is $840. How many of each kind does he have?

10. In a cash register there are 87 bills, $5 and $1 bills only. If their value is $179, how many of each kind of bill are there?

11. Mrs. Chavis has twice as much money invested at 7.5% as she has at 6.0%. The yearly income from both investments is $840. How much has she invested at each rate?

12. In a laboratory a pharmacist has 30% and 80% alcohol solutions. She needs 100 mL of a 50% alcohol solution. How many milliliters of the 30% and 80% solutions should she mix?

13. A dairy has milk that is 4% butterfat and cream that is 40% butterfat. To make 36 gal of a mixture that is 20% butterfat, how many gallons of milk and cream must be used?

14. The number of $2 raffle tickets printed is $3\frac{1}{2}$ times the number of $5 tickets. If all the tickets are sold, receipts from the $5 tickets will be $500 less than those from the $2 tickets. How many $5 tickets were printed?

15. The attendance at a school football game was 350. Tickets for adults cost $2.25, compared to $1.00 for children. If the total receipts were $600, how many children and adults attended?

16. In order to get a thicker sauce, a restaurant owner combines a sauce that is 70% tomato paste with the original sauce, which is 40% tomato paste. How much of each should be used to make 5 L of the new 60% tomato paste sauce?

17. Walnuts cost $9.95/lb, while peanuts cost $6.50/lb. If there are 3 less lb of peanuts than there are walnuts, how many lb of each are there in a box which costs $62.75?

18. Rosa has $3.10 in nickels and dimes. She has 10 fewer nickels than she has dimes. How many dimes and nickels does she have?

19. A collection of 90 coins, pennies, nickels, and dimes, has a value of $2.85. If there are twice as many pennies as there are nickels and dimes combined, how many pennies, dimes, and nickels are there?

20. Miquel invested $10,000 in 3 banks. He invested the same amount of money in two banks, which gave him a 6% yearly interest rate. The third bank gave him a rate of 7% for one year. If he made $640 in interest after 1 year, how much did he invest in each bank?

Applications

Computer Bill does not like to carry more than $1.50 in change. More than eight coins tear holes in his pockets. The computer program at the right will generate a chart of the possible combinations of dimes and quarters that Bill can carry.

```
10 PRINT "QUARTERS","DIMES","TOTAL"
20 D = 0
30 FOR Q = 8 TO 0 STEP -1
40 QT = Q * .25:DT = D * .10
50 PRINT Q;"   $";QT,D;"   $";DT,"   $";QT + DT
60 D = D + 1: NEXT Q
70 END
```

Solve.

21. What is the least number of quarters and dimes that Bill can carry and have $1.25? What are the coins?

22. If Bill carried all dimes, how many would he need to carry to have as much money as if he carried all quarters?

23. How many of each coin should Bill carry if he wants to have an equal amount of money for each coin?

BIOGRAPHY: Sonja Corvin-Kowalewski

Sonja Corvin-Kowalewski was born in Moscow in 1850. By the time she was 18, her progress in the study of mathematics was so rapid that she enrolled in the University of Heidelberg for advanced studies. At age 20 she became a student of Karl Weierstrass, the father of modern analysis, with whom she did considerable work on the refraction of light in crystalline media. In 1888, her paper "On the Rotation of a Solid Body About a Fixed Point" won the coveted Bordin Prize of the French Academy of Sciences. She taught at the University of Stockholm from 1889 until she died, at 41, during a flu epidemic.

An Investigation

Kowalewski's interest in crystals may have been generated by the fact that crystals, including snowflakes, provide striking examples of symmetry and other important mathematical principles. Do some research on crystals, and prepare a report on this natural phenomenon.

11.9 Problem Solving: Wind and Water Current Problems

Objective: To use systems of equations to solve uniform-motion problems involving winds and water currents

Cal and Marty were about to start a round trip cycling race on a windy day. "Because of the wind, our times won't be as good as they are on a calm day," Cal said.

"You're wrong," said Marty. "It will slow us down when we head into the wind, but it will help us when the wind is at our backs, and so it won't make any difference."

Which boy is right?

EXAMPLE 1 Supply numbers to consider Cal's and Marty's opinions. Assume that they are making a 60-mi round trip, that with no wind they can average 15 mi/h on their bicycles, and that there is a constant wind of 5 mi/h.

Understand the Problem For half the round trip they head directly into the wind. Their rate is 15 − 5, or 10 mi/h. For half the trip the wind blows in the same direction they are traveling. Their rate then is 15 + 5, or 20 mi/h.

Plan Your Approach Use the distance formula, distance = rate × time, to find the average rate of speed with the wind blowing. Compare this with the given rate of speed with no wind.

Make a rate-time-distance table to determine the time it would take to cycle each part of the trip when the wind is blowing. Then use the total time and the total distance to find the average rate of speed for the round trip.

Let t_1 = time against wind.
Let t_2 = time with wind.
$t_1 + t_2$ = total time.

	r	× t =	d
Against wind	15 − 5	t_1	30
With wind	15 + 5	t_2	30

Write equations to find t_1 and t_2.

Complete the Work

$(15 - 5)t_1 = 30$ $(15 + 5)t_2 = 30$
$10t_1 = 30$ $20t_2 = 30$
$t_1 = \frac{30}{10} = 3$ $t_2 = \frac{30}{20}$, or $\frac{3}{2}$

Find the total time: $t_1 + t_2 = 3 + \frac{3}{2}$, or $\frac{9}{2}$

Substitute the total time, $\frac{9}{2}$, to find the rate of speed for the total distance with the wind blowing: $r \cdot \frac{9}{2} = 60$, or $r = 13.\overline{3}$

Interpret the Results

$13.\overline{3}$ mi/h is less than 15 mi/h for the round trip. Cal was right. On a windy day the average rate of speed for a round trip is less than on a day with no wind.

The distance formula and systems of equations can be used to solve water-current problems.

EXAMPLE 2   It takes 2 h for a boat to travel 28 mi downstream. The same boat can travel 18 mi upstream in 3 h. Find the rate of speed of the boat in still water and the rate of the current.

Understand the Problem

Draw a diagram.   downstream ←——— boat ———→ upstream
                              ←——— current

The boat travels with the current when it goes downstream. Thus, the rate the boat travels downstream is the rate of the boat in still water plus the rate of the current. What is the rate of speed of the boat when it travels upstream?

Plan Your Approach

Make a table to show a system of equations. Then solve.
Let r = rate of speed of the boat in still water.
Let c = rate of current.

	r	$\times$	t	$=$	d
Downstream	$r + c$		2		28
Upstream	$r - c$		3		18

Complete the Work

$\begin{cases} (r + c)2 = 28 \to r + c = 14 \\ (r - c)3 = 18 \to r - c = 6 \end{cases}$

                      $2r = 20$ Add the equations.
                        $r = 10$

Substitute 10 for r, and solve for c: $(r + c)2 = 28$
                                             $c = 4$

> **Interpret the Results**
>
> Conclusion: The rate of the boat in still water is 10 mi/h. The rate of the current is 4 mi/h.
>
> The check is left for you.

CLASS EXERCISES

Complete the table. Write a system of equations. Then solve.

1. An airplane averages 255 mi/h with no wind. How long will it take the airplane to travel 2160 mi flying against a 15 mi/h wind? How long does it take the airplane to travel the same distance flying with the wind? Find the average rate of speed for the round trip.

	Rate	× Time	= Distance
Against wind	?	t_1	2160
With wind	?	t_2	2160

2. A boat takes 3 h to go 24 mi upstream and 3 h to go 36 mi downstream. Find the speed of the current and the rate of the speed of the boat in still water.

	Rate	× Time	= Distance
Downstream	?	?	36
Upstream	?	?	24

PRACTICE EXERCISES

1. A small aircraft flies a 2160-mi round trip in a 45 mi/h wind. Find the average rate of speed for the trip if the aircraft can fly 225 mi/h with no wind.

2. A cyclist travels a 90-mi round trip in a 3 mi/h wind. Find the average rate of speed for the trip if the cyclist can travel 12 mi/h with no wind.

3. A crew rowing with the current traveled 16 mi in 2 h; against the current, the crew rowed 8 mi in 2 h. Find the rate of rowing in still water and the rate of the current.

4. Traveling with the wind, a plane flew 4000 km in 5 h. Against the wind the plane only flew 3000 km in the same time. Find the rate of speed of the plane in calm air and the speed of the wind.

5. A boat travels 12 km upstream in 2 h. If the return trip takes 1 h, find the rate of speed of the boat in calm water.

11.9 Problem Solving: Wind and Water Current Problems

6. A plane travels 1200 mi with the wind in 2 h. The return trip against the wind takes 3 h. Find the rate of speed of the plane in calm air.

7. A plane traveled 2400 km in 4 h with a tailwind. However, returning against the same wind took 1 h longer. Find the plane's airspeed and the wind speed.

8. Flying a propeller plane against a headwind for 700 mi, Sandy took 5 h to reach the city. He flew with the tailwind when he made the return trip in $3\frac{1}{2}$ h. What was the wind speed?

9. Natasha rowed upstream 28 mi in 7 h and traveled the same distance downstream in 4 h. What was her rate without the current?

10. Pam swims downstream for 10 mi in 2 h. The return trip upstream takes four times as long. What is the rate of the current?

11. Traveling at 14 mi/h, Noah makes a 140-mi bike trip against the wind. He covers the same distance at 20 mi/h with the wind. Find his total time traveled.

12. Bjorn traveled 100 mi on his motorcycle in 2 h with the wind. It took him 3 hours to return the same distance against the wind. Find the rate of the wind.

13. Against prevailing headwinds, a pilot calculates his flight as 6 h for a distance of 2100 mi. The return trip with the wind would require 1 h less. Find the rate of speed of the plane with no wind.

14. With a tailwind, a plane flew 300 mi in 40 min. With no change in wind, the return trip took 5 min more. Find the rate of the wind in mi/h.

15. A riverboat travels 10 km upstream in 50 min. The riverboat travels 15 km downstream in 45 min. Find the rate of speed of the boat in calm water in km/h.

16. A swimmer swims upstream at 40 m/min and downstream at twice the rate. How fast can she swim in still water?

17. A steamboat travels 8.4 mi downstream in 3 h. Traveling upstream, it can travel $\frac{3}{7}$ of the distance in the same time. What is the rate of the current and the rate of the steamboat in still water?

18. A canoe covers a distance of 40 mi downstream in 6 h. Returning upstream, it takes the canoe 3 times as long to cover $\frac{3}{5}$ of the distance. Find the rate of the current and the rate of the canoe in still water.

19. Flying for 5 h against the wind, François makes a journey of 180 km in his ultralight. With the wind the ultralight makes the same trip in 3 h. Find the rate of speed of the wind.

20. An eagle flies 300 mi in 8 h with the wind. It flies only $\frac{1}{3}$ of the distance in 7 h against the wind. Find the rate of the eagle in the air against the wind.

21. It takes a salmon the same time to swim a distance upstream that it takes to swim twice that distance downstream. If the salmon swims at a rate r and the rate of the current is c, find the relationship between the two rates.

22. A plane travels with the wind at a mi/h and travels against the wind at b mi/h. Find the rate of the wind and the rate of speed of the plane in still air in terms of a and b.

CAREER: Geophysicist

A **geophysicist** is a person who specializes in the branch of science that deals with the physics of the Earth. This includes weather, winds, tides, earthquakes, volcanoes, and their effect on Earth. The data gathered and studied by geophysicists are often used to improve the environment in which we live, and to help us live more safely and comfortably in an environment over which we may have little control.

Most geophysicists specialize in one area of geophysics, such as climatology (the study of climate and climatic phenomena), meteorology (the study of the atmosphere and atmospheric phenomena), and seismology (the study of earthquakes and related phenomena).

If you are curious about the physical environment, enjoy collecting and analyzing data, and like mathematics and science, then you may be interested in becoming a geophysicist.

An Investigation

An earthquake sends shock waves through the earth. A primary wave travels at a rate of 5 mi/s, and a secondary wave, which starts at the same time, travels at a rate of 3 mi/s. A seismologist working in a seismic station notes that the time between the primary and secondary waves of an earthquake was 16 s. Using the distance formula $d = rt$, the scientist writes and solves two equations to determine how far from the station the earthquake occurred. What two equations did the scientist write? How far from the station was the earthquake?

11.10 Graphing Systems of Linear Inequalities

Objective: To graph the solution set of a system of linear inequalities

In order to analyze ways of maximizing profits and minimizing costs, some companies use graphs of *systems of linear inequalities*. A **system of linear inequalities** involves two or more linear inequalities. Such a system can often be solved by graphing.

Capsule Review

To graph an inequality, it is important to graph the related equality first.

Graph each inequality.

1. $y \leq -3x$ 2. $2x + y < 6$ 3. $2x + 3y > 12$

To solve a system of linear inequalities in two variables x and y, first solve each inequality for y.

EXAMPLE Graph the solution of the system $\begin{cases} 2x + y > 2 \\ x - y \geq 3 \end{cases}$

$2x + y > 2$ $\qquad$ $x - y \geq 3$
$\quad y > -2x + 2$ $\qquad -y \geq -x + 3$ $\quad$ *Reverse the direction of the*
$\qquad\qquad\qquad\qquad\qquad\;\; y \leq x - 3$ $\quad$ *inequality when you multiply or divide by a negative number.*

Graph $y > -2x + 2$.
First, graph $y = -2x + 2$. Use a broken line since the graph of $y > -2x + 2$ does not include the line. Shade the region above the line to represent the graph of $y > -2x + 2$.

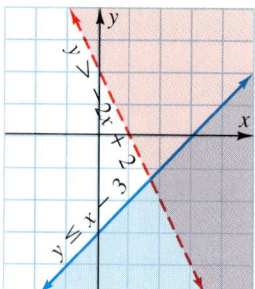

Graph $y \leq x - 3$ in the same coordinate plane. First graph $y = x - 3$. Use a solid line since $y \leq x - 3$ includes points on the line. Shade the region below the line to represent the graph of $y \leq x - 3$.

The graph of the solution is shaded in both colors and is the intersection of the inequalities. The solution includes points on the boundary line $y = x - 3$, but not on the boundary line $y = -2x + 2$.

484 Chapter 11 Systems of Linear Equations

CLASS EXERCISES

Tell which region, *A*, *B*, *C*, or *D*, is the graph of the system of inequalities.

1. $\begin{cases} x < 0 \\ y < 0 \end{cases}$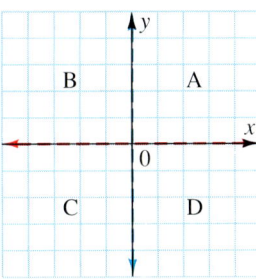

2. $\begin{cases} x \geq -1 \\ y \geq 1 \end{cases}$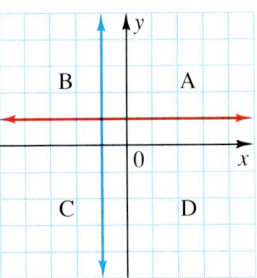

3. $\begin{cases} x \geq 2 \\ y < 3 \end{cases}$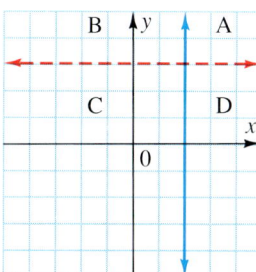

4. $\begin{cases} x < -2 \\ y > -2 \end{cases}$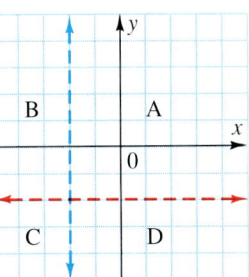

Determine whether the ordered pair is a solution of the system $x + y \leq 6$ and $x - y < 1$.

5. $(1, -3)$
6. $(1, 2)$
7. $(0, 0)$
8. $(6, 0)$
9. $(-2, 1)$
10. $(7, 1)$
11. $(8, 6)$
12. $(-3, -1)$

For Discussion

13. In the Example on page 484, each equation divides the coordinate plane into three sets.
    s_1: The set of points on the line
    s_2: The set of points for which the *greater than* relation holds
    s_3: The set of points for which the *less than* relation holds
    Use these three sets to describe the solution of the system $2x + y > 2$ and $x - y \geq 3$.

11.10 Graphing Systems of Linear Inequalities

PRACTICE EXERCISES

Graph the system of inequalities, and indicate the solution of the system by shading.

1. $\begin{cases} x \leq 1 \\ y \geq 4 \end{cases}$
2. $\begin{cases} x \geq 2 \\ y \leq 3 \end{cases}$
3. $\begin{cases} y \geq 0 \\ x > 0 \end{cases}$
4. $\begin{cases} y < -2 \\ x \leq 2 \end{cases}$
5. $\begin{cases} y \leq 2x \\ x > 1 \end{cases}$
6. $\begin{cases} y > x \\ x < -2 \end{cases}$
7. $\begin{cases} y > 1 - x \\ x \leq -1 \end{cases}$
8. $\begin{cases} y < 2 - x \\ x > 4 \end{cases}$
9. $\begin{cases} y < x + 3 \\ y < 5 - 2x \end{cases}$
10. $\begin{cases} y < 5x - 1 \\ y \geq 7 - 3x \end{cases}$
11. $\begin{cases} y - x \leq 5 \\ y > 2x \end{cases}$
12. $\begin{cases} y < -3x \\ x > 4 \end{cases}$
13. $\begin{cases} x + y > 5 \\ 2x - y < 4 \end{cases}$
14. $\begin{cases} x + y > 2 \\ 2x - y \leq 1 \end{cases}$
15. $\begin{cases} y > 4x + 2 \\ x \leq 4 \end{cases}$
16. $\begin{cases} y \leq 4x - 3 \\ x > 3 \end{cases}$
17. $\begin{cases} 3x + y \geq 0 \\ 2x - y \leq -4 \end{cases}$
18. $\begin{cases} 2x + y \leq 4 \\ 3x - y > 6 \end{cases}$
19. $\begin{cases} 4x - y < 4 \\ x + 2y < 2 \end{cases}$
20. $\begin{cases} x + y > 0 \\ x - 3y > 3 \end{cases}$
21. $\begin{cases} x - 2y < 3 \\ 2x + y > 8 \end{cases}$
22. $\begin{cases} x + 2y > 4 \\ 2x - y > 6 \end{cases}$
23. $\begin{cases} 2x \geq y + 3 \\ x < 3 - 2y \end{cases}$
24. $\begin{cases} 3 < 2x - y \\ x - 3y \leq 4 \end{cases}$
25. $\begin{cases} 2x - y \leq 3 \\ x - \frac{1}{3}y > 1 \end{cases}$
26. $\begin{cases} 5x - y > 6 \\ 2x - \frac{1}{2}y \leq 3 \end{cases}$
27. $\begin{cases} 3y - x + 1 \leq 0 \\ x - y - 1 > 0 \end{cases}$

A system of inequalities may contain more than two inequalities. Graph the system.

28. $\begin{cases} 3y - x + 1 \leq 0 \\ x + 2y - 3 \geq 0 \\ x - y - 1 > 0 \end{cases}$
29. $\begin{cases} 4x + 5 > y \\ x - 4 \leq 2y \\ 4y \leq -5x + 20 \end{cases}$
30. $\begin{cases} 3x + 2 \geq 2y \\ 2x - 8 > 4y \\ -4y < x - 4 \end{cases}$
31. $\begin{cases} \frac{1}{2}x - y > 3 \\ 3x - \frac{3}{2}y \geq -1 \\ y < 4 \end{cases}$
32. $\begin{cases} x - 4y < 4 \\ \frac{1}{3}x - 2y \geq 2 \\ -x + \frac{1}{2}y \geq -2 \end{cases}$
33. $\begin{cases} x > \frac{1}{2}y + 1 \\ -\frac{2}{3}x \leq 2y - 5 \\ \frac{5}{6}y \geq -4x + 2 \end{cases}$

Applications

Write and solve a system of two inequalities.

34. **Office Administration** Two typists are working on a statistical report. To finish on time, together they need to type between 20 and 30 pages per day. This is the first time typist A has typed material that contains mostly numbers, so she may be expected to type no more than $\frac{2}{3}$ as many pages as typist B. On the graph, find at least two acceptable combinations of typing output that the typists might produce in a day.

Chapter 11 Systems of Linear Equations

35. Agriculture Some land on a farm is to be fenced as a feeder lot for cattle. The farmer wants the distance around the lot to be no more than 2600 ft. The length should be greater than 800 ft. From the graph, name two sets of dimensions for the feeder lot that satisfy the farmer's requirements. Must the graph be shown in all four quadrants of the coordinate plane?

TEST YOURSELF

Find the original number described in the problem.   11.6

1. The sum of the digits is 11. Reversing the digits gives a new number that is 9 more than the original.

2. The tens digit is twice the units digit. Reversing the digits gives a new number that is 27 less than the original.

Write and solve a system of linear equations to find the present ages.   11.7

3. Robert is four years older than Dawn. Seven years ago Dawn was $\frac{3}{4}$ as old as Robert.

4. Greg is three times as old as Jane. Five years ago he was five times as old.

Solve by first writing a system of equations.   11.8–11.9

5. Anne Marie has 52 coins in dimes and quarters, which are worth $6.25. How many of each coin does she have?

6. John has a total of 9 stamps, which consist of 25¢ and 2¢ stamps. If his stamps have a value of $1.10, how many of each stamp does he have?

7. A car travels a 240 mi round trip in a 15 mi/h wind. Find the average rate of speed for the trip if the car travels 45 mi/h with no wind.

8. It takes 4 h for a boat to travel 56 mi downstream. The same boat can travel 36 mi upstream in 6 h. Find the rate of speed of the boat in still water and the rate of the current.

Graph the system of inequalities, and indicate the solution of the system by shading.   11.10

9. $\begin{cases} x \leq 3 \\ y \geq 1 \end{cases}$

10. $\begin{cases} x \geq 2 \\ y \leq 3 \end{cases}$

11. $\begin{cases} y \geq 1 - 2x \\ x > y \end{cases}$

12. $\begin{cases} y > -4 - x \\ 3y - x \leq 3 \end{cases}$

CHAPTER 11 SUMMARY AND REVIEW

Vocabulary

consistent system (444)
dependent system (445)
equivalent systems (463)
inconsistent system (445)
independent system (444)
simultaneous equations (444)
system of linear equations (444)
system of linear inequalities (484)

The Graphing Method To solve a system of linear equations by graphing, graph both equations in the same coordinate plane. The coordinates of any point of intersection of the lines are a solution of the system. **11.1**

Solve by graphing. State the number of solutions and describe the system.

1. $\begin{cases} x + y = 3 \\ 2x - y = 3 \end{cases}$
2. $\begin{cases} x = 2y + 2 \\ 2x - 4y = 4 \end{cases}$
3. $\begin{cases} 2x - y = 10 \\ 2x - y = 8 \end{cases}$

The Substitution Method To solve a system of linear equations by substitution: **11.2**

- In either equation solve for one variable in terms of the other.
- Substitute for that variable in the other equation. Solve.
- Find the corresponding value of the other variable and check the solution.

Solve by the substitution method.

4. $\begin{cases} y = 2x \\ x + 2y = 8 \end{cases}$
5. $\begin{cases} 2r + 3s = 27 \\ 4r - s = 9 \end{cases}$
6. $\begin{cases} \frac{1}{4}c - d = -1 \\ 4c + 3d = 22 \end{cases}$

Mixed Problems To solve problems that involve two unknowns and two conditions, use the four problem solving steps. Assign two variables to the unknowns, translate the given statements into a system of equations, and solve the system. **11.3**

7. The perimeter of a triangular sign is 21 ft. If the lengths of two sides are equal and the third side is 3 ft longer than one of the equal sides, find the lengths of the three sides.

The Addition/Subtraction Method To solve a system of linear equations using the addition/subtraction method: **11.4**

- add or subtract the equations to eliminate one of the variables.
- solve the resulting equation for the remaining variable.
- replace that value in either equation, and solve for the unknown variable.
- check the solution.

Solve by the addition/subtraction method.

8. $\begin{cases} 2a + b = 8 \\ a - b = 4 \end{cases}$

9. $\begin{cases} 3p = 13 - q \\ 2p - q = 2 \end{cases}$

10. $\begin{cases} \frac{1}{2}m - \frac{1}{2}n = 10 \\ \frac{3}{4}m + \frac{1}{2}n = 20 \end{cases}$

The Multiplication-Addition/Subtraction Method To solve a system of equations by the multiplication-addition/subtraction method, multiply one or both equations to produce an equivalent system in which the coefficients of one variable are the same or additive inverses of one another. Then use the addition/subtraction method. **11.5**

Solve by the multiplication-addition/subtraction method.

11. $\begin{cases} x + 4y = 17 \\ 3x + 2y = 6 \end{cases}$

12. $\begin{cases} 6a + 7b = -33 \\ 5a - 4b = 2 \end{cases}$

13. $\begin{cases} \frac{1}{4}c + d = \frac{7}{2} \\ \frac{1}{2}c - \frac{1}{4}d = 1 \end{cases}$

Digit Problems To solve problems involving two-digit integers, use two variables to represent the digits, write a system of equations to describe the relationships between them, and solve the system. **11.6**

14. The sum of the digits of a two-digit number is 5. If the digits are interchanged, the new number is 13 more than twice the original. Find the original number.

Age Problems, Money and Mixture Problems, Wind and Water-Current Problems To solve these types of problems, translate the statements into a system of equations, and solve the system. **11.7–11.9**

15. The ages of two sisters, Sue and Amy, total 35 yr. In 5 yr, Sue will be twice as old as Amy. How old is each now?

16. Ed spent $3.40 for 22 stamps. If he bought only 25-cent and 10-cent stamps, how many of each kind did he buy?

17. It takes a boat $1\frac{1}{2}$ hr to go 12 mi downstream and 6 hr to return. Find the rate of speed of the boat in still water and the speed of the current.

Solving Systems of Inequalities To solve a system of inequalities, graph each inequality on the same coordinate plane. The solution of the system is the double-shaded area, which is the intersection of the inequalities. **11.10**

Solve by graphing.

18. $\begin{cases} x < -1 \\ y > 3 \end{cases}$

19. $\begin{cases} y \leq x + 4 \\ x + 2y \geq 5 \end{cases}$

20. $\begin{cases} x + 2y < -4 \\ 2x - y < -3 \end{cases}$

CHAPTER 11   TEST

Solve the system of equations by graphing.

1. $\begin{cases} 3x - y = 4 \\ -9x + 3y = -12 \end{cases}$
2. $\begin{cases} x + 2y = -2 \\ x - y = 4 \end{cases}$
3. $\begin{cases} x + 2y = 6 \\ x + y = -1 \end{cases}$

Solve the system by any algebraic method you prefer.

4. $\begin{cases} x - y = 1 \\ y = 2x - 3 \end{cases}$
5. $\begin{cases} x + y = 2 \\ 3x - 2y = -9 \end{cases}$
6. $\begin{cases} 2x + y = 0 \\ y = \dfrac{x}{3} \end{cases}$
7. $\begin{cases} 2x - y = -1 \\ 3x = y + 2 \end{cases}$
8. $\begin{cases} \dfrac{x}{3} - \dfrac{y}{2} = 6 \\ \dfrac{x}{2} + \dfrac{y}{8} = 2 \end{cases}$
9. $\begin{cases} 5x = 4y - 7 \\ 8y = 6x + 2 \end{cases}$

Solve the system of linear inequalities by graphing.

10. $\begin{cases} 2x + y < -4 \\ x \geq 1 \end{cases}$
11. $\begin{cases} y \leq -2x \\ x > -2y + 6 \end{cases}$
12. $\begin{cases} 2x + y < 0 \\ 2x - y < -6 \end{cases}$

Use a system of equations to solve the problem.

13. The sum of the digits of a two-digit number is 9. If the digits are reversed, the new number is 45 more than the original number. Find the two-digit number.

14. A store sells cashews for $4.40/lb and peanuts for $1.20/lb. How many lb of each can be bought to get exactly 3 lb of nuts for $6?

15. A barge travels 16 mi upstream in 4 h. On the return trip it takes 2 h to travel the same distance with the current. Find the rate of speed of the barge in still water and the speed of the stream's current.

16. Allen is 4 yr older than his brother Randy. Three years ago, Allen was twice as old as Randy. How old is each now?

Challenge

1. The jetstream is the wind that blows across the country from west to east. A jet plane flying directly against the jetstream makes an 800-mi nonstop trip from Kennedy Airport to O'Hare Airport in 2 h. The jet plane makes the return trip in 1 h 36 min with the jetstream directly behind it. Find the speed of the jetstream.

PREPARING FOR STANDARDIZED TESTS

Select the best choice for each question.

1. Find the value of $2ab(3a^2 - 5b)^2$ when $a = -3$ and $b = 5$.
   A. -1470
   B. -120
   C. -100
   D. 100
   E. 120

2. What is the value of x in the system of equations:
   $$\begin{cases} 2x - 7y = 8 \\ x - 4y = 3 \end{cases}$$
   A. 2 B. 5 C. 11 D. 17 E. $21\frac{1}{4}$

3. In his small motorboat it takes Sam 5 h to go 30 mi up the river, but he can return down the river in only 3 h. What is the rate, in mi/h, of the river?
   A. 8 B. 6 C. 5 D. 4 E. 2

4. Solve the equation:
   $3(x + 7) + 5(2x - 4) = 11(x - 3)$
   A. -17
   B. -16
   C. 16
   D. 17
   E. 33

5. Three years ago, Martin's age was one half his sister's age. If the sum of their ages today is 21, how old is Martin now?
   A. 4 B. 5 C. 8 D. 10 E. 13

6. $\frac{54.81}{2.7} = \frac{?}{}$
   A. 2.03 B. 20.3 C. 203
   D. 2.3   E. 23

7. An equation of a line through P(4, 3) and Q(5, 5) is:
   A. $y = 2x - 11$
   B. $y = -2x + 11$
   C. $y = 2x - 5$
   D. $y = -2x + 5$
   E. $y = 2x + 11$

8. Sara purchased 3 cassette tapes at $7.90 each, 1 video tape for $22.50, and 4 blank cassette tapes for $1.95 each. How much sales tax, at 6%, was added to the total?
   A. $0.32 B. $0.33 C. $3.24
   D. $3.27 E. $5.40

9. The sum of the digits of a two-digit number is 11. If the digits are reversed, the new number is 27 more than the original number. What is the units digit of the original number?
   A. 3 B. 4 C. 5 D. 6 E. 7

10. Some boys wanted to make shelves for their clubhouse, and after measuring the space, decided each shelf should be a board 18 in. long and 6 in. wide. They found three boards, one measuring 2 ft × 12 in., another 20 in. × 6 in., and the third one 38 in. × 8 in. Using a saw and these boards, how many shelves can they make?
    A. 0 B. 2 C. 3 D. 4 E. 5

11. Find the coordinates of the point of intersection of the graphs of $y = 3x - 1$ and $2x + y = 9$.
    A. (1, 2) B. (-1, -4) C. (0, 9)
    D. (2, 5) E. (3, 3)

Preparing for Standardized Tests   **491**

MAINTAINING SKILLS

Perform the indicated operations.

Example 1 $1.2 \cdot 2 + 5.3 - 4.2 \div 7 + 3 \cdot 5$

$\underline{1.2 \cdot 2} + 5.3 - \underline{4.2 \div 7} + \underline{3 \cdot 5}$ *Do all multiplications and divisions from left to right.*

$= 2.4 + 5.3 - 0.6 + 15$ *Now do all additions and subtractions from left to right.*
$= 22.1$

1. $729.58 + 3.46$
2. $345.15 + 5.98$
3. $238.98 - 23.07$
4. $765.39 - 36.09$

5. $5.24 \cdot 2 \div 2.62 + 4$
6. $25.6 \div 8 \cdot 3.2 + 2$
7. $50.5 - 4 \cdot 1.2 \div 6 + 3$

Simplify.

Example 2 $\sqrt{900}$
$= \sqrt{9} \cdot \sqrt{100}$
$= 3 \cdot 10 = 30$

8. $\sqrt{25}$
9. $\sqrt{36}$
10. $\sqrt{400}$
11. $\sqrt{625}$
12. $\sqrt{144}$
13. $\sqrt{225}$
14. $\sqrt{0.04}$
15. $\sqrt{0.09}$

Multiply.

Example 3 $(a + 3)(a - 2)$

$(a + 3)(a - 2) = (a + 3)(a - 2) = a^2 - 2a + 3a - 6 = a^2 + a - 6$

16. $(a - 5)(a + 5)$
17. $(y - 3)(y - 4)$
18. $(2x + 6)(x - 3)$
19. $(3b + 5)(b - 2)$
20. $(2m + 4)(3m - 2)$
21. $(3x + 2)(4x - 5)$

Solve.

22. Sam is 185 cm tall. Otis is 1.5 m tall. What is the difference in their heights?

23. Box A weighs 3 lb 8 oz. Box B weighs 4 lb 9 oz. What is the combined weight of the boxes?

12 Radicals

Knowledge gained through technology is changing how we live and work in many ways. Mathematics plays an important role in the application of this knowledge to solve practical problems.

12.1 Square Roots

Objective: To find the square roots of numbers with rational square roots

A landscape architect wants to use 121 tiles, each one foot square, to create a square patio. Since $11^2 = 11 \times 11 = 121$, the architect can make a square patio that has 11 tiles on each side.

Subtracting a number is the inverse of adding the number and dividing by a nonzero number is the inverse of multiplying by the number. The inverse of squaring a number is finding the *square root*, $\sqrt{121} = 11$.

Capsule Review

Simplify.

1. 4^2
2. $(-2)^2$
3. $\left(\dfrac{1}{2}\right)^2$
4. 12^2

Write in exponential form.

5. $\left(-\dfrac{1}{3}\right)\left(-\dfrac{1}{3}\right)$
6. $(-4)(-4)$
7. $s \cdot s$
8. $a \cdot a \cdot b \cdot b$

If $x^2 = y$, then x is called a **square root** of y. Since $11^2 = 121$, then 11 is a square root of 121. However, $(-11)^2 = 121$, and -11 is also a square root of 121. Every positive real number has two square roots, one positive and one negative. Does every negative real number have two square roots?

To indicate the **positive** (or **principal**) **square root** of a number y, write $\sqrt{y}$. The symbol $\sqrt{}$ is called a **radical sign**. An expression like $\sqrt{25}$ is a **radical**, and the number under the radical sign is a **radicand**.

To indicate the negative square root of a number y, write $-\sqrt{y}$. The expression $\pm\sqrt{y}$ indicates both square roots of y. For example, $\sqrt{9} = 3$, $-\sqrt{9} = -3$, and $\pm\sqrt{9} = \pm 3$. When using your calculator, pressing the $\sqrt{}$ key will give the positive (or principal) square root.

494 Chapter 12 Radicals

EXAMPLE 1 Find the indicated square root: a. $\sqrt{64}$ b. $-\sqrt{\frac{1}{9}}$ c. $\pm\sqrt{0.81}$

a. $\sqrt{64} = 8$ b. $-\sqrt{\frac{1}{9}} = -\frac{1}{3}$ c. $\pm\sqrt{0.81} = \pm 0.9$

An important property of square roots is suggested by the following: $\sqrt{100} = 10$ and $\sqrt{25} \cdot \sqrt{4} = \sqrt{5^2} \cdot \sqrt{2^2} = 5 \cdot 2 = 10$. Therefore, $\sqrt{100} = \sqrt{25} \cdot \sqrt{4}$.

Product Property of Square Roots

For all real numbers m and n, where $m \geq 0$ and $n \geq 0$,
$$\sqrt{mn} = \sqrt{m} \cdot \sqrt{n}$$

A similar property for division of square roots is suggested by the following: $\sqrt{\frac{36}{9}} = \sqrt{4} = 2$ and $\frac{\sqrt{36}}{\sqrt{9}} = \frac{6}{3} = 2$. Therefore, $\sqrt{\frac{36}{9}} = \frac{\sqrt{36}}{\sqrt{9}}$.

Quotient Property of Square Roots

For all real numbers m and n, where $m \geq 0$ and $n > 0$, $\sqrt{\frac{m}{n}} = \frac{\sqrt{m}}{\sqrt{n}}$

Use one of the two properties of square roots to find the square root of a number.

EXAMPLE 2 Simplify. Do not use a calculator.

a. $\sqrt{1156}$ b. $\sqrt{\frac{1296}{169}}$ c. $-\sqrt{2.56}$

a. $\sqrt{1156}$
$= \sqrt{4 \cdot 289}$
$= \sqrt{4} \cdot \sqrt{289}$
$= \sqrt{2^2} \cdot \sqrt{17^2}$
$= 2 \cdot 17$
$= 34$

b. $\sqrt{\frac{1296}{169}}$
$= \frac{\sqrt{1296}}{\sqrt{169}}$
$= \frac{\sqrt{16} \cdot \sqrt{81}}{\sqrt{169}}$
$= \frac{\sqrt{4^2} \cdot \sqrt{9^2}}{\sqrt{13^2}}$
$= \frac{4 \cdot 9}{13}$
$= \frac{36}{13}$

c. $-\sqrt{2.56}$
$= -\sqrt{\frac{256}{100}}$
$= -\frac{\sqrt{256}}{\sqrt{100}}$
$= -\frac{\sqrt{16} \cdot \sqrt{16}}{\sqrt{10} \cdot \sqrt{10}}$
$= -\frac{4 \cdot 4}{10}$
$= -\frac{16}{10}$, or -1.6

CLASS EXERCISES

Find the indicated square roots.

1. $\sqrt{49}$
2. $-\sqrt{121}$
3. $\pm\sqrt{144}$
4. $\pm\sqrt{289}$
5. $\sqrt{5^2}$
6. $\sqrt{9^2}$
7. $\sqrt{\frac{1}{49}}$
8. $-\sqrt{\frac{1}{16}}$
9. $\pm\sqrt{1600}$
10. $\pm\sqrt{576}$

PRACTICE EXERCISES

Find the indicated square roots.

1. $\sqrt{25}$
2. $-\sqrt{81}$
3. $-\sqrt{256}$
4. $\sqrt{361}$
5. $\sqrt{\frac{4}{9}}$
6. $\pm\sqrt{\frac{16}{25}}$
7. $-\sqrt{\frac{1}{144}}$
8. $\sqrt{\frac{1}{121}}$
9. $\pm\sqrt{\frac{144}{225}}$
10. $\pm\sqrt{\frac{196}{361}}$
11. $-\sqrt{10,000}$
12. $-\sqrt{90,000}$
13. $\sqrt{1089}$
14. $\sqrt{784}$
15. $\sqrt{0.01}$
16. $\sqrt{0.04}$
17. $\sqrt{0.64}$
18. $\sqrt{0.25}$
19. $-\sqrt{1.96}$
20. $\sqrt{2.89}$
21. $\sqrt{0.0009}$
22. $\sqrt{0.0025}$
23. $\pm\sqrt{0.0625}$
24. $\pm\sqrt{0.0196}$
25. $\sqrt{0.1296}$
26. $\sqrt{0.1089}$
27. $\sqrt{\frac{0.25}{0.09}}$
28. $\sqrt{\frac{0.01}{0.81}}$
29. $\sqrt{\frac{1.44}{2.25}}$
30. $\sqrt{\frac{1.21}{3.61}}$
31. $\sqrt[4]{81}$
32. $\sqrt[4]{10,000}$
33. $\sqrt[3]{125}$
34. $\sqrt[3]{64}$

Applications

35. **Physics** The approximate time t (in seconds) that it takes an object to fall a distance d (in ft) under the influence of gravity is given by the formula $t = \sqrt{\frac{d}{16}}$. Find the time it takes an object to fall 144 ft.

HISTORICAL NOTE

Figurate numbers are a subset of the counting numbers.

1. Diagram the next three triangular numbers.
2. Diagram the next three square numbers.
3. The Greek mathematician Pythagoras and his followers suggested that any square number greater than 1 is the sum of two successive triangular numbers. Draw diagrams to show that this is true for the square numbers shown here.

Triangular Numbers

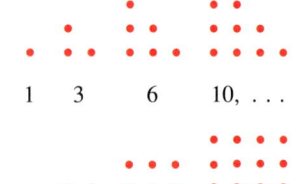

1   3   6   10, ...

Square Numbers

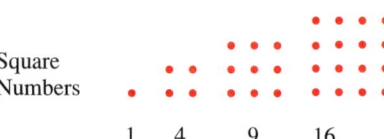

1   4   9   16, ...

12.2 Irrational Square Roots

Objective: To approximate irrational square roots

An artist has 68 square marble tiles. The artist wants to use all of the tiles to create a square mosaic design. Assuming that no open spaces are left and that no tiles are broken, is this possible?

Since there is no integer a such that $a^2 = 68$, the artist cannot use all 68 tiles to make a square design.

Because 68 is not the square of an integer, its square root is not a rational number. The square root of 68 is an *irrational number*. An **irrational number** is a real number that cannot be written in the form $\frac{m}{n}$, where m and n are integers and $n \neq 0$. Some examples are $\sqrt{2}$, $\sqrt{5}$, $\sqrt{13}$, and π.

Capsule Review

Graph each of these sets of numbers on a number line.

1. $\{-5, 5, 1, -1, 0\}$
2. $\{0, 1, 2, 3, \ldots\}$
3. {positive integers}
4. $\left\{2.5, -1.5, 3\frac{3}{4}, -0.5, -\frac{1}{4}\right\}$
5. {odd integers}

The set of real numbers is made up of rational numbers and irrational numbers. If you were to graph all rational numbers and irrational numbers on the number line, the graph would be continuous, or complete. This property of the real numbers is called the *completeness property*.

The **completeness property** states that there is a one-to-one correspondence between the real numbers and the points of a number line.

Irrational square roots are often left in radical form, but for some problems it may be necessary to approximate the square roots. You probably will use a calculator to find decimal approximations of irrational square roots. You can, however, use a square-root table such as the one on page 670. It lists decimal values for the square roots of integers from 1 through 150. A calculator gives a closer approximation of the square root.

EXAMPLE 1 Approximate $\sqrt{68}$ to the nearest hundredth using a square-root table. Verify your answer using a calculator.

n	n^2	$\sqrt{n}$
67	4489	8.185
68 →	4624 →	8.246
69	4761	8.307

$\sqrt{68} \approx 8.246$, or 8.25 to the nearest hundredth

A calculator will display the result as 8.2462113. Therefore, $\sqrt{68} \approx 8.25$ to the nearest hundredth.

Another method for approximating irrational square roots is called the *divide-and-average method*.

EXAMPLE 2 Approximate $\sqrt{75}$ using the divide-and-average method.

Step 1   75 is between 64 and 81; therefore, $\sqrt{75}$ is between $\sqrt{64}$ and $\sqrt{81}$.

$64 < 75 < 81$
$8 < \sqrt{75} < 9$

Step 2   75 is closer to 81 than to 64, and so 9 is a closer approximation of $\sqrt{75}$. Therefore, divide 75 by 9. Divide until the quotient has one more decimal place than the divisor.

$9\overline{)75.0} = 8.3$

Step 3   Find the average of 9 and 8.3. The average should have the same number of decimal places as the quotient.

$\dfrac{9 + 8.3}{2} = \dfrac{17.3}{2} \approx 8.7$

Step 4   Repeat steps 2 and 3 using the average 8.7 as the divisor.

$8.7\overline{)75.000} = 8.62 \qquad \dfrac{8.7 + 8.62}{2} = \dfrac{17.32}{2} = 8.66$

So, $\sqrt{75} \approx 8.66$. If the steps were repeated would you get a closer approximation? Each successive repetition yields a number closer to, but never equal to, the irrational number. This illustrates the **density property** of real numbers, which states that between any two real numbers there is another real number.

CLASS EXERCISES

Use the square-root table on page 670 to approximate the square root. Then, compare your answers using a calculator.

1. $\sqrt{3}$ 2. $\sqrt{2}$ 3. $\sqrt{27}$ 4. $\sqrt{58}$ 5. $\sqrt{31}$
6. $\sqrt{97}$ 7. $\sqrt{85}$ 8. $\sqrt{19}$ 9. $\sqrt{45}$ 10. $\sqrt{76}$

Use the divide-and-average method to approximate the square root.

11. $\sqrt{50}$ 12. $\sqrt{79}$ 13. $\sqrt{18.5}$ 14. $\sqrt{5.9}$ 15. $\sqrt{39.5}$

PRACTICE EXERCISES

Approximate the square root to the nearest tenth using the divide-and-average method. Then compare your answers using a calculator.

1. $\sqrt{5}$
2. $\sqrt{8}$
3. $\sqrt{22}$
4. $\sqrt{17}$
5. $\sqrt{83}$
6. $\sqrt{67}$
7. $\sqrt{95}$
8. $\sqrt{59}$
9. $\sqrt{27}$
10. $\sqrt{32}$
11. $\sqrt{15}$
12. $\sqrt{12}$
13. $\sqrt{99}$
14. $\sqrt{86}$
15. $\sqrt{19.5}$
16. $\sqrt{17.7}$
17. $\sqrt{149}$
18. $\sqrt{165}$
19. $\sqrt{127}$
20. $\sqrt{105}$
21. $\sqrt{71.5}$
22. $\sqrt{9.5}$
23. $\sqrt{4.51}$
24. $\sqrt{58.9}$

Use the divide-average-method to compute the following to two decimal places of accuracy.

25. $\sqrt{40{,}098.52}$
26. $\sqrt{95{,}000}$
27. $\sqrt{0.432}$
28. $\sqrt{0.915}$

Applications

29. **Geometry** The area of a square rug is 48 ft². What is the approximate length of one side of the rug?

30. **Geometry** The length of a rectangular poster is 3 times the width. The area of the rectangle is 192 in². Find the dimensions of the poster.

ALGEBRA IN INDUSTRY

In the printing industry, printers often use a measurement unit called an *em* to measure the available type space on a page. This unit is helpful when computing printing costs. An *em* is the area occupied by the letter M in any type size. The size of the em, of course, varies with the size of the type. In 6-point type there are 144 ems in the square inch; in 12-point type only 36. The formula for finding the cost of composition is:

$$C = \frac{T}{E} \times H$$

where C = cost of composition, E = total ems set per hour, T = total ems set, and H = hourly wage. Use the formula to find the cost of setting type for the following book.

A total of 16,800,000 ems are needed to set the type for an encyclopedia. The compositor works at a rate of 3500 em/h and charges $42.50 per hour.

12.3 Decimal Forms of Rational Numbers

Objective: To write rational numbers in decimal or fraction form

A carpenter has two boards. One is $\frac{3}{4}$ in. thick. The other is $1\frac{5}{16}$ in. thick. If the carpenter uses a calculator to find the total thickness, she changes the fractions to decimals.

The rational numbers $\frac{3}{4}$ and $1\frac{5}{16}$ can be expressed as **terminating decimals**.

$\frac{3}{4} = 0.75$ and $1\frac{5}{16} = 1.3125$. The division results in a zero remainder.

Capsule Review

Show that the number is a rational number.

1. 8.3 2. 1.5 3. 4 4. 0
5. $-2\frac{1}{2}$ 6. $-1\frac{3}{4}$ 7. 6.01 8. 5.99

Not every rational number can be expressed as a terminating decimal. Some rational numbers are represented by **repeating decimals**. However, *every* rational number can be expressed as either a terminating or a repeating decimal.

EXAMPLE 1 Divide to express $\frac{5}{6}$ as a decimal.

$$\frac{5}{6} \longrightarrow \begin{array}{r} 0.833 \\ 6\overline{)5.000} \\ \underline{4\ 8} \\ 20 \\ \underline{18} \\ 20 \\ \underline{18} \\ 2 \end{array}$$

Notice that the digit 3 repeats in the quotient and that the remainder 2 also repeats.

So, $\frac{5}{6} = 0.8333\ldots,\ = 0.8\overline{3}$.   *The bar shows the repeating digit or block of digits.*

500 Chapter 12 Radicals

EXAMPLE 2 Write $0.\overline{45}$ in the form $\dfrac{m}{n}$, where m and n are integers, $n \ne 0$.

Let $x = 0.\overline{45}$.
$100x = 45.\overline{45}$ *Multiply each side by 10^2, or 100, because two digits repeat.*
$x = 0.\overline{45}$ *Subtract $x = 0.\overline{45}$ from $100x = 45.\overline{45}$.*
$99x = 45.00$

$x = \dfrac{45}{99}$, or $\dfrac{5}{11}$ *Solve for x.*

So, $0.\overline{45} = \dfrac{5}{11}$.

EXAMPLE 3 Write $0.5\overline{3}$ in the form $\dfrac{m}{n}$, where m and n are integers, $n \ne 0$.

Let $x = 0.5\overline{3}$.
$10x = 5.\overline{3}$ *Multiply each side by 10^1, or 10, because one digit repeats.*
$x = 0.5\overline{3}$
$9x = 4.80$ *Subtract $x = 0.5\overline{3}$ from $10x = 5.\overline{3}$.*

$x = \dfrac{4.8}{9} = \dfrac{48}{90} = \dfrac{8}{15}$ *Solve for x.* $\dfrac{4.8}{9} = \dfrac{4.8(10)}{9(10)} = \dfrac{48}{90}$

So, $0.5\overline{3} = \dfrac{8}{15}$.

CLASS EXERCISES

Express each rational number as a terminating or repeating decimal.

1. $\dfrac{5}{8}$ 2. $-\dfrac{5}{2}$ 3. $-\dfrac{1}{6}$ 4. $\dfrac{2}{9}$

Write each decimal in the form $\dfrac{m}{n}$, where m and n are integers, $n \ne 0$.

5. 0.75 6. 1.8 7. $0.\overline{6}$ 8. $0.8\overline{3}$

PRACTICE EXERCISES

Express each rational number as a decimal.

1. $\dfrac{1}{8}$ 2. $\dfrac{7}{2}$ 3. $\dfrac{7}{3}$ 4. $\dfrac{2}{3}$ 5. $-\dfrac{5}{4}$ 6. $\dfrac{8}{5}$

7. $\dfrac{4}{9}$ 8. $-\dfrac{5}{11}$ 9. $-\dfrac{51}{4}$ 10. $\dfrac{37}{8}$ 11. $\dfrac{5}{33}$ 12. $-\dfrac{8}{11}$

Write each decimal in the form $\dfrac{m}{n}$, where m and n are integers, $n \ne 0$.

13. 2.33 14. -1.6 15. $0.\overline{3}$ 16. $0.\overline{1}$

. $0.\overline{27}$ **18.** $0.\overline{45}$ **19.** $1.\overline{1213}$ **20.** $-5.2\overline{3}$

21. $0.3\overline{25}$ **22.** $0.\overline{125}$ **23.** $0.\overline{142857}$ **24.** $0.\overline{857142}$

Write each number in the form $\dfrac{m}{n}$, and then find their sum.

25. $\dfrac{2}{3}; 0.\overline{3}$ **26.** $0.\overline{36}; \dfrac{7}{11}$ **27.** $\dfrac{1}{18}; 0.3\overline{8}$ **28.** $\dfrac{5}{7}; 0.\overline{285714}$

29. The rational numbers $\dfrac{1}{20}$ and $\dfrac{3}{125}$ can be expressed as terminating decimals. Write the prime factorization of each denominator. What pattern do you observe?

30. The rational numbers $\dfrac{1}{6}, \dfrac{5}{12},$ and $\dfrac{1}{11}$ can be expressed as repeating decimals. Write the prime factorization of each denominator. Do factors other than 2 or 5 occur?

31. Use the information in Exercises 29–30 to make a general rule for determining, without dividing, whether a rational number in simplest form will be a terminating or a repeating decimal.

Applications

32. Electricity To find how long a series circuit runs (in s), divide the total charge of the circuit (in amp × s) by the number of amps. Find how long a student runs a series circuit at 0.12 amp for a total charge of 20.2 amp × s.

EXTRA

Newton's Method, named after Sir Isaac Newton (1642–1727), provides a way to approximate the square root of any positive number using only addition and division. This computer program applies Newton's method five times and prints each approximation.

```
10 INPUT "APPROXIMATE THE SQUARE ROOT OF
      WHAT POSITIVE NUMBER?   ";N: PRINT
20 INPUT "FIRST APPROXIMATION:   ";W: PRINT
30 FOR I = 1 TO 5
40 X = (W + N / W) / 2
50 PRINT X
60 W = X
70 NEXT I
80 END
```

Tell how the outcome of the program is changed if line 40 is replaced by

1. 40 X = INT(((W + N/W)/2) * 10 + 0.05)/10

2. 40 X = INT(((W + N/W)/2) * 100 + 0.005)/100

3. What is the relationship between this method and the divide and average method? Is one more accurate than the other? If so, which one?

12.4 Simplifying Square Roots

Objectives: To simplify square roots
To simplify square roots containing variable expressions

Electricians use the formula $V = \sqrt{WR}$, where V is voltage, W is power in watts, and R is resistance in ohms. If an appliance uses 150 watts and has a resistance of 20 ohms, what voltage does the appliance need?

Substituting the values for W and R:

$$V = \sqrt{150 \cdot 20}$$
$$= \sqrt{3000}$$

However, 3000 is not the square of an integer.

When a radicand is not the square of an integer, you may use the product and quotient properties of square roots to *simplify* a square-root radical.

Capsule Review

Find the indicated square root.

1. $\sqrt{900}$
2. $\sqrt{3600}$
3. $\sqrt{100}$
4. $\sqrt{144}$
5. $-\sqrt{6400}$
6. $-\sqrt{8100}$
7. $\sqrt{\frac{1}{400}}$
8. $\sqrt{\frac{1}{900}}$
9. $\sqrt{\frac{2500}{49}}$
10. $\sqrt{\frac{9}{1600}}$

An expression that contains a square-root radical is in *simplest form* if

- the radicand contains no square factor other than 1 and no fractions, and
- no denominator contains a radical.

EXAMPLE 1 Simplify: $\sqrt{3000}$

$\sqrt{3000} = \sqrt{100 \cdot 30}$ *The greatest square factor of 3000 is 100.*
$\phantom{\sqrt{3000}} = \sqrt{100} \cdot \sqrt{30}$ *Use the product property of square roots.*
$\phantom{\sqrt{3000}} = 10\sqrt{30}$

From Example 1, you see that the voltage needed by the appliance described at the beginning of the lesson is $10\sqrt{30}$ volts.

12.4 Simplifying Square Roots   **503**

You know that $\sqrt{6^2} = 6$, but $\sqrt{x^2} = x$ is true only when x is positive or zero. For example: $\sqrt{(-6)^2} \ne -6$, since $\sqrt{(-6)^2} = \sqrt{36} = 6$. However, the following property is true for *all* real numbers.

> For any real number n, $\sqrt{n^2} = |n|$.

EXAMPLE 2 Simplify: **a.** $\sqrt{72y^2}$ **b.** $\sqrt{27x^3}$ **c.** $\sqrt{50a^6}$

a. $\sqrt{72y^2} = \sqrt{36 \cdot 2 \cdot y^2}$ The greatest perfect square factor of $72y^2$ is $36y^2$.
$\phantom{\sqrt{72y^2}} = \sqrt{36y^2} \cdot \sqrt{2}$ Product property of square roots
$\phantom{\sqrt{72y^2}} = 6|y|\sqrt{2}$ Absolute value symbols ensure that $\sqrt{y^2}$ is positive.

b. $\sqrt{27x^3} = \sqrt{9x^2 \cdot 3x}$ Find the greatest square factor of $27x^3$.
$\phantom{\sqrt{27x^3}} = \sqrt{9x^2} \cdot \sqrt{3x}$ Product property of square roots
$\phantom{\sqrt{27x^3}} = 3x\sqrt{3x}$ No absolute value symbols are needed. If x were negative, $\sqrt{27x^3}$ would be undefined.

From now on in this text, unless stated otherwise, you can assume that all variables in a radicand are *greater than* or *equal* to zero. Therefore, absolute value notation may be omitted.

c. $\sqrt{50a^6} = \sqrt{25a^6 \cdot 2}$ Find the greatest square factor of $50a^6$.
$\phantom{\sqrt{50a^6}} = \sqrt{25a^6} \cdot \sqrt{2}$ Product property of square roots.
$\phantom{\sqrt{50a^6}} = 5a^3\sqrt{2}$

EXAMPLE 3 Evaluate $\sqrt{2n - 10}$:   **a.** for $n = 3$ **b.** for $n = 9$

a.
$\sqrt{2n - 10}$
$= \sqrt{2 \cdot 3 - 10}$ Replace n with 3.
$= \sqrt{6 - 10}$
$= \sqrt{-4}$

For $n = 3$, $\sqrt{2n - 10} = \sqrt{-4}$, which is not a real number.

b.
$\sqrt{2n - 10}$
$= \sqrt{2 \cdot 9 - 10}$ Replace n with 9.
$= \sqrt{18 - 10}$
$= \sqrt{8}$
$= \sqrt{4} \cdot \sqrt{2}$, or $2\sqrt{2}$

For $n = 9$, $\sqrt{2n - 10} = 2\sqrt{2}$.

To find values of y that make a radical expression like $\sqrt{-2y - 5}$ a real number, find values of y that make the radicand nonnegative.

EXAMPLE 4 For what values of y will $\sqrt{-2y - 5}$ be a real number?

$-2y - 5 \ge 0$ The radicand must be nonnegative.
$-2y \ge 5$
$y \le -\frac{5}{2}$

For $y \le -\frac{5}{2}$, $\sqrt{-2y - 5}$ is a real number.

504   Chapter 12 Radicals

CLASS EXERCISES

Assume that the variable may be positive or negative, then tell why each radical is not in the simplest form and simplify.

1. $\sqrt{20}$
2. $\sqrt{27}$
3. $\sqrt{18x^3}$
4. $\sqrt{45b^4}$

Evaluate for $a = 3$. Then simplify, if possible.

5. $\sqrt{a + 9}$
6. $\sqrt{2a - 3}$
7. $\sqrt{-5a - 10}$
8. $\sqrt{3a - 1}$

For what values of x will each radical represent a real number?

9. $\sqrt{3x - 9}$
10. $\sqrt{2x + 5}$
11. $\sqrt{-2x - 7}$
12. $\sqrt{-x + 4}$

PRACTICE EXERCISES

Evaluate for the given value of the variable. Then simplify, if possible.

1. $\sqrt{c - 7}$, $c = 15$
2. $\sqrt{5x - 6}$, $x = -6$
3. $\sqrt{3a + 6}$, $a = 7$

Find the values of x that make each radical expression a real number.

4. $\sqrt{5x - 10}$
5. $\sqrt{2x + 7}$
6. $\sqrt{-4x + 2}$
7. $\sqrt{-3x - 9}$

Simplify and assume all variables are greater than or equal to zero.

8. $\sqrt{24}$
9. $\sqrt{48}$
10. $\sqrt{98}$
11. $\sqrt{54}$
12. $\sqrt{288}$
13. $\sqrt{150}$
14. $\sqrt{784}$
15. $\sqrt{1000}$
16. $\sqrt{200x^2}$
17. $\sqrt{75x^4}$
18. $\sqrt{63x^3}$
19. $\sqrt{300x^6}$
20. $\sqrt{8b^8}$
21. $\sqrt{12y^{12}}$
22. $\sqrt{5c^5}$
23. $\sqrt{15d^7}$
24. $-\sqrt{120m^6}$
25. $-\sqrt{125x^{16}}$
26. $\sqrt{100b^9}$
27. $\sqrt{144p^8}$
28. $\sqrt{a^2b}$
29. $\sqrt{x^4y}$
30. $\sqrt{m^4n^6}$
31. $\sqrt{p^{12}q^{10}}$
32. $\sqrt{320x^5}$
33. $\sqrt{243x^{13}}$
34. $-\sqrt{128n^9}$
35. $-\sqrt{150a^{11}}$
36. $\sqrt{88a^4}$
37. $\sqrt{500r^7}$
38. $\sqrt{720a^{12}}$
39. $\sqrt{\frac{36}{49}x^5y}$
40. $-\sqrt{\frac{100}{64}a^3}$
41. $\sqrt{0.09c^5}$
42. $\sqrt{0.04p^9}$
43. $\sqrt{0.16m^3}$
44. $\sqrt{\sqrt{144}}$
45. $\sqrt{\sqrt{324}}$
46. $\sqrt{\sqrt{625}}$
47. $\sqrt{\sqrt{10,000}}$
48. $\sqrt{x^2 + 4x + 4}$
49. $\sqrt{a^2 - 6a + 9}$
50. $\sqrt{9x^2 + 6x + 1}$

Applications

51. Physics A formula in physics that can be used to determine the velocity v in feet per second of an object (neglecting air resistance) after it has fallen a certain distance is $v = \sqrt{64h}$, where h is the distance the object has fallen in feet. Find the velocity of an object after it has fallen 128 ft.

52. Geometry Find the length of a side of a square whose area is 125 in.²

53. Physics The formula $T = 2\pi\sqrt{\dfrac{L}{32}}$ gives the time T (in seconds) required for a pendulum to make one complete swing back and forth. If L is 8 ft (the length of the pendulum), how long does it take the pendulum to make one complete swing?

ALGEBRA IN POLICE SCIENCE

Did you know that the approximate speed of a car prior to an accident can be determined from the skid marks left by the car? An investigating officer will use the formula $S = 2\sqrt{5f}$, where S represents the speed in mi/h and f represents the length of the skid mark in ft.

EXAMPLE An automobile involved in an accident left skid marks 135 ft in length. What was the approximate speed of the car prior to the accident?

$S = 2\sqrt{5f}$
$S = 2\sqrt{5(135)}$ *Substitute the value that you know.*
$S = 2\sqrt{675}$ *Simplify.*
$ = 2\sqrt{3(9)(25)}$
$ = 30\sqrt{3}$
$S \approx 30(1.73) \approx 51.9$ *Use an approximation for $\sqrt{3}$.*

So, the approximate speed of the car was 51.9 mi/h.

Give the approximate speed of the car, to the nearest whole mi/h, for each of the following skid marks.

1. 25 ft   2. 80 ft   3. 125 ft   4. 175 ft   5. 200 ft

APPLICATION:
Image Formation

Did you know that photography, the art of taking quality photographs, involves many mathematical computations that people often take for granted? One such calculation concerns the focal length of a camera lens. The focal length of a camera is the distance between the center of the lens and the focal point. The size of the image formed on the film depends upon the focal length of the lens. The shorter the focal length of the lens, the smaller the image.

A formula can be used to find the focal length of a camera lens:

$$\frac{1}{p} + \frac{1}{q} = \frac{1}{f}$$

where p is the object distance, q the image distance, and f the focal length.

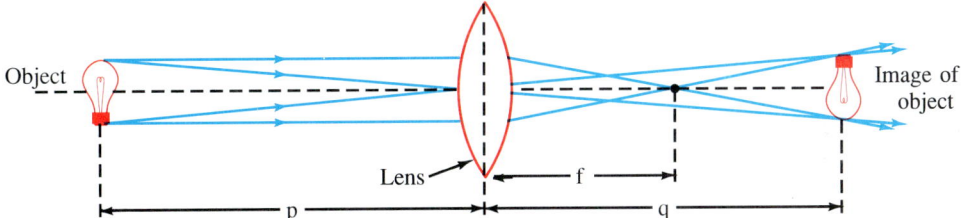

When an object is placed on one side of a converging lens beyond the principal focus, a real image will be formed on the opposite side of the lens. If the object is moved closer to the focal point, the image will be formed farther away from the lens and will be bigger. One way of accurately determining the position of an image is to use the lens formula.

If a camera with an image distance of 4 cm is held 64 cm from an object, what is the focal length of the camera lens?

$$\frac{1}{64} + \frac{1}{4} = \frac{1}{f}; \quad f = \frac{64}{17} \approx 3.76$$

The focal length is approximately 4 cm.

Another way to solve problems of the form $\frac{1}{a} + \frac{1}{b} = \frac{1}{c}$ is to use a **nomogram**. A nomogram is a graph which usually contains three scales graduated for different variables.

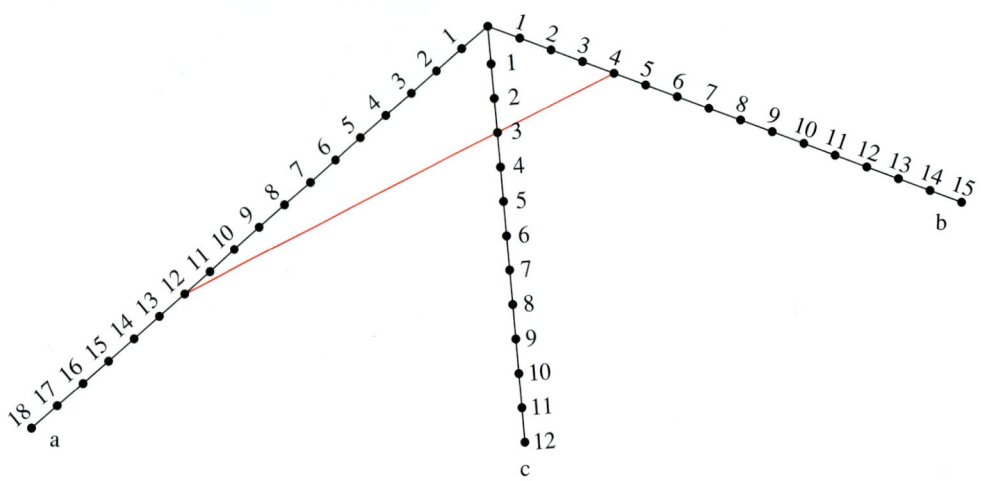

When a straight line connects values of any two points, the related value may be read directly from the third at the point intersected by the line. For example, to solve $\frac{1}{12} + \frac{1}{4} = \frac{1}{c}$, follow these steps:

- Copy the nomogram above.
- Locate the point 12 on the a scale and the point 4 on the b scale.
- Connect them with a straight line.

Notice that this line intersects the c scale at the point 3. Thus, it is clear from the nomogram that $\frac{1}{12} + \frac{1}{4} = \frac{1}{3}$. How would you check this?

Use your Nomogram in the manner shown to solve each of the following. Check your answers algebraically.

1. $\frac{1}{6} + \frac{1}{3} = \frac{1}{c}$
2. $\frac{1}{15} + \frac{1}{b} = \frac{1}{6}$
3. $\frac{1}{a} + \frac{1}{9} = \frac{1}{6}$

4. A camera with an image distance of 5 cm is held 70 cm from an object. What is the focal length of the camera lens?

5. Mike took a picture of an object that was 66 cm away. His camera has a focal length of $\frac{66}{23}$ cm. What is the image distance of the picture?

12.5 Addition and Subtraction of Radicals

Objective: To simplify sums and differences of radicals

Adding and subtracting radical expressions is very similar to adding and subtracting *similar terms* of polynomials. Like radicals are like similar terms.

Like Radicals	Unlike Radicals
$-\sqrt{3}$ and $-2\sqrt{3}$	$\sqrt{3}$ and $\sqrt{7}$
$5\sqrt{x}$ and $6\sqrt{x}$	$-\sqrt{x}$ and $2\sqrt{y}$

Capsule Review

EXAMPLES a. $-x^2y + 5x^2y = 4x^2y$ — Like terms

b. $4mn^2 + 3m^2n$ — Unlike terms cannot be combined.

Simplify. If not possible, explain why.

1. $12m^3 - 9m^3$
2. $3a^2b + 6a^2b$
3. $r^2s - 4rs^2$
4. $5x + 2x^2 - 2x + x^2$
5. $3y^3 - y + 2y^3 + 8y$
6. $9a^2b^3 - 8b^3 + 2a^2$

You add or subtract radicals in the same way you combine like terms; that is, you use the distributive property.

EXAMPLE 1 Simplify: $\sqrt{3} - 2\sqrt{5} + 5\sqrt{3}$.

$\sqrt{3} - 2\sqrt{5} + 5\sqrt{3}$
$= (\sqrt{3} + 5\sqrt{3}) - 2\sqrt{5}$ *Group like radicals.*
$= (1 + 5)\sqrt{3} - 2\sqrt{5}$ *Use the distributive property.*
$= 6\sqrt{3} - 2\sqrt{5}$

The expression $6\sqrt{3} - 2\sqrt{5}$ is in its simplest radical form since $6\sqrt{3}$ and $2\sqrt{5}$ are unlike radicals and therefore, cannot be combined. As with other expressions like $2x - 3x$, you can add or subtract radicals if they are similar. Two or more radicals are similar if their radicands are identical. In all other cases the radicals are dissimilar or unlike.

In Example 2, to add or subtract some radicals, you must write each one in simplest form. Then, you can combine the like radicals.

EXAMPLE 2 Simplify: $\sqrt{54} + 2\sqrt{40} - 3\sqrt{96}$.

$\sqrt{54} + 2\sqrt{40} - 3\sqrt{96}$
$= \sqrt{9 \cdot 6} + 2\sqrt{4 \cdot 10} - 3\sqrt{16 \cdot 6}$ *Write each radical in*
$= 3\sqrt{6} + 2 \cdot 2\sqrt{10} - 3 \cdot 4\sqrt{6}$ *simplest form.*
$= 3\sqrt{6} + 4\sqrt{10} - 12\sqrt{6}$
$= (3\sqrt{6} - 12\sqrt{6}) + 4\sqrt{10}$ *Group like radicals.*
$= (3 - 12)\sqrt{6} + 4\sqrt{10}$ *Use the distributive property.*
$= -9\sqrt{6} + 4\sqrt{10}$

Recall that, you may assume all radicals with variables represent nonnegative numbers unless stated otherwise.

EXAMPLE 3 Simplify: $\sqrt{18x^2y} - \sqrt{8x^2y} + \sqrt{50x^2y}$.

$\sqrt{18x^2y} - \sqrt{8x^2y} + \sqrt{50x^2y}$
$= \sqrt{9x^2 \cdot 2y} - \sqrt{4x^2 \cdot 2y} + \sqrt{25x^2 \cdot 2y}$
$= 3x\sqrt{2y} - 2x\sqrt{2y} + 5x\sqrt{2y}$
$= (3x - 2x + 5x)\sqrt{2y}$
$= 6x\sqrt{2y}$

CLASS EXERCISES

Simplify. Assume that all variables represent nonnegative real numbers.

1. $5\sqrt{2} + \sqrt{2}$
2. $\sqrt{3} - 4\sqrt{3}$
3. $-5\sqrt{a} + 2\sqrt{a}$
4. $\sqrt{x} - 6\sqrt{x}$
5. $\sqrt{6} - 4\sqrt{6}$
6. $-\sqrt{10} + \sqrt{7} - 3\sqrt{7}$
7. $\sqrt{16y} - \sqrt{y}$
8. $-5\sqrt{b} + \sqrt{4b}$
9. $\sqrt{75} + 2\sqrt{27} - \sqrt{12}$
10. $-\sqrt{80} - 3\sqrt{45} + \sqrt{20}$
11. $\sqrt{12x^3} + 2\sqrt{27x^3}$
12. $2\sqrt{28y^3} + \sqrt{63y^3}$

PRACTICE EXERCISES

Perform the indicated operations. Assume that all variables represent nonnegative real numbers and simplify.

1. $8\sqrt{5} - 3\sqrt{5}$
2. $-2\sqrt{2} - \sqrt{2}$
3. $6\sqrt{10} + \sqrt{10}$
4. $-4\sqrt{7} + 3\sqrt{7}$
5. $2\sqrt{27} + 3\sqrt{16}$
6. $3\sqrt{18} - \sqrt{25}$
7. $6\sqrt{3} - 4\sqrt{5} - \sqrt{3}$
8. $-\sqrt{7} - 7\sqrt{3} + 4\sqrt{7}$
9. $\sqrt{8} + 2\sqrt{50} - \sqrt{18}$
10. $2\sqrt{12} - \sqrt{48} + 3\sqrt{27}$
11. $8\sqrt{72} - 3\sqrt{8} - \sqrt{98}$
12. $-5\sqrt{80} + \sqrt{125} + 10\sqrt{45}$
13. $\sqrt{9x} - \sqrt{16x} + \sqrt{25x}$
14. $3\sqrt{36a} + \sqrt{100a} - 2\sqrt{64a}$

15. $\sqrt{12x} - \sqrt{27x} + \sqrt{48x}$
16. $\sqrt{18y} - 2\sqrt{32y} - \sqrt{50y}$
17. $4\sqrt{54} - \sqrt{6} + 5\sqrt{24}$
18. $2\sqrt{250} - 3\sqrt{640} - 5\sqrt{10}$
19. $\sqrt{100x^2y} - \sqrt{144x^2y} - 5\sqrt{x^2y}$
20. $-3\sqrt{81ab^2} + 2\sqrt{ab^2} - 5\sqrt{9ab^2}$
21. $\sqrt{72xy^2} - 2\sqrt{98xy^2}$
22. $-3\sqrt{125c^4d} + \sqrt{80c^4d}$
23. $4\sqrt{320a^3} - 2\sqrt{180a^3}$
24. $\sqrt{44x^5} + \sqrt{99x^5}$
25. $\sqrt{175a^2b} - 3\sqrt{112a^2b}$
26. $-5\sqrt{192rs^3} + 2\sqrt{300rs^3}$
27. $5\sqrt{176} - 3\sqrt{99} + 4\sqrt{162}$
28. $-3\sqrt{200} + 2\sqrt{147} - \sqrt{338}$
29. $\sqrt{\frac{3}{4}} - 2\sqrt{\frac{3}{4}}$
30. $-3\sqrt{\frac{5}{16}} - 5\sqrt{\frac{5}{16}}$
31. $\sqrt{\frac{5xy^2}{9}} - \sqrt{\frac{5xy^2}{4}}$
32. $-\sqrt{\frac{5t^3}{16}} + \sqrt{\frac{5t^3}{36}}$
33. $a\sqrt{ab^3} + ab\sqrt{ab} - b\sqrt{a^3b}$
34. $3x\sqrt{25x^3} - x\sqrt{4x^3} + 3x\sqrt{9x^3}$
35. $y\sqrt{8x^5} + xy\sqrt{18x^3} - x\sqrt{50x^3y^2}$
36. $a\sqrt{27a^3} - b\sqrt{48b^3} - a^2\sqrt{75a}$

Applications

37. **Geometry** Find the perimeter of a square whose side has a length of $\sqrt{125}$ cm.

38. **Geometry** Find the perimeter of a triangle whose sides measure $4\sqrt{2}$ cm, $4\sqrt{6}$ cm, and $8\sqrt{2}$ cm.

BIOGRAPHY: Amalie (Emmy) Noether

Amalie (Emmy) Noether, born in Erlanger, Germany, on March 23, 1882, is considered to be the most creative abstract algebraist of modern times. In 1907, she earned her Ph.D. degree *summa cum laude* from the University of Erlanger. In 1915 she went to the University of Göttingen, where she was persuaded to remain by the eminent mathematicians *David Hilbert* and *Felix Klein*.

The real extent of her work cannot be accurately judged from her papers, because much of her work during this period appeared in the publications of her students and colleagues. In 1933 Noether left Germany for the United States, where she became a visiting professor of mathematics at Bryn Mawr College in Pennsylvania. She also lectured and conducted research at the Institute for Advanced Study in Princeton, New Jersey. The names of Noether, Hilbert, and Klein are among the most prominent of those active in establishing new frontiers in modern mathematics. Research one of these mathematicians and be prepared to present your findings in a group discussion.

12.6 Multiplication of Radicals

Objective: To simplify products of radicals

To simplify expressions containing square roots, you used the product property of square roots: $\sqrt{mn} = \sqrt{m} \cdot \sqrt{n}$ for any real numbers where $m \geq 0$, $n \geq 0$.

According to the *symmetric property of equality*, $\sqrt{m} \cdot \sqrt{n} = \sqrt{mn}$. This means that you can use the product property to multiply radicals.

Capsule Review

Simplify. Assume that the variables represent nonnegative real numbers.

1. $\sqrt{625}$ 2. $\sqrt{900}$ 3. $\sqrt{54}$ 4. $\sqrt{63}$

5. $\sqrt{r^4 s^5}$ 6. $\sqrt{a^3 b^{10}}$ 7. $\sqrt{8x^3 y^7}$ 8. $\sqrt{27c^2 d}$

The *commutative* and *associative properties* for multiplication are also used when multiplying radicals.

EXAMPLE 1   Multiply and simplify: $2\sqrt{6} \cdot 3\sqrt{2}$

$2\sqrt{6} \cdot 3\sqrt{2} = (2 \cdot 3) \cdot (\sqrt{6} \cdot \sqrt{2})$ *Commutative and associative properties*
$= 6 \cdot (\sqrt{6} \cdot \sqrt{2})$ $\sqrt{m} \cdot \sqrt{n} = \sqrt{mn}$
$= 6\sqrt{12}$
$= 6\sqrt{4 \cdot 3}$ *Simplify:* $\sqrt{12}$
$= 6\sqrt{4} \cdot \sqrt{3}$ $\sqrt{mn} = \sqrt{m} \cdot \sqrt{n}$
$= 6 \cdot 2\sqrt{3} = 12\sqrt{3}$

It is sometimes easier to simplify a radical before you multiply.

EXAMPLE 2   Simplify: $-5\sqrt{8x} \cdot 4\sqrt{3x}$

$-5\sqrt{8x} \cdot 4\sqrt{3x} = (-5 \cdot 4) \cdot (\sqrt{8x} \cdot \sqrt{3x})$
$= -20 \cdot (\sqrt{4 \cdot 2x} \cdot \sqrt{3x})$ *Simplify:* $\sqrt{8x}$
$= -20 \cdot 2\sqrt{2x} \cdot \sqrt{3x}$ $\sqrt{4 \cdot 2x} = \sqrt{4} \cdot \sqrt{2x} = 2\sqrt{2x}$
$= -40\sqrt{6x^2}$ $\sqrt{2x} \cdot \sqrt{3x} = \sqrt{6x^2}$
$= -40\sqrt{x^2} \cdot \sqrt{6}$ *Simplify:* $\sqrt{6x^2}$
$= -40x\sqrt{6}$

Chapter 12 Radicals

You can also use the distributive property to simplify products of expressions that contain radicals.

EXAMPLE 3 Simplify: $\sqrt{2}(5 + 3\sqrt{6})$

$$\begin{aligned}\sqrt{2}(5 + 3\sqrt{6}) &= \sqrt{2} \cdot 5 + \sqrt{2} \cdot 3\sqrt{6} &&\text{Use the distributive property.}\\ &= 5\sqrt{2} + 3\sqrt{12}\\ &= 5\sqrt{2} + 3\sqrt{4} \cdot \sqrt{3} &&\text{Simplify: } 3\sqrt{12}\\ &= 5\sqrt{2} + 3 \cdot 2\sqrt{3}\\ &= 5\sqrt{2} + 6\sqrt{3}\end{aligned}$$

In this example, notice that $5\sqrt{2} + 6\sqrt{3}$ is in simplest form because $\sqrt{2}$ and $\sqrt{3}$ are unlike radicals.

Recall that you used the FOIL method to multiply two binomials. You can also use FOIL to multiply binomials that contain radicals.

EXAMPLE 4 Simplify: $(2 + 3\sqrt{7})(3 - 2\sqrt{7})$

$(2 + 3\sqrt{7})(3 - 2\sqrt{7}) = (2 + 3\sqrt{7})(3 - 2\sqrt{7})$
with FOIL: O, F, I, L

$$\begin{aligned}&= 2 \cdot 3 - 2 \cdot 2\sqrt{7} + 3\sqrt{7} \cdot 3 - 3\sqrt{7} \cdot 2\sqrt{7}\\ &= 6 - 4\sqrt{7} + 9\sqrt{7} - 6\sqrt{7} \cdot \sqrt{7}\\ &= 6 + 5\sqrt{7} - 6(\sqrt{7})^2 &&\text{Combine like radicals.}\\ &= 6 + 5\sqrt{7} - 6 \cdot 7 &&\text{Simplify: } (\sqrt{7})^2\\ &= -36 + 5\sqrt{7}\end{aligned}$$

In this example, notice that $\sqrt{7} \cdot \sqrt{7} = 7$, since $\sqrt{7} \cdot \sqrt{7} = (\sqrt{7})^2$. This suggests the following property.

> For all nonnegative real numbers n, $(\sqrt{n})^2 = n$.

EXAMPLE 5 Simplify: $(2 + \sqrt{3})(2 - \sqrt{3})$

$$\begin{aligned}(2 + \sqrt{3})(2 - \sqrt{3}) &= 2^2 - (\sqrt{3})^2 &&\text{Recall: } (m + n)(m - n) = m^2 - n^2\\ &= 4 - 3\\ &= 1\end{aligned}$$

12.6 Multiplication of Radicals

CLASS EXERCISES

Multiply and simplify.

1. $\sqrt{5} \cdot \sqrt{10}$
2. $(\sqrt{2})(-\sqrt{6})$
3. $(3\sqrt{2})^2$
4. $(-5\sqrt{3})^2$
5. $-2\sqrt{7y} \cdot 3\sqrt{8y}$
6. $\sqrt{2x} \cdot \sqrt{24x}$
7. $-\sqrt{3}(5 + 2\sqrt{3})$
8. $\sqrt{2}(2\sqrt{6} - 3)$
9. $(5 - 2\sqrt{6})(2 + 3\sqrt{6})$
10. $(4 + 3\sqrt{7})(3 - 4\sqrt{7})$
11. $(3 - \sqrt{5})(3 + \sqrt{5})$
12. $(\sqrt{3} - 7)(\sqrt{3} + 7)$

For Discussion

13. In Example 5, $(2 + \sqrt{3})(2 - \sqrt{3}) = 1$, which is a rational number. Explain why the product of two binomials of the form $a + \sqrt{b}$ and $a - \sqrt{b}$ is always rational.

14. Explain and show how multiplying radical expressions is like multiplying polynomials.

PRACTICE EXERCISES

Multiply and simplify.

1. $\sqrt{2} \cdot \sqrt{8}$
2. $\sqrt{12} \cdot \sqrt{3}$
3. $\sqrt{4} \cdot \sqrt{3}$
4. $\sqrt{25} \cdot \sqrt{2}$
5. $(\sqrt{7})^2$
6. $(-\sqrt{5})^2$
7. $3\sqrt{50} \cdot \sqrt{2}$
8. $2\sqrt{3} \cdot \sqrt{30}$
9. $-5\sqrt{6} \cdot 2\sqrt{3}$
10. $3\sqrt{3} \cdot 5\sqrt{15}$
11. $2\sqrt{7} \cdot 3\sqrt{3}$
12. $-5\sqrt{5} \cdot 4\sqrt{6}$
13. $3\sqrt{5} \cdot 4\sqrt{70}$
14. $7\sqrt{14} \cdot 2\sqrt{7}$
15. $\sqrt{\frac{3}{5}} \cdot \sqrt{\frac{5}{3}}$
16. $\sqrt{\frac{9}{2}} \cdot \sqrt{\frac{2}{3}}$
17. $3\sqrt{12x} \cdot 2\sqrt{2x}$
18. $6\sqrt{5a} \cdot 4\sqrt{10a}$
19. $(3\sqrt{n})^2$
20. $(-2\sqrt{m})^2$
21. $\sqrt{3}(4 + \sqrt{3})$
22. $\sqrt{5}(\sqrt{5} - 3)$
23. $\sqrt{6}(\sqrt{6} - 2)$
24. $\sqrt{2}(2 + \sqrt{2})$
25. $(3\sqrt{2} + 1)(2\sqrt{2} - 3)$
26. $(5 - 4\sqrt{3})(4 + 5\sqrt{3})$
27. $(6 - \sqrt{7})(6 + \sqrt{7})$
28. $(\sqrt{3} - 4)(\sqrt{3} + 4)$
29. $\sqrt{2ab} \cdot 3\sqrt{6ab}$
30. $-2\sqrt{8xy} \cdot \sqrt{3xy}$
31. $\sqrt{c^2d} \cdot \sqrt{de^3}$
32. $\sqrt{a^5b} \cdot \sqrt{bc^5}$
33. $5a\sqrt{\frac{a}{b}} \cdot \sqrt{\frac{a}{b}}$
34. $3r\sqrt{\frac{4r}{s}} \cdot \sqrt{\frac{r}{s}}$
35. $(3\sqrt{7} + 2)^2$
36. $(5 - 2\sqrt{6})^2$
37. $(3 - \sqrt{3})(4 + \sqrt{2})$
38. $(\sqrt{5} + 2)(\sqrt{6} - 4)$
39. $(\sqrt{x} + 2\sqrt{y})^2$
40. $(5\sqrt{u} + 2\sqrt{v})^2$
41. $(2\sqrt{m} - 3\sqrt{n})(4\sqrt{m} + \sqrt{n})$
42. $(\sqrt{x} + 4\sqrt{y})(5\sqrt{x} - 2\sqrt{y})$

Applications

Geometry Find the area of each of the following figures. Express answers in simplest form.

43. A triangle whose base is $2\sqrt{3}$ m and whose height is $2\sqrt{2}$ m.
44. A rectangle whose length measures $\frac{5\sqrt{2} + 3}{3}$ cm and whose width measures $\frac{3\sqrt{2} - 2}{5}$ cm.
45. A square whose side is $4\sqrt{x}$ ft.
46. A trapezoid whose height and upper base each are $\sqrt{x}$ m and whose lower base is $2\sqrt{y}$ m. $\left[\text{Hint: } A = \frac{1}{2}h(b_1 + b_2)\right]$

Number Problems Write an equation for each word sentence. Do not solve.

47. The square root of a number squared is 12.
48. The product of 20 and the square root of the sum of 5 and some number is equal to 100.
49. When 3 is subtracted from the square root of the product of 2 and some number the result is 7.
50. The product of 4 and the square root of the product of 3 and some number is 24.

TEST YOURSELF

Approximate the square root using the divide-and-average method.   12.2

1. $\sqrt{45}$
2. $\sqrt{80}$

Write each rational number as a common fraction.   12.3

3. $0.\overline{3}$
4. $0.\overline{12}$

Simplify.   12.1, 12.4

5. $\sqrt{196}$
6. $-\sqrt{90,000}$
7. $\sqrt{\frac{25}{16}}$
8. $\pm\sqrt{\frac{4}{81}}$
9. $-\sqrt{48}$
10. $\sqrt{63y^2}$
11. $\sqrt{25x^3}$
12. $\sqrt{\frac{64}{9}}$

Perform the indicated operations.   12.5–12.6

13. $6\sqrt{2} - 4\sqrt{7} - \sqrt{2}$
14. $\sqrt{8} - 2\sqrt{50} + \sqrt{18}$
15. $3\sqrt{6} \cdot 4\sqrt{2}$
16. $\sqrt{3}(2 - 2\sqrt{6})$

12.7 Division of Radicals

Objective: To simplify quotients of radicals

Recall the quotient property of square roots; that is, for any real numbers $m \geq 0$ and $n > 0$,

$$\sqrt{\frac{m}{n}} = \frac{\sqrt{m}}{\sqrt{n}}$$

You have already used this property to simplify radical expressions that contain perfect square numbers.

Capsule Review

Simplify.

1. $\sqrt{\frac{4}{9}}$
2. $-\sqrt{\frac{16}{25}}$
3. $\sqrt{\frac{64}{49}}$
4. $\sqrt{\frac{1}{144}}$
5. $-\sqrt{\frac{1}{100}}$
6. $\sqrt{\frac{36}{121}}$
7. $\sqrt{\frac{225}{81}}$
8. $\sqrt{\frac{4z^8}{100}}$

The quotient property of square roots also applies when you simplify irrational square roots. Recall that an expression in simplest form cannot have a radical in the denominator.

EXAMPLE 1 Simplify: $\sqrt{\frac{2}{3}}$

$$\sqrt{\frac{2}{3}} = \frac{\sqrt{2}}{\sqrt{3}} = \frac{\sqrt{2}}{\sqrt{3}} \cdot \frac{\sqrt{3}}{\sqrt{3}} \qquad \text{Multiply by 1 in the form } \frac{\sqrt{3}}{\sqrt{3}}.$$

$$= \frac{\sqrt{6}}{\sqrt{9}}$$

$$= \frac{\sqrt{6}}{3}, \text{ or } \frac{1}{3}\sqrt{6}$$

Changing a fraction such as $\frac{\sqrt{2}}{\sqrt{3}}$ (with an irrational denominator) into a fraction such as $\frac{\sqrt{6}}{3}$ (with a rational denominator) is called **rationalizing the denominator.** Examples 2 and 3 show how to rationalize a denominator that contains variables.

EXAMPLE 2 Simplify: $\sqrt{\dfrac{1}{x^7}}$

$$\sqrt{\dfrac{1}{x^7}} = \dfrac{\sqrt{1}}{\sqrt{x^7}}$$

$$= \dfrac{\sqrt{1}}{\sqrt{x^7}} \cdot \dfrac{\sqrt{x}}{\sqrt{x}}$$

$$= \dfrac{\sqrt{x}}{\sqrt{x^8}} = \dfrac{\sqrt{x}}{x^4}$$

Notice that multiplying by $\dfrac{\sqrt{x}}{\sqrt{x}}$ gives an equivalent rational expression without a radical in the denominator.

Sometimes it is easier to simplify the expression before rationalizing the denominator.

EXAMPLE 3 Simplify: $\dfrac{\sqrt{24x^2y}}{\sqrt{3xy^2}}$

$$\dfrac{\sqrt{24x^2y}}{\sqrt{3xy^2}} = \sqrt{\dfrac{24x^2y}{3xy^2}} = \sqrt{\dfrac{8x}{y}} = \dfrac{\sqrt{4}\cdot\sqrt{2x}}{\sqrt{y}} = \dfrac{2\sqrt{2x}}{\sqrt{y}} \cdot \dfrac{\sqrt{y}}{\sqrt{y}}$$ *Multiply by 1 in the form $\dfrac{\sqrt{y}}{\sqrt{y}}$.*

$$= \dfrac{2\sqrt{2xy}}{\sqrt{y^2}} = \dfrac{2\sqrt{2xy}}{y}$$

The radical expression $3 - 2\sqrt{2}$ is called the *conjugate* of $3 + 2\sqrt{2}$. In general, if m and n are nonnegative, then the binomials $a\sqrt{m} + b\sqrt{n}$ and $a\sqrt{m} - b\sqrt{n}$ are **conjugates** of each other. Notice that conjugates are the sum and the difference of the same two terms.

You can use conjugates to rationalize denominators containing binomial radical expressions.

EXAMPLE 4 Rationalize the denominator: $\dfrac{2\sqrt{5} - 3}{3 + 2\sqrt{2}}$

$$\dfrac{2\sqrt{5} - 3}{3 + 2\sqrt{2}} = \dfrac{2\sqrt{5} - 3}{3 + 2\sqrt{2}} \cdot \dfrac{3 - 2\sqrt{2}}{3 - 2\sqrt{2}}$$ *Multiply the numerator and the denominator by the conjugate of $3 + 2\sqrt{2}$.*

$$= \dfrac{(2\sqrt{5} - 3)(3 - 2\sqrt{2})}{(3 + 2\sqrt{2})(3 - 2\sqrt{2})}$$

$$= \dfrac{6\sqrt{5} - 4\sqrt{10} - 9 + 6\sqrt{2}}{9 - 4\cdot 2}$$ *Use FOIL to multiply.*

$$= \dfrac{6\sqrt{5} - 4\sqrt{10} - 9 + 6\sqrt{2}}{1}$$

$$= 6\sqrt{5} - 4\sqrt{10} - 9 + 6\sqrt{2}$$

CLASS EXERCISES

Rationalize the denominator and simplify.

1. $\sqrt{\dfrac{1}{5}}$
2. $\sqrt{\dfrac{1}{2}}$
3. $\dfrac{\sqrt{2}}{\sqrt{5}}$
4. $\dfrac{\sqrt{3}}{\sqrt{7}}$
5. $\dfrac{2\sqrt{2}}{\sqrt{18}}$
6. $\dfrac{3\sqrt{2}}{4\sqrt{32}}$
7. $\sqrt{\dfrac{4}{x}}$
8. $\sqrt{\dfrac{9}{b^3}}$

Write the conjugate of each binomial.

9. $1 - \sqrt{2}$
10. $2 + \sqrt{3}$
11. $5 + 3\sqrt{5}$
12. $4 - 2\sqrt{7}$

PRACTICE EXERCISES

Rationalize the denominator and simplify.

1. $\sqrt{\dfrac{2}{11}}$
2. $\sqrt{\dfrac{3}{10}}$
3. $-2\sqrt{\dfrac{1}{2}}$
4. $-3\sqrt{\dfrac{1}{3}}$
5. $\dfrac{3\sqrt{8}}{\sqrt{3}}$
6. $\dfrac{2\sqrt{12}}{-\sqrt{5}}$
7. $\dfrac{-\sqrt{90}}{\sqrt{10}}$
8. $\dfrac{\sqrt{75}}{2\sqrt{3}}$
9. $\dfrac{5\sqrt{2}}{3\sqrt{6}}$
10. $\dfrac{4\sqrt{3}}{2\sqrt{8}}$
11. $\dfrac{-3\sqrt{5}}{4\sqrt{15}}$
12. $\dfrac{6\sqrt{6}}{-2\sqrt{30}}$
13. $\sqrt{\dfrac{16}{y}}$
14. $\sqrt{\dfrac{25}{a}}$
15. $\sqrt{\dfrac{8}{x}}$
16. $\sqrt{\dfrac{12}{b}}$
17. $\dfrac{3}{\sqrt{a^3}}$
18. $\dfrac{-5}{\sqrt{x^5}}$
19. $\dfrac{-2}{3\sqrt{y^3}}$
20. $\dfrac{\sqrt{3}}{\sqrt{12m}}$
21. $\sqrt{\dfrac{24c^3}{6c}}$
22. $\sqrt{\dfrac{27x^5}{3x}}$
23. $\dfrac{\sqrt{a^3b^4}}{\sqrt{ab}}$
24. $\dfrac{\sqrt{cd}}{\sqrt{c^3d^3}}$
25. $\dfrac{5}{3 - \sqrt{2}}$
26. $\dfrac{-6}{5 + \sqrt{7}}$
27. $\dfrac{-3}{\sqrt{5} - 1}$
28. $\dfrac{2}{\sqrt{6} + 3}$
29. $\dfrac{\sqrt{5} - 1}{\sqrt{5} + 3}$
30. $\dfrac{\sqrt{6} + 3}{\sqrt{6} - 4}$
31. $\dfrac{2 + \sqrt{3}}{1 - \sqrt{5}}$
32. $\dfrac{4 - \sqrt{3}}{2 + \sqrt{7}}$
33. $\dfrac{-5}{2\sqrt{11} + 2}$
34. $\dfrac{-7}{3\sqrt{10} - 5}$
35. $\dfrac{1 - 3\sqrt{7}}{3\sqrt{3} + 2}$
36. $\dfrac{3\sqrt{2} + 2}{3 - 2\sqrt{5}}$

37. $\dfrac{2\sqrt{7} + 3\sqrt{5}}{3\sqrt{7} - 2\sqrt{5}}$

38. $\dfrac{5\sqrt{2} - \sqrt{3}}{\sqrt{2} + 3\sqrt{3}}$

39. $\dfrac{\sqrt{x} - \sqrt{y}}{\sqrt{x}}$

40. $\dfrac{\sqrt{x} + 2\sqrt{y}}{5\sqrt{x}}$

41. $\dfrac{2\sqrt{x} - 3}{\sqrt{x} + 1}$

42. $\dfrac{3\sqrt{x} + 4\sqrt{y}}{3\sqrt{x} - 4\sqrt{y}}$

Applications

Solve. Express answers in simplest form.

43. **Chemistry** The ratio of the rates of diffusion of two gases is given by the formula $\dfrac{r_1}{r_2} = \dfrac{\sqrt{m_1}}{\sqrt{m_2}}$ where m_1 and m_2 are masses of the molecules of the gases. Find $\dfrac{r_1}{r_2}$ if $m_1 = 25$ units and $m_2 = 80$ units.

44. **Physics** The period T in seconds of a pendulum of length L in feet is given by the formula $T = 2\pi\sqrt{\dfrac{L}{32}}$. Find the period of a pendulum 4 ft long.

45. **Number Problem** The square root of a number divided by the square root of the product of 4 and the number is $\dfrac{1}{2}$.

46. **Number Problem** The square root of the product of 5 and some number divided by the square root of the product of 20 and the number squared is $\dfrac{1}{2}$.

EXTRA

Tell whether each statement is true or false. If false, tell why.

1. $\sqrt{36n} = \sqrt{36} \cdot \sqrt{n}$

2. $\dfrac{5}{6} = 0.\overline{83}$

3. $\sqrt{36} = -6$

4. $\sqrt{20} = 4\sqrt{5}$

5. $(3 + \sqrt{2})^2 = 11 + 6\sqrt{2}$

6. $2\sqrt{5} \cdot 3\sqrt{6} = 6\sqrt{11}$

7. The simplest form of $\sqrt{\dfrac{1}{8}}$ is $\dfrac{\sqrt{8}}{8}$.

8. The simplest form of $\dfrac{\sqrt{25}}{\sqrt{7}}$ is $\dfrac{5\sqrt{7}}{7}$.

12.7 Division of Radicals

12.8 Solving Radical Equations

Objective: To solve simple radical equations

Many scientific formulas contain radicals. For example, the velocity (in ft/s) of an object (disregarding air resistance) after it has fallen a certain distance is given by the formula $v = \sqrt{2gh}$, where g is the acceleration of gravity and h is the distance (in ft) the object has fallen. The equation $v = \sqrt{2gh}$ is called a *radical equation*.

Capsule Review

For what values of x is the radical a real number?

EXAMPLE $\sqrt{2x + 6}$

$2x + 6 \geq 0$ $\sqrt{2x + 6}$ is a real number if $2x + 6$ is nonnegative.
$2x \geq -6$ Solve for x.
$x \geq -3$

So, $\sqrt{2x + 6}$ is a real number when $x \geq -3$.

1. $\sqrt{3x - 9}$ **2.** $\sqrt{-2x + 5}$ **3.** $\sqrt{18 - 5x}$ **4.** $\sqrt{6 + 3x}$

A **radical equation** is an equation that contains radicals with variables in the radicand. To solve such an equation:

- Isolate the radical on one side of the equation and combine any like terms.
- Square both sides of the equation to eliminate the radical.
- Repeat steps 1 and 2, if necessary.

EXAMPLE 1 Solve and check: $3 + \sqrt{a - 1} = 5$

$3 + \sqrt{a - 1} = 5$
$\sqrt{a - 1} = 5 - 3$ *Isolate the radical by subtracting 3 from each side.*
$(\sqrt{a - 1})^2 = (2)^2$ *Combine like terms: $5 - 3 = 2$; then square each side.*
$a - 1 = 4$ $(\sqrt{n})^2 = n, n \geq 0$
$a = 5$

Check: $3 + \sqrt{a - 1} = 5$
$3 + \sqrt{5 - 1} \stackrel{?}{=} 5$ *Replace a with 5.*
$3 + \sqrt{4} \stackrel{?}{=} 5$
$3 + 2 \stackrel{?}{=} 5$
$5 = 5$ ✔

The solution is 5.

You can square each side of an equation anytime it is convenient to do so, as long as you check all solutions in the original equation.

EXAMPLE 2 Solve and check: $0 = 8 + 4\sqrt{x}$

$0 = 8 + 4\sqrt{x}$
$-8 = 4\sqrt{x}$ *Isolate the term that contains the radical.*
$-2 = \sqrt{x}$ *Divide each side by 4.*
$(-2)^2 = (\sqrt{x})^2$ *Square each side.*
$4 = x$ $(\sqrt{n})^2 = n, n \geq 0$

Check: $0 \stackrel{?}{=} 8 + 4\sqrt{4}$
$0 \stackrel{?}{=} 8 + 4 \cdot 2$
$0 \neq 16$

The equation $0 = 8 + 4\sqrt{x}$ has no real number solution.

In this example, the simplified equation $-2 = \sqrt{x}$ is obtained. Explain why there is no real number solution for this equation.

When an equation is solved by squaring, an *extraneous solution* may be introduced. Recall from Chapter 8 an extraneous solution will not check in the original equation.

EXAMPLE 3 Solve and check: $\sqrt{5t^2 - 16} = t$

$\sqrt{5t^2 - 16} = t$
$(\sqrt{5t^2 - 16})^2 = (t)^2$ *Square each side.*
$5t^2 - 16 = t^2$
$4t^2 = 16$
$t^2 = 4$
$t = 2$ or $t = -2$

Check: $\sqrt{5(2)^2 - 16} \stackrel{?}{=} 2$ $\sqrt{5(-2)^2 - 16} \stackrel{?}{=} -2$
$\sqrt{5(4) - 16} \stackrel{?}{=} 2$ $\sqrt{5(4) - 16} \stackrel{?}{=} -2$
$\sqrt{20 - 16} \stackrel{?}{=} 2$ $\sqrt{20 - 16} \stackrel{?}{=} -2$
$\sqrt{4} \stackrel{?}{=} 2$ $\sqrt{4} \stackrel{?}{=} -2$
$2 = 2$ ✔ $2 \neq -2$

The solution is 2. -2 is an extraneous solution.

CLASS EXERCISES

Solve and check.

1. $\sqrt{x} = 5$
2. $\sqrt{a} = 12$
3. $\sqrt{2b} = 16$
4. $\sqrt{3y} = 9$
5. $2\sqrt{3t} = 4$
6. $3\sqrt{5p} = 6$

State each equation in the form needed before squaring each side. Then give the equation obtained by squaring.

7. $\sqrt{2x + 1} = 3$
8. $\sqrt{3a - 1} = 2$
9. $\sqrt{4y + 1} + 3 = 2$
10. $\sqrt{3c + 1} - 4 = 1$
11. $0 = 3 - 2\sqrt{y}$
12. $6 - 3\sqrt{2x} = 0$

PRACTICE EXERCISES

Solve and check.

1. $\sqrt{y} = 8$
2. $\sqrt{a} = 9$
3. $\sqrt{3m} = 18$
4. $\sqrt{4x} = 8$
5. $\sqrt{n} = \frac{1}{4}$
6. $\sqrt{m} = \frac{2}{3}$
7. $\sqrt{\frac{x}{3}} = 2$
8. $\sqrt{\frac{b}{7}} = 3$
9. $\sqrt{y - 2} = 3$
10. $\sqrt{a + 4} = 6$
11. $\sqrt{4x} - 4 = 0$
12. $0 = \sqrt{2z} - 9$
13. $6 + \sqrt{m} = 5$
14. $10 + 2\sqrt{b} = 0$
15. $6 - 3\sqrt{y} = 0$
16. $8 - \sqrt{a} = 1$
17. $\sqrt{5x - 1} + 3 = 7$
18. $\sqrt{3x + 4} + 1 = 4$
19. $\sqrt{6m - 1} - 5 = -2$
20. $\sqrt{2x - 3} - 2 = 4$
21. $7 - 3\sqrt{2b} = 1$
22. $5 - 2\sqrt{y} = 2$
23. $\sqrt{\frac{2m}{3}} + 5 = 7$
24. $9 = \sqrt{\frac{5n}{4}} - 1$
25. $\sqrt{x} = 3\sqrt{2}$
26. $\sqrt{t} = 2\sqrt{5}$
27. $\sqrt{5y^2 - 7} = 2y$
28. $\sqrt{3n^2 + 12} = 3n$
29. $\sqrt{y^2 + 3} = y + 1$
30. $t - 5 = \sqrt{t^2 - 35}$
31. $\sqrt{2x - 3} = \sqrt{3x - 2}$
32. $\sqrt{3y - 5} = \sqrt{5y - 3}$
33. $\sqrt{2b + 1} = \sqrt{4b + 7}$
34. $2\sqrt{5a - 1} = 3\sqrt{2a + 4}$
35. $\sqrt{3y + 4} = y - 2$
36. $m - 4 = \sqrt{2m - 5}$
37. $\sqrt{y} + 2 = \sqrt{y + 16}$
38. $\sqrt{x - 1} = 2 - \sqrt{x}$

Cube both sides. Solve for a.

Example: $\sqrt[3]{7a + 5} = 3 \longrightarrow 7a + 5 = 27 \longrightarrow a = \frac{22}{7}$

39. $\sqrt[3]{2x + 3} = 7$
40. $\sqrt[3]{3y + 5} = \sqrt[3]{5 - 2y}$

41. Explain why cubing both sides of cubic-root equations will not produce extraneous solutions.

Applications

42. **Number Problem** The square root of $\frac{1}{5}$ of a number is 5. Find the number.

43. **Number Problem** When 5 times a number is decreased by 2, the square root of the result is 6. Find the number.

44. **Physics** In $v = \sqrt{2gh}$, find the value of h if $v = 18$ ft/s and $g = 32$ ft/s^2.

45. **Physics** The time t in seconds that it takes an object to fall a distance of d ft from rest is given by the formula $t = \sqrt{\frac{2d}{g}}$ where g is the acceleration due to gravity. Solve for d. Find the value of d when $g = 32$ ft/s^2 and $t = 3$ s.

46. **Physics** The period (T) of a pendulum is the time needed to swing side to side and back. The period is given by the formula $T = 2\pi\sqrt{\frac{l}{980}}$, where l is the length of a pendulum in centimeters. Find the length of the pendulum given a period of 1 s and $\pi \approx 3.14$.

EXTRA

When you were younger, you discovered that you could see farther if you climbed a tree. The higher you climbed, the farther you could see. There is a formula for this:

$$V = 3.5\sqrt{h}$$

where h is your height, in meters, above the ground, and V is the distance, in kilometers, that you can see.

Solve. Round your answer to the nearest whole number.

1. How far could you see if you climbed 7 m to the top of a tree?

2. Suppose that you wanted to see a distance of 21 km. How tall a tree should you climb to do this?

3. Suppose that you could see a distance of 55 km if you stood on the roof of a building. To the nearest meter, how tall is the building?

4. Suppose that you are flying in an airplane at an altitude of 8 km. How far could you see from the window of the plane?

5. Suppose that the plane in Exercise 4 loses altitude and that you could then see a distance of about 200 km. How much altitude did the plane lose?

12.9 The Pythagorean Theorem

Objective: To use the Pythagorean theorem to find the length of a leg or the length of the hypotenuse of a right triangle

In physics, two forces that pull at right angles and the resultant force are represented by a rectangle with a diagonal, as shown at the right. The magnitude of this resultant force can be calculated by thinking of the rectangle as being divided into two right triangles. The calculations involve powers and square roots.

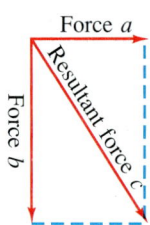

Capsule Review

Simplify.

1. 6^2
2. $3^2 + 2^2$
3. $10^2 - 8^2$
4. 18^2
5. $\sqrt{36}$
6. $\sqrt{81}$
7. $\sqrt{13^2 - 5^2}$
8. $\sqrt{3^2 + 4^2}$
9. $(\sqrt{3})^2$
10. $(\sqrt{11})^2$
11. $(x + 1)^2$
12. $(y - 2)^2$

In a right triangle, the side opposite the right angle is the longest side. It is called the **hypotenuse**. The other two sides are called the **legs** of the triangle.

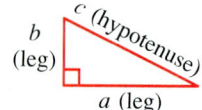

From the work of a Greek named Pythagoras and the members of his secret society came an important right-angled triangle property called the *Pythagorean theorem*. This theorem relates the squares on the sides of a right triangle and thus describes the relationship of the lengths of those sides.

The Pythagorean Theorem

In a right triangle, the square of the length of the hypotenuse is equal to the sum of the squares of the lengths of the two legs.

$$c^2 = a^2 + b^2$$

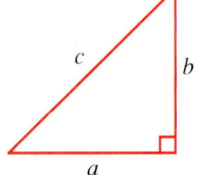

The *Pythagorean theorem* is an important mathematical tool for solving problems in surveying, carpentry, and navigation. It also provides mathematicians with insight in finding the square and square roots of a number.

EXAMPLE 1 Two forces, $a = 6$ lb and $b = 8$ lb, pull at right angles. Find the resultant force, c.

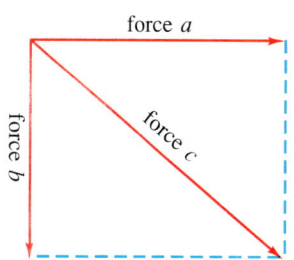

$c^2 = a^2 + b^2$
$c^2 = 6^2 + 8^2$ *Replace a with 6 and b with 8.*
$c^2 = 36 + 64$
$c^2 = 100$ *Since $c^2 = 100$, $c = 10$ or $c = -10$.*
$c = 10$ *The length must be positive.*

So, the resultant force is 10 lb.

In geometry, two special right triangles are often used. They are referred to as the 45-45-90 triangle and the 30-60-90 triangle. The numbers are the measures, in degrees, of the angles of the triangles.

EXAMPLE 2 Find the length of the unknown side to the nearest tenth. Use the table on page 670 or a calculator.

a.
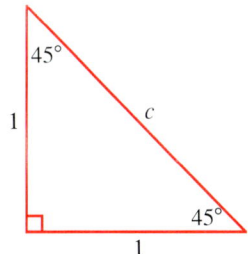

b.
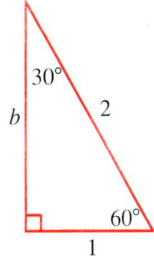

a. $c^2 = a^2 + b^2$
$c^2 = 1^2 + 1^2$ *Solve for c^2*
$c^2 = 2$ *Since $c^2 = 2$, $c = \sqrt{2}$ or $c = -\sqrt{2}$.*
$c = \sqrt{2}$ *The length, c, cannot be negative.*
$c \approx 1.4$ *Use a table or calculator.*

b. $c^2 = a^2 + b^2$
$2^2 = 1^2 + b^2$ *Solve for b^2.*
$4 = 1 + b^2$
$3 = b^2$
$\sqrt{3} = b$
$1.7 \approx b$

When you know the lengths of the three sides of a triangle, you can determine whether it is a right triangle by using the *converse* of the Pythagorean theorem: If $a^2 + b^2 = c^2$, then the triangle is a right triangle.

EXAMPLE 3 The lengths of the sides of a triangle are 4, 5, and 8. Determine whether the triangle is a right triangle.

$a^2 + b^2 = c^2$
$4^2 + 5^2 \stackrel{?}{=} 8^2$ *Substitute: $a = 4$, $b = 5$, $c = 8$*
$16 + 25 \stackrel{?}{=} 64$
$41 \neq 64$ So, the triangle is not a right triangle.

CLASS EXERCISES

Use the Pythagorean theorem, $c^2 = a^2 + b^2$, to find the missing length.

1. $a = 6, b = 8, c = ?$
2. $a = 10, b = 24, c = ?$
3. $a = 5, c = 13, b = ?$
4. $b = 4, c = 5, a = ?$
5. $a = 2, c = \sqrt{13}, b = ?$
6. $a = \sqrt{5}, b = \sqrt{5}, c = ?$

PRACTICE EXERCISES

Use the Pythagorean theorem, $c^2 = a^2 + b^2$, to find the missing length to the nearest tenth. Use the table on page 670 or a calculator.

1. $a = 3, b = 4, c = ?$
2. $a = 5, b = 12, c = ?$
3. $a = 10, c = 19, b = ?$
4. $b = 8, c = 24, a = ?$
5. $a = 4, b = 9, c = ?$
6. $a = 7, b = 5, c = ?$
7. $b = 10, c = 13, a = ?$
8. $a = 40, c = 41, b = ?$

The lengths of three sides of a triangle are given. Is it a right triangle?

9. 3, 4, 6
10. 18, 15, 23
11. 1, 1, $\sqrt{2}$
12. 1, $\sqrt{3}$, 2

Use $c^2 = a^2 + b^2$ to find the missing length to the nearest tenth. Use the table on page 670 or a calculator.

13. $a = 0.8, b = 0.6, c = ?$
14. $a = 2.4, b = 1.0, c = ?$
15. $a = \frac{1}{5}, c = \frac{1}{3}, b = ?$
16. $a = 2\frac{1}{2}, c = 6\frac{1}{2}, b = ?$

The diagrams below suggest a proof of the Pythagorean theorem. Each figure shows a square with side of length $(a + b)$.

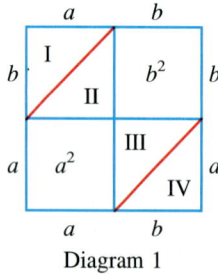
Diagram 1

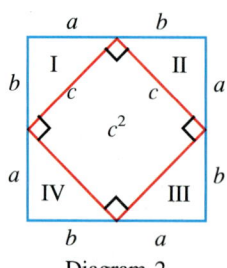
Diagram 2

17. In Diagram 1, show that $(a + b)^2 = a^2 + b^2 + 4\left(\frac{1}{2}ab\right)$.

    In Diagram 2, show that $(a + b)^2 = c^2 + 4\left(\frac{1}{2}ab\right)$.

18. Use the equations in Exercise 17 to verify the Pythagorean theorem.

Applications

Solve. Round the answer to the nearest tenth.

19. **Construction** A carpenter braces an 8 × 15 ft wall by nailing a board diagonally across the wall. How long is the bracing board?

20. **Physics** Two jeeps at a 90° angle to each other try to pull a third car out of the snow. If one jeep exerts a force of 600 lb and the other exerts a force of 800 lb, what is the resulting force on the car stuck in snow?

21. **Geometry** The diagonal of a square is $8\sqrt{2}$ cm. Find the length of a side of the square.

22. **Sports** A hiker leaves her camp in the morning. How far is she from camp after walking 5 mi due west and then 7 mi due north?

23. **Geometry** The lengths of the sides of a right triangle are given by three consecutive integers. Find the lengths of the sides.

24. **Construction** A wire is stretched from the top of a 4-ft pole to the top of a 9-ft fence. If the pole and fence are 10 ft apart, how long is the wire?

DID YOU KNOW?

The gaps between the rails on a railroad track are deliberately placed there for the safety of the train. These are called *expansion gaps* and allow the rails to expand when the temperature rises. Without such gaps, the track would buckle and rise off the ground.

1. Suppose a 20-ft piece of rail expands 1 in. during a hot spell. If there were no expansion gaps between the rails, about how high off the ground would the rail rise? (*Hint:* Use the Pythagorean theorem and right triangles to estimate the height.) Before expansion, the rail is 240 in. long. After expansion, it is 241 in. long. Find h.

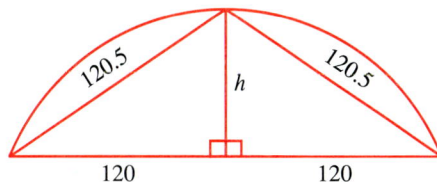

2. To make the ride smoother for passengers riding a train, it is desirable not to have too many gaps. Suppose rails are 50 ft long. If a rail expands 1 in. during a temperature change, how far off the ground would it rise if there were no expansion gaps?

12.10 The Distance Formula

Objective: To find the distance between any two points in the coordinate plane

A city planner is making a map for a new subdivision that will be laid out in square blocks. Various locations are labeled with ordered pairs, just as points in the coordinate plane are labeled. If two locations are labeled $A(3, 5)$ and $B(9, 2)$, what is the distance between the points?

In Lesson 5.6, you found the distance between two points on a number line. This is important to recall before studying the *distance formula*.

Capsule Review

Find the distance between the given points on a number line.

EXAMPLE $A(-5)$ and $B(2)$

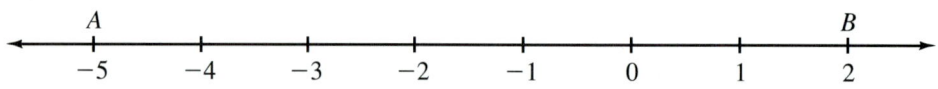

$|2 - (-5)| = |2 + 5| = 7$, or $|-5 - 2| = |-7| = 7$

The distance between A and B is 7 units.

1. $R(2)$ and $S(5)$
2. $C(7)$ and $D(3)$
3. $X(8)$ and $Y(-1)$
4. $E(-5)$ and $F(1)$
5. $A(-5)$ and $D(-4)$
6. $P(-2)$ and $Q(-13)$

In the coordinate plane, you can think of the axes and the lines parallel to the axes as number lines.

For the lines parallel to the x-axis:

$DE = |1 - 5| = |5 - 1| = 4$
$FG = |1 - 5| = |5 - 1| = 4$

For the lines parallel to the y-axis:

$DF = |4 - 1| = |1 - 4| = 3$
$EG = |4 - 1| = |1 - 4| = 3$

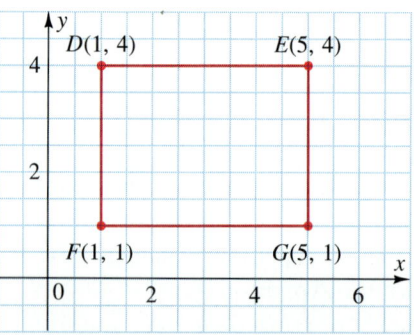

528   Chapter 12 Radicals

To find the distance between two points A and B not on the x-axis, y-axis, or a line parallel to either axis, you can use the Pythagorean theorem.

$(AB)^2 = (AC)^2 + (CB)^2$
$AB = \sqrt{(AC)^2 + (CB)^2}$
$= \sqrt{(5-2)^2 + (9-3)^2}$
$= \sqrt{3^2 + 6^2}$
$= \sqrt{9 + 36} = \sqrt{45} = 3\sqrt{5}$

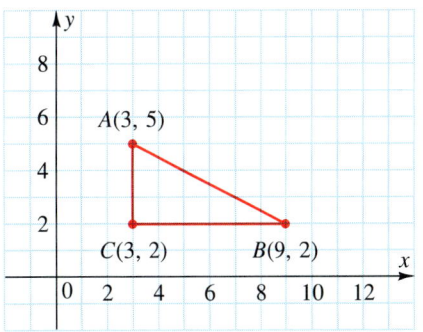

So, the distance between A and B on the city planner's map is $3\sqrt{5}$ units. This method can be generalized as follows:

Distance Formula

For any points $P_1(x_1, y_1)$ and $P_2(x_2, y_2)$ in the coordinate plane, the distance d between the points is given by

$$d = \sqrt{(x_2 - x_1)^2 + (y_2 - y_1)^2}$$

EXAMPLE Find the distance between $E(-4, 2)$ and $F(5, 4)$.

$d = \sqrt{(x_2 - x_1)^2 + (y_2 - y_1)^2}$
$d = \sqrt{(-4 - 5)^2 + (2 - 4)^2}$
$d = \sqrt{(-9)^2 + (-2)^2}$
$d = \sqrt{81 + 4}$
$d = \sqrt{85}$

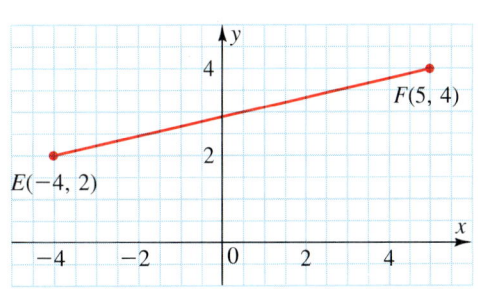

The distance between $E(-4, 2)$ and $F(5, 4)$ is $\sqrt{85}$.

CLASS EXERCISES

State the values of x_1, x_2, y_1, and y_2 for each pair of points. Then find the distance.

1. $A(0, 0)$, $B(-3, 4)$
2. $P(5, 12)$, $Q(0, 0)$
3. $R(2, 2)$, $T(5, -2)$
4. $C(1, 4)$, $D(6, 9)$
5. $X(3, 1)$, $Y(-1, 6)$
6. $E(5, 6)$, $F(4, -1)$

For Discussion

7. When using the distance formula, does it matter which point has coordinates (x_1, y_1) or (x_2, y_2)? Explain.

PRACTICE EXERCISES

Find the distance between each pair of points.

1. $A(1, 2)$, $B(4, -2)$
2. $R(4, 2)$, $T(-2, 10)$
3. $P(4, -2)$, $Q(-2, 4)$
4. $X(-6, -2)$, $Y(-5, 4)$
5. $L(-5, 1)$, $M(7, 6)$
6. $E(5, 13)$, $F(5, -1)$
7. $E(3, 1)$, $F(-2, -1)$
8. $A(-5, 0)$, $B(-9, 6)$
9. $C(10, 8)$, $D(2, -3)$
10. $P(7, -9)$, $Q(4, -3)$
11. $S(8, -10)$, $T(3, 2)$
12. $A(-2, 4)$, $B(7, -8)$

Find the lengths of the sides of the triangle whose vertices are given.

13. $A(0, 0)$, $B(8, 0)$, $C(4, 3)$
14. $R(-1, 7)$, $S(0, 0)$, $T(8, 4)$
15. $X(-4, 2)$, $Y(-1, 6)$, $Z(5, 4)$
16. $D(1, 5)$, $E(5, 5)$ $F(5, 1)$
17. $A(3, 6)$, $B(-1, 3)$, $C(5, -5)$
18. $R(7, -3)$, $S(0, 4)$, $T(8, -1)$
19. $M(-7, 5)$, $N(-7, -7)$, $P(2, -7)$
20. $J(6, 3)$, $K(9, 7)$, $L(10, 0)$

21. The distance between two points with coordinates $(1, 1)$ and $(4, y)$ is 5. Find all possible values for y.

22. The distance between two points with coordinates $(2, -1)$ and $(x, 3)$ is 5. Find all possible values for x.

23. Can the point with coordinates $(-2, 4)$ be the center of a circle that passes through the points with coordinates $(1, 0)$, $(-5, 0)$, and $(1, 8)$?

Applications

24. **Computer** The distance formula has many applications in coordinate geometry. Use it to prove that quadrilateral $ABCD$ is a rhombus. $A(6, 5)$, $B(2, 2)$, $C(9, 1)$ and $D(5, -2)$. This computer program will find the lengths of each side of the figure.

```
10  PRINT "ENTER THE COORDINATES FOR"
20  INPUT "POINT A:   ";AX,AY
30  INPUT "POINT B:   ";BX,BY
40  INPUT "POINT C:   ";CX,CY
50  INPUT "POINT D:   ";DX,DY
60  S1 =   SQR ((AX - BX) ^ 2 + (AY - BY) ^ 2)
70  S2 =   SQR ((BX - CX) ^ 2 + (BY - CY) ^ 2)
80  S3 =   SQR ((CX - DX) ^ 2 + (CY - DY) ^ 2)
90  S4 =   SQR ((DX - AX) ^ 2 + (DY - AY) ^ 2)
100 PRINT : PRINT "SIDE AB IS ";S1
110 PRINT "SIDE BC IS ";S2
120 PRINT "SIDE CD IS ";S3
130 PRINT "SIDE DA IS ";S4
140 END
```

LOGICAL REASONING

A geometric figure is formed by the four lines whose equations are:

$$y = x \qquad y = -x$$
$$y = x - 4 \qquad y = -x + 4$$

What kind of geometric figure is it? What is its area?

12.11 Problem Solving Strategy: Use Coordinate Geometry

Coordinate geometry combines concepts from algebra and geometry by using the coordinate plane. Certain kinds of problems are more easily solved by this approach than by either algebra or geometry alone.

Given a line segment, how can the midpoint of the segment be found without measurement?

EXAMPLE Find the midpoint M of segment RP given the endpoints of your choice; for this example, use the points $(-4, -1)$ and $(6, 5)$.

Understand the Problem

What are the given facts?
Segment RP has endpoints of $(-4, -1)$ and $(6, 5)$.

What are you asked to find?
The coordinates of the midpoint $M(x, y)$.

Plan Your Approach

Choose a strategy.
The strategy is to use a right triangle and the lengths of the sides of the triangle to locate the x- and y-coordinates for the midpoint M.

- Plot the points R and P, then draw a right triangle RSP. The coordinates of S are $(6, -1)$. Why?
- Draw lines parallel to the x- and y-axes from M to sides RS and SP.
- Then A is the midpoint of RS and B is the midpoint of SP.

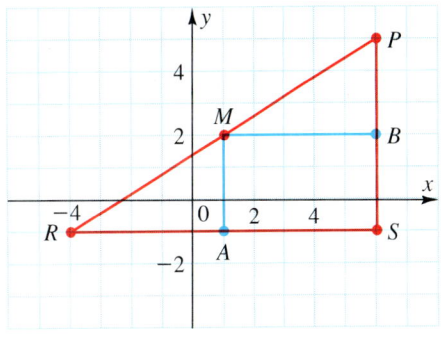

Complete the Work

Locate the values of A and B.
Since A is the midpoint of RS, then $RA = AS$. Since B is the midpoint of SP, then $SB = BP$.

$$\text{length of } RA = \text{distance from } R(-4, -1) \text{ to } A(x, -1)$$
$$= |x - (-4)|$$

12.11 Problem Solving Strategy: Use Coordinate Geometry **531**

Since x is the larger value, the statement can be written as

$$x - (-4)$$

The lengths of each segment can be determined by subtracting the smaller coordinate from the larger coordinate.

$RA = x - (-4)$ and $AS = 6 - x$ $SB = y - (-1)$ and $BP = 5 - y$
$x + 4 = 6 - x$ Solve for x. $y + 1 = 5 - y$ Solve for y.
$2x = 2$ $2y = 4$
$x = 1$ $y = 2$

Interpret the Results

State your answer.
The coordinates of midpoint M are $(1, 2)$. Notice that given $R(-4, -1)$ and $P(6, 5)$, the midpoint coordinates, $(1, 2)$, are the averages of the x- and y-coordinates. How would you show this?

For this example, the endpoints of segment RP were given. If the endpoints were any coordinates (x_1, y_1) and (x_2, y_2), then the midpoint could be found using the average of the x- and y-coordinates.

$$M\left(\frac{x_1 + x_2}{2}, \frac{y_1 + y_2}{2}\right)$$

CLASS EXERCISES

Find the coordinates of the midpoints of the segments with the following endpoints.

1. $(2, 4)$ and $(8, -4)$
2. $(-4, 3)$ and $(6, -11)$
3. $(-5, 2)$ and $(7, 7)$
4. $(-6, 0)$ and $(0, 6)$
5. $(x, 0)$ and $(0, y)$
6. $(2a, b)$ and $(6a, -4b)$

For Discussion

7. How would you use the midpoint formula to show that the midpoint of the hypotenuse of a right triangle is equidistant from the three vertices?

PRACTICE EXERCISES

Find the midpoints of the segments with the following endpoints.

1. $(-6, 4)$ and $(-4, 6)$
2. $(-4, -1)$ and $(-1, -4)$
3. $(-7, 3)$ and $(-5, 3)$
4. $(5, -2)$ and $(8, -2)$
5. $(1, 7)$ and $(-8, -3)$
6. $(-5, 2)$ and $(-7, 5)$
7. $(2a, b)$ and $(a, 2b)$
8. $(3a, -3b)$ and $(-2a, -3b)$

Chapter 12 Radicals

Write the missing coordinates of the labeled points using the given points.

9.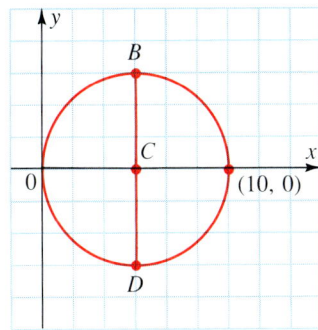

10.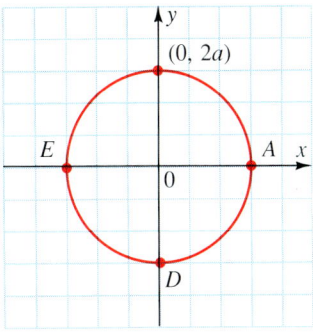

Complete each statement in order to show that the diagonals are equal in length.

11.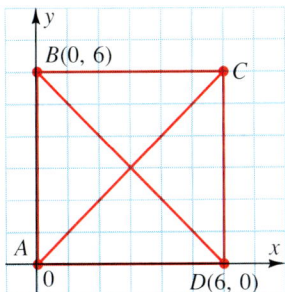

12.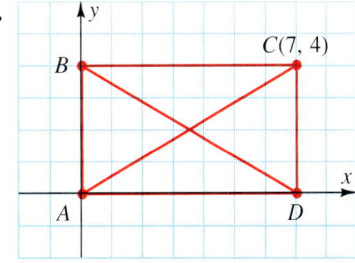

a. The coordinates of C are _?_.
b. The length of segment AC is _?_.
c. The length of segment BD is _?_.
d. What statement can be made concerning the length of the diagonals of a square?

a. The coordinates of B are _?_.
b. The coordinates of D are _?_.
c. The length of segment AC is _?_.
d. The length of segment BD is _?_.
e. What statement can be made about the diagonals of a rectangle?

13. Use the square in Exercise 11 to show that the diagonals of a square bisect each other. That is, show that the diagonals have the same midpoint.

14. For Exercise 12, show that the diagonals of the rectangle have the same midpoint.

15. Triangle ABC has vertices $A(-2, 1)$, $B(5, 2)$, and $C(1, -2)$. Find the length of each side of the triangle. Is it a right triangle? Explain.

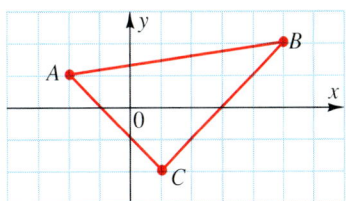

12.11 Problem Solving Strategy: Use Coordinate Geometry 533

For Exercises 16–17, find:

16.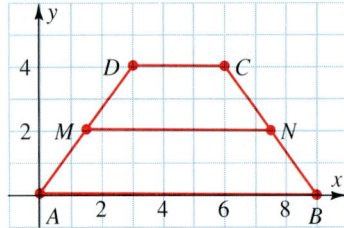

17.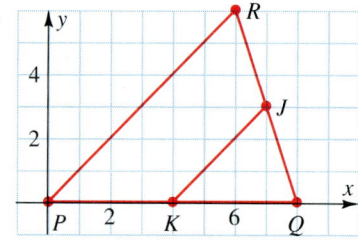

a. the midpoint M of $\overline{AD}$
b. the midpoint N of $\overline{BC}$
c. the length of $\overline{MN}$
d. the length of $\overline{CD}$
e. the length of $\overline{AB}$
f. a statement that can be made concerning MN and $AB + CD$

a. the midpoint K of $\overline{PQ}$
b. the midpoint J of $\overline{QR}$
c. the length of $\overline{JK}$
d. the length of $\overline{PR}$
e. a statement that can be made concerning JK and PR

18. The perpendicular bisector of a line segment intersects the segment's midpoint at right angles. Any point lying in the perpendicular bisector is equidistant from the endpoints of the segment. In the figure line PQ is the perpendicular bisector of segment AB.

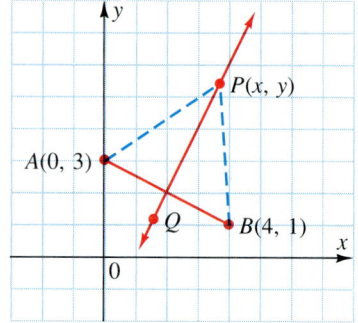

a. Using the coordinates of $A(0, 3)$ and $P(x, y)$, find the length of segment AP.
b. Using the coordinates of $B(4, 1)$ and $P(x, y)$, find the length of segment BP.
c. Using the lengths of segments AP and BP, write an equation that expresses the fact that P is equidistant from A and B.
d. Square both sides of the equations, and simplify. What is your resulting equation?

Mixed Problem Solving Review

1. How many kilograms of pinto beans costing $1.29/kg and how many kilograms of garbanzos at $1.59/kg should be mixed to make 27 kg of a mixture valued at $38.43?

2. Suki's age is now 75% of her brother's. Seven years ago his age was 50% greater than hers. How old is each one now?

3. Canoeing downstream for 2 h Josh travels 11 mi. Traveling upstream for the same time he goes only 7 mi. How fast does he go in still water?

4. How many milliliters of water must be mixed with 25 ml of a 15% solution of alcohol in water to get a 10% solution.

PROJECT

René Descartes, a mathematician of the 1600s, provided the foundation for coordinate geometry, and Sir Isaac Newton was famous for his development of calculus. Newton wrote to a friend, "If I have seen farther than Descartes, it is because I have stood on the shoulders of giants." Research why he might have made such a statement.

TEST YOURSELF

Simplify. 12.7

1. $\sqrt{\dfrac{4}{8}}$

2. $\dfrac{5}{3 - \sqrt{2}}$

3. $\sqrt{t^2 + 6t + 9}$

4. $\dfrac{\sqrt{48x^3y^2}}{\sqrt{3xy}}$

5. $\dfrac{-3\sqrt{5}}{4 - \sqrt{15}}$

6. $\dfrac{5}{2 - \sqrt{3}}$

Find the solution to each of the following equations. 12.8

7. $\sqrt{10t^2 - 16} = t$

8. $\sqrt{9a^2 - 6a + 1} = 0$

9. $\sqrt{12x - 35} = x$

10. $\sqrt{-5h + 3} = \sqrt{2h}$

Determine the value of x for each of the following triangles. 12.9

11.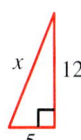

12.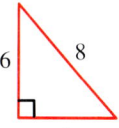

13.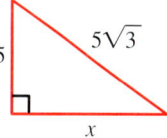

Find the distance and midpoint for each of the following pairs of points. 12.10–12.11

14. $A(10, 7)$, $B(-3, 4)$

15. $A(8, -10)$, $B(3, 5)$

12.11 Problem Solving Strategy: Use Coordinate Geometry 535

CHAPTER 12 SUMMARY AND REVIEW

Vocabulary

completeness property (497)
conjugate (517)
density property (498)
hypotenuse (524)
irrational number (497)
leg (524)
principal square root (494)
radical (494)

radical equation (520)
radical sign (494)
radicand (494)
rationalizing the denominator (516)
repeating decimal (500)
square root (494)
terminating decimal (500)

Square Roots If $x^2 = y$, then x is called a square root of y. — 12.1

Product and Quotient Properties of Square Roots
For all real numbers m and n where $m \geq 0$ and $n \geq 0$, $\sqrt{mn} = \sqrt{m} \cdot \sqrt{n}$

For all real numbers m and n where $m \geq 0$ and $n > 0$, $\sqrt{\dfrac{m}{n}} = \dfrac{\sqrt{m}}{\sqrt{n}}$.

Find the principal square root.

1. $\sqrt{25}$
2. $\sqrt{\dfrac{4}{9}}$
3. $\sqrt{\dfrac{81}{36}}$
4. $\sqrt{900}$

Irrational Square Roots An irrational number is a real number that cannot be written in the form $\dfrac{m}{n}$, where m and n are integers and $n \neq 0$. — 12.2

Use the square-root table on page 670 to approximate the square root.

5. $\sqrt{5}$
6. $\sqrt{10}$
7. $\sqrt{89}$
8. $\sqrt{60}$

Use the divide-and-average method to the nearest hundredth to approximate the square root.

9. $\sqrt{8}$
10. $\sqrt{17}$
11. $\sqrt{27}$
12. $\sqrt{86}$

Decimal Form of Rational Numbers A rational number can be written in the form $\dfrac{m}{n}$, where m and n are integers, $n \neq 0$. A rational number can also be written as a terminating or repeating decimal. — 12.3

Write each rational number as a terminating or repeating decimal.

13. $-\dfrac{7}{2}$
14. $\dfrac{1}{9}$
15. $\dfrac{5}{6}$

536   Chapter 12 Radicals

Write each decimal in the form $\frac{m}{n}$, where m and n are integers, $n \neq 0$.

16. 0.25 **17.** $0.\overline{6}$ **18.** $0.\overline{54}$

Simplifying Square Roots An expression that contains a square root is in *simplest form* when: The radicand contains no square factors other than 1, and no fractions. No denominator contains a radical. **12.4**

Simplify.

19. $\sqrt{20}$ **20.** $\sqrt{45}$ **21.** $\sqrt{98x^2}$ **22.** $\sqrt{12y^3}$

Combining Square Roots Only like square roots can be simplified using addition or subtraction. Use the Product and Quotient Properties to multiply and divide. **12.5–12.7**

Perform the indicated operations.

23. $\sqrt{5} + 2\sqrt{6} - 4\sqrt{5}$ **24.** $-5\sqrt{6x} \cdot 2\sqrt{3x}$ **25.** $\sqrt{3}(2\sqrt{2} - 5)$

26. $(2 - 3\sqrt{5})(3 + 2\sqrt{5})$ **27.** $\dfrac{\sqrt{3}}{\sqrt{5}}$ **28.** $\dfrac{1 + \sqrt{5}}{2 - \sqrt{5}}$

Solving Radical Equations To solve a radical equation, isolate the radical and then square each side of the equation. **12.8**

Solve.

29. $\sqrt{3x - 1} = 4$ **30.** $0 = 6 + 2\sqrt{x}$ **31.** $\sqrt{2y + 1} + 2 = 5$

Pythagorean Theorem In a right triangle, the square of the hypotenuse is equal to the sum of the squares of the legs: $c^2 = a^2 + b^2$. **12.9**

If c is the hypotenuse, find the length of the unknown side.

32. $a = 10$, $b = 24$, $c = ?$ **33.** $a = 1$, $c = \sqrt{2}$, $b = ?$

Distance Formula $d = \sqrt{(x_2 - x_1)^2 + (y_2 - y_1)^2}$ **12.10**

Find the distance between the given points.

34. $A(0, 9)$ and $B(0, -12)$ **35.** $X(-3, 7)$ and $Y(-9, -6)$

Use Coordinate Geometry Many problems in geometry can be solved using algebraic techniques. **12.11**

36. A rectangle has vertices at $(0, 0)$, $(-3, 3)$, $(3, 9)$, and $(6, 6)$. Find the length and the midpoint of each diagonal.

Summary and Review

CHAPTER 12 TEST

Simplify.

1. $\sqrt{\frac{100}{49}}$
2. $\sqrt{900}$
3. $\sqrt{450}$
4. $\sqrt{27y^3}$

5. Use the divide-and-average method to approximate $\sqrt{24}$.

Write each decimal in the form $\frac{m}{n}$ where m and n are integers, $n \neq 0$.

6. $0.\overline{3}$
7. $1.\overline{45}$

Perform the indicated operations.

8. $2\sqrt{3} - 4\sqrt{3} + 5\sqrt{4}$
9. $\sqrt{12b^2} - \sqrt{27b^2}$
10. $3\sqrt{6} \cdot 2\sqrt{3}$
11. $\sqrt{2}(2\sqrt{5} - 3)$
12. $\frac{\sqrt{5}}{\sqrt{8}}$
13. $\sqrt{\frac{2}{3}}$

Solve.

14. $2\sqrt{2x + 1} = 8$
15. $\sqrt{3x - 1} + 2 = 7$

16. If c is the hypotenuse of a right triangle, find b when $a = 2$ and $c = 4$.

17. Find the distance between $A(-2, 4)$ and $B(7, -8)$.

18. A triangle has vertices at $(-4, 0)$, $(4, 0)$, and $(0, 8)$. Find the midpoint of each side of the triangle.

19. A rectangle has sides with lengths of 6 and 9. Find the lengths of the diagonals of this rectangle.

Solve. Round to the nearest tenth, if necessary.

20. Find the perimeter and area of a rectangle whose length is $5\sqrt{y}$ ft and whose width is $2\sqrt{y}$ ft.

21. When 5 is subtracted from the square root of twice a number, the result is 61. Find the number.

22. What length of wire is needed to reach from the top of a 3.5 m pole to a point 1.5 m from the base of the pole?

Challenge

In dealing with the theory of vibratory motion in physics, the equation $a^2 - 2al + k^2 = 0$ is derived. Show that $a = l \pm \sqrt{l^2 - k^2}$ are both solutions of the equation.

PREPARING FOR STANDARDIZED TESTS

Select the best choice for each question.

1. When simplified, $\sqrt{18x^3y^2} =$
   A. $2xy\sqrt{3x}$ B. $3xy\sqrt{2x}$
   C. $2xy\sqrt{3}$ D. $3xy\sqrt{2}$
   E. $9xy\sqrt{x}$

2. In 1987, the towns in one county spent the following amounts for education:

$6,472,891	$4,989,300
$1,473,600	$12,394,000
$3,456,863	$947,965

   To the nearest million, what was the total spent for education by these six towns?
   A. $27 million   B. $28 million
   C. $29 million   D. $30 million
   E. $31 million

3. $\sqrt{2809}$ equals
   A. 43   B. 47   C. 53
   D. 57   E. 63

4. $\sqrt{162} + \sqrt{50} - 3\sqrt{8} =$
   A. $6 - \sqrt{2}$ B. $2\sqrt{5} - 3\sqrt{2}$
   C. $20\sqrt{2}$ D. $9\sqrt{2}$
   E. $8\sqrt{2}$

5. Find the better price: a bicycle advertised at 30% off on a list price of $348 or one advertised at 25% off on a list price of $295?
   A. $73.75   B. $104.40
   C. $209.97   D. $221.25
   E. $243.60

6. Solve for x: $\sqrt{2x - 5} = 3$
   A. 3   B. 4   C. 5
   D. 6   E. 7

7. What fraction is equivalent to the decimal 1.125?
   A. $\frac{45}{4}$ B. $\frac{9}{8}$ C. $\frac{6}{5}$
   D. $\frac{5}{6}$ E. $\frac{9}{80}$

8. Paul has been assigned a term paper which is to be 20 pages long. He typed his first draft and found he had only 15 pages. What per cent of the term paper did he have left to write?
   A. 5%   B. 20%   C. 25%
   D. $33\frac{1}{3}$%   E. 75%

9. Solve for x:
   $$\frac{3}{x+1} - \frac{2}{x} = \frac{2}{x+1}$$
   A. -2 B. -1 C. 0
   D. 1   E. 2

10. Use $a^2 + b^2 = c^2$ to find b when $a = 24$ and $c = 25$.
    A. 7   B. 8   C. 51
    D. 49   E. 51

11. If the circle has a radius of 14 cm and one side of the square is 16 cm, what is the area of the shaded portion? (Use $\frac{22}{7}$ as an approximation for π.)
    A. 1232 sq cm
    B. 976 sq cm
    C. 616 sq cm
    D. 360 sq cm
    E. 106 sq cm

Preparing for Standardized Tests

CUMULATIVE REVIEW (CHAPTERS 1–12)

Write the numbers in order from least to greatest.

1. $\sqrt{4}, 6, 3, -2$

2. $-\frac{1}{3}, \sqrt{\frac{4}{9}}, 0, -\sqrt{\frac{9}{16}}, \frac{1}{2}$

Match the equation or inequality in exercises 3–7 with its solution set a–e.

3. $\frac{1}{x^2} - \frac{2}{3x} = \frac{5}{3}$ a. $\{x: -1 < x < 1\}$

4. $|x| - 3 = -2$ b. $\{-1, \frac{3}{5}\}$

5. $2 < 5 - 3x < 8$ c. {all real numbers greater than 1}

6. $2x - 3 > -1$ d. $\{(-1, 1)\}$

7. $\begin{cases} -7y + 6 = x \\ 3x + y = -2 \end{cases}$ e. $\{-1, 1\}$

Solve each equation, inequality, or system.

8. $15x = 5(x - 4)$

9. $6 - \frac{3}{4}a > 3$

10. $|2y - 1| = 5$

11. $m - \frac{3}{4} = \frac{5m}{8}$

12. $\frac{2}{3}t - 1 = 7$

13. $\begin{cases} 2c + 3d = 14 \\ 4c - 3d = 10 \end{cases}$

14. $z^2 - 9z = 0$

15. $\frac{3}{b+2} = \frac{1}{b-2}$

16. $6 + \sqrt{n - 3} = 2$

17. $4r - 5(r + 1) = -(3r - 2) + 5$

18. $-9 \leq -3n$ and $-3n < 6$

19. $g^2 + 2g - 15 = 0$

20. $\sqrt{3w + 1} = 3\sqrt{2w - 1}$

Solve. Use the table on the right.

21. ABC Roofing charges $25 plus a certain amount per hour. How much does ABC Roofing charge per hour?

22. Tin Roofing charges by the hour. How much does Tin Roofing charge per hour?

23. How long must a job be for both to charge the same amount?

24. Complete the table.

Cost of Job

Time worked	1 h	2 h	3 h	4 h
ABC Roofing	$40	$55	?	?
Tin Roofing	$20	$40	?	?

Chapter 12 Radicals

Simplify. Assume that no variable equals zero.

25. $|-\sqrt{36}|$

26. $-\left[-\left(-\frac{1}{3}\right)\right]$

27. $\sqrt{\frac{9}{144}}$

28. $(1.2 \times 10^{-3})^2$

29. $(3p^2q)(2pq^3)$

30. $(c-d)^2 - d^2$

31. $\dfrac{-6a^5b^2}{9a^3b^3}$

32. $\dfrac{-2}{3\sqrt{2x}}$

33. $\dfrac{3-m}{m^2 - m - 6}$

34. $\dfrac{z^2 - 4}{3z + 6}$

35. $\dfrac{s+t}{\frac{3}{s} + \frac{3}{t}}$

36. $\sqrt{12x^3y}$

37. $2(\sqrt{3^2 + 1})^2 - 3(\sqrt{2^2 - 1})$

38. $2[(h-3)^2] + 2[h - (-3)^2]$

Write the equation for each line in slope-intercept form.

39. The line represented by $x - 2y = 6$.

40. The line that has slope -2 and contains the point $P\,(1, 1)$.

41. The line that contains the points $A\,(-1, -2)$ and $B\,(1, 4)$.

Factor completely. If a polynomial cannot be factored, write prime.

42. $m^2 - 11m + 18$

43. $12x^2y^2 - 9xy^2$

44. $4a^6 + 12a^3 + 9$

45. $a^3 + 23$

46. $2b^2 - 9b - 18$

47. $3p^3 + 9p^2 + 3p$

48. $xy - 3y - 2x + 6$

49. $3m + 6n - 2m^2 - 4mn$

The fee for renting a computer is $45 plus $25 for each day you keep the machine. Use this information for Exercises 50–53.

50. Complete this table.

Number of days	1	2	3
Rental fee	$70	?	?

51. Use the table to graph three points for this relationship. Draw a line through the points.

52. Use the graph to determine the rental fee for 5 days.

53. Jan can spend $220. For how many days can he rent a computer?

Evaluate each expression if $a = -\frac{1}{2}$, $b = 4$, and $c = -\frac{3}{4}$.

54. $[(a+c)b]^2 - 1$

55. $\sqrt{b} \div a$

56. $b + c \div a$

57. $\sqrt{-b} \div a$

58. $(a^2 + c)(a^2 - c)$

59. $\sqrt{(a-c)b}$

60. $\dfrac{a^2b}{a+b}$

61. $\sqrt{\dfrac{-a\sqrt{b}}{(a-c)^2}}$

62. $\sqrt{b} + \dfrac{c - a^2}{b}$

Cumulative Review

Perform the indicated operation.

63. $(0.6 \div 0.3)^2$
64. $\sqrt{0.25} - (0.2)^2$
65. $-0.8 - (-\sqrt{0.36})$
66. $(-2a^2d)(5a^3d^2)$
67. $(-3m^2n^3)^2$
68. $(3t - 2)^2$
69. $\dfrac{p+2}{p+1} \cdot \dfrac{p^2-1}{p^2-4}$
70. $\dfrac{12x^3y^2}{4x^5yz}$
71. $\dfrac{c^2-9}{c-1} \div c - 3$
72. $\dfrac{2h}{h^2-1} - \dfrac{1}{h+1}$
73. $\sqrt{2}(3 + 2\sqrt{10})$
74. $\dfrac{-3\sqrt{14}}{\sqrt{63}}$
75. $(5r^2 + 2r) - (3r^2 - 4r + 1)$
76. $(-b^3 + 3b^2 - 2)(-3b^2)$
77. $3x(y^2 + 2y) - 2y(xy + 2x)$
78. $\sqrt{25x^2y} - \sqrt{9x^2y} + \sqrt{36x^2y}$

Determine the range for each given domain $\{-2, 0, 1\}$.

79. $y = 2x - 3$
80. $y = x^2 + 1$
81. $\tfrac{1}{2}y = \tfrac{1}{4}x + 3$
82. $\tfrac{1}{3}y = x + \tfrac{1}{6}$

Solve.

83. The cost of apples c varies directly with their weight w. If $c = \$1.29$ when $w = 3$, find c when $w = 5$.

84. Find three consecutive odd integers such that the first times the second is 1 more than 2 times the third.

85. A crew rowing with the current traveled 18 mi in 2 h and against the current, the crew rowed 10 mi in 2 h. Find the rate of the crew in still water and the rate of the current.

86. Ben earned $127.50 in annual simple interest on an investment of $1500. What annual interest rate was paid?

87. The length of a rectangle is 4 less than twice its width. If the perimeter is 52 ft, what are the dimensions of the rectangle?

88. If a bus traveling at 50 mi/h takes 6 h to travel from San Francisco to Fresno, and a train takes 4 h to travel the same distance, does the rate vary indirectly as the time? If so, what is the constant of variation?

89. The width of a rectangle is $\dfrac{2w+8}{(w+2)(w-2)}$ in., and the area is $\dfrac{4}{w^2-4}$ in². Find the length of the rectangle.

90. Find two positive numbers that differ by 3 and whose reciprocals have a sum of $\tfrac{8}{9}$.

13 Quadratic Equations and Functions

Designing and constructing uniquely shaped buildings involves solving various kinds of problems. For example, quadratic equations can be used to find the strengths and weaknesses of circular and rectangular beam configurations.

13.1 Quadratic Equations with Perfect Squares

Objective: To solve quadratic equations involving perfect-square expressions

A contractor has 225 square paving stones to cover a square outdoor area. Each stone is one square foot. The contractor can find the number of paving stones for one side by solving the quadratic equation, $x^2 = 225$. The property of square roots can be useful in solving quadratic equations of the form $x^2 = k$ or $(ax + b)^2 = k$, where $k \geq 0$.

Capsule Review

Recall that if $x^2 = y$, then x is called the square root of y.

Find the square root. Assume all variables are nonnegative.

1. $4x^2$
2. $16y^2$
3. $25m^2n^2$
4. $(3y - 1)^2$

An algebraic expression like $4x^2$ or $(3y - 1)^2$ is called a **perfect square**. When an equation contains a perfect square on one side and a nonnegative constant on the other, you can use the square-root property.

> **Square-Root Property**
>
> If $x^2 = k$, then $x = +\sqrt{k}$ or $x = -\sqrt{k}$ for any real number k, $k \geq 0$.

EXAMPLE 1 Solve: $4x^2 = 20$

$4x^2 = 20$
$x^2 = 5$ *Divide each side by 4.*
$x = \pm\sqrt{5}$ *Apply the square-root property.*

Check:
$4x^2 = 20$ $4x^2 = 20$
$4(\sqrt{5})^2 \stackrel{?}{=} 20$ $4(-\sqrt{5})^2 \stackrel{?}{=} 20$
$4(5) \stackrel{?}{=} 20$ $4(5) \stackrel{?}{=} 20$
$20 = 20$ ✓ $20 = 20$ ✓

The solutions are $\pm\sqrt{5}$. Use your calculator to find decimal approximations for the solutions.

EXAMPLE 2 Solve: $(y - 7)^2 = 64$

$(y - 7)^2 = 64$ Take the square root of each side of the equation.
$y - 7 = \pm 8$

$y - 7 = 8 \quad | \quad y - 7 = -8$
$y = 15 \quad | \quad y = -1$

Check: $(y - 7)^2 = 64 \qquad\qquad (y - 7)^2 = 64$
$(15 - 7)^2 \stackrel{?}{=} 64 \qquad\qquad [(-1) - 7]^2 \stackrel{?}{=} 64$
$8^2 \stackrel{?}{=} 64 \qquad\qquad\qquad (-8)^2 \stackrel{?}{=} 64$
$64 = 64 \checkmark \qquad\qquad\quad 64 = 64 \checkmark$

The solutions are 15 and -1.

EXAMPLE 3 Solve: $7(2n - 1)^2 + 10 = 3$.

$7(2n - 1)^2 + 10 = 3$
$7(2n - 1)^2 = -7$
$(2n - 1)^2 = -1$ No real solution

If you try to compute $\sqrt{-1}$ on your calculator, the display will indicate an error. There is no real number solution since the square of any real number is always a nonnegative real number.

CLASS EXERCISES

Solve. Express all radicals in simplest form.

1. $5x^2 = 30$
2. $n^2 = \frac{9}{16}$
3. $3(x - 5)^2 = 75$
4. $(x - 2)^2 = 7$
5. $5(m + 3)^2 = 25$
6. $4(x + 4)^2 = 256$

PRACTICE EXERCISES

Solve. Express all radicals in simplest form.

1. $p^2 = \frac{4}{25}$
2. $n^2 = \frac{1}{100}$
3. $x^2 = 49$
4. $y^2 = 64$
5. $3x^2 = 18$
6. $5m^2 = 35$
7. $2n^2 = 16$
8. $4y^2 = 48$
9. $x^2 - 15 = 0$
10. $y^2 - 10 = 0$
11. $2n^2 - 6 = 0$
12. $3m^2 - 15 = 0$
13. $4x^2 + 9 = 0$
14. $6y^2 + 24 = 0$
15. $3p^2 - 5 = 7$
16. $2n^2 - 13 = -1$
17. $8m^2 - 40 = 0$
18. $6z^2 - 42 = 0$
19. $(x - 1)^2 = 9$
20. $(y - 3)^2 = 36$
21. $(2x + 1)^2 = 16$
22. $(3z + 2)^2 = 4$
23. $5(2x - 1)^2 = 45$
24. $6(3y + 2)^2 = 24$

13.1 Quadratic Equations with Perfect Squares

25. $\frac{1}{9}p^2 - 1 = -\frac{3}{4}$ 26. $\frac{4}{25}n^2 + \frac{6}{16} = \frac{15}{16}$ 27. $(x + 1)^2 = 72$

28. $(n - 9)^2 = 27$ 29. $\left(m + \frac{2}{3}\right)^2 = \frac{1}{9}$ 30. $\left(n + \frac{3}{4}\right)^2 = \frac{1}{16}$

31. $\left(t + \frac{2}{3}\right)^2 = \frac{5}{9}$ 32. $\left(y + \frac{3}{4}\right)^2 = \frac{3}{16}$ 33. $2(5x + 1)^2 = 50$

34. $3(2s + 4)^2 = 36$ 35. $5(z - 3)^2 = 35$ 36. $7(m + 4)^2 = 70$

37. $(x - 5)^2 + 3 = 27$ 38. $(z + 3)^2 - 5 = 49$ 39. $2(3x - 5)^2 + 3 = 7$

40. $3(2t + 7)^2 + 1 = 37$ 41. $3(2p + 1)^2 + 5 = 59$

42. $5(6m - 5)^2 - 3 = 57$ 43. $6(2r + 3)^2 - 5 = 85$

44. $\left(x - \frac{3}{5}\right)^2 - \frac{7}{9} = -\frac{1}{3}$ 45. $\left(y - \frac{1}{3}\right)^2 - \frac{5}{8} = -\frac{1}{2}$

46. $5\left(x - \frac{1}{3}\right)^2 + \frac{1}{2} = \frac{3}{5}$ 47. $5\left(z - \frac{1}{2}\right)^2 + \frac{1}{3} = \frac{3}{4}$

48. $x^2 + 2x + 1 = 4$ 49. $y^2 - 4y + 4 = 25$

50. $25x^2 - 10x + 1 = 16$ 51. $16m^2 + 24m + 9 = 9$

52. $x^2 + x + \frac{1}{4} = 4$ 53. $y^2 - 18y + 81 = 25$

54. $y^2 - 12y + 36 = 84$ 55. $b^2 + 14b + 42 = -7$

Applications

Solve by the square-root property.

56. **Coordinate Geometry** The quadratic equation whose graph is a circle with the center at the origin is $x^2 + y^2 = r^2$, $r > 0$. Solve for x.

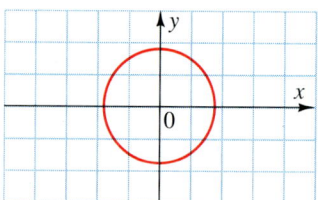

HISTORICAL NOTE

The Palantine Anthology is a book of problems dating back to 500 A.D. The following problem was found in this book. Try to solve it.

> Brickmaker: I must finish this house soon. Today is cloudless and I need only a few more bricks. I have all I need but three hundred. I know you could make that many by yourself in one day. Your son will quit work when he makes 200. Your son-in-law will quit when he makes 250. The three of you working together, how soon can you have my bricks ready?

13.2 Completing the Square

Objective: To solve quadratic equations by completing the square

The product of two numbers is 40. The larger number exceeds the smaller one by 6. What equation can you use to determine the two numbers?

The equation $x(x + 6) = x^2 + 6x = 40$ is different from those of the form you solved in the preceding lesson because $x^2 + 6x$ is not a perfect square.

Capsule Review

Recall that an expression of the form $ax^2 + bx + c$ is a perfect-square trinomial if it can be factored into two identical binomial factors.

EXAMPLE
a. $x^2 + 6x + 9 = (x + 3)(x + 3) = (x + 3)^2$
b. $x^2 - \frac{2}{3}x + \frac{1}{9} = \left(x - \frac{1}{3}\right)\left(x - \frac{1}{3}\right) = \left(x - \frac{1}{3}\right)^2$

Tell which of the following are perfect-square trinomials and factor them.

1. $x^2 + 4x + 4$
2. $4m^2 + 12m + 9$
3. $y^2 + y + 1$
4. $z^2 - \frac{2}{5}z + \frac{1}{25}$
5. $n^2 - n + \frac{1}{4}$
6. $49p^2 - 28p + 4$
7. $x^2 - x + 1$
8. $n^2 - \frac{4}{5}n + \frac{4}{25}$
9. $c^2 - \frac{3}{2}c + \frac{9}{16}$

Study the perfect square at the right. In the expanded form of the square, notice that the coefficient of the first term is 1 and that the constant term is the square of one-half the coefficient of the term of the first degree.

$(n - 6)^2 = n^2 - 12n + 36$

$36 = \left(\frac{1}{2} \cdot -12\right)^2$

This fact can be used to develop a procedure for finding the value of c that makes $ax^2 + bx + c$ a perfect square.

To **complete the square** for $ax^2 + bx + \underline{\;?\;}$:
- If $a \neq 1$, then divide the expression by a.
- Find one-half of the coefficient of x.
- Square the value obtained.
- Add the squared result to $x^2 + \frac{b}{a}x$.

Completing the square is one method of solving quadratic equations that works whether or not the equation can be factored. That is, it works for all quadratic equations.

EXAMPLE 1 Find the value of c to complete the square for $x^2 - 8x + c$.

Using the procedure given, the coefficient of x^2 is 1, and $\frac{1}{2}(-8) = -4$. The term to be added to $x^2 - 8x$ is $(-4)^2$ or 16. Therefore, $x^2 - 8x + 16$ is equal to $(x - 4)^2$, which is a perfect square.

EXAMPLE 2 Solve $x^2 + 6x = 40$ by completing the square.

$x^2 + 6x = 40$	Complete the square.
$x^2 + 6x + 9 = 40 + 9$	Take one-half of the coefficient of x, square it, and add the result to both sides of the equation. $\frac{1}{2}(6) = 3; 3^2 = 9$
$(x + 3)^2 = 49$	Write $x^2 + 6x + 9$ as a perfect square.
$x + 3 = 7 \quad \mid \quad x + 3 = -7$	Solve the equation.
$x = 4 \quad \quad \mid \quad x = -10$	

The solutions are 4 and -10. The check is left for you.

Example 2 shows the solution to the problem posed at the beginning of this lesson. If $x = 4$, then $x + 6 = 10$, so 4 and 10 is one answer. If $x = -10$, then $x + 6 = -4$, so -10 and -4 is another answer. Both pairs of numbers, 4, 10 and $-4, -10$, have a product of 40, and the larger number of each pair is 6 more than the smaller.

EXAMPLE 3 Solve $y^2 - y - \frac{3}{4} = 0$ by completing the square.

$y^2 - y - \frac{3}{4} = 0$	
$y^2 - y = \frac{3}{4}$	Add $\frac{3}{4}$ to each side.
$y^2 - y + \frac{1}{4} = \frac{3}{4} + \frac{1}{4}$	Complete the square: $\frac{1}{2}(-1) = -\frac{1}{2}$, and $\left(-\frac{1}{2}\right)^2 = \frac{1}{4}$. Add $\frac{1}{4}$ to each side of the equation.
$\left(y - \frac{1}{2}\right)^2 = 1$	Write $y^2 - y + \frac{1}{4}$ as a perfect square.
$y - \frac{1}{2} = 1 \quad \mid \quad y - \frac{1}{2} = -1$	Solve the equation.
$y = \frac{3}{2} \quad \quad \mid \quad y = -\frac{1}{2}$	

The solutions are $\frac{3}{2}$ and $-\frac{1}{2}$. The check is left for you.

Although the method of completing the square always works for solving quadratic equations, it is sometimes faster to solve equations by factoring. So, try factoring first, and if necessary try completing the square.

Before completing the square to solve a quadratic equation, make sure the equation is in the form $ax^2 + bx = c$ and then divide through by a.

EXAMPLE 4 Solve $2t^2 + 3t - 4 = 0$ by completing the square.

$$2t^2 + 3t - 4 = 0$$
$$2t^2 + 3t = 4 \quad \text{Add 4 to each side.}$$
$$t^2 + \frac{3}{2}t = 2 \quad \text{Divide each term by 2.}$$
$$t^2 + \frac{3}{2}t + \frac{9}{16} = 2 + \frac{9}{16} \quad \text{Complete the square. Since } \frac{1}{2}\left(\frac{3}{2}\right) = \frac{3}{4} \text{ and}$$
$$\left(\frac{3}{4}\right)^2 = \frac{9}{16}, \text{ add } \frac{9}{16} \text{ to each side of the equation.}$$
$$\left(t + \frac{3}{4}\right)^2 = \frac{41}{16} \quad \text{Write as a perfect square.}$$

$$t + \frac{3}{4} = \frac{1}{4}\sqrt{41} \qquad \qquad t + \frac{3}{4} = -\frac{1}{4}\sqrt{41}$$
$$t = \frac{-3 + \sqrt{41}}{4} \qquad \qquad t = \frac{-3 - \sqrt{41}}{4}$$

The solutions are $\frac{-3 + \sqrt{41}}{4}$ and $\frac{-3 - \sqrt{41}}{4}$. The check is left for you.

The computer program below can also be used to solve Example 4.

```
10 INPUT "FOR AT^2 + BT + C = 0, ENTER A,B,C:  ";A,B,C: PRINT
20 B = B / A:C =  - C / A
30 PRINT "T^2 + ";B;"T = ";C
40 N = ((1 / 2) * B) ^ 2
50 PRINT "T^2 + ";B;"T + ";N;" = ";C;" + ";N
60 PRINT "(T + ";(1 / 2) * B;") ^ 2 = ";C + N
70 PRINT : PRINT "THE SOLUTIONS ARE "; SQR (C + N) - (1 / 2) * B;
   " AND "; - SQR (C + N) - (1 / 2) * B
80 END
```

Compare these solutions with the results obtained using a calculator or the table on page 670.

CLASS EXERCISES

Find the value of c to complete the square.

1. $y^2 + 10y + c$ **2.** $x^2 - 6x + c$ **3.** $p^2 - \frac{2}{3}p + c$

Solve by completing the square.

4. $x^2 - 8x = -15$ **5.** $z^2 + 18z + 56 = 0$ **6.** $4n^2 = 4n + 1$

PRACTICE EXERCISES

Find the value of c to complete the square.

1. $m^2 + 8m + c$ **2.** $t^2 + 12t + c$ **3.** $y^2 - 10y + c$

13.2 Completing the Square **549**

Solve by completing the square. Express all radicals in simplest form.

4. $y^2 + 3y = 3$
5. $b^2 - 4b = 5$
6. $x^2 - 2x = 15$
7. $z^2 + 12z = 45$
8. $h^2 + 12h = 28$
9. $t^2 + 4t = 12$
10. $t^2 - 5t = 2\frac{3}{4}$
11. $y^2 + 7y = 7\frac{3}{4}$
12. $x^2 + 3x = -1$
13. $m^2 - 3m = -1$
14. $x^2 + 3x = 0$
15. $c^2 + 5c = 0$
16. $r^2 - 6r + 5 = 0$
17. $p^2 + 8p - 9 = 0$
18. $t^2 - 3t - 10 = 0$
19. $p^2 + 5p + 6 = 0$
20. $n^2 - 3n + \frac{5}{4} = 0$
21. $r^2 - r - 3\frac{3}{4} = 0$
22. $x^2 - 4x + 2 = 0$
23. $4y^2 - 6y - \frac{1}{2} = 0$
24. $3b^2 - 12b - 1 = 0$
25. $5p^2 - 10p = -4$
26. $2t^2 - 5t = 4$
27. $2c^2 + c = 5$
28. $2z^2 - 10z = 3$
29. $2n^2 + 10n - 3 = 0$
30. $2m^2 - m - 5 = 0$
31. $3n^2 - 8n + 4 = 0$
32. $2x^2 - 5x - 12 = 0$
33. $2y^2 + 1 = -5y$
34. $3z^2 - 1 = 4z$
35. $3y^2 = 10 + 5y$
36. $6b^2 = 10b - 3$
37. $1 + \frac{2}{y^2} = \frac{7}{2y}$
38. $\frac{3}{y-2} - \frac{1}{y-1} = 2$
39. $y^2 + by = -1$
40. $x^2 + bx = -2$
41. $x^2 + bx + c = 0$
42. $ax^2 + bx + c = 0$

Applications

Number Problems Solve each of the following.

43. When twice a number is added to the square of the number, the sum is 3. Find the number(s).

44. When 4 is subtracted from 9 times a number, the result is equal to twice the square of the number. Find the number.

MATH CLUB ACTIVITY

When Kevin was 25, he began saving $1800 a year at 7% interest compounded annually. After 10 years, he stopped saving but left all the deposits and compound interest in his account until he was 62.

Marie didn't begin saving until she was 35. She saved $1800 a year at 7% interest compounded annually until she was 62. She also left the deposits and compound interest in her account.

Who had more in their account at age 62? If you deposit P dollars at r percent interest (expressed as a decimal), the amount A in the account after t years is given by the formula $A = P(1 + r)^t$. You may wish to use a calculator.

13.3 The Quadratic Formula

Objective: To develop the quadratic formula and to use it in solving quadratic equations

In the preceding lessons you learned that quadratic equations of the form $ax^2 + bx + c = 0$ can be solved by factoring, by using the square root property, or by completing the square. These same techniques can be used to solve any equation of the form $ax^2 + bx + c = 0$ where $a \neq 0$.

Capsule Review

Write the following equations in the form $ax^2 + bx + c = 0$.

1. $3x^2 - 5 = 7x$
2. $4x^2 = x$
3. $-7 = x^2$
4. $-2(x + 5)^2 = 3x$

Solve $ax^2 + bx + c = 0$ by completing the square.

$ax^2 + bx + c = 0$

$ax^2 + bx = -c$ Subtract c from each side.

$x^2 + \dfrac{b}{a}x = \dfrac{-c}{a}$ Divide each term by a.

$x^2 + \dfrac{b}{a}x + \dfrac{b^2}{4a^2} = \dfrac{-c}{a} + \dfrac{b^2}{4a^2}$ Complete the square. $\left(\dfrac{1}{2} \cdot \dfrac{b}{a}\right)^2 = \dfrac{b^2}{4a^2}$, so add $\dfrac{b^2}{4a^2}$ to each side of the equation.

$\left(x + \dfrac{b}{2a}\right)^2 = \dfrac{b^2 - 4ac}{4a^2}$ Write the left side as a perfect square, and the right side as a single fraction.

$x + \dfrac{b}{2a} = \pm\sqrt{\dfrac{b^2 - 4ac}{4a^2}}$ Take the square root of each side.

$x = -\dfrac{b}{2a} \pm \dfrac{\sqrt{b^2 - 4ac}}{2a}$ Solve for x.

$x = \dfrac{-b \pm \sqrt{b^2 - 4ac}}{2a}$ Simplify.

The solutions are $\dfrac{-b + \sqrt{b^2 - 4ac}}{2a}$ and $\dfrac{-b - \sqrt{b^2 - 4ac}}{2a}$.

The solutions for the quadratic equation $ax^2 + bx + c = 0$ lead to the *quadratic formula*.

> **Quadratic Formula**
>
> If $ax^2 + bx + c = 0$, where a, b, and c are real numbers and $a \neq 0$, then
> $$x = \frac{-b \pm \sqrt{b^2 - 4ac}}{2a}.$$

EXAMPLE Solve $5x^2 = 10x - 4$ using the quadratic formula.

$5x^2 = 10x - 4 \rightarrow 5x^2 - 10x + 4 = 0$

$x = \dfrac{-b \pm \sqrt{b^2 - 4ac}}{2a}$

$x = \dfrac{-(-10) \pm \sqrt{(-10)^2 - 4(5)(4)}}{2(5)}$ $a = 5$, $b = -10$, and $c = 4$.

$x = \dfrac{10 \pm \sqrt{100 - 80}}{10}$ Simplify.

$x = \dfrac{10 \pm \sqrt{20}}{10}$

$x = \dfrac{10 \pm 2\sqrt{5}}{10}$ Use the Product Property of Square Roots.

$x = \dfrac{5 \pm \sqrt{5}}{5}$ Reduce.

The solutions are $\dfrac{5 + \sqrt{5}}{5}$ and $\dfrac{5 - \sqrt{5}}{5}$. The check is left for you.

CLASS EXERCISES

State the values of a, b, and c. Then use the quadratic equation to solve.

1. $x^2 + 6x + 1 = 0$ 2. $3x^2 - 4x = 7$ 3. $5y = y^2$

For Discussion

4. What will happen if $4ac > b^2$ in the quadratic formula?

PRACTICE EXERCISES

Solve by using the quadratic formula.

1. $x^2 + 5x + 6 = 0$ 2. $y^2 - y - 6 = 0$ 3. $c^2 - 3c - 10 = 0$
4. $m^2 + 6m + 8 = 0$ 5. $p^2 - 9p = -18$ 6. $t^2 - 3t = -2$
7. $3y^2 - 3y - 1 = 0$ 8. $2n^2 - 5n - 12 = 0$ 9. $4x^2 - 12x = -9$

10. $5b^2 - 4b = 33$
11. $x^2 - 2x = 10$
12. $3x^2 - 8x = -2$
13. $6x^2 + 7x - 5 = 0$
14. $2p^2 + 5p + 3 = 0$
15. $2y^2 + 3y - 1 = 0$
16. $3n^2 - 4n - 2 = 0$
17. $3z^2 - 8z = -4$
18. $6y^2 - y = 2$
19. $m^2 + 6m - 10 = 0$
20. $m^2 + 4m - 6 = 0$
21. $c^2 = -7c - 5$
22. $2p^2 = 5 - 4p$
23. $2b^2 - 5 = 2b$
24. $5n^2 - 2 = 8n$
25. $(2t + 3)(t + 4) = 1$
26. $(2x - 5)(x + 1) = 2$
27. $\dfrac{2y^2}{3} - y = \dfrac{-1}{6}$
28. $\dfrac{m^2}{3} - m = \dfrac{-1}{2}$
29. $\dfrac{x^2}{3} - \dfrac{3}{2} = \dfrac{x}{2}$
30. $\dfrac{c^2}{3} - \dfrac{1}{2} = \dfrac{5c}{6}$
31. $\dfrac{x - 1}{2x} = \dfrac{x + 1}{x - 2}$
32. $\dfrac{x - 4}{x} = \dfrac{3}{x + 2}$
33. $1 + \dfrac{2}{c^2} = \dfrac{7}{2c}$
34. $\dfrac{1}{y} - \dfrac{2}{y^2} = -6$
35. $\dfrac{x^2}{x - 2} + 3 = \dfrac{2x}{x - 2}$
36. $\dfrac{3}{m - 2} - 2 = \dfrac{1}{m - 1}$

Use the quadratic formula to solve the equation for x.

37. $4x - x^2 + k^2 = 0$
38. $x^2 + 2xk + k^2 = 0$
39. $\dfrac{x^2}{a} + \dfrac{x}{b} = -\dfrac{1}{c}$
40. $\dfrac{x^2}{a} - \dfrac{1}{c} = \dfrac{5x}{b}$
41. $(2x + a)^2 = (x + a) + 6$
42. $(x - b)^2 = (x - b) + 4$

Applications

43. **Geometry** If the length of a rectangle is one foot less than the width and the area is 12 ft², find the dimensions.

44. **Number Problem** Find two consecutive integers such that 3 times the square of the first is equal to 7 more than 5 times the second.

DID YOU KNOW?

Projectile motion makes a parabolic path of any object fired and maintaining a constant speed. The height y of the path and the horizontal distance x the projectile travels can be determined by a quadratic equation. For example, $y = -\dfrac{1}{16}x^2 + 2x$ represents the path of a fired projectile. Assuming that the starting point is $(0, 0)$, the horizontal distance traveled of 32 ft is found by solving the equation.

Solve for the horizontal distances of these projectiles if their paths are represented by the following quadratic equations.

1. $y = -10x^2 + 10x$
2. $y = -2x^2 + 36x$
3. $y = -4x^2 + 75x$

13.4 Mixed Practice: Solving Quadratic Equations by Any Method

Objective: To choose the best method for solving a quadratic equation

This table summarizes the procedures for solving quadratic equations. It can help you decide which is the best method for solving a particular quadratic equation.

METHODS FOR SOLVING QUADRATIC EQUATIONS

Method	Can Be Used	Best to Use
Factoring	sometimes	If the constant term is 0, or if $ax^2 + bx + c$ can be factored.
Square-Root Property	sometimes	For equations of the form $x^2 = k$ or $(ax + b)^2 = k$, $k \geq 0$.
Completing the square	always	For equations of the form $x^2 + bx + c = 0$, where b is even.
Quadratic Formula	always	For any equation of the form $ax^2 + bx + c = 0$.

EXAMPLE Choose a method for solving each equation. Explain your choice. Then solve.

a. $3m^2 + 2m = 8$ b. $y^2 - 8y = 2$ c. $(2x - 1)^2 = 9$ d. $3x^2 + 5x - 6 = 0$

a. $\quad 3m^2 + 2m = 8$
$\quad\quad 3m^2 + 2m - 8 = 0$ Write in the form $ax^2 + bx + c = 0$.
$\quad\quad (3m - 4)(m + 2) = 0$ Factor the left side.
$\quad\quad 3m - 4 = 0 \mid m + 2 = 0$ Solve for m.
$\quad\quad\quad m = \dfrac{4}{3} \quad\quad\mid\quad m = -2$

b. $y^2 - 8y = 2$
$\quad (y - 4)^2 = 18$ Since b is even, use completing the square.
$\quad y - 4 = \pm\sqrt{18}$ Solve for y.
$\quad y = 4 \pm 3\sqrt{2}$

c. $(2x - 1)^2 = 9$ It is in the form $(ax + b)^2 = k$, $k \geq 0$.
$\quad 2x - 1 = \pm 3$ Use the square-root property.
$\quad 2x - 1 = 3 \mid 2x - 1 = -3$
$\quad\quad x = 2 \quad\mid\quad x = -1$

Chapter 13 Quadratic Equations and Functions

d. $3x^2 + 5x - 6 = 0$

$$x = \frac{-5 \pm \sqrt{5^2 - 4(3)(-6)}}{2(3)}$$

$$x = \frac{-5 \pm \sqrt{97}}{6}$$

Use the quadratic formula.

$$x = \frac{-b \pm \sqrt{b^2 - 4ac}}{2a}$$

The check for the solution of each equation is left for you.

CLASS EXERCISES

Choose an appropriate method for solving each equation. Explain.

1. $x^2 = \frac{5}{9}$
2. $(3y + 7)^2 = 1$
3. $x^2 + 5x + 6 = 0$
4. $t^2 - 4t + 1 = 0$
5. $a^2 - 6a = -8$
6. $2m^2 + 5m = 7$
7. $n^2 - n - 1 = 0$
8. $x^2 + 7x - 16 = 0$
9. $3m^2 - 5m = 0$
10. $2r^2 + 3r = 9$
11. $4x^2 = \frac{9}{25}$
12. $(5x - 4)^2 = 25$

13. Use a calculator to solve this equation: $175x^2 - 250x - 225 = 0$

For Discussion

14. Why can both completing the square and the quadratic formula be used to solve all quadratic equations (assuming $b^2 - 4ac \geq 0$)?

15. Why do you think it is stated, in the table at the beginning of the lesson, that the square root property can only *sometimes* be used?

PRACTICE EXERCISES

Solve the quadratic equation by using an appropriate method.

1. $x^2 - x - 2 = 0$
2. $y^2 - 6y + 9 = 0$
3. $6m^2 = 72$
4. $4x^2 = 80$
5. $z^2 - 4z = -3$
6. $x^2 + 8x = 20$
7. $4x^2 + 3x - 1 = 0$
8. $2x^2 + 5x + 3 = 0$
9. $3t^2 - 2t = 0$
10. $4x^2 - 20 = 0$
11. $x^2 + \frac{1}{6}x = \frac{1}{6}$
12. $m^2 - \frac{7}{6}m = \frac{1}{2}$
13. $n^2 = 45 + 12n$
14. $15 - 14t = t^2$
15. $2x^2 - 9x = -8$
16. $2x^2 + 7x = 9$
17. $(3x - 2)^2 = 10$
18. $(4t - 1)^2 = 15$
19. $4x^2 - 4x + 1 = 45$
20. $9y^2 + 42y + 49 = 32$
21. $2x^2 = 6x - 3$
22. $3 - 2m = 3m^2$
23. $5 - \frac{3}{x} = \frac{2}{x^2}$
24. $3 + \frac{2}{m} = \frac{4}{m^2}$
25. $(n - 4)^2 - 3(n - 4) = 10$
26. $(2y + 5)^2 = -7(2y + 5) - 6$
27. $(3x - 8)^2 = (2x - 5)^2$
28. $(4x + 6)^2 = (2x + 4)^2$

13.4 Mixed Practice: Solving Quadratic Equations by Any Method

Solve for x in terms of the other variable.

29. $2b^2x^2 - 3bx = -1$
30. $x^2 - cx = 2c^2$
31. $(x - b)^2 + 5(x - b) + 4 = 0$
32. $(x - a)^2 + (x + a)^2 = 5a$
33. $\left(\frac{1}{4}x - \frac{1}{16}b\right)^2 = \frac{1}{36}$
34. $\left(\frac{1}{3}x + \frac{1}{4}a\right)^2 = \frac{1}{9}$

Applications

35. **Physics** The distance s traveled by an object with acceleration a in time t is given by the equation $s = \frac{1}{2}at^2$. Solve for t, when $s = 64$ ft and $a = 32$ ft/s².

36. **Physics** A projectile is shot vertically up in the air. Its distance s, in ft, after t seconds is given by the equation $s = 96t - 16t^2$. Find the values of t, to the nearest hundredth of a second, when $s = 96$ ft.

37. **Construction** The length of a rectangular floor is twice the width. The area of the floor is 32 ft². What are the dimensions of the room?

38. **Number Problem** Find two consecutive even integers whose product is 224.

39. **Number Problem** Find an integer such that the square of the integer is 81 less than 18 times the integer.

40. **Landscaping** The length of a rectangular garden exceeds three times the width by 4 m. The area of the garden is 24 m². Find the length and width to the nearest hundredth of a meter.

41. **Number Problem** Three times the square of a number equals 2 times the number. Find the number(s).

TEST YOURSELF

Solve by using the square-root property. 13.1

1. $5x^2 = 40$
2. $6(3z - 2)^2 = 24$

Solve by completing the square. 13.2

3. $m^2 - 2m - 3 = 0$
4. $2x^2 - 4x = -1$

Solve by using the quadratic formula. 13.3

5. $x^2 - 2x = 5$
6. $2n^2 - 8n = -5$

Solve by using an appropriate method. 13.4

7. $4t^2 - 9 = 0$
8. $y^2 - 2y = 3$
9. $3x^2 + 5x - 2 = 0$
10. $m^2 + 4m = 9$

13.5 Graphing Quadratic Functions

Objectives: To graph quadratic functions
To use graphs of quadratic functions to solve quadratic equations

Arrow Company finds that its profit y for producing x units is given by the equation $y = x^2 + 2x - 3$.

(Number of units, Profit)
(x, y)

Profit changes as the number of units change. That is, the amount of profit is a function of the number of units.

A function f given by the equation $f(x) = ax^2 + bx + c$, where a, b, and c are real numbers, and $a \neq 0$, is a **quadratic function**.

$y = f(x) = x^2 + 2x - 3$ is a quadratic function. The graph of the function will enable Arrow Company to determine for what values of x its profit y will be zero.

Capsule Review

You can use the slope and y-intercept to graph equations.

Give the slope and y-intercept of the graph of each of the following linear functions. Then draw its graph.

1. $f(x) = 2x + 1$

x	−2	−1	0	1	2
f(x)	?	?	?	?	?

2. $g(x) = -x + 1$

x	−4	−2	0	2	4
g(x)	?	?	?	?	?

3. $h(x) = 5x$ 4. $f(x) = -x$ 5. $h(x) = 3$ 6. $g(x) = -\dfrac{1}{2}$

One way to graph a quadratic equation is to find several ordered pairs that satisfy the equation, then graph the ordered pairs on a coordinate plane and connect the points with a smooth curve.

13.5 Graphing Quadratic Functions   **557**

EXAMPLE 1 Use graphing to find the values of x for which $y = f(x) = x^2 + 2x - 3$ is zero.

A calculator can be used to evaluate the function for each value of x.

x	$x^2 + 2x - 3$	y
-4	$(-4)^2 + 2(-4) - 3$	5
-3	$(-3)^2 + 2(-3) - 3$	0
-2	$(-2)^2 + 2(-2) - 3$	-3
-1	$(-1)^2 + 2(-1) - 3$	-4
0	$(0)^2 + 2(0) - 3$	-3
1	$(1)^2 + 2(1) - 3$	0
2	$(2)^2 + 2(2) - 3$	5

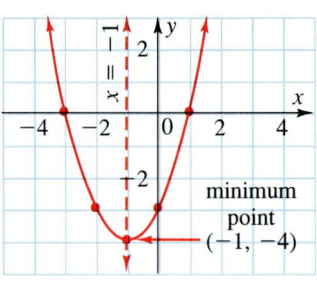

So, $y = f(x) = x^2 + 2x - 3$ is zero when $x = -3$ and $x = 1$.

The graph in Example 1 shows that the values of x for which the company's profit is 0 are -3 and 1. Since manufacturing -3 units is not a reasonable answer, -3 is not a reasonable solution. The company can conclude that it makes a profit on every unit after the first.

The curve of $y = f(x) = x^2 + 2x - 3$ is a parabola. A **parabola** is a graph of a quadratic function. The x-coordinate of a point where the curve intersects the x-axis, the point where $y = f(x) = 0$, is called an *x-intercept*.

EXAMPLE 2 Graph $y = g(x) = -x^2 - 2x$.

x	y
-3	-3
-2	0
-1	1
0	0
1	-3

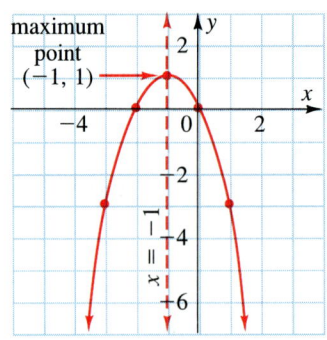

The parabola in Example 1 opens upward and has a **minimum** (lowest) **point** at $(-1, -4)$. The parabola in Example 2 opens downward and has a **maximum** (highest) **point** at $(-1, 1)$. These examples suggest the following:

For any quadratic function $f(x) = ax^2 + bx + c$, $a \neq 0$:

- If a is positive, the parabola opens upward.
- If a is negative, the parabola opens downward.

The minimum or maximum point of a parabola is called the **vertex**. In Examples 1 and 2, notice that the x-coordinate of the vertex of each parabola is -1. In each example, it is the average of the x-coordinates of the points where the parabola intersects the x-axis, that is where $y = f(x) = 0$.

For $y = f(x) = x^2 + 2x - 3$
$(-3, 0), (1, 0)$
$\frac{-3 + 1}{2} = \frac{-2}{2} = -1$

For $y = g(x) = -x^2 - 2x$
$(-2, 0), (0, 0)$
$\frac{-2 + 0}{2} = -1$

But in each example, $-\frac{b}{2a}$ also equals -1. This leads to the following:

The x-coordinate of the vertex of a parabola, the graph of a quadratic function $f(x) = ax^2 + bx + c$, $a \neq 0$, is $-\frac{b}{2a}$.

If you folded the curves in Example 1 and Example 2 along the line $x = -1$, the two halves of the parabola would coincide. In each example, the vertical line, $x = -1$, is called the *axis of symmetry*. Notice that the axis of symmetry passes through the vertex, or *turning point*, of the parabola.

For a parabola with the equation $f(x) = ax^2 + bx + c$, $a \neq 0$, the equation of the **axis of symmetry** is $x = -\frac{b}{2a}$.

It is helpful to use the vertex and the axis of symmetry to graph a quadratic function.

EXAMPLE 3 Find the vertex, the axis of symmetry, and six points on the parabola $h(x) = x^2 + 6x + 8$. Graph the equation.

a. $x = -\frac{b}{2a} = -\frac{6}{2} = -3$ *The x-coordinate of the vertex of the parabola.*

$y = (-3)^2 + 6(-3) + 8$
$= -1$ *Substitute -3 for x to find the y-coordinate of the vertex.*

Therefore the vertex is $(-3, -1)$.

b. $x = -\frac{b}{2a} = -\frac{6}{2} = -3$ *The axis of symmetry passes through the vertex.*

Therefore, $x = -3$ is the axis of symmetry.

13.5 Graphing Quadratic Functions

c. Make a table. Since the points, except the vertex, occur in pairs that have the same $h(x)$-coordinate, choose three values of x that are greater than -3 and three values of x that are less than -3.

x	$h(x)$
-6	8
-5	3
-4	0
-3	-1
-2	0
-1	3
0	8

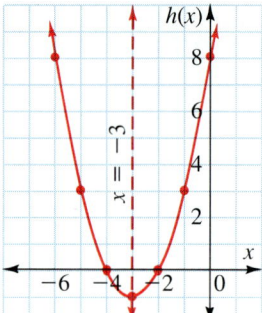

CLASS EXERCISES

Tell whether the graph of each function opens upward or downward. Then tell whether there is a minimum or a maximum point.

1. $g(x) = 2x^2$
2. $h(x) = 2x - x^2$
3. $g(x) = x^2 + 3x + 4$

Find the vertex and the axis of symmetry of each function.

4. $g(x) = x^2 - x - 12$
5. $f(x) = 3 - 2x - x^2$
6. $h(x) = 3x^2$
7. $h(x) = 4 - x^2$
8. $g(x) = x^2 - 9$
9. $f(x) = x^2 + x$

For Discussion

10. Graph the following on the same set of axes. You may use a graphing calculator or a computer graphing program.

$$y = x^2 + 2x \qquad y = x^2 + 2x + 4 \qquad y = x^2 + 2x - 3$$

How does the value of c in $y = ax^2 + bx + c$ affect the graph?

PRACTICE EXERCISES

Graph each equation. State whether the parabola opens upward or downward and if it has a minimum or maximum point.

1. $y = x^2$
2. $y = 2x^2$
3. $y = x^2 - 2x - 3$
4. $y = x^2 - 5x - 6$
5. $y = x^2 + 4x$
6. $y = x^2 - 6x$
7. $y = 4 - x^2$
8. $y = 1 - x^2$
9. $y = x^2 - 1$
10. $y = x^2 - 9$
11. $y = 3 - x^2$
12. $y = 5 + x^2$

Find the vertex, the axis of symmetry, and the *x*-intercepts. Use each to graph the following equations.

13. $s(x) = x^2 + 3x + 2$
14. $g(x) = x^2 - 2x + 1$
15. $r(x) = x^2 + 5x + 8$
16. $f(x) = x^2 - 4x + 5$
17. $h(x) = 7 - 6x - x^2$
18. $f(x) = 5 - 4x - x^2$
19. $g(x) = 5 + 3x - 2x^2$
20. $h(x) = 10 - x - 3x^2$
21. $f(x) = -\frac{1}{2}x^2 - 5$
22. $g(x) = -2 + \frac{1}{2}x^2$
23. $m(x) = \frac{3}{4}x^2 + 2$
24. $r(x) = \frac{1}{5}x^2 + 3$
25. $g(x) = (x + 2)^2$
26. $f(x) = (x + 3)^2$
27. $h(x) = (x + 1)^2 - 5$
28. $r(x) = (x + 4)^2 - 3$
29. $g(x) = -2(x + 3)^2 - 4$
30. $h(x) = -4\left(x + \frac{1}{2}\right)^2 - 4$

Graph each equation. State whether the parabola opens upward or downward and if it has a minimum or maximum point.

31. $y = 5(x - 2)^2 + 3$
32. $y = 3(x - 4)^2 + 5$
33. $y = -2(x - 5)^2 - 15$

Applications

Solve graphically.

34. **Business** The total profit *p* made by an engineering firm is given by the equation $p = x^2 - 25x + 5000$. Graph the equation. Find the minimum profit made by the company.

35. **Physics** A projectile is propelled upward. Its distance *s*, in ft, after *t* seconds is given by the function $s = f(t) = 96t - 16t^2$. Graph the function. Find the maximum height reached by this projectile.

MATH CLUB ACTIVITY

Can This Hiker Be Saved?

On a hiking trip, you are $\frac{2}{3}$ the distance across a stream when you spot a bear approaching. You assume it is approaching at 0.75 km/h. You can just escape by running at a uniform speed to either side of the stream. What must your speed be?

APPLICATION:
Physics

Did you know that today computers are used to track projectiles of many forms, such as rockets and satellites?

Applications of projectile motion often occur in physics and are greatly aided by the use of computers. When a rocket is launched it is necessary to know how long it will take to land, how high it goes, and how much time it takes the rocket to go a specified distance.

It was Galileo who first accurately described projectile motion. He often used mathematics to help him analyze physical science. Galileo proved in the 1600's that when a frictionless object is projected from the ground, it takes the same amount of time to go up as it does to come down.

The path that the object traces is a parabola, represented by the function:

$$H(t) = -\frac{1}{2}gt^2 + vt$$

where $H(t)$ = height
t = time
g = force of gravity
v = initial speed

The force of gravity for the earth is 32 ft/s², while our moon has a force of gravity of 5.32 ft/s². The forces of gravity for other planets in our solar system include:

Planet	Gravity (ft/s²)
Venus	28.8
Mars	12.2
Jupiter	82.6
Saturn	35.5

The following computer program enables you to find the answer to the three questions involved with tracking a rocket launched from earth with an initial speed of 100 ft/sec. Since the program is quite simple and straightforward, it will be necessary to refine the input in repeated runs to give more precise information.

```
10 INPUT "ENTER THE BEGINNING TIME, THE ESTIMATED TIME AND
      THE TIME INCREMENT:   '';B,E,I: PRINT
20 PRINT "TIME","DISTANCE": PRINT "_____","_____"
30 FOR X = B TO E STEP I
40 PRINT X, - 16 * X ^ 2 + 100 * X
50 NEXT X
60 END
```

Use the following steps when running the program.

1. Estimate how long it will take the rocket to land. Enter the beginning time (0), the estimated time to land (say 7 or 8) and the time increment (start with 1 and refine it in later RUNs.) Repeat this step until the rocket reaches a distance of 0 again.

2. Choose a beginning and ending time that you think surround the peak of the rocket height. Enter the time and a decimal increment. Do repeated RUNs until you get the peak height. You can tell which value is the peak because the values around it are symmetric and smaller.

3. Use this program to determine how much time is needed for the rocket to cover a certain distance.

EXERCISES

1. A rocket is propelled upward at a speed of 120 ft/sec. Change line 40 to be

    $$40 \text{ PRINT } X, -.5*32*X^2 + 120*X$$

    in the above program to answer these questions:

    a. How high does it go?
    b. How long is it in the air?
    c. How many seconds does it take for the rocket to reach 100 ft?

2. Revise line 40 so that the same rocket can be tracked when it is launched from the surface of the moon ($g = 5.32$).

    a. How high does it go?
    b. How long will it take before it reaches the surface again?
    c. How does this data compare with the data of the earth launch?

3. Investigate the force of gravity for other planets and revise the computer program to track a rocket launched from there.

13.6 The Discriminant

Objective: To use the discriminant to determine the nature and number of solutions of a quadratic equation

You have solved quadratic equations by several methods and found that there may be two distinct solutions (or real roots), one solution, or no real solution. Without actually solving, you can find this information by using a part of the quadratic formula.

Capsule Review

To solve $ax^2 + bx + c = 0$, $a \neq 0$, use the quadratic formula:
$$x = \frac{-b \pm \sqrt{b^2 - 4ac}}{2a}$$

Use the quadratic formula to solve.

1. $t^2 - 3t - 10 = 0$
2. $2y^2 + 4y = 3$
3. $12x^2 - 4x - 3 = 0$
4. $m^2 - 10m + 25 = 0$
5. $6t^2 - 1 = 0$
6. $z^2 + 4 = 0$

In the quadratic formula, the algebraic expression $b^2 - 4ac$ is called the **discriminant**. The *discriminant* determines the nature and number of solutions of a quadratic equation.

EXAMPLE 1 Solve $x^2 + 3x - 5 = 0$, and graph $y = x^2 + 3x - 5$.

$x^2 + 3x - 5 = 0$

$x = \dfrac{-b \pm \sqrt{b^2 - 4ac}}{2a}$

$= \dfrac{-3 \pm \sqrt{3^2 - 4(1)(-5)}}{2(1)}$

$= \dfrac{-3 \pm \sqrt{9 + 20}}{2}$

$= \dfrac{-3 \pm \sqrt{29}}{2}$

$y = x^2 + 3x - 5$

x	y
-4	-1
-3	-5
-2	-7
-1.5	-7.25
-1	-7
0	-5
1	-1

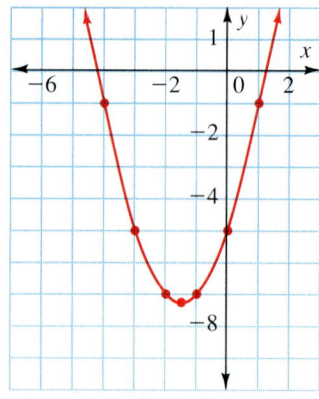

There are two x-intercepts.

There are two real solutions, $\dfrac{-3 + \sqrt{29}}{2}$ and $\dfrac{-3 - \sqrt{29}}{2}$. Notice that $b^2 - 4ac > 0$.

564 Chapter 13 Quadratic Equations and Functions

EXAMPLE 2 Solve $x^2 - 2x + 1 = 0$, and graph $y = x^2 - 2x + 1$.

$x^2 - 2x + 1 = 0$

$x = \dfrac{-b \pm \sqrt{b^2 - 4ac}}{2a}$

$= \dfrac{-(-2) \pm \sqrt{(-2)^2 - 4(1)(1)}}{2(1)}$

$= \dfrac{2 \pm \sqrt{4 - 4}}{2} = \dfrac{2 \pm 0}{2}$, or 1

$y = x^2 - 2x + 1$

x	y
-1	4
0	1
1	0
2	1
3	4

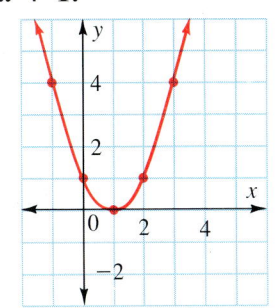

There is one real solution. Notice that $b^2 - 4ac = 0$.   There is one x-intercept.

EXAMPLE 3 Solve $x^2 + 2x + 3 = 0$, and graph $y = x^2 + 2x + 3$.

$x^2 + 2x + 3 = 0$

$x = \dfrac{-b \pm \sqrt{b^2 - 4ac}}{2a}$

$= \dfrac{-2 \pm \sqrt{(2)^2 - 4(1)(3)}}{2(1)}$

$= \dfrac{-2 \pm \sqrt{4 - 12}}{2}$

$= \dfrac{-2 \pm \sqrt{-8}}{2}$, or $-1 \pm \sqrt{-2}$

$y = x^2 + 2x + 3$

x	y
-3	6
-2	3
-1	2
0	3
1	6

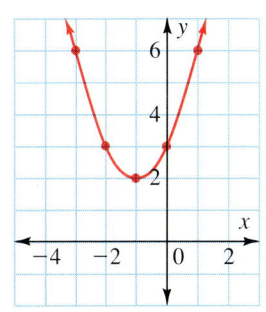

There are no real solutions. Why? Notice that $b^2 - 4ac < 0$. There are no x-intercepts.

Look for a relationship between the value of the discriminant and the nature of the solutions of the equation.

$b^2 - 4ac$	Nature of the solutions of $ax^2 + bx + c = 0$	No. of x-intercepts
Positive	Two distinct, real solutions	2
Zero	One distinct, real solution (double solution)	1
Negative	No real solutions	0

EXAMPLE 4 Use the discriminant to determine the nature of the solutions of the equation:
   a. $-x^2 + 3 = 0$ b. $-x^2 + 4x - 5 = 0$

a. $-x^2 + 3 = 0$
   $b^2 - 4ac = 0^2 - 4(-1)(3) = 12$ *Substitute: $a = -1$, $b = 0$, $c = 3$*
   Since $12 > 0$, there are two distinct real solutions.

b. $-x^2 + 4x - 5 = 0$
   $b^2 - 4ac = (4)^2 - 4(-1)(-5) = -4$ *Substitute: $a = -1$, $b = 4$, $c = -5$*
   Since $-4 < 0$, there are no real solutions.

13.6 The Discriminant   **565**

CLASS EXERCISES

Find the value of the discriminant. Then state the nature of the solutions.

1. $x^2 - 1 = 0$
2. $x^2 + 4 = 0$
3. $x^2 - 6x + 9 = 0$
4. $x^2 + 4x + 3 = 0$
5. $-2x^2 + 3x - 2 = 0$
6. $4x^2 - 20x + 25 = 0$

PRACTICE EXERCISES

Find the value of the discriminant, and state the nature of the solutions.

1. $x^2 + 2x - 15 = 0$
2. $x^2 - 5x + 4 = 0$
3. $x^2 - 8x + 16 = 0$
4. $x^2 - 10x + 25 = 0$
5. $x^2 + 4x - 6 = 0$
6. $x^2 - x - 3 = 0$
7. $2x^2 - 3x + 2 = 0$
8. $3x^2 - 4x + 5 = 0$
9. $-2x^2 + 7x - 3 = 0$
10. $-2x^2 - 3x + 2 = 0$
11. $-6x^2 + 5x - 4 = 0$
12. $4x^2 - 3x + 3 = 0$
13. $4x^2 + 12x = -9$
14. $9x^2 - 30x = -25$
15. $\frac{3}{2}x^2 = 4x - \frac{1}{2}$
16. $x^2 + 3x = 3$
17. $2x^2 + \frac{1}{2}x + \frac{2}{3} = 0$
18. $\frac{1}{5}x^2 - \frac{5}{4}x - 1 = 0$

Without graphing, find the number of x-intercepts for each function.

19. $f(x) = 9 - 12x - 4x^2$
20. $f(x) = -16 - 40x + 25x^2$
21. $f(x) = x^2 + \frac{1}{2}x - \frac{17}{4}$
22. $f(x) = -\frac{1}{2}x^2 - x - \frac{11}{6}$
23. $f(x) = 4x^2 - kx + 625$ if $|k| > 100$
24. $f(x) = kx^2 - 8x + 3$ if $k > 5\frac{1}{3}$

Applications

25. **Engineering** In a certain electrical circuit an engineer found that the power p can be expressed by the function $p = 2i^2 - 60i + 480$ where i is the current. Can the power in this circuit ever be zero?

26. **Finance** A rental agency uses the formula $T = 5400 + 300x - 50x^2$ to find its total income when x units are rented. Can they expect their total ever to equal $7000?

LOGICAL REASONING

The formula $2D = n(n - 3)$ tells how many diagonals D in a polygon of n sides. An architect designs a building with 14 diagonals. How many sides does the building have?

Chapter 13 Quadratic Equations and Functions

13.7 Problem Solving Strategy: Use a Function

Martin's monthly income is $500 plus 5% of his sales. His income i can be written as a function of his sales s (in dollars).

$$i = f(s) = 0.05s + 500 \quad \text{Do you recognize this as a linear function?}$$

In this situation, the slope of 0.05 is the rate at which Martin's income increases or decreases for a change in sales. There is a $0.05 income rise for every dollar increase in sales. The i-intercept, 500, refers to the $500 he earns if he makes no sales (if $s = 0$).

EXAMPLE 1 The table shows the prices and net weights of boxes of Apple Wheat Crackles. Recall that in a linear function, the slope is constant.

Weight (oz) w	8	20	40
Price ($) p	1.55	3.35	6.35

a. Confirm that the price p is a linear function of the weight w.
b. Write this function for the price $f(w)$.

Using the slope and intercepts, what can you tell about the function?

■ **Understand the Problem** **What are the given facts?**
From a chart of ordered pairs of weight and prices, you should show that the ordered pairs belong to a linear function, and then write the function. You must then interpret the meaning of the slope and intercepts for this situation.

■ **Plan Your Approach** **Choose a strategy.**
a. Show that the slope is constant.
b. Use the slope and one ordered pair to find the p-intercept. Write the function in the form $f(w) = p = mx + b$.

■ **Complete the Work** **Solve.**
a. For (8, 1.55) and (20, 3.35):
$$m = \frac{3.35 - 1.55}{20 - 8} = \frac{1.80}{12} = 0.15$$

For (20, 3.35) and (40, 6.35):
$$m = \frac{6.35 - 3.35}{40 - 20} = \frac{3.00}{20} = 0.15$$

b. $f(w) = p = 0.15w + b$
Use (20, 3.35) to find b.
$3.35 = 0.15(20) + b$
$3.35 = 3.00 + b$
$b = 0.35$
$f(w) = p = 0.15w + 0.35$

Interpret the Results

State your conclusion.
The slope 0.15 is the increase in price per 1 oz increase in weight. The w-intercept has no meaning since weight cannot be negative. The price intercept $0.35 is the price of zero ounces; it could be the packaging cost.

Quadratic functions, as well as linear functions, frequently serve as mathematical models. Such a model enables you to make predictions and often provides a better understanding of the situation.

EXAMPLE 2 Lia Design Company produces a special drawing tool. As the company produces a larger number of tools, the profit per tool increases. However, beyond a certain number, complications in the manufacturing procedure reverse this advantage. The company finds that the following quadratic function is a formula that approximates the profit per tool p for the number made n.

$$p = f(n) = -0.00025n^2 + 0.105n - 1.025$$

a. Find the profit per tool if 145 tools are made.
b. Find the number of tools the Lia Design Company made if the profit was $7.50 per tool.
c. How many tools should be produced to make the most profit? What is the maximum profit?
d. At what point does the Lia Design Company start to lose money on this venture?

Understand the Problem

What are the given facts?
Analyze the given quadratic function to answer questions **(a)** through **(d)**.

Plan Your Approach

Choose a strategy.
Use your understanding of quadratic functions.
a. Substitute 145 for n, and find $f(145)$.
b. Substitute 7.5 for p [or $f(n)$], and find n.
c. Since the coefficient of n^2 is negative, the function will have a maximum value. You must find the vertex point.
d. For what value of n does $p = 0$? Set $f(n) = 0$, and solve.

◼ **Complete the Work**

Solve.
a. $f(145) = 8.94$
b. Solving the quadratic equation for $p = 7.5$, $n = 110$ or 310.
c. $-\dfrac{b}{2a} = -\dfrac{0.105}{2(-0.00025)} = 210 = n$
   Substitute 210 for n and find the p-coordinate for the vertex.
   $f(210) = 10.00$
   The vertex is (210, 10).
d. If $f(n) = 0$, $n = 10$ or 410.

◼ **Interpret the Results**

State your conclusion.
a. If 145 tools are made, the profit per tool will be $8.94.
b. If the profit was $7.50 per tool, the number made was 110 or 310.
c. Making 210 tools permits the maximum profit, $10 per tool.
d. The company loses money if they make fewer than 10 or more than 410 tools.

CLASS EXERCISES

1. Write a function to express Susan's income if she receives a base salary of $12,000 and a 15% commission on her sales.

A State highway department gives estimates for the cost of constructing a particular kind of highway.

No. miles of highway m	6	9	15
Cost (millions of $) c	8.3	12.2	20.0

2. Show that cost is a linear function of the number of miles constructed.

3. Write the cost c as a function of the number of miles m.

4. What are the meaning of the slope and intercepts in this situation?

PRACTICE EXERCISES

In the relationships shown, first determine whether each is a linear function. If so, then write the function with the upper variable as a function of the lower. Tell the meaning of slopes and intercepts.

1. The costs of different weights of pure silver

Cost ($) c	58	70	78
No. oz w	3	18	28

2. The temperature at various times on the same day

Temp. (°F) t	65	67	70
Time h	2 pm	5 pm	9 pm

13.7 Problem Solving Strategy: Use a Function

3. The cost of lighting a field for different lengths of time

Cost ($) c	1.88	3.02	4.35
Hours h	1.5	6.3	11.9

4. The price of different weights of a rare perfume

Price ($) p	4.56	6.12	9.63
Grams g	0.5	2.1	5.7

Write a linear function to describe each of the following situations.

5. The total cost of belonging to a health club if there is an initial fee of $75 in addition to the monthly charge of $24.50

6. The cost of consulting a lawyer if there is a base charge of $35 in addition to the hourly charge of $55

7. The number of gallons of solvent left in a 55 gal drum if it suddenly develops a leak through which liquid escapes at the rate of one qt/h

8. The volume of alcohol in a flask, if it initially held 3.00 L and is filling at the rate of 2.5 mL/min

Assume that a linear relationship exists between the variables. Use the appropriate function to solve the problem.

9. If a designer charged $150 for $2\frac{1}{2}$ h consultation and $325 for 6 h, how long was your consultation, if you were charged $200?

10. If a biologist measures 80 cricket chirps/min at 60°F and 144 chirps/min at 76°F, at what temperature should he expect 180 chirps/min?

One would expect that the more walnut trees planted per acre, the higher the yield of walnuts. However, beyond a certain number of trees per acre, the trees crowd one another, and the yield drops. A state agricultural department estimates that the yield in bushels of walnuts per tree can be given as a function of the number of trees per acre t by the formula: $f(t) = -0.01t^2 + 0.8t$.

11. What yield should 30 trees per acre give?

12. How many trees per acre should produce a yield of 16 bu/tree?

13. What number of trees per acre gives the greatest yield? What is this maximum yield?

14. What does the t-intercept represent in this situation? What does the $f(t)$ intercept represent?

A student newspaper currently has 500 subscriber's who pay $6.00/yr. The editors assume that for each $0.25 price decrease, they would sell 50 more subscriptions.

15. On this assumption, write a function for their annual receipts in terms of the number x of $0.25 price decreases.

16. What number of $0.25 decreases produces the maximum receipts? What is this maximum?

Sponsors of a design show believe 600 people will attend if the price per ticket is $6.00. They assume that 25 fewer people will attend for each $0.50 increase in ticket price.

17. On this assumption, write a function for the ticket sales in terms of the number x of $0.50 price increases.

18. What number of $0.50 increases produces the maximum ticket sales? What is this maximum?

Mixed Problem Solving Review

1. If one pair of opposite sides of a square is made 10% longer and the other pair is made 15% shorter, the resulting rectangle will have a perimeter of 234. What is the perimeter of the original square?

2. One printer can do a job alone in 3 h. With a second printer also working on this job, it takes 2 h. How long would it take the second printer alone?

3. Write a linear function for the total cost of cable television if the installation charge is $35 and the monthly fee is $15. If you have paid a total of $125, how many months have you been connected?

4. Find a two-digit number if its tens' digit is 2 more than its units' digit and the number with the digits reversed is $\frac{4}{7}$ of the original.

PROJECT

The quadratic function $h(t) = 150t - 5t^2$ can be used to predict the height in meters of a small rocket launched with an initial velocity of 150 m/sec. t is the number of seconds after launching. Construct a table as shown below for $t = 0$ to 20. Graph the ordered pairs, $(t, h(t))$.

t	$h(t)$
0	0
1	145
2	280
3	405

1. What kind of curve is the graph?
2. What is the maximum height reached by the rocket?
3. After how many seconds does the rocket hit the ground?

13.8 The Sum and Product of the Solutions

Objective: To use the sum and product of the solutions of a quadratic equation to write the equation
To determine, without solving, if two numbers are the solutions of a quadratic equation

Suppose you know that the solutions of a quadratic equation are 3 and −5. Could you work backwards to write the equation? Think of the two factors from which the solutions must come. $x = 3$ comes from $(x - 3)$, and $x = -5$ from $(x + 5)$. The equation is $(x - 3)(x + 5) = 0$ or $x^2 + 2x - 15 = 0$.

Capsule Review

Recall that factoring is helpful in determining the solutions of an equation.

Find the solutions of these equations.

1. $x^2 - 5x + 6 = 0$ 2. $x^2 + 9x + 14 = 0$ 3. $x^2 + 11x = 0$

The equation $x^2 + 2x - 15 = 0$ has solutions 3 and −5. Examine the sum and product of these solutions.

The sum of the solutions is $3 + (-5) = -2$.
The product of the solutions is $3(-5) = -15$.

opposite of the sum of solutions
↓
$x^2 + 2x - 15 = 0$
↑
product of the solutions

In general, if the solutions of a quadratic equation are s_1 and s_2, then the equation can be written as $(x - s_1)(x - s_2) = 0$, or as:

$$x^2 - (s_1 + s_2)x + s_1 s_2 = 0$$

opposite of the sum of the solutions   product of the solutions

Consider the standard form of a quadratic equation, $ax^2 + bx + c = 0$. Divide each side by a: $x^2 + \dfrac{b}{a}x + \dfrac{c}{a} = 0$

Compare this equation with the equation $x^2 - (s_1 + s_2)x + s_1 s_2 = 0$. The coefficients of corresponding terms must be the same. This means:

$$\frac{b}{a} = -(s_1 + s_2), \text{ and } \frac{c}{a} = s_1 s_2.$$

These relationships can be used to write the quadratic equation whose solutions are given.

EXAMPLE 1 Write a quadratic equation whose solutions are $\frac{4}{5}$ and -1.

$\frac{b}{a} = -\left[\frac{4}{5} + (-1)\right] = \frac{1}{5}$ *The opposite of the sum of the solutions*

$\frac{c}{a} = \left(\frac{4}{5}\right)(-1) = -\frac{4}{5}$ *Product of the solutions*

$x^2 + \frac{1}{5}x - \frac{4}{5} = 0$ $x^2 + \frac{b}{a}x + \frac{c}{a} = 0$

How else can the equation $x^2 + \frac{1}{5}x - \frac{4}{5} = 0$ be written?

The sum and product relationship provide a method of checking the solutions of a quadratic equation without solving.

EXAMPLE 2 Determine whether $\frac{7}{4}$ and -2 are solutions of $4x^2 + x - 14 = 0$.

$-(s_1 + s_2) = -\left[\frac{7}{4} + (-2)\right] = \frac{1}{4}$

Does $\frac{1}{4} = \frac{b}{a}$ from the equation?

$\frac{1}{4} = \frac{1}{4}$ ✓

$s_1 s_2 = \frac{7}{4}(-2) = -\frac{7}{2}$

Does $-\frac{7}{2} = \frac{c}{a}$ from the equation?

$-\frac{7}{2} \stackrel{?}{=} -\frac{14}{4}$

$-\frac{7}{2} = -\frac{7}{2}$ ✓

$\frac{7}{4}$ and -2 are solutions of $4x^2 + x - 14 = 0$

CLASS EXERCISES

Without solving, find the sum and product of the solutions of each equation.

1. $x^2 + 4x - 12 = 0$ 2. $y^2 - 6y + 8 = 0$ 3. $p^2 - 3p = 10$

13.8 The Sum and Product of the Solutions

4. $3m^2 + 9m + 6 = 0$ **5.** $x^2 + 5x = 0$ **6.** $2n^2 - 10n = 12$

Write a quadratic equation that has the given solutions.

7. $1, -2$ **8.** $-2, -3$ **9.** $-\frac{1}{3}, 1$ **10.** $3, -\frac{3}{2}$

For Discussion

11. If the only solution to a quadratic equation is -4, what are the sum and product of its solutions?

PRACTICE EXERCISES

Write a quadratic equation in standard form that has the given solutions.

1. $1, 3$ **2.** $3, 6$ **3.** $-2, 6$ **4.** $-3, 2$
5. $-5, -1$ **6.** $-1, -2$ **7.** $-5, 0$ **8.** $0, 3$
9. $-3, \frac{4}{5}$ **10.** $-\frac{1}{2}, 2$ **11.** $\frac{2}{3}, 1$ **12.** $-\frac{3}{2}, 3$

Determine if the numbers are solutions of the given equation.

13. $-4, 4; \; x^2 - 16 = 0$ **14.** $3, -3; \; x^2 + 9 = 0$
15. $-2, -1; \; x^2 - 3x + 2 = 0$ **16.** $-3, -4; \; x^2 + 7x + 12 = 0$

Write a quadratic equation that has the given solutions.

17. $-\frac{5}{2}, \frac{5}{2}$ **18.** $\frac{2}{3}, -\frac{2}{3}$ **19.** $-\frac{3}{2}, \frac{4}{3}$ **20.** $\frac{3}{2}, -\frac{1}{2}$
21. $1 + \sqrt{6}, \; 1 - \sqrt{6}$ **22.** $-2 + \sqrt{3}, \; -2 - \sqrt{3}$

Determine if the numbers are solutions of the given equation.

23. $-\frac{7}{2}, 5; \; 2x^2 - 3x = 35$ **24.** $1, -\frac{2}{3}; \; 3 = 6x^2 - x$
25. $1 \pm \sqrt{2}; \; z^2 - z - 1 = 0$ **26.** $5 \pm \sqrt{5}; \; y^2 - 10y + 20 = 0$
27. $\frac{-3 \pm \sqrt{5}}{2}; \; p^2 + 3p = -1$ **28.** $\frac{1 \pm \sqrt{7}}{3}; \; 3r^2 - 2r = 2$

Show that the following statements are true. (*Hint:* The solutions of a quadratic equation are $\dfrac{-b \pm \sqrt{b^2 - 4ac}}{2a}$.)

29. Sum of the solutions is $-\dfrac{b}{a}$.   **30.** Product of the solutions is $\dfrac{c}{a}$.

Applications

31. Engineering An engineer thinks that $1 \pm \sqrt{5}$ are solutions of the equation $x^2 - 2x = 4$. Determine if these are solutions of the given equation.

32. Business In 1990 a financial officer projects from the equation $p = -x^2 - 20x + 300$, that his company will show a 0 profit in the year 2000. Determine if he is correct.

33. Engineering A research engineer proposes that a rocket which follows a path represented by the equation $y = -16t^2 - 256t + 4096$ has solutions of $-8 \pm 8\sqrt{5}$. Determine if this is correct.

TEST YOURSELF

Find the vertex and the axis of symmetry of each equation. 13.5

1. $y = x^2$ **2.** $y = 5 - x^2$ **3.** $y = x^2 - 2x$

Tell whether the graph of each function opens upward or downward. Then tell whether there is a minimum or a maximum point.

4. $g(x) = \frac{1}{4}x^2$ **5.** $f(x) = 3x - x^2$ **6.** $h(x) = x^2 + 2x + 3$

Find the value of the discriminant, and state the nature of the solutions. 13.6

7. $x^2 - 3x + 4 = 0$ **8.** $4y^2 + 12y + 9 = 0$ **9.** $2z^2 - 5z = 12$

The Villo Travel Agency offers a vacation package to the Carribean at a discount rate. They figure that the amount of profit per person can be determined by the function: $f(x) = 40x - x^2$, where x is the number of people.

Use the quadratic function to solve these problems. 13.7

10. What is Villo's profit/person if 15 people were to go to the Carribean?

11. If the profit was $351/person, how many people went to the Carribean?

12. Find the number of people which will give Villo's the largest profit/person. What is the maximum profit?

13. What is the least or greatest number of people they can accept and not lose any money?

For each equation, find the sum and product of its solutions. 13.8

14. $m^2 + 7m + 9 = 0$ **15.** $2n^2 - 4n = 5$ **16.** $5r^2 = 4r$

CHAPTER 13 SUMMARY AND REVIEW

Vocabulary

axis of symmetry (559)
completing the square (547)
discriminant (564)
maximum point (558)
minimum point (558)
parabola (558)
perfect square (544)
quadratic formula (552)
quadratic function (557)
vertex (559)

Using Perfect Square Quadratics To solve quadratic equations of the form $x^2 = k$ and $(ax + b)^2 = k$, where $k \geq 0$, use the square-root property: if $x^2 = k$, then $x = +\sqrt{k}$ or $x = -\sqrt{k}$.   **13.1**

Solve by using the square-root property.

1. $4p^2 = 80$
2. $(y - 1)^2 = 9$
3. $(m + 3)^2 - 4 = 32$

Using Completing the Square To complete the square for $ax^2 + bx + \underline{\ ?\ }$:   **13.2**
- If $a \neq 1$, then divide the equation by a.
- Find one-half of the coefficient of x obtained.
- Square the value.
- Add the squared result to $x^2 + \dfrac{b}{a}x$.

Solve by completing the square.

4. $x^2 + 4x - 12 = 0$
5. $y^2 - 6y - 5 = 0$
6. $2z^2 - 4 = 5$

Using the Quadratic Formula To solve any quadratic equation of the form $ax^2 + bx + c = 0$, you can use the quadratic formula: **13.3**

$$x = \dfrac{-b \pm \sqrt{b^2 - 4ac}}{2a}.$$

Solve by using the quadratic formula.

7. $3x^2 + 7x + 3 = 0$
8. $4z^2 = 20z - 5$
9. $4p^2 - 6p = -9$

Solving Quadratic Equations by Any Method You can use factoring, the square-root property, completing the square, or the quadratic formula to solve quadratic equations.   **13.4**

Solve each quadratic equation by using an appropriate method.

10. $4t^2 - 8t = 0$
11. $5x^2 = 30$
12. $(y + 3)^2 = 16$
13. $3p^2 + 1 = 10p$
14. $k^2 + 10k + 30 = 0$
15. $m^2 + 5m - 24 = 0$

Graphing a Quadratic Function, $f(x) = ax^2 + bx + c$, $a \neq 0$ One way to graph a quadratic function is to find several ordered pairs that satisfy the equation, plot the ordered pairs, and then connect the points with a smooth curve.

13.5

Graph the function. Find its x-intercepts. Tell whether there is a maximum or a minimum point. Determine the axis of symmetry.

16. $y = x^2$
17. $y = 1 - x^2$
18. $y = x^2 - 2x + 1$

Determining the Nature of the Solutions In the quadratic formula, $b^2 - 4ac$ is called the discriminant. There is a relationship between the value of the discriminant and the nature of the solutions of a quadratic equation.

13.6

Find the value of the discriminant, and state the nature of the solutions.

19. $y^2 + 4y - 12 = 0$
20. $x^2 = 14x - 49$
21. $c^2 - 2c + 25 = 0$

Using the Sum and Product of the Solutions of a Quadratic Equation To write a quadratic equation in the general form $ax^2 + bx + c = 0$, when the solutions, s_1 and s_2, of the equation are given, use these facts:

13.8

- When the equation is written as $x^2 + \frac{b}{a}x + \frac{c}{a} = 0$, the coefficient of the x term, $\frac{b}{a}$, is the opposite of the sum of the solutions. That is $-(s_1 + s_2) = \frac{b}{a}$.
- The constant term, $\frac{c}{a}$, is the product of the solutions. That is $s_1 s_2 = \frac{c}{a}$.

Write a quadratic equation in standard form with the given solutions.

22. $8, -1$
23. $-6, -5$
24. $\frac{4}{5}, -1$
25. $\frac{10}{3}, -\frac{5}{2}$

Using a Function to Solve a Problem Quadratic functions, as well as linear functions, frequently serve as mathematical models. Such models enable you to make predictions and often provide better understanding of the situation.

13.7

Use the function to solve the problem.

26. The Sundries Company uses the function $f(x) = -0.6x^2 + 15x - 4$ to determine their profits/sale. Find the amount of sales (x) that will give the Sundries Company the largest profit/sale. What is the maximum profit per sale?

Summary and Review

CHAPTER 13 TEST

Solve by using the square-root property.

1. $2x^2 = 24$
2. $2(y - 1)^2 = 50$

Solve by completing the square.

3. $c^2 - 2c = 15$
4. $2m^2 + 10m - 3 = 0$

Solve by using the quadratic formula.

5. $6x^2 + 7x - 5 = 0$
6. $2n^2 - 5 = 2n$

Solve each quadratic equation by using an appropriate method.

7. $m^2 - 6m - 7 = 0$
8. $p^2 + \dfrac{p}{6} - \dfrac{1}{6} = 0$

Graph the function. Find its *x*-intercepts. Tell whether there is a minimum or a maximum point. Determine the axis of symmetry.

9. $y = x^2 - x - 12$
10. $y = -x^2 - 2$

Find the value of the discriminant, and state the nature of the solutions.

11. $p^2 - 8p + 16 = 0$
12. $2y^2 + 7y - 3 = 0$
13. $3x^2 = 2x - 5$

Write a quadratic equation in standard form with the given solutions.

14. 3, 2
15. $\dfrac{5}{2}, \dfrac{9}{4}$
16. $-\dfrac{2}{3}, 0$

Use the appropriate function to solve.

17. The length of a rectangle is 5 ft more than twice its width. Find the dimensions, to the nearest tenth of a foot, if the rectangle has an area of 20 ft².

18. The ABC Tile Company uses the function $f(x) = 115x - x^2$ to determine their profit/sale. How many sales (*x*) did they have if they had a profit of $3300 per sale?

Challenge

Solve graphically: $y > x^2$
$y \leq 4$

PREPARING FOR STANDARDIZED TESTS

Select the best choices for each question.

1. If $4x^2 + kx + 16$ is a perfect square trinomial then k could equal:
   A. 0   B. 4   C. 8
   D. 16   E. 32

2. What value of x will make the following three ratios equal?
   $$\frac{64}{144}, \frac{x}{108}, \frac{12}{27}$$
   A. 40   B. 48   C. 52
   D. 56   E. 60

3. $\quad 10.08$
   $\times\; 0.05$
   A. 0.504   B. 5.04   C. 0.054
   D. 0.540   E. 5.40

4. Solve the equation:
   $2x - 3 = 5x + 18$
   A. −7   B. −5   C. −3
   D. 5   E. 7

5. Find the value of $|3x + 2y|$ when $x = -4$ and $y = 5$.
   A. 0   B. 2
   C. 3   D. 7
   E. 22

6. Solve for x: $2x + 5 = \sqrt{7}$
   A. 1
   B. 22
   C. $\dfrac{5 - \sqrt{7}}{2}$
   D. $\dfrac{5 + \sqrt{7}}{2}$
   E. $\dfrac{\sqrt{7} - 5}{2}$

7. Solve for x: $5x - 7 < 3x + 11$.
   A. $x < \frac{1}{2}$ B. $x < 2$
   C. $x > 2$ D. $x < 9$
   E. $x > 9$

8. What is the prime factorization of 180?
   A. $2^3 \cdot 3 \cdot 5$ B. $2^2 \cdot 3^2 \cdot 5$
   C. $2 \cdot 3^3 \cdot 5$ D. $2^2 \cdot 9 \cdot 5$
   E. $2 \cdot 3 \cdot 6 \cdot 5$

9. Solve for x: $x^2 - 7x - 18 = 0$
   A. −2 or −9
   B. −2 or 9
   C. 2 or −9
   D. 2 or 9
   E. $\pm 3\sqrt{2}$

10. Find the value of $5a^3b + 2a^2b^2 - 3ab^3$ when $a = -1$ and $b = 2$.
    A. 42   B. 26   C. 22
    D. −22   E. −26

11. Tom started typing his history paper at 6:48 p.m. and finished at 10:15 p.m. that evening. How long did it take him to type the paper?
    A. 3 h 17 min
    B. 3 h 27 min
    C. 4 h 3 min
    D. 4 h 27 min
    E. 4 h 33 min

12. The product of two positive consecutive odd integers is 195. Find the smaller one.
    A. 41   B. 39   C. 17   D. 15
    E. None of the above

Preparing for Standardized Tests

MAINTAINING SKILLS

Order each set of numbers from least to greatest.

Example 1   Write in order: $-5, 0, -2, 3, 1$
A number line can help.
Answer: $-5, -2, 0, 1, 3$

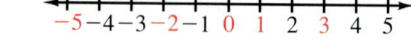

1. $4, -1, -3, 2, -4$
2. $0, -1.5, -1, -1.75, -2$
3. $-3.0, 2.8, 2.08, 2.88, -3.1$
4. $\frac{1}{4}, \frac{1}{3}, -\frac{3}{4}, -\frac{2}{3}, 0$
5. $\frac{5}{6}, \frac{7}{8}, -\frac{3}{5}, -\frac{2}{3}, -\frac{4}{7}$
6. $\frac{3}{4}, 0.7, -\frac{2}{3}, -0.6, -\frac{1}{2}$

Use the bar graph to solve each word problem.

7. How many records were sold in December?
8. In which month were the fewest number of records sold?
9. How many more records were sold in March then in January?
10. How many records were sold from November through March?

Simplify.

Example 2 $(-3.5)^2$
$(-3.5)^2 = (-3.5)(-3.5) = 12.25$

11. $(1.8)^2$
12. $(0.7)^2$
13. $(-2.6)^2$
14. $-(3.2)^2$
15. $-(-4.9)^2$
16. $(1.2)^2$
17. $(-5.1)^2$
18. $-(6.5)^2$
19. $-(-8.3)^2$
20. $(7.4)^2$

Find the probability of choosing at random a marble of the given color.

Example 3 $P(\text{red}) = \dfrac{\text{number of red marbles}}{\text{total number of marbles}}$
$= \dfrac{2}{10} = \dfrac{1}{5}$

21. $P(\text{blue})$
22. $P(\text{green})$
23. $P(\text{not blue})$
24. $P(\text{not green})$

14 Statistics and Probability

The collection and analysis of information concerning population is called demography. This type of information provides insight into many areas such as voting patterns, career development, marriage trends, or academic achievement.

14.1 Statistics: Measures of Central Tendency

Objective: To find the mode, median, and mean from data and from frequency distributions

Statistics is the collection, organization, analysis, and interpretation of numerical information, called *data*. Before attempting to analyze a set of data, it is a common practice to order the numbers, either from least to greatest or greatest to least.

Capsule Review

A number line shows the order of real numbers.

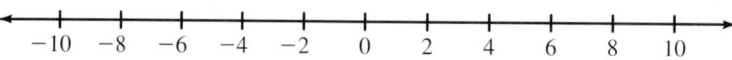

List the numbers in order from least to greatest.

1. 4, −5, 6, −1, −2, 0, 7, 9, −11
2. 1.2, −3.2, −4, 5.8, 3
3. $3\frac{1}{2}$, 3, $-3\frac{1}{2}$, $-\frac{1}{2}$, 0, −3
4. −6.08, 6.1, 6.0, 6.12, −6.10

Some specific characteristics of a set of data can be obtained by using *measures of central tendency*. A **measure of central tendency** is a statistic (a number) that is in some way representative or typical of a set of data. The *mode*, *median*, and *mean* are measures of central tendency.

A **mode** of a set of data is a number in the set that occurs most often.

A **median** of a set of data is the middle number in the set when the numbers are arranged in order from least to greatest.

The **mean** of a set of data consisting of n numbers is the sum of the numbers in the set divided by n.

For a given set of data, there is exactly one mean and exactly one median. However, the set may have one, more than one, or no modes.

EXAMPLE 1 Find the mode, mean, and median for the prices of 10 brands of vegetable soup packed in $10\frac{3}{4}$ oz cans.

Brand	A	B	C	D	E	F	G	H	I	J
Price in cents	68	59	73	89	68	94	69	89	68	89

Mode: Order the data   59, 68, 68, 68, 69, 73, 89, 89, 89, 94
Both 68 and 89 appear three times; there are two modes.

Mean: $\dfrac{59 + 68 + 68 + 68 + 69 + 73 + 89 + 89 + 89 + 94}{10} = \dfrac{766}{10} \approx 77$

In the statistics mode your calculator keeps track of the number of entries. You may wish to use a calculator to display the mean.

Median: For 10 values (an even number), there is no one middle number. Find the mean of the two middle numbers: $\dfrac{69 + 73}{2} = 71$

Data can be summarized in a table called a **frequency distribution.** Each number is matched with its frequency, the number of times it occurs.

EXAMPLE 2 The distribution of scores earned by 60 students on a 10-point quiz are shown below. Find the mode, median, and mean.

Score	Frequency	Frequency × Score
1	1	1 × 1 = 1
2	2	2 × 2 = 4
3	2	2 × 3 = 6
4	5	5 × 4 = 20
5	4	4 × 5 = 20
6	5	5 × 6 = 30
7	13	13 × 7 = 91
8	11	11 × 8 = 88
9	9	9 × 9 = 81
10	8	8 × 10 = 80
Total	60 (students)	421 (all scores)

Mode: The mode is 7, because its frequency, 13, is the highest.

Median: In a list of 60, the middle scores are the 30th and 31st score. Add down the frequency column until the sum is greater than 31: 1 + 2 + 2 + 5 + 4 + 5 + 13 = 32. The 30th and 31st scores are both 7 so the median is 7.

Mean: $\dfrac{\text{Total of all scores}}{\text{Number of students}} \longrightarrow \dfrac{421}{60} = 7.0$, to the nearest tenth

14.1 Statistics: Measures of Central Tendency

CLASS EXERCISES

Find the mode, median, and mean.

1. Number of eggs found in Mallard duck nests at the Tenafly Nature Center: 7, 9, 9, 8, 11, 9, 8, 10, 10

This table shows the frequency distribution of gas mileage for some new car models.

2. What is the mode?
3. What is the median to the nearest tenth?
4. What is the mean to the nearest tenth?

mi/gal	× Frequency	= Total
24	3	72
25	2	50
26	3	78
27	2	54
Sum	10	254

PRACTICE EXERCISES

Find the mode, median, and mean. Round to the nearest tenth.

1. The price in cents of 10 brands of a pint-size orange juice: 69, 71, 68, 70, 65, 71, 67, 71, 69, 67

2. The ages in years of players on the Oakwood Acorns Basketball Team: 22, 23, 28, 33, 24, 27, 24, 26, 23, 26, 25, 29, 21, 30

3. The weights in pounds of linemen on the Ferndale High School football team: 159, 167, 171, 162, 155, 183, 168, 153, 164, 148

4. The number of students attending the high schools in Wexford County: 539, 495, 517, 525, 400, 560, 478

5. Margot's weekly weight losses in pounds following the Doctors Diet: 2.1, 1.8, 1.1, 1.5, 1.4, 1.4, 1.3, 1.5, 1.2, 1.7, 1.2

6. Lengths in centimeters of mature Canada geese found in the Meadville Town Park: 35.2, 42.6, 41.0, 37.2, 34.5, 36.8, 41.0, 37.9, 42.1, 41.5

7. This table shows the number of children per family.

Children	× Frequency	= Total
0	9	0
1	10	10
2	14	28
3	10	30
4	4	16
5	2	10
6	1	6
Sum	50	100

8. This table shows runs scored this season by a little league team.

Score	× Frequency	= Total
0	2	0
1	8	8
2	4	8
3	5	15
4	2	8
7	2	14
10	1	10
12	1	12
Sum	25	75

9. Change one number in the set {7, 12, 19, 16} so that the median is 12.
10. Insert a number in the set {2, 9, 11, 16} so that the median of the new set is the same as that of the original.
11. Find n so that the mean of the set {6, 8, n, 10, 16} is 13.
12. Find n so that the mean of the set {18, 35, n, 9, 15} is 17.

Tell how the mean, median, and mode are affected by these changes.

13. Each number in a set of numbers is decreased by 8.
14. Each number in a set of numbers is increased by 4.
15. Each number in a set of numbers is halved.
16. Each number in a set of numbers is squared.
17. The mean of 30 bowling scores is 145. If the four lowest and the four highest scores are removed, the mean of the remaining scores is 148. What is the mean of the scores removed?

Applications

Identify the measure of central tendency that is being used.

18. **Geography** Half the continents in the world have an area that is less than South America.
19. **Business** The most popular shirt size at Jay's Stash is large.
20. **Education** The average test score in English for the district is 82.

ALGEBRA IN DEMOGRAPHY

Measures of central tendency are used in forecasting population growths, weather prediction, and analysis of data.

The population of San Antonio, Texas for 1900–1980 is given at the right.

1. How has the population changed since 1900?
2. What was the mean population for the years 1950, 1960, 1970, and 1980?
3. Based on the information in the table, would you expect the population to increase or decrease from 1980–2000?

Year	Population
1900	53,321
1950	408,442
1960	587,718
1970	654,153
1980	785,880

4. Research how the population where you live has changed.
5. What was the average population for your town or city since 1900?

14.2 Statistics: Graphing Data

Objectives: To make a frequency distribution table from data
To make a histogram from a frequency distribution table

To communicate information about data quickly, statisticians frequently display the data in a graph. Graphs provide a visual interpretation of the data. One type of graph commonly used is called a *histogram*. A **histogram** is a bar graph of a frequency distribution.

Capsule Review

Recall that in a set of data the mode is the number occurring most frequently, the median is the middle number, and the mean is the average of the numbers.

For this set of data find:
70, 70, 71, 71, 72, 72, 72, 72, 73, 73, 73, 74, 74, 75, 75, 76, 77, 78.

1. the mode.   **2.** the median.   **3.** the mean.

For certain data, in order to construct a histogram, you must first organize the data into a frequency distribution table which shows the data grouped into intervals.

EXAMPLE 1 The speeds of cars as they passed a checkpoint are shown. Make a frequency distribution table for the data. Locate the median from the frequency distribution table.

```
31  46  37  45  30  51  41  38
44  47  26  32  38  39  44  25
37  46  51  34  35  33  28  37
38  41  33  48  36  31  36  39
43  46  33  42  40  34  34  41
```

- Order the speeds in convenient intervals.
- Tally the data.
- Count the tallies and write the frequencies.
- Total the frequencies.

Intervals (mi/h)	Tally	Frequency
25–29	III	3
30–34	JHT JHT	10
35–39	JHT JHT I	11
40–44	JHT III	8
45–49	JHT I	6
50–54	II	2
	Total	40

The median speed is the mean of the 20th and 21st values, which are in the 35–39 mi/h speed interval.

586   Chapter 14 Statistics and Probability

EXAMPLE 2   Make a histogram for the data in the frequency table of Example 1.

- List frequencies on the vertical axis.
- List speeds, in intervals, on the horizontal axis.
- Draw bars with no space between. The height of the bar is determined by the frequency.
- Note that, on this graph, only the lower values of each interval are labeled on the axis. For example, for 50 to 54 only 50 is written.

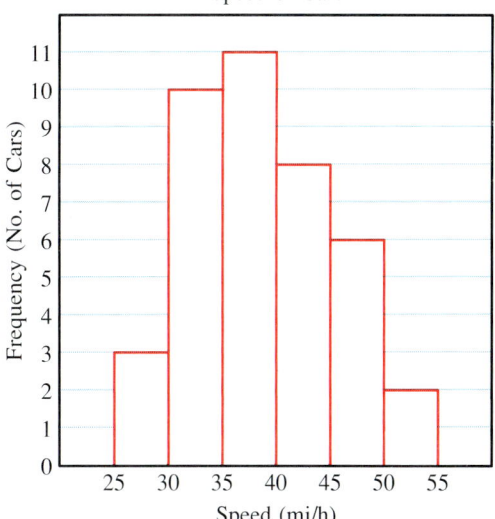

a. **Which interval represents the speed at which most cars were driven?**

The tallest bar shows the mode of the interval. Most cars were driven from 35 to 39 mi/h.

b. **What percent of the cars were driven at speeds less than 35 mi/h?**

The graph shows 13 cars at less than 35 mi/h. $\frac{13}{40} = 0.325$ or 32.5%

Graphs such as histograms are often preferred to tables of numbers because they are more interesting and easier to interpret. Can you tell the precise mean, mode, or median from the graph? Why?

CLASS EXERCISES

The Algebra I midterm test scores for 30 students at the Martin Luther King High School are listed at the right.

71	85	70	75	76
83	61	69	80	70
88	67	72	73	90
99	72	64	82	76
100	75	68	96	84
94	81	87	74	79

1. Make a frequency distribution table that shows the scores. Use intervals 60–64, 65–69, ...

2. Make a histogram from your frequency distribution table.

3. About what percent of the students earned a 70 or more?

At an airport parking lot the following number of cars were parked overnight during a 25 day period (numbers are in hundreds).

1.5	1.0	1.2	1.5	1.7
2.3	1.8	1.9	1.3	1.5
1.6	2.2	1.8	1.9	1.9
1.7	1.4	1.3	2.0	2.2
2.5	2.0	1.9	1.4	1.4

4. Make a histogram from the data. Use intervals 1.0–1.4, 1.5–1.9, ...

5. From the histogram state the mode and median, then calculate the mean using all the data.

For Discussion

6. What other graphical displays can be used to present statistical data? Use examples from magazines or newspapers.

PRACTICE EXERCISES

The 30 students whose midterm scores are listed in the Class Exercises took a year-end test. Their scores are at the right.

```
76  88  82  92  68
75  68  58  62  55
77  52  70  74  66
60  67  48  65  77
79  69  70  57  63
92  84  77  88  70
```

1. Make a frequency distribution table that shows the scores in 5-point intervals, use 45–49, 50–54, ….

2. Find the interval that contains the median.

3. Use your calculator to find the mean score.

4. Make a histogram of the year-end scores.

5. From the histogram in Exercise 4, how many students earned at least a 70?

The ABC Manufacturing Company employs 35 people. The age of each employee is given at the right.

```
18  20  17  20  24
28  25  19  20  26
34  38  40  45  44
36  42  50  52  54
38  40  42  48  62
51  49  34  37  41
63  64  60  35  45
```

6. Make a frequency distribution table that shows the ages in 10-year intervals of 10–19, 20–29, ….

7. Find the interval that contains the median.

8. Use your calculator to find the mean age.

9. Make a histogram of the employees' ages.

10. What percent of the employees are between 20 and 39 years old, inclusive?

In January 98 students worked on a class project to earn money for the class trip. This histogram displays the results.

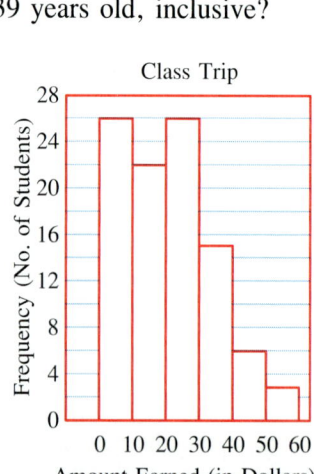

11. What percent of those in the entire class earned less than $30 for the trip?

12. In which interval is the median earnings?

13. What percentage of students earned above the median interval? below the median interval?

14. Give the interval(s) that contains the mode(s) for the data.

For exercises 15 and 16, 25 students recorded the number of hours a television news program was viewed in their homes.

6	10	1	4	5
0	7	9	3	11
7	4	2	6	8
4	2	5	3	9
5	8	12	15	6

15. Make a frequency distribution and find the mode, median, and mean.

16. Make a histogram to show the data.

17. Construct a histogram from which the mean, mode, and median are all in the same interval.

18. Construct a histogram with a mode, median, and mean in different intervals. Explain your work.

Applications

19. Geography The 13 greatest known depths in the seas and oceans of the world, to the nearest 100 ft are:

12,000; 12,300; 12,400; 12,400; 15,200; 15,700; 16,500; 18,500; 22,800; 22,800; 24,500; 30,200; 35,800

Show the data in a frequency distribution table. Then construct a histogram from the table.

20. Demography It has been projected that in the year 2100 the population, in millions, of 20 countries will be:

632, 571, 509, 376, 356, 316, 309, 297, 293, 196, 173, 168, 164, 139, 128, 125, 120, 116, 112, 111

Show the data in a frequency distribution table. Then construct a histogram from the table.

EXTRA

The mayor of a town with a population of 10,000 people wants to know how many of the people will support a bill she plans to propose at a council meeting next week. Since it is impossible to ask each person, she plans to conduct a survey. A reasonable estimate for support of the bill can be obtained if the people are selected in a random manner.

The mayor's staff conducts a telephone survey of 300 people chosen at random from the town's telephone directory. An analysis of the responses reveals that 228 of the 300 people support the bill. The mayor concludes that it is reasonable to expect 7600 people to support her bill. Is she right? Can you justify the position? Describe at least two methods the mayor's staff might have used to make the selections random.

14.3 Statistics: Measures of Variability

Objective: To compute the range, variance, and standard deviation of a set of data

The **range** of a set of data is the difference between the greatest number and the least number in the set. The range gives you a measure of the spread, or *variability,* of the numbers. Since the range is computed from only two values, it is not always representative of the entire set of data.

To get a better picture of variability, you can use other statistical measures. These measures involve the use of square roots.

Capsule Review

Recall the Square-Root Property, if $x^2 = k$, then $x = \pm\sqrt{k}$, for any real number k, $k \geq 0$.

EXAMPLES $\sqrt{36} = 6$ $\sqrt{4^2} = 4$ $\sqrt{(-3)^2} = 3$ $\sqrt{0.49} = 0.7$

Find the principal (positive) square root.

1. 100   2. $\frac{1}{64}$ 3. 1.21   4. $(-144)^2$ 5. $\left(\frac{-9}{16}\right)^2$

The frequency distributions for two sets of data are shown below. They have the same range and mean, but the data are distributed differently.

Weekly Wages for Department A

Wages	Frequency	Freq. × Wages
$100	4	$ 400
200	0	0
300	1	300
400	1	400
500	0	0
600	4	2400
Total	10	$3500

Range: $600 − $100 = $500
Mean: $3500 ÷ 10 = $350

Weekly Wages for Department B

Wages	Frequency	Freq. × Wages
$100	1	$ 100
200	0	0
300	4	1200
400	4	1600
500	0	0
600	1	600
Total	10	$3500

Range: $600 − $100 = $500
Mean: $3500 ÷ 10 = $350

How does the distribution of the two sets of data differ? Which department has a greater extreme of wages? In which department do more people receive wages close to the mean?

One way to describe how much a set of data varies from the mean is by its *standard deviation* (symbol σ, pronounced "sigma"). The **standard deviation** measures how much each value in the data differs from the mean of the data. A larger standard deviation indicates that most of the values in the data do not cluster around the mean. A smaller standard deviation indicates that most of the values in the data do cluster around the mean.

To find a standard deviation:

- Compute the deviation (difference) of each number from the mean of the set.
- Compute the square of each deviation.
- Multiply each squared deviation by the corresponding frequency.
- Compute the sum of the squared deviations.
- Compute the mean of the sum; this is called the **variance**.
- Find the principal square root of the variance. The result is the standard deviation.

EXAMPLE 1 Find the standard deviation from the mean ($350) for the wages earned in Department A and in Department B, as listed in the frequency distribution tables on page 590.

Department A

Income	Deviation from Mean	Freq. × (Dev.)²
$100	−250	4(62,500)
300	−50	1(2500)
400	50	1(2500)
600	250	4(62,500)
	Sum of squares	505,000

$$\sigma = \sqrt{\frac{\text{sum of squares}}{\text{frequency}}}$$

$$\sigma = \sqrt{\frac{505,000}{10}}$$

$\sigma = \sqrt{50,500}$, or $225 to nearest dollar

Variance = $50,500
Standard deviation = $225

Department B

Income	Deviation from Mean	Freq. × (Dev.)²
$100	−250	1(62,500)
300	−50	4(2500)
400	50	4(2500)
600	250	1(62,500)
	Sum of squares	145,000

$$\sigma = \sqrt{\frac{\text{sum of squares}}{\text{frequency}}}$$

$$\sigma = \sqrt{\frac{145,000}{10}}$$

$\sigma = \sqrt{14,500}$, or $120 to nearest dollar

Variance = $14,500
Standard deviation = $120

In Example 1, the smaller standard deviation for Department B shows that more incomes in Department B, than in Department A, cluster around the mean. Note that the range was the same for both departments ($500). The range is computed by using only the greatest and least numbers, whereas the standard deviation is computed by using all of the data. This makes the standard deviation a more precise and informative measure of variability than the range.

EXAMPLE 2 In the 10 Major League All-Star baseball games, the winning teams scored the following numbers of runs: 3, 3, 4, 4, 5, 6, 7, 7, 7, 13. Find the measures of central tendency: mode, median, and mean. Find the measures of variability: range, variance, and standard deviation.

Mode: 7 runs per game

Median: $\dfrac{5 + 6}{2} = 5.5$ runs per game

Mean: $\dfrac{3 + 3 + 4 + 4 + 5 + 6 + 7 + 7 + 7 + 13}{10} = \dfrac{59}{10}$, or 5.9 runs/game

Range: $13 - 3 = 10$

You can use a computer program to generate the table of computations for the variance and the standard deviation from the mean of 5.9. The statistics mode on a calculator will also provide a useful means for computing the measures of variability.

This table shows computations for the variance and the standard deviation from the mean of 5.9.

Variance: $\dfrac{78.90}{10} = 7.89$

Standard Deviation:
$\sigma = \sqrt{7.89} = 2.81$, to the nearest hundredth

Runs	Deviation from Mean	Freq. × (Dev.)²
3	3 − 5.9 = −2.9	2(8.41)
4	4 − 5.9 = −1.9	2(3.61)
5	5 − 5.9 = −0.9	1(0.81)
6	6 − 5.9 = 0.1	1(0.01)
7	7 − 5.9 = 1.1	3(1.21)
13	13 − 5.9 = 7.1	1(50.41)
	Sum of squares	78.90

CLASS EXERCISES

The 10 families in an apartment building have these numbers of pets: 1, 2, 3, 0, 1, 0, 1, 6, 2, 4.

1. Find the measures of central tendency.   **2.** Find the measures of variability.

PRACTICE EXERCISES

To the nearest dollar, find the variance and the standard deviation from the mean for the wages in each department.

1. Department C

Wages	100	150	200	250
Frequency	5	2	1	2

2. Department D

Wages	100	150	200	250
Frequency	4	5	0	1

For each set of data, find the measures of central tendency: mode, median, and mean. Find the measures of variability: range, variance, and standard deviation.

3. The noontime temperature readings in degrees Celsius for 7 days:
   9, 12, 8, 10, 5, 8, 11.

4. The hourly wages of Company XYZ:
   $7, $10, $10, $11, $12, $12, $14, $20.

5. The noontime barometer readings for 7 days:
   30.1, 29.0, 29.4, 29.6, 29.9, 29.0, 29.4.

6. The population of a city during the past 10 years:
   23,000; 26,000; 32,000; 45,000; 52,000; 50,000; 47,000; 46,000; 42,000; 37,000.

7. The earnings per share of a stock during a 10-month period:
   $4.29, $4.78, $5.06, $6.45, $6.32, $5.12, $3.25, $2.86, $2.08, $1.76.

Applications

8. **Health** The amount of protein in grams in 3-oz servings of several types of fish and seafood are: bluefish, 22; clams, 11; crabmeat, 24; salmon, 17; sardines, 20; shrimp, 17; tuna, 24. What is the mean and the standard deviation for this set of data?

9. **Business** From 1981 through 1985, XYZ Communications earned $21,000, $24,000, $30,000, $25,000, and $30,000. Find the mean and standard deviation.

TEST YOURSELF

1. The mean of 24 students' test scores is 74. If the two lowest scores of 60 and 62 along with the two highest scores of 97 and 92 were removed, find the mean of the remaining scores. **14.1**

The ages of the employees at the HELP Manufacturing Company are: 18, 25, 35, 28, 20, 21, 41, 38, 30, 25, 19, 24, 19, 22, 28, 33, 37, 38, 22, 21, 25, 19, 29, 40.

2. Find the mode, the median, and the mean for the data.

3. Draw a histogram of the data using five-year intervals. What percentage of the employees are under 25? **14.2**

4. Find the range, the variance, and the standard deviation for the data. **14.3**

5. The number of cars passing through a busy intersection each hour over a 24 h period are given as follows:

80	60	60	20	10	10	40	70
140	100	70	120	130	80	70	40
50	85	100	95	70	80	75	70

   Find the mean and standard deviation for the data.

APPLICATION: Scattergrams

A scattergram is a dot or point graph of data. A scattergram can show how two sets of data vary relative one to the other.

The table shows the population of New York State in millions. The scattergram plots the population count against the year.

Year	1790	1820	1850	1880	1910	1940
No.	0.3	1.4	3.1	5.1	9.1	13.5

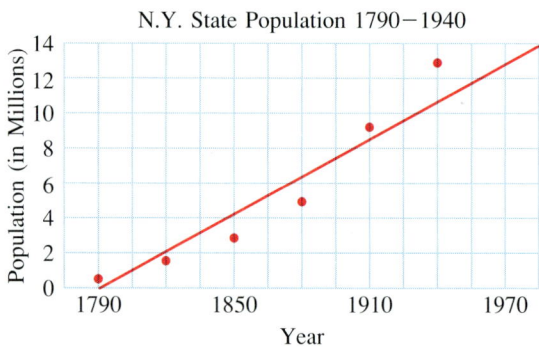

Statisticians try to draw a *line of best fit*, a line having the same number of points above and below it.

Here as the years increase, the population also increases. The upward slant or positive slope of the line of best fit indicates a *positive correlation*. A positive correlation might indicate that one variable will have a positive influence on the other variable.

The scattergram on the right shows the relationship between years and death rates. As the years have passed, the number of deaths per thousand has decreased in the United States. The downward slant or negative slope indicates a *negative correlation*.

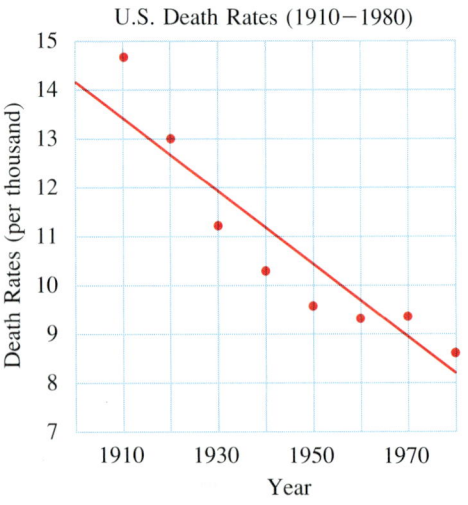

In some cases, no line of best fit can be drawn. In such cases there may be no correlation between the two sets of data.

1. From the scattergram below, what can you conclude about the data? Find the slope of the line of best fit.

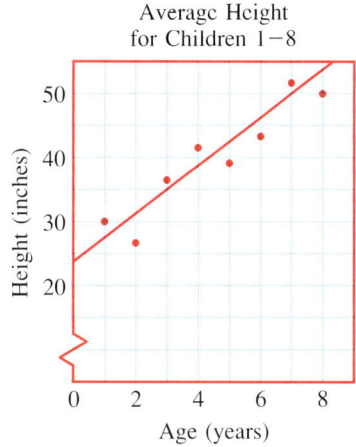

2. Draw a scattergram of the data for workers and number of hours to complete a job. Can a line of best fit be drawn? What is the correlation between the two sets of data?

Workers	0–1	1–2	2–3	3–4	4–5	5–6	6–7	7–8	8–9	9–10
Hours	860	345	262	168	105	88	68	56	42	6

3. From the following frequency distribution table, draw a scattergram of the data. Can a line of best fit be drawn? Is there a positive or negative correlation between the two sets of data?

Height	62"	64"	66"	68"	70"	72"	74"	76"
Weight	123	130	136	145	154	162	171	181

4. Draw a scattergram from the following table. What type of correlation can be found between the two sets of data? Can the number of immigrants from Spain to the United States be estimated for 1981–90? Why or why not?

Years	1931–40	1941–50	1951–60	1961–70	1971–80
Immigration	3,258	2,898	7,894	44,659	34,141

5. Refer to the scattergram on page 594 which showed the relationship between years and death rates. Make a frequency distribution table representing the two sets of data.

6. Work with a partner and design your own statistical experiment. Use the techniques you have learned to describe the data, and explain how you and your partner can use data to predict trends.

Application: Scattergrams

14.4 Simple Probability

Objectives: To determine the probability of an event and of the complementary event for a random experiment
To solve problems involving probability

The Italian mathematician Girolamo Cardano (1501–1576) helped to develop the field of *probability*. **Probability** measures the likelihood that a particular event will occur. The measures are expressed as ratios.

Capsule Review

EXAMPLE Write the ratio 4 out of 10 as a fraction in simplest form and as a percent.

$$\frac{4}{10} = \frac{2}{5}$$ *Write as a fraction. Simplify.*

$$= 0.40 \text{ or } 40\%.$$ *Divide to find the percent.*

Write each ratio as a fraction in simplest form and as a percent.

1. 15 to 25
2. 18 out of 48
3. 28 to 49
4. 48 to 64

In probability, each repetition of an experiment is a **trial.** A possible result of each trial is called an **outcome.** When you toss a fair coin, the two possible outcomes are *heads (H)* or *tails (T)*. The probability of tossing heads is 1 out of 2 possible outcomes.

$$\text{Probability (heads)} = \frac{1}{2} \text{ or } P(H) = \frac{1}{2}$$

$$\text{Probability (of an event)} = \frac{\text{number of favorable outcomes}}{\text{total number of possible outcomes}}$$

When the possible outcomes have the same chance of occurring, they are described as **equally likely;** they occur at *random*. The set of all possible outcomes is called the **sample space.** The sample space for tossing a coin is heads or tails. An **event** is any of the possible outcomes, including all or none.

596 Chapter 14 Statistics and Probability

EXAMPLE 1 A box contains 12 buttons, identical in size and shape but not in color. There are 2 blue, 3 yellow, 6 black, and 1 white. Find the probability of selecting:
 a. a blue button b. a blue or yellow button
 c. a green button d. a blue, yellow, black, or white button

a. $P(\text{blue}) = \dfrac{2}{12}$ ← number of blue buttons
                   ← total number of buttons

  $= \dfrac{1}{6}$ *Simplify.*

b. $P(\text{blue or yellow}) = \dfrac{5}{12}$ ← *There are 2 blues and 3 yellows*

c. $P(\text{green}) = 0$. There are no green buttons.

d. $P(\text{blue, yellow, black, or white}) = \dfrac{12}{12} = 1$

The probability of an event that is impossible is 0. The probability of an event that is certain to happen is 1. All other probabilities are between 0 and 1. In general, for any probability $P(E)$,

$$0 \leq P(E) \leq 1$$

In a random experiment, the two situations—that an event does occur and that the event does not occur—are **complementary events**.

In Example 1, $P(\text{blue})$ and $P(\text{not blue})$ are complementary events.

$P(\text{blue}) = \dfrac{1}{6}$ $P(\text{not blue}) = \dfrac{5}{6}$ Note: $\dfrac{1}{6} + \dfrac{5}{6} = 1$

The sum of the probability of an event $P(E)$ and the probability of its complement written as $P(\overline{E})$, is 1. $P(E) + P(\overline{E}) = 1$

Odds are a ratio that compares the probability of an event to the probability of its complement. *Odds of 2 to 1 for* means that the probability of the event occurring is $\dfrac{2}{3}$. *Odds of 2 to 1 against* means that the probability of the event not occurring is $\dfrac{2}{3}$.

EXAMPLE 2 The table below shows the SAT math scores for 186 seniors at Garrison High School. If one senior is chosen at random, find

a. the probability that the student had a score between 501 and 600

b. the odds that the student scored between 301 and 500

Score	201–300	301–400	401–500	501–600	601–700	701–800
Students	2	15	35	62	42	30

14.4 Simple Probability   **597**

a. $P(\text{scored } 501\text{–}600) = \dfrac{62}{186}$ ← Number of students scoring 501–600
← Total number of students

$= \dfrac{1}{3}$ *Simplify.*

$= 33\dfrac{1}{3}\%$ *Write as a percent.*

There is a probability of $\dfrac{1}{3}$ or a $33\dfrac{1}{3}\%$ chance that a senior chosen at random scored between 501–600.

b. Odds of scoring 301–500 $= \dfrac{\frac{50}{186}}{\frac{136}{186}}$ ← probability of scoring 301–500
← probability of not scoring 301–500

$= \dfrac{50}{136} = \dfrac{25}{68}$

The odds in favor of picking a senior who scored between 301–500 are 25 to 68.

CLASS EXERCISES

The numbers 1–5 are written on five red and five yellow tags (one number on each tag). The 10 tags are placed in a bag and are thoroughly mixed. One tag is picked at random. Find each probability.

1. $P(\text{red})$
2. $P(5)$
3. $P(\text{yellow } 4)$
4. $P(\text{red or yellow})$
5. $P(1 \text{ or } 2)$
6. $P(\text{even number})$
7. $P(8)$
8. $P(3)$
9. $P(\text{not a red } 3)$

For Discussion

10. Does $P(\overline{E}) = 1 - P(E)$? Explain.
11. Are the Odds of E equal to $\dfrac{P(E)}{1 - P(E)}$? Explain.

PRACTICE EXERCISES

The numbers 1–8 are written on eight red, eight white, and eight blue tags (one number on each tag). The tags are placed in a bag and are thoroughly mixed. One tag is picked at random. Find each probability.

1. $P(\text{blue})$
2. $P(\text{white})$
3. $P(6)$
4. $P(2)$
5. $P(\text{red or blue})$
6. $P(\text{blue or white})$
7. $P(3 \text{ or } 4)$
8. $P(9 \text{ or } 0)$
9. $P(\text{greater than } 6)$
10. $P(\text{less than } 6)$
11. $P(\overline{7})$
12. $P(\overline{5})$

Use the table for Exercises 13–18.

13. P(rainy day in July)
14. P(rainy day in May)
15. Odds for rainy day in April

No. of Days of Rain Fall

Apr.	May	June	July	Aug.	Sep.
14	10	9	5	7	9

16. Odds against rainy day in September
17. P(clear day in August)
18. Odds against rainy day in June

A letter is chosen at random from the word *favorable*. Find the probability of each event.

19. P(consonant)
20. P(vowel)
21. P(the letter **a**)
22. P(the letter **v**)
23. P(the letters **a** or **e**)
24. P(the letters **b** or **l**)
25. P(the letter **y**)
26. P(the letter **z**)

Use the frequency table. Find each probability.

Color	Frequency
red	12
green	8
blue	4

27. P(red)
28. P(red or green)
29. P(green)
30. P(red or blue)

Applications

31. **Testing** A question on a multiple-choice test has four possible answers. What are the odds against guessing the correct answer?

32. **Finance** Rufus has four $1 bills, two $5 bills, and one $10 bill in his wallet. He takes out a bill without looking. What is the probability that it will be a $5 bill? a $10 bill?

MATH CLUB ACTIVITY

This type of question was asked on the Annual High School Mathematics Examination. See if you can solve it.

There are two cards with the same size and shape. One is red on both sides and the other is red on one side and blue on the other. The cards have the same probability of being selected. In other words, the probability is $\frac{1}{2}$. One of the cards is selected and placed on a table. The side of the card on the table that you can see is red. What is the probability that the other side of the card is also red?

14.5 Problem Solving Strategy: Draw a Diagram

A **Venn diagram** is a pictorial representation of sets. It is used for counting purposes. In this counting technique, each person or object is counted once.

EXAMPLE A survey was taken of the types of magazines people read. 90 people read news magazines; 70 people read sports magazines; 40 people read business magazines; 30 people read news and sports magazines; 20 people read news and business magazines; 10 people read sports and business magazines; and 10 people read all three. If one of the people surveyed is selected at random, what is the probability the person reads only a news magazine?

Understand the Problem

What are the given facts?
Of the 90 people who read a news magazine, 30 people read a news magazine and a sports magazine; 20 people read a news magazine and a business magazine; and 10 people read all three.

What are you asked to find?
The probability that a person selected at random from the survey reads *only* a news magazine.

Plan Your Approach

Choose a strategy.
Draw a Venn diagram. Let each circle represent people who read news, sports, and business magazines.

Complete the Work

Draw a diagram. Work backwards.
Place the 10 people who read all three magazines in the intersection of the three circles. Since 10 people who read news and business magazines have already been placed in the intersection of the three sets, write 10 in the intersection of news and business only. Explain how the other numbers were placed.

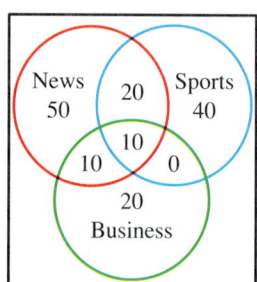

The diagram shows that 50 people of the 150 surveyed read only news magazines.

$$P(\text{only news}) = \frac{50}{150} \text{ or } \frac{1}{3}$$

600   Chapter 14 Statistics and Probability

> **Interpret the Results**

State your answer.
The probability that a person selected at random from the survey reads *only* a news magazine is $\frac{1}{3}$ or $33\frac{1}{3}\%$.

Check your answer.
Find the probability of the complement.

$P(\text{not only news}) = \frac{100}{150}$ ← $40 + 20 + 20 + 10 + 10$

$= \frac{2}{3}$

Note: $\frac{1}{3} + \frac{2}{3} = \frac{3}{3}$ or 1 $P(E) + P(\overline{E}) = 1$

Since the sum is 1, the probability of $\frac{1}{3}$ is correct.

CLASS EXERCISES

Use a Venn diagram to solve.

1. The following number of students studied Spanish, French, and/or German in high school: 20 studied Spanish; 15 studied French; 10 studied German; 8 studied Spanish and French; 6 studied Spanish and German; 4 studied French and German; 1 studied all three. If a student is selected at random, what is the probability the student studied only Spanish and French.

2. In a recent poll, several parents were asked what sport(s) they preferred to watch: a total of 8 preferred football, 7 liked basketball, 3 chose golf, 7 preferred football and basketball, and 2 liked to watch all of them. If one of the parents is chosen at random, what is the probability the parent preferred only golf?

PRACTICE EXERCISES

Use a Venn diagram to solve.

1. Students were asked what kind of pet they had:
   28 had a dog.
   15 had a cat.
   8 had a dog and a cat.
   What is the probability that a randomly selected student will have only a cat?

2. Students were asked whether they drank milk or juice with lunch:
   50 drank milk.
   30 drank juice.
   20 drank milk and juice.
   What is the probability that a randomly selected student will only drink milk?

3. Here are the results of a luncheon survey:
   83 people had soup.
   72 people had chicken.
   54 people had salad.
   27 people had soup and chicken.
   23 people had soup and salad.
   12 people had chicken and salad.
   10 people had all three.
   A person at the luncheon is selected at random. What is the probability the person had soup, chicken, and salad?

4. Here is a list of track meet participations:
   42 students ran.
   28 students jumped.
   8 students pole vaulted.
   12 students ran and jumped.
   2 students ran and pole vaulted.
   6 students jumped and pole vaulted.
   2 students did all three.
   A member of the team is selected at random. What is the probability the member only jumped and pole vaulted?

5. The following number of campers chose these activities: 32 chose swimming; 32 chose boating; 20 chose hiking; 15 chose swimming and boating; 12 chose swimming and hiking; 11 chose boating and hiking; and 8 chose all three. What is the probability that a camper selected at random chose swimming and boating or boating and hiking?

Mixed Problem Solving Review

1. The length of a room is 3 ft longer than its width. What are the dimensions of the room if the area is 108 ft^2?

2. Al's father is 3 times as old as Al. Five years ago he was 4 times as old as Al. How old is each now?

3. A 20-kg object is 15 ft from the fulcrum, and balances a 30 kg object. How far is the 30 kg object from the fulcrum?

4. George can paint a room in 12 h. Joe can paint a room in 15 h. How long will it take them if they work together?

5. Erin has 45 coins worth $9.65. There are some nickels, the rest are quarters. Find how many of each kind of coin Erin has.

6. It takes 4 h for a boat to travel 64 mi downstream. The same boat can travel 35 mi upstream in 5 h. Find the rate of the boat in still water.

PROJECT

Conduct a survey. Choose any three sports. Ask students in your school to name the sport(s) they would participate in from your list. Draw a Venn diagram of the results and calculate the various probabilities.

14.6 Probability: Compound Events

Objective: To solve problems involving independent, dependent, mutually exclusive, and inclusive events

A **compound event** is made up of two (or more) events. Probability in compound events is influenced by the relationship between the separate events. A *tree diagram* can be used to show these relationships.

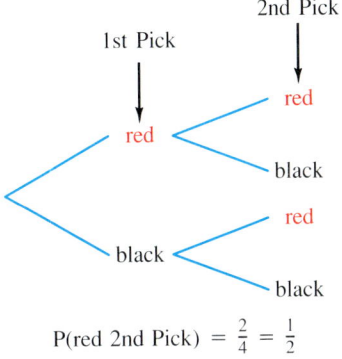

Suppose there are two checkers, red and black, in a box. Luanne picks one, looks at it, and puts it back. In this case does the probability of picking a red checker on the second pick depend upon whether she picked a red or a black checker the first time? Explain.

$P(\text{red 2nd Pick}) = \frac{2}{4} = \frac{1}{2}$

What happens if Luanne does not put that checker back after the first pick? Does the probability of picking a red checker the second time depend upon whether she picked a red or a black checker the first time? Explain.

Capsule Review

In a box of colored pencils there are 4 red (R), 3 yellow (Y), 2 green (G), and 1 blue (B). Pick 1 pencil at random.

EXAMPLE $P(B) = \dfrac{\text{number of favorable outcomes}}{\text{total number of possible outcomes}} = \dfrac{1}{10}$

Find each probability.

1. $P(R)$
2. $P(G)$
3. $P(Y)$
4. $P(\overline{Y})$
5. $P(B \text{ or } G)$
6. $P(R \text{ or } G)$
7. $P(\text{a pencil})$
8. $P(\text{a pen})$

Independent events are events that do not influence one another. That is, each event occurs without changing the probability of the other event.

EXAMPLE 1 Jason and Carl each own four pairs of jeans, one blue, white, tan, and gray pair. Find the probability that both will wear gray jeans today. Find $P(J \text{ and } C)$, where J and C are independent events.

14.6 Probability: Compound Events **603**

List the sample space to show all the possible combinations. Each boy's jeans are represented by blue, white, tan, and gray. The ordered pairs will be in the form Carl's choice, Jason's choice.

In the diagram, each dot stands for a possible pairing. The dots circled in black stand for pairings in which the boys choose the same color: (B, B), (W, W), (T, T), and (G, G). The dot circled in red, (G, G), stands for the pairing in which each boy chooses gray.

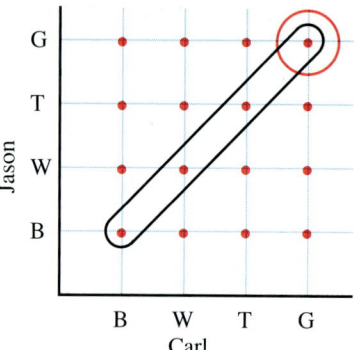

Of the 16 possible pairings, only one, (G, G), means that both Jason and Carl wear gray jeans on the same day. The probability of both selecting gray is $\frac{1}{16}$.

The probability for independent events can be considered another way:
Let $P(J)$ = the probability of Jason's wearing gray jeans today.
Let $P(C)$ = the probability of Carl's wearing gray jeans today.
$P(J) = \frac{1}{4}$, $P(C) = \frac{1}{4}$, $P(J \text{ and } C) = \frac{1}{4} \cdot \frac{1}{4} = \frac{1}{16}$

This suggests the following generalization.

For two independent events A and B, $P(A \text{ and } B) = P(A) \cdot P(B)$.

Dependent events are events that influence each other. If one of the events occurs, it changes the probability of the other event.

EXAMPLE 2 Beth has 2 red socks and 4 white socks in her sock drawer. Find the probability that she will randomly select one red sock and then, without replacing it, randomly select another red sock. Find $P(\text{red and red})$ for dependent events.

First pick (6 socks) | Second pick (5 socks)
$P(\text{red}) = \frac{2}{6} = \frac{1}{3}$ | $P(\text{red}) = \frac{1}{5}$, why?

$P(\text{1st red and 2nd red}) = \frac{1}{3} \cdot \frac{1}{5} = \frac{1}{15}$

For two dependent events A and B, where B is influenced by A,
$$P(A \text{ and } B) = P(A) \cdot P(B, \text{ given } A)$$

Chapter 14 Statistics and Probability

Mutually exclusive events are events that cannot happen at the same time.

EXAMPLE 3 Fred has 2 green, 4 blue, and 3 red shirts in a drawer. He chooses a shirt at random and does not want a red shirt. Find the probability that Fred will pick a green or blue shirt. Find P(g or b), for mutually exclusive events.

$$P(\text{green or blue}) = \frac{6}{9} = \frac{2}{3}$$

The probability can be calculated another way.
$P(\text{green}) = \frac{2}{9}$, $P(\text{blue}) = \frac{4}{9}$; $P(\text{green or blue}) = \frac{2}{9} + \frac{4}{9}$ or $\frac{2}{3}$.

> For two mutually exclusive events A and B: $P(\text{A or B}) = P(\text{A}) + P(\text{B})$

Events which can occur at the same time are called **inclusive events**.

EXAMPLE 4 Of the 20 members of the bicycle club, 7 were in Race A, 8 were in Race B, and 3 were in both races. Find the probability of selecting at random a club member who was in Race A or in Race B. Find $P(\text{A or B})$ for mutually inclusive events.

Draw a diagram to find how many members were racing.

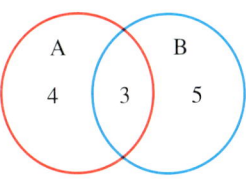

Begin with the 3 members who were in both races.
If 7 were in Race A, 7 − 3 or 4 were in Race A only.
If 8 were in Race B, 8 − 3 or 5 were in Race B only.
Count 4 + 3 + 5 = 12.

12 out of 20 members were in Race A or in Race B. $P(\text{A or B}) = \frac{12}{20}$, or $\frac{3}{5}$.

It is not necessary to find out how many individual members were racing. The probability can be calculated in another way.

$$P(A) = \frac{7}{20} \quad P(B) = \frac{8}{20} \quad P(\text{both A and B}) = \frac{3}{20}$$

$$P(\text{A or B}) = \frac{7}{20} + \frac{8}{20} - \frac{3}{20} = \frac{12}{20}, \text{ or } \frac{3}{5}$$

> For two inclusive events A and B:
> $P(\text{A or B}) = P(\text{A}) + P(\text{B}) - P(\text{A and B})$

CLASS EXERCISES

A bag contains 5 red, 3 green, and 4 white marbles. One is selected randomly, replaced, and another is selected. Draw a tree diagram and find the probability.

1. P(1st R and 2nd W)
2. P(1st R and 2nd R)
3. P(1st R and 2nd $\overline{R}$)

A bag contains 5 red, 3 green, and 4 white marbles. One is selected randomly and not replaced, then another is selected. Draw a tree diagram and find the probability.

4. P(1st R and 2nd W)
5. P(1st R and 2nd R)
6. P(1st R and 2nd $\overline{R}$)

7. Of the 15 members in the stage band, 5 play saxophone, 7 play trumpet and 2 play both. If a member is chosen at random, what is the probability the member plays saxophone or trumpet?

PRACTICE EXERCISES

A bag contains cards labeled with numbers. There are three 5s, one 4, two 3s, and two 2s. One card is randomly selected, noted, and replaced. Then another card is randomly selected. Find the probability.

1. P(1st 5 and 2nd 3)
2. P(1st 4 and 2nd 2)
3. P(1st 5 and 2nd 5)
4. P(1st 3 and 2nd 3)
5. P(1st even and 2nd odd)
6. P(1st odd and 2nd odd)

One card is randomly selected, noted, and not replaced. Then another card is randomly selected. Draw a tree diagram and find the probability.

7. P(1st 5 and 2nd 3)
8. P(1st 4 and 2nd 2)
9. P(1st 4 and 2nd 4)
10. P(1st 2 and 2nd 2)
11. P(1st even and 2nd even)
12. P(1st odd and 2nd even)

One card is randomly selected; find the probability that the card chosen is:

13. P(4 or 5)
14. P(2 or 3)
15. P(3 or 5)
16. P(2 or 4)
17. P(2 or 5)
18. P(3 or 4)

A bag contains number cards and color cards. There are five 5s, three 3s, four red, and three green cards. One card is randomly selected, noted, and replaced. Then another card is randomly selected. What is the probability for each situation?

19. P(1st 5 and 2nd red)
20. P(1st 3 and 2nd green)
21. P(1st number and 2nd color)
22. P(1st color and 2nd color)
23. P(1st 3 and 2nd 3)
24. P(1st red and 2nd red)

Chapter 14 Statistics and Probability

25. P(1st color and 2nd number) **26.** P(1st number and 2nd number)

27. Of 20 students, 6 read mysteries, 10 read science fiction, and 2 read both. If two students are picked at random, what is the probability they both read only mysteries?

28. Of 28 students eating lunch, 8 eat salads, 12 eat sandwiches, and 5 eat both. If two students are randomly selected, what is the probability that each student chosen ate only a sandwich or only a salad?

29. Of the 30 students in the class, 12 study Biology, 10 study Geometry, and 11 study neither. If two students are chosen at random, what is the probability that they both study Biology and not Geometry?

30. Suppose 30% of couples surveyed purchased both a dishwasher and a microwave oven, and 10% purchased neither. If there were twice as many couples who purchased only a dishwasher as there were couples who purchased only a microwave, what is the probability that a couple selected at random purchased only a dishwasher?

Applications

31. Marketing Five people are chosen for product testing by lot from 15 men and 12 women. What is the probability, as a decimal to the nearest hundredth, that all will be men? that all will be women? that all will be men or all will be women? $\left(\textit{Hint: } \text{Probability that 2 women are selected is } \frac{12}{27} \cdot \frac{11}{26}.\right)$

32. Finance A wallet contains three $1 bills and four $5 bills. Two bills are taken from the wallet at random. What is the probability that $2 or $10 is taken?

TEST YOURSELF

One box contains color tags: 3 red and 5 green. Another box contains number tags: four 2s and six 3s. A tag is selected at random from each box. Find each probability.

14.4, 14.6

1. P(red) **2.** P(2) **3.** P(blue)

4. P(2 or 3) **5.** P(1st green and 2nd 2) **6.** P(1st red and 2nd 3)

Solve.

7. Jan selects a red tag at random from the first box and keeps it. Find the probability of her selecting a second red tag.

8. Of 30 people surveyed, 15 take a train to work, 20 drive a car to work, and 8 take both. If a person is selected at random, find the probability that the person only drives a car to work.

CHAPTER 14 SUMMARY AND REVIEW

Vocabulary

complementary events (597)
compound event (603)
dependent events (604)
equally likely (596)
event (596)
frequency distribution (583)
histogram (586)
inclusive events (605)
independent events (603)
mean (582)
measure of central tendency (582)
median (582)
mode (582)
mutually exclusive events (605)
odds (597)
outcome (596)
probability (596)
range (590)
sample space (596)
standard deviation (591)
statistics (582)
trial (596)
variance (591)
Venn diagram (600)

Finding Measures of Central Tendency A **mode** of a set of data is a number in the set that occurs most frequently. A **median** of a set of data is the middle number in a set, or the mean of the two middle numbers, when the numbers are arranged in order from least to greatest. The **mean** of a set of data consisting of n numbers is the sum of the numbers in the set divided by n. **14.1**

Find the mode, median, and mean.

1. 7, 10, 24, 12, 7, 12
2. −2, 0, 1.5, 0.5, −5.5, 3.5, 3.8, 3.5

A **Histogram** is a bar graph of a frequency distribution that plots data against their frequencies. **14.2**

3. Construct a histogram to picture the distribution of the following 32 test scores. Group the scores in 5-point intervals. 75, 55, 69, 78, 88, 63, 73, 89, 66, 36, 44, 62, 57, 67, 70, 74, 58, 45, 95, 33, 65, 40, 80, 63, 69, 71, 57, 53, 67, 48, 69, 81

Finding Measures of Variability The **range** of a set of data is the difference between the greatest number and the least number in the set. The **variance** is the mean of the sum of the squares of the deviations. The **standard deviation** is the square root of the variance. **14.3**

4. Find the range of the data: 4.0, 5.0, 6.0, 4.5, 4.0, 6.0, 5.5.
5. To the nearest hundredth, find the variance and the standard deviation of the data: 4.0, 5.0, 6.0, 4.5, 4.0, 6.0, 5.5.

Determining Simple Probability and Odds The **probability** $P(E)$ of an   **14.4**
event is the ratio of the number of favorable outcomes to the total number of
possible outcomes, $0 \leq P(E) \leq 1$. The **complement** $P(\overline{E})$ of an event is that
situation in which the event does not occur, $P(E) + P(\overline{E}) = 1$. Odds are a ratio
that compares the probability of an event to the probability of its complement.

Four packs of tags, gold, red, white, and blue, each numbered 1–20, make 80 tags in all. One tag is picked at random. Find the probability.

6. $P(8)$ **7.** $P(\text{red})$ **8.** $P(\text{blue})$ **9.** $P(\text{blue 5})$

10. Odds in favor of rain on a day in June if the average number of rain days in June over a 40-year period is 20.

Determining the Probability of Two Events Independent events   **14.5–14.6**
do not influence each other, but dependent events do. Mutually exclusive
events cannot occur at the same time, whereas inclusive events can.

For independent events A, B: $P(A \text{ and } B) = P(A) \cdot P(B)$

dependent events A, B: $P(A \text{ and } B) = P(A) \cdot P(B, \text{ given } A)$

mutually exclusive events A, B: $P(A \text{ or } B) = P(A) + P(B)$

inclusive events A, B: $P(A \text{ or } B) = P(A) + P(B) - P(A \text{ and } B)$.

A box contains decorative disks: 3 gold, 6 red, 4 white, and 5 blue. For Exercises 11 and 12 a disk is picked at random and replaced, and then another disk is picked at random. For Exercises 13 and 14 only one disk is picked. Find each probability.

11. $P(\text{1st gold and 2nd red})$ **12.** $P(\text{1st gold and 2nd gold})$

13. $P(\text{red or white})$ **14.** $P(\text{white or blue})$

The decorative disks, 3 gold, 6 red, 4 white, and 5 blue are mixed in a bag. A disk is picked at random and not replaced. Another disk is picked at random. Find each probability.

15. $P(\text{1st gold and 2nd red})$ **16.** $P(\text{1st white and 2nd white})$

17. $P(\text{1st blue and 2nd gold})$ **18.** $P(\text{1st gold and 2nd gold})$

Solve.

19. Of the 20 students in class, 8 like science, 5 like math, and 3 like both. If a student from class is randomly selected, what is the probability the student will like science or math?

Summary and Review

CHAPTER 14 TEST

1. Find the mode, median, and mean, to the nearest tenth, for the set of data: 56, 58, 75, 70, 53, 90, 93, 70.

2. Construct a histogram of the following data: The heights (greater than 30 m) of redwood trees in a California Park. Use 10 m intervals.

52	64	79	39	52	41
33	46	78	67	58	44
61	75	37	55	57	61
46	80	86	53	58	69
75	74	81	63	68	64

3. Find the range, the variance, and the standard deviation, to the nearest tenth, for the set of data: 8, 10, 10, 11, 12, 14, 14, 16, 16, 19.

There are 4 sets of model boats. Each set is a different color: red (R), blue (B), yellow (Y), and green (G). There are 5 different models in each set—sailboat (S), paddle boat (P), motorboat (M), tugboat (T), and fireboat (F). Find each probability for models selected at random and put back each time.

4. $P(B)$

5. $P(T)$

6. $P(\overline{Y})$

7. $P(\text{any model})$

8. Pick two: $P(\text{1st G and 2nd G})$

9. Pick two: $P(\text{1st S and 2nd F})$

10. Pick only one: $P(R \text{ or } B)$

11. Pick only one: $P(M \text{ or } T)$

12. Find the odds for rain on any day in April if there is an average of 12 rain days in the month.

13. Jane and Sylvia each have 5 blouses. Each has 2 white ones. If both girls choose a blouse, what is the probability that they both will wear a white blouse today?

14. Of the 60 people surveyed, 20 people travel to work by car, 37 people use public transportation, and 12 people use both. If a person from the survey is selected at random, what is the probability the person uses a car or public transportation?

Challenge

In a group of coaches, there are 8 track coaches, 3 soccer coaches, and 1 football coach. If 3 coaches are selected at random to be on the Central Athletic Committee, what is the probability of selecting at least 1 soccer coach?

PREPARING FOR STANDARDIZED TESTS

Directions: In each item you are to compare a quantity in Column 1 with a quantity in Column 2. Write the letter of the correct answer from these choices.

A. The quantity in Column 1 is greater than the quantity in Column 2.
B. The quantity in Column 2 is greater than the quantity in Column 1.
C. The quantity in Column 1 is equal to the quantity in Column 2.
D. The relationship cannot be determined from the given information.

Notes: Information centered over both columns refers to one or both of the quantities being compared. A symbol that appears in both columns has the same meaning in each column. All variables represent real numbers. Most figures are not drawn to scale.

	Column 1	Column 2
1.	0.035	$\frac{7}{20}$
2.	80% of 45	36

$a = 0, b = -1$

	Column 1	Column 2
3.	$a^2b + ab^2$	$5b + 4a$

$\frac{4}{7} = \frac{x}{21}$

	Column 1	Column 2
4.	x	10
5.	$|5 - 4(3 - 6)|$	17
6.	$\sqrt{16 + 9}$	$4 + 3$

$9x + 1 \geq 2x - 6$

	Column 1	Column 2
7.	x	-2

Use this equation for questions 8–9:

$$2x^2 + 5x - 5 = 0$$

	Column 1	Column 2
8.	Discriminant	50
9.	Sum of the roots.	Product of the roots.

	Column 1	Column 2
10.	$\frac{11}{12} \div \frac{7}{18}$	$2\frac{2}{7}$

This triangle is isosceles.

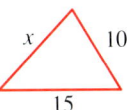

	Column 1	Column 2
11.	x	10
12.	$\frac{2}{3} + \frac{1}{2}$	$\frac{3}{5}$

$f(x) = 12x^2 - 3$

	Column 1	Column 2
13.	$f(-1)$	$f(1)$

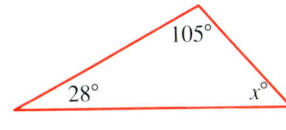

	Column 1	Column 2
14.	x	57

Use this information for questions 15–16:

x and y are measures of complementary angles.

	Column 1	Column 2
15.	$x + y$	180
16.	x	y

CUMULATIVE REVIEW   CHAPTERS 1–14

Replace each ? with >, <, or = to make a true statement.

1. $-\frac{1}{2}$? $-\frac{1}{4}$
2. $\left(-\frac{1}{2}\right)^2$? $\frac{1}{4}$
3. $\sqrt{\frac{1}{4}}$? $\frac{1}{4}$
4. $\frac{2}{3}$? $\frac{16}{24}$
5. $\left|-\frac{2}{3}\right|$? $\sqrt{\frac{4}{9}}$
6. $4\sqrt{5}$? $\sqrt{20}$
7. $-\sqrt{\frac{9}{16}}$? $-\frac{3}{5}$
8. $\frac{2}{\sqrt{3}}$? $\frac{1}{3}\sqrt{6}$
9. $2\sqrt{16} - 3\sqrt{18}$? $\sqrt{2}$
10. $\sqrt{2}(3 + \sqrt{2})$? $3\left(\frac{2}{3} + \sqrt{2}\right)$

Graph.

11. $y = -x^2 + 4x + 3$
12. $2x + y = -3$
13. $y > -\frac{1}{2}x + 1$

Perform the indicated operation. Simplify if possible.

14. $\sqrt{81} - (-\sqrt{36})$
15. $-\sqrt{18} + \sqrt{50}$
16. $3\sqrt{8} \cdot 2\sqrt{3}$
17. $\frac{2\sqrt{3}}{1 + \sqrt{2}}$
18. $\frac{2\sqrt{6}}{6\sqrt{3}}$
19. $\frac{(-2\sqrt{2})^2}{\sqrt{6}\sqrt{2}}$

Solve.

20. The diagonal of a square is $7\sqrt{2}$ cm. Find the length of the side of the square.

21. Al's father is $2\frac{1}{4}$ times as old as Al. Nine years ago he was 3 times as old as Al. How old is each one now?

22. The number of days needed to do a job varies inversely as the number of people working on the job. It takes 12 days for 4 people to complete the job. If the job has to be finished in 8 days, how many people are needed?

23. The monthly normal precipitation (in inches) that falls in Houston, Texas is recorded in the chart below. These normals are based on records for the 30-year period 1951–1980 inclusive.

Jan.	Feb.	March	April	May	June	July	Aug.	Sept.	Oct.	Nov.	Dec.
3.2	3.3	2.7	4.2	4.7	4.0	3.3	3.7	4.9	3.7	3.4	3.7

  a. Find the mean, mode, and median for the data.
  b. Draw a histogram to show the data.
  c. Find the variance and standard deviation of the data.

15 | Right Triangle Relationships

Triangular and rectangular shapes are used in many modern buildings. Architects arrange such shapes so that they are in logical relation to each other and are pleasing to the eye. The perspective of this photograph gives the appearance of a change in the shape of the structure.

15.1 Basic Geometric Figures

Objectives: To identify basic geometric figures, and to classify angles according to their measures
To identify complementary and supplementary angles, and to solve word problems about them

Geometry is a branch of mathematics that deals with sets of points. You have used dots to represent points on a number line as well as points in the coordinate plane. A point, however, is an abstract idea that has no size or shape, merely position. A set of points can represent a line or a geometric figure such as a triangle, a rectangle, or a circle.

Capsule Review

EXAMPLE Graph $-2 \leq x \leq 5$.

Graph each sentence.

1. $-3 \leq x \leq 3$ 2. $2x \leq 8$ 3. $x \geq -3$ 4. $-10 \geq 5x$

A **line** consists of infinitely many points extending without end in both directions. A *line* determined by points A and B is denoted by $\overleftrightarrow{AB}$ or $\overleftrightarrow{BA}$. The arrowheads show that the line extends infinitely in both directions.

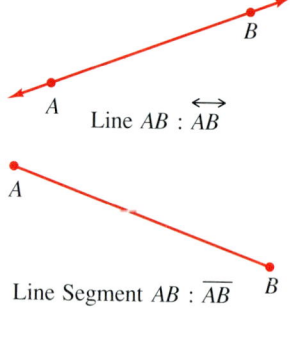

Line AB : $\overleftrightarrow{AB}$

A set of points on a line is a **line segment** if it consists of two points A and B called *endpoints* and all points in between them. Line segment AB is written as $\overline{AB}$ or $\overline{BA}$. The length or measure of a line segment is denoted AB.

Line Segment AB : $\overline{AB}$

The part of $\overleftrightarrow{AB}$ that starts at point A and extends infinitely passing through point B is called **ray** AB and is denoted $\overrightarrow{AB}$. A is called the endpoint of $\overrightarrow{AB}$.

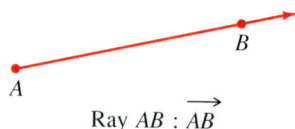

Ray AB : $\overrightarrow{AB}$

EXAMPLE 1 Graph and tell whether the figure is a point, a line, a line segment, or a ray.

a. $-2 \leq x \leq 3$ b. $x \geq 1$ c. $x = 0$

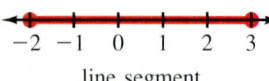
line segment

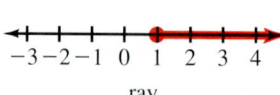
ray

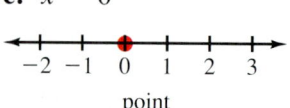
point

614 Chapter 15 Right Triangle Relationships

An **angle** is the union of two non-collinear rays with a common endpoint. The rays are the sides of the angle, and the common endpoint is the **vertex** of the angle. The angle at the right can be denoted as ∠A, ∠CAB, or ∠BAC.

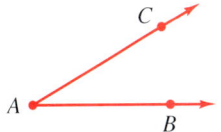

Angles are measured with a **protractor.** Using a protractor, you can see that the measure of ∠CAB is 30°. This is written as $m\angle CAB = 30$.

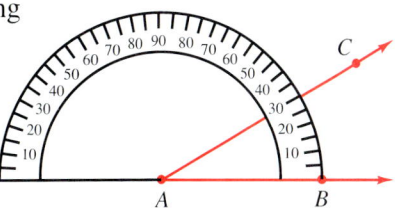

Angles are classified according to their measures.

A **right angle** has a measure of 90°.
An **acute angle** has a measure between 0° and 90°.
An **obtuse angle** has a measure between 90° and 180°.

Some pairs of angles have special names.

Complementary angles are two angles whose measures have a sum of 90°. Each angle is a complement of the other.
Supplementary angles are two angles whose measures have a sum of 180°. Each angle is a supplement of the other.

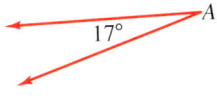
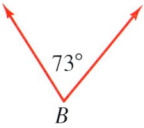
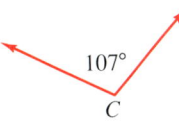

EXAMPLE 2 In the figure above, which pair of angles is complementary? Which pair of angles is supplementary?

$m\angle A + m\angle B = 17 + 73$ $m\angle B + m\angle C = 73 + 107$
$ = 90$ $ = 180$

So ∠A and ∠B are complementary, and ∠B and ∠C are supplementary.

EXAMPLE 3 Angles *M* and *N* are supplementary.
 a. Find $m\angle N$ if $m\angle M = 35$.
 b. Classify each angle as acute or obtuse.

 a. $m\angle M + m\angle N = 180$ *The sum of the measures of supplementary angles is 180.*
 $ 35 + m\angle N = 180$
 $ m\angle N = 145$

 b. ∠M is acute and ∠N is obtuse.

EXAMPLE 4 The measure of an angle is 20 less than the measure of its supplement.
   a. Find the measure of the angle.
   b. Find the measure of the supplement of the angle.
   c. Find the measure of the complement of the angle.

   a. Let x = the measure of the angle.
   Then $180 - x$ = measure of the supplement.
   $x = (180 - x) - 20$
   $2x = 160$
   $x = 80$

   b. The measure of the supplement is $180 - x$.
   $180 - x = 180 - 80 = 100$

   c. The measure of the complement is $90 - x$.
   $90 - x = 90 - 80 = 10$

   So, the angle measures 80. The measure of the complement is 10 and the measure of the supplement is 100.

CLASS EXERCISES

Use the figure to the right for Exercises 1 and 2.

1. Name three different line segments.
2. Name five different angles in the figure.

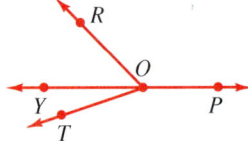

Tell if each angle is acute, right, or obtuse.

3. 38°   4. 100°   5. 90°   6. 145°

Find the complement of each angle.

7. 15°   8. 75°   9. 45°   10. 89°

Find the supplement of each angle.

11. 140°   12. 10°   13. 90°   14. 150°

PRACTICE EXERCISES

Graph each sentence on a number line, and state whether the figure represents a point, a line, a line segment, or a ray.

1. $x \leq 2$ 2. $x \geq 4$ 3. $x = 7$ 4. $x = -3$
5. $4 \leq x \leq 7$ 6. $0 \geq x \geq -5$ 7. $6 > x$ or $x > 3$ 8. $-1 \leq x$ or $x \leq 4$

State whether the given angles are complementary or supplementary.

9. 47°, 43°   10. 12°, 78°   11. 2°, 178°   12. 154°, 26°

In Exercises 13–16, ∠M and ∠N are complementary, ∠A and ∠B are supplementary. Find the measure of the missing angle and classify the angle.

13. $m\angle M = 30$ **14.** $m\angle N = 75$ **15.** $m\angle A = 78$ **16.** $m\angle B = 25$

17. The measure of an angle is 40 more than that of its complement. Find the measure of the angle and its complement.

18. The measure of an angle is 30 less than that of its complement. Find the measure of the angle and its complement.

19. The measure of an angle is 50 less than that of its complement. Find the measure of the angle and its supplement.

20. The measure of an angle is 10 more than that of its supplement. Find the measure of the angle and its supplement.

21. The measure of the complement of an angle is 10 less than three times the measure of the angle. Find the measure of the angle and its supplement.

22. The measure of the supplement of an angle is 45 less than four times the measure of the angle. Find the measure of the angle.

23. Two times the measure of the supplement of an angle is five times the measure of the complement of the angle. Find the measure of the angle.

Applications

24. Construction A builder rests his 8-m ladder against the side of a building. The ladder forms a 28°-angle between itself and the ground. Find the complement and supplement of the angle formed.

25. Travel An airplane heading from California takes off and flies 65° NE. If the plane turns 90° to the right, in what direction will the plane be heading?

GEOMETRY IN ARCHITECTURE

Geometry plays a key role in architecture through the incorporation of different shapes into various designs. Ludwig Mies Van der Rohe and Frank Lloyd Wright were probably two of the most famous architects of the twentieth century. Some basic geometric shapes in architecture include arches, circles, triangles, ellipses, parabolas, and domes. Many famous structures incorporate these shapes, including the White House and the Rose Bowl.

Investigate an unusual architectural design of your choice. Research the history of the structure and prepare a report of your findings for your class.

15.2 Triangles

Objectives: To classify triangles according to the measures of their angles and according to the lengths of their sides
To find the measures of the angles of a triangle and to solve word problems involving the angles of a triangle

The sides of the Transamerica Building in San Francisco are triangular in shape. A **triangle** is the figure formed by three segments joining three noncollinear points.

An architect can design a triangular side of a building by determining two angles that would be formed at the base of the building. Is it possible for the architect to find the measure of the third angle? Solving a problem of this type involves setting up the proper equation.

Capsule Review

To solve an equation in which the same variable occurs more than once:

- First combine like terms on each side.
- Use the addition or subtraction property for equations so that all variables are on one side and all constant terms are on the other.

EXAMPLE Solve: $3s + 7s = 15 - 5s$

$$10s = 15 - 5s$$
$$15s = 15$$
$$s = 1$$

Solve for the unknown.

1. $7b + 4b - 2b - 13 = 26 - 4b$
2. $\frac{1}{2}x - 12 = \frac{3}{4}x - 36$
3. $7w + 2w + 5 = w + 16$
4. $3x - 15 = x + 18 - 4x$

Triangle TRN can be written $\triangle TRN$.
Sides of $\triangle TRN$: $\overline{TR}, \overline{RN}, \overline{NT}$
Vertices of $\triangle TRN$: T, R, N
Angles of $\triangle TRN$: $\angle T\ (\angle NTR)$
$\angle R\ (\angle TRN)$
$\angle N\ (\angle RNT)$

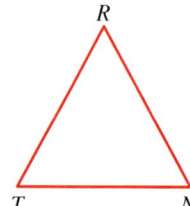

618 Chapter 15 Right Triangle Relationships

A triangle may be classified according to its angles.

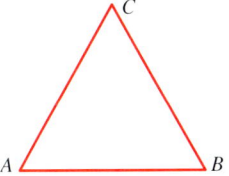 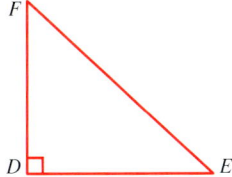 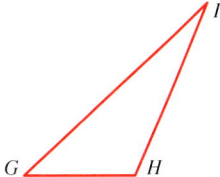

△ABC is an **acute triangle** because each of its three angles is acute.

△DEF is a **right triangle** because it has one right angle.

△GHI is an **obtuse triangle** because it has one obtuse angle.

A triangle may also be classified according to the lengths of its sides.

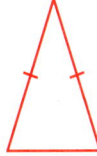 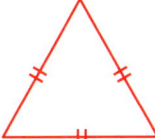 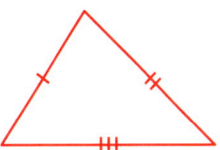

An **isosceles triangle** has at least two sides of equal length.

An **equilateral triangle** has all sides of equal length.

A **scalene triangle** has no sides of equal length.

In an isosceles triangle, the measures of the angles opposite the equal sides are equal. In an equilateral triangle, the measures of all the angles are equal. In a scalene triangle, the measures of the angles are not equal.

EXAMPLE 1 Classify each triangle according to the measures of the angles and according to the lengths of the sides.

a. 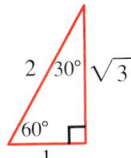   b. 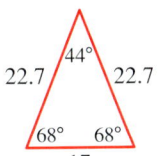   c. 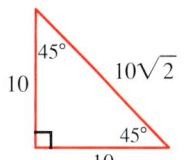   d.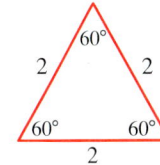

a. right triangle
   scalene triangle

b. acute triangle
   isosceles triangle

c. right triangle
   isosceles triangle

d. acute triangle
   equilateral triangle

The sum of the measures of the angles of a triangle is 180.

$$m\angle A + m\angle B + m\angle C = 180$$

15.2 Triangles **619**

EXAMPLE 2 In $\triangle DES$, $m\angle D = 12$ and $m\angle E = 40$. Find $m\angle S$.

$m\angle D + m\angle E + m\angle S = 180$ *The sum of the measures of the*
$12 + 40 + m\angle S = 180$ *angles of a triangle is 180.*
$52 + m\angle S = 180$
$m\angle S = 128$

EXAMPLE 3 If $\triangle RAN$ is isosceles, $RA = AN$, and $m\angle A = 32$, find $m\angle R$ and $m\angle N$.

$m\angle A + m\angle R + m\angle N = 180$
Let $x = m\angle R$, then $m\angle R = m\angle N$.   *Angles opposite equal sides are*
So $32 + x + x = 180$ *equal and $RA = AN$.*
$32 + 2x = 180$
$2x = 148$
$x = 74$
So $m\angle R = 74$ and $m\angle N = 74$.

EXAMPLE 4 In $\triangle QRS$, the measure of $\angle Q$ is twice the measure of $\angle R$, and the measure of $\angle S$ is 15 more than the measure of $\angle Q$. Find the measure of each angle of the triangle.

Let $r = m\angle R$.
Then $m\angle Q = 2r$ and $m\angle S = 2r + 15$.
$r + 2r + (2r + 15) = 180$
$5r + 15 = 180$
$5r = 165$
$r = 33$
$2r = 66$
$2r + 15 = 81$
So $m\angle R = 33$, $m\angle Q = 66$, and $m\angle S = 81$.

CLASS EXERCISES

Classify each triangle according to the measures of the angles and according to the lengths of the sides.

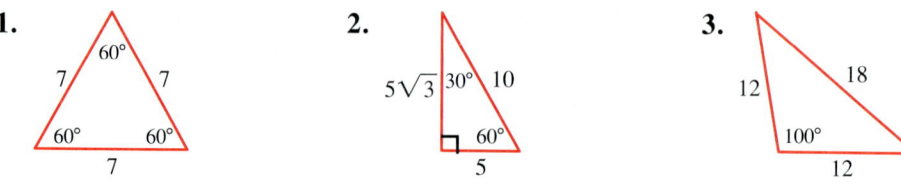

Given $\triangle MNO$, find $m\angle O$.

4. $m\angle M = 45$, $m\angle N = 30$ **5.** $m\angle M = 90$, $m\angle N = 13$

Use a protractor for Exercises 6 and 7.

6. Draw a large triangle. Find the measure of each angle with a protractor. Then find the sum of the measures.

7. Repeat Exercise 6 with two more types of triangles. What can you conclude about your results?

PRACTICE EXERCISES

Classify each triangle according to the measures of the angles and according to the lengths of the sides.

1.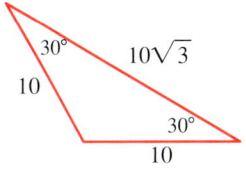
2.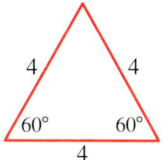
3.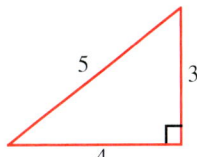
4.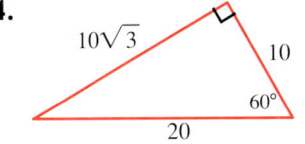
5.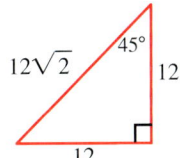
6.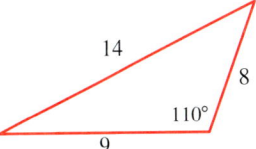

Given $\triangle JKL$, find $m\angle L$.

7. $m\angle J = 70$, $m\angle K = 40$
8. $m\angle J = 30$, $m\angle K = 40$
9. $m\angle J = 33$, $m\angle K = 80$
10. $m\angle J = 50$, $m\angle K = 72$
11. $m\angle J = 82$, $m\angle K = 73$
12. $m\angle J = 42$, $m\angle K = 64$

13. If $\triangle MNO$ is isosceles, $MN = NO$, and $m\angle N = 80$, find $m\angle M$.
14. If $\triangle WXY$ is isosceles, $m\angle Y = 24$, and $WY = XY$, find $m\angle X$.
15. If $\triangle ABC$ is a right isosceles triangle and $m\angle A = 90$, find $m\angle B$.
16. If $\triangle DEF$ is an equilateral triangle, find $m\angle E$.
17. In an isosceles triangle, one of the equal angles measures 33°. Find the measure of the remaining angles in the triangle.
18. In an isosceles triangle, one of the equal angles measures 47°. Find the measure of the remaining angles.
19. The measures of the angles of a triangle are consecutive integers. Find all the angles.
20. The measures of the angles of a triangle are consecutive even integers. Find all the angles.

15.2 Triangles   **621**

21. The measures of the angles of a triangle are in the ratio 2:3:4. Find all the angles.

22. The measures of the angles of a triangle are in the ratio 3:4:5. Find all the angles.

23. In $\triangle QRS$ the measure of $\angle Q$ is three times the measure of $\angle R$. The measure of $\angle S$ is 9 more than 5 times the measure of $\angle R$. Find the angles.

24. In $\triangle ABC$ the measure of $\angle C$ is four times the measure of $\angle A$. The measure of $\angle B$ is 2 less than twice the measure of $\angle A$. Find the angles.

25. In $\triangle MNO$ the measure of $\angle N$ is 11 more than seven times the measure of $\angle M$. The measure of $\angle O$ is 5 times that of $\angle M$. Find the angles.

26. In $\triangle XYZ$ the measure of $\angle X$ is 53 and the measure of $\angle Z$ is 22 more than 6 times that of $\angle Y$. Find the angles.

27. In $\triangle SUM$ the measure of $\angle M$ is 4 less than the measure of the supplement of $\angle U$. The measure of $\angle S$ is 30 less than $\frac{1}{2}$ the measure of $\angle U$. Find the angles.

28. In $\triangle FUN$ the measure of $\angle U$ is 8 more than $\frac{1}{3}$ the measure of the supplement of $\angle F$. The measure of $\angle N$ is 31 more than the complement of $\angle F$. Find the angles.

Applications

29. **Hobbies** Jason's kite is 30 ft above the ground. Draw a right triangle so that the angle formed between the kite string and the ground is 35°. Which side of the triangle represents the kite string?

30. **Forestry** A forest ranger wishes to determine the height of a redwood tree. Draw a right triangle so that the line drawn from the ground to the top of the tree forms a 60° angle when the ranger is standing 50 ft from the base of the tree.

ALGEBRA IN CONSTRUCTION

Becca and Steve work summers for the city recreation department. Their first "you're on your own" job was to design and build a sandbox for a playground. They made it in the shape of a right triangle with sides of one unit and b units, and hypotenuse AB.

The sandbox is very popular, so Becca was given the job of enlarging it. Because of the contour of the ground and location of other play equipment, she decided to build another right triangle, using the same hypotenuse AB. Another side is 2 units. How long is the third side?

15.3 Congruence

Objective: To identify congruent figures and name the corresponding parts of congruent triangles

In the early 1940s, R. Buckminster Fuller introduced the idea of constructing buildings and other structures with a design based on equilateral triangles rather than rectangles. Note that the equilateral triangles are the same size and shape. The properties of equality for real numbers can also be applied to corresponding parts of triangles that are the same size and shape.

Capsule Review

For all real numbers a, b, and c, the following properties are true:
 Reflexive property: $a = a$
 Symmetric property: If $a = b$, then $b = a$.
 Transitive property: If $a = b$ and $b = c$, then $a = c$.

Which property of real numbers justifies each statement?

1. If $5 + x = -2$, then $-2 = 5 + x$
2. $15 - 5x = 15 - 5x$
3. If $a + 2 = 9$ and $9 = 7 + 2$, then $a + 2 = 7 + 2$.

Angles that have the same measure are called **congruent angles.** Line segments that have the same length are **congruent line segments.** **Congruent figures** are figures that have the same size and shape.

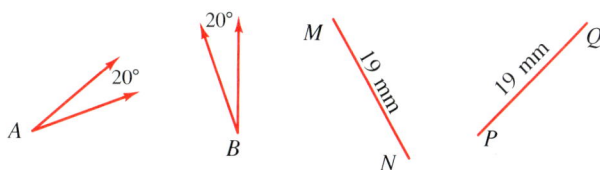

The symbol for "is congruent to" is ≅. Therefore, in the figure above $\angle A \cong \angle B$ and $\overline{MN} \cong \overline{PQ}$.

EXAMPLE 1 Find the length of each segment and tell whether the segments are congruent.

a. $\overline{BE}$ and $\overline{DH}$ b. $\overline{FD}$ and $\overline{CA}$

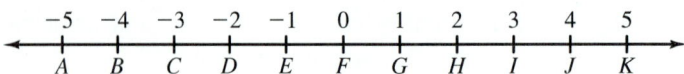

a. $BE = 3$ and $DH = 4$. So $\overline{BE}$ and $\overline{DH}$ are not congruent.

b. $FD = 2$ and $CA = 2$. So $\overline{FD} \cong \overline{CA}$.

The properties for equality will also be true for congruent angles and line segments. For example, the symmetric property for congruent angles would state

"If $\angle A \cong \angle B$, then $\angle B \cong \angle A$."

For congruent line segments, the symmetric property would state

"If $\overline{AB} \cong \overline{CD}$, then $\overline{CD} \cong \overline{AB}$."

EXAMPLE 2 Name the property of congruence that illustrates each statement.
a. If $\angle WXY \cong \angle XYW$, then $\angle XYW \cong \angle WXY$.
b. $\angle BND \cong \angle BND$
c. If $\overline{PQ} \cong \overline{QR}$ and $\overline{QR} \cong \overline{ST}$, then $\overline{PQ} \cong \overline{ST}$.

a. symmetric property   b. reflexive property   c. transitive property

If two triangles such as $\triangle ABC$ and $\triangle XYZ$ are congruent, you could mentally superimpose one over the other. In congruent triangles there are three types of **corresponding parts.**

Corresponding Vertices
$A \leftrightarrow X$ $B \leftrightarrow Y$ $C \leftrightarrow Z$

Corresponding Angles
$\angle A \leftrightarrow \angle X$ $\angle B \leftrightarrow \angle Y$ $\angle C \leftrightarrow \angle Z$

Corresponding Sides
$\overline{AB} \leftrightarrow \overline{XY}$ $\overline{BC} \leftrightarrow \overline{YZ}$ $\overline{AC} \leftrightarrow \overline{XZ}$

In congruent figures, corresponding parts are named in the same order. So, $\triangle ABC \cong \triangle XYZ$.

EXAMPLE 3 $\triangle RAT \cong \triangle MOP$
a. Name the corresponding sides.
b. If $\triangle RAT \cong \triangle MOP$ then is it true that $\triangle ART \cong \triangle OPM$?

a. If $\triangle RAT \cong \triangle MOP$, the following vertices correspond:
$R \leftrightarrow M \quad A \leftrightarrow O \quad T \leftrightarrow P$
Therefore, the following sides correspond:
$\overline{RA} \leftrightarrow \overline{MO} \quad \overline{AT} \leftrightarrow \overline{OP} \quad \overline{TR} \leftrightarrow \overline{PM}$

b. If $\triangle ART \cong \triangle OPM$, then the following vertices correspond:
$A \leftrightarrow O \quad R \leftrightarrow P \quad T \leftrightarrow M$
But this is not the same as the correspondence defined by $\triangle RAT \cong \triangle MOP$. So, $\triangle ART$ is not necessarily congruent to $\triangle OPM$.

CLASS EXERCISES

Refer to the number line below. Find the length of each line segment given and tell whether the segments are congruent.

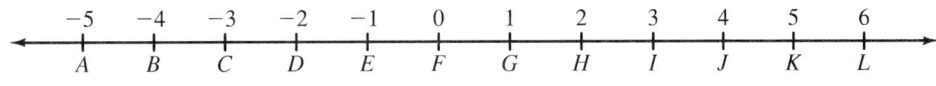

1. $\overline{AE}$ and $\overline{EI}$
2. $\overline{DF}$ and $\overline{JL}$
3. $\overline{CG}$ and $\overline{EJ}$

Use the indicated property of equality or congruence to complete each statement.

4. reflexive property: $\overline{AC} \cong$ _____

5. symmetric property: If $m\angle G = 90$, then _____ $= m\angle G$.

6. transitive property: If $m\angle K = m\angle M$ and _____ $= m\angle N$, then $m\angle K = m\angle N$.

7. transitive property: If $\overline{XY} \cong$ _____ and $\overline{ZA} \cong \overline{YZ}$, then $\overline{XY} \cong \overline{YZ}$.

8. symmetric property: If $\overline{TP} \cong \overline{AB}$, then _____ $\cong \overline{TP}$.

Given $\triangle ABC \cong \triangle TOD$:

9. Name the corresponding vertices.
10. Name the corresponding sides.
11. Name the corresponding angles.

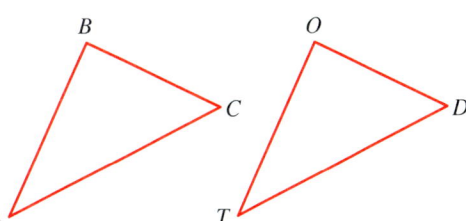

PRACTICE EXERCISES

Refer to the number line below. Find the length of each segment and then tell whether the segments are congruent.

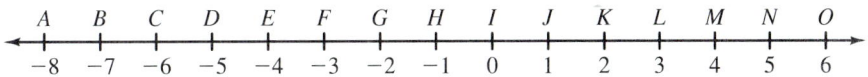

1. $\overline{BF}$ and $\overline{KO}$
2. $\overline{CH}$ and $\overline{IN}$
3. $\overline{EI}$ and $\overline{KM}$
4. $\overline{AF}$ and $\overline{CE}$
5. $\overline{JM}$ and $\overline{DG}$
6. $\overline{BE}$ and $\overline{KN}$

Name the property of equality or congruence that illustrates each statement.

7. $m\angle M = m\angle M$

8. If $\angle J \cong \angle K$ and $\angle K \cong \angle L$, then $\angle J \cong \angle L$.

9. If $\overline{AC} \cong \overline{BC}$, then $\overline{BC} \cong \overline{AC}$

10. The length of $\overline{UV}$ is equal to itself.

11. If the complement of $\angle A$ is congruent to $\angle B$ and $\angle B$ is congruent to $\angle C$, then the complement of $\angle A$ is congruent to $\angle C$.

12. If the supplement of $\angle M$ is congruent to $\angle M$, then $\angle M$ is congruent to its supplement.

For Exercises 13–16, $\triangle TOP \cong \triangle CAR$.

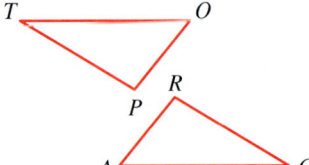

13. Name the corresponding angles.
14. Name the corresponding sides.
15. *True* or *false*? $\triangle ARC \cong \triangle OPT$
16. *True* or *false*? $\triangle RCA \cong \triangle OTP$

For Exercises 17–20, $\triangle PAL \cong \triangle DAN$.

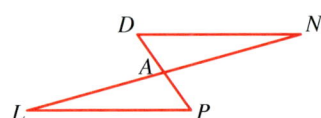

17. Name the corresponding sides.
18. Name the corresponding angles.
19. *True* or *false*? $\triangle LAP \cong \triangle AND$
20. *True* or *false*? $\triangle APL \cong \triangle ADN$

21. $RECT$ is a rectangle. $\overline{RC}$ and $\overline{ET}$ are diagonals of the rectangle. Name all triangles you think may be congruent to $\triangle RCT$.

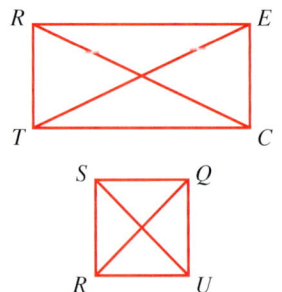

22. $SQUR$ is a square. $\overline{US}$ and $\overline{RQ}$ are diagonals of the square. Name all triangles you think may be congruent to $\triangle RSQ$.

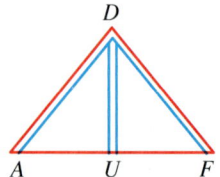

Applications

23. **Construction** A truss for a garage has two congruent sides and a vertical tie rod as shown in the diagram. Name two triangles that you think are congruent and their corresponding sides and angles.

24. **Construction** The diagram at the right shows the cross section of a truss for a bridge. The vertical tie rods, ($\overline{TF}$, $\overline{UD}$, $\overline{SG}$), are equal in length and $\overline{BF}$, $\overline{FD}$, $\overline{DG}$, and $\overline{GE}$ are congruent. Name all triangles you think may be congruent.

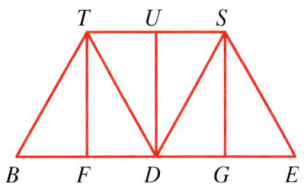

25. **Construction** A carpenter writes these six congruence statements. Complete correctly the statement about congruent triangles.

$\angle P \cong \angle Q$ $\overline{SP} \cong \overline{RQ}$
$\angle S \cong \angle R$ $\overline{PT} \cong \overline{QU}$
$\angle T \cong \angle U$ $\overline{ST} \cong \overline{RU}$
$\triangle \underline{\ ?\ } \cong \triangle \underline{\ ?\ }$

TEST YOURSELF

Graph the inequality and tell whether it is a point, a line, a line segment, a ray or none of these.   **15.1**

1. $x \geq -2$
2. $-3 \leq x \leq 0$
3. $x \geq -2$ or $x \leq 0$

4. If 10 more than $m\angle T$ is twice its complement, find $m\angle T$.

5. If the complement of $\angle X$ is $\frac{1}{4}$ its supplement, find the measure of $\angle X$.

Classify each triangle according to the measures of the angles and according to the lengths of the sides.   **15.2**

6.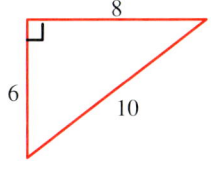
7.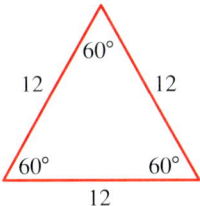
8.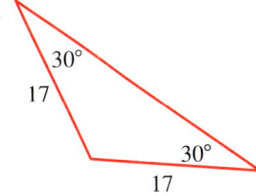

9. In $\triangle HIJ$, $\overline{HI} \cong \overline{JI}$, $m\angle I = 48$, find $m\angle H$ and $m\angle J$.

Given $\triangle DEF \cong \triangle PAT$ **15.3**

10. $\angle D$ corresponds to $\angle P$, true or false?
11. $\angle T$ corresponds to $\angle E$, true or false?
12. $\overline{FD}$ corresponds to $\overline{AP}$, true or false?
13. $\triangle FDE \cong \triangle TPA$, true or false?
14. If $\angle D \cong \angle E$ and $\angle E \cong \angle A$ then $\angle D \cong \angle A$, true or false?

15.4 Similar Figures

Objectives: To identify similar figures and name the corresponding vertices and sides of similar triangles
To use proportions to solve problems concerning similar triangles

When a photograph is enlarged, each object in the enlargement has the same shape as in the original photograph, but a different size.

The photographs are *similar* and the corresponding dimensions of each object are in proportion.

Capsule Review

A proportion is an equation that states that two ratios are equal. To solve a proportion, you use the fact that the product of the means equals the product of the extremes.

EXAMPLE   3:x = 5:9 (means are x and 5; extremes are 3 and 9)

$\frac{3}{x} = \frac{5}{9}$; $5x = 27$; $x = \frac{27}{5} = 5.4$

Solve these proportions.

1. $\frac{5}{6} = \frac{10}{x}$
2. $\frac{2}{3} = \frac{x}{18}$
3. $\frac{51}{9} = \frac{n}{15}$
4. $\frac{22}{7} = \frac{11}{x}$
5. $3:4 = x:12$
6. $5:7 = 15:x$
7. $21:7 = 3:x$
8. $1:6 = x:3$

Similar figures have the same shape, but not necessarily the same size. Two triangles are **similar** if and only if their corresponding angles are congruent. Corresponding sides of similar triangles are in proportion.

The symbol for "is similar to" is ∼. In the figure, △ABC ∼ △XYZ. Note that the corresponding vertices are named in the same order, and that the angles are congruent.

∠A ≅ ∠X   ∠B ≅ ∠Y   ∠C ≅ ∠Z

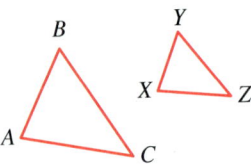

Corresponding sides are opposite congruent angles in similar triangles. For △ABC and △XYZ above, the corresponding sides are $\overline{AB}$ and $\overline{XY}$, $\overline{BC}$ and $\overline{YZ}$, and $\overline{AC}$ and $\overline{XZ}$.

EXAMPLE 1 Is it true that △ABD ~ △VWY? If so, name the corresponding sides.

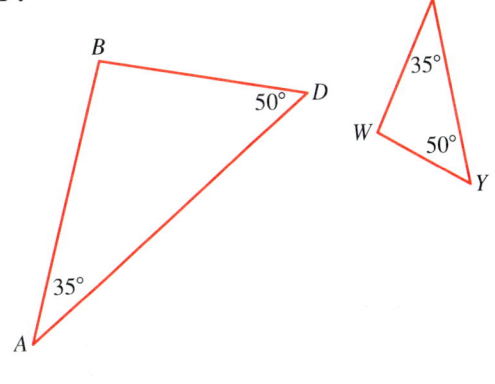

Since the sum of the angles of a triangle is 180°, the missing angle in both △ABD and △VWY is 180 − (35 + 50) = 95. Therefore, the corresponding angles are congruent,

∠A ≅ ∠V, ∠B ≅ ∠W, and ∠D ≅ ∠Y

and the triangles are similar. The corresponding sides are $\overline{AB}$ and $\overline{VW}$, $\overline{BD}$ and $\overline{WY}$, and $\overline{AD}$ and $\overline{VY}$.

△TRS ~ △XVW, so the lengths of corresponding sides are in the same ratio or proportion.

$\frac{ST}{WX} = \frac{4}{6} = \frac{2}{3}$; $\frac{SR}{WV} = \frac{6}{9} = \frac{2}{3}$; $\frac{TR}{XV} = \frac{8}{12} = \frac{2}{3}$

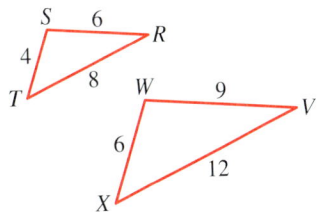

To find an unknown side using similar triangles, first find a pair of corresponding sides whose lengths are known. Then find the side that corresponds to the unknown side. Solve by writing a proportion.

EXAMPLE 2 △WIN ~ △DOS, WI = 10, IN = 8, WN = 6, and DO = 5. Find the lengths of $\overline{DS}$ and $\overline{OS}$.

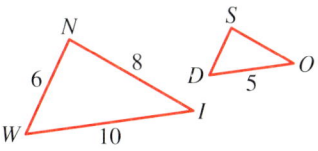

Use the lengths of corresponding sides $\overline{WI}$ and $\overline{DO}$ to write a known ratio, 10:5. Then show that the lengths of corresponding sides are in proportion.

$\frac{WI}{DO} = \frac{WN}{DS}$ $\frac{10}{5} = \frac{6}{DS}$ and $\frac{WI}{DO} = \frac{IN}{OS}$ $\frac{10}{5} = \frac{8}{OS}$

 30 = 10(DS) 40 = 10(OS)

 3 = DS 4 = OS

15.4 Similar Figures

CLASS EXERCISES

For Exercises 1–7, $\triangle GFE \sim \triangle MNO$.

1. Name the corresponding angles.
2. Name the corresponding sides.
3. *True* or *false*? $\triangle FGE \sim \triangle MNO$.
4. *True* or *false*? $\triangle EGF \sim \triangle OMN$.
5. Complete the proportion: $\dfrac{FE}{NO} = \dfrac{FG}{?}$

Write a proportion involving the following lengths.

6. *EG* and *MN*
7. *NO* and *GF*

Tell whether the triangles are similar, not similar, or not possible to determine.

8.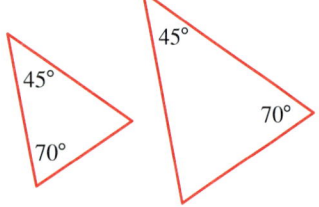

9.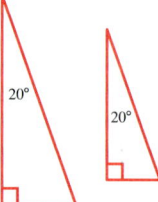

10. Two isosceles triangles, each with one of their angles equal to 20°.
11. Two right triangles, each with a 30° angle.

PRACTICE EXERCISES

Tell whether or not the triangles are similar. If similar, name the corresponding sides.

1.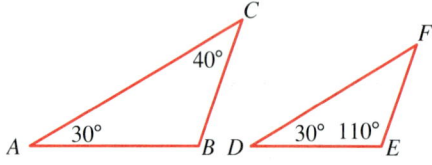

2.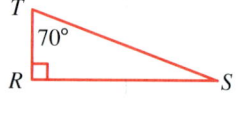

    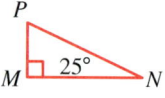

3.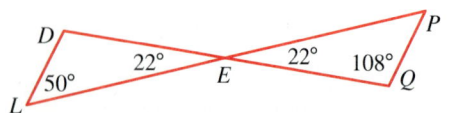

4.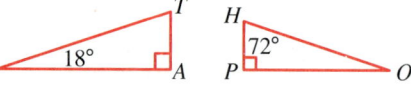

630   Chapter 15 Right Triangle Relationships

If △ABC ~ △XYZ, find the lengths of the two missing sides.

5. $AB = 4$, $BC = 7$; $AC = 5$; $XY = 8$
6. $AB = 18$; $BC = 36$; $AC = 21$; $XY = 6$
7. $XY = 4$, $YZ = 8$; $XZ = 10$; $AC = 7.5$
8. $XY = 3$, $YZ = 2.5$, $XZ = 4$; $BC = 7.5$

In Exercises 9–12, △DEF ~ △MNO. Find the lengths of the two missing sides.

9. $DE = 7$, $EF = 4$; $MO = 18$; $NO = 12$.
10. $EF = 18$, $DF = 54$; $MN = 45$; $NO = 15$.
11. $DF = 4$, $DE = 2.4$; $NO = 5$; $MN = 6$.
12. $MO = 3.6$, $MN = 7.2$, $EF = 8$; $DE = 12$.

For Exercises 13–16, △BAC ~ △PAR. Find the lengths of the missing sides.

13. $AB = 12$, $BC = 21$, $AC = 15$, and $AP = 20$. Find PR and AR.
14. $AP = 30$, $PR = 50$, $AR = 40$, and $AB = 20$. Find BC and AC.
15. $AC = 8$, $AP = 16$, $BC = 12$, and $AR = 20$. Find AB and PR.
16. $AB = 28$, $AR = 63$, $BC = 54$, and $PR = 81$. Find AC and AP.

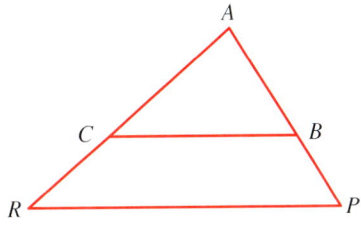

For Exercises 17–18, △JAR ~ △TOP. Find the lengths of the missing sides if the triangles are isosceles. ($JA = AR$ and $TO = OP$)

17. $JR = 3.6$, $OP = 5.6$, $\dfrac{JA}{TO} = \dfrac{3}{4}$
18. $TP = 11$, $AR = 33$, $\dfrac{JA}{TO} = \dfrac{3}{22}$

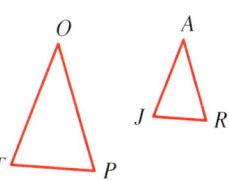

Find x. Assume that each figure shows a pair of similar triangles.

19.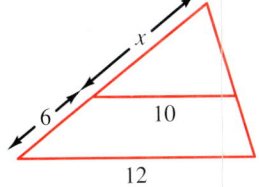

20.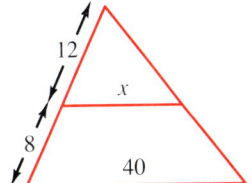

15.4 Similar Figures **631**

21.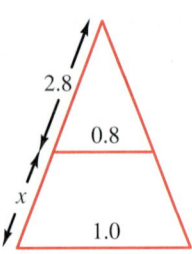

22.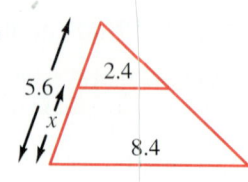

Applications

Surveying Find *h*. Assume the triangles in each exercise are similar.

23.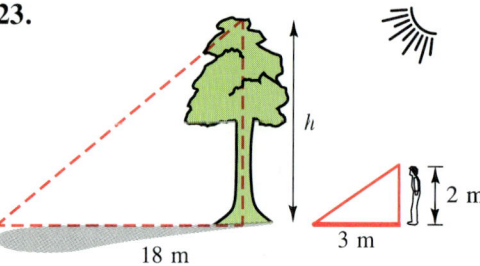

24.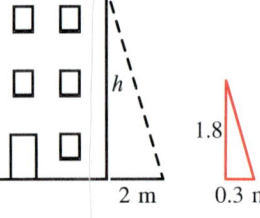

25.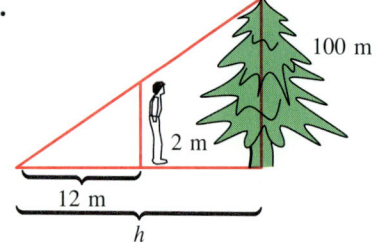

26.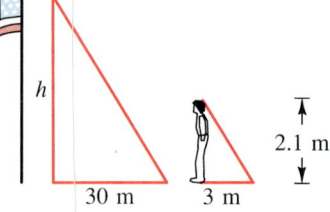

27. **Gardening** A triangular flower garden in a park has a sidewalk around it. The sides of the triangle around the outside of the sidewalk are 60 ft, 45 ft, and 20 ft. The smallest side of the garden is 14.5 ft long. How much will it cost to put a fence around the garden if the fencing costs $15.00 per yard?

EXTRA

Barry has a summer job as a carpenter's helper at Kid World, a local amusement park. The work crew is building picnic shelters. Barry's job is to cut rafters so that the ends make a vertical join at the peak of the shelter. What size angle will he measure before cutting? *Hint:* If two rafters are joined vertically, the angle formed between them is 90°.

632 Chapter 15 Right Triangle Relationships

15.5 Trigonometric Ratios

Objective: To identify and compute the sine, cosine, and tangent ratios in any right triangle

The word **trigonometry** is based on Greek words that mean "triangle measurement." In similar triangles, ratios of corresponding sides are equal. In trigonometry, some of these ratios are given special names when the triangles are right triangles. In such triangles the measures of the acute angles are related to the ratios of the lengths of the sides.

Capsule Review

A ratio with a radical in the denominator can be changed to simplest form by rationalizing the denominator.

EXAMPLE Write $\dfrac{2}{\sqrt{3}}$ in simplest form with a rational denominator.

$$\frac{2}{\sqrt{3}} = \frac{2}{\sqrt{3}} \cdot \frac{\sqrt{3}}{\sqrt{3}} = \frac{2\sqrt{3}}{3}$$

Write each ratio in simplest form with a rational denominator.

1. $\dfrac{1}{\sqrt{5}}$
2. $\dfrac{3}{\sqrt{2}}$
3. $\dfrac{1}{3\sqrt{2}}$
4. $\dfrac{5}{\sqrt{10}}$
5. $\dfrac{\sqrt{3}}{\sqrt{6}}$
6. $\dfrac{\sqrt{2}}{\sqrt{5}}$
7. $\dfrac{\sqrt{3}}{\sqrt{50}}$
8. $\dfrac{\sqrt{7}}{\sqrt{21}}$

Recall that in right triangle ABC, the side opposite the right angle $\angle C$, is called the *hypotenuse* and is labeled c. The other two sides are called *legs*. Notice that a is the leg opposite $\angle A$, and b is opposite $\angle B$. You can also say that b is *adjacent to* $\angle A$ and a is adjacent to $\angle B$.

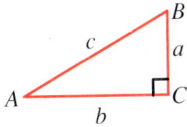

For any right triangle, there are three **trigonometric ratios** of the lengths of its sides.

The trigonometric ratios for right triangle ABC with acute angles A and B and right angle C:

sine of $\angle A = \dfrac{\text{length of leg opposite } \angle A}{\text{length of hypotenuse}}$; $\sin A = \dfrac{a}{c}$

cosine of $\angle A = \dfrac{\text{length of leg of adjacent } \angle A}{\text{length of hypotenuse}}$; $\cos A = \dfrac{b}{c}$

tangent of $\angle A = \dfrac{\text{length of leg opposite } \angle A}{\text{length of leg adjacent } \angle A}$; $\tan A = \dfrac{a}{b}$

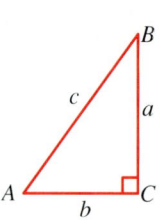

EXAMPLE 1 For $\triangle ABC$, $m\angle C = 90$. Find $\sin A$, $\cos A$, $\tan A$, $\sin B$, $\cos B$, and $\tan B$.

$\sin A = \dfrac{6}{10} = \dfrac{3}{5}$ $\sin B = \dfrac{8}{10} = \dfrac{4}{5}$

$\cos A = \dfrac{8}{10} = \dfrac{4}{5}$ $\cos B = \dfrac{6}{10} = \dfrac{3}{5}$

$\tan A = \dfrac{6}{8} = \dfrac{3}{4}$ $\tan B = \dfrac{8}{6} = \dfrac{4}{3}$

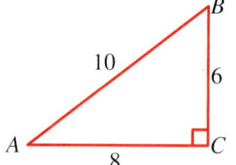

EXAMPLE 2 In $\triangle XYZ$, $YX = 3$, $YZ = 1$, and $m\angle Z = 90$. Use the Pythagorean theorem to find XZ. Then find the sine, cosine, and tangent of $\angle X$.

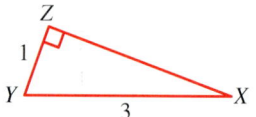

$(YX)^2 = (YZ)^2 + (XZ)^2$
$3^2 = 1^2 + (XZ)^2$
$9 = 1 + (XZ)^2$
$8 = (XZ)^2$
$2\sqrt{2} = XZ$

Therefore, $\sin X = \dfrac{1}{3}$; $\cos X = \dfrac{2\sqrt{2}}{3}$; $\tan X = \dfrac{1}{2\sqrt{2}} = \dfrac{1 \cdot \sqrt{2}}{2\sqrt{2} \cdot \sqrt{2}} = \dfrac{\sqrt{2}}{4}$.

CLASS EXERCISES

Find the value of each trigonometric ratio. Express radicals in simplest form.

1. $\sin A$
2. $\cos A$
3. $\tan A$
4. $\sin B$
5. $\cos B$
6. $\tan B$
7. $\sin M$
8. $\cos M$
9. $\tan M$
10. $\sin N$
11. $\cos N$
12. $\tan N$

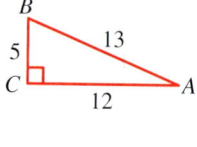

PRACTICE EXERCISES

Find the value of each trigonometric ratio.

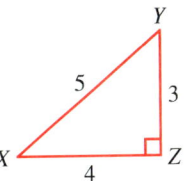

1. sin X
2. cos X
3. tan X
4. sin Y
5. cos Y
6. tan Y

Find the values for the sine, cosine, and tangent of ∠A and ∠B.

7. 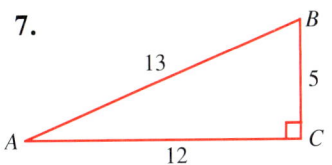 8.  9.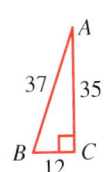

10. 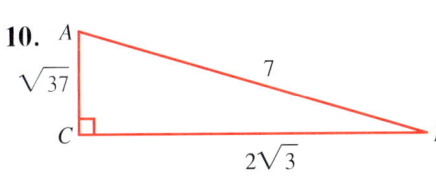 11. 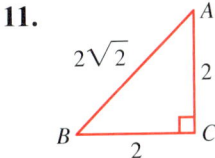 12.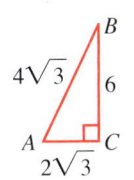

Use the Pythagorean Theorem to find the length of the missing side. Then find the sine, cosine, and tangent of ∠Q and ∠T.

13. 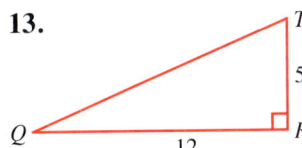 14. 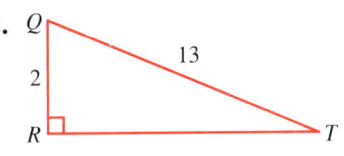 15.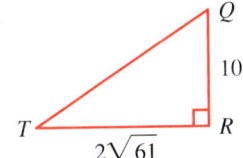

16. 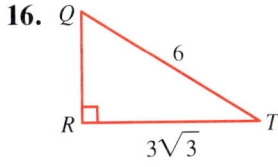 17. 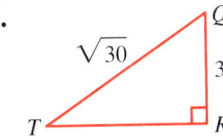 18.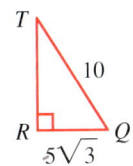

△JKL is a right triangle with m∠L = 90. Show that each of the following is true.

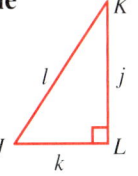

19. cos J = sin K
20. cos K = sin J
21. $(\sin J)^2 + (\cos J)^2 = 1$
22. $\tan K = \frac{\sin K}{\cos K}$

15.5 Trigonometric Ratios **635**

Evaluate each expression. Use the table on page 671.

23. $\dfrac{\sin 30° + \cos 60°}{\sin 60° + \cos 30°}$

24. $\dfrac{2 \times \sin 15°}{2 \times \cos 15°}$

25. $\dfrac{1 - \sin 45°}{1 + \tan 45°}$

26. $\dfrac{\tan 60° - \tan 30°}{\tan 45°}$

27. $\dfrac{(\sin 75°)^2 + (\cos 75°)^2}{1 - (\tan 35°)^2}$

28. $\dfrac{\tan 30° - \cos 60° + \sin 45°}{\tan 45° + \cos 30° - \sin 60°}$

Applications

29. **Navigation** A submarine travels 36 miles diving at an angle of 32°. Which ratio would be used to find out how deep the submarine is?

30. **Geometry** A rectangle has a length of 8 m and a width of 6 m. Find the length of the diagonal. What is the sine, cosine, and tangent for the angle formed by the diagonal and the length?

31. **Navigation** A boat sails 15 mi east, then 10 mi north. Find the distance from the boat to its starting point. Which trigonometric ratio is used to find the angle of the boat's straight-line course?

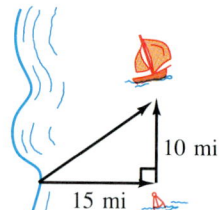

32. **Architecture** A 35-ft roof slopes at an angle 35° to the floor. Which trigonometric ratio would be used to find the width of the floor?

EXTRA

Making a Tent

Celia has a large rectangular piece of tent cloth 5 m long. She wants to build a tent with a front and back opening in the shape of a triangle and with sides of equal length. So that she can stand comfortably in the tent, she decides to have a height of 1.7 m. Celia needs enough floor space to place two sleeping bags side by side with a walkway between them around the poles. Is this possible?

Rule of Thumb

Hold your thumb vertically, at arm's length. You now have a tool for estimating the height of most buildings. You are missing one fact. What is it?

APPLICATION: Networks

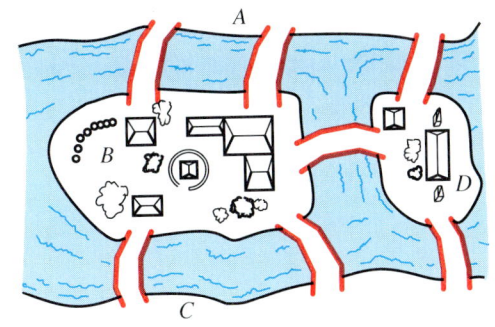

Did you know that geometry has been applied to problems in designing telephone and communications systems, transportation connections, and linking computer stations? These design problems often can be represented by a *network*. A **network** consists of a set of vertices and the connecting paths between them.

One of the most famous network problems occurred in the sixteenth century in Königsberg, Prussia. The town was on two islands in the middle of a river. The town was connected by seven bridges which crossed the river as shown in the sketch. The question was whether a person could plan a walk, starting anywhere, so that all seven bridges could be crossed without crossing the same bridge twice.

Leonhard Euler (pronounced *OY luhr*), a Swiss mathematician, first studied this problem in 1735. Here is Euler's approach to the problem:

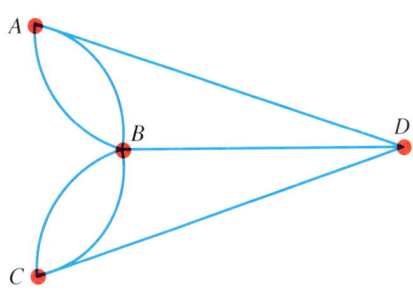

Euler began by redrawing the map so that the four areas of land were represented by four points A, B, C, and D. He then connected the four points with seven lines to represent the bridges. The Königsberg problem can now be restated as follows: Given four distinct vertices, try to draw a continuous path from one vertex to another, without retracing a previously drawn path.

Euler discovered that such a network can only be drawn if one of two conditions are true:

1. There are exactly 2 vertices with an odd number of paths leading from them.
2. All vertices have an even number of paths leading from them.

Application: Networks **637**

Since vertex *B* has 5 paths, and vertices *A*, *C*, and *D* all have three paths, it is impossible to cross each bridge only once.

Trace each one of these figures on a separate piece of paper. Then try to draw a copy of each figure without picking your pencil up from the paper and without retracing a previously drawn path.

1.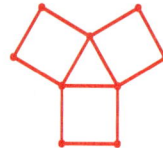
2.
3.
4.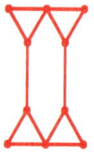
5.
6.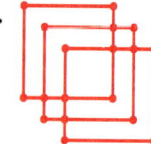

A set of problems similar to the Königsberg Bridge problem concerns closed geometric figures. The object is to cross each segment of the figure exactly once without picking the pencil up from the paper. Copy each figure on a separate piece of paper. Then draw a continuous path through each segment. Keep track of the following information:

The number of segments in the figure or figures.
Starting on the outside, does the path end on the inside or outside?
Starting on the inside, does the path end on the inside or outside?

7.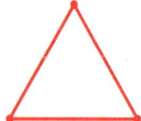
8.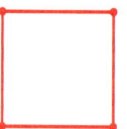
9.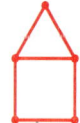
10.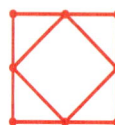
11.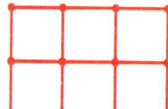
12.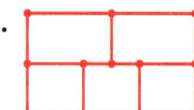

13. Using the problem solving strategies studied in this book, develop a set of rules concerning the figures that can be solved in Exercises 7–12.

14. Research network problems further and investigate a branch of mathematics known as topology.

15.6 Using Trigonometric Ratios

Objective: To use a trigonometric table to find the lengths of sides and the measures of angles of a right triangle

When architects design access ramps, the angle of inclination must be small enough to allow easy access. If a ramp rises 5 ft vertically over a horizontal distance of 100 ft, the architect can use a trigonometric table to determine the angle of inclination.

Capsule Review

In right triangle ABC, $m\angle C = 90$.

$\sin A = \dfrac{a}{c}$, $\cos A = \dfrac{b}{c}$, and $\tan A = \dfrac{a}{b}$.

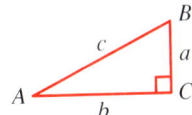

For right triangle ABC, use the Pythagorean theorem to find the length of the missing side. Then find the sine, cosine, and tangent of $\angle A$ and $\angle B$.

1. $a = 4$, $b = 3$
2. $b = 2$, $c = 3$
3. $a = 2\sqrt{5}$, $b = 2\sqrt{6}$

The three right triangles shown are similar. If the measures of angles A, D and G are equal to 60° then the sine of 60° will be the same since the ratios of the length of the corresponding sides are equal.

$\triangle ABC$	$\triangle DEF$	$\triangle GHI$
$\dfrac{\sqrt{3}}{2}$	$\dfrac{2\sqrt{3}}{4} = \dfrac{\sqrt{3}}{2}$	$\dfrac{3\sqrt{3}}{6} = \dfrac{\sqrt{3}}{2}$

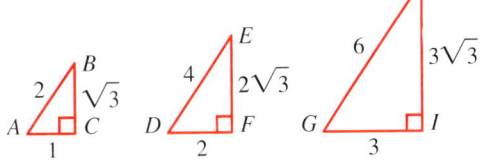

The equivalent of $\dfrac{\sqrt{3}}{2}$ is approximately 0.8660, so $\sin 60° = 0.8660$.

Trigonometric tables have been developed for angles from 0° to 90°. Most values listed are approximations. For convenience, you can use = when you write a trigonometric ratio. The following is a portion of the table found on page 671.

Angle	sin	cos	tan
58°	0.8480	0.5299	1.6003
59°	0.8572	0.5150	1.6643
60°	0.8660	0.5000	1.7321

15.6 Using Trigonometric Ratios

To find a trigonometric ratio of an angle, first find the angle in the left-hand column. Then look across the row until you are under the column with the appropriate heading. When using a calculator, enter the degree measurement and press either the sine, cosine or tangent key.

EXAMPLE 1 Find cos 62°.

cos 62° = 0.4695

Angle	sin	cos	tan
61°	0.8746	0.4848	1.8040
62°	0.8829	0.4695	1.8807
63°	0.8910	0.4550	1.9626

To find an angle when given its trigonometric ratio, first find the column headed with the trigonometric ratio. Then look down the column until you find the closest decimal. On some calculators you can find the angle by entering the decimal and pressing the $\sin^{-1}$, $\cos^{-1}$, or $\tan^{-1}$ key.

EXAMPLE 2 Find the measure of $\angle x$ if sin $x = 0.8572$.

$x = 59°$

Angle	sin	cos	tan
58°	0.8480	0.5299	1.6003
59°	0.8572	0.5150	1.6643
60°	0.8660	0.5000	1.7321

When you know one acute angle and one side in a right triangle, you can use a trigonometric table to find the other two sides.

EXAMPLE 3 In right triangle JET, $m\angle E = 90$, $JE = 28$, and $m\angle J = 78$. Find TE and JT to the nearest whole number.

$\tan 78° = \frac{TE}{28}$

$4.7046 = \frac{TE}{28}$

$TE = 4.7046(28)$
$TE = 131.7288$
$TE \approx 132.$

$\cos 78° = \frac{28}{JT}$

$0.2079 = \frac{28}{JT}$

$JT = \frac{28}{0.2079}$

$JT = 134.6801$
$JT \approx 135.$

EXAMPLE 4 In right triangle ABC, $m\angle C = 90$, $AC = 100$, and $BC = 5$. Find $m\angle A$ and $m\angle B$ to the nearest whole number.

$\tan A = \frac{5}{100} = 0.05$

By referring to the trigonometric table, you find that $m\angle A = 3$, to the nearest whole number. So, $m\angle B = 90 - 3 = 87$.

CLASS EXERCISES

Use the table on page 671 to find the value of each trigonometric ratio.

1. cos 45°
2. sin 15°
3. tan 30°
4. cos 53°
5. sin 41°
6. tan 17°
7. sin 81°
8. cos 71°

Find the measure of ∠x.

9. cos x = 0.5000
10. sin x = 0.7071
11. tan x = 0.4663
12. sin x = 0.9336
13. tan x = 1.8807
14. cos x = 0.8387

Tell which trigonometric ratio you would use to find x.

15.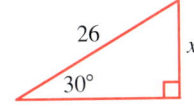
16.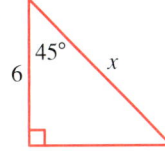
17.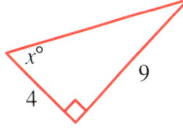

PRACTICE EXERCISES

Use the table on page 671 to find the values of each.

1. cos 5°
2. sin 23°
3. tan 57°
4. tan 16°
5. sin 73°
6. cos 82°
7. cos 66°
8. sin 2°
9. tan 88°
10. tan 65°
11. sin 55°
12. cos 43°

Find the measure of ∠x to the nearest degree.

13. cos x = 0.8660
14. sin x = 0.3420
15. tan x = 1.000
16. tan x = 3.7321
17. sin x = 0.8192
18. cos x = 0.2924
19. cos x = 0.9511
20. sin x = 0.6561
21. tan x = 0.0349

Refer to right △ABC with m∠C = 90. Find the measures of the other sides to the nearest whole number.

22. m∠A = 40, BC = 5
23. m∠A = 32, AB = 42
24. m∠B = 71, AC = 17
25. m∠B = 5, BC = 50

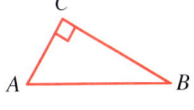

Refer to right △XYZ, with m∠Y = 90. Find the measures of the other sides to the nearest whole number.

26. m∠X = 50, XY = 50
27. m∠X = 22, YZ = 32
28. m∠Z = 71, XZ = 34
29. m∠Z = 14, XY = 15

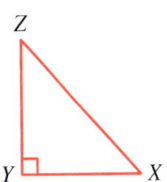

15.6 Using Trigonometric Ratios **641**

Refer to right △JKL, with m∠L = 90. Find the other angle measures to the nearest whole number.

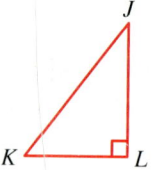

30. JK = 13, JL = 12
31. JK = 5, KL = 3
32. KL = 7, JL = 12
33. KL = 9, JL = 3

△ABC is a right triangle with m∠A = 90. Find the length of the requested side to the nearest tenth.

34. cos ∠B = 0.4226, BC = 18. Find AC.
35. sin ∠B = 0.6157, BC = 25. Find AB.
36. tan ∠C = 0.2867, AB = 12. Find BC.
37. tan ∠C = 4.7046, AC = 22. Find BC.

Applications

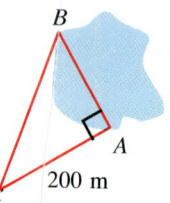

38. **Surveying** A surveyor wants to find the width of a lake. She places her transit at point L, which is 200 m from one side of the lake (point A). With a transit, she measures ∠L to be 37°. How wide is the lake?

39. **Architecture** An architect is designing an access ramp which forms an angle of 5° with the ground. The ramp will rise 5 ft, how long will the ramp be?

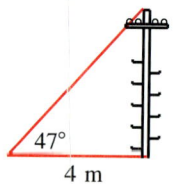

40. **Engineering** To support a pole, a cable is drawn tight between the pole and the ground. The cable is 4 m from the base of the pole. When the cable is attached it makes a 47° angle with the ground. How tall is the pole?

41. **Hobbies** Benjamin is flying a kite. If he knows that the kite string is 60 m and that the angle the string makes with the ground is 65°, how high above the ground is his kite?

EXTRA

Walking Distance

Elise walks by several large buildings as she goes from home to school to her after-school job, so she can't take a direct route. She travels 500 m east, 400 m north, 400 m east, and then 800 m north.

If Elise can take a straight path home from her job, how long is her return route?

15.7 Problem Solving Strategy: Check for Hidden Assumptions

When solving mathematical problems, it is important to check for any concepts that might be hidden within a problem. These are known as *hidden assumptions*, and they may or may not be true.

Before proceeding with a problem, be sure you clearly state what your assumptions are. This is important in order not to make an incorrect assumption about the problem. Such faulty reasoning can deter you from obtaining the correct solution.

In this lesson, certain assumptions are being made in order to apply the mathematics you know. These assumptions include:

1. Any triangle formed will be a right triangle.
2. When solving a problem involving navigation or surveying, vertical lines are perpendicular to the horizon or ground.

EXAMPLE 1 At a point 100 ft from a tree a forest ranger measures the angle of elevation to the top of the tree to be 62°. How tall is the tree?

Understand the Problem

What are the given facts?

You are told that the forest ranger is 100 ft from the tree and that the angle of elevation is 62°.

What is the hidden assumption?

The triangle formed is a right triangle and the ground is level.

Plan Your Approach

Choose a strategy.

You are asked to determine the height of the tree. Draw a diagram. Fill in all the given information. The **angle of elevation** is the angle formed by the line of sight and the horizon.

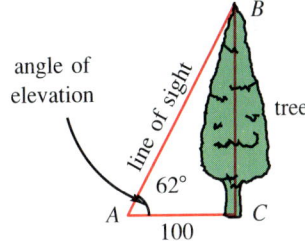

There are several approaches to take in solving this problem. For example use trigonometric functions and the Pythagorean theorem.

Complete the Work

Determine the measures of the angles of the triangle.

$\angle A + \angle B + \angle C = 180°$ *The sum of the angles in a triangle is 180°.*
$62° + \angle B + 90° = 180°$
$\angle B = 28°$

To find side BC two approaches may be used.

Approach One	Approach Two
$\tan A = \dfrac{\text{opp}}{\text{adj}}$	$\sin B = \dfrac{\text{opp}}{\text{hyp}}$
$\tan 62° = \dfrac{BC}{100}$	$\sin 28° = \dfrac{100}{AB}$
$\tan 62°(100) = BC$	$\sin 28°(AB) = 100$
$1.881(100) = BC$	$0.4695(AB) = 100$
Therefore, $BC = 188$.	$AB = 213$
	Use the Pythagorean theorem, $a^2 + b^2 = c^2$, to find BC.
	$a^2 + 100^2 = 213^2$
	$a = 188$
	Therefore, $BC = 188$.

Interpret the Results

If the ground between the ranger and the tree is level and a right triangle is formed, then the tree is 188 ft tall.

It is also important to make certain assumptions and create an accurate drawing when solving the problem. The following example illustrates how to calculate the distance to an object using trigonometry.

EXAMPLE 2   An observer in a lighthouse 225 ft above sea level spots a ship in the harbor. The angle of depression from the observer to the ship is 48°. Determine how far the ship is from the shore.

Understand the Problem

What are the given facts?

You are told that the lighthouse is 225 ft above sea level, and that the angle of depression is 48°.

What are the hidden assumptions?

1. The line of center of the lighthouse forms a 90° angle with the ground.
2. A right triangle can be formed by the ship, the observer, and the base of the lighthouse.

Chapter 15 Right Triangle Relationships

- **Plan Your Approach**

    Choose a strategy.

    The problem asks to determine the distance between the ship and the lighthouse. Draw a diagram.

    Fill in all the given information keeping in mind that the **angle of depression** is the angle formed by the line of sight and the horizontal plane. Notice that the angle of depression to the ship is equal to the angle of elevation from the ship.

    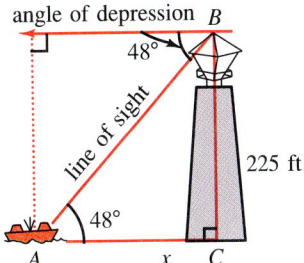

- **Complete the Work**

    $\angle A + \angle B + \angle C = 180°$ The sum of the angles in a triangle is 180°.
    $48° + \angle B + 90° = 180°$
    $\angle B = 42°$

    Find the length of $\overline{AC}$ using tan B.

    $\tan B = \dfrac{AC}{BC}$

    $\tan 42° = \dfrac{x}{225}$

    $x = 203$

- **Interpret the Results**

    The distance between the lighthouse and the boat is 203 ft.

CLASS EXERCISES

For Discussion

1. What is the meaning of the word *assumption*?
2. Do problem solvers make hidden assumptions intentionally or unintentionally?
3. Can a hidden assumption make a solution to a problem incorrect?
4. Why are hidden assumptions not helpful to a problem solver?
5. Why must a problem solver be careful in using drawings to help with the solution of a problem?

PRACTICE EXERCISES

For each situation, does the assumption made change the conditions of the problem? Write *yes* or *no*.

1. In a problem about any positive real numbers a, b, c, and d, it is assumed that $a < 0 < b$.

2. In a problem about any two lines in a coordinate plane having slopes m_1 and m_2, it is assumed that $m_1 \cdot m_2 = -1$.

3. In a problem about any system of linear equations, it is assumed that the system is an independent system.

4. In a problem involving a circle of diameter d it is assumed that d_1 is another diameter and that $d = d_1$.

To write each equation, state an assumption that has been made about the drawing.

5. $a^2 + b^2 = c^2$

6. $P = 4s$

For each problem, draw a diagram of the situation. State what assumptions are made and find each answer to the nearest whole number.

7. The length of a tree's shadow is 13 m. The angle of elevation of the sun is 37°. What is the height of the tree?

8. José is standing at a point 75 m from a base of the building. The angle of elevation from where he stands to the top of the building is 77°. How high is the building?

9. A jet flying at an altitude of 8700 m spotted another aircraft flying at the same altitude. A pedestrian who was standing on the ground directly under the jet spotted the other aircraft at an angle of elevation of 38°. How far was the jet from the other aircraft?

10. A sailboat was passing directly under a bridge where Lloyd was standing to watch his friend windsurf. If the angle of depression from Lloyd to his friend was 68° and he was 380 m above the water, how far was the sailboat from his friend?

11. A plane is flying at an altitude of 10,000 ft. The airport is 112 miles ahead. What is the angle of depression from the plane to the airport?

12. A truck traveled 1600 m down a hill. If it is now 64 m below the starting point, what is the angle of depression from the starting point to the truck?

13. An air-traffic controller is in a 78-meter-high tower and sees a plane, which is 7500 m away, at an angle of elevation of 38°. What is the altitude of the airplane?

Mixed Problem Solving Review

1. The length of a rectangle is four times its width. The number of square feet in its area is four times the number of feet in its perimeter. What are the dimensions of the rectangle?

2. The difference between two integers is 12. If twice the first integer is added to the second the sum is 84. Find the two integers.

3. Rob received grades of 85, 78, 80, and 94 on his first four exams. What is the lowest possible test score he can receive on his fifth exam to have an average greater than 85?

PROJECT

A definition of a term in mathematics gives a precise meaning to the term. A mathematical term is a name for a concept, such as a triangle, solution, or irrational number. Make a list of three definitions given in this chapter. Then change the definitions. Can an unintentional change made in a definition introduce a hidden assumption into a problem?

TEST YOURSELF

1. Given $\triangle ABC \sim \triangle AJN$, $AC = 10$, $AB = 12$, and $AJ = 4$, then $AN = \underline{\ ?\ }$ **15.4**

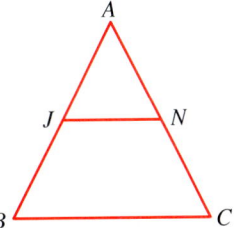

2. In right triangle XYZ, $YX = 3$, $YZ = 5$ and $m\angle Y = 90$. Use the Pythagorean theorem to find XZ. Then find sin X, cos X and tan X. **15.5**

3. In right triangle ABC $m\angle C = 90$, $AC = 12$ and $BC = 4$. Find $m\angle A$ and $m\angle B$ to the nearest whole number. **15.6**

4. A lighthouse is 246 ft tall. Find the angle of depression to the nearest degree if a ship is 3 mi from the base of the lighthouse. **15.7**

CHAPTER 15 SUMMARY AND REVIEW

Vocabulary

acute angle (615)
acute triangle (619)
angle (615)
angle of depression (645)
angle of elevation (643)
complementary angles (615)
congruent angles (623)
congruent line segments (623)
corresponding parts (624)
cosine (634)
equilateral triangle (619)
geometry (614)
isosceles triangle (619)
line (614)
line segment (614)
obtuse angle (615)
obtuse triangle (619)
protractor (615)
ray (614)
right angle (615)
right triangle (618)
scalene triangle (619)
similar triangles (628)
sine (634)
supplementary angles (615)
tangent (634)
trigonometry (633)
trigonometric ratios (633)
vertex (615)

Basic Geometric Figures Angles can be classified as follows: A **right angle** has a measure of 90°. An **acute angle** has a measure between 0° and 90°. An **obtuse angle** has a measure between 90° and 180°. **Complementary angles** are two angles whose measures have a sum of 90°. **Supplementary angles** are two angles whose measures have a sum of 180°. — 15.1

1. If $\angle M$ and $\angle N$ are supplementary, find $m\angle N$ if $m\angle M = 67$.

2. The measure of an angle is 20 more than the measure of its complement. Find the measure of the angle and its complement.

Triangles Triangles can be classified as follows: A **right triangle** has one right angle. An **acute triangle** has all acute angles. An **obtuse triangle** has one obtuse angle. An **isosceles triangle** has at least two sides of equal length. An **equilateral triangle** has all sides of equal length. A **scalene triangle** has no sides of equal length. — 15.2

The sum of the measures of the angles of a triangle is 180.

3. Classify $\triangle RST$ according to the measures of its angles and according to the lengths of its sides.

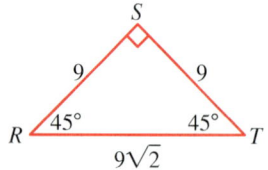

4. Given $\triangle MOP$, find $m\angle M$ if $m\angle O = 55$ and $m\angle P = 63$.

Congruence **Congruent angles** are angles that have the same measure. **Congruent line segments** are line segments that have equal length. **Congruent figures** are figures that have the same size and shape. — 15.3

$\triangle PAD \cong \triangle CAR$.

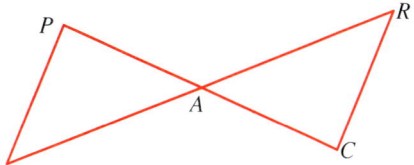

5. Name the corresponding angles.
6. Name the corresponding sides.
7. *True* or *false*? $\triangle DAP \cong \triangle ARC$

Similar Figures In **similar triangles**, corresponding angles are congruent and corresponding sides are proportional.   15.4

8. $\triangle DOG \sim \triangle CAT$. Find the lengths of the two missing sides if $DO = 4$, $GO = 3$, $CA = 10$, and $CT = 15$.
9. $\triangle ABC \sim \triangle ADE$, $AB = 15$, $AD = 25$, $DE = 50$, and $AC = 32$. Find AE and BC.

Trigonometric Ratios Trigonometric ratios for right $\triangle ABC$ are as follows: 15.5

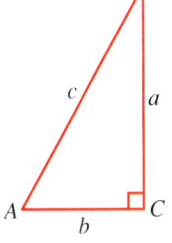

sine of $\angle A = \dfrac{\text{length of leg opposite } \angle A}{\text{length of hypotenuse}}$; $\sin A = \dfrac{a}{c}$

cosine of $\angle A = \dfrac{\text{length of leg adjacent } \angle A}{\text{length of hypotenuse}}$; $\cos A = \dfrac{b}{c}$

tangent of $\angle A = \dfrac{\text{length of leg opposite } \angle A}{\text{length of leg adjacent } \angle A}$; $\tan A = \dfrac{a}{b}$

10. Use the Pythagorean theorem to find the length of $\overline{AB}$. Find the values for $\sin A$, $\cos A$, $\tan A$, $\sin B$, $\cos B$, and $\tan B$.

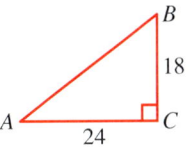

Use the table on page 671.   15.6

11. Find $\cos 29°$.
12. Find x if $\tan x = 0.5774$.
13. Given right $\triangle XYZ$, $m\angle Y = 90$. Find the lengths of the other sides to the nearest whole number if $m\angle X = 34$ and $XY = 40$.

For Exercises 14–15, refer to the problem below.   15.7

An 8-meter ladder rests against the side of a house. If the foot of the ladder is 2 m from the house, find the angle the ladder makes with the ground.

14. Draw a diagram of a right triangle to represent the situation.
15. Give the trigonometric ratio that relates the known sides of the right triangle to the unknown angle. Find the angle to the nearest degree.

CHAPTER 15 TEST

1. Graph $x \leq 2$ on a number line and state whether the figure represents a point, a line, a line segment, or a ray.

2. *True* or *false*? Angles with measures of 72 and 18 are complementary.

3. If $\angle M$ and $\angle N$ are complementary, find $m\angle N$ if $m\angle M = 71$.

4. The measure of an angle is 18 more than the measure of its supplement. Find the measure of the angle and its supplement.

5. Classify $\triangle PRQ$ (a) according to the measures of its angles and (b) according to the lengths of its sides.

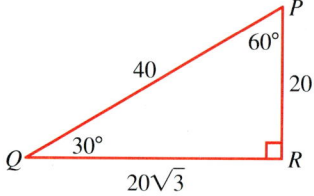

6. Given $\triangle BOG$, find $m\angle B$ if $m\angle O = 35$ and $m\angle G = 21$.

7. $\triangle MNO$ is isosceles, $m\angle N = 100$, and $MN = NO$. Find $m\angle M$.

For Exercises 8–9, $\triangle RIG \cong \triangle TIN$.

8. Name the corresponding angles and sides.

9. *True* or *false*? $\triangle GIR \cong \triangle NIT$.

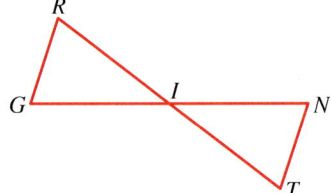

10. $\triangle HIT \sim \triangle POS$. Find the lengths of the missing sides if $HI = 6$, $PO = 9$, $HT = 5$, and $SO = 15$.

11. $\triangle JAM \sim \triangle JFT$, $JA = 12$, $JF = 18$, $AM = 15$, and $JT = 10$. Find FT and JM.

12. Use the Pythagorean theorem to find the length of $\overline{AB}$. Find the values for $\sin A$ and $\sin B$.

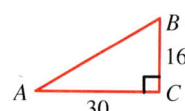

Use the table on page 671 for Exercises 13–14.

13. Find $\cos 71°$.

14. Find x if $\tan x = 2.7475$.

15. Given right $\triangle XYZ$, $m\angle Y = 90$, find the other sides to the nearest whole number if $m\angle X = 68$ and $XY = 10$.

Challenge

Find the acute angle that is formed by the line $y = \frac{1}{4}x - 3$ and the *x*-axis.

PREPARING FOR STANDARDIZED TESTS

Select the best choice for each question.

1. Solve the equation:
   $x^2 - 6x - 5 = 0$.
   A. 5, 1   **B.** $-1, -5$
   C. $3 \pm \sqrt{14}$ **D.** $-3 \pm \sqrt{14}$
   E. $\dfrac{6 \pm \sqrt{58}}{2}$

2. Find $f(-2)$ if $f(x) = 3x^2 + 5x + 1$.
   A. 23   **B.** 21   **C.** 3   **D.** 2   **E.** 1

3. The graph of $y = x^2 - 4x + 3$ is a parabola intersecting the y-axis at a point whose coordinates are:
   A. $(3, 0)$ **B.** $(0, 3)$ **C.** $(0, 2)$
   D. $(1, 0)$ **E.** $(0, 1)$

4. Factor $6x^2 + 5x - 6$.
   A. $(3x - 2)(2x + 3)$
   B. $(3x + 2)(2x - 3)$
   C. $(3x + 3)(2x - 2)$
   D. $(6x + 1)(x - 6)$
   E. $(6x - 1)(x + 6)$

Use this information for questions 5–8.

The heights of 11 students were recorded in inches as 57, 52, 61, 57, 49, 54, 61, 57, 53, 51, and 53.

5. The mean is:
   A. 54   **B.** 55   **C.** 56   **D.** 57   **E.** 61

6. The median is:
   A. 54   **B.** 55   **C.** 56   **D.** 57   **E.** 61

7. The mode is:
   A. 54   **B.** 55   **C.** 56   **D.** 57   **E.** 61

8. The range is:
   A. 4   **B.** 6   **C.** 8   **D.** 10   **E.** 12

9. What is the conjunction of these two inequalities?
   $$x < 5,\ x > -2$$
   A. $-2 < x < 5$
   B. $-2 < x$ or $x < 5$
   C. $-2 > x > 5$
   D. $x > 5$ or $x < -2$
   E. $x > 5$ and $x < -2$

10. What is $\dfrac{3}{4}\%$ of 64?
    A. 48   **B.** 4.8   **C.** 0.48
    D. 0.048   **E.** 0.0048

11. Ira bought two pens at $1.66 each, a box of pencils for $1.89, and a package of paper for $2.99. If the sales tax rate is 5%, what was the total Ira paid for these items?
    A. $7.56   **B.** $7.57   **C.** $7.61
    D. $8.61   **E.** $8.62

12. What is the probability that in a throw of a die, a multiple of two will appear?
    A. $\dfrac{1}{12}$ **B.** $\dfrac{1}{6}$ **C.** $\dfrac{1}{3}$ **D.** $\dfrac{1}{2}$ **E.** $\dfrac{2}{5}$

13. What is the slope of a line that is the graph of $5x - 2y = 8$?
    A. $\dfrac{5}{2}$ **B.** $-\dfrac{5}{2}$ **C.** $\dfrac{2}{5}$ **D.** $-\dfrac{2}{5}$ **E.** -4

14. Solve for x:
    $(x - 2)(x + 3) - 5 = (x - 1)(x + 1)$
    A. -12 **B.** -10 **C.** 8
    D. 10   **E.** 12

15. The probability of a particular event happening is $\dfrac{2}{7}$. What are the odds *against* it occurring?
    A. $7:5$ **B.** $7:2$ **C.** $5:7$
    D. $2:5$ **E.** $5:2$

CUMULATIVE REVIEW (CHAPTERS 1–15)

Find the distance between each pair of points.

1. $A(2, 3)$ and $B(5, 7)$
2. $X(-1, -2)$ and $Y(7, 4)$
3. $M(3, -2)$ and $N(6, 1)$
4. $C(5, -4)$ and $D(-3, 1)$
5. $R(-1, 2)$ and $T(3, 6)$
6. $J(-2, -5)$ and $K(2, 1)$

Simplify. Assume that no variable equals zero.

7. $|2 - (3\sqrt{2})^2|$
8. $\sqrt{3}(2 + \sqrt{6})$
9. $3\sqrt{12} + 3\sqrt{18} - 2\sqrt{27}$
10. $a - 3(a - 2)$
11. $\sqrt{20c^2d^3}$
12. $(2b^3 - 1)^2$
13. $(2z - 6)(3z + 2)$
14. $(2\sqrt{3} - 1)^2$
15. $(2r^2t)^3(3rt^3)$
16. $\dfrac{(8m^3n^5)^2}{(-4m^4n^3)^2}$
17. $\dfrac{2\sqrt{3}}{3 + \sqrt{2}}$
18. $\dfrac{3x - 7}{5} - \dfrac{3 - 2x}{5}$
19. $\dfrac{24d^6}{\sqrt{5}} \cdot \dfrac{45}{\sqrt{6d^2}}$
20. $\dfrac{4x^2 - 9}{x + 2} \div \dfrac{2x + 3}{x^2 - 4}$
21. $\dfrac{2x^2 + 2x}{x^2 - 9} + \dfrac{x + 1}{3 - x}$
22. $(2.5 \times 10^6)(0.3 \times 10^{-2})$
23. $7(3c - 2) + 6 - 5c$
24. $(2 \div 3)2^2 + 8 \div 3$
25. $(2\sqrt{3} + 3\sqrt{2})(3\sqrt{3} - 2\sqrt{2})$
26. $-\dfrac{2}{3}g^2h(12gh^3 - 3h)$
27. $(8x^3 + 27) \div (2x + 3)$
28. $(9p^2 - 3p + 1)(3p + 1)$
29. $(2a^2b - 3c)(2ab^2 + 3c)$
30. $\dfrac{36w^5 - 48w^3 + 12w^2}{-12w^2}$
31. $\dfrac{8m^2 - 6mn - 9n^2}{12m^2 + mn - 6n^2}$

Write an equation, inequality, or system for each problem. Solve.

32. The sum of the digits of a two-digit number is 13. If the digits are reversed then new number is 27 more than the original. Find the original number.

33. Find the first of three consecutive integers if the product of the first and the second is 34 less than the square of the third.

34. Thirty mL of a 15% acid solution is to be changed to a 10% acid solution by adding water. How much water is to be added?

35. Mae's scores on three math tests are 81, 92, and 86. What is the lowest she can get on the fourth test to have an average greater than 85?

Factor completely. If the polynomial cannot be factored, write "prime."

36. $16x^2 - 9y^2$
37. $t^2 + 2t - 35$
38. $c^2 - 13c + 36$
39. $2m^2 + m - 15$
40. $6m^2 + 5m - 6$
41. $6a^3 - 24ab^2$
42. $8a^3b + 12a^2b^2 - 4a^2b$
43. $rt^2 - 9r + t^3 - 9t$

Solve each equation, inequality, or system. If there is no solution, state that fact.

44. $0.2x + 3.9 = 1.5$
45. $\frac{3}{5}r + 10 = 19$
46. $2 - 3(a - 4) = 2a - 1$
47. 18% of $m = 90$
48. $2y^2 - 3y = 27$
49. $c^2 + 5c = 5c + 4$
50. $8 - 5t > 13$
51. $|d - 7| = 3$
52. $\sqrt{3n + 1} = 10$
53. $|p| + 6 = 5$
54. $|4 - 3x| < 10$
55. $\sqrt{b^2} + 2 = b - 2$
56. $\frac{u}{u+1} = 2 - \frac{3}{u+1}$
57. $\begin{cases} 2x + 3y = 5 \\ 5x + 3y = 11 \end{cases}$
58. $\frac{2h}{h^2 - 4} + 3 = \frac{3h}{h+2}$
59. $|w - 5| > 1$
60. $9x^2 = 18$
61. $|a - 2| \leq 3$
62. $\begin{cases} b = 2a + 1 \\ a + 3b = 30 \end{cases}$
63. $(m - 2)^2 = 36$
64. $\begin{cases} 6r + 7t = 24 \\ 2r - 3t = -8 \end{cases}$
65. $d^2 - 6d = 16$
66. $3x^2 = 2x + 5$
67. $5a^2 = 20(a + 3)$
68. $p - 5(p - 2) + (p - 3) = 2$
69. $5 + \sqrt{3a - 2} = 9$

Evaluate $b^2 - 4ac$ for each equation, then state the nature of the solutions.

70. $2x^2 - x + 1 = 0$
71. $x^2 + 3x - 2 = 0$
72. $x^2 - 2x - 1 = 0$

Write a quadratic equation with the given roots.

73. 2, 5
74. $\frac{1}{3}$, 2
75. $-3, \frac{1}{2}$

Find the mode, median, mean, and range for each set of data.

76. 6, 8, 8, 9, 14, 15
77. 21, 32, 63, 21, 35, 32

78. Gloria scored the following number of runs in 5 games: 4, 2, 0, 4, 5. Find the range, variance, and standard deviation.

There are 4 green, 5 red, and 3 yellow cards in a box. Find each probability. One card is picked at random.

79. $P(\text{green})$
80. $P(\text{green or red})$
81. $P(\text{blue})$

Cumulative Review **653**

One card is picked, replaced, and another card is picked.

82. P(1st green and 2nd green)
83. P(1st green and 2nd not green)

One card is picked, not replaced, and another card is picked.

84. P(1st green and 2nd green)
85. P(1st red and 2nd green)
86. The measure of an angle is 20 less than its complement. Find the measure of the angle and its complement.
87. In triangle ABC, the measure of angle B is twice $m\angle A$. The measure of angle C is 20 more than $m\angle A$. Find the measure of each angle.
88. The length of a tree's shadow is 8 ft. The angle of elevation of the sun is 33°. What is the height of the tree to the nearest whole number?
89. Find the area of a rectangle whose length is $3\sqrt{x}$ cm and width is $2\sqrt{x}$ cm.
90. A square lot has an area of 625 square feet. How many feet of fencing is needed to enclose the lot?

Find the slope of a line that contains the given points.

91. $A(3, 2); B(4, 6)$
92. $C\left(-\frac{1}{2}, 0\right); D\left(\frac{3}{4}, 1\right)$

Through the given point, draw a line with the given slope.

93. $m = \frac{1}{2}; T(4, 5)$
94. $m = -\frac{3}{5}; S(10, 2)$

Find the solution set of each sentence. The replacement set is $\{-2, 0, 1, 3, 4\}$.

95. $3x - 1 = 8$
96. $z + 5 = -3$
97. $2m - 5 = 3$
98. $y - 1 < 5$
99. $s + 2 > 3$
100. $3 > 2y + 1$

101. A department store finds that its profit for selling x units of a certain product is given by the equation $p = x^2 - 30x + 125$. For which values of x is the profit zero?
102. The length of a rectangle is 5 m longer than its width. What are the dimensions if its area is 36 m²?

EXTRA PRACTICE

Chapter 1 Real Numbers

Evaluate each expression if $u = -5$, $w = \frac{4}{3}$, $x = 3$, $y = -\frac{1}{2}$, $z = -4.1$ 1.1–1.8

1. $w + 6$
2. $(wx) - 1$
3. $(9w) \div 4$
4. $3 + z$
5. $y + z$
6. $|x + y|$
7. $y - x$
8. $5 - (y - z)$
9. $|z| - |y|$
10. $(xyw) + z$
11. $w[x + (6y)]$
12. $\left(\dfrac{y}{w}\right) - (2z)$
13. $u + z - 11.3$
14. $(uw) - z + 0.3$
15. $\dfrac{|(6w) - (4y)|}{-0.5}$

Write an algebraic expression for each phrase.   1.1

16. The sum of a number i and five
17. 55 decreased by a number n
18. The number of miles traveled m divided by the number of hours h elapsed

For Exercises 19–26, refer to the number line below.   1.2

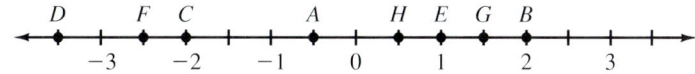

Give the coordinate of each point.

19. D
20. A
21. B
22. G

Name the graph of each number.

23. 1
24. -2
25. $-2\frac{1}{2}$
26. 0.5

Show that each number is a rational number.   1.2

27. -7
28. 0
29. $3\frac{2}{3}$
30. -1.26

Simplify.   1.3, 1.7

31. $-(-3)$
32. $-[-(-8)]$
33. $(-7)(-9)\left(\frac{1}{3}\right)$
34. $\frac{2}{5}(10)(-x)$

Replace each __?__ with >, <, or = to make a true statement.   1.3–1.6

35. -2 __?__ 0
36. -4 __?__ -3
37. $-\frac{1}{2}$ __?__ $-\frac{1}{3}$
38. -2.1 __?__ -2.1

39. $-11.1 + (-4.6) + 2.3$ __?__ $11.1 + (-4.6) - (+2.3)$

40. $0.56 + 2.12 + (-2.12)$ __?__ $-0.56 + (-2.21) + 2.21$

41. The temperatures at sunrise for 1 week were: $-1.5°F$, $-3.1°F$, $2.2°F$, $0.4°F$, $-1.1°F$, $3.6°F$, and $0.5°F$; what was the average temperature at sunrise?

Chapter 2 Algebraic Expressions

Evaluate each expression if $a = 4$, $b = -2$, $c = 0$, $d = -5$, $e = 8$. 2.1–2.2

1. $2b - e$
2. $e - \dfrac{a}{c}$
3. $3[a - 2(4b - 3e) + 3c]$
4. $\dfrac{-c + d}{2ae}$
5. $5b^3 - 2d^2$
6. $(2 - ab)(e - c) \div (2d + a)$
7. $\dfrac{e^2 - a^2}{(e - a)^4}$
8. $4ac(2e + b)$
9. $-4(2d^2 + b) \div (c^2 - e^2)$

Which property of real numbers justifies each statement? 2.3

10. $-7(a + 11) = -7(a) + (-7)(11)$
11. $(-3 + 9) + 7 = -3 + (9 + 7)$
12. $5(j + k) = 5(k + j)$
13. $-2(3x + y) = (3x + y)(-2)$

Simplify. 2.4–2.5

14. $5r^2 - 2r + r^2 - r^3 - r$
15. $j^2k^2 + 3jk^2 - j^2k^2 + 5j^2k + jk$
16. $5m^2 - [2m^2 - m(2m + 1)]$
17. $3x - 2[x^2 - x - x^2 + 3x(x + 2)]$
18. $2g - [(-h + g) - (g^2 - h) - g^2 + (h + g)]$

Translate each word expression to an algebraic expression. 2.6

19. Three less than a number
20. The product of five and a number squared
21. Eight times the sum of a number squared and one

Write an algebraic expression for each phrase.

22. Let s be Sarah's age now.
    a. What was Sarah's age 5 years ago?
    b. Greg is 6 years younger than Sarah. Represent Greg's age now.
    c. What was Greg's age 5 years ago?

Write an algebraic equation for each sentence. 2.9

23. Thirteen reduced by a number n is -24.
24. The product of a number and -6 reduced by 2 is 10.
25. Thirty-eight increased by the quotient of 7 and some number is 50.
26. Adult tickets cost \$10 and children's \$7.50. The total value of the tickets is \$1520.

Find each solution set for a replacement set $\left\{-3, 1, 0, \dfrac{1}{3}\right\}$. 2.7

27. $-2 > 3y + 5$
28. $-2y = 1$
29. $y + 2 > -4$

Chapter 3 Equations in One Variable

Solve each equation. 3.1–3.3

1. $a + 26 = 31$
2. $98 = 14m$
3. $\frac{1}{2}y = -17$
4. $4.5 + b = 11.2$
5. $\frac{f}{8} = -3$
6. $-\frac{2}{3} = \frac{5}{3} + c$
7. $-2.23 + s = 0$
8. $0 = -\frac{9}{4}r$
9. $3.45 - (-t) = 2.82$
10. $d - 4.2 = -3.1$
11. $-9.6 = -h$
12. $-8.21 + e = -8.21$
13. $4k + 3 = 11$
14. $-4 + \frac{m}{6} = 2$
15. $6.4p - 1.5 = -0.22$

16. Find a number n if this number reduced by 11 is -15. 3.4, 3.7, 3.8

17. Find a number n if $\frac{7}{8}$ of this number is 49.

18. Find a number n if 0.25 of this number decreased by 7 is -9.

19. Jan's father is 56 years old. If he is $3\frac{1}{2}$ times as old as Jan, how old is Jan?

20. Prove that for all real numbers a, b, and c, if $a = b$, then $c - a = c - b$. 3.5

    Proof:

Statements	Reasons
1. $c = c$	1.
2. $c - a = c - a$	2.
3. $a = b$	3.
4. $c - a = c - b$	4.

21. Prove that for all real numbers a, b, and c, $c \neq 0$, if $\frac{a}{c} = \frac{b}{c}$, then $a = b$.

    Proof:

Statements	Reasons
1. $\frac{a}{c} = \frac{b}{c}$	1.
2. $\frac{a}{c} = \frac{1}{c} \cdot a;\ \frac{b}{c} = \frac{1}{c} \cdot b$	2.
3. $\frac{1}{c} \cdot a = \frac{1}{c} \cdot b$	3.
4. $c \cdot \frac{1}{c} \cdot a = c \cdot \frac{1}{c} \cdot b$	4.
5. $a = b$	5.

22. If $P = 2l + 2w$, find l when $P = 48$ and $w = 11$. 3.6

23. $V = \pi r^2 h$. Find h if $r = 4$, $\pi \approx 3.14$, and $V = 251.2$.

24. $A = \frac{1}{2}h(a + b)$. Find b if $A = 60.4$, $h = 2.0$, and $a = 3.8$.

Chapter 4 More Equations in One Variable

Solve and check. 4.1, 4.2, 4.4

1. $-3a + 5a = 14$
2. $-15 = 4b - 7b + 3$
3. $3c + 2(c + 2) = -6$
4. $-4(5d + 6) = 2$
5. $5e + 1 = 7e + 7$
6. $3(4f + 6) = -2f + 18$
7. $2g + 4(3 - g) = 5g + 19$
8. $-h + 2(h + 8) = -2(1 - 3h) + 3$
9. $-3(2a + 5) = 9$
10. $11k - 2k + 4 = 15 - 2k$
11. $-0.5l + 0.7 = 1.4 - 0.4l$
12. $0.03(2m - 3) + 0.05 = 5$

13. Write three consecutive odd integers starting with odd integer i.   4.3
14. Write the two consecutive integers preceding integer c.
15. Find five consecutive integers if the sum of the first and the third is -28.
16. What is 45% of 350?
17. What percent of 108 is 24?   4.5
18. 0.15 is 30% of what number?
19. What percent of $50,000 is $5?
20. What is the new price of a $12.50 item reduced by 30%?   4.6
21. If a haircut cost $12 last year and now costs $15.60, by what percent has the price increased?
22. After a 20% weight loss, Gemma now weighs 124 lb. What was her original weight?
23. How many mL of water should be added to 15 mL of a 20% solution of formic acid in water to produce a 5% solution?   4.7
24. How many grams of ester must be added to 25 g of a 10% ester solution in alcohol in order to make a 25% solution?

Solve each equation for the underlined variable. 4.8

25. $p = r + \underline{s} + t$
26. $5\underline{a}c = 3d$
27. $2(\underline{p} - 3q) = 5\underline{p}$

Make a sketch and a table to solve these problems. 4.9–4.10

28. Two planes traveling at 540 mi/h and at 630 mi/h start from the same place at the same time and fly in opposite directions. How long will it take for the two planes to be 1755 mi apart?
29. Robin can ride at an average rate of 25 km/h. She sets out to overtake a cyclist averaging 20 km/h. If the second cyclist has a 1 h headstart, how long will it take Robin to overtake the cyclist?
30. A truck and a car leave the same town and travel in the same direction. The car travels at 40 mi/h and the truck at 55 mi/h. How many minutes will it take before they are 10 mi apart?

Chapter 5 Inequalities in One Variable

Draw the graph of each inequality. 5.1

1. $x > -2$ 2. $y \leq 3$ 3. $z = 0$ 4. $v < -0.5$ 5. $w \neq -1$

Solve. Then draw the graph of each inequality. 5.2–5.5

6. $a + 11 > -6$
7. $4b \leq -15$
8. $-2c + 5 + 3c < 7$
9. $-\frac{2}{3} \geq d + \frac{1}{6}$
10. $-\frac{5}{3} > -\frac{5}{4}e$
11. $3(2f - 1) - 5f \leq 7$
12. $-2g \geq 0$
13. $-0.12 + j \leq -1.002$
14. $h - 2h < -1.5$
15. $5l - 10 > 16 - 8l$
16. $4 - (1 - t) \geq 2t - 3$
17. $3(1 - m) > -9m$
18. $-6n > 3$ or $-5 \geq -5n + 20$
19. $8(2 - p) > 0$ and $p \leq -2$
20. $-5 + 3q \geq 4$ or $6 - 5q > -9$
21. $-7 < 5 - 3r < 2$

Solve. Then draw the graph of each equation or inequality. 5.6–5.7

22. $|x| = 1.1$
23. $|z - 2| = 3$
24. $9 = |3v + 6|$
25. $-3 + |2w - 1| = 1$
26. $-6|1 - 2p| = -3$
27. $-\left|\frac{2g}{3}\right| = 19$
28. $|m| < 0.5$
29. $14 \leq |n + 8|$
30. $|3l - 5| \leq 0$
31. $-3|2h + 1| < 0$
32. $-2(3 - |1 - 2e|) < 3$
33. $5 + |2 - g| \geq 6$

Write the inequality represented by each graph. 5.2–5.7

34.    35.

36.    37.

38.    39.

Write and solve an inequality for each problem. 5.8

40. If Mark must not spend more than $15 for stamps and the stamps cost $0.35 each, what is the greatest number he can buy?

41. If the circumference of a round plate must be less than 36 in., what can be said about the plate's radius?

42. To meet design specifications, the perimeter of a photograph must be at least 150 cm. If the width is 19 cm, what restrictions must be placed upon the length?

Chapter 6 Polynomials

Simplify. Assume that no variable equals zero.   6.1–6.3

1. $-2h^4 \cdot 11h^3$
2. $(1.5 \times 10^8)(2.0 \times 10)$
3. $5jk^2 \cdot (-7j^2k^2)$
4. $\dfrac{b}{b^4}$
5. $\dfrac{-14c^3d^2}{7c^2d^3}$
6. $\dfrac{1.8 \times 10^4}{0.3 \times 10}$
7. $-f^0$
8. $(a^2)^4$
9. $(-3bc^4)^2$
10. $\dfrac{5ef^7}{-15e^4f^2g}$
11. $\left(\dfrac{2c^2}{d}\right)^3$
12. $\dfrac{(-5gh^3)^2}{20g^2h}$

Evaluate if $x = 3$, $y = -2$, and $z = -\dfrac{1}{2}$.

13. $5x^3(yz^5)$
14. $\dfrac{-7x^2 \cdot (4z^3)^2}{9y^2}$
15. $\dfrac{-5y^5}{(-z^7)^2}$

Rewrite each in scientific notation.   6.4

16. 0.00814
17. $23{,}100{,}000$
18. 0.012×10^{-3}
19. $(0.7 \times 10^2)^2$
20. $(3.4 \times 10^{-2})(4.2 \times 10)$
21. $\dfrac{1.9 \times 10^2}{38 \times 10^2}$

Simplify. Then write in descending order with respect to x.   6.5–6.6

22. $-xy^2 + 2x^2y^2 + 3xy^2 - 7$
23. $-2(ax^2 - 3a^2x^3) + 5ax^2$
24. $\begin{array}{r} -8x^3 + 5x^2 + x - 11 \\ + \; 5x^3 + x^2 - x + 6 \end{array}$
25. $\begin{array}{r} 6x^4 - 7x^3 - 2x^2 + x - 9 \\ -(-2x^4) + 9x^3 + 2x^2 + 2x + 9 \end{array}$
26. $(2d - 7x) + (d - 5x)$
27. $(x^4 + 2x^3 - x + 5) - (3x^3 + 2x^2)$

Multiply. Simplify, if possible.   6.7–6.8

28. $3y(-2y^2 + y - 3)$
29. $(7z^3 - z^2 + 10)(-2z^3)$
30. $-2a^2b(8ab^2 - 7a^3 + b)$
31. $w(6w^2 - 5w) + 2w^2(7w - 8)$
32. $(2x + 3)(x + 5)$
33. $(c - 7)(c + 11)$
34. $(5t - 2)(2t - 5)$
35. $(4p + 7)(3p^2 - p + 2)$
36. $(2a + b)(3a - 7b)$
37. $(5d^3 - 2d^2 - d)(3d^2 - 4)$

38. If the width of a rectangle is 6 cm shorter than its length l, write a polynomial to express its area.

Write as a polynomial.   6.9

39. $(12x + 7)^2$
40. $(4a - 1)(4a + 1)$
41. $(-2b + c)^2$
42. $(5c^2 - 2d^3)^2$
43. $(13x^2 - y)(13x^2 + y)$
44. $(11 - 3e^3)^2$

45. Use a pattern to find an algebraic expression that can be used to find the sum of the first n positive even integers.   6.10

Chapter 7 Factoring Polynomials

Give the prime factorization of each number. 7.1

1. 48
2. 720
3. 41
4. 960
5. 4375
6. 1119

Factor each polynomial completely, if possible. 7.2–7.8

7. $12m^2 + 27m^3 - 3m - 6$
8. $8r^5 - 4r^4 + 32r^3 + 6r$
9. $-3a^2b + 15a^3b^2 - 18ab^3$
10. $14xy^2 - 21x^4 + 35y^2 - 28$
11. $y^2 + 7y + 12$
12. $b^2 - 6b + 8$
13. $x^2 - 25x + 144$
14. $n^2 + 17n + 72$
15. $j^2 - 22j + 121$
16. $s^2 + 7sk + 12k^2$
17. $t^2 + t - 2$
18. $r^2 - 5r - 150$
19. $m^2 + 11m - 42$
20. $b^2 - 9b - 6$
21. $x^2 - 5xy + 6y^2$
22. $m^2 + 14mn - 72n^2$
23. $2h^2 + 3h + 1$
24. $5m^2 - 6m + 1$
25. $2t^2 - 5t + 2$
26. $3n^2 + 2n - 1$
27. $4k^2 + 3k - 1$
28. $7l^2 + 12l + 5$
29. $2x^2 + 5x - 12$
30. $6y^2 - 13y + 5$
31. $4z^2 + 11z + 6$
32. $10x^2 - x + 6$
33. $16q^2 + q - 15$
34. $4g^2 + 13gh - 12h^2$
35. $39k^2 + 37k + 8$
36. $6a^2b^2 - ab - 40$
37. $5c^2 - 7c - 15$
38. $25y^2 - 9$
39. $4w^2 + 4w + 1$
40. $9u^2 - 6ur + r^2$
41. $1 - 49a^2$
42. $4m^2 + 2mn - 2n^2$
43. $a^2b^4 - c^6$
44. $196 - 121b^2$
45. $36 - 60v + 25v^2$
46. $81x^2 - 234x + 169$
47. $16q^2 + 48$
48. $3a^4 - 3a^2$
49. $7x^2 + 14xy + 28y^2$
50. $1 - x - y + xy$
51. $2p^3 + 3p^2 - 20p$
52. $5 - 40w^2 + 80w^4$
53. $a^2 + 2a + 1 - b^2$
54. $(3x - 2)^2 - 4y^2$
55. $162 - 2d^4$

Solve and check. 7.9

56. $(2y + 3)(3y - 2) = 0$
57. $5m - 6m^2 = 0$
58. $v^2 + 10v + 25 = 0$
59. $j^2 - 6j = 7$
60. $3x^2 - 2x - 1 = 0$
61. $(2a + 4)(a - 1) = 20$
62. $3a^3 + 6a^2 + 3a = 0$
63. $36x^2 - 49 = 0$

64. The square of a number reduced by three times the number gives 28. Find the number. 7.10

65. The height of a triangle is 5 m less than its base. If its area is 33 m², find its dimensions.

66. Find three consecutive integers such that the square of twice the first is 30 more than three times the product of the second and third.

Chapter 8 Rational Expressions

Give the values of the variable for which each expression is undefined. 8.1

1. $\dfrac{5}{4x}$
2. $\dfrac{-3y}{y+1}$
3. $\dfrac{a+3}{a^2+3a}$
4. $\dfrac{z^2-1}{z^2-2z+1}$

Simplify each expression if possible. Assume that no denominator is zero.

5. $\dfrac{15v}{27v^3}$
6. $\dfrac{d+3}{d-3}$
7. $\dfrac{a-b}{b-a}$
8. $\dfrac{2x^2-5x-3}{x-3}$

Multiply or divide. Express your answer in simplest form. 8.2–8.3

9. $-\dfrac{5m^2}{2n} \cdot \dfrac{6n^4}{10m}$
10. $\dfrac{x-3}{x+5} \cdot \dfrac{2x+10}{3-x}$
11. $\dfrac{2v+3}{2v^3} \cdot \dfrac{v^3+4v^2}{3+2v}$
12. $\dfrac{2y^2-3y-2}{3y^2} \cdot \dfrac{12y^2-6y}{y-2}$
13. $\dfrac{-14c^2}{d} \div \dfrac{21c^3}{d^2}$
14. $\dfrac{9h^2+3h}{h+1} \div (6h+2)$
15. $\dfrac{5j-15}{6j} \div \dfrac{6-2j}{9j^2}$
16. $\dfrac{x^3-4x}{3x} \cdot \dfrac{x-5}{x+2} \div \dfrac{x-2}{x+5}$

Add or subtract. Express your answer in simplest form. 8.4–8.6

17. $\dfrac{2}{3ab} - \dfrac{5}{2a^2b} + \dfrac{6}{a^2b^2}$
18. $\dfrac{y}{x-y} - \dfrac{x}{y-x}$
19. $\dfrac{d^2+3d+2}{d^2-4} - \dfrac{1}{2-d}$
20. $\dfrac{2n^2+2mn}{m^2-n^2} + 1$
21. $\dfrac{\dfrac{1}{x}-\dfrac{3}{x^2}}{2x-\dfrac{1}{x^3}+1}$
22. $\dfrac{1-\dfrac{2}{c}-\dfrac{3}{c^2}}{\dfrac{2}{c}+\dfrac{2}{c^2}}$

Divide. 8.7

23. $(5j^2 - 10jk^2l + 25j^2kl^2 - 30kl) \div 20jkl$
24. $(3x^3 - 2x^2 + 3x - 5) \div (x-2)$
25. $(y^3 - 27) \div (y-3)$

Express as a ratio in simplest form. 8.8

26. $39:169$
27. $3 \text{ m}:20 \text{ cm}$
28. $3.25 \text{ h}:20 \text{ min}$

Solve and check. 8.9

29. $\dfrac{1}{6}x^2 + \dfrac{1}{3}x = \dfrac{5}{2}$
30. $\dfrac{8}{t^2-t} + \dfrac{t}{1-t} = \dfrac{7}{2}$

31. If Sue walks 2 km/h faster, she can cover 16 km in the same time it takes to walk 12 km at her present rate. How fast is she walking? 8.10

Chapter 9 Linear Equations

Plot each point in the coordinate plane. Tell its quadrant or axis. 9.1

1. $A(3, 5)$
2. $B(-2, 3)$
3. $C(4, 0)$
4. $D(-4, -1)$
5. $E(0, -2)$

Determine whether the ordered pairs are solutions of the equation. 9.2

6. $3y - 2x = 0$: $(3, 2)$, $\left(-\frac{3}{2}, -\frac{2}{3}\right)$, $(-2, -3)$, $\left(\frac{5}{4}, \frac{5}{6}\right)$, $\left(-4, -\frac{8}{3}\right)$

From the graph of $x + y = -1$, determine whether the ordered pairs given are solutions of the equation $x + y = -1$. Check your answers.

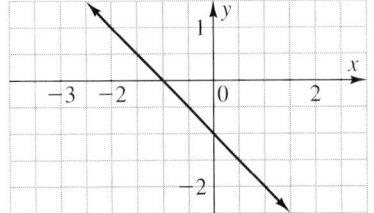

7. $(2, -1)$
8. $(-1, 0)$
9. $(0, -1)$
10. $(1, 1)$

The cost c ($) of w (oz) of milk is given by $c = 0.015w + 0.20$. Graph this equation and answer the following questions. 9.3

11. Approximately how much does 24 oz of milk cost?
12. Approximately how much milk can be purchased for $2?

Find the slope of a line containing the given points. 9.4

13. $(-4, 3)$; $(2, -5)$
14. $(7, -3)$; $(7, 1)$
15. $\left(-\frac{3}{2}, -1\right)$; $\left(\frac{4}{3}, -\frac{1}{5}\right)$

Through each given point, draw a line with the given slope.

16. $(-1, -3)$; 2
17. $(-2, 0)$; $-\frac{3}{4}$
18. $(5, -3)$; no slope

Find the slope and y-intercept of each equation. Then graph the equation. 9.5

19. $3x + y = -2$
20. $2x - 5y = 0$
21. $-\frac{1}{3}x + \frac{1}{2}y = -\frac{3}{2}$

Determine if the graphs of each pair of equations are parallel lines.

22. $y = -2x + 4$; $4x + 2y = 0$
23. $y = -\frac{4}{3}x + \frac{3}{5}$; $15y - 20x = 0$

Write an equation in standard form for each line described. 9.6

24. The line contains points $P(-2, 6)$ and $Q(3, -4)$.
25. The line has a zero slope and contains point $S(-6, 4)$.

Graph each inequality in a coordinate plane. 9.7

26. $y > -2x + 3$
27. $5x + 2y \leq -10$
28. $3x + 12 < 0$

Chapter 10 Relations, Functions, and Variation

1. Given the mapping shown: 10.1–10.3
   a. State the ordered pairs of the relation.
   b. Give the domain and range.
   c. Graph the relation.
   d. Tell whether the relation is a function.

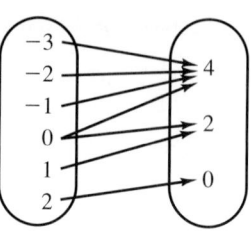

2. Given the graphed relation:
   a. State the ordered pairs of the relation.
   b. Give the domain and range.
   c. Tell whether the relation is a function.

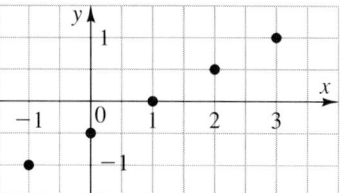

3. Given the relation $y = |x + 2|$. Domain: $\{-3, -2, -1, 0, 1, 2\}$.
   a. Graph the relation.
   b. Tell whether the relation is a function.

4. Given the graphed relation:
   a. Give three ordered pairs of the relation.
   b. Give the domain and range.
   c. What kind of function is this relation?
   d. Give the equation of this function.

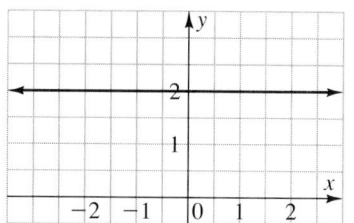

If $f(x) = -3x + 5$ and $g(x) = \frac{1}{10}x$, evaluate the following.   10.2

5. $f(0)$
6. $f\left(-\frac{1}{3}\right)$
7. $g(-20)$
8. $g(0.2)$
9. $f(g(9))$
10. $g(f(0))$
11. $f(1) \cdot g\left(\frac{1}{2}\right)$
12. $g(5) + f\left(-\frac{5}{3}\right)$

For the data shown in the given tables, tell whether y varies directly   10.4, 10.5
as x, inversely as x, or neither of these. If there is direct or inverse
variation, give the constant of proportionality and the equation.

13.
x	4	18	6	10
y	6	27	9	15

14.
x	3	12	2	8
y	16	4	24	6

15. If y varies directly as x and $y = 69$ when $x = 3$, find y when $x = 5$.

16. If y varies inversely as x and $y = 12$ when $x = 4$, find y when $x = 6$.

If $C = \frac{5}{9}(F - 32)$ is the formula for converting degrees F to degrees C;   10.6
find:

17. The temperature in degrees Celsius equivalent to 5°F

18. The temperature in degrees Fahrenheit equivalent to 100°C

Chapter 11 Systems of Linear Equations

Graph each system and give its solution set.   11.1

1. $\begin{cases} x - 2y = 0 \\ 3x + y = -7 \end{cases}$
2. $\begin{cases} 3y - 2x = 6 \\ y = \frac{2}{3}x + 1 \end{cases}$
3. $\begin{cases} 5x + 2y = 4 \\ 5x - 4 = -2y \end{cases}$

Solve each system by an appropriate algebraic method.   11.2, 11.4, 11.5

4. $\begin{cases} x + 5y = -4 \\ -3x + 2y = 12 \end{cases}$
5. $\begin{cases} 2a - 3b = 18 \\ -4a + 3b = -24 \end{cases}$
6. $\begin{cases} -c = 10 - 5d \\ -2c + 5d = -5 \end{cases}$

7. $\begin{cases} 5x + 7y = 11 \\ 14y = 13 - 10x \end{cases}$
8. $\begin{cases} 4x = 3y - 4 \\ 5y - 2x = 9 \end{cases}$
9. $\begin{cases} 3 = 4j - 9k \\ 5j + 12k = -4 \end{cases}$

10. $\begin{cases} 3u - 4v = 6 \\ u = 2 + \frac{4}{3}v \end{cases}$
11. $\begin{cases} 8(r - t) = \frac{7}{2} - 5t \\ t = -0.5 \end{cases}$
12. $\begin{cases} -\frac{1}{2}q + \frac{1}{3}p = p + 2 \\ \frac{p - q}{2} = p + 1 \end{cases}$

Write a system of linear equations and solve.

13. The sum of two numbers is 22 and one is 1 more than twice the other. Find the two numbers.   11.3

14. Belinda invested $1200 in two accounts. One paid 7.5% simple annual interest and the other 8.5%. At the end of one year, the total interest was $96.50. How much was invested in each account?

15. The sum of the digits of a two-digit number is 10. The number with the digits reversed is 18 less than the original. Find the original number.   11.6

16. Lonnie is 6 yr younger than his sister. In 13 yr his age will be $\frac{4}{5}$ of hers. What are their present ages?   11.7

17. A combination of forty $1 and $5 bills is worth $144. How many of each kind of bill are there?   11.8

18. Sunflower seeds priced at $2.35/lb are mixed with peanuts at $1.85/lb to make a 16 lb mixture valued at $32.40. How many pounds of each are used?

19. A boat going downstream traveled 15 mi in 3 h but took $3\frac{3}{4}$ h for the return trip. What was the speed of the current?   11.9

Graph each system of inequalities on a coordinate plane.   11.10

20. $\begin{cases} 2y + 5x > -2 \\ y < 2x + 3 \end{cases}$
21. $\begin{cases} x + 3y \leq 0 \\ y \geq \frac{1}{2}x \end{cases}$
22. $\begin{cases} x < -1 \\ y \geq 1 \end{cases}$

Extra Practice

Chapter 12 Radicals

Express in simplest form. Assume variables are positive. 12.1, 12.4–12.7

1. $-\sqrt{225}$
2. $\sqrt{\dfrac{5}{9}}$
3. $\sqrt{96}$
4. $\sqrt{50x^3}$
5. $3\sqrt{12} - \sqrt{3} + 2\sqrt{5}$
6. $(3\sqrt{3a})(5\sqrt{a})$
7. $\sqrt{0.0001b^4}$
8. $\dfrac{1}{\sqrt{27}}$
9. $\dfrac{\sqrt{2}}{2} + 3\sqrt{8} - \sqrt{32}$
10. $\dfrac{\sqrt{40c^3d}}{\sqrt{10cd}}$
11. $\sqrt{288xy^2z^3}$
12. $(2 + 3\sqrt{5})(4\sqrt{5})$
13. $(5\sqrt{3} + 3\sqrt{2})^2$
14. $(7\sqrt{5} - 1)(7\sqrt{5} + 1)$
15. $\dfrac{-6}{3\sqrt{3} - 2}$

Simplify. Use the square root table on p. 670 when necessary. 12.2

16. $\sqrt{51}$
17. $\sqrt{94}$
18. $\sqrt{7}$
19. $\sqrt{145}$

Use the divide-and-average method to approximate each square root to the nearest tenth.

20. $\sqrt{11}$
21. $\sqrt{241}$
22. $\sqrt{87.9}$
23. $\sqrt{459.6}$

Express each rational number as a decimal. 12.3

24. $-\dfrac{3}{8}$
25. $\dfrac{25}{3}$
26. $-\dfrac{7}{25}$
27. $\dfrac{5}{7}$

Express each decimal as a rational number in the form $\dfrac{m}{n}$, where m and n are integers and $n \neq 0$.

28. 0.13
29. $-0.\overline{4}$
30. 9.01
31. $3.\overline{12}$

Solve and check. 12.8

32. $7 + 3\sqrt{x} = 9$
33. $\sqrt{5 - 2x} = -2$
34. $2\sqrt{1 - 2v} = 3\sqrt{v}$

Given right $\triangle ABC$ with right $\angle C$. Use the table on p. 670 or a calculator to find the remaining side to the nearest tenth. 12.9

35. $a = 12$, $b = 16$
36. $a = 7$, $c = 25$
37. $b = 11.4$, $c = 14.5$

The lengths of three sides of a triangle are given. Is it a right triangle? Write *yes* or *no*.

38. $1.0, 2.4, 2.6$
39. $6\sqrt{3}, 3\sqrt{3}, 9$
40. $7, 12, 13$

Find the distance between each pair of points. 12.10

41. $A(-1, 3)$, $B(2, 7)$
42. $C(2, -5)$, $D(-3, -10)$

Find the lengths of the sides of a triangle with the following vertices.

43. $P(-2, 1)$, $Q(-2, 13)$, $R(-7, 1)$
44. $H(5, 12)$, $K(0, 0)$, $L(-3, -4)$

Chapter 13 Quadratic Equations and Functions

Solve.   **13.1**

1. $z^2 = 225$
2. $3y^2 = 81$
3. $(x - 5)^2 = 98$
4. $(2b + 1)^2 = 4$

Solve each equation by completing the square.   **13.2**

5. $r^2 - 4r - 96 = 0$
6. $q^2 - q = 2$
7. $2p^2 + 4p = 7$

Use the quadratic formula to solve each equation.   **13.3**

8. $x^2 - 6x + 4 = 0$
9. $6y^2 + 11y = 10$
10. $9z^2 - 12z = 1$

Use the most appropriate method to solve each equation.   **13.4**

11. $t^2 - t - 110 = 0$
12. $(3m - 4)^2 = 50$
13. $16m^2 = 49$
14. $10b^2 + 21b = 10$
15. $c^2 + 2c - 1 = 0$
16. $d^2 - 6d + 10 = 0$

17. If the product of two consecutive even integers is 288, find the two integers.   **13.2–13.4**

Find the vertex point and graph each function.   **13.5**

18. $f(x) = -x^2$
19. $y = 2x^2 + 3$
20. $f(x) = (x + 2)^2$
21. $y = x^2 - 6x + 6$
22. $y = -3x^2 + 12x + 1$
23. $f(x) = 2x^2 + 8x$

Find the value of the discriminant and state the nature of the solutions.   **13.6**

24. $m^2 + 12m + 36 = 0$
25. $4n^2 - 8n + 1 = 0$
26. $3k^2 - 2k + 5 = 0$

Find the number of x-intercepts of each function without graphing.

27. $f(x) = x^2 + 2x + 1$
28. $y = -2x^2 + x - 2$
29. $f(x) = 3x^2 - x - 1$

Use the sum and product of the solutions to write quadratic equations with the given solutions.   **13.8**

30. $-7, 11$
31. $-\frac{2}{3}, 5$
32. $\frac{4}{5}, -\frac{1}{2}$
33. $-3 \pm \sqrt{2}$

34. A ski lodge cook keeps track of the number of people at breakfast and the amount of oatmeal cooked in three days.   **13.7**

No. people n	12	20	35
No. cups oatmeal c	4.8	8	14

    a. Show that this is a linear function, and write the function.

    b. Tell the meaning of the slope and intercepts.

Chapter 14 Statistics and Probability

Find the mode, median, and mean for each set of data. 14.1

1. −3, −3, −2, 0, 1, 3, 4
2. −4.0, −2.3, −1.1, 2.7, 3.8, 4.2

Distances from home to school were compiled for a class of 30, and a histogram was constructed. 14.2

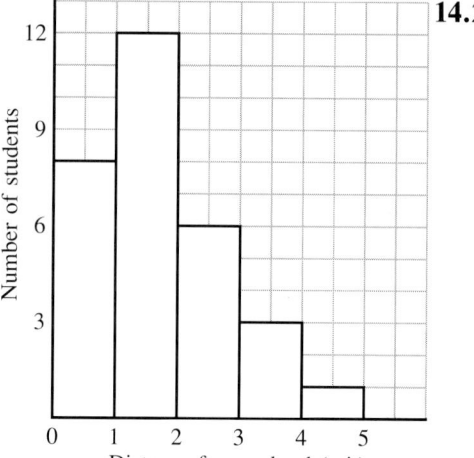

3. What percent of students live more than 2 mi away?
4. In what interval is the median distance?
5. What is the mean distance to the nearest tenth of a mile? *Hint:* Use the middle of the interval as the distance; e.g., 0.5 mi, 1.5 mi, etc.

The number of stray dogs taken by an animal shelter over a 12-mo period is shown below. 14.3

Jan	Feb	Mar	Apr	May	Jun	Jul	Aug	Sep	Oct	Nov	Dec
3	2	2	4	3	6	9	13	11	5	0	2

For the animals taken per month, find:

6. The mean
7. The range
8. The variance
9. The standard deviation

A letter is chosen at random from the word PROBABILITY. Find the probability of choosing: 14.4

10. P(vowel)
11. P(B)
12. P(J)
13. P(B or I)

Of a class of 30 students, 5 have motor scooters, 19 have bicycles, and 4 have both. Make a Venn diagram and find the probability that a student chosen at random has: 14.5

14. Just a motor scooter
15. A motor scooter or a bicycle
16. Neither a bicycle nor a motor scooter

A bag contains lettered, colored cards of the following type: 5 red As, 4 blue As, 3 red Bs, and 2 blue Bs. One card is randomly chosen, noted, and replaced. Then a second card is randomly selected. What is the probability of each event? 14.6

17. P(both red Bs)
18. P(both blues)
19. P(both cards same letter)

Chapter 15 Right Triangle Relationships

Given the plane figure shown, find:

15.1

1. $m\angle BFE$ 2. $m\angle BFA$ 3. $m\angle AFD$
4. Two right angles 5. Two obtuse angles
6. Three pairs of supplementary angles
7. If an angle measures 15 more than half its supplement, find the angle and its supplement.

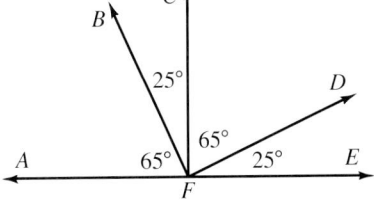

Complete.

15.2

8. In an isosceles triangle if the unequal angle has a measure of 74, then the angles opposite the congruent sides have measures of ? . This is a(n) ? triangle.

9. In right $\triangle ABC$, $m\angle C = 90$ and the measure of $\angle B$ is $\frac{1}{5}$ that of $\angle A$. Find the measures of $\angle A$ and $\angle B$.

Given $\triangle JKL \cong \triangle PQR$:

15.3

10. Name the corresponding angles. 11. Name the corresponding sides.
12. True or false? $\triangle KJL \cong \triangle QPR$?
13. In $\triangle LMN$, $m\angle M = 75$ and $m\angle N = 33$. In $\triangle RST$, $m\angle S = 72$ and $m\angle T = 33$. Show that the triangles are similar. Write the similarity.

15.4

14. $\triangle ABC \sim \triangle DEF$, $AB = 12$, $BC = 16$, and $EF = 12$. Find DE.

In $\triangle ABC$, $\angle B$ is a right angle. In terms of a, b, and c, find:

15.5

15. $\sin C$ 16. $\cos C$ 17. $\tan A$

In right triangle GHK, $m\angle K = 90$, $GH = 4$, and $HK = 2\sqrt{3}$. Find:

18. GK 19. $\sin G$ 20. $\tan H$

Use the trigonometric table on p. 671 to find:

15.6

21. If $m\angle A = 23$ **(a)** $\sin A$ **(b)** $\tan A$ **(c)** $\cos A$
22. If $\cos B = 0.9903$ **(a)** $m\angle B$ **(b)** $\tan B$ **(c)** $\sin B$
23. In $\triangle ABC$, $m\angle C = 90$, $AB = 25$, and $BC = 15$. To the nearest whole number, find: **(a)** $m\angle A$ **(b)** $m\angle B$ **(c)** AC

24. From a lighthouse 350 ft high, the angle of a depression to a ship at sea is 12°. How far is the ship from shore? Are there hidden assumptions in this problem?

15.7

Extra Practice **669**

Squares and Approximate Square Roots

Number n	Square n^2	Positive Square Root $\sqrt{n}$	Number n	Square n^2	Positive Square Root $\sqrt{n}$	Number n	Square n^2	Positive Square Root $\sqrt{n}$
1	1	1.000	51	2,601	7.141	101	10,201	10.050
2	4	1.414	52	2,704	7.211	102	10,404	10.100
3	9	1.732	53	2,809	7.280	103	10,609	10.149
4	16	2.000	54	2,916	7.348	104	10,816	10.198
5	25	2.236	55	3,025	7.416	105	11,025	10.247
6	36	2.449	56	3,136	7.483	106	11,236	10.296
7	49	2.646	57	3,249	7.550	107	11,449	10.344
8	64	2.828	58	3,364	7.616	108	11,664	10.392
9	81	3.000	59	3,481	7.681	109	11,881	10.440
10	100	3.162	60	3,600	7.746	110	12,100	10.488
11	121	3.317	61	3,721	7.810	111	12,321	10.536
12	144	3.464	62	3,844	7.874	112	12,544	10.583
13	169	3.606	63	3,969	7.937	113	12,769	10.630
14	196	3.742	64	4,096	8.000	114	12,996	10.677
15	225	3.873	65	4,225	8.062	115	13,225	10.724
16	256	4.000	66	4,356	8.124	116	13,456	10.770
17	289	4.123	67	4,489	8.185	117	13,689	10.817
18	324	4.243	68	4,624	8.246	118	13,924	10.863
19	361	4.359	69	4,761	8.307	119	14,161	10.909
20	400	4.472	70	4,900	8.367	120	14,400	10.954
21	441	4.583	71	5,041	8.426	121	14,641	11.000
22	484	4.690	72	5,184	8.485	122	14,884	11.045
23	529	4.796	73	5,329	8.544	123	15,129	11.091
24	576	4.899	74	5,476	8.602	124	15,376	11.136
25	625	5.000	75	5,625	8.660	125	15,625	11.180
26	676	5.099	76	5,776	8.718	126	15,876	11.225
27	729	5.196	77	5,929	8.775	127	16,129	11.269
28	784	5.292	78	6,084	8.832	128	16,384	11.314
29	841	5.385	79	6,241	8.888	129	16,641	11.358
30	900	5.477	80	6,400	8.944	130	16,900	11.402
31	961	5.568	81	6,561	9.000	131	17,161	11.446
32	1,024	5.657	82	6,724	9.055	132	17,424	11.489
33	1,089	5.745	83	6,889	9.110	133	17,689	11.533
34	1,156	5.831	84	7,056	9.165	134	17,956	11.576
35	1,225	5.916	85	7,225	9.220	135	18,225	11.619
36	1,296	6.000	86	7,396	9.274	136	18,496	11.662
37	1,369	6.083	87	7,569	9.327	137	18,769	11.705
38	1,444	6.164	88	7,744	9.381	138	19,044	11.747
39	1,521	6.245	89	7,921	9.434	139	19,321	11.790
40	1,600	6.325	90	8,100	9.487	140	19,600	11.832
41	1,681	6.403	91	8,281	9.539	141	19,881	11.874
42	1,764	6.481	92	8,464	9.592	142	20,164	11.916
43	1,849	6.557	93	8,649	9.644	143	20,449	11.958
44	1,936	6.633	94	8,836	9.695	144	20,736	12.000
45	2,025	6.708	95	9,025	9.747	145	21,025	12.042
46	2,116	6.782	96	9,216	9.798	146	21,316	12.083
47	2,209	6.856	97	9,409	9.849	147	21,609	12.124
48	2,304	6.928	98	9,604	9.899	148	21,904	12.166
49	2,401	7.000	99	9,801	9.950	149	22,201	12.207
50	2,500	7.071	100	10,000	10.000	150	22,500	12.247

Trigonometric Ratios

Angle	sin	cos	tan	Angle	sin	cos	tan
0°	0.0000	1.0000	0.0000	45°	0.7071	0.7071	1.0000
1	0.0175	0.9998	0.0175	46	0.7193	0.6947	1.0355
2	0.0349	0.9994	0.0349	47	0.7314	0.6820	1.0724
3	0.0523	0.9986	0.0524	48	0.7431	0.6691	1.1106
4	0.0698	0.9976	0.0699	49	0.7547	0.6561	1.1504
5	0.0872	0.9962	0.0875	50	0.7660	0.6428	1.1918
6	0.1045	0.9945	0.1051	51	0.7771	0.6293	1.2349
7	0.1219	0.9925	0.1228	52	0.7880	0.6157	1.2799
8	0.1392	0.9903	0.1405	53	0.7986	0.6018	1.3270
9	0.1564	0.9877	0.1584	54	0.8090	0.5878	1.3764
10	0.1736	0.9848	0.1763	55	0.8192	0.5736	1.4281
11	0.1908	0.9816	0.1944	56	0.8290	0.5592	1.4826
12	0.2079	0.9781	0.2126	57	0.8387	0.5446	1.5399
13	0.2250	0.9744	0.2309	58	0.8480	0.5299	1.6003
14	0.2419	0.9703	0.2493	59	0.8572	0.5150	1.6643
15	0.2588	0.9659	0.2679	60	0.8660	0.5000	1.7321
16	0.2756	0.9613	0.2867	61	0.8746	0.4848	1.8040
17	0.2924	0.9563	0.3057	62	0.8829	0.4695	1.8807
18	0.3090	0.9511	0.3249	63	0.8910	0.4540	1.9626
19	0.3256	0.9455	0.3443	64	0.8988	0.4384	2.0503
20	0.3420	0.9397	0.3640	65	0.9063	0.4226	2.1445
21	0.3584	0.9336	0.3839	66	0.9135	0.4067	2.2460
22	0.3746	0.9272	0.4040	67	0.9205	0.3907	2.3559
23	0.3907	0.9205	0.4245	68	0.9272	0.3746	2.4751
24	0.4067	0.9135	0.4452	69	0.9336	0.3584	2.6051
25	0.4226	0.9063	0.4663	70	0.9397	0.3420	2.7475
26	0.4384	0.8988	0.4877	71	0.9455	0.3256	2.9042
27	0.4540	0.8910	0.5095	72	0.9511	0.3090	3.0777
28	0.4695	0.8829	0.5317	73	0.9563	0.2924	3.2709
29	0.4848	0.8746	0.5543	74	0.9613	0.2756	3.4874
30	0.5000	0.8660	0.5774	75	0.9659	0.2588	3.7321
31	0.5150	0.8572	0.6009	76	0.9703	0.2419	4.0108
32	0.5299	0.8480	0.6249	77	0.9744	0.2250	4.3315
33	0.5446	0.8387	0.6494	78	0.9781	0.2079	4.7046
34	0.5592	0.8290	0.6745	79	0.9816	0.1908	5.1446
35	0.5736	0.8192	0.7002	80	0.9848	0.1736	5.6713
36	0.5878	0.8090	0.7265	81	0.9877	0.1564	6.3138
37	0.6018	0.7986	0.7536	82	0.9903	0.1392	7.1154
38	0.6157	0.7880	0.7813	83	0.9925	0.1219	8.1443
39	0.6293	0.7771	0.8098	84	0.9945	0.1045	9.5144
40	0.6428	0.7660	0.8391	85	0.9962	0.0872	11.4301
41	0.6561	0.7547	0.8693	86	0.9976	0.0698	14.3007
42	0.6691	0.7431	0.9004	87	0.9986	0.0523	19.0811
43	0.6820	0.7314	0.9325	88	0.9994	0.0349	28.6363
44	0.6947	0.7193	0.9657	89	0.9998	0.0175	57.2900
45	0.7071	0.7071	1.0000	90	1.0000	0.0000	

SYMBOLS

Symbol	Meaning	Page
( )	parentheses—a grouping symbol	3
...	"and so on"	7
{ , }	encloses the numbers of a set	7
−	negative	7
+	positive	7
π	pi, a number approximately equal to $\frac{22}{7}$ or 3.14	8
=	equals, is equal to	12
>	is greater than	12
<	is less than	12
$-a$	opposite or additive inverse of a	13
$\|a\|$	absolute value of a	14
$\frac{1}{n}$	reciprocal of a number	35
[ ]	brackets—a grouping symbol	37
$\times$, $\cdot$	multiply	42
a^n	exponent, a is used as a factor n times; a to the nth power	55
$\approx$	is approximately equal to	66
$\neq$	is not equal to	78
$\emptyset$, { }	empty set, null set	79
$\stackrel{?}{=}$	is this statement true?	137
%	percent	152
a_b	subscript, for descriptive discrimination	160
$\geq$	is greater than or equal to	187
$\leq$	is less than or equal to	187
$a:b$	ratio $\frac{a}{b}$	346

Symbol	Meaning	Page
(a, b)	ordered pair whose first component is a and second component is b	370
$P(x, y)$	point P with coordinates x, y	370
$f(x)$	f of x, the value of f at x	418
$F: x \rightarrow x + 1$	the function F that takes x into $x + 1$	418
$\pm$	positive or negative	494
$\sqrt{}$	principal square root	494
$0.6\overline{6}$	repeating decimal	500
σ	sigma, standard deviation	591
$P(A)$	probability of event A	596
$P(\overline{A})$	probability of the complement of A	597
$\overleftrightarrow{AB}$	a line containing points A and B	614
$\overline{AB}$	a line segment with two endpoints A and B	614
$\overrightarrow{AB}$	a ray with endpoint A passing through point B	614
AB	length or measure of line segment AB	614
$\angle A$	angle A	615
$m\angle A$	measure of angle A	615
$\triangle ABC$	triangle ABC	618
$\llcorner$	right angle, 90°	619
$\cong$	is congruent to	623
$\sim$	is similar to	628
$\cos A$	$\dfrac{\text{length of leg adjacent to } \angle A}{\text{length of hypotenuse}}$	634
$\sin A$	$\dfrac{\text{length of leg opposite } \angle A}{\text{length of hypotenuse}}$	634
$\tan A$	$\dfrac{\text{length of leg opposite } \angle A}{\text{length of leg adjacent to } \angle A}$	634

ANSWERS TO SELECTED EXERCISES

Chapter 1 Real Numbers

Practice Exercises, pages 5–6 **1.** 27 **3.** 6
5. 2 **7.** 24 **9.** 15 **11.** $\frac{1}{300}$ **13.** 6
15. 7 **17.** $3 + x$ **19.** $t - 1.07$
21. $\frac{9}{8}$ **23.** $\frac{1}{4}$ **25.** 1 **27.** 0 **29.** $\frac{17}{6}$
31. $\frac{9}{2}$ **33.** $\frac{3}{8}j$ **35.** $\frac{6.4}{b}$ **37.** 23
39. 5 **41.** 8 **43.** $\frac{m}{g}$

Practice Exercises, pages 10–11 **1.** -1.0
3. 0.2 **5.** -1.2 **7.** 1.2 **9.** G
11. A **13.** I **15.** D **17.** $\frac{7}{1}$
19. $-\frac{3012}{1000}$ **21.** $\frac{47}{4}$ **23.** $\frac{-1}{1}$
25. [number line: $-1, 0, 1, 2, 4$]
27. [number line: $-4, -1.5, 0, 2, 2.5$]
29. [number line: $-1\frac{1}{4}, -\frac{1}{2}, 0, \frac{3}{4}$]
31. natural numbers; whole number; integer; rational number; real number **33.** whole number; integer; rational number; real number **35.** rational number; real number **37.** Yes. The number line extends indefinitely in both directions, and all real numbers, rational and irrational, can be graphed on it. **39.** 71
41. 4 **43.** 15 **45.** when X is an integer

Practice Exercises, pages 15–16
1. [number line: $-4, -1, 0, 2, 3$]
3. [number line: $-2\frac{1}{2}, -\frac{1}{2}, 0, 2, 2\frac{1}{2}$]
5. < **7.** > **9.** = **11.** > **13.** 3.1
15. $-\frac{2}{5}$ **17.** 6.2 **19.** $-3\frac{1}{5}$ **21.** 13 **23.** 5
25. $\{\frac{1}{7}, \frac{1}{6}, \frac{1}{5}\}$ **27.** $\{-1\frac{3}{4}, -1\frac{2}{3}, -\frac{5}{4}\}$
29. $\{-0.1041, -0.104, -0.1\}$ **31.** -14 **33.** $-\frac{3}{8}$
35. -21 **37.** 12 **39.** 7 **41.** -7
43. -6.1 **45.** $3\frac{7}{20}$ **47.** false **49.** true
51. -20; 20°C below freezing
53. $+3\frac{1}{4}$; $3\frac{1}{4}$ lb weight gain
55. $+55$; 55 ft above sea level
57. -785.95; loss of $785.95

Practice Exercises, pages 20–21 **1.** -5 **3.** -6
5. 1 **7.** -7 **9.** -8 **11.** -3.5
13. -8.0 **15.** -6 **17.** 1 **19.** 6
21. -6 **23.** -17 **25.** -6.5 **27.** $-2\frac{1}{2}$
29. $\frac{7}{3}$ **31.** < **33.** <
35. $5 + (-17) = -12$; -12°C
37. $-3 + (-3) + (-3) + (-3) = -12$; 12 yd loss
39. $-3\frac{1}{8} + 4\frac{5}{8} = 1\frac{1}{2}$; up $1\frac{1}{2}$ points

Practice Exercises, pages 24–26 **1.** -500
3. -453 **5.** 4.61 **7.** -1.709 **9.** $-\frac{4}{7}$
11. $-\frac{5}{8}$ **13.** $\frac{1}{14}$ **15.** 0 **17.** 8.6
19. -8 **21.** 8 **23.** $\frac{17}{24}$ **25.** $1\frac{49}{75}$
27. 23.98 **29.** $3\frac{1}{3}$ **31.** -13 **33.** -25
35. 59 **37.** 15 **39.** 7 **41.** 1
43. $19 + 6 + (-8) = 17$; 17°C
45. $1000 + (-60) + 2200 + (-200) + 1700 = 4640$; 4640 m **47.** 15 **49.** 5.07 **51.** 72

Practice Exercises, pages 29–30 **1.** 4.1
3. -0.75 **5.** -8 **7.** -13 **9.** 3.7
11. 9.5 **13.** 17 **15.** 18 **17.** $\frac{7}{9}$
19. 46 **21.** 27 **23.** 0.3 **25.** 1.9
27. -3.6 **29.** $-\frac{13}{8}$ **31.** -5.65 **33.** $-14\frac{1}{3}$
35. $30\frac{9}{10}$ **37.** -17 **39.** 64 **41.** -24.3
43. -7.2 **45.** 3 **47.** -5 **49.** 0.4

Test Yourself, page 30 **1.** 22.3 **3.** 76
5. [number line: $-4, -3, -2, -1, 0, 1$] **7.** > **9.** -1
11. $4\frac{1}{2}$ **13.** -4.42

Practice Exercises, pages 33–34 **1.** -30
3. -0.468 **5.** 65 **7.** 2.22 **9.** 30

11. −45   13. 120   15. 200   17. −1200
19. −900   21. −18   23. −6
25. 55,555.5   27. −11,111.1111
29. −1558.788   31. 15   33. 17
35. −12   37. −18   39. −3   41. 4
43. $(-7) \times 2 = -14: -14°$

Practice Exercises, pages 38–40   1. $\frac{1}{5}$ 3. $\frac{1}{75}$

5. −5   7. $-\frac{10}{9}$ 9. −7   11. −12

13. $\frac{15}{2}$ 15. $-\frac{1}{2}$ 17. 0

19. not possible   21. 2   23. $-\frac{1}{6}$

25. $-\frac{1}{16}$ 27. −7   29. 3   31. −1

33. 2   35. 3   37. −2.909088   39. −587.22

41. 1800   43. −8   45. $-7\frac{1}{2}$ 47. true

49. true   51. 80   53. 190 cm
55. 1.5 km; no   57. 88.8°F
59. 0.01   61. −2.5

Practice Exercises, pages 44–45   1. $m - 2$

3. $s + 500$ 5. $p + 0.06p$ 7. $\frac{a + b + c}{3}$

9. $2x - 1$ 11. Yes, since one can be a simplified form of the other.   13. $-|2x + 1|$
15. True. $x - y$ and $y - x$ are opposites, but the absolute values of opposites are equal.

Test Yourself, page 45   1. −34   3. 8   5. 0

7. −6   9. 8   11. $-\frac{1}{9}$ 13. −9   15. $\frac{3}{2}$

Summary and Review, pages 46–47   1. 3.5

3. 12   5. 32   7. $\frac{1}{8}$ 9. $x + 18$

11. $a - b$
13. [number line from −4½ to 2½]
15. [number line from −2 to 3]
17. >   19. −4.31   21. −9   23. −1.2
25. −11.7   27. −5.92   29. 150   31. −120
33. 0   35. +2.5 yd

Maintaining Skills, page 50   1. 3.3147   3. 42.807
5. 0.9   7. 3.7   9. 14.94   11. 280.5
13. 2.8   15. 8.2   17. 0.452   19. −7
21. −28   23. 14   25. −15   27. −4
29. −42   31. 456 mi   33. 65 packages

Chapter 2   Algebraic Expressions

Practice Exercises, pages 53–54   1. −5   3. 130
5. 50   7. 23   9. 7   11. 13

13. $-\frac{39}{31}$ 15. 54   17. −33   19. $\frac{76}{13}$

21. −90   23. $6.25

Practice Exercises, pages 57–58   1. 25   3. $\frac{1}{4}$

5. −16   7. 16   9. −16   11. −16   13. 98
15. 2209   17. 363   19. 44944   21. 216
23. −33
25. 0.09   27. 0.000000001   29. 100   31. 100
33. 5   35. 56   37. 64   39. 400   41. −15
43. −9   45. −226   47. 7776   49. −231.3441

Practice Exercises, pages 61–62   1. 68   3. 134

5. $\frac{1}{7}$ 7. 5130   9. 2325   11. −24.75   13. 2

15. Distributive property   17. Commutative property for multiplication   19. Symmetric property
21. $(2)(3x) - (2)(11)$

23. $\left(-\frac{3}{4}\right)\left(\frac{2}{3}x\right) - \left(-\frac{3}{4}\right)(8)$ 25. $(12 - 3)a$

27. False   29. False   31. False   33. $13.47

Practice Exercises, pages 64–66   1. like
3. unlike   5. like   7. unlike   9. unlike
11. like   13. $8x$ 15. $-12t$ 17. $2t + 15$
19. $12g^2$ 21. $7x^2 + 7x$ 23. $12g - h + 2$
25. $9ab$ 27. $5ab^2$ 29. $4m + 13$
31. $-6x + 7y - 4$ 33. $3x + 11$ 35. $5r - 2$
37. $6x + 7$ 39. $-15c^2 + 16c$ 41. $8x^3 - x^2$
43. $155r + 295$ 45. $7m + 7n$ 47. $6x^2 - 6y$
49. $3t^2 + 7t$ 51. $6x^2y^2 + 3x^2y - 9xy^2$

53. $4u$ 55. $\frac{5}{8}y$ 57. $-23c + 24d$

59. $1.518b - 0.69$ 61. $n + (n + 1) + (n + 2)$; $3n + 3$ 63. $5(7w) + 2(7w)$; $49w$

Test Yourself, page 66   1. −9   3. 1   5. 8
7. 192   9. 56.52
11. Associative property for multiplication
13. Commutative property for addition
15. $11hk - 5hk^2 - 5h^2k$ 17. $2x^2 + 7$

Practice Exercises, pages 69–71   1. $g - 8$
3. $-2a + 13$ 5. $3t^2 - 9$ 7. $-2x - 1$
9. $ab^2c - 3abc^2$ 11. −27   13. −3   15. 26
17. 5   19. $2x^2yz + x^2yz^2$ 21. $-6x + 14$
23. $21r - 153$ 25. 21   27. 41   29. $-5m - 1$
31. $a - 2$ 33. $-36x + 37$ 35. 18 cm^2
37. $2(a + 6) + 4(3a)$; $14a + 12$ 39. 13.19
41. 5   43. 40

674   Answers to Selected Exercises

Practice Exercises, pages 76–77   **1.** $n - 7$
3. $n + 5$ **5.** $8n - 1$ **7.** $5n^2$ **9.** $\dfrac{n^2}{3}$
11. $6(n + 3)$
17. a. $m - 4$ **b.** $m - 4$ **c.** $m - 8$
19. a. l **b.** $l - 7$ **c.** $l^2 - 7l$ **d.** $4l - 14$
21. $9t + 36$ dollars; $2t - 36$ dollars
23. $p + 1 = 3c$

Practice Exercises, pages 80–81   **1.** $\{1\}$ **3.** $\{0\}$
5. $\{-1, 0, 1\}$ **7.** $\{-1, 1\}$ **9.** $\{0\}$ **11.** $\{-1, 0, 1\}$
13. $\{-1\}$ **15.** $\{-1, 0\}$ **17.** $\emptyset$ **19.** $\left\{\dfrac{1}{2}\right\}$
21. $\{-1\}$ **23.** $\{-1\}$ **25.** $\{-1\}$ **27.** $\emptyset$
33. Any negative number.
35. Negative five is greater than negative seven.
37. Six subtracted from a is greater than zero.

Practice Exercises, pages 83–84
1. no solution   **3.** $\{0\}$ **5.** $\{-2\}$ **7.** $\{-3\}$
9. $\{1\}$ **11.** 15   **13.** 33   **15.** $m + 50 = 2m$; $50, $100   **17.** $n + (n + 6) = -8$; -7
19. $l(32 - l) = 252$; 18 ft × 14 ft

Practice Exercises, pages 87–89   **1.** $x - 3 = 18$
3. $x + 4 = 27$ **5.** $5x - 2 = 18$ **7.** $8x^2 = 56$
9. $\dfrac{5}{3x} = 10$ **11.** $6(x + 9) = 132$
13. a. $b - 6$ **b.** $b - 1$ **c.** $b - 7$
d. $(b - 6) + (b - 7) = 35$
15. a. $45n$ **b.** $45n + 35$ **c.** $45n + 35 = 485$
21. $3j + 5 = 2(j + 5)$
23. $25e + 10(6 + e) = 235$
25. $13{,}250 = 2c + 150$; What was the cost of the old car?   **27.** $110 = 35 + 5p$; How much was a monthly payment?   **29.** $s = 100 + 2(1200)$; How much tuition did Sheila pay?
31. $55 = \dfrac{1}{2}b(4b + 2)$; What are the base and height of the triangle?

Test Yourself, page 89   **1.** $-2r - 12$
3. $-m^2 + n$; -11 **5.** $\dfrac{n}{9} - 8$ **9.** $\emptyset$
11. $n - (-2) = 11$

Summary and Review, pages 90–91   **1.** 12
3. -0.0001 **5.** 7   **7.** Distributive property
9. Associative property for addition
11. Associative property for multiplication
13. $-2c + 5 + 4c^2$ **15.** $4mn^2 - 3mn$

17. $2h^2$ **19.** $-24 + 26p$; 28   **21.** $2[n + (-2)]$
23. $5n$ **25.** $\{2\}$ **27.** $\{-2\}$ **29.** $\{-1, 1\}$
31. $2x - 2 = 1$

Cumulative Review, page 94
1. $-1.5, -1, 0, 2, 2.5$ **3.** [number line with points at $-2, -\tfrac{1}{2}, 0, 1, 3$]
5. 0.8   **7.** $1\dfrac{1}{2}$ **9.** 14   **11.** -4.1 **13.** -4
15. $\dfrac{1}{4}$ **17.** 16   **19.** 18   **21.** $6x - 3y$
23. $-2a + 6$ **25.** $2c + 8$ **27.** -3 **29.** 8
31. 44   **33.** $\dfrac{1}{3}$ **35.** 14   **37.** $2n - 3$ **39.** $2w + 1$

Chapter 3   Equations in One Variable

Practice Exercises, pages 98–99   **1.** 13.01   **3.** 8.3
5. 8.1   **7.** $6\dfrac{2}{3}$ **9.** $2\dfrac{2}{3}$ **11.** $4\dfrac{2}{3}$ **13.** -30 **15.** 5
17. 44   **19.** -3.6 **21.** -6.1 **23.** -7
25. 19.019
27. 1.062   **29.** $2\dfrac{2}{3}$ **31.** 5   **33.** 27   **35.** -5.8
37. no solution   **39.** 5   **41.** no solution   **43.** 26
45. 12   **47.** $\{3\}$ **49.** $\{9\}$

Practice Exercises, pages 104–105   **1.** -8 **3.** -7
5. -5 **7.** -3 **9.** -7 **11.** -21 **13.** -12
15. 55   **17.** 20   **19.** 24   **21.** 35
23. $m = 10.5 \times 15$ **25.** $t = \dfrac{55.5 \times 11}{5}$
27. $r = \dfrac{24.6 \times 6}{-5}$ **29.** -18 **31.** -9 **33.** $\dfrac{2}{3}$
35. -0.118 **37.** 9.2   **39.** 20   **41.** $\{-12, 12\}$
43. $\left\{-\dfrac{2}{3}, \dfrac{2}{3}\right\}$ **45.** $m = 4.5 \times 33$; 148.5
47. $y = 6 \times 0.89$; 5.34
49. $m = 62.4 \div 81.23$; 0.77

Practice Exercises, pages 108–109   **1.** $\dfrac{8}{3}$ **3.** 3
5. $-\dfrac{11}{3}$ **7.** 17   **9.** 76   **11.** $-\dfrac{15}{2}$ **13.** $\dfrac{15}{2}$
15. $\dfrac{1}{4}$ **17.** $-\dfrac{7}{9}$ **19.** 113   **21.** 117   **23.** $\dfrac{3}{2}$
25. $-4, 4$ **27.** $-2, 2$ **29.** $-1, 1$ **31.** 7   **33.** 6
35. $r = \dfrac{2.79 - 23.58}{583}$; -0.04
37. $x = (-0.14 + 0.03)0.24$; -0.03

Answers to Selected Exercises   **675**

Test Yourself, page 109 1. -7 3. $\frac{13}{4}$ 5. -60
7. 3 9. 0

Practice Exercises, pages 111–112
1. $7x + 3 = 24$; 3 albums
3. $5x = 40$; books; 0 left over
5. $4e + 5 = 25$; $5 per hour

Practice Exercises, pages 115–116

1. Statements: 2. $\frac{a}{c}$ 3. $\frac{b}{c}$
   Reasons: 1. Given 3. Substitution property

3. | Statements | Reasons |
    |---|---|
    | 1. $a = b$ | 1. Given |
    | 2. $a - c = a - c$ | 2. Reflex. prop. |
    | 3. $a - c = b - c$ | 3. Substit. prop. |

5. | Statements | Reasons |
    |---|---|
    | 1. $a + [b + (-a)]$ | |
    | $\quad = a + [(-a) + b]$ | 1. Comm. prop. add. |
    | 2. $\quad = [a + (-a)] + b$ | 2. Assoc. prop. add. |
    | 3. $\quad = 0 + b$ | 3. Add. inv. prop. |
    | 4. $\quad = b$ | 4. Ident. prop. add. |

7. | Statements | Reasons |
    |---|---|
    | 1. $a = b$ | 1. Given |
    | 2. $a - d = a - d$ | 2. Reflex. prop. |
    | 3. $\quad = b - d$ | 3. Substit. prop. |
    | 4. $\quad = b + (-d)$ | 4. Def. of subtrac. |
    | 5. $\quad = -d + b$ | 5. Commut. prop. of add. |
    | 6. $\quad = -d - (-b)$ | 6. Def. of subtrac. |
    | 7. $\quad = (-1)(d) - (-1)(b)$ | 7. Prop. of -1 for mult. |
    | 8. $\quad = (-1)(d - b)$ | 8. Dist. prop. |
    | 9. $\quad = -(d - b)$ | 9. Prop. of -1 for mult. |

9. transitive, symmetric 11. symmetric

Practice Exercises, pages 118–119 1. 3.5 units
3. $h = 5.9$ 5. $r = 7$ 7. $h = 3$ 9. $w = 110$
13. 95.5 km/h 15. 71.02 m^2

Practice Exercises, pages 122–123 1. $10 = \frac{1}{3}x$
3. $2w + 3 = 7$ 5. $n - 92 = -28$; 64
7. $11x = -165$; -15
9. $n + 7.13 = 2.09$; -5.04
11. $\frac{2}{5}x = 16$; 40 yr 13. $3n + 11 = 50$; 13
15. $21 + 2(12) = 54$; 15 cm 17. $\frac{3}{5}p = 245.97$; $409.95 19. $4.50x + 2.50(117) = 733.50$; 98
21. $\frac{1}{2}x + \frac{1}{10}x + \frac{1}{20}x + 35{,}000 = x$; $100,000

Practice Exercises, pages 128–129 1. 3 3. -5
5. 9 7. 75 9. $21.98 11. 22 yr 13. -425
15. 7.7 17. 18 m 19. 3.2 h 21. 11 in.
23. $3000 25. $-4°$F

Test Yourself, page 129 1. -216

3. | Statements | Reasons |
    |---|---|
    | 1. $m = r; n \neq 0$ | 1. Given |
    | 2. $\frac{m}{n} = \frac{m}{n}$ | 2. Reflex. prop. |
    | 3. $\frac{m}{n} = \frac{r}{n}$ | 3. Subst. prop. |

5. $h = 10$

Summary and Review, pages 130–131 1. 21
3. -32 5. 5 7. $\frac{36}{5}$ 9. 5 11. -12 13. 1
15. 8 17. $w = 8$ 19. $r = 0.04$ 21. For all real numbers x, y, and z, if $x = y$, then $x + z = y + z$
23. $3x + 500 = 12{,}200$; 3900 lb 25. $2x + 14.95 = 809.45$; $397.25 27. $54 = 2(12) + 2(x)$; 15 cm 29. $2x = 210$; 105 lb

Maintaining Skills, page 134 1. $-2b + 35$
3. $y + 5$ 5. $-4m + 4$ 7. 0.20 9. 0.06
11. 0.1 13. 75% 15. 7% 17. 38.5%
19. 25% 21. 70% 23. 37.5% 25. $\frac{7}{20}$ 27. $\frac{3}{8}$
29. $\frac{1}{12}$ 31. $1\frac{1}{2}$ h 33. 95°F

Chapter 4 More Equations in One Variable

Practice Exercises, pages 138–139 1. 2 3. 4
5. 2 7. -3 9. 0 11. 7 13. 7 15. 9
17. -2 19. 10 21. 1 23. -2 25. -6 27. 5
29. 0.2 31. -1.5 33. $-9\frac{1}{2}$ 35. $-5\frac{9}{11}$ 37. 7
39. -10 41. -20 43. $-4\frac{3}{4}$ 45. 2 47. 7
49. $2w + 2(2w - 10) = 118$; $w = 23$ ft, $l = 36$ ft

Practice Exercises, pages 142–143 1. -3 3. 1
5. 0 7. 10 9. $\frac{10}{3}$ 11. 22 13. -2
15. 4 17. $1\frac{2}{3}$ 19. 5 21. $-\frac{5}{4}$ 23. $-1\frac{3}{10}$
25. $\frac{5}{4}$ 27. 1 29. $-\frac{1}{29}$ 31. 9 33. -3 35. $\frac{5}{2}$
37. $\frac{11}{15}$ 39. $-\frac{6}{13}$ 41. $\frac{1}{2}$ 43. $\frac{3}{4}$ 45. $9\frac{1}{4}$
47. $5\frac{1}{3}$ sq units

Practice Exercises, pages 146–147
1. $x + (x + 1) + (x + 2) = 99$; 32, 33, 34
3. $x + (x + 2) + (x + 4) + (x + 6) = -124$; $-34, -32, -30, -28$
5. $x + (x + 8) = 3(x + 6) - 2$; $-8, -6, -4, -2, 0$
7. $x + (x + 1) = 105$; 52, 53   9. $x + (x + 1) + (x + 2) = -354$; $-119, -118, -117$
11. $x + (x + 2) = -54$; $-28, -26$
13. $x + (x + 2) + (x + 4) = -45$; $-17, -15, -13$
15. $(x + 1) + (x + 3) = 48$; 22, 23, 24, 25
17. $2(x + 2) + x = 85$; 27, 29
19. $x + (x + 1) + (x + 2) = 39 - x$; 9, 10, 11
21. $x + (x + 1) + (x + 2) = 4x - 9$; 12, 13, 14
23. $x + (x + 5) + (x + 10) + (x + 15) = 90$; 15, 20, 25, 30   25. $2(x + 1) = x + (x + 2)$; any three consecutive integers   27. $3[x + (x + 1)] - [(x + 2) + (x + 3)] = 70$, 18, 19, 20, 21

Practice Exercises, pages 150–151   1. 9.5   3. 24
5. 18.72   7. $\frac{3}{2}$ 9. 3   11. -10 13. $-\frac{1}{5}$ 15. 1
17. 2   19. 6   21. 0   23. 105   25. 1400
27. -1.077 29. -0.4 31. 960   33. $-\frac{16}{5}$
35. $-\frac{11}{2}$ 37. 0   39. $2\frac{2}{3}$ 41. 19   43. 0.2
45. ≈ -0.46 47. ≈ 0.37
49. 3   51. $l = 26$ m; $w = 16$ m

Test Yourself, page 151   1. 14   3. 2   5. $-1\frac{2}{3}$
7. $i + 2, i + 4$ 9. 11, 13, 15, 17   11. 34.95

Practice Exercises, pages 154–155   1. 0.98
3. 375   5. 20%   7. $66\frac{2}{3}\%$ 9. 15   11. 35
13. $3500   15. 2.5%   17. 54   19. 250%
21. 7.2   23. $133\frac{1}{3}\%$ 25. $90   27. $25   29. 8%
31. 100   33. 275   35. 8.25%

Practice Exercises, pages 157–159   1. $147; 25% (dec.)   3. $82.86; $111.86   5. $26 (inc.); 2.9% (inc.)   7. $5.25 (dec.); $78.75   9. $22,425
11. 12.6%   13. 11.1%   15. 1.4%   17. old price: 60¢; new price: 70¢   19. $10,398.40   21. 33.7%
23. 66¢   25. 17.1%   27. 96.9¢   29. 74
31. $22.92

Practice Exercises, pages 163   1. $16\frac{2}{3}\%$
3. $11.\overline{3}\%$ 5. 6.7 mL   7. 2.9 g

Practice Exercises, pages 166–167   1. $x = y - 10$
3. $c = d + 3$ 5. $d = \frac{2y}{a}$ 7. $n = \frac{8}{3}k$

9. $m = \frac{2j}{kl}$ 11. $y = \frac{12 - 3x}{4}$ 13. $y = 4t$
15. $a = \frac{-11b - 7}{3}$ 17. $m = \frac{2l - h}{4}$
19. $w = \frac{A}{l}$; 7   21. $b = p - 2a$; 20   23. $b = \frac{V}{h}$; $346\frac{2}{3}$ 25. $y = \frac{-7x - 14z}{4}$ 27. $r = \frac{158 - q}{115}$
29. $w = \frac{16}{3}t$ 31. $q = \frac{-5v - 9}{2}$
33. $h = \frac{16i - l}{3}$ 35. $C = \frac{5F - 160}{9}$; -20
37. $t = \frac{A - p}{pr}$; 5   39. $n = \frac{L - a + d}{d}$
41. $l = \frac{2S - na}{n}$ 43. $c = \frac{a + 2b}{12b}$
45. $d = rt$; $r = \frac{d}{t}$; 5.14 km/h

Practice Exercises, pages 170–171
1. $45t = 55(t - 1)$ 3. $550(t + 2) = 625t$
5. $3(r + 20)$; $4r$ 7. $5\frac{1}{2}$ 9. $14\frac{2}{3}$ 11. 90
13.

	Rate	× Time	= Distance
Car 1	40	$t + 1$	$40(t + 1)$
Car 2	55	t	$55t$

15.

	Rate	× Time	= Distance
Going	50	t	$50t$
Return	55	$t - \frac{1}{2}$	$55\left(t - \frac{1}{2}\right)$

17. $2\frac{2}{3}$ h; hiker 1: 7 mi/h, hiker 2: 4 mi/h; 275 mi

Practice Exercises, pages 175–177   1. 9h   3. 4h
5. $3\frac{1}{2}$ h   7. 1st bus 50 mi/h; 2nd bus 45 mi/h
9. 60 km   11. $4\frac{2}{3}$ mi   13. Bus 24 mi/h; Car 36 mi/h   15. $9\frac{2}{3}$ h   17. 2:15 AM   19. 10:00 AM
21. $3\frac{5}{7}$ h   23. 1000 m

Test Yourself, page 177   1. 0.75   3. 300%
5. 5%   7. 6.25%   9. $a = 2b + 3$ 11. 3h

Summary and Review, pages 178–179   1. -9
3. $-12\frac{3}{4}$ 5. -2.82 7. $z + 8, z + 10, z + 12, z + 14$ 9. 2.85   11. $1350   13. 108 students
15. 37.5 mL   17. $n = \frac{15m - p}{5}$

19. $a = \dfrac{p-b}{2}; \dfrac{11}{2}$ **21.** 1650 mi   **23.** $2\dfrac{2}{3}$ km

Cumulative Review, pages 182–184   **1.** $-\dfrac{1}{2}$
3. $\dfrac{3}{4}$ **5.** -2 **7.** $\dfrac{1}{12}$ **9.** $-\dfrac{29}{24}$ **11.** $-\dfrac{7}{6}$
13. $\dfrac{7}{10}$ **15.** $\dfrac{2}{3}$ **17.** 7   **19.** -20 **21.** $2x - 6$
23. $2h + 9$ **25.** $5mn - 1$ **27.** 0   **29.** 56
31. 16   **33.** 8   **35.** -1 **37.** $m - 2$
39. $m + (m - 2) = 30$ **41.** 5   **43.** -9 **45.** $\dfrac{1}{9}$
47. -14 **49.** $-\dfrac{1}{10}$ **51.** -1 **53.** 2   **55.** 4
57. 32   **59.** 8   **61.** 25   **63.** $42 = 2l + 2(6)$; 15 cm
65. $n + (n + 1) = 49$; 24, 25
67. $a = 0.15(40)$; 6   **71.** $48 = 0.12c$; 400
75. $4 = 0.08c$; 50   **79.** $15 = r(50)$; 30%
83. $5 = 0.20x$; 25   **87.** $x = 0.375(96)$; 36
91. $30 = 1.50x$; 20   **95.** $x = 0.005(200)$; $1
97. $y = x - 9$ **99.** $g = 2h - 6$ **103.** $c = 3d$
105. $q = \dfrac{2p - 400}{3}$ **107.** 20%   **109.** $8h$

Chapter 5   Inequalities in One Variable

Practice Exercises, pages 188–189
1. -3;
3. 1;
9.
11.
17.
19. $\{0\}$ **21.** {all real numbers}
23. {all real numbers except -1}
25. $\{-5\}$;
27. $\left\{\dfrac{5}{2}\right\}$;
33.
35.

37. all real numbers   **39.** all real numbers where a and b are additive inverses

Practice Exercises, pages 192–193
1. $m > -2$
3. $a < 1\dfrac{1}{2}$;
9. $x < 12$
11. $a > 4$
17. $\left\{\text{all integers less than } -\dfrac{5}{2}\right\}$ **19.** {all integers less than 0}   **21.** {all integers less than -2}   **23.** $\emptyset$
25. $t < -0.2$
27. $x > 10\dfrac{1}{2}$
29. $\{7, 8, 9, \ldots\}$ **31.** $\{3, 4, 5, \ldots\}$
33. Answers may vary. An example: $5 > 4$, $-2 > -4, 5 - (-2) \underline{?} 4 - (-4), 7 < 8$
35. False   **37.** $k - 6 < -6; k < 0$

Practice Exercises, page 197
1. $x < 5$
3. $a < 7$
9. $a > -7$
11. $b < 32$
17. $x \geq 5$
19. $z \leq 10$
25. $x \leq -40$
29. $m \geq 3$
33. $k < 9.6$
37. $\dfrac{3}{4}y > -18; y > -24$

Practice Exercises, pages 200–201
1. $y > -3$
3. $x \geq -2$
7. $\{x: x > 3\}$ **9.** $\{b: b \leq 9\}$ **11.** $\{y: y \leq -5\}$
13. $\left\{a: a < \dfrac{2}{3}\right\}$ **15.** $\{d: d \geq 4\}$ **17.** $\left\{y: y < -\dfrac{20}{3}\right\}$
19. $\{y: y > -15\}$ **21.** $\left\{k: k > -\dfrac{1}{6}\right\}$
23. $\{n: n \geq -23\}$ **25.** $\{m: m \geq 1\}$
27. $\{w: w \text{ is a real number}\}$
29. $\left\{t: t > \dfrac{7}{6}\right\}$ **31.** $\left\{x: x \neq -\dfrac{1}{4}\right\}$ **33.** $\{w: w \leq 3\}$
35. $\left\{y: y > \dfrac{1}{4}\right\}$ **37.** $\left\{a: a \geq -\dfrac{19}{11}\right\}$
39. $\{m: m \text{ is a real number}\}$ **41.** $\dfrac{2}{3}y + 14 \geq 8$; {all real numbers greater than or equal to -9}

Test Yourself, page 201
1. [number line with open circle at -2] 3. [number line with closed circle at 0]
5. {all real numbers greater than -1}
9. {a: a < -12} [number line with open circle at -12]

Practice Exercises, page 205
1. [number line 34 to 35] 3. [number line -1 to 1]
9. [number line -4 to 4] 11. [number line at 0]
17. [number line -1 to 0 to 3] 19. [number line -1 to 2]
25. no solution
27. [number line $\frac{3}{5}$ to 2]
31. $23 < c < 23.5$ 33. $l \leq 107$

Practice Exercises, pages 208–209
1. $\{-5, 3\}$
3. $\emptyset$ 5. $\{-9, -3\}$ 7. $\{-8, 1\}$ 9. $\{-\frac{7}{3}, 1\}$
11. $\{-\frac{11}{5}, 3\}$ 13. $\{-\frac{13}{7}, 3\}$ 15. $\{\frac{16}{3}, 8\}$
17. $\{-2, 2\}$ 19. $\{-\frac{1}{4}, \frac{5}{4}\}$ 21. $\{-3, 1\}$
23. $\{-10, 8\}$ 25. $\{-4, 2\}$ 27. $\{-9, 11\}$
29. $\{-11, 19\}$ 31. $\{\frac{3}{2}, \frac{9}{2}\}$ 33. $\{\frac{3}{5}, 3\}$ 35. $\{\frac{4}{3}\}$
37. $\{\frac{3}{5}, 1\}$ 39. $|4 + x| = 16; -20, 12$
41. $|2x - x| = 33; -33, 33$
43. $2|2x - 2| - 4 = 24; -6, 8$

Practice Exercises, pages 212–213
1. $y < -2$ or $y > 10$ [number line -2 and 10]
5. $-12 < c < 6$ [number line -12 to 6]
9. $-1 < x < 7$ [number line -1 to 7]
13. $m < 4$ or $m > 5$ [number line 4 and 5]
17. $-\frac{7}{5} \leq a \leq \frac{11}{5}$ [number line $-\frac{7}{5}$ to $\frac{11}{5}$]
21. $\frac{3}{2} \leq x \leq \frac{5}{2}$ [number line $\frac{3}{2}$ to $\frac{5}{2}$]
25. $\frac{1}{2} \leq z \leq \frac{5}{2}$ [number line $\frac{1}{2}$ to $2\frac{1}{2}$]
33. all real numbers   35. $\{d: d < 1 \text{ or } d > 5\}$
37. $\{t: \frac{1}{3} \leq t \leq 3\}$ 39. $\{p: p \text{ is a real number}\}$
41. $\{z: z \text{ is a real number}\}$ 43. $\{g: g > \frac{1}{4}\}$
45. $\{y: y > 2\}$ 47. $\emptyset$ 49. $|t - 0| < 5$

Practice Exercises, pages 215–216
1. at most $30,000   3. 90   5. at least 24 h   7. 5 m
9. no more than 5 mi
11. bonds: at least $2500; stocks: at most $7500

Test Yourself, page 217
1. $p < -3$ or $p > -1$ [number line -3 and -1]
5. $-1, 7$ 7. $-\frac{5}{3}, \frac{1}{3}$
9. $y < -\frac{1}{2}$ or $y > \frac{1}{2}$ [number line $-\frac{1}{2}$ and $\frac{1}{2}$]
13. $11 - \frac{1}{3}n > 5$; number less than 18
15. $\frac{13.1 + 12.8 + 13.0 + 13.3 + 12.7 + x}{6} <$ 12.8; less than 11.9 s

Summary and Review, pages 220–221
1. [number line -3 to 0]   3. [number line -2 to 0]
9. $g < -3$ [number line -3 to 0]
13. $\{-5, 9\}$ 15. $\{-5, 4\}$
17. $\{y: y \leq 2 \text{ or } y \geq 6\}$ [number line 2 and 6]
19. $\frac{1}{2}x + 3 > 2; x > -2$
21. $\frac{94 + x}{6} \geq 20$; at least $26

Maintaining Skills, page 224
1. 100,000   3. 1
5. -64 7. 729   9. 1   11. 256   13. $\frac{1}{8}$ 15. $\frac{3}{5}$
17. $\frac{5}{6}$ 19. $-5x^2 + 5xy$ 21. $-5cd$
23. $2p^2 - 4pq + 6q^2$ 25. $3920   27. $378

Chapter 6   Polynomials

Practice Exercises, pages 228–229
1. $3 \cdot 3 \cdot 3 \cdot z \cdot z$ 3. $-7 \cdot a \cdot a \cdot a \cdot b \cdot b \cdot b \cdot b$
5. yes   7. no   9. yes   11. yes   13. $15x^7y^3$
15. $18r^7s^6$ 17. $20g^8$ 19. 15×10^9
21. 3×10^9 23. 2.1×10^8 25. k^{m+4}
27. b^{x+3} 29. 2^m 31. $-2,476,099$ 33. $130,321$
35. $2r^{5x}d^x$ 37. $-27z^{5a}y^{6m}$ 39. 135 41. -27
43. 288 45. $\frac{1}{2}x^2$

Practice Exercises, pages 232–233
1. x^3
3. $\frac{1}{y^8}$ 5. 1   7. 8   9. $\frac{1}{10^4}$ 11. $\frac{3b^3}{a^4}$ 13. $\frac{3c^5}{7a^2}$
15. $\frac{4h}{t}$ 17. 4×10^3 19. 3×10 21. 1   23. -1
25. $\frac{1}{2}$ 27. 3   29. $-\frac{a}{2b}$ 31. $-\frac{5y^2}{3x}$ 33. $\frac{7y^3z^2}{3x^2}$
35. $-2q^2r^3$ 37. 1.4×10^9 39. $-\frac{3}{5}$ 41. x^{a-2}

43. $\dfrac{1}{y^{2k-1}}$ **45.** 6 **47.** 1 **49.** x^2

Practice Exercises, page 236 **1. a.** x^{30} **b.** z^{27}
3. -10^{10} **5.** $(-3)^4$ or 81 **7.** $\dfrac{-a^5}{b^{10}}$
9. 16×10^{12} **11.** $24x^{10}$ **13.** $\dfrac{-27a^3}{b^6}$ **15.** $\dfrac{16y^2}{25x^2}$
17. $\dfrac{-8k^{12}}{27j^9}$ **19.** $\dfrac{1}{81b^4}$ **21.** $4a^2b^6$ **23.** $-250m^6n^7$
25. $-3a^8$ **27.** $\dfrac{-1}{32x^5y^{15}}$ **29.** z^{4k} **31.** $-32k^{5m}$
33. a^{2m+4} **35.** $5184a^{12}b^4$ **37.** $46^2 = 2116$;
$47^2 = 2209$; $48^2 = 2304$; $49^2 = 2401$; $50^2 = 2500$;
$51^2 = 2601$; $52^2 = 2704$; $53^2 = 2809$; $54^2 = 2916$;
$55^2 = 3025$; $56^2 = 3136$; $57^2 = 3249$; $58^2 = 3364$;
$59^2 = 3481$; $60^2 = 3600$

Practice Exercises, page 239 **1.** no; 27 is greater
than 10 **3.** no; 0.009 is less than 1 **5.** 2×10^6
7. 7.65×10^{-3} **9.** 3.98×10^9
11. 2.092×10^{11} **13.** 9×10^6 **15.** 6×10^9
17. $1.\overline{3} \times 10^9$ **19.** 9.216×10^{12} **21.** 3.2×10^5
23. 4.8×10^{10}; 6.912×10^{13}; 2.52288×10^{16}

Test Yourself, page 239 **1.** no; expression has
2 terms **3.** yes **5.** $-12x^4$ **7.** -81 **9.** $\dfrac{-2}{x^3y}$
11. 3.12×10^{-4} **13.** 2.3×10^{-4}

Practice Exercises, pages 243–244 **1.** 1 **3.** 7
5. 0 **7.** 4 **9.** 5 **11.** $-4x^3y + 6x^2y^2 + 2xy^3$
13. $5x^2y^2 - 2xy^3$ **15.** $-17x^3 - 25x^2 + 10$
17. $4x^2z - \dfrac{15}{8}xz^2$ **19.** $-\dfrac{23}{5}x^2z + 2xz^2$
21. $-\dfrac{1}{2}cd^2$ **23.** $21x^2 - 17$ **25.** $-2xy^2 + 8x^2y$
27. $-x^{3a} + x^{2a} + 2x^a$ **29.** $x^{2a} - x^a$ **31.** $-\dfrac{13}{4}$
33. $8w + 8$ **35.** $6x - 3$

Practice Exercises, pages 246–247 **1.** $5x^3 + 2x^2 - 9x + 2$ **3.** $7r + 6s - 5t$ **5.** $2x^3 + 9x^2 - x + 2$ **7.** $5x^3 + 3x - 16$ **9.** $7r - 2s + t$
11. $2x^3 + 3x^2 + 3x - 2$ **13.** $5x^3 + 4x - 17$
15. $m^3 + 3$ **17.** $5c^3 - 4c$ **19.** $13a + b + 2c$
21. $9r + 6s + 10$ **23.** $j + k + 6m$ **25.** $x^2 + 5x$
27. $3x^4 - 13x^3 - 7x^2 - 3x$ **29.** $14a^2 + a + 7$
31. $10x^3 - x^2 + 6x - 1$ **33.** $4n + 6$

Practice Exercises, pages 249–250 **1.** $-5x^5 + 30x^4 - 40x^3 + 25x^2$ **3.** $-6p^7 + 4p^5 - 10p^3$
5. $12p^3q + 6p^2q^2 - 3p^2qr$ **7.** $2x^5 - 2x^3 + 6x^2$
9. $15m^4n^2 - 5m^2n^3 - 10mn^4 + 5mn^2p$

11. $49h^2k - 7hk^2 + 56hk$ **13.** $-a^2b^3 - 11a^3b^2 + 40a^2b$ **15.** $-2y^2 + 7y - 10$
17. $6m^3 - 6m^2 + 2m$ **19.** $6a^2b^3 - 4a^3b^2$
21. $15r^4s^4 - 10r^2s^6 + 5rs^5$ **23.** $-12a^3bc^4 - 3a^2b^2c^3 + 6ab^3c^6$ **25.** $-5m^3 + 7m^2 + 2m^2n$
27. $24a^{2x} - 8a^{x+2} + 16a^x$ **29.** $j^2 + 2j$
31. $w^2 + 2w$ **33.** $3l^2 - 10l$

Practice Exercises, pages 253–255 **1.** $a^2 - 3a - 40$ **3.** $h^2 - 6h - 7$ **5.** $3a^2 + 11a - 20$
7. $10m^2 - 33m - 54$ **9.** $7r^2 - 38r + 15$
11. $10x^2 - 31x + 24$ **13.** $4y^2 - 7yz - 2z^2$
15. $8j^2 - 10jk - 3k^2$ **17.** $-6b^2 - 5b + 21$
19. $-24y^4 + 18y^3 - 15y^2 + 6y$ **21.** $-10z^4 - 13z^3 + 11z^2 + 12z$ **23.** $x^3 + 5x^2 + x - 10$
25. $z^3 - 8z^2 + 3z + 36$ **27.** $c^3 - 3c^2 - 9c - 5$
29. $-k^3 + 4k^2 + 29k + 24$ **31.** $y^3 - 5y^2 + 7y - 3$ **33.** $9x^2 - 24x + 16$ **35.** $4a^2 - 12ab + 9b^2$ **37.** $6a^3 + a^2 + 13a + 10$ **39.** $-2c^3 + 9c^2 + c - 2$ **41.** $10g^3 - 9g^2 + 8g^2 - 3$
43. $x^4 - 6x^3 + 8x^2 - 15x$ **45.** $4y^4 - 8y^3 + 7y^2 - 6y$ **47.** $5t^4 + 4t^3 - 14t^2 - 4t$
49. $a^3 + b^3$ **51.** $y^4 + y^3 - 6y^2 + 3y + 9$
53. $x^3 + 9x^2 + 27x + 27$ **55.** $64b^3 - 48b^2 + 12b - 1$ **57.** $3x^5 - 4x^3 + 6x^2 + x - 2$
59. $10a^5 - 9a^4 + 4a^3 + 14a^2 - 6a + 3$
61. $n^2 + 3n + 2$ **63.** $2x^2 + 9x + 7$
65. $e^2 + 6e + 8$

Practice Exercises, pages 258–259
1. $9x^2 + 12xy + 4y^2$ **3.** $64m^2 + 32mn + 4n^2$
5. $16m^4 - 48m^2 + 36$ **7.** $49s^4 - 42s^2 + 9$
9. 576 **11.** 144 **13.** $x^2 - 16$ **15.** $y^2 - 25$
17. $16x^6 - 9$ **19.** $4t^2 - 4tu + u^2$
21. $d^4 + 2d^2e^2 + e^4$ **23.** $h^4 - j^4$
25. $1 - 24g^3h^2 + 144g^6h^4$ **27.** $25a^2b^6 + 60ab^3c^2d^4 + 36c^4d^8$ **29.** $3721x^6y^4z^2$
31. $16 - a^{2x}$ **33.** $x^{2a+2} + 2x^{a+2} + x^2$
35. $3^{4y+2} - 4(3^{2y+1}) + 4$ **37.** $a^2 + 2a + 1$
39. $4x^2 - 9$

Test Yourself, page 259 **1.** $3x^3 - 4x$ **3.** $2y^2 - 4$
5. $6y^3 - 2y^2$ **7.** $55a^5b^3 - 33a^4b^4 + 77a^3b^5$
9. $3r^2 + 40r + 77$ **11.** $24n^2 - 5n - 75$
13. $81y^2 + 54y + 9$ **15.** $25a^2 - 90ab + 81b^2$

Practice Exercises, pages 262–263 **1.** 100,000,
1,000,000, 10,000,000 **3.** 35, 48, 63 **5.** 4558
7. 6, 12, 18, 30, 48, 78, 126 **9.** $\dfrac{n(n+1)}{2}$
11. It is twice as large. **13.** 233, 377, 610, 987,
1597, 2584, 4181, 6765, 10,946, 17,711
15. 1, 1, 2, 3, 5

Summary and Review, pages 264–265 1. a^7
3. $-15a^4c^5$ 5. x^2 7. 1 9. 3×10^4 11. 1
13. $\dfrac{25a^2}{c^8}$ 15. 2.89×10^5 17. 1.6×10^6
19. 9.0×10^3 21. 1; 1 23. $3x^3 - x^2 + 5$
25. $9x + 2y$ 27. $-y + 5$ 29. $3x^3 + 6xy - 3x$
31. $-35x^3y^2 + 5x^2y^3$ 33. $3b^2 - 17b + 24$
35. $d^2 - 18d + 81$ 37. $25x^2 - 16$ 39. 30, 39, 49

Cumulative Review, page 268 1. -0.5 3. 7
5. $-5x + 1$ 7. $9m^2 - m + 2$ 9. $4b^3 - 6ab^2 + 2b^2$ 11. $6n^2 - 11n + 4$ 13. 15 15. 8
17. $-4xy$ 19. $6n^5$ 21. $8p^6$ 23. $\dfrac{m^2}{4}$ 25. -14
27. -8 29. 20 31. 5 33. 11 35. 4
37. All real numbers less than -2 or greater than or equal to zero 39. 20 km/h, 25 km/h

Chapter 7 Factoring Polynomials

Practice Exercises, page 272 1. $2 \cdot 13$ 3. $2^2 \cdot 13$
5. $2 \cdot 7 \cdot 11$ 7. $3 \cdot 5 \cdot 13$ 9. $3 \cdot 5 \cdot 7$
11. $11 \cdot 13$ 13. $5^2 \cdot 11 \cdot 13$ 15. $5^3 \cdot 7 \cdot 11$
17. $3^2 \cdot 5 \cdot 7^2$ 19. $2^5 \cdot 5 \cdot 11$ 21. $2 \cdot 11^2$
23. $2^5 \cdot 3^2$ 25. prime 27. $2^3 \cdot 5^3$ 29. 59
31. 109 33. yes; $3x$ is the product of two primes
35. yes; $x \cdot x = x^2$ is the product of two primes.
41. yes; the maximum area is obtained when the rectangle is a square.

Practice Exercises, pages 276–277 1. 7 3. 18
5. 25 7. 8 9. $2x^2y(x - 6y^3)$
11. $7cd^3(1 + 2c^2d^2)$ 13. $13x^2y^2(y + 2)$
15. $6mn^3(2m^3n^2 - 3)$ 17. $2jk(2j^2 - 3k + 4)$
19. $4ab(a + 2ab + 3)$ 21. $12x^3y^4(1 + 3x - 5y)$
23. $3m^2n^2(8m + 7n - 13n^2)$
25. $x^3y^3(13x^2y - 11y + 17x)$
27. $6l^2m(7lm - 6m^4 - 9l^2)$
29. $33x^5y^7(1 - 3xy^2 - 2y)$
31. $r^2s^3(89s^6 + 113rs^4 + 73r^2)$
33. $-5a^2b^4(3a + 7a^2b + 11)$
35. $-5x^4y^4(23 + 45xy + 5x^2y^2)$
37. $7c^4d^2e^3(11d^4 + 4ce)$ 39. $18x^2y^2z^2(3y^3z - 2x)$
41. $r^2\left(\dfrac{\pi}{2} + 4\right)$ 43. $2r^2(8 - \pi)$

Practice Exercises, pages 280–281
1. $(x + 9)(x + 1)$ 3. $(a + 11)(a + 1)$
5. $(z + 29)(z + 1)$ 7. $(y + 11)(y + 3)$
9. $(s - 6)(s - 1)$ 11. $(m - 41)(m - 1)$
13. $(y - 2)(y - 11)$ 15. $(z - 17)(z - 3)$
17. $(13 - b)(5 - b)$ 19. $(12 - z)(3 - z)$
21. $(3 + y)(16 + y)$ 23. $(9 - f)(8 - f)$
25. $(a + 9b)(a + 3b)$ 27. $(x - 7y)(x - y)$
29. $(r + 3t)(r + 2t)$ 31. $(m - 25n)(m - n)$
33. $(s + 9t)(s + 2t)$ 35. $(y - 8z)(y - 2z)$
37. $(x - 25y)(x - 2y)$ 39. $(r + 16t)(r + 4t)$
41. $(m - 24n)(m - 3n)$ 43. $(m - 8n)(m + 2n)$
45. $(c^2 + 6)(c^2 + 2)$ 47. $(y^2 - 3)(y^2 - 8)$
49. $(a + 2)(a + 8)$ 41. x^2 53. $(2 - z)(11 - z)$
55. $(a^x + 1)(a^x + 2)$ 57. $(x^{2n} - 3)(x^{2n} - 4)$; $(x^{2n} - 3)(x^n + 2)(x^n - 2)$
59. length: $x + 4$; width: $x + 1$
61. length: $m + 3$; width: $m + 1$
63. length: $x + 2a$; width: $x + a$

Practice Exercises, page 284
1. $(x + 5)(x - 1)$ 3. $(m - 10)(m + 1)$
5. $(x - 2)(x + 4)$ 7. $(m - 6)(m + 2)$
9. $(x - 1)(x + 3)$ 11. $(a - 1)(a + 6)$
13. $(m - 4)(m + 3)$ 15. $(k - 15)(k + 2)$
17. $(x - 3y)(x + 2y)$ 19. $(r - 7s)(r + 3s)$
21. $(x - 10y)(x + 4y)$ 23. $(k - 9j)(k + 4j)$
25. $(x - 2y)(x + y)$ 27. $(r - 2t)(r + 7t)$
29. $(y - 16z)(y + 2z)$ 31. $(x - 10y)(x + 2y)$
33. $(r - 6t)(r + 9t)$ 35. $(y - 20z)(y + 5z)$
37. $(y^2 - 2)(y^2 + 25)$ 39. $(a - 9)(a + 3)$
41. $(x^k - 2)(x^k + 14)$
43. $(a + 35b)(a - 10b)$; 550 ft $\times$ 100 ft

Practice Exercises, pages 287–288
1. $(3x + 2)(x - 8)$ 3. $(5z + 2)(z - 3)$
5. $(2m + 5)(m + 1)$ 7. $(5r + 3)(3r + 7)$
9. $(9l + 5)(9l + 3)$ 11. $(6z + 5)(z - 1)$
13. $(2m - 9)(2m + 1)$ 15. $(2x + 11)(x - 1)$
17. $(7d - 5)(2d + 3)$ 19. $(8x - 5)(3x - 4)$
21. $(4z + 5)(5z + 6)$ 23. $(13c - 5)(2c + 3)$
25. $(3x + 2y)(x - y)$ 27. $(2r + t)(r + 7t)$
29. $(3a - b)(a - 5b)$ 31. $(3n - 2p)(2n + p)$
33. $(5x - 2y)(x + 4y)$ 35. prime
37. $(4x + 9y)(3x + 2y)$ 39. $(7k - 3m)(2k - 11m)$
41. $(9a + 8b)(3a - 4b)$ 43. $(4x + 11)(x + 4)$
45. $(4a - 9)(3a - 16)$ 47. $(3x^k + 2)(2x^k + 7)$
49. $(5x^{2k+3} + 4)(2x^{2k+3} - 3)$
51. $(y - 12)(6y + 1)$ 53. $5c + 17$

Test Yourself, page 288 1. $2 \cdot 5 \cdot 11$
3. $2^2 \cdot 3^2 \cdot 5$ 5. $3 \cdot 7^2$ 7. $14x(3 - x^2)$
9. $(a + 5)(a + 7)$ 11. $(m - 10)(m - 6)$
13. $(y - 15)(y + 4)$ 15. $(2x + 1)(x + 12)$
17. $(3x - 2)(x + 2)$

Practice Exercises, pages 290–291 1. yes 3. no
5. yes 7. $(x - 6)^2$ 9. $(2d + 9)^2$
11. $(k + 10)^2$ 13. $(9y - 2)^2$ 15. $(5t + 1)^2$

Answers to Selected Exercises **681**

17. $(10k^5 + 1)^2$ **19.** $(5y^4 + 1)^2$
21. $(a^5b^2 - 4)(a^5b^2 + 4)$ **23.** $(x^9y^5 - 6)(x^9y^5 + 6)$
25. yes **27.** yes **29.** no **31.** yes
33. $(12x^2y - 25)(12x^2y + 25)$ **35.** $(a^3bc^2 - d)$
$(a^3bc^2 + d)$ **37.** $(2p^2q^2r^4 - 9)(2p^2q^2r^4 + 9)$
39. $(7rs + 1)^2$ **41.** $9(2x^3y^2 - 1)^2$
43. $(e^{32}f^{50} - g^{72}h^{18})(e^{32}f^{50} + g^{72}h^{18})$
45. $2a + 1$ **47.** x^2
49. $(a + b - c)(a + b + c)$
51. $(3x + 4)^2$ **53.** $t^2 - rs$

Practice Exercises, pages 294–295
1. $5(a + 3)$ **3.** $4(x - 3)$ **5.** $(5x - 4)(x - 2)$
7. $19(j - 2)$ **13.** $-2(a - b)$ **15.** $21(a - b)$
17. $(6 - h^3)(h - k)$ **19.** $(r - 3)(s + t)$
21. $(3r + s)(t + w)$ **23.** $(c - 2)(a + 3b)$
25. $(6y - 3x - 4)(6y + 3x + 4)$
27. $(x + y)(x + z)$
29. $(3b + 4c)(5a - 3c)$
31. $(3r + 1)(2t - 5s)$
33. $(r + 4s + t)(r - 4s - t)$
35. $(a + 3 + c)(a + 3 - c)$
37. $(5y + 2x - 3)(5y - 2x + 3)$
39. $(x + 3z^{3r})(x^{2r} + y)$
41. $(m^r + n)(p + n^{2r})$
43. $(x^a + y^b + 1)(x^a + y^b - 1)$
45. $3(8y - 3x)(2y + x)$

Practice Exercises, pages 298–299
1. $5(x + 2)(x - 2)$ **3.** $3(k + 7)(k - 7)$
5. $4y(y - 3z)(y + 3z)$ **7.** $6x(x + 2y)(x - 2y)$
9. $2(x + 5)(x - 2)$ **11.** $6(k + 1)^2$
13. $5(2k + 1)(k + 3)$ **15.** $-4(x + 3)(x - 2)$
17. $-10(m + 3)(m - 7)$ **19.** $3x(5x - 1)^2$
21. $12x(x + 1)^2$ **23.** $4x^2y(2x + 3)(x - 1)$
25. $6r^2s^2(r + 1)(2r - 1)$
27. $3x^3y^3(3x + 2y)(2x - 3y)$
29. $4x^4y^4(2x - 3y)^2$ **31.** $-3m^5p^2(2m + 5p)^2$
33. $2x(x^4 - 5)(x^4 + 5)$
35. $2x(x^3 - 4)(x^3 + 4)$
37. $3m(m + 4)(m - 4)(m + 2)(m - 2)$
41. $3x^4y^5(8x - 3)^2$ **43.** $3(2a - 1)(a - 4)$
45. $5(a + 2)(a - 1)$
47. $(a + 3)(a - 3)(a + 2)^2$
49. $3a(a^8 + 5)^2$ **51.** $x^{k+7}(x^{k+7} + 1)^2$
53. $x(x + 2)(x + 1)$ **55.** $2a(a + 3)(a + 1)$

Practice Exercises, pages 302–303
1. $\frac{2}{3}, -4$
3. $3, -6$ **5.** $-\frac{3}{5}, 2$ **7.** $0, -4$ **9.** $0, 4$
11. $0, -3$ **13.** $-4, 2$ **15.** $-\frac{1}{3}, -1$ **17.** $-\frac{1}{4}, 3$

19. $10, -9$ **21.** $5, 3$ **23.** $4, -\frac{5}{3}$ **25.** $\frac{5}{3}, -3$
27. $-\frac{5}{2}, \frac{2}{3}$ **29.** $-\frac{3}{7}, \frac{5}{2}$ **31.** $-1, \frac{1}{12}$
33. $0, -7, 3$ **35.** $0, -\frac{4}{3}, \frac{5}{2}$ **37.** $0, -\frac{9}{2}, \frac{7}{4}$
39. $0, -\frac{9}{5}, \frac{5}{2}$ **41.** $0, -\frac{3}{2}, \frac{11}{6}$ **43.** $-2, -1$
45. $-7, -3, -1$ **47.** $0, -3, 3, -1, 1$
49. $n^2 + 5n = 24$ **51.** $2n^2 = n - 10$

Test Yourself, page 303
1. $2(y - 7)(y + 7)$ **3.** $4a(a - 2)(a - 1)$
5. $3x(1 - y)$ **7.** $16(x^2 + 4)$ **9.** $-1, 10$

Practice Exercises, page 306
1. 14 ft × 10 ft **3.** 14 cm × 6 cm
5. $-8, -6$ or $6, 8$ **7.** $-9, -7$ or $7, 9$
9. 6 **11.** 27 m × 8 m **13.** 30 in. × 30 in.
15. 8 in. × 4 in.

Summary and Review, pages 310–311
1. prime **3.** composite **5.** $2^4 \cdot 3$ **7.** $2^3 \cdot 3 \cdot 5$
9. 5 **11.** $3ab^2(1 - 3a)$ **13.** $(x + 3)(x + 2)$
15. $(a + 11b)(a + b)$ **17.** $(y - 12)(y + 2)$
19. $(2m + 1)(m + 7)$ **21.** $(3x - 5)(2x + 3)$
23. $4(5m - 2n)(5m + 2n)$ **25.** $7(m - 2)(m + 2)$
27. $3a(2a^2 + 4a + 3)$ **29.** $5(x - y)$
31. $4x(3x - 1)^2$ **33.** $9(2x^2 - 5)$ **35.** $-8, 10$
37. $-28, -26$ or $26, 28$

Maintaining Skills, page 314
1. $\frac{1}{2}$ **3.** $\frac{1}{10}$ **5.** $\frac{6}{7}$
7. $\frac{25}{14}$ **9.** $\frac{5}{9}; \frac{6}{9}$ **11.** $\frac{10}{36}; \frac{21}{36}$ **13.** $\frac{25}{24}$ **15.** $\frac{43}{60}$
17. 40 in.^{2}

Chapter 8 Rational Expressions

Practice Exercises, pages 318–319 **1.** -4
3. $-\frac{2}{5}$ **5.** $-\frac{1}{3}, \frac{1}{3}$ **7.** $-3, 3$ **9.** $\frac{1}{4c}; c \neq 0$
11. $\frac{2a + 3}{4}$ **13.** $7; a \neq 2$ **15.** $-\frac{1}{3}; m \neq \frac{5}{2}$
17. $\frac{1}{2x - 1}; x \neq \frac{1}{2}, -3$
19. $\frac{1}{5x - 3}; x \neq \frac{3}{5}, -2$
21. $-\frac{c + 6}{3 + c}; c \neq -3, 3$

23. $\dfrac{y+3}{y-2}$; $y \neq -2, 2$

25. $\dfrac{m-4}{m-5}$; $m \neq -5, 5$ **27.** $\dfrac{5c^2}{3a^2}$; 0; 0

29. $\dfrac{3+2n}{7m^3n^3}$; 0; 0 **31.** $\dfrac{7z+2}{z-1}$; 1; -3

33. $\dfrac{2c+3}{c-7}$; $-\dfrac{3}{2}$; 7 **35.** $\dfrac{4a^2+8a-5}{15-a-2a^2}$; -3; $\dfrac{5}{2}$

37. $\dfrac{9+2x}{x-11}$; -1; 11 **39.** $\dfrac{2r-5}{2r-1}$; $\dfrac{1}{2}$; $-\dfrac{1}{3}$

41. $\dfrac{a-3b}{a+4b}$; $2b$; $-4b$ **43.** $\dfrac{x-y}{x+3y}$; $-y$; $-3y$

45. $\dfrac{3r+4s}{2r-s}$; $\dfrac{4}{3}s$; $\dfrac{1}{2}s$ **47.** $\dfrac{6}{s}$

Practice Exercises, pages 322–323 **1.** $\dfrac{6}{5x^2y}$

3. $\dfrac{12}{5x^2y}$ **5.** $\dfrac{5x^2}{y^2}$ **7.** $\dfrac{6r^2}{5st}$ **9.** $\dfrac{6q^2}{pr^2}$

11. $24m^2n^3p$ **13.** $\dfrac{1}{2}$ **15.** $\dfrac{c-5}{3c-4}$ **17.** $\dfrac{b+2}{3(2b-3)}$

19. $-\dfrac{3r+2}{r}$ **21.** -1 **23.** $\dfrac{15r^2s^2}{2q^2t^3}$

25. $\dfrac{2(2i-3)}{i^3}$ **27.** 1 **29.** $\dfrac{l+5}{2l-9}$ **31.** -1

33. $-\dfrac{2a+1}{7a+2}$ **35.** $\dfrac{(s-1)(s-4)}{2(s+2)(s+3)}$ **37.** 1

39. $\dfrac{7t-3u}{2t+9u}$ **41.** $\dfrac{(r-2e)(r+2e)}{(2r-e)(2r+e)}$ **43.** 36

45. -125 **49.** $\dfrac{2}{x+5}$

Practice Exercises, pages 326–327 **1.** $\dfrac{10}{b}$

3. $\dfrac{4}{d}$ **5.** $\dfrac{d}{6e}$ **7.** $\dfrac{3}{2z+5}$ **9.** $\dfrac{1}{x+1}$

11. $\dfrac{7(r-7)}{5(r+2)}$ **13.** $\dfrac{x+2}{x+6}$ **15.** $\dfrac{4a^2}{15b^2}$

17. 1 **19.** $\dfrac{3ac}{10b}$ **21.** $\dfrac{x}{5(y+6)}$ **23.** $6(4-c)$

25. $\dfrac{5h-2}{3h-4}$ **27.** $\dfrac{2t+3}{2t-2}$ **29.** $\dfrac{5(2x-5)}{x-5}$

31. $\dfrac{7y+3}{3(3y-2)}$ **33.** $\dfrac{3y-1}{4y+5}$ **35.** $\dfrac{x}{y+5}$

37. $\dfrac{(g-3h)(g-4h)}{3(h+g)(2h+g)}$ **39.** $\dfrac{(3c+2d)(c-3d)}{(3c-d)(2c+3d)}$

41. $3x-1$

Practice Exercises, pages 331–332 **1.** $10ab$
3. $42xy$ **5.** $42a^3b^2$ **7.** $18x^2y^3$
9. $3(n+1)(n+5)$ **11.** $6(r-9)(2r+5)$

13. $(t+6)(t-6)$ **15.** $\dfrac{5}{12mn^2}$; $\dfrac{9m^2n}{12mn^2}$

17. $\dfrac{5(y+3)}{7(y+1)(y+3)}$; $\dfrac{21}{7(y+1)(y+3)}$

19. $\dfrac{3y}{y^2-9}$; $\dfrac{2(y+3)}{y^2-9}$ **21.** $(x+3)(x+2)(x+5)$

23. $(3c+7)(4c-5)(c-6)$
25. $(x-3)(16x+5)(2x+1)(x+3)$

27. $\dfrac{9b^2}{6a^2b^2}$; $\dfrac{18a^2b}{6a^2b^2}$; $\dfrac{a(a+b)}{6a^2b^2}$

29. $\dfrac{-8n(3n+5)}{(3n-5)(3n+5)}$; $\dfrac{5n^2}{(3n-5)(3n+5)}$

31. $\dfrac{3d(d+1)}{2d(d-2)(d+1)}$; $\dfrac{2d(d+3)}{2d(d-2)(d+1)}$

33. $\dfrac{c(c+2)}{3(c+1)(c+2)}$; $\dfrac{3c^2}{3(c+1)(c+2)}$; $\dfrac{9c(c+1)}{3(c+1)(c+2)}$

35. $(5x-7y)(3x+4y)(4x-3y)$
37. $72(a-b)^2(a+b)(a^2+b^2)$
39. $6(5m-4)(3m+2)(m-4)$
41. $6x^3$

Practice Exercises, pages 335–336 **1.** $\dfrac{4}{m}$

3. $\dfrac{1}{2x}$ **5.** 2 **7.** -1 **9.** $\dfrac{a+1}{2a^2}$ **11.** $\dfrac{1-5m}{3m^2}$

13. $\dfrac{10-a}{5(a+4)}$ **15.** $\dfrac{8l+7}{(l+4)(l-1)}$

17. $\dfrac{24}{(x-5)(x+5)}$ **19.** $\dfrac{11}{(r-3)(r+3)}$

21. $\dfrac{8y+4}{(y-1)(y+3)}$ **23.** $\dfrac{a-2}{a-1}$

25. $\dfrac{g^3+g^2+2g-4}{g(g+2)}$ **27.** $\dfrac{5b^2-b+2}{6b^2}$

29. $\dfrac{5u^2-11u+42}{6(u+2)(u-2)}$ **31.** $\dfrac{a^2+9a+7}{(a-4)(a+3)}$

33. $-\dfrac{d+3}{d+4}$ **35.** $\dfrac{t^2-t-9}{(2t+3)(4t-1)(t+5)}$

37. $\dfrac{4n^2+3n+18}{(n+3)(n-2)(3n+4)}$

39. $\dfrac{14k^2+67k+45}{k(k-3)(k+3)(k+5)}$ **41.** $-\dfrac{8}{a(a+4)}$;

$\dfrac{6}{(h+1)(h-1)}$; $\dfrac{-3d-15}{(d-3)(d+3)}$

Practice Exercises, pages 339–340 **1.** $\dfrac{2x+3}{x}$

3. $\dfrac{3x-10}{x}$ **5.** $\dfrac{2z^2-z-1}{z}$ **7.** $\dfrac{4y^2-10y+3}{4y}$

9. $\dfrac{2d^2+2d-3}{2d+1}$ **11.** $\dfrac{u^2+4u-1}{u+2}$ **13.** 2

15. $\dfrac{2}{5}$ 17. $\dfrac{mn+2}{2n^2-m}$ 19. $\dfrac{a}{a-4}$ 21. $\dfrac{3}{2}$ 23. $u-1$

25. $\dfrac{3x^2-4x-3}{9x^2-15x-5}$ 27. $\dfrac{3z^2-13z-10}{3z^2-5}$

29. $\dfrac{x^2+2x-3}{x+1}$ 31. $\dfrac{a^2-a+1}{a+1}$

33. $\dfrac{2m^2-2m-23}{2m+5}$ 35. $c+3$ 37. $\dfrac{a-2}{a+4}$

39. $\dfrac{n+6}{n-3}$ 41. $-\dfrac{2u}{u^2+1}$ 43. $\dfrac{x^2+2}{4x-1}$

45. $\dfrac{4a^2-2a-6}{3a^2-3a-18}$ 47. $\dfrac{3a+9}{a(3a^2+a+3)}$

Test Yourself, page 340

1. $\dfrac{m+6}{3m}$ 3. $\dfrac{3}{10ab^2}$; $\dfrac{4a^2b}{10ab^2}$

5. $\dfrac{3m+10}{6m}$ 7. $\dfrac{3y-1}{3}$

Practice Exercises, page 344 1. $2x+1-\dfrac{1}{3x}$

3. $x^2-18x+3-\dfrac{7}{x}$ 5. y^3-5y^2+

$7y+2-\dfrac{3}{y}$ 7. $n-5$. 9. $k+5-\dfrac{12}{k+6}$

11. $3j+16+\dfrac{43}{j-3}$ 13. $5m+12$

15. $d-4-\dfrac{8}{d-3}$ 17. $2b^2+2b+10+\dfrac{10}{b-1}$

19. $3x^2-3x+3-\dfrac{6}{x+1}$ 21. $6s-17$

23. $2p^4q^2-6p^2q+3pq^3-1$ 25. p^2+p-

$7+\dfrac{10}{p+2}$ 27. $3h^2+h-5-\dfrac{2}{2h+1}$

29. $x^2+4x+13+\dfrac{57}{x-4}$

31. $3h^2+h-1+\dfrac{1}{3h-1}$ 33. $9t^2+21t+49$

35. $3x+2y$ 37. $2p+2q+\dfrac{2q^2}{3p+2q}$

39. $p=x-1$; $100

Practice Exercises, pages 348–349 1. $\dfrac{3}{4}$ 3. $\dfrac{1}{6}$

5. $\dfrac{1}{5}$ 7. $\dfrac{1}{2x}$ 9. $\dfrac{5}{6}$ 11. $\dfrac{4}{1}$ 13. false 15. true

17. true 19. 36 21. 17 23. $\dfrac{3}{2}$ 25. 4

27. $\dfrac{7}{2}$ 29. $\dfrac{1}{3}$ 31. -3 33. $-2, \dfrac{7}{6}$ 35. $-6, 2$

37. $-6, 4$ 39. a. $\dfrac{32}{7}$ b. $\dfrac{5}{12}$ c. $\dfrac{3}{4}$ 41. 17 km

43. 8.5 h

Practice Exercises, pages 352–353 1. $\dfrac{9}{28}$ 3. $\dfrac{4}{35}$

5. 4 7. no solution 9. 7 11. $\dfrac{2}{5}$ 13. -2

15. $-\dfrac{1}{3}$ 17. $\dfrac{1}{2}$ 19. 5 21. no solution 23. $2, -3$

25. $-\dfrac{3}{2}, 4$ 27. $\dfrac{2}{3}, -\dfrac{3}{5}$ 29. $1, 4$ 31. -14

33. $3, -3$ 35. $3, -1$ 37. $-\dfrac{6}{5}, -1$ 39. $260

Practice Exercises, pages 356–357

1. $\dfrac{12}{7}$ h 3. $2\dfrac{2}{9}$ h

5. Rhoda: 55 mi/h; Van: 48 mi/h

7. $62\dfrac{1}{2}$ mi/h 9. 12 h 11. 11 h 20 min

Practice Exercises, pages 359–360 1. $9276.69

3. $p = \dfrac{15 \times (500)^2 - 7500u - ut}{500}$; $5835.00

5. $t = r(5.50) + s(4.50)$ 7. $t = \dfrac{10v+7u}{uv}$

9. $r = (n-3)\left(\dfrac{200}{n}+2\right)$

11. $p = 15n - 15u - uc$

Test Yourself, page 361 1. $6 - \dfrac{1}{y} + \dfrac{5}{y^2}$ 3. $\dfrac{1}{25}$

5. 1300 m 7. 14, 2 9. b = number bought,

n = number sold; $p = n\left(\dfrac{25}{b}+1.50\right) - 25$

Summary and Review, pages 362–363

1. $\dfrac{b}{3}$, $b \neq 0$ 3. $\dfrac{1}{m+2}$; $m \neq -2, 1$ 5. $\dfrac{16x}{5z}$

7. $\dfrac{a-1}{a-2}$ 9. $\dfrac{3n}{24m^3n^2}$; $\dfrac{10m^2}{24m^3n^2}$

11. $\dfrac{2x}{(x+3)(x-3)}$; $\dfrac{3(x-3)}{(x+3)(x-3)}$

13. $\dfrac{2c-13}{(c+1)(c-2)}$ 15. $\dfrac{u^2+4u+3}{u+2}$

17. $y - x$ 19. $2r^2 + 9r - 5$ 21. $\dfrac{5}{2}$ 23. 10

25. 18 min

Cumulative Review, pages 366–368 1. -1

3. -2 5. $2\dfrac{1}{2}$ 7. $-3\dfrac{1}{2}$ 9. $\{x: x \geq -1\}$

11. $\{x: x \neq -2\}$ 13. $\{x: x \leq -1 \text{ or } x > 0\}$

15. -3 17. $t \geq 6$ 19. no solution 21. 8

23. 2 **25.** $-2 \leq p \leq 3$ **27.** 12 **29.** 3 **31.** 7
33. 6×10^6 **35.** 1.28×10^4 **37.** -5 **39.** 35
41. -1 **43.** -8 **45.** 4 **47.** $-\frac{2}{3}$ **49.** -2
51. -4 **53.** $4b^6$ **55.** $\frac{x}{3y}$ **57.** -1 **59.** $\frac{1}{z-2}$
61. $3a^2 + 5a$ **63.** $3r - 15$ **65.** $(h+3)(h+1)$
67. $\frac{(k-5)(k-2)}{(k+2)(k+5)}$ **69. a.** reflexive property
b. substitution property **c.** commutative property
for addition **71.** 24 oz **73.** $4(a - 2b)$
75. $3c(2c - 3)$ **77.** $7p(3p^2 + 2p - 4)$
79. $(d+4)^2$ **81.** $(g+5)(g-3)$
83. $(r-6)(r-4)$ **85.** $6a^3$ **87.** $3x(x+1)$
89. $-\frac{2}{9}$ **91.** $\frac{5b+6}{4b^2}$ **93.** $\frac{5h^2 + 11h + 3}{h+2}$
95. 1 **97.** $4x^3y - 8x^2y^3 + 4x^2y$
99. $a^2 + 7a - 4$ **101.** $55.25 **103.** 3, 5, 7 or
$-3, -1, 1$ **105.** $66\frac{2}{3}$ min. **107.** $41\frac{2}{3}$ mi/h

Chapter 9   Linear Equations

Practice Exercises, pages 372–374
1.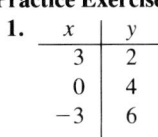
7.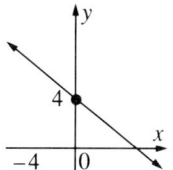
13.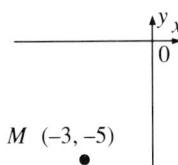
15.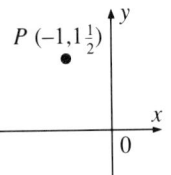

17. $(-3, -7)$ **19.** $(3, 1)$ **29.** y-axis
31. Answers may vary; possible solutions are
$y = \frac{1}{2}x - \frac{7}{2}$, $y = 3x - 6$ **35.** $(-1, 4)$
37. parallel; perpendicular **39.** one; $(-2, 4)$
41. 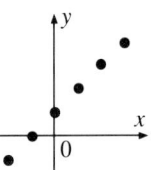 y-coordinate is one more than x-coordinate; $y = x + 1$

43. a. ordered pairs of (time, distance)
b. All are solutions except (0.5, 14).

Practice Exercises, pages 378–379
1.

x	y
3	2
0	4
-3	6

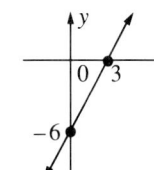

5. 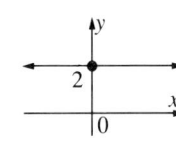 **9.**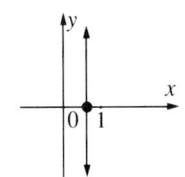

13. 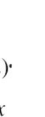 **15.**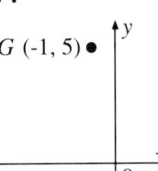

21. no **23.** yes **33.** 3 **35.** -3 **37.** $6a^2$
39. $8d^4 + 6$ **41.**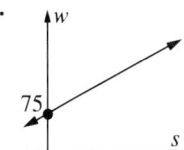

Practice Exercises, pages 381–383 **1.** $55; $70
3. $115 **5.** $102; $204 **7.** $306 **9.** 30; 36
11.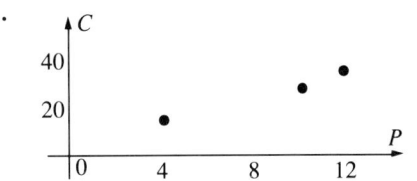

13. 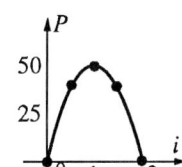 **15.**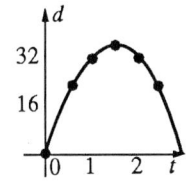

Practice Exercises, pages 387–388 **1.** 2 **3.** $\frac{5}{3}$
5. $-\frac{3}{2}$ **7.** $-\frac{6}{5}$ **9.** -1 **11.** $-\frac{4}{5}$
13. 0; horizontal **15.** no slope; vertical
17. 0; horizontal

19. 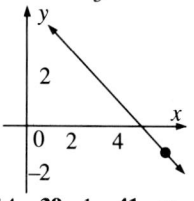 **21.**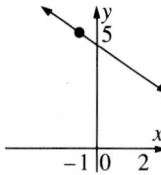

23. yes **25.** no **27.** no **29.** yes
31. $m_{AC} = -\dfrac{4}{5}$; $m_{BC} = -\dfrac{4}{3}$; $m_{AB} = 0$

33. 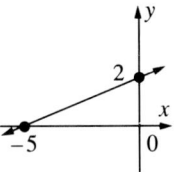 **35.**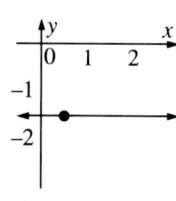

37. 14 **39.** 1 **41.** $m_{AB} = 3$
43. $m_{RS} + \dfrac{1}{2}$; $m_{ST} = -\dfrac{4}{5}$; $m_{TU} = \dfrac{2}{7-x}$; $m_{RU} = \dfrac{4}{-2-x}$

Test Yourself, page 388 **1.** yes
3.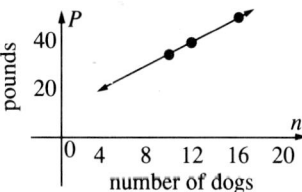

7.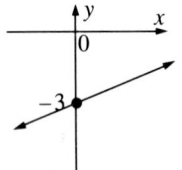

9. 66 lb **11.** $\dfrac{1}{2}$

Practice Exercises, pages 394–395 **1.** $-\dfrac{3}{2}$; 3
3. $\dfrac{1}{3}$; -3 **5.** -2; 0 **7.** 3; 6 **9.** $y = -2x$
11. $y = x + 3$ **13.** $y = \dfrac{2}{3}x + 4$ **15.** $y = -1$
17. 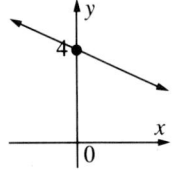 **19.**

25. not parallel **27.** parallel
29. $y = 2x$; $m = 2$, $b = 0$

31. $y = \dfrac{3}{5}x - 3$; $m = \dfrac{3}{5}$, $b = -3$
33. $y = \dfrac{5}{2}x$; $m = \dfrac{5}{2}$, $b = 0$
35. $y = 2x$; $m = 2$, $b = 0$
37. 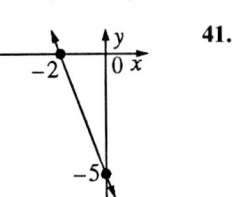 **41.**

45. -8 **47.** parallel **49.** perpendicular
51. No; parallel lines do not intersect.
53. $F = \dfrac{9}{5}C + 32$
55. a. 167°F **b.** 212°F **c.** 405°F

Practice Exercises, pages 398–400
1. $y = \dfrac{3}{2}x - 3$ **3.** $-x + 2y = 0$
5. $2x + y = -1$ **7.** $5x + 4y = 8$
9. $-14x + 6y = -1$ **11.** $y = 5x - 2$
13. $y = 3x - 9$ **15.** $y = -3x + 8$
17. $y = -\dfrac{2}{3}x + 5$ **19.** $-3x + y = -9$
21. $2x + 5y = -4$ **23.** $-2x + y = -8$
25. $-x + y = -9$ **27.** $-2x + 2y = -1$
29. $-2x + 3y = 9$ **31.** $-3x + 5y = 15$
33. $-x + 3y = -17$ **35.** $y = 2gx + 7 - 4g$
37. $y = \dfrac{1}{n}x + \dfrac{2}{n} + 1$

Practice Exercises, pages 403–405 **1.** $(-6, 0)$
3. $(0, 0)$, $(3, 5)$ **5.** $y < -x + 2$ **7.** $y \leq x$
9. $x \leq 2$

13. 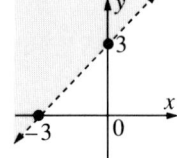 **15.**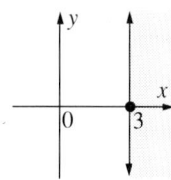

23. 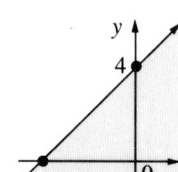 **25.**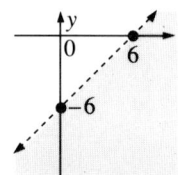

33. $y < \dfrac{5}{2}x - 3$ **35.** $y \geq \dfrac{1}{4}x + \dfrac{3}{4}$ **37.** $y \geq 1$

41. 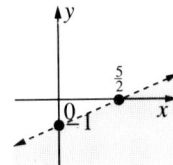   **43.**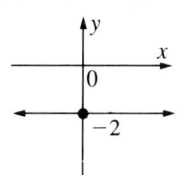

Maintaining Skills, page 410   **1.** (2, 1)
3. (1, −3)   **5.** (0, −2)
13.   **15.**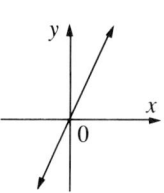

Test Yourself, page 405
3. $2x + y = 8$ **7.**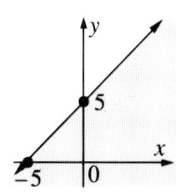

17. −5   **19.** 12   **21.** −11   **23.** $\frac{5}{2}$ **25.** 0
27. 60 students

Chapter 10   Relations, Functions, and Variation

Summary and Review, pages 406–407
1. and 3. 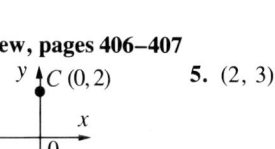   **5.** (2, 3).

Practice Exercises, pages 414–416
1. {(−3, 5), (−1, 4), (1, 4), (3, −3)};
$D = \{-3, -1, 1, 3\}$; $R = \{5, 4, -3\}$
3. $\{(\frac{1}{3}, -1), (0, 0), (\frac{1}{3}, 2), (\frac{1}{2}, 2\frac{1}{2})\}$;
$D = \{\frac{1}{3}, 0, \frac{1}{2}\}$; $R = \{-1, 0, 2, 2\frac{1}{2}\}$
5. {(−3, −4), (1, 6), (4, 30), (5, 24)};
$D = \{-3, 1, 4, 5\}$; $R = \{-4, 6, 24, 30\}$
7. {(−2, 0), (−1, 1), (0, 2), (1, 1), (2, 0), (3, −1), (4, −2)}; $D = \{-2, -1, 0, 1, 2, 3, 4\}$;
$R = \{-2, -1, 0, 1, 2\}$
9. $R = \{2, 5, 11\}$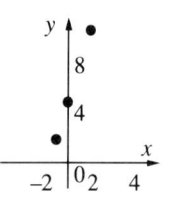

7.

x	y
−2	2
0	1
2	0

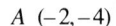
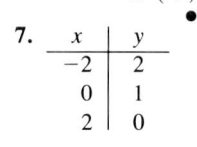   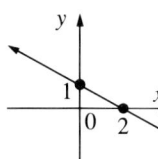

11.

t	F
0	39
40	79
80	119

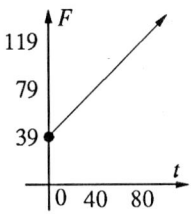

13. $\frac{1}{2}$

13. $R = \{\frac{1}{2}, 1, 2\}$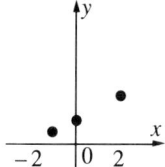

17. $m = \frac{1}{2}$; $b = -4$ **19.** $m = -\frac{1}{2}$; $b = 3$
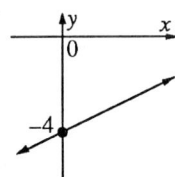   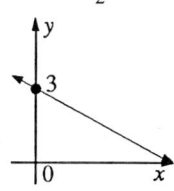

17. {(−1, 2), (−1, 8), (0, 4), (1, 4), (2, 6), (3, 4)};
$D = \{-1, 0, 1, 2, 3\}$; $R = \{2, 4, 6, 8\}$
21. $R = \{-7, 0, 6\}$
23. $D = \{1, 2, 3, 4, 5\}$; $R = \{5000, 5200, 5400, 5600, 5800\}$; {(1, 5000), (2, 5200), (3, 5400), (4, 5600), (5, 5800)}   **25.** $D = \{1, 2, 3, 4, 5\}$;
$R = \{4, 5, 6, 7, 8\}$; {(1, 4), (2, 5), (3, 6), (4, 7), (5, 8)}

21. not parallel   **23.** $-2x + 3y = -12$
25. 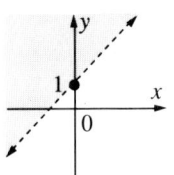   **27.**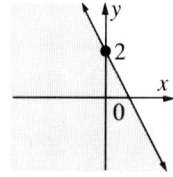

Practice Exercises, pages 419–421   **1.** a function
3. a function   **5.** a function   **7.** a function
9. not a function   **11.** a function   **13.** 5   **15.** −2

17. $R = \{-8, -5, 4, 13\}$; a function

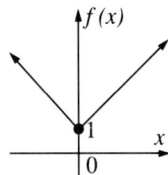

21. $R = \{$all real numbers greater than or equal to $-1\}$; a function

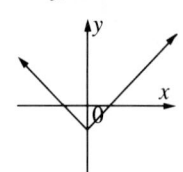

23. 2 **25.** -15 **27.** 3

Practice Exercises, page 424
3. yes; no; no **7.** not a function

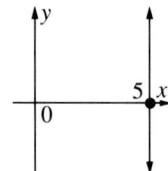

9. yes; yes; yes

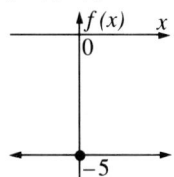

13. -1 **15.** -3 **17.** -1 **19.** 23
23. yes; no; no

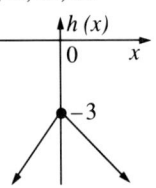

27. -8 **29.** 19 **31.** 2 **33.** -1 **35.** $-\dfrac{29}{15}$

Test Yourself, page 424
3. $R = \left\{-\dfrac{7}{4}, -1, 0\right\}$

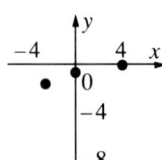

5. yes; no; no
9. yes; no; no

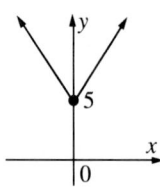

11. -7 **13.** 23

Practice Exercises, pages 428–429 **1.** yes **3.** no
5. yes; 4; $y = 4x$ **7.** yes; $\dfrac{1}{3}$; $y = \dfrac{1}{3}x$ **9.** 11
11. $5\dfrac{1}{3}$ **13.** 35 **15.** 12.8 **17.** 11.88 **19.** 52
21. 9 **23.** 20 **25.** 7.5 **27.** no **29.** $560

Practice Exercises, pages 432–433 **1.** yes; $xy = 12$
3. yes; $xy = -1.44$ **5.** 8 **7.** 4 **9.** 24 **11.** 6.3
13. 5.29 **15.** $\dfrac{36}{121}$ **17.** 21.03 **19.** 10 in.^{3}
21. 42 kg **23.** yes; $r_1 t_1 = r_2 t_2$ **25.** 35 mi/h
27. yes

Practice Exercises, pages 435–436 **1.** $A = s^2$
3. $V = C\left(1 - \dfrac{n}{N}\right)$ **5.** $V = \dfrac{4}{3}\pi r^3$ **7.** $40.32
9. $1.30 **11.** 18 **13.** 8 **15.** $4000 **17.** $0
19. $12,500 **21.** 6.79 **23.** 0.0218 **25.** 2.4

Test Yourself, page 439
1. yes; $y = -3x$; in the form $y = kx$
3. yes; $y = -\dfrac{1}{2}x$; in the form $y = kx$
5. 6 **7.** yes; $xy = 14$ **9.** no **11.** 10
13. $44.90 **15.** $3000

Summary and Review, pages 438–439
1. $\{(2, -2), (5, -5), (2, 0), (-2, 1)\}$;
$D = \{-2, 2, 5\}$; $R = \{-2, -5, 0, 1\}$; not a function
3. a function;   **7.** a function;
a linear function   not a linear function

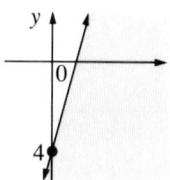

9. directly; $y = 4x$; 4 **11.** 11 **13.** $480 **15.** 8 m

Cumulative Review, page 442 **1.** -3 **3.** $-2, -6$
5. $3xy^2$ **7.** $\dfrac{7m + n}{7mn}$ **9.** 1 to 2 **11.** 1 to 5
13. $\dfrac{7x^2 + 5x - 3}{x}$ **15.** $\dfrac{3x}{x^2 + 4y}$
17. $(2x + 3)(2x - 5)$ **19.** $-3x + y = -3$
21. $x - 2y = 24$ **25.**

27. $R = \left\{0, -\dfrac{63}{4}, -7\right\}$ **29.** $\dfrac{2}{3} < t < \dfrac{14}{3}$
31. 11 h **33.** $l = 30$ cm; $w = 25$ cm

Chapter 11 Systems of Linear Equations

Practice Exercises, pages 446–449
1. 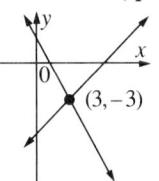 **3.**

13. independent and consistent; different graphs; one solution **15.** independent and inconsistent; different graphs; no solution **17.** dependent and consistent; same graph; more than one solution **19.** independent and consistent; different graphs; one solution **21.** dependent and consistent; same graph; more than one solution **23.** independent and inconsistent; different graphs; no solution **25.** intersecting lines; one solution **27.** parallel lines; no solution **29.** parallel lines; no solution
31. 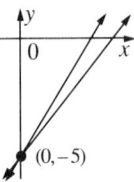 **33.**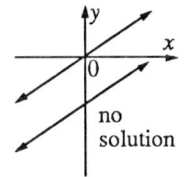

37. parallel lines; no solution **39.** parallel lines; no solution **41.** intersecting lines; one solution
47. $m_1 = m_2$; $b_1 \neq b_2$
49. $m_1 \neq m_2$ or $m_1 = m_2$ and $b_1 = b_2$

Practice Exercises, pages 453–454 **1.** (1, 1)
3. (5, −2) **5.** (3, 0) **7.** (−2, −1) **9.** (5, 2)
11. $\left(-\dfrac{13}{3}, \dfrac{56}{3}\right)$ **13.** (0, 4) **15.** $\left(-\dfrac{2}{3}, 2\right)$
17. $\left(11, \dfrac{13}{4}\right)$ **19.** no solution **21.** $\left(-\dfrac{4}{7}, \dfrac{6}{35}\right)$
23. $\left(\dfrac{1}{2}, -2\right)$ **25.** $\left(\dfrac{55}{9}, \dfrac{38}{9}\right)$ **27.** $\left(\dfrac{2}{5}, -\dfrac{3}{5}\right)$
29. (−4, −9) **31.** $\left(\dfrac{1}{3}, -6\right)$ **33.** $\left(\dfrac{3}{4}, \dfrac{1}{2}\right)$
35. (2, −3) **37.** $\left(-4, \dfrac{1}{7}\right)$
39. $\{(a, b): a - b = 2\}$ **41.** $\left(-\dfrac{7}{2}, -7\right)$
43. $\left(-\dfrac{10}{3}, -\dfrac{8}{3}\right)$ **45.** (1, 1) **47.** $\left(-\dfrac{23}{41}, -\dfrac{50}{41}\right)$ **49.** (−2, 3)

Practice Exercises, pages 457–458
1. $l = 3w - 2$ **3.** $l + s = 48$
$2l + 2w = 68$ $l - s = 12$
$l = 25$ ft; $w = 9$ ft   30, 18
5. $f = 13 + 5s$ **7.** $x + y = 27$
$s + f = 247$ $x = 3 + 2y$
Said 39; father 208   8, 19
9. $x = \dfrac{2}{3}y$ **11.** $x + y = 6000$
$y - x = 5$ $0.075x + 0.08y = 469.75$
15, 10 $3950 at 8%; $2050 at 7.5%
17. $x + y = 18$
$x = 3 + \dfrac{1}{4}y$
6 blocks

Practice Exercises, pages 461–462 **1.** (2, 10)
3. (−3, 3) **5.** (−6, 4) **7.** (5, −6) **9.** (30, 6)
11. (−3, 1) **13.** (2, −3)
15. $\{(a, b): 2a - b = 1\}$
17. $\left(-\dfrac{5}{3}, -13\right)$ **19.** $\left(\dfrac{9}{2}, 7\right)$ **21.** (1, 0)
23. (−4, −5) **25.** $\left(-2, \dfrac{17}{2}\right)$ **27.** $\left(\dfrac{3}{4}, \dfrac{1}{2}\right)$
29. $\left(\dfrac{3}{2}, \dfrac{1}{2}\right)$ **31.** (−2, 8.5) **33.** (3, 8)
35. (4.1, 3.2) **37.** $\left(1, \dfrac{7}{9}\right)$ **39.** $\left(4, -\dfrac{3}{2}\right)$
41. $\left(\dfrac{75}{4}, -\dfrac{3}{2}\right)$ **43.** $\left(2, \dfrac{1}{3}\right)$ **45.** (−5, 1)
47. 25 acres **49.** $-\dfrac{1}{3}$ amp; 1 amp

Practice Exercises, pages 465–466 **1.** (−1, −1)
3. (3, 2) **5.** (−2, −3) **7.** (−22, −33) **9.** (4, 3)
11. (−3, −4) **13.** $\left(\dfrac{1}{2}, -\dfrac{1}{3}\right)$ **15.** (1, 2)
17. $\left(2, \dfrac{1}{2}\right)$ **19.** $\left(\dfrac{1}{2}, \dfrac{1}{3}\right)$ **21.** $\left(-\dfrac{4}{5}, -\dfrac{7}{5}\right)$
23. (1400, 450) **25.** no solution **27.** (−1, −2)
29. $\left(\dfrac{1}{4}, -\dfrac{1}{2}\right)$ **31.** $x = \dfrac{c(1 - b)}{1 - ab}, y = \dfrac{c(1 - a)}{1 - ab}$
33. $\{(x, y): ax + y = c \text{ and } c = 0\}$ **35.** 25

Test Yourself, page 466 **1.** intersecting lines; one solution **3.** intersecting lines; one solution
5. (6, −4) **7.** (−2, 14) **9.** $22,000; $11,000

Practice Exercises, pages 468–469 **1.** 24 **3.** 39
5. 28 **7.** 82 **9.** 94 **11.** 83 **13.** 21
15. $t + u = 1$

Practice Exercises, pages 471–473 **1.** $c = 2b$; $c - 8 = 3(b - 8)$; Cordell 32 yr; Beth 16 yr
3. $m = 2 + 2n$; $n + 9 = \frac{2}{3}(m + 9)$; Nadia 5 yr; Mario 12 yr **7.** $s + 4 = a$; $a - 14 = 2(s - 14)$; Akhil 22 yr; Sarat 18 yr **11.** $(h - 6) + (c - 6) = 71$; $h = 27 + c$; Harvey 55 yr; Carol 28 yr
13. $m = 9 + s$; $m + 1 = 3(s + 1)$; Seth $3\frac{1}{2}$ yr; Mary $12\frac{1}{2}$ yr **15.** $s = 3r$; $\frac{1}{4}(s + r) = 8$; Randi 8 yr; Susan 24 yr **17.** Liz 1909; Bill 1927 **19.** Bill 59 yr; Ann 41 yr **21.** Spero $22\frac{1}{2}$ yr; Chris $27\frac{1}{2}$ yr
23. 20 yr **25.** Emma 20; Kim 32; Vangie 25
27. Mary 28 yr; Jim 53 yr; Grace 52 yr

Practice Exercises, pages 476–477
1. $n + d = 28$; $5n + 10d = 260$; 24 dimes; 4 nickels **3.** $c + a = 175$; $2c + 6a = 750$; 75
5. $p + r = 50$; $1.20p + 2.10r = 147(50)$; 35 lb peanuts; 15 lb raisins **7.** $x + y = 32,000$; $0.075x + 0.09y = 2670$; $18,000 **9.** $f + t = 124$; $5f + 10t = 840$; 44 tens; 80 fives **11.** $y = 2x$; $0.075y + 0.06x = 840$; $4000 at 6%; $8000 at 7.5%
13. $x + y = 36$; $0.04x + 0.4y = 0.2(36)$; 16 gal cream; 20 gal milk **15.** $x + y = 350$; $2.25x + 1.00y = 600$; 150 children; 200 adults
17. $w = p + 3$; $9.95w + 6.50p = 62.75$; 5 lb walnuts; 2 lb peanuts **19.** $p + n + d = 90$; $p + 5n + 10d = 285$; $p = 2(n + d)$; 60 pennies; 15 nickels; 15 dimes

Practice Exercises, pages 481–483 **1.** 216 mi/h
3. 6 mi/h; 2 mi/h **5.** 9 km/h **7.** 540 km/h; 60 km/h **9.** $5\frac{1}{2}$ mi/h **11.** 17 h **13.** 385 mi/h
15. 16 km/h **17.** 2 mi/h; 0.8 mi/h **19.** 12 km/h
21. $\frac{2d}{t} - c = \frac{d}{t} + c$

Practice Exercises, pages 486–487
1. 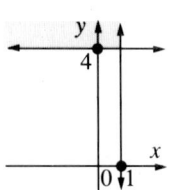 **7.**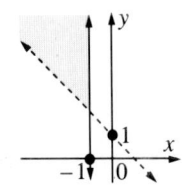

11. 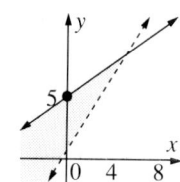 **15.**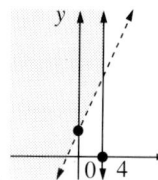

17. 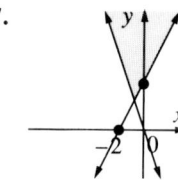 **21.**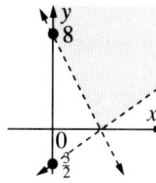

23. 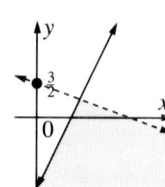 **29.**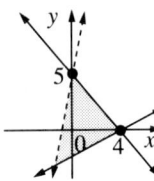

33. 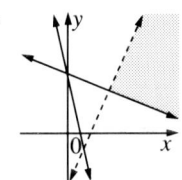 **35.**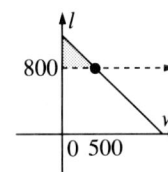

$l = 900$ ft; $w = 400$ ft
or
$l = 1000$ ft; $w = 300$ ft

Test Yourself, page 487 **1.** 56 **3.** Robert 23 yr; Dawn 19 yr **5.** 7 quarters; 45 dimes **7.** 40 mi/h
11.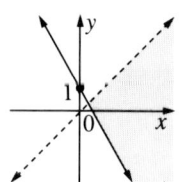

Summary and Review, pages 488–489
1. 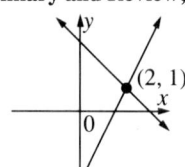 **3.**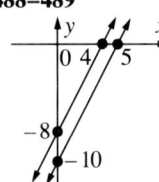

(2, 1); 1 solution; lines intersect

no solution; parallel lines

5. $\left(\frac{27}{7}, \frac{45}{7}\right)$ **7.** 6 ft; 6 ft; 9 ft **9.** (3, 4)
11. $\left(-1, \frac{9}{2}\right)$ **13.** $\left(\frac{10}{3}, \frac{8}{3}\right)$
15. Sue 25 yr; Amy 10 yr **17.** 5 mi/h; 3 mi/h

19.

Maintaining Skills, page 492   **1.** 733.04
3. 215.91 **5.** 8 **7.** 52.7 **9.** 6 **11.** 25
13. 15 **15.** 0.3 **17.** $y^2 - 7y + 12$
19. $3b^2 - b - 10$ **21.** $12x^2 - 7x - 10$
23. 8 lb 1 oz

Chapter 12   Radicals

Practice Exercises, page 496   **1.** 5 **3.** -16
5. $\frac{2}{3}$ **7.** $-\frac{1}{12}$ **9.** $\pm\frac{4}{5}$ **11.** -100 **13.** 33
15. 0.1 **17.** 0.8 **19.** -1.4 **21.** 0.03 **23.** ± 0.25
25. 0.36 **27.** $\frac{5}{3}$ **29.** $\frac{4}{5}$ **31.** 3 **33.** 5 **35.** 3 s

Practice Exercises, page 499   **1.** 2.2 **3.** 4.7
5. 9.1 **7.** 9.7 **9.** 5.2 **11.** 3.9 **13.** 9.9
15. 4.4 **17.** 12.2 **19.** 11.3 **21.** 8.5 **23.** 2.1
25. 200.25 **27.** 0.66 **29.** 6.93 ft

Practice Exercises, pages 501–502   **1.** 0.125
3. $2.\overline{3}$ **5.** -1.25 **7.** $0.\overline{4}$ **9.** -12.75 **11.** $0.1\overline{5}$
13. $\frac{233}{100}$ **15.** $\frac{1}{3}$ **17.** $\frac{3}{11}$ **19.** $\frac{11212}{9999}$ **21.** $\frac{325}{999}$
23. $\frac{1}{7}$ **25.** $\frac{2}{3}, \frac{1}{3}; 1$ **27.** $\frac{1}{18}, \frac{7}{18}; \frac{4}{9}$
29. The prime factors are all 2's or 5's.
31. A rational number represents a terminating decimal if the prime factors of its denominator consist of only the factors 2 and/or 5. A rational number represents a repeating decimal if the prime factors of its denominator include factors other than 2 and/or 5.

Practice Exercises, pages 505–506   **1.** $2\sqrt{2}$
3. $3\sqrt{3}$ **5.** $x \geq -\frac{7}{2}$ **7.** $x \leq -3$ **9.** $4\sqrt{3}$
11. $3\sqrt{6}$ **13.** $5\sqrt{6}$ **15.** $10\sqrt{10}$ **17.** $5x^2\sqrt{3}$
19. $10x^3\sqrt{3}$ **21.** $2y^6\sqrt{3}$ **23.** $d^3\sqrt{15d}$
25. $-5x^8\sqrt{5}$ **27.** $12p^4$ **29.** $x^2\sqrt{y}$ **31.** p^6q^5
33. $9x^6\sqrt{3x}$ **35.** $-5a^5\sqrt{6a}$ **37.** $10r^3\sqrt{5r}$
39. $\frac{6}{7}x^2\sqrt{xy}$ **41.** $0.3c^2\sqrt{c}$ **43.** $0.4m\sqrt{m}$
45. $3\sqrt{2}$ **47.** 10 **49.** $a - 3$ **51.** $64\sqrt{2}$ ft/s
53. π s

Practice Exercises, pages 510–511   **1.** $5\sqrt{5}$
3. $7\sqrt{10}$ **5.** $6\sqrt{3} + 12$ **7.** $5\sqrt{3} - 4\sqrt{5}$
9. $9\sqrt{2}$ **11.** $35\sqrt{2}$ **13.** $4\sqrt{x}$ **15.** $3\sqrt{3x}$
17. $21\sqrt{6}$ **19.** $-7x\sqrt{y}$ **27.** $11\sqrt{11} + 36\sqrt{2}$
29. $-\frac{1}{2}\sqrt{3}$ **31.** $-\frac{y}{6}\sqrt{5x}$ **33.** $ab\sqrt{ab}$
35. 0 **37.** $20\sqrt{5}$ cm

Practice Exercises, pages 514–515   **1.** 4 **3.** $2\sqrt{3}$
5. 7 **7.** 30 **9.** $-30\sqrt{2}$ **11.** $6\sqrt{21}$ **15.** 1
17. $12x\sqrt{6}$ **19.** $9n$ **21.** $4\sqrt{3} + 3$
23. $6 - 2\sqrt{6}$ **25.** $9 - 7\sqrt{2}$ **27.** 29
29. $6ab\sqrt{3}$ **31.** $cde\sqrt{e}$ **33.** $\frac{5a^2}{b}$
35. $67 + 12\sqrt{7}$ **37.** $12 + 3\sqrt{2} - 4\sqrt{3} - \sqrt{6}$
39. $x + 4\sqrt{xy} + 4y$ **41.** $8m - 10\sqrt{mn} - 3n$
43. $2\sqrt{6}$ m^2 **45.** $16x$ ft^2
47. $\sqrt{x^2} = 12$ **49.** $\sqrt{2x} - 3 = 7$

Test Yourself, page 515   **1.** 6.71 **3.** $\frac{1}{3}$ **5.** 14
7. $\frac{5}{4}$ **9.** $-4\sqrt{3}$ **11.** $5x\sqrt{x}$ **13.** $5\sqrt{2} - 4\sqrt{7}$

Practice Exercises, pages 518–519   **1.** $\frac{\sqrt{22}}{11}$
3. $-\sqrt{2}$ **5.** $2\sqrt{6}$ **7.** -3 **9.** $\frac{5}{9}\sqrt{3}$ **11.** $-\frac{\sqrt{3}}{4}$
13. $\frac{4\sqrt{y}}{y}$ **15.** $\frac{2\sqrt{2x}}{x}$ **17.** $\frac{3\sqrt{a}}{a^2}$ **19.** $\frac{-2\sqrt{y}}{3y^2}$
21. $2c$ **23.** $ab\sqrt{b}$ **25.** $\frac{15 + 5\sqrt{2}}{7}$
27. $\frac{-3\sqrt{5} - 3}{4}$ **29.** $-2 + \sqrt{5}$
31. $\frac{2 + 2\sqrt{5} + \sqrt{3} + 15}{-4}$ **33.** $\frac{-\sqrt{11} + 1}{4}$
35. $\frac{3\sqrt{3} - 2 + 9\sqrt{21} + 6\sqrt{7}}{23}$ **39.** $\frac{x - \sqrt{xy}}{x}$
41. $\frac{2x - 5\sqrt{x} + 3}{x - 1}$ **43.** $\frac{\sqrt{5}}{4}$
45. all positive real numbers

Practice Exercises, pages 522–523   **1.** 64 **3.** 108
5. $\frac{1}{16}$ **7.** 12 **11.** 4 **13.** no solution **15.** 4
17. $\frac{17}{5}$ **19.** $\frac{5}{3}$ **21.** 2 **23.** 6 **25.** 18 **27.** $\sqrt{7}$
29. 1 **31.** no solution **33.** no solution **35.** 7
37. 9 **39.** 170 **41.** Cubing a number does not change its sign. **43.** $\frac{38}{5}$ **45.** 144 ft

Practice Exercises, pages 526–527 **1.** 5 **3.** 16.2
5. 9.8 **7.** 8.3 **9.** no **11.** yes **13.** 1
15. $\frac{4}{15}$ **19.** 17 ft **21.** 8 cm **23.** 3, 4, 5

Practice Exercises, page 530 **1.** 5 **3.** $6\sqrt{2}$
5. 13 **7.** $\sqrt{29}$ **9.** $\sqrt{185}$ **11.** 13
13. $AB = 8, BC = 5, AC = 5$
15. $XY = 5, YZ = 2\sqrt{10}, XZ = \sqrt{85}$
17. $AB = 5, BC = 10, AC = 5\sqrt{5}$
19. $MN = 12, NP = 9, MP = 15$
21. 5 or -3 **23.** yes

Practice Exercises, pages 532–534 **1.** $(-5, 5)$
3. $(-6, 3)$ **5.** $\left(-3\frac{1}{2}, 2\right)$ **7.** $\left(\frac{3a}{2}, \frac{3b}{2}\right)$
9. $B(5, 5), C(5, 0), D(5, -5)$
11. a. (6, 6) **b.** $6\sqrt{2}$ **c.** $6\sqrt{2}$ **d.** same length
13. $M_{BD} = M_{AC} = (3, 3)$
15. $AB = 5\sqrt{2}; AC = 3\sqrt{2}; BC = 4\sqrt{2}$; yes because $(3\sqrt{2})^2 + (4\sqrt{2})^2 = (5\sqrt{2})^2$
17. a. (4, 0) **b.** (7, 3) **c.** $3\sqrt{2}$ **d.** $6\sqrt{2}$
f. $3\sqrt{2} = \frac{1}{2}(6\sqrt{2})$

Test Yourself, page 535 **1.** $\frac{\sqrt{2}}{2}$ **3.** $t + 3$
5. $-12\sqrt{5} - 15\sqrt{3}$ **7.** $\frac{4}{3}$ **9.** 7, 5
11. 13 **13.** $5\sqrt{2}$ **15.** $5\sqrt{10}$; $\left(5\frac{1}{2}, -2\frac{1}{2}\right)$

Summary and Review, pages 536–537 **1.** 5
3. $\frac{3}{2}$ **5.** 2.236 **7.** 9.434 **9.** 2.83 **11.** 5.20
13. -3.5 **15.** $0.8\overline{3}$ **17.** $\frac{2}{3}$ **19.** $2\sqrt{5}$
21. $7x\sqrt{2}$ **23.** $-3\sqrt{5} + 2\sqrt{6}$ **25.** $2\sqrt{6} - 5\sqrt{3}$
27. $\frac{\sqrt{15}}{5}$ **29.** $\frac{17}{3}$ **31.** 4 **33.** 1 **35.** $\sqrt{205}$

Cumulative Review, pages 540–542
1. $-2, \sqrt{4}, 3, 6$ **3.** b **5.** a **7.** d
9. $a < 4$ **11.** 2 **13.** (4, 2) **15.** 4 **17.** 6
19. $-5, 3$ **21.** $15 **23.** 5 h **25.** 6
27. $\frac{1}{4}$ **29.** $6p^3q^4$ **31.** $-\frac{2a^2}{3b}$
33. $-\frac{1}{m+2}$ **35.** $\frac{st}{3}$ **37.** $20 - 3\sqrt{3}$
39. $y = \frac{1}{2}x - 3$ **41.** $y = 3x + 1$

43. $3xy^2(4x - 3)$ **45.** prime
47. $3p(p^2 + 3p + 1)$ **49.** $(m + 2n)(3 - 2m)$
51.

53. 7 days **55.** -4 **57.** $2\sqrt{2}$ **59.** 1 **61.** 4
63. 4 **65.** -0.2 **67.** $9m^4n^6$ **69.** $\frac{p-1}{p-2}$
71. $\frac{c+3}{c-1}$ **73.** $3\sqrt{2} + 4\sqrt{5}$ **75.** $2r^2 + 6r - 1$
77. $xy^2 + 2xy$ **79.** $R = \{-7, -3, -1\}$
81. $R = \left\{5, 6, 6\frac{1}{2}\right\}$ **83.** $2.15
85. crew: 7 mi/h; current: 2 mi/h
87. $l = 16$ ft; $w = 10$ ft **89.** $l = \frac{2}{w + 4}$

Chapter 13 Quadratic Equations and Functions

Practice Exercises, pages 545–546
1. $\pm\frac{2}{5}$ **3.** ±7 **5.** $\pm\sqrt{6}$ **7.** $\pm2\sqrt{2}$
9. $\pm\sqrt{15}$ **11.** $\pm\sqrt{3}$ **13.** no real solution
15. ±2 **17.** $\pm\sqrt{5}$ **19.** $-2, 4$
21. $-\frac{5}{2}, \frac{3}{2}$ **23.** $-1, 2$ **25.** $\pm\frac{3}{2}$
27. $-1 \pm 6\sqrt{2}$ **29.** $-1, -\frac{1}{3}$ **31.** $\frac{-2 \pm \sqrt{5}}{3}$
33. $-\frac{6}{5}, \frac{4}{5}$ **35.** $3 \pm \sqrt{7}$ **37.** $5 \pm 2\sqrt{6}$
39. $\frac{\pm\sqrt{2} + 5}{3}$ **41.** $\frac{\pm3\sqrt{2} - 1}{2}$ **43.** $\frac{\pm\sqrt{15} - 3}{2}$
45. $\frac{1}{3} \pm \frac{\sqrt{2}}{4}$ **47.** $\frac{1}{2} \pm \frac{\sqrt{3}}{6}$ **49.** $-3, 7$
51. $-\frac{3}{2}, 0$ **53.** 4, 14 **55.** -7

Practice Exercises, pages 549–550
1. 16 **3.** 25 **5.** $-1, 5$ **7.** $-15, 3$
9. $-6, 2$ **11.** $-\frac{7}{2} \pm 2\sqrt{5}$

13. $\dfrac{3 \pm \sqrt{5}}{2}$ 15. $-5, 0$ 17. $-9, 1$
19. $-3, -2$ 21. $-\dfrac{3}{2}, \dfrac{5}{2}$ 23. $\dfrac{3 \pm \sqrt{11}}{4}$
25. $1 \pm \dfrac{\sqrt{5}}{5}$ 27. $\dfrac{-1 \pm \sqrt{41}}{4}$
29. $\dfrac{-5 \pm \sqrt{31}}{2}$ 31. $\dfrac{2}{3}, 2$ 33. $\dfrac{-5 \pm \sqrt{17}}{4}$
35. $\dfrac{5 \pm \sqrt{145}}{6}$ 37. $\dfrac{7 \pm \sqrt{17}}{4}$
39. $\dfrac{-b \pm \sqrt{b^2 - 4}}{2}$ 41. $\dfrac{-b \pm \sqrt{b^2 - 4ac}}{2}$
43. $-3, 1$

Practice Exercises, pages 552–553

1. $-3, -2$ 3. $-2, 5$ 5. $3, 6$ 7. $\dfrac{3 \pm \sqrt{21}}{6}$
9. $\dfrac{3}{2}$ 11. $1 \pm \sqrt{11}$ 13. $-\dfrac{5}{3}, \dfrac{1}{2}$
15. $\dfrac{-3 \pm \sqrt{17}}{4}$ 17. $\dfrac{2}{3}, 2$ 19. $-3 \pm \sqrt{19}$
21. $\dfrac{-7 \pm \sqrt{29}}{2}$ 23. $\dfrac{1 \pm \sqrt{11}}{2}$
25. $\dfrac{-11 \pm \sqrt{33}}{4}$ 27. $\dfrac{3 \pm \sqrt{5}}{4}$ 29. $-\dfrac{3}{2}, 3$
31. $\dfrac{-5 \pm \sqrt{33}}{2}$ 33. $\dfrac{7 \pm \sqrt{17}}{4}$ 35. $2, -3$
37. $-2 \pm \sqrt{4 + k^2}$ 39. $\dfrac{-ac \pm \sqrt{ac(ac - 4b^2)}}{2bc}$
41. $\dfrac{-4a + 1 \pm \sqrt{8a + 97}}{8}$ 43. $w = 4$ ft; $l = 3$ ft

Practice Exercises, pages 555–556

1. $-1, 2$ 3. $\pm 2\sqrt{3}$ 5. $1, 3$ 7. $-1, \dfrac{1}{4}$
9. $0, \dfrac{2}{3}$ 11. $-\dfrac{1}{2}, \dfrac{1}{3}$ 13. $-3, 15$
15. $\dfrac{9 \pm \sqrt{17}}{4}$ 17. $\dfrac{\pm\sqrt{10} + 2}{3}$
19. $\dfrac{1 \pm 3\sqrt{5}}{2}$ 21. $\dfrac{3 \pm \sqrt{3}}{2}$ 23. $-\dfrac{2}{5}, 1$
25. $2, 9$ 27. $\dfrac{13}{5}, 3$ 29. $\dfrac{1}{b}, \dfrac{1}{2b}$
31. $b - 1, b - 4$ 33. $\dfrac{3b \pm 8}{12}$ 35. 2 s
37. $l = 8$ ft; $w = 4$ ft   39. 9 41. $0, \dfrac{2}{3}$

Test Yourself, page 556

1. $\pm 2\sqrt{2}$ 3. $-1, 3$ 5. $1 \pm \sqrt{6}$
7. $\pm \dfrac{3}{2}$ 9. $-2, \dfrac{1}{3}$

Practice Exercises, pages 560–561

1. 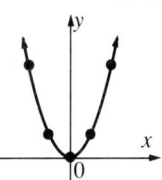   upward; minimum

5. 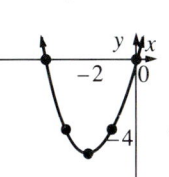   upward; minimum

7. 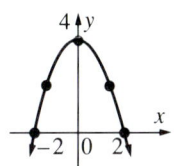   downward; maximum

15. $\left(-\dfrac{5}{2}, \dfrac{7}{4}\right)$; $x = -\dfrac{5}{2}$; none
17. $(-3, 16)$; $x = -3$; $-7, 1$
19. $\left(\dfrac{3}{4}, \dfrac{49}{8}\right)$; $x = \dfrac{3}{4}$; $\dfrac{5}{2}, -1$
21. $(0, -5)$; $x = 0$; none   23. $(0, 2)$; $x = 0$; none
25. $(-2, 0)$; $x = -2$; -2
27. $(-1, -5)$; $x = -1$; $-1 \pm \sqrt{5}$
29. $(-3, -4)$; $x = -3$; none
31. 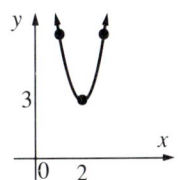   upward; minimum

35. 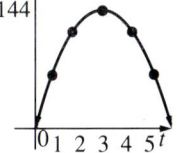   144 ft

Practice Exercises, page 566

1. 64; 2 real solutions   3. 0; 1 real solution
5. 40; 2 real solutions   7. -7; no real solutions
9. 25; 2 real solutions   11. -71; no real solutions
13. 0; 1 real solution   15. 13; 2 real solutions
17. $-\dfrac{61}{12}$; no real solutions   19. 2   21. 2
23. 2   25. no

Practice Exercises, pages 569–571

1. linear function; $c = 0.80w + 55.6$; c-intercept 55.6 is the cost of producing zero ounces of silver; slope 0.80 is the increase ($) for each additional ounce of silver 3. linear function; $c = 0.24h + 1.52$; c-intercept: 1.52 is the cost ($) for zero hours, perhaps a maintenance charge; h-intercept: no meaning—number of hours cannot be negative; slope: 0.24 is the increase ($) for each additional hour.

5. $c = 24.5m + 75$ 7. $g = -\frac{1}{4}h + 55$
9. 3.5 h 11. 15 bu per tree
13. 40 trees; 16 bushels per tree
15. $R(x) = (6 - 0.25x)(500 + 50x)$
17. $S(x) = (6 + 0.5x)(600 - 25x)$

Practice Exercises, pages 574–575

1. $x^2 - 4x + 3 = 0$ 3. $x^2 - 4x - 12 = 0$
5. $x^2 + 6x + 5 = 0$ 7. $x^2 + 5x = 0$
9. $5x^2 + 11x - 12 = 0$ 11. $3x^2 - 5x + 2 = 0$
13. yes 15. no 17. $4x^2 - 25 = 0$
19. $6x^2 + x - 12 = 0$ 21. $x^2 - 2x - 5 = 0$
23. yes 25. no 27. yes
29. $\frac{-b + \sqrt{b^2 - 4ac}}{2a} + \frac{-b - \sqrt{b^2 - 4ac}}{2a} =$
$\frac{-2b}{2a} = -\frac{b}{a}$ 31. yes 33. yes

Test Yourself, page 575

1. $(0, 0)$; $x = 0$
3. $(1, -1)$; $x = 1$ 5. downward; maximum
7. -7; no real solutions 9. 121; 2 real solutions
11. 27 or 13 people 13. 0 or 40 people
15. 2; -2.5

Summary and Review, pages 576–577

1. $\pm 2\sqrt{5}$
3. $-9, 3$ 5. $3 \pm \sqrt{14}$ 7. $\frac{-7 \pm \sqrt{13}}{6}$
9. no real solutions 11. $\pm\sqrt{6}$ 13. $\frac{5 \pm \sqrt{22}}{3}$
15. $-8, 3$
17. ± 1; maximum; $x = 0$

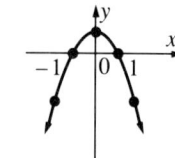

19. 64; 2 real solutions 21. -96; no real solutions
23. $x^2 + 11x + 30 = 0$ 25. $6x^2 - 5x - 50 = 0$

Maintaining Skills, page 580

1. $-4, -3, -1, 2, 4$
3. $-3.1, -3.0, 2.08, 2.8, 2.88$
5. $-\frac{2}{3}, -\frac{3}{5}, -\frac{4}{7}, \frac{5}{6}, \frac{7}{8}$ 7. 450 9. 200
11. 3.24 13. 6.76 15. -24.01 17. 26.01
19. -68.89 21. $\frac{3}{10}$ 23. $\frac{7}{10}$

Chapter 14 Statistics and Probability

Practice Exercises, pages 584–585

1. mode: 71; median: 69; mean: 68.8
3. mode: none; median: 163; mean: 163
5. mode: 1.2, 1.4, 1.5; median: 1.4; mean: 1.5
7. mode: 2; median: 2; mean: 2
9. Change 16 to 12, or 19 to 12, or 12 to 8
11. 25
13. All three measures decrease by 8.
15. All three measures are halved.
17. 136.75 19. mode

Practice Exercises, pages 588–589

1.
Test Score	Tally	Frequency
45–49	I	1
50–54	I	1
55–59	III	3
60–64	III	3
65–69	⊬⊬ I	6
70–74	IIII	4
75–79	⊬⊬ I	6
80–84	II	2
85–89	II	2
90–94	II	2

3. 70.97 5. 16 students 7. 40–49 11. 75.5%
13. 24.5%; 49%
17.

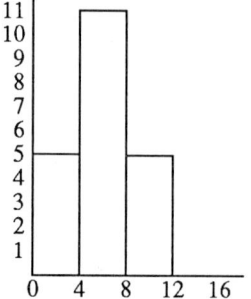

694 Answers to Selected Exercises

19.

Interval	Tally	Frequency
10,000–14,900	IIII	4
15,000–19,900	IIII	4
20,000–24,900	III	3
25,000–29,900		0
30,000–34,900	I	1
35,000–39,900	I	1

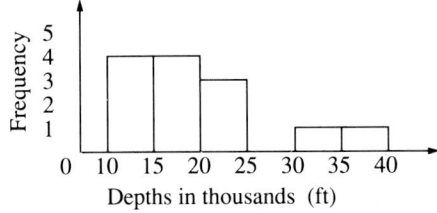

Tree diagram for Exercises 7–12

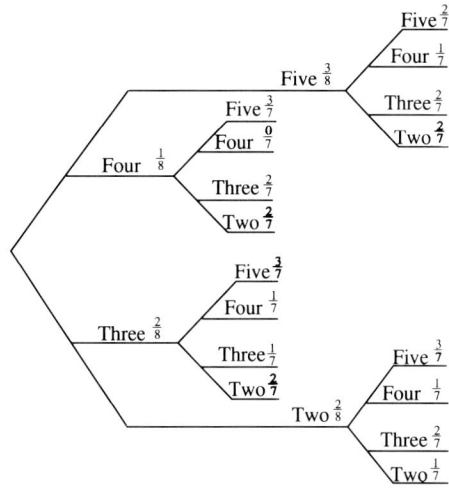

Practice Exercises, pages 592–593
1. variance: $3500; std. deviation: $59.16
3. mode: 8; median: 9; mean: 9; range: 7; variance: 4.6; std. deviation: 2.1   **5.** mode: 29.0, 29.4; median: 29.4; mean: 29.5; range: 1.1; variance: 0.2 std. deviation: 0.4   **7.** mode: none; median: $4.54; mean: $4.20; range: $4.69; variance: $2.46; std. deviation: $1.57   **9.** mean: $26,000; std. deviation: $3521.36

Test Yourself, page 593   **1.** 73.25
3.

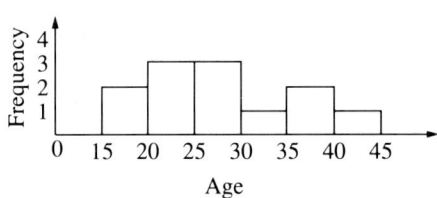

Percentage under 25: 42%
5. mean: 71.9; std. deviation: 32.9

Practice Exercises, pages 598–599   **1.** $\frac{1}{3}$ **3.** $\frac{1}{8}$
5. $\frac{2}{3}$ **7.** $\frac{1}{4}$ **9.** $\frac{1}{4}$ **11.** $\frac{7}{8}$ **13.** $\frac{5}{31}$
15. 7 to 8 for   **17.** $\frac{24}{31}$ **19.** $\frac{5}{9}$
21. $\frac{2}{9}$ **23.** $\frac{1}{3}$ **25.** 0   **27.** $\frac{1}{2}$ **29.** $\frac{2}{3}$
31. 3 to 1 against

Practice Exercises, pages 601–602
1. $\frac{1}{5}$ **3.** $\frac{10}{157}$ **5.** $\frac{1}{3}$

Practice Exercise, pages 606–607
1. $\frac{3}{32}$ **3.** $\frac{9}{64}$ **5.** $\frac{15}{64}$

7. $\frac{3}{8} \times \frac{2}{7} = \frac{3}{28}$ **9.** $\frac{1}{8} \times 0 = 0$
11. $\frac{1}{8} \times \frac{0}{7} + \frac{1}{8} \times \frac{2}{7} = \frac{1}{28}$; $\frac{2}{8} \times \frac{1}{7} + \frac{2}{8} \times \frac{1}{7} = \frac{1}{14}$; $\frac{1}{28} + \frac{1}{14} = \frac{3}{28}$ **13.** $\frac{1}{2}$
15. $\frac{5}{8}$ **17.** $\frac{5}{8}$ **19.** $\frac{4}{25}$ **21.** $\frac{56}{225}$ **23.** $\frac{1}{25}$
31. 0.04; 0.01; 0.05

Test Yourself, page 607   **1.** $\frac{3}{8}$ **3.** 0   **5.** $\frac{1}{4}$ **7.** $\frac{3}{28}$

Summary and Review, pages 608–609
1. mode: 7, 12; median: 11; mean: 12
3.

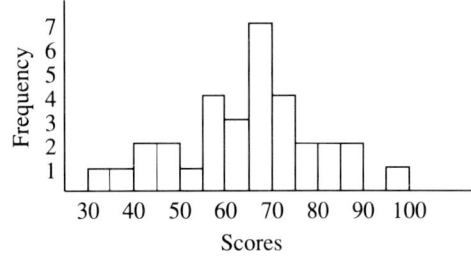

5. variance: 0.64; std. deviation: 0.80
7. $\frac{1}{4}$ **9.** $\frac{1}{80}$ **11.** $\frac{1}{18}$ **13.** $\frac{5}{9}$ **15.** $\frac{1}{17}$ **17.** $\frac{5}{102}$
19. $\frac{1}{2}$

Cumulative Review, page 612
1. <   **3.** >   **5.** =   **7.** <   **9.** <

11. 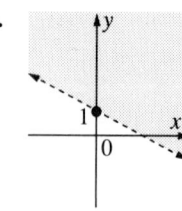 13.

15. $2\sqrt{2}$ 17. $2\sqrt{3}(-1 + \sqrt{2})$ 19. $\frac{4}{3}\sqrt{3}$
21. Father: 54 yr; Al: 24 yr 23. a. mean: 3.7;
mode: 3.7; median: 3.7 c. variance: 0.37; standard deviation: 0.61

Chapter 15 Right Triangle Relationships

Practice Exercises, pages 616–617
1. ray 3. point 5. line segment 7. line
9. complementary 11. supplementary
13. $m\angle N = 60$; acute 15. $m\angle B = 102$; obtuse
17. 25°; 65° 19. 20°; 160° 21. 25°; 155°
23. 30° 25. 155° SE

Practice Exercises, pages 621–622
1. obtuse triangle; isosceles triangle
3. right triangle; scalene triangle
5. right triangle; isosceles triangle
7. 70 9. 67 11. 25 13. 50 15. 45
17. 33°, 114° 19. 59°; 60°; 61° 21. 40°; 60° 80°
23. $m\angle Q = 57$; $m\angle R = 19$; $m\angle S = 104$
25. $m\angle M = 13$; $m\angle N = 102$; $m\angle O = 65$
27. $m\angle S = 4$; $m\angle U = 68$; $m\angle M = 108$
29. the hypotenuse

Practice Exercises, pages 625–627
1. 4; 4; congruent 3. 4; 2; not congruent
5. 3; 3; congruent 7. reflexive
9. symmetric 11. transitive
13. $\angle T \leftrightarrow \angle C$; $\angle O \leftrightarrow \angle A$; $\angle P \leftrightarrow \angle R$ 15. true
17. $\overline{DA} \leftrightarrow \overline{PA}$; $\overline{AL} \leftrightarrow \overline{AN}$; $\overline{PL} \leftrightarrow \overline{DN}$ 19. false
21. $\triangle ETC$; $\triangle TER$; $\triangle CRE$
23. $\triangle AUD \cong \triangle FUD$; $\angle A \leftrightarrow \angle F$;
$\angle AUD \leftrightarrow \angle FUD$; $\angle ADU \leftrightarrow \angle FDU$; $\overline{AU} \leftrightarrow \overline{FU}$;
$\overline{UD} \leftrightarrow \overline{UD}$; $\overline{AD} \leftrightarrow \overline{FD}$

Test Yourself, page 627 1. ray 3. line
5. $m\angle X = 60$ 7. acute; equilateral
9. 66°, 66° 11. false 13. true

Practice Exercises, pages 630–632
1. similar; $\overline{AB} \leftrightarrow \overline{DE}$; $\overline{BC} \leftrightarrow \overline{EF}$; $\overline{AC} \leftrightarrow \overline{DF}$
3. similar; $\overline{EQ} \leftrightarrow \overline{ED}$; $\overline{EP} \leftrightarrow \overline{EL}$; $\overline{PQ} \leftrightarrow \overline{LD}$
5. $YZ = 14$; $XZ = 10$ 7. $AB = 3$; $BC = 6$

9. $MN = 21$; $DF = 6$ 11. $MO = 10$; $EF = 2$
13. $PR = 35$; $AR = 25$ 15. $AB = 6.4$; $PR = 30$
17. $AR = AJ = 4.2$; $TP = 4.8$ 19. 30 21. 0.7
23. 12 m 25. 600 m 27. $453.13

Practice Exercises, pages 635–636 1. $\frac{3}{5}$ 3. $\frac{3}{4}$
5. $\frac{3}{5}$ 7. $\sin A = \frac{5}{13}$; $\sin B = \frac{12}{13}$; $\cos A = \frac{12}{13}$;
$\cos B = \frac{5}{13}$; $\tan A = \frac{5}{12}$; $\tan B = \frac{12}{5}$
9. $\sin A = \frac{12}{37}$; $\sin B = \frac{35}{37}$; $\cos A = \frac{35}{37}$;
$\cos B = \frac{12}{37}$; $\tan A = \frac{12}{35}$; $\tan B = \frac{35}{12}$
11. $\sin A = \frac{\sqrt{2}}{2}$; $\sin B = \frac{\sqrt{2}}{2}$; $\cos A = \frac{\sqrt{2}}{2}$;
$\cos B = \frac{\sqrt{2}}{2}$; $\tan A = \frac{\sqrt{2}}{2}$; $\tan B = 1$
13. $\sin Q = \frac{5}{13}$; $\cos Q = \frac{12}{13}$; $\tan Q = \frac{5}{12}$;
$\sin T = \frac{12}{13}$; $\cos T = \frac{5}{13}$; $\tan T = \frac{12}{5}$
15. $\sin Q = \frac{\sqrt{5246}}{86}$; $\cos Q = \frac{5\sqrt{86}}{86}$;
$\tan Q = \frac{\sqrt{61}}{5}$; $\sin T = \frac{5\sqrt{86}}{86}$; $\cos T = \frac{\sqrt{5246}}{86}$;
$\tan T = \frac{5\sqrt{61}}{61}$
17. $\sin Q = \frac{\sqrt{70}}{10}$; $\sin T = \frac{\sqrt{30}}{10}$; $\cos Q = \frac{\sqrt{30}}{10}$;
$\cos T = \frac{\sqrt{70}}{10}$; $\tan Q = \frac{\sqrt{21}}{3}$; $\tan T = \frac{\sqrt{21}}{7}$
23. 0.5774 25. 0.1465 27. 1.9617
31. tangent ratio

Practice Exercises, pages 641–642
1. 0.9962 3. 1.539 5. 0.9563 7. 0.4067
9. 28.6363 11. 0.8192 13. 30° 15. 45°
17. 55° 19. 18° 21. 2° 23. $BC = 22$; $AC = 36$
25. $AB = 50$; $AC = 4$ 27. $XY = 79$; $XZ = 85$
29. $XZ = 62$; $YZ = 60$ 31. $m\angle K = 53$;
$m\angle J = 37$ 33. $m\angle K = 18$; $m\angle J = 72$ 35. 19.7
37. 105.8 39. 57 ft 41. 54 m

Practice Exercises, pages 646–647
1. yes 3. yes 5. $m\angle C = 90$
7.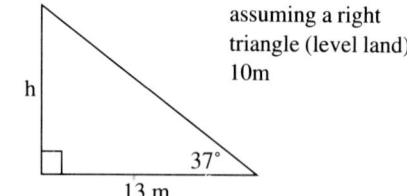
assuming a right triangle (level land)

9. 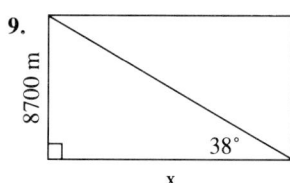 assuming a right triangle (level land) 11,135 m

11. 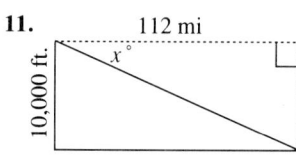 assuming a flat earth; assuming 112 mi ahead means horizontal distance; 1°

13. 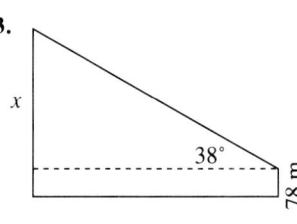 assuming a flat earth; 5938 m

Test Yourself, page 647
1. $3\frac{1}{3}$ **3.** $m\angle A = 18; m\angle B = 72$

Summary and Review, pages 648–649
1. 113   **3.** right triangle, isosceles triangle
5. $\angle P \leftrightarrow \angle C; \angle PAD \leftrightarrow \angle CAR; \angle D \leftrightarrow \angle R$
7. false   **9.** $AE = 53\frac{1}{3}; BC = 30$
11. 0.8746   **13.** $XZ = 48; YZ = 27$
15. $\cos A = \frac{2}{8}; m\angle A = 76$

Cumulative Review, pages 652–654
1. 5   **3.** $3\sqrt{2}$ **5.** $4\sqrt{2}$ **7.** 16   **9.** $9\sqrt{2}$
11. $2cd\sqrt{5d}$ **13.** $6z^2 - 14z - 12$ **15.** $24r^7t^6$
17. $\frac{6\sqrt{3} - 2\sqrt{6}}{7}$ **19.** $36d^4\sqrt{30}$ **21.** $\frac{x+1}{x+3}$
23. $8(2c - 1)$ **25.** $6 + 5\sqrt{6}$ **27.** $4x^2 - 6x + 9$
29. $4a^3b^3 + 6a^2bc - 6ab^2c - 9c^2$ **31.** $\frac{2m - 3n}{3m - 2n}$
33. $x(x + 1) = (x + 2)^2 - 34;\ 10$
35. 82   **37.** $(t + 7)(t - 5)$
39. $(2m - 5)(m + 3)$ **41.** $6a(a - 2b)(a + 2b)$
43. $(t - 3)(t + 3)(r + t)$ **45.** 15   **47.** 500
49. $-2, 2$ **51.** 4, 10   **53.** no solution
55. no solution   **57.** $\left(2, \frac{1}{3}\right)$ **59.** $w < 4$ or $w > 6$
61. $-1 \leq a \leq 5$ **63.** $-4, 8$ **65.** $-2, 8$
67. $-2, 6$ **69.** 12   **71.** 17; 2 real solutions
73. $x^2 - 7x + 10 = 0$ **75.** $2x^2 + 5x - 3 = 0$
77. mode: 21, 32; median: 32; mean: 34; range: 42
79. $\frac{1}{3}$ **81.** 0   **83.** $\frac{2}{9}$ **85.** $\frac{5}{33}$
87. $m\angle A = 40; m\angle B = 80; m\angle C = 60$
89. $6x\ \text{cm}^2$ **91.** 4
93.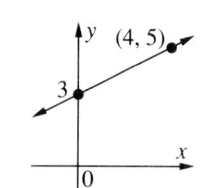
95. $\{3\}$ **97.** $\{4\}$ **99.** $\{3, 4\}$ **101.** 5, 25

GLOSSARY

The explanations given in this Glossary include definitions and brief descriptions of the key terms used in this book.

abscissa (p. 371) The abscissa is the x-coordinate of an ordered pair.

absolute value (p. 14) The absolute value of a number is its distance from zero on a number line.

acute angle (p. 615) An angle that has a measure between 0° and 90°.

acute triangle (p. 619) A triangle that has all acute angles.

addition property for equations (p. 97) If a, b, and c are any real numbers, and $a = b$, then $a + c = b + c$.

addition property for inequalities (p. 190) For all real numbers a, b, and c: If $a > b$, then $a + c > b + c$; if $a < b$, then $a + c < b + c$.

additive identity (p. 59) The sum of 0 and any real number a is equal to the number a. That is, $a + 0 = a$, so 0 is called an additive identity.

additive inverse (p. 27) The opposite of a number is its additive inverse.

additive inverse property (p. 27) For every real number n, there is exactly one real number $-n$ such that $n + (-n) = 0$ and $-n + n = 0$.

algebraic expression (p. 3) An algebraic expression contains variable and/or numerical expressions.

angle (p. 615) The union of two non-collinear rays with a common endpoint.

angle of depression (p. 645) The angle formed by the line of sight and below the line of the horizon.

angle of elevation (p. 643) The angle formed by the line of sight and above the line to the horizon.

ascending order (p. 242) The degree of each term in a polynomial increases from left to right.

associative property for addition (p. 60) For all real numbers a, b, and c, $(a + b) + c = a + (b + c)$.

associative property for multiplication (p. 60) For all real numbers a, b, and c, $(a \cdot b) \cdot c = a \cdot (b \cdot c)$.

average (p. 39) The sum of n numbers divided by n.

axis of symmetry (p. 559) For a quadratic function, $f(x) = ax^2 + bx + c$, the axis of symmetry of the graph is $x = -\dfrac{b}{2a}$.

base (p. 55) In the expression, a^n, a is the base.

binomial (p. 242) A binomial is a polynomial with two terms.

boundary of two half-planes (p. 401) The line which separates two half-planes is the boundary of each half-plane.

closed half-plane (p. 402) If the inequality uses $\leq$ or $\geq$, the boundary line is part of the graph and is drawn as a solid line. The graph is called a closed half-plane.

closure property for addition (p. 24) For all real numbers m and n, $m + n$ is a unique real number.

closure property for multiplication (p. 33) For all real numbers m and n, $m \cdot n$ is a unique real number.

coefficient (p. 63)   In the term $6x^2$, 6 is the coefficient or numerical coefficient.

collinear points (p. 386)   A set of points that lie on the same line are collinear points.

combined inequality (p. 202)   A conjunction or disjunction expressed using $>$, $<$, $\geq$, and/or $\leq$.

common factor (p. 274)   The same factor for two or more integers is a common factor.

commutative property for addition (p. 59)   For all real numbers a and b, $a + b = b + a$.

commutative property for multiplication (p. 59)   For all real numbers a and b, $a \cdot b = b \cdot a$.

comparison property (p. 186)   For all real numbers a and b, one and only one of the following is true: $a = b$, $a < b$, $a > b$.

complementary angles (p. 615)   Two angles whose measures have a sum of 90°.

complementary events (p. 597)   In a random experiment, the two situations—that an event does occur and that the event does not occur—are complementary events.

completeness property (p. 497)   A one-to-one correspondence between the real numbers and the points of a number line.

completing the square (p. 547)   A method used to form a perfect square trinomial.

complex rational expression (p. 337)   A rational expression whose numerator or denominator contains one or more rational expressions.

composite function (p. 423)   A composite function combines two or more functions.

composite number (p. 270)   A positive integer that has more than two positive integral factors is a composite number.

compound event (p. 603)   A compound event is made up of two (or more) events.

conclusion (p. 114)   The final statement of a proof or theorem is the conclusion.

conjugates (p. 517)   The sum and difference of the same two terms are conjugates. For example, $a + \sqrt{b}$ and $a - \sqrt{b}$.

conjunction (p. 202)   A sentence formed by joining two sentences with the word *and*.

congruent angles (p. 623)   Angles that have the same measure are congruent angles.

congruent figures (p. 623)   Figures that have the same size and shape.

congruent line segments (p. 623)   Line segments that have the same length.

consecutive integers (p. 144)   Integers that differ by one are consecutive integers.

consistent systems (p. 444)   A system of equations with at least one solution.

constant function (p. 423)   A linear function whose range contains only one element.

constant of proportionality (p. 426)   The constant, k, in a direct variation, $y = kx$, or inverse variation, $xy = k$.

coordinate of a point on a number line (p. 8)   The graph of a real number is a point on a number line.

coordinate of a point in a plane (p. 370)   The two numbers paired with a given point.

corresponding parts (p. 624)   The corresponding parts of congruent triangles are corresponding vertices, corresponding angles, and corresponding sides.

coordinate system (p. 370)   Determined when two number lines in a plane are drawn so they intersect at right angles.

cosine (p. 634)   In a right triangle, the cosine of an acute angle is the ratio of the length of the leg adjacent to the angle to the length of the hypotenuse.

data (p. 582)   A collection of numerical information.

degree of a monomial (p. 241)   The sum of the exponents of its variables.

degree of a polynomial (p. 242)   The highest degree of any of the polynomial's terms.

density property (p. 498) The density property states that between any two real numbers, there is another real number.

dependent events (p. 604) If one event influences the other, they are dependent events.

dependent system (p. 445) A linear system of equations where the graph of each equation is the same.

descending order (p. 242) The degree of each term in a polynomial decreases from left to right.

difference of two squares (p. 290) The product of two binomials that are the sum and difference of the same two numbers; $a^2 - b^2 = (a + b)(a - b)$.

direct proof (p. 114) The reasoning that takes you from the hypothesis to the conclusion.

direct variation (p. 426) A function in the form $y = kx$, $k \neq 0$, where it is said that y varies directly as x, or y is directly proportional to x.

discriminant (p. 564) Determines the nature and number of solutions of a quadratic equation.

disjunction (p. 203) A sentence formed by joining two sentences with the word *or*.

distance formula (p. 529) For two points $P_1(x_1, y_1)$ and $P_2(x_2, y_2)$ in the coordinate plane, the distance d between the points is given by $d = \sqrt{(x_2 - x_1)^2 + (y_2 - y_1)^2}$.

distributive property (p. 60) For all real numbers a, b, and c, $a(b + c) = ab + ac$ and $(b + c)a = ba + ca$.

divide-and-average method (p. 498) A method for approximating irrational square roots.

division property for equations (p. 101) For all real numbers a, b, and c, $c \neq 0$; if $a = b$, then $\dfrac{a}{c} = \dfrac{b}{c}$.

division property for inequalities (p. 195) For all real numbers a, b, and c:

If $a > b$ and $c > 0$, then $\dfrac{a}{c} > \dfrac{b}{c}$,

If $a > b$ and $c < 0$, then $\dfrac{a}{c} < \dfrac{b}{c}$.

domain of a relation (p. 413) The set of all the first elements or *x*-coordinates of a relation is called the domain.

empty set (p. 79) The set with no members. Used when no member of the replacement set makes an open sentence true.

equally likely (p. 596) The possible outcomes of an experiment which have the same chance of occurring.

equation (p. 78) A mathematical sentence in which the symbol = (equals) connects two numerical or variable expressions.

equilateral triangle (p. 619) A triangle with three sides of equal length.

equivalent equations (p. 96) Equations that have the same solution(s) for the same replacement set are equivalent equations.

equivalent systems (p. 463) Linear systems that may look different but have the same solution are equivalent systems.

evaluate (p. 3) To substitute a given number for each variable and simplify.

even integers (p. 144) Integers that are exactly divisible by two are even integers.

event (p. 596) Any of the possible outcomes for an experiment is an event.

exponents (p. 55) Used to show how many times the factor is multiplied. In a^n, n is the exponent.

extraneous solution (p. 351) An apparent solution which does not satisfy the original equation is an extraneous solution.

extremes (p. 346) In the ratio $a:b = c:d$, a and d are the extremes.

factor (p. 55) When two or more numbers are multiplied, each number is called a factor of the product.

factored completely (p. 296) A polynomial is factored completely when it is written as a product of prime polynomials.

Fibonacci sequence (p. 261) The sequence of numbers, 1, 1, 2, 3, 5, 8, 13, 21, . . . , where each number, except the first two, is the sum of the two preceding numbers.

finite solutions (p. 82) The number of possible solutions can be counted.

FOIL (p. 251) Method used for multiplying two binomials: add the products of the First, Inside, Outside, and Last terms.

formula (p. 117) An equation that expresses a relationship among two or more quantities.

frequency distribution (p. 583) A summary of data where each number is matched with the number of times it occurs.

function (p. 417) A relation in which each element of the domain is paired with exactly one element of the range.

functional notation (p. 418) Notation used to specify a function, for example, the arrow notation, $f: x \rightarrow 2x + 1$, or the f of x notation, $f(x) = 2x + 1$.

Geometry (p. 614) A branch of mathematics that deals with sets of points.

graph of a number (p. 8) Drawings which show relations between numbers or sets of numbers.

greatest common factor (GCF) (p. 274) The greatest integer that is a common factor of two or more integers.

half-plane (p. 401) The regions formed when a boundary line divides a plane.

histogram (p. 586) A bar graph of a frequency distribution.

hypotenuse (p. 524) In a right triangle, the side opposite the right angle is the longest side and is called the hypotenuse.

hypothesis (p. 114) The given statement of a theorem or proof.

identity property for addition (p. 18) For every real number n, there is exactly one real number 0 such that $n + 0 = n$ and $0 + n = n$.

identity property for multiplication (p. 32) For every real number n, $n \cdot 1 = n$ and $1 \cdot n = n$.

inclusive events (p. 605) Two events which can occur at the same time.

inconsistent system (p. 445) A system of equations that has no solution.

independent events (p. 603) Two events that do not influence one another.

independent system (p. 444) A linear system of equations where the graph of each equation is different.

inequality (p. 78) A mathematical sentence in which $<$ (less than), $>$ (greater than), or $\neq$ (is not equal to) is used.

integers (p. 7) The set of numbers, $\{. . . , -2, -1, 0, 1, 2, . . .\}$.

intersection (p. 203) The graph of a combined inequality containing "and." The overlap of the graphs of the two inequalities.

inverse operations (p. 96) An inverse operation is used as an efficient method to solve an algebraic equation, addition and subtraction are inverse operations, as are multiplication and division.

inverse variation (p. 430) A function in the form $xy = k$, $k \neq 0$, where it is said that y varies inversely as x.

irrational numbers (p. 8) Numbers that cannot be written in the form $\frac{m}{n}$, where m and n are integers and $n \neq 0$.

isosceles triangle (p. 619) A triangle that has at least two congruent sides.

least common denominator (LCD) (p. 328) The smallest possible common denominator of two or more fractions.

leg (p. 524) A leg of a right triangle is one of the sides which form the right angle.

like terms (p. 63) Terms that are exactly the same or differ only in their numerical coefficients.

linear equation (p. 375) An equation whose graph is a straight line.

linear function (p. 422) A function whose graph is a straight line.

linear inequality (p. 401) A linear expression with inequality signs.

line (p. 614) A line consists of infinitely many points extending in both directions without end.

line segment (p. 614) A set of points on a line consisting of A and B, and all points between A and B, is called the line segment AB.

literal equation (p. 164) An equation in which constants or coefficients of the variables are expressed by letters.

mathematical model (p. 110) Represents the known parts of a problem.

maximum point (p. 558) The highest point of a parabola that opens downward.

mean (p. 582) In a set of n numbers, the sum of the numbers divided by n.

measure of central tendency (p. 582) A statistic (a number) that is in some way representative or typical of a set of data.

median (p. 582) The middle number in the set, or the mean of the two middle numbers, when the numbers are arranged in order from least to greatest.

minimum point (p. 558) The lowest point of a parabola that opens upward.

mixed expression (p. 337) The sum or difference of a polynomial and a rational expression.

mode (p. 582) A number in a set of data that occurs most often.

monomials (p. 226) An expression that is either a real number, a variable, or a product of a real number and one or more variables.

multiplication property for equations (p. 101) For all real numbers a, b, and c; if $a = b$ then $ac = bc$.

multiplication property for inequalities (p. 195) For all real numbers a, b, and c, if $a > b$ and $c > 0$, then $ac > bc$. If $a > b$ and $c < 0$, then $ac < bc$.

multiplicative identity (p. 59) The product of 1 and any real number a is equal to the number a. That is, $a \cdot 1 = a$, so 1 is called the multiplicative identity.

multiplicative inverse (p. 35) For every nonzero real number n, there is exactly one real number $\frac{1}{n}$, such that $n \cdot \frac{1}{n} = 1$ and $\frac{1}{n} \cdot n = 1$.

mutually exclusive events (p. 605) Two events that cannot happen at the same time.

negative exponents (p. 237) For all real numbers a, $a \neq 0$, and for all positive integers n, $a^{-n} = \frac{1}{a^n}$.

negative numbers (p. 7) Numbers to the left of the zero (0) on a number line.

numerical coefficient (p. 63) The numeral part of a term. In the term $6x^2$, 6 is the numerical coefficient.

numerical expression (p. 3) An expression that may contain one or more of the operations of addition, subtraction, multiplication, or division.

obtuse angle (p. 615) An angle that has a measure between 90° and 180°.

obtuse triangle (p. 619) A triangle that has one obtuse angle.

odd integers (p. 145) Integers that are not exactly divisible by 2.

odds (p. 597) A ratio that compares the probability of an event to the probability of its complement.

open half-plane (p. 402) If the inequality uses $<$ or $>$, the boundary line is not part of the graph and is drawn as a dashed line. The graph is called an open half-plane.

open sentences (p. 79) Equations and inequalities that contain variables.

opposites (p. 13) Numbers whose graphs are the same distance from the origin but in opposite directions.

order of an inequality (p. 194) The direction of the inequality.

ordered pairs (p. 379) Each point in a coordinate plane is assigned a unique ordered pair of real numbers, (x, y).

ordinate (p. 371) The ordinate is the y-coordinate of an ordered pair.

origin (p. 7) The 0 (zero) point on a number line or coordinate plane.

outcome (p. 596) The possible result of each trial or experiment.

parabola (p. 558) A graph of a quadratic function $f(x) = ax^2 + bx + c, a \neq 0$.

parallel (p. 393) Any lines that have the same slope.

percent (p. 152) A percent means hundredth or per hundred.

percent decrease (p. 156) The ratio of an amount of decrease to the previous amount, expressed as a percent.

percent increase (p. 156) The ratio of an amount of increase to the previous amount, expressed as a percent.

perfect square (p. 544) An algebraic expression which contains squares of integers and square algebraic expressions.

perfect square trinomial (p. 289) The square of a binomial equals a perfect square trinomial.
$(m + n)^2 = m^2 + 2mn + n^2$, and
$(m - n)^2 = m^2 - 2mn + n^2$.

point (p. 186) The graph of a real number on a number line is a point.

polynomial (p. 242) A sum of monomials.

polynomial equation (p. 300) An equation whose sides are both polynomials.

positive numbers (p. 7) Numbers to the right of zero (0) on a number line.

postulates (p. 113) Statements that are accepted as true without requiring a proof.

prime factorization (p. 271) A composite number written as the product of prime numbers.

prime number (p. 263) An integer greater than 1, whose only positive factors are itself and 1, is a prime number.

prime polynomial (p. 286) A prime polynomial is a polynomial that has no polynomial factors with integral coefficients except itself and 1.

principal square root (p. 494) The positive square root indicated by $\sqrt{y}$.

probability (p. 596) Measures the likelihood that a particular event will occur.

product property of square roots (p. 495) For all real numbers m and n, where $m \geq 0$ and $n \geq 0$, $\sqrt{mn} = \sqrt{m} \cdot \sqrt{n}$.

properties of equality (p. 61) For all real numbers a, b, c:
Reflexive: $a = a$
Symmetric: If $a = b$, then $b = a$.
Transitive: If $a = b$ and $b = c$, then $a = c$.

properties of real numbers (p. 59) Properties of real numbers are statements that are true for all real numbers.

property of exponents for division (p. 230) For all real numbers $a, a \neq 0$, and for all positive integers m and n:
If $m > n$, then $\dfrac{a^m}{a^n} = a^{m-n}$.
If $m < n$, then $\dfrac{a^m}{a^n} = \dfrac{1}{a^{n-m}}$.
If $m = n$, then $\dfrac{a^m}{a^n} = a^0 = 1$.

property of exponents for multiplication (p. 227) For all real numbers a, and for positive integers m and n:
$a^m \cdot a^n = a^{m+n}$.

property of −1 for multiplication (p. 67) For every real number a: $-1 \cdot a = -a$ and $a \cdot -1 = -a$.

property of proportions (p. 346) In a proportion, the product of the means equals the product of the extremes.

property of zero for multiplication (p. 32) For all real numbers n, $n \cdot 0 = 0$ and $0 \cdot n = 0$.

proportion (p. 345) A statement that ratios are equal is called a proportion.

protractor (p. 345) A measuring device used for finding the size of angles.

Pythagorean theorem (p. 524) In a right triangle, the square of the length of the hypotenuse is equal to the sum of the squares of the lengths of the sides.

quadrants (p. 370) The four regions formed by the axes in a coordinate plane.

quadratic equation (p. 300) An equation of the form $ax^2 + bx + c = 0$, where a, b, and c are real numbers and $a \neq 0$.

quadratic formula (p. 552) If $ax^2 + bx + c = 0$, where a, b, and c are real numbers and $a \neq 0$, then $x = \dfrac{-b \pm \sqrt{b^2 - 4ac}}{2a}$.

quadratic function (p. 557) A function given by the equation $f(x) = ax^2 + bx + c$, where a, b, and c are real numbers and $a \neq 0$.

radical equation (p. 520) An equation that contains radicals with variables in the radicand.

radical expression (p. 494) An expression in the form $\sqrt{a}$.

radical sign (p. 494) The radical sign is a symbol $\sqrt{}$ used to express a radical.

radicand (p. 494) The expression under the radical sign.

ratio (p. 345) The comparison of two quantities by division.

range (p. 590) The difference between the greatest number and the least number in a set.

range of a relation (p. 413) The set of all the second elements or y-coordinates of a relation is called the range.

rational expression (p. 316) An expression that can be written in the form $\dfrac{p}{q}$, where p and q are polynomials, $q \neq 0$.

rational number (p. 8) A number that can be written in the form $\dfrac{m}{n}$, where m and n are integers, and $n \neq 0$.

rationalize a denominator (p. 516) To write an equivalent rational expression so the denominator is changed from an irrational expression to a rational expression.

ray (p. 614) A ray is a part of a line which starts at a point and extends infinitely passing through a second point.

real numbers (p. 7) The set of irrational numbers together with the set of rational numbers form the set of real numbers.

reciprocal (p. 35) Two numbers whose product is 1 are called reciprocals.

reflexive property of equality (p. 61) For all real numbers a, $a = a$.

relation (p. 413) A set of one or more pairs of numbers is a relation.

repeating decimal (p. 500) A decimal which repeats endlessly the same digit, or the same set of digits.

replacement set (p. 79) The set of numbers that may be substituted for the variable.

right angle (p. 615) An angle that has a measure of 90°.

right triangle (p. 619) A triangle that has one right angle.

sample space (p. 596) The set of all possible outcomes in an experiment.

scalene triangle (p. 619) A triangle with three sides of unequal length.

scientific notation (p. 238) A number expressed in the form $n \times 10^m$, where n is a real number such that $1 \leq n < 10$ and m is an integer.

sequence (p. 260) A sequence is a set of numbers arranged in a pattern.

set-builder (p. 198) Set-builder notation is a way of expressing a solution set, such as $\{x: x < -12\}$, read as: the set of all real numbers x such that x is less than -12.

similar triangles (p. 628) Two triangles are similar if and only if the measures of their corresponding angles are equal. The measures of their corresponding sides are proportional.

simplest form of an expression (p. 317) When the numerator and the denominator have no common factors other than 1.

simplify (p. 13) To replace the expression with its simplest name.

simultaneous equations (p. 444) Simultaneous equations are two (or more) linear equations using the same variables.

sine (p. 634) In a right triangle, the sine of an acute angle is the ratio of the length of the leg opposite the angle to the length of the hypotenuse.

slope of a line (p. 384) The slope of a line is the ratio of the change in y to the corresponding change in x. For any two points on a line, (x_1, y_1) and (x_2, y_2); $m = \frac{y_2 - y_1}{x_2 - x_1}$.

slope-intercept form (p. 391) A linear equation in the form $y = mx + b$, where m is the slope of the line and b is the y-intercept.

solution (p. 79) Any value of the variable that makes the open sentence true.

solution set (p. 79) The set of all the numbers from the replacement set that make the open sentence true.

standard deviation (p. 591) Measures how much each value in the data differs from the mean of the data.

standard form of linear equations (p. 375) An equation in the form $Ax + By = C$, where A, B, and C are integers and A and B are not both zero.

statistics (p. 582) Statistics is the collection, organization, analysis, and interpretation of numerical information.

substitution (p. 114) If $a = b$, then a may replace b and b may replace a.

subtraction property for equations (p. 97) For all real numbers a, b, and c, if $a = b$, then $a - c = b - c$.

subtraction (p. 28) For all real numbers m and n: $m - n = m + (-n)$.

supplementary angles (p. 615) Two angles whose measures have a sum of 180°.

square of a binomial (p. 257) Square the first term, double the product of the two terms, square the last term, and write the sum as a perfect square trinomial.

square root (p. 494) If $x^2 = y$, then x is called a square root of y.

square root property (p. 544) If $x^2 = k$, then $x = +\sqrt{k}$ or $x = -\sqrt{k}$, for any real number k, $k \geq 0$.

symmetric property of equality (p. 114) For all real numbers a and b, if $a = b$ then $b = a$.

systems of linear equations (p. 444) Two (or more) linear equations using the same variables.

tangent (p. 634) The tangent of an angle in a right triangle is the ratio of the length of the leg opposite the angle to the length of the leg adjacent to the angle.

terminating decimal (p. 500) A decimal which ends at a finite number of places.

theorem (p. 114) A general conclusion that is shown to be true by using postulates, definitions, given facts, and other proved theorems.

transitive property of equality (p. 114) For all real numbers a, b, and c, if $a = b$ and $b = c$, then $a = c$.

tree diagram (p. 603) A diagram used to show relationships in compound events.

trial (p. 596) A repetition of a probability experiment.

triangle (p. 618) A figure formed by three segments joining three noncollinear points.

trigonometric ratios (p. 634) Ratios of the lengths of the sides of a right triangle, sine, cosine, and tangent.

trigonometry (p. 633) A Greek word for triangle measurement.

trinomial (p. 242) A trinomial is a polynomial with three terms.

uniform motion (p. 172) An object that moves at a constant speed, or rate, is said to be in uniform motion.

unlike terms (p. 63) Terms that are not exactly the same.

union (p. 204) For a combined inequality involving *or*, the union contains both graphs of the inequalities.

variable (p. 2) A symbol used to represent one or more numbers.

variable expression (p. 2) An expression that contains one or more variables.

variability (p. 590) A measure of the spread of the numbers in a set of data.

variance (p. 591) The mean of the sum of the squares of the deviations from the mean is called the variance.

Venn diagram (p. 600) A pictorial representation of sets.

vertex of an angle (p. 615) The common endpoint of the two rays which form the sides of an angle.

vertex of a parabola (p. 559) The minimum or maximum point of a parabola.

vertical line test (p. 418) A method for determining whether a relation is a function.

***x*-axis** (p. 370) The horizontal number line on the coordinate plane.

***x*-coordinate** (p. 371) The first element in a set of ordered pairs.

***x*-intercept** (p. 376) The *x*-coordinate of the point where a graph intercepts the *x*-axis.

***y*-axis** (p. 370) The vertical number line on the coordinate plane.

***y*-coordinate** (p. 371) The second element in a set of ordered pairs.

***y*-intercept** (p. 376) The *y*-coordinate of the point where a graph intercepts the *y*-axis.

zero-product property (p. 300) For all real numbers a and b, if $ab = 0$, then $a = 0$ or $b = 0$ or both a and $b = 0$.

INDEX

Abscissa, 371
Absolute value, 14, 213
  equations, 206–208
  inequalities, 210–212
  on a number line, 14
Accounting, algebra in, 429
Acute angle, 615
Addition
  additive identity property, 59
  additive inverse property, 27
  associative property for, 60
  closure property for, 24
  commutative property for, 59
  distributive property, 60
  of fractions, 328, 333
  identity property for, 18
  on a number line, 17–19
  of polynomials, 245, 337
  of radicals, 509–510
  of rational expressions, 333–334, 337
  of real numbers, 22–24
  rules for, 22–23
  solving systems of equations by, 459–460
Addition/subtraction method, 459–460
Addition property
  for equations, 97
  for inequalities, 190
Additive identity, 18
Additive inverse, 27
Algebra in
  accounting, 429
  aviation, 167
  bookkeeping, 469
  construction, 622
  demography, 585
  engineering, 159
  geometry, 71, 205, 299 (see also Geometry)
  health, 247, 344
  industry, 499
  mechanics, 213
  meteorology, 34
  physics, 336
  police science, 506
  recreation, 284

Algebra in (cont.)
  space technology, 255
  taxation, 416
  transportation, 143
Algebraic expression(s)
  evaluating, 3–4, 52–53, 55–57, 67–69
  involving exponents, 55–57
  numerical, 3
  parentheses in, 3–4, 52–53, 67–69
  simplifying, 13, 52–64
  translating word phrases into, 5, 74–75, 454
  value of, 3–4, 52
  variable, 2–5
  writing, 6, 105
Algebraic notation, 42 (see also Symbol)
Algebraic proof, 113–114
Angle(s), 615–616
  acute, 615
  complementary, 615–616
  congruent, 623
  corresponding, 624, 628–629
  of depression, 643–645
  of elevation, 643–644
  measure of, 615
  obtuse, 615
  right, 615
  sides of, 615
  supplementary, 615–616
  of a triangle, 619
  vertex of, 615
Applications (see also Problem solving)
  agriculture, 308–309
  break-even point, 450
  cost analysis, 425
  image formation, 507–508
  interest, 341
  marathon running, 218–219
  mean, 39
  messenger service, 160
  meteorology, 41
  networks, 637–638
  physics, 562–563
  radio waves, 389–390

Applications (cont.)
  scattergrams, 594–595
  technology, 240 (see also Technology)
Area
  of a circle, 276, 284, 299
  greatest, 272, 309
  of a rectangle, 71, 281, 304, 353
  of a shaded region, 275–276, 299
  of a square, 256, 276, 299
  of a trapezoid, 70
  of a triangle, 118, 276, 462
Arrow notation, 418
Associative property
  for addition, 60
  for multiplication, 60
Average (see Mean)
Aviation, algebra in, 167
Axes, coordinate, 370–371
Axis of symmetry, 559

Base, 55
Binomial(s), 242
  difference of two squares, 290
  dividing by, 342–343
  factoring products of, 289–290
  multiplying, 251–252, 256–258
  opposites, 318
  squaring, 256–257
Binomial factor, 275
Biography
  Gauss, Carl Friedrich, 21
  Hypatia, 273
  Noether, Amalie, 511
  Poncelet, Jean-Victor, 349
  Ramanujan, Srinivasa, 58
  Turing, Alan, 81
Bookkeeping, algebra in, 469
Boundary line of a half-plane, 401–402
Brackets, 37, 52, 69

Calculator, 6
  calculation-ready form, 103–105, 109, 341

Index 707

Calculator (cont.)
  change sign key, 26
  evaluating expressions with, 323
  factoring with, 277, 291
  Fibonacci sequence and, 263
  graphing, 419–420, 448, 454, 560
  input sequence, 323
  interest and, 341
  memory keys, 71, 317
  power key, 58, 229
  reciprocals and, 40
  scientific notation, 255, 238
  simplifying expressions with, 234
  solve equations with, 103, 556
  square numbers with, 236
  square root, 494, 497–498, 544–545
  statistics, 583, 592
  trigonometry and, 525–526, 640
Capacity, formulas for, 308
Careers
  air traffic controller, 449
  computer designer, 81
  computer software developer, 77
  geophysicist, 483
  meteorologist, 41
  seismologist, 483
Challenge, 48, 92, 132, 180, 222, 226, 312, 364, 408, 440, 490, 538, 578, 610
Chapter Review (see Reviews)
Chapter Test (see Tests)
Circle, area of, 276, 284, 299
Closed half-plane, 402
Closure property
  for addition, 24
  for multiplication, 33
Coefficient, 63
Collinear points, 386
Combined inequalities, 202–204, 209, 213
Combining like terms, 63–64, 136–138
Common factors, 274–275
Commutative property
  for addition, 59
  for multiplication, 59
Comparison property, 186
Complementary angles, 615–616
Complementary events, 597
Completeness property, 497

Completing the square, 547–549
Complex rational expression, 337–338
Composite function, 423
Composite number, 270–271
Compound events, 603–605
Computer (see also Technology)
  ABS function, 30
  computer applications, 51, 54, 72–73, 77, 81
  graphing, 189, 376, 395
  INT function, 11
  SUM function, 125
  using computer programs, 54, 72–73, 125–126, 189, 197, 217, 236, 240, 247, 272, 277, 309, 376–377, 390, 395, 405, 469, 478, 502, 530, 549, 562–563, 592
Conclusion, 114, 332
Conditional, 332
Congruence, properties of, 624
Congruent angles, 623
Congruent line segments, 623
Congruent triangles, 624–625
Conjugates, 517
Conjunction, 202
Consecutive integers, 144–145, 305
Consistent systems, 444
Constant, 164
  function, 425
  of proportionality, 426–428
  of variation, 426–428
Construction, algebra in, 622
Coordinate
  axes, 370–371
  plane, 370–372
  system, 370
Coordinate geometry, 531–532
Coordinate(s) of a point, 8, 370–371
Coriolis force, 379
Corresponding angles, 624, 628–629
Corresponding sides, 624, 628–629
Corresponding vertices, 624, 628
Cosine, 634, 639–640
Counting numbers (see Natural numbers)
Critical Thinking
  analysis, 40
  classifying, 244

Critical Thinking (cont.)
  generalizing, 353
  predicting consequences, 400
Cross-ratio, 349
Cubes, sum of two, 58
Cubic-root equations, 522
Cumulative Review (see Reviews)
Cylinder, volume of, 118

Decimal(s)
  coefficients, 148–149
  in equations, 98, 103
  expressed as fractions, 500–501
  as irrational numbers, 497–498
  on a number line, 8
  and percent, 152–154
  repeating, 500–501
  terminating, 500–501
Degree
  of a monomial, 241
  of a polynomial, 242
Demography, algebra in, 585
Denominator(s)
  adding or subtracting expressions with like and unlike, 334
  fractions with irrational, 516–517
  least common, 328–330
  rationalizing, 516–517
Density property, 498
Dependent events, 604
Dependent systems, 445
Depreciation
  annual rate of, 429
  linear, 434–435
Did You Know?, 81, 163, 209, 233, 281, 379, 527, 553
Difference of two squares, 290
Diophantine problem, 273
Direct proof, 114
Direct variation, 426–428
  graph of, 427
Discriminant, 564–565
Disjunction, 203
Distance formula, 528–529
  uniform motion, 119, 389, 479–480, 483
Distributive property, 60
Divide-and-average method, 498
Divisible, 270, 469

708 Index

Division
  definition of, 36
  fraction form of, 4, 42, 52
  of fractions, 324
  of monomials, 230–231
  of polynomials, 342–343
  property of exponents for, 230
  of radicals, 516–517
  of rational expressions, 324–325, 327
  of real numbers, 35–37
  by zero, 37, 40, 291, 351
Division property
  for equations, 101
  for inequalities, 195
Domain
  of a function, 417
  of a relation, 413

Empty set, 79
Endpoint(s), 614
Engineering, algebra in, 159
Enrichment (*see* Challenge; Critical Thinking; Extra; Logical Reasoning; Math Club Activities; Projects; Puzzles)
Equal sign, 12, 42
Equality, properties of, 61, 114
Equally likely, 596
Equation(s), 78 (*see also* Solving equations)
  absolute value, 206–208
  addition property for, 97
  cubic-root, 522
  with decimal coefficients, 148
  decimals in, 98, 103
  division property for, 101
  equivalent, 96, 460
  fractions in, 97, 102-103, 107, 507–508
  and functions, 418–419, 422–423
  graphing, 186
  of inverse variations, 430–431
  of a line, 375, 396–398
  linear, 375 (*see also* Linear equation)
  literal, 164–165
  multiplication property for, 101
  with percent, 152–154

Equation(s) *(cont.)*
  polynomial, 300–302, 304–305
  as proportion, 345–347
  quadratic, 300 (*see also* Quadratic equation)
  radical, 520–521
  rational, 350–359
  simultaneous, 444
  solutions of (*see* Solution)
  standard forms of polynomial, 300
  subtraction property for, 97
  systems of linear, 444–446
  translating word statements to, 85–87, 139
  in two variables, 371–372
  writing, 110–111, 120–121
Equilateral triangle, 619
Equivalent equations, 96, 460
Equivalent expressions, 329–330
Equivalent systems of equations, 460, 463–464
Estimating from graphs, 380
Euclid's Algorithm, 277
Evaluating algebraic expressions, 3–4, 52–53
  involving exponents, 55–57
  parentheses in, 67–69
  rule for order of operations, 52
Evaluating formulas, 55–57, 117–118
Evaluating functions, 418–419, 423
Even integers, 144
Event, probability of, 596–597, 600–601
Exponent(s), 55
  ascending order of, 242
  descending order of, 242
  division with, 230–232
  evaluating expressions with, 55–57
  factored form of expressions with, 226
  multiplication with, 226–227
  negative, 237–238, 240
  order of operations with, 56
  positive integers as, 55
  in prime factorization, 271
  properties of, 227, 230, 234–235
  scientific notation, 237–238, 240
  simplifying expressions with, 55–56

Exponent(s) *(cont.)*
  in simplifying polynomials, 242–243
  in simplifying radicals, 504
  zero, 232
Exponential form, 227
Exponential notation, 240
Expressions (*see* Algebraic expression, Radical expression, Rational expression)
Extra, 155, 197, 250, 277, 319, 395, 433, 462, 473, 502, 519, 523, 589, 632, 636, 642
Extra Practice, 655–669
Extraneous
  solutions, 351–352, 521
Extremes of a proportion, 346

Factor(s), 270
  binomial, 275
  common, 274–275
  greatest common, 274–275, 277
  integral, 270–271
  missing, 274
  monomial, 275
  prime, 271
  square, 503
  variable, 226
Factored form, 226
Factoring
  applications of, 299, 308–309
  completely, 296–297
  a difference of two squares, 290
  by grouping, 292–293
  perfect square trinomials, 289
  polynomials, 274–276
  by removing a common binomial factor, 292–293
  solving polynomial equations by, 300–302, 304–305
  trinomials, 278–287
False statement, 82, 519
Fibonacci Sequence, 260–261
Figurate numbers, 496
FOIL method, 251
Formula(s), 117
  amount in savings account, 118

Formula(s) *(cont.)*
  area of a circle, 276
  area of a rectangle, 304
  area of a square, 256
  area of a trapezoid, 70
  area of a triangle, 118
  aspect ratio, 167
  capacity of a rectangular bin, 308
  capacity of a round bin, 308
  circular motion, 55
  cost of composition, 499
  depreciation, annual rate of, 429
  depreciation, linear, 434–435
  diagonals in a polygon, 566
  distance, 528–529
  evaluating, 55–57, 117–118
  focal length, 507
  height of a rocket, 307
  interest, compound, 550
  interest, simple, 54, 154
  midpoint, 532
  monthly payments, 341
  perimeter, 117–118, 304
  in problem solving, 358–359, 434–435
  profit, 358–359
  quadratic, 551–552
  resistance, 336
  slope, 384
  speed, 506
  sum of n terms of arithmetic series, 118
  temperature conversion, 118
  uniform motion, 168–169, 172–174, 479–481
  vision, 523
  volume of a cylinder, 118
  volume of a rectangular prism, 118
  volume of a sphere, 233
  writing, 358–359, 361
Fraction(s) *(see also* Rational expression)
  adding and subtracting, 328, 333
  decimal form of, 500–501
  dividing, 324
  in equations, 97, 102–103, 107, 507–508
  equivalent, 328
  with irrational denominator, 516–517
  least common denominator of, 328
  mixed numbers, 337

Fraction(s) *(cont.)*
  multiplying, 320
  on a number line, 8
  ratio expressed as, 345
  reciprocal of, 324
  simplifying, 230
Frequency distribution, 583, 586–587, 590
Function(s), 417
  applications of, 425, 429, 562–563, 571
  arrow notation, 418
  composite, 423
  constant, 423
  defined by equations, 418–419, 422–423
  described in tables, graphs, and mappings, 417–419
  direct variation, 426–428
  domain of, 417
  evaluating, 418–419, 423
  graphing, 418, 423, 427, 430, 557–560
  inverse variation, 430–431
  linear, 422–423, 567–568
  quadratic, 557–560, 568–569
  range of, 417
  trigonometric, 634
  vertical line test for, 418
Functional notations, 418
$f(x)$ notation, 418

Gauss, Carl Friedrich, 21
Geometry, 614
  algebra in, 71, 205, 299
  angles, 615–616 *(see also* Angle)
  applications of, 116, 281, 353, 622, 637–638
  in architecture, 617
  area *(see* Area)
  capacity, 308
  congruent figures, 623–625
  coordinate, 531–532
  formulas *(see* Formula)
  lines, 614 *(see also* Line; Line segment)
  networks, 637–638
  perimeter, 71, 117–118, 304, 353, 455–456
  points, 614 *(see also* Point)
  projective, 349
  similar triangles, 628–629
  triangles, 618–620 *(see also* Triangle)
  volume *(see* Volume)

Graph(s)
  and absolute value, 14, 206–207, 210–212
  bar, 586
  with a calculator, 419–420, 448, 454, 560
  with computer programs, 189, 376, 395
  of a direct variation, 427
  of an equation, 186
  estimating from, 380–381
  of functions, 418, 423, 427, 430
  of a horizontal line, 377, 385–386
  of a hyperbola, 430
  of inequalities, 187–188, 205
  of intersecting lines, 444
  of an inverse variation, 430
  of a line, 375–378
  of linear equations, 375–378, 384–386, 391–393
  of linear inequalities, 401–403
  on a number line, 7–9, 186–188
  of opposites, 13
  of ordered pairs, 371–372
  of a parabola, 558
  of parallel lines, 393
  of quadratic functions, 557–560
  of relations, 412–414
  scattergram, 594–595
  of solution set of an equation, 186
  of solution sets of inequalities, 187–188, 191, 195–196, 203–204
  of systems of linear equations, 444–445
  of systems of linear inequalities, 484
  of a vertical line, 377, 385–386
Greater than (>), 12, 42,
Greater than or equal to (≥), 187, 215
Greatest common factor, 274–275, 277
Grouping
  factoring by, 292–293
  symbols, 52

Half-planes, 401–402
Health, algebra in, 247, 344

Histogram, 586–587
History, 77, 81, 163, 209, 233, 295, 496, 546
Horizontal number line, 370
Hypatia, 273
Hyperbola, 430
Hypotenuse, 524
Hypothesis, 114, 332

Identity
    additive, 59
    multiplicative, 59
Identity property
    for addition, 18
    for multiplication, 32
If-then statements, 332
Inclusive events, 605
Inconsistent systems, 445
Independent events, 603–604
Independent systems, 444
Industry, algebra in, 499
Inequality(ies), 78–80
    absolute value, 210–212
    addition property for, 190
    applications of, 209, 213
    combined, 202–204, 209, 213
    division property for, 195
    graphing, 187–188, 205
    linear, 401–403
    multiplication property for, 195
    solution sets of, 79–80, 187–188
    subtraction property for, 190
    systems of linear, 484
    transitive property of order for, 189
    writing, 214–215
Inequality symbols, 12, 42, 78, 187, 215
Integer(s), 7
    absolute value of, 14
    addition of, 17–19, 21
    consecutive, 144–145, 305
    even, 144
    negative, 7–8
    on a number line, 7–9
    odd, 145
    positive, 7–8
Integral factor, 270–271
Integral solution, 273
Interest
    compound, 550
    simple, 54, 154
Intersecting lines, 444
Intersection

Intersection (cont.)
    graph of combined inequality, 203
    point of, 444
    of sets, 600
Inverse(s)
    additive, 27
    multiplicative, 35
Inverse operations, 35, 96
Inverse variation, 430–431
    graphs of, 430
Irrational numbers, 8, 497–498
    on a number line, 9
Isosceles triangle, 619

Königsberg Bridge problem, 637–638

Least common denominator, 328–330
Less than ($<$), 12, 42,
Less than or equal to ($\leq$), 187, 215
Lever, principle of, 431
Like radicals, 509–510
Like terms, 63
    combining, 64
    in equations, 136–138
Line(s), 614
    equation of, 375, 396–398
    graphing, 375–378
    horizontal, 385–386
    intersecting, 444
    parallel, 393, 445
    slant of, 385–386
    slope of, 384–386
    vertical, 385–386
    x-intercept of, 376
    y-intercept of, 376
Line segment(s), 614
    congruent, 623
    finding midpoint of, 531–532
Linear equation(s), 375
    applications of, 433, 458
    graphing, 375–378, 384–386, 391–393
    slope-intercept form of, 391–393
    standard form of, 375
    systems of (see Systems of linear equations)
Linear functions, 422–423, 567–568
Linear inequality(ies), 401–403

Linear inequality (cont.)
    graphing, 402–403
    systems of, 484
Literal equation(s), 164–165
    application of, 167
Logical reasoning, 26, 54, 116, 189, 236, 291, 327, 332, 530, 566

Maintaining Skills (see Reviews)
Mappings, 412–413, 417
Math Club Activities, 11, 147, 193, 229, 357, 374, 550, 561, 599
Mathematical model, 110
Maximum point, 558
Mean, 39, 41, 582–583, 585
Means of a proportion, 346
Measure of an angle, 615
Measures of central tendency, 582–583
Measures of variability, 590–592
Mechanics, algebra in, 213
Median, 582–583
Members of a set, 7
Mental computation, 238, 257
Meteorology
    algebra in, 34
    application, 41
Midpoint formula, 532
Minimum point, 558
Mixed expression, 337
Mode, 582–583
Monomial(s), 226
    degree of, 241
    dividing, 230–231
    division of a polynomial by, 342
    and exponents, 234–235
    multiplying, 227
    multiplying a polynomial by, 248–249
    terms of a polynomial, 242
Monomial factor, 275
Multiplication
    associative property for, 60
    of binomials, 251–252, 256–258
    closure property for, 33
    commutative property for, 59
    distributive property, 60
    of fractions, 320
    identity property for, 32
    of monomials, 227

Multiplication (cont.)
  of a polynomial by a monomial, 248–249
  of polynomials, 248–258
  property of exponents for, 227
  property of −1, 67
  property of zero for, 32
  of radicals, 512–513
  of rational expressions, 320–321
  of real numbers, 31–33
  rules for, 31–32
  symbols for, 3, 42
  zero-product property, 300
Multiplication-addition/subtraction method, 463–464
Multiplication property
  for equations, 101
  for inequalities, 195
Multiplicative identity, 59
Multiplicative inverse, 35
Mutually exclusive events, 605

Natural (counting) numbers, 7
  on a number line, 9
Negative direction on a number line, 17–18
Negative exponents, 237–238, 240
Negative numbers, 7–9
Negative slope, 385
Newton's method, 502
Noether, Amalie, 511
Nomogram, 508
Number(s)
  absolute value of, 14
  comparing, 12
  composite, 270–271
  consecutive, 144–145
  corresponding to a point on a number line, 7–9
  decimals (see Decimal)
  Egyptian, 163
  even, 43, 144
  graphing, 7–9, 11
  integers, 7 (see also Integer)
  irrational, 8–9, 497–498
  natural, 7
  negative, 7–9
  odd, 145
  opposite of, 13
  ordered pairs of, 370–372
  patterns, 236, 260–263

Number(s) (cont.)
  positive, 7–9
  prime, 270–271
  prime factorization of, 271
  rational, 8–9, 316
  real, 7–9 (see also Real numbers)
  reciprocal of, 35
  square, 496
  square root of, 494–495
  squaring, 494
  triangular, 496
  whole, 7
Number line
  absolute value and, 14
  addition on, 17–19
  coordinate of a point on, 8
  direction on, 17–19
  distance on, 17–19, 206–208
  graphing on, 7–9, 11, 186–188
  horizontal, 370
  ordering numbers on, 12
  origin on, 7
  real numbers on, 7–9, 12–13
  vertical, 370
Numerical coefficient, 63
Numerical expression, 3

Obtuse Angle, 615
Obtuse triangle, 619
Odd integers, 145
Odds, 597
One as multiplicative identity, 59
Open half-plane, 402
Open sentence(s), 79 (see also Equation; Inequality)
  replacement set of, 79–80, 82–83
  solution of, 79–80
  solution set of, 79–80
Operations
  inverse (opposite), 35, 96
  order of, 52
Opposite(s)
  additive inverse, 27
  of a binomial, 318
  of a number, 13
  operations, 35, 96
Order
  of operations, 52
  of real numbers, 12
  transitive property of, 189
Ordered pair(s), 370–372

Ordered pair(s) (cont.)
  graph of, 371–372
  set of, 412–413
Ordinate, 371
Origin
  on a number line, 7
  in a plane, 370–371
Outcome, 596

Palantine Anthology problem, 546
Parabola, 558–560
  axis of symmetry, 559
  graph of, 558
  maximum point, 558
  minimum point, 558
  vertex, 559
  x-coordinate of the vertex of, 559
Parallel lines, 393, 445
Parentheses
  adding negative numbers with, 18
  algebraic expressions with, 3–4, 52–53, 67–69
  equations with, 98, 137–138
Pattern(s), number
  Fibonacci sequence, 260–261
  look for, 260–261, 263
  Pascal's triangle, 263
  with squares of numbers, 236
Percent, 152–154
  applications of, 159, 160, 357, 416
  of change, 156–157
Perfect square, 544
Perfect square trinomial, 289
Perimeter, 71, 117–118, 304, 353, 455–456
Physics
  algebra in, 336
  application of, 562–563
Pi (π), 8
Plane, coordinate, 370–372
  coordinates of points in, 370–371
  distance between two points in, 528–529
  origin in, 370–371
  plotting points in, 371–372
  quadrants in, 370
  x-axis in, 370
  y-axis in, 370
Point(s), 614
  collinear, 386

Point(s) (cont.)
  coordinate(s) of, 8, 370–371
  distance between two, 206–208, 528–529
  graph of, 186
  of intersection, 444
  maximum, 558
  minimum, 558
  on a number line, 7–9
  plotting, 371–372
Police science, algebra in, 506
Polynomial(s), 241–242
  adding, 245
  ascending order of exponents in, 242–243
  binomial factors of, 278–287
  classifying, 244
  degree of, 242
  descending order of exponents in, 242
  dividing, 342–343
  equations, 300–302, 304–305
  factoring, 274–276 (see also Factoring)
  monomial factors of, 274–276
  multiplying, 248–258
  prime, 286
  simplifying, 242–243
  subtracting, 246
  terms of, 242
Polynomial equation(s), 300 (see also Linear equation; Quadratic equation)
  applications of, 307, 308–309
  solutions (roots) of, 300–302
  solving, by factoring, 300–302, 304–305
  standard forms of, 300
Poncelet, Jean-Victor, 349
Positive direction on a number line, 17–18
Positive numbers, 7–9
Postulates, 113
Power(s)
  multiplying by a power of 10, 148–149
  of a number, 55
  of a product, 234–235
  of a quotient, 235
Preparing for Standardized Tests (see Tests)
Prime factorization, 271

Prime numbers, 263, 270–271, 295
Prime polynomial, 286
Principal square root, 494
Probability, 596
  of two dependent events, 604
  of an event, 596–597, 600–601
  of two inclusive events, 605
  of two independent events, 603–604
  of two mutually exclusive events, 605
  odds, 597
Problem solving (see also Application)
  age, 470–471
  angle of depression, 643–645
  angle of elevation, 643–644
  commission, 380
  consecutive integers, 144–145, 305
  digit, 467–468
  linear depreciation, 434–435
  mixed types, 127–129 (see also Reviews)
  mixture, 161–163, 474–475
  money, 474–475
  percent of change, 156–157
  perimeter, 455–456
  probability, 600–601
  profit, 358–359, 567–568
  reminders, 121
  steps, 43
  trigonometry, 643–644
  uniform motion, 168–169, 172–174, 355–356, 479–481
  wind and water current, 479–481
  work, 354–355
Problem solving strategies
  account for all possibilities, 82–83
  check for hidden assumptions, 643–645
  draw a diagram, 600–601
  estimate from graphs, 380
  look for a pattern, 260–261
  make a drawing or a table, 168–169
  make a model, 110–111
  select appropriate notation, 42–44
  solve a simpler problem, 358–359

Problem solving strategies (cont.)
  write an equation, 120–121
  write an inequality, 214–215
  write a linear system, 455–456
  use an appropriate formula, 434–435
  use coordinate geometry, 531–532
  use a function, 567–569
  use polynomial equations, 304–305
Product(s)
  of binomials, 251–252, 256–257, 289–290
  of means and extremes, 346
  of monomials, 227
  power of a, 235
  of powers, 227
  property of square roots, 495
  of radicals, 512–513
  of rational expressions, 320–321
  of signed numbers, 31–33
  of sum and difference of two terms, 257–258
Product property of square roots, 495
Projectile motion, 553
Projective geometry, 349
Projects, 45, 84, 112, 124, 171, 217, 263, 307, 361, 383, 436, 458, 535, 571, 602, 647
Proof, 113–114
Property(ies)
  additive inverse, 27
  associative, for addition, 60
  associative, for multiplication, 60
  closure, for addition, 24
  closure, for multiplication, 33
  commutative, for addition, 59
  commutative, for multiplication, 59
  comparison, 186
  of completeness, 497
  of congruence, 624
  density, 598
  distributive, 60
  of equality, 61, 114
  for equations, 97, 101
  of exponents, 234–235

Property(ies) *(cont.)*
    of exponents for division, 230
    of exponents for multiplication, 227
    identity, for addition, 18
    identity, for multiplication, 32
    for inequalities, 190, 195
    multiplicative inverse, 35
    of parallel lines, 393
    product, of square roots, 495
    of proportions, 346
    quotient, of square roots, 495
    reciprocal, 35
    reflexive, 61, 624
    square-root, 544
    substitution, 114
    symmetric, 61, 624
    transitive, 61, 624
    transitive, of order for $<$ and $>$, 189
    of zero for multiplication, 32
    zero-product, 300
Proportion(s), 345–347
    property of, 346
    and similar triangles, 628–629
Proportionality, constant of, 426–428
Protractor, 615
Pythagorean theorem, 524–525
    applications of, 527, 622
    converse of, 525
Puzzles, 16, 54, 100, 155, 473

Quadrants, 370
Quadratic equation(s), 300
    application of, 553
    involving perfect-square expressions, 544–545
    solutions (roots) of, 564–565, 572–573
    solving, by completing the square, 547–549
    solving, by factoring, 300–302
    solving, by using the quadratic formula, 551–552, 564–565
    solving, by using the square-root property, 544–545
    standard form of, 300

Quadratic formula, 551–552
Quadratic function(s), 557–560, 568–569
    applications of, 562–563, 571
    graphing, 557–560
Quotient(s)
    power of a, 235
    of powers, 230–231
    property of square roots, 495
    of radicals, 516–517
    of signed numbers, 35–37
Quotient property of square roots, 495

Radical(s), 494
    like, 509
    unlike, 509
Radical equation(s), 520–521
    applications of, 506, 523
Radical expression(s)
    adding, 509–510
    application of, 506
    conjugates, 517
    dividing, 516–517
    in equations, 520–521
    multiplying, 512–513
    rationalizing the denominator of, 516–517
    simplifying, 503–504, 509–517
    subtracting, 509–510
    with variables, 504
Radical sign, 494
Radicand, 494
Radio signals, 255, 389–390
Ramanujan, Srinivasa, 58
Random experiment, 596–597
Random survey, 124, 589, 600–602
Range
    of a function, 417
    of a relation, 413
    of a set of data, 590
Ratio(s), 345
    cross-ratio, 349
    expressed as a fraction, 345
    probability, 596–597
    and proportion, 345–347
    in simplest form, 345–346
    slope of a line expressed as, 384
    trigonometric, 633–634
Rational equations, 350–359
Rational expression(s), 316
    adding, 333–334

Rational expression(s) *(cont.)*
    complex, 337–338
    dividing, 324–325, 327
    in equations, 350–359
    equivalent, 329–330
    least common denominator of, 328–330
    mixed expression, 337
    multiplying, 320–321
    in simplest form, 317
    simplifying, 317–318, 338
    subtracting, 333–334
    undefined, 317
Rational number(s), 316
    decimal forms for, 500–501
    on a number line, 8–9, 11
Rationalizing the denominator, 516–517
Ray, 614
Reading in Algebra, 421
Real number(s), 7–9
    adding, 22–24
    application of, 41
    dividing, 35–37
    multiplying, 31–33
    on a number line, 7–9, 12–13
    opposite of, 13
    order of, 12
    properties (*see* Property)
    rules for, 22–23, 31–32
    subtraction of, 27–29
Reciprocal(s), 35
    of a fraction, 324
Reciprocal property, 35
Recreation, algebra in, 284
Rectangle, area of, 71, 281, 304, 353
Rectangular prism, volume of, 118
Reflexive property, 61, 114, 624
Relation(s), 413 (*see also* Function)
    domain of, 413
    graphing, 414
    range of, 413
Repeating decimal, 500–501
Replacement set, 79–80
    finite, 82
    infinite, 82
Reviews
    Chapter Summary and, 46–47, 90–91, 130–131, 178–179, 220–221, 264–265, 310–311, 362–363, 406–407, 438–439, 488–489,

Reviews *(cont.)*
 536–537, 576–577, 608–609, 648–649
 Cumulative, 94, 182–184, 268, 366–368, 442, 540–542, 612, 652–654
 Extra Practice, 655–669
 Maintaining Skills, 50, 134, 224, 314, 410, 492, 580
 Mixed Problem Solving, 84, 112, 124, 171, 216, 263, 307, 361, 383, 436, 458, 571, 602, 647
Right angle, 615
Right triangle(s), 619
 applications of, 527, 622
 hypotenuse of, 524
 Pythagorean theorem, 524–525
  converse of, 525
 trigonometric ratios in, 633–634, 639–640
Root(s), 302, 564
 extraneous, 521
Rule for order of operations, 52

Sample space, 596
Scalene triangle, 619
Scattergram, 383, 594–595
Scientific notation, 237–238, 240
 application of, 240
Segment, line, 614
Self-test (*see* Tests)
Sentence(s)
 conjunction, 202
 disjunction, 203
 open, 78–80
Set(s)
 empty, 79
 intersection of, 600
 members of, 7
 of ordered pairs, 412–413
 of rational numbers, 8
 of real numbers, 7–9
 replacement, 79–80, 82–83
 solution, 79–80
Set-builder notation, 198–199
Set notation, 7, 42
Sides
 of an angle, 615
 corresponding, 624, 628–629
 of a triangle, 618
Similar terms, 63, 509
Similar triangles, 628–629

Simplest form
 of radical expressions, 503–504
 of rational expressions, 317
 of ratios, 345–346
Simplifying
 algebraic expressions, 13, 52–64
 expressions with parentheses, 67–69
 fractions, 230
 polynomials, 242–243
 products of radicals, 512–513
 quotients of radicals, 516–517
 rational expressions, 317–318, 338
 square roots, 503–504
 sums and differences of radicals, 509–510
Simultaneous equations, 444 (*see also* System of linear equations)
Sine, 634, 639–640
Slant of a line, 385–386
Slope of a line, 384–386, 400
Slope-intercept form of a linear equation, 391–393
Solution(s)
 of equations, 79, 96–98, 186, 300–302, 371–372
 extraneous, 351–352, 521
 integral, 273
 of open sentences, 79–80
 of quadratic equations, 564–565, 572–573
Solution set(s)
 of equations, 79, 97
 graphing (*see* Graph)
 of inequalities, 79–80, 187–188
 of open sentences, 79–80
 of systems of linear equations, 444–445
 of systems of linear inequalities, 484
Solving equations, 96–108
 by combining like terms, 136–138
 with decimal coefficients, 148–149
 by factoring, 300–302
 fractional, 507–508
 involving absolute value, 206–208

Solving equations *(cont.)*
 involving percent, 152–154, 156–157
 involving proportion, 346–347
 literal, 164–165
 parentheses in, 98, 137–138
 polynomial, 300–302, 304–305
 quadratic, 544–555, 564–565, 572–573
 radical, 520–521
 rational, 350–359
 shortcuts for, 148–149
 steps for, 141
 with the variable on both sides, 140–141
Solving inequalities, 190–199
 application of, 218–219
 combined, 202–204
 involving absolute value, 210–212
 systems of linear, 484
 using a computer program, 197
Solving systems of linear equations
 by the addition/subtraction method, 459–460
 by graphing, 444–446
 by the multiplication-addition/subtraction method, 463–464
 in problems, 455–456, 467–483
 by the substitution method, 451–452
Space technology, algebra in, 255
Sphere, volume of, 233
Square, area of, 256, 276, 299
Square(s)
 of a binomial, 256–257
 factoring the difference of two, 290
 factoring a perfect square trinomial, 289
 of a number, 494
 perfect, 544
Square factor, 503
Square numbers, 496
Square root(s), 494–495
 approximating, 497–498, 502
 irrational, 497–498, 516–517
 negative, 494

Square root(s) (cont.)
  positive, 494
  principal, 494
  product property of, 495
  property, 544
  quotient property of, 495
  simplifying, 503–504
  with variable expressions, 503–504
Standard deviation, 591–592
Standard form
  of a linear equation, 375
  of a polynomial equation, 300
  of a quadratic equation, 300
Standardized Tests (see Tests)
Statistics, 582
  applications of, 585, 594–595
  frequency distribution, 583, 586–587, 590
  graphing data, 586–587, 594–595
  measures of central tendency, 582–583
  measures of variability, 590–592
Substitution method, 451–452
Substitution property, 114
Subtraction
  definition of, 28
  distributive property, 60
  of a polynomial and a rational expression, 337
  of polynomials, 246
  of radicals, 509–510
  of rational expressions, 333–334
  of real numbers, 27–29
  solving systems of equations by, 459–460
Subtraction property
  for equations, 97
  for inequalities, 109
Summary and Review (see Reviews)
Supplementary angles, 615–616
Symbol(s)
  absolute value, 14, 42
  congruence, 623
  division, 4, 42, 52
  empty set, 79
  equality, 12, 42
  greater than, 12, 42
  greater than or equal to, 187, 215

Symbol(s) (cont.)
  grouping, 3–4, 37, 52
  inequality, 78
  less than, 12, 42
  less than or equal to, 187, 215
  multiplication, 3, 42
  parentheses, 3–4, 52
  positive and negative signs, 7, 42
  radical sign, 494
  for repeating digit, 500
  for a set, 7, 42
  is similar to, 628
  square root, 494
  table of, xii
Symmetric property, 61, 114, 624
Symmetry, axis of, 559
Systems of linear equations, 444–446 (see also Solving systems of linear equations)
  applications of, 450, 458, 473
  consistent, 444
  dependent, 445
  equivalent, 460, 463–464
  graphing, 444–445
  inconsistent, 444
  independent, 445
  solution sets of, 444–445
  writing, 455–456
Systems of linear inequalities, 484

Table(s)
  frequency distribution, 583, 586–587, 590
  of squares and square roots, 670
  of symbols, xii
  of trigonometric ratios, 671
Tangent, 634, 639–640
Taxation, algebra in, 416
Technology
  applications of, 240, 255, 562–563
  spreadsheets, 72–73, 125–126, 217
Term(s), 63
  combining like, 64, 136–138
  like, 63
  of a polynomial, 242
  unlike, 63–64
Terminating decimal, 500–501

Tests
  Chapter, 48, 92, 132, 180, 222, 266, 312, 364, 408, 440, 490, 538, 578, 610, 650
  Preparing for Standardized, 49, 93, 133, 181, 223, 267, 313, 365, 409, 441, 491, 539, 579, 610, 651
  Test Yourself, 45, 66, 89, 109, 129, 151, 177, 217, 239, 259, 288, 303, 340, 361, 388, 405, 424, 437, 466, 487, 515, 535, 556, 575, 593, 607, 627, 647
Theorem, 114
  Pythagorean, 524–525
Tolerance, 213
Transitive property
  of congruence, 624
  of equality, 61, 114
  of order for $<$ and $>$, 189
Translating word phrases into algebraic expressions, 5, 74–75, 454
Translating word statements to equations, 85–87, 139
Transportation, algebra in, 143
Trapezoid, area of, 70
Tree diagram, 603
Trial, 596
Triangle(s), 618–620
  acute, 619
  angles of, 619
  area of, 118, 276, 462
  congruent, 624–625
  corresponding parts, 624–625
  equilateral, 619
  isosceles, 619
  obtuse, 619
  right, 619 (see also Right triangle)
  scalene, 619
  sides of, 618
  similar, 628–629
  in trigonometry, 633–634, 639–640
  vertices of, 618
Triangular numbers, 496
Trigonometric ratios
  cosine, 634, 639–640
  sine, 634, 639–640
  tangent, 634, 639–640
Trigonometry, 633
  using a calculator in, 640
  using trigonometric ratios, 639–640, 643–645

Trigonometry *(cont.)*
   using trigonometric tables, 639–640
Trinomial(s), 242
   factoring, 278–287
   perfect square, 289
True statement, 79, 82, 319, 519

Undefined
   rational expression, 317
   slope, 386
Uniform motion, 168–169, 172–174, 355–356, 479–481
   applications of, 255, 389–390, 483, 561
Union, graph of combined inequality, 204
Unit distance, 17–19
Units of measure
   bushel, 308
   cubit, 383
   em, 499
   foot, 383
   megahertz, 389
Unlike radicals, 509
Unlike terms, 63–64

Value(s)
   of an expression, 3–4, 52
   of a variable, 3–4, 317
Variability, 590

Variable(s), 2–5
   on both sides of an equation, 140–141
   value of, 3–4, 317
Variable expression, 2–4
Variable factor, 226
Variance, 591
Variation
   constant of, 426–428
   direct, 426–428
   inverse, 430–431
Velocity, 55, 143, 307, 370, 506, 571
Venn diagram, 600–602
Vertex (vertices)
   of an angle, 615
   corresponding, 624, 628
   of a parabola, 559
   of a triangle, 618
Vertical line test, 418
Vertical number line, 370
Vocabulary, 46, 90, 130, 178, 220, 264, 310, 362, 406, 438, 488, 536, 576, 608, 648
Volume
   of a cylinder, 118
   of a rectangular prism, 118
   of a sphere, 233

Whole numbers, 7
   on a number line, 9

Writing in Algebra, 6, 62, 105, 119, 139, 323, 454

x-axis, 370
x-coordinate, 370
x-coordinate of the vertex of a parabola, 559
x-intercept, 376

y-axis, 370
y-coordinate, 370
y-intercept, 376, 396–397

Zero
   absolute value of, 14
   as additive identity, 59
   division by, 37, 40, 291, 351
   as an exponent, 232
   as origin, 7
   product property, 300
   property for multiplication, 32
Zero-product property, 300

Photo Credits

Chapter 1 1: Arjen Verkaik/The Stock Market. 2: Ken Karp. 21: The Bettmann Archive. 26: Chris Sorensen/The Stock Market. 34: Ted Kaufman/The Stock Market. 41: Russ Kinne/Comstock. Chapter 2 51: Brian Brake/Science Source/Photo Researchers. 55: Chris Jones/The Stock Market. 58: The Granger Collection. 77: Ed Kashi. 78: Arjen Verkaik/The Stock Market. 81: The Science Museum. Chapter 3 95: Mark E. Gibson/The Stock Market. 96: Seaworld of Florida. 100: Michal Heron/Woodfin Camp. 118: Tom King/The Image Bank. Chapter 4 135: Hank Morgan/Photo Researchers. 136: Frank Whitney/The Image Bank. 143: Cliff Feulner/The Image Bank. 160: Roy Morsch/The Stock Market. 161: DPI. 167: Peter F. Runyon/The Image Bank. Chapter 5 185: Bill Ross/Westlight/Woodfin Camp. 189: Ken Karp. 194: David Madison/Duomo. 202: D. Strohmeyer/Focus West. 208: Rutger's New Jersey Bowl/NJN. 209: Harvard College Observatory/Science Photo Library/Photo Researchers. 214: Richard Steedman/The Stock Market. 218: Paul Sutton/Duomo. Chapter 6 225: NASA. 255: NASA. Chapter 7 269: DPI. 273: The Bettmann Archive. 308: Gary Cralle/The Image Bank. Chapter 8 315: Henley & Savage/The Stock Market. 327: Jay Freis/The Image Bank. 341: Barry O'Rourke/The Stock Market. 345: Obremski/The Image Bank. 354: Gregory Heisler/The Image Bank. Chapter 9 369: Brownie Harris/The Stock Market. 389: Co. Rentmeester/The Image Bank. Chapter 10 411: Robert Kristofik/The Image Bank. 419: John Lei/Omni–Photo Communications, Inc. 425: Jay Freis/The Image Bank. Chapter 11 443: Michal Heron/Woodfin Camp. 449: Walter Bibikow/The Image Bank. 455: William Strode/Woodfin Camp. 478: The Bettmann Archive. 479: William Strode/Woodfin Camp. 481: (t) Phillip Wallick/The Stock Market. 481: (b) Tom Leigh/Rainbow. 482: Eddie Hironaka/The Image Bank. 483: David Austen/Stock Boston. 487: Joe Bator/The Stock Market. Chapter 12 493: Gary Gladstone/The Image Bank. 494: Sonya Jacobs/The Stock Market. 497: Miguel/The Image Bank. 500: Dan McCoy/Rainbow. 503: Roger Ressmeyer/Starlight Photo Agency. 506: George Anderson. 507: Tom Stack & Associates. 520: Duomo. Chapter 13 543: Wes Thompson/The Stock Market. 544: Charles West/The Stock Market. 557: Jay Freis/The Image Bank. 562: NASA. Chapter 14 581: Jeff Lowenthal/Woodfin Camp. 582: Robert Frerck/Odyssey Prod. 594: John M. Roberts/The Stock Market. 596: The Bettmann Archive. Chapter 15 613: John Blaustein/Woodfin Camp. 618: Paul Barton/The Stock Market. 623: Murray & Assoc., Inc./The Stock Market. 628: Russ Kinne/Comstock. 639: Ken Karp.

Demon Pie